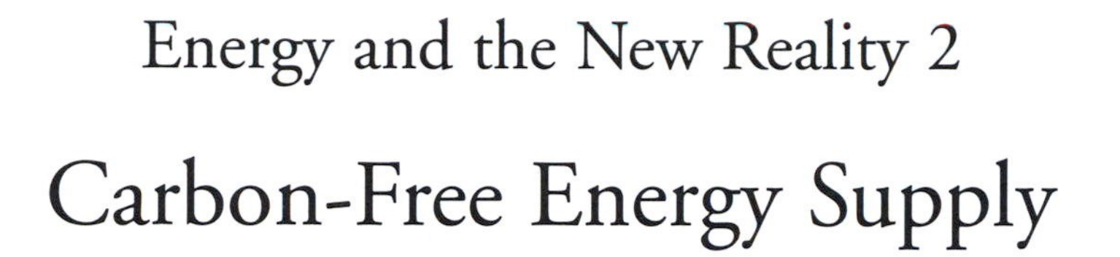

Energy and the New Reality 2

Carbon-Free Energy Supply

Energy and the New Reality 2

Carbon-Free Energy Supply

L. D. Danny Harvey

London • Washington, DC

First published in 2010 by Earthscan

Earthscan Ltd, Dunstan House, 14a St Cross Street, London EC1N 8XA, UK
Earthscan LLC, 1616 P Street, NW, Washington, DC 20036, USA

Earthscan publishes in association with the International Institute for Environment and Development

For more information on Earthscan publications, see www.earthscan.co.uk or write to earthinfo@earthscan.co.uk

ISBN: 978-1-84407-913-1 hardback
978-1-84971-073-2 paperback

Typeset by Domex e-Data, India
Cover design by Susanne Harris

A catalogue record for this book is available from the British Library

Library of Congress Cataloging-in-Publication Data
Harvey, Leslie Daryl Danny, 1956-
Energy and the new reality 2 : carbon-free energy supply / L.D. Danny Harvey.
p. cm.
Includes bibliographical references and index.
ISBN 978-1-84407-913-1 (hardback) – ISBN 978-1-84971-073-2 (pbk.) 1. Renewable energy resources. 2. Carbon dioxide mitigation. 3. Energy conservation. 4. Energy consumption. 5. Climatic changes–Prevention. I. Title.
TJ163.3.H3825 2010
333.79–dc22

2009053713

At Earthscan we strive to minimize our environmental impacts and carbon footprint through reducing waste, recycling and offsetting our CO_2 emissions, including those created through publication of this book. For more details of our environmental policy, see www.earthscan.co.uk.

This book was printed in the UK
by The Cromwell Press Group.
The paper used is FSC certified.

To all those dedicated to changing the world for the better

Contents

List of Figures, Tables and Boxes

Figures

Tables

Boxes

Preface

This book and the accompanying Volume 1 (*Energy Efficiency and the Demand for Energy Services*) are an attempt to objectively, comprehensively and quantitatively examine what it would take to limit the atmospheric CO_2 concentration to no more than 450ppmv. A 450ppmv (parts per million by volume) concentration limit is chosen not because I consider it to be a safe level (it most decidedly is not), but because it is the lowest concentration limit that I dare imagine that we could stay below (we are currently at 390ppmv and increasing by about 2ppmv per year). The first step in assessing how we could stay below this limit is to assess how small the global energy demand could be in the future through the assiduous application of all known and foreseeable energy efficiency measures. Thus, Volume 1 examines the technical potential for dramatic improvements in energy efficiency in a vast array of end uses, along with costs, co-benefits and practical considerations. Human population and the relentless pursuit of ever more material consumption are strong drivers of increasing greenhouse gas emissions, so alternative scenarios for the growth of regional population and GDP (gross domestic product) per person are also considered in Volume 1.

Volume 2 critically assesses the potential contribution, economics, implementation timescale and environmental and social issues related to the deployment of various forms of C-free energy supply. Here, the term 'C-free' includes potentially C-neutral energy sources such as biomass and fossil fuels equipped with CO_2 capture and storage. This book provides all the information needed for a solid understanding of the issues and complexities associated with all of the major and many of the minor potential C-free energy sources, and equips the more technically inclined reader with the tools needed to perform his or her own rough quantitative analysis of the C-free energy potential and cost. Two key conclusions are that solar, wind and biomass together are a viable alternative to expanded use of nuclear energy and use of carbon capture and storage, and that the transition to an energy system with very low to zero CO_2 emissions in time to limit atmospheric CO_2 to no more than 450ppmv can be achieved at an affordable cost if future population follows a trajectory near the low end of current scenarios and if there is substantial moderation in the rate of economic growth per person in the rich countries. For scenarios of high population and economic growth it will be substantially more difficult to satisfy a 450ppmv limit, and may not be possible at all.

These two books are a comprehensive blueprint concerning what needs to be done to solve the global warming problem (that is, which will stabilize climate at a warming that will still preserve much that is valuable and beautiful in the world). Nothing less than a complete and rapid transformation of our energy system, and indeed, of our deep-seated ways of thinking is required. However, the political and (in some cases) business response so far has been to consider incremental changes – adjustments – to what is still fundamentally a business-as-usual trajectory. There is still little evidence of a political acceptance of the nature and the magnitude of the changes needed. Global warming changes all the old rules about energy, economic growth and the jostling for perceived comparative advantage in international negotiations. There is a new 'reality', but it is a new reality that we, by and large, have not yet faced up to. It is high time that we did.

L. D. Danny Harvey

Department of Geography
University of Toronto
harvey@geog.utoronto.ca
March 2010

Online Supplemental Material

The following supplemental material can be accessed by visiting www.earthscan.co.uk/resources and selecting the link for this book:

- Powerpoint presentations for each chapter, containing figures, bullet points suitable for teaching purposes, selected tables and supplemental photographs;
- Excel-based problem sets for several chapters;
- supplemental tables;
- Excel spreadsheets used to generate the supply scenarios presented here.

Acknowledgements

I would like to thank the people listed below for kindly reviewing parts of (or, in some cases, all of) the indicated chapters, in many cases also providing additional information that I would not otherwise have obtained. As well, numerous people responded to questions about their work and in so doing contributed greatly to the final product. I accept responsibility, however, for any errors, misconceptions or important omissions that may remain.

Chapter 2	Steve Hegedus (University of Delaware), S. A. Kalogirou (Higher Technical Institute, Nicosia)
Chapter 3	Al Cavallo (Department of Homeland Security, New York), Gregor Giebel (Risø National Laboratory, Denmark), Tariq Iqbal (Memorial University, St John's, Canada)
Chapter 4	Andrew Knox (University of Toronto), Stefan Wirsenius (Chalmers University of Technology, Sweden)
Chapter 5	John Lund (Oregon Institute of Technology)
Chapter 6	Patrick McCully (International Rivers, Berkeley)
Chapter 7	Wilfried van Sark (Utrecht University)
Chapter 8	Dean Abrahamson (Professor Emeritus, University of Minnesota), Andrew Knox (University of Toronto), Storm van Leeuwen (Ceedata Consultancy, The Netherlands)
Chapter 9	John Davison (IEA Greenhouse Gas R&D Programme), Derek Paul (Professor Emeritus, University of Toronto)
Chapter 10	Ulf Bossel (European Fuel Cell Forum, Switzerland)

Special thanks is also extended to Richard Harrod of 4COffshore Ltd, UK, for preparing Figure 3.28 specifically for this book.

Chapter Highlights

Chapter 1 Introduction and key points from Volume 1

Significant and widespread negative impacts can be expected in association with continued emissions of greenhouse gases. Reducing these risks to a meaningful degree requires that human emissions be eliminated before the end of this century. Such a rapid reduction in CO_2 emissions will probably be sufficient to limit the atmospheric CO_2 concentration to no more than 450ppmv (parts per million by volume), compared to a pre-industrial concentration of 280ppmv and a concentration in 2010 of 390ppmv. Depending on how quickly we act, how large the climate sensitivity is and how sensitive the Greenland and West Antarctic ice sheets are to warming, it may or may not be possible to avoid triggering an eventual 10m or more sea level rise.

Assuming full implementation between now and 2050 of the energy savings potential identified in Volume 1, and taking into account structural shifts in national economies as wealth increases, the primary energy intensity of the global economy (primary energy use per unit of GDP) would decrease at an average annual compounded rate of 2.7 per cent/year between 2005 and 2050. Over the period 2005–2100, the primary energy intensity of the global economy could improve at an average compounded rate of 1.8 per cent/year. Depending on the growth in the human population and in average GDP per person, this implies that 1–5TW of additional C-free primary power will be needed by 2050 if the atmospheric CO_2 concentration is to peak at no more than 450ppmv. The total C-free primary power in 2050 would be 4–8TW. By comparison, the world primary power demand in 2005 was 15.3TW, of which 3.3TW were from C-free energy sources. To the extent that a 2.7 per cent/year average annual improvement in energy intensity is not achieved between now and 2050, more than 4–8TW of new C-free power will be needed by 2050.

Chapter 2 Solar energy

The amount of solar radiation intercepted by the earth is over 11,000 times present world primary energy demand. Solar energy can be used passively (to provide heat, ventilation and light) as well as actively (to produce electricity and hot water). Electricity can be generated from solar energy through large, centralized arrays of photovoltaic (PV) modules in sunny areas, using either flat collectors or concentrating collectors; through PV panels that are incorporated into the roofs and façades of buildings (a system called building-integrated PV or BiPV); and by concentrating solar energy with mirrors to make steam for use in a steam turbine (a system called concentrating solar thermal power or CSTP). PV electricity is currently expensive (25–30 cents/kWh in sunny locations), but costs can probably be reduced by a factor of two over the next decade or two. There are a number of different and promising PV technologies that could achieve these costs. Many PV technologies rely on rare elements, but by concentrating sunlight onto the PV modules by a factor of 100 or so, the required use of rare elements can be reduced by a similar factor, thereby allowing many different PV technologies to supply a significant fraction of future world electricity demand. BiPV is particularly attractive because it would generate electricity in the centre of the major demand centres at times of peak demand, thereby alleviating transmission bottlenecks. BiPV alone could provide 15–60 per cent of total present-day electricity demand in individual OECD (Organisation for Economic Co-operation and Development) countries.

CSTP is also attractive. A number of different CSTP technologies are under development, and it is reasonable to expect that electricity costs of 5–10 cents/kWh can be achieved in two decades. CSTP requires direct-beam solar radiation and so can be applied only in arid and semi-arid regions. However, by storing solar heat in hot molten salt or other media, electricity can be generated 24 hours per day and shorter-term intermittency in the supply of electricity from sunlight can be largely eliminated.

Solar energy can be used to produce hot water for household consumption, space heating and to drive various air conditioning and dehumidification systems. The first application is already economically attractive in many parts of the world, while solar air conditioning and dehumidification are still expensive. Solar heat can be provided at the range of temperatures needed for a wide range of industrial processes (including processing of metal ores and fixation of atmospheric nitrogen to produce nitrogen fertilizer), crop drying and cooking. The time required for the energy saved with various solar technologies to pay back the energy invested in building solar energy systems is two years or less in moderately sunny locations (southern Europe/northern US) and in the near future should drop to less than one year.

Chapter 3 Wind energy

Wind energy is a rapidly growing C-free energy source for the generation of electricity, the installed capacity having grown at an average compounded rate of 25 per cent/year from 1994 to the end of 2008. This rapid growth has been due to its relatively low cost among renewable-energy options, with electricity costs reaching a highly competitive 5–8 cents/kWh in many jurisdictions. The potential wind resource at a cost of up to 7 cents/kWh (taking into account expected future cost reductions) is probably greater than present world electricity demand, and larger still with higher cost ceilings. The difficulties with regard to wind relate to its intermittency and to the fact the some of the best wind regions are 1000–3000km from major electricity demand centres.

There are a number of strategies available for dealing with wind variability. Fluctuations at a timescale of seconds can be reduced by aggregating the output of several wind turbines in a wind farm, or by using short-term storage such as flywheels, supercapacitors and superconducting magnetic storage. Fluctuations at longer timescales will be reduced when geographically dispersed wind farms are linked together. Underground compressed air energy storage (CAES) is a proven technology that could be used for storing excess wind energy on a large scale and generating electricity when needed. Current CAES systems require supplemental fuel (natural gas now, biogas in the future), but advanced adiabatic CAES systems under development would store the heat that is generated when air is compressed and use this, along with the compressed air, to generated electricity later without the use of supplemental fuel. Flexible loads, including heat pumps with thermal energy storage, can also be employed to match fluctuations in wind energy. Parked plug-in hybrid vehicles could provide important regulation (voltage and frequency control) and storage functions.

The average output of a wind turbine is typically only 20–40 per cent that of the peak output, so it is possible to oversize a wind farm compared to the transmission capacity with little wasted electricity-generation potential. Although wasting some potential generation due to transmission limitations would increase the generation component of the electricity cost, it would reduce the transmission component of the cost because more energy would be transmitted through the given transmission link. By locating oversized wind farms in the regions with the best winds and transmitting electricity with high-voltage DC power lines, average power output as a fraction of peak power in the order of 70 per cent can be achieved at a cost in the order of 8 cents/kWh, which is comparable to that expected from wind farms located next to major demand centres but with much less favourable winds.

Chapter 4 Biomass energy

Biomass energy is perhaps the most complicated of the (net) C-free energy options. Complications arise from the many different possible kinds of biomass that can be used, the different forms in which biomass can be used (as solid, liquid or gaseous fuels), the variety of energy conversion processes available, the many potential end-use applications, the potential adverse and beneficial environmental impacts, social considerations through competition with land for food and impacts on the price of food crops, and uncertainties related to long-term sustainability and the impacts of climatic change during the coming century and beyond.

One of the most difficult issues with regard to biomass energy is the determination of the net energy gain using biomass. In spite of the resulting uncertainties, there are clear differences in the relative benefits of using biomass in different applications. Use of biomass as a solid fuel is superior to its conversion to liquid fuels. However, liquid fuels from biomass represent one of the few options for displacing petroleum-based fuels in the transportation sector.

The use of corn (maize) to produce ethanol is the least effective of the possible uses of biomass, and is also the least effective of the possible ways of reducing transportation energy use. Ethanol from sugarcane is substantially more attractive, as would be ethanol from lignocellulosic materials (grasses such as switchgrass or *Miscanthus*, corn stover and wood) if this technology (which is still under development) performs as projected. Direct use of woody biomass as a fuel for heating, or for subsequent gasification and use in the generation of electricity or for cogeneration, provides the greatest net energy benefit. If biomass used for electricity generation displaces coal, this would provide the largest CO_2 emission reduction per unit of biomass grown.

The use of plantation-scale biomass energy is fraught with numerous potential environmental and social problems. Its development should therefore proceed in a gradual, well-planned manner and with minimal government subsidies so as to avoid distortions that create unexpected and undesirable side-effects. Enforceable restrictions in the development of bioenergy resources will be needed so as to prevent destruction of forests or other natural areas that are better left intact.

Biomass energy could supply a significant fraction of future energy needs and at attractive costs in the long run (generally \$3–6/GJ for solid fuels, less than \$15/GJ for liquid fuels and about 5 cents/kWh for electricity). Most of this biomass will have to come from bioenergy plantations, but if existing forests are to be protected, the plantations will have to be limited to surplus agricultural and grazing lands. Whether or not there are surplus lands in the future will depend on (1) the future human population, (2) future diet (in particular, the proportion of food energy provided by meat and the kinds of meat consumed), (3) future agricultural productivity, and (4) the future efficiency in converting animal feed into animal food products. Diet emerges as a significant factor in determining both the future potential of biomass energy and the environmental impacts associated with the food system.

Chapter 5 Geothermal energy

Geothermal energy occurs in several different forms: hydrothermal, geopressurized, hot dry rock and magma. Geothermal heat can be used directly to provide domestic hot water and space heating, or can be used indirectly to produce electricity by first generating steam that is used in a steam turbine. Hot dry rock systems (also known as enhanced geothermal systems (EGSs) involve (1) drilling an injection borehole, down which water will be injected; (2) fracturing the rock in the region at the bottom of the injection borehole, so that water can flow through the rock and become hot in the process; and (3) drilling an extraction borehole in the vicinity of the injection borehole, from which the heated water will be drawn. The key for geothermal energy to make a significant contribution to world energy needs is the development of hot dry rock technology. If this technology can be developed according to industry projections, geothermal energy could provide one to several 100EJ/yr for many centuries. Electricity costs of 8–10 cents/kWh might be feasible. In heating applications, the geothermal resource could be significantly extended through the use of heat pumps to extract additional heat from the already-used water that is returned to the ground.

Chapter 6 Hydroelectric power

Existing hydroelectric powerplants already constitute a large renewable source of electricity – about 16 per cent of current global electricity demand. Projects under construction or planned will increase electricity production by about 50 per cent, while the economic potential is more than three times current global hydroelectric electricity supply under current conditions. However, not all of the economic potential can be developed, because of adverse environmental or social impacts. Further hydro development should be critically scrutinized to make sure that there is a large benefit to cost ratio, including minimal greenhouse gas emissions from decomposition of organic matter in flooded terrain. Hydropower projects should only go forward based on negotiated and legally binding agreements with the affected people based on their free and informed consent. Displaced peoples should be fully and fairly compensated, and the cost of such compensation should be included in the projected cost of hydroelectric energy and taken into account, along with the cost of other renewable or C-free supply alternatives and energy efficiency measures, when deciding whether or not to proceed. Hydroelectric developments are particularly advantageous when they serve to compensate for variable wind or solar electricity production, as they then serve to leverage greater renewable energy contributions without requiring greater fossil fuel spinning reserve.

Chapter 7 Ocean energy

Ocean energy in the form of waves, tides, tidal currents and energy from vertical thermal and salinity gradients can be converted to electricity. All of these forms of energy are at a very early stage of development, and intensive research would be needed to bring them to commercial viability. The greatest challenge is to build systems that work reliably for 10–20 years in a continuously harsh environment. The smallest technical glitch can undermine otherwise promising designs. The various forms of renewable energy from the oceans are not likely to form a large fraction of the renewable energy mix, except locally where there are particularly good and economical energy resources. The technical potential is largest (up to several times current global electricity demand) for wave energy, but the realizable potential is likely to be a very small fraction of the technical potential. Ocean thermal energy conversion (OTEC) could be broadly applicable in low-latitude regions and could make an important contribution but requires significant further research, development and testing before a clearer picture of its potential emerges.

Chapter 8 Nuclear energy

Nuclear energy is beset with a number of adverse environmental impacts and risks, concerns over nuclear weapons proliferation and terrorism and the continuing inability to find an acceptable long-term method for the isolation of high-level nuclear waste. The probability of a serious accident during the operation of nuclear powerplants is thought to be extremely low as long as proper procedures are followed and there is a culture of safety. However, these conditions may break down somewhere at some point in time.

It will be difficult for nuclear energy to play a significant role in reducing greenhouse gas emissions at a global scale. Indeed, the industry will be hard pressed to maintain the current electricity production over the coming two decades. Currently, nuclear energy provides 16 per cent of world electricity demand. However, the current fleet is old (average age is 25 years), and a new reactor would have to be built once every five weeks over the next ten years and once every 22 days over the following decade just to maintain the current capacity.

If, in spite of the above, the nuclear power capacity were to increase to 1500GW (almost four times the current capacity of 370GW) by 2050, known and speculative uranium supplies available at a cost of $130/kg or less would last about 40 years with once-through use of uranium fuel. The supply could be expanded by up to a factor of 75 with reprocessing of spent fuel to separate and use plutonium in fast breeder reactors, but this would create enormous risks of nuclear materials sufficient to build a crude bomb getting into the hands of terrorists. Alternative reprocessing schemes that would provide some resistance to proliferation of nuclear weapons and that would stretch the existing uranium supply (to perhaps the end of this century for a nuclear capacity of 1500GW) have been proposed. However, these would require significant technological development.

The savings in primary energy when nuclear-generated electricity displaces fossil fuel-generated electricity at 40 per cent efficiency, divided by the primary energy inputs throughout the nuclear lifecycle from mining to decommissioning and isolation of wastes, averages about 15 at present. This ratio, referred to as the energy return over energy invested (EROEI), is quite uncertain. However, the EROEI decreases as the grade of ore used decreases because the mass of ore that must by mined and processed, and the mass of tailing that must be treated for land reclamation, increases faster than the inverse of the ore grade. By the time uranium prices have reached $130/kg, the grade of much of the remaining ore will probably have dropped to about 0.02 per cent and the EROEI will have dropped to about three to five (except for uranium that is co-mined with other minerals). Further decreases in ore grade would see an accelerating drop in the EROEI.

The cost of new nuclear powerplants is uncertain. Recent assessments indicate a cost of about $4000/kW to over $10,000/kW. Decommissioning costs are also uncertain but are likely to be at least as expensive as the lower estimates of the cost of constructing a nuclear powerplant.

Development of a new generation of nuclear reactors will require another 15–20 years or more to reach the pilot demonstration stage, and then perhaps ten years of operation before such reactors could begin to be produced on a large scale. Another 20 years or more would be required before the new generation could provide a significant fraction of the total nuclear fleet. Thus, a major change in the technology of the operating fleet cannot be expected before 2050 at the earliest.

Chapter 9 Carbon capture and storage

Capture of carbon dioxide from new fossil fuel powerplants using existing technologies would increase fuel requirements by 11–40 per cent according to various estimates, while retrofitting existing coal-fired plants is estimated to increase fuel requirements by 43–77 per cent. With future technologies the energy penalty in new plants might be reduced to 2–12 per cent. Additional energy would be required to compress or liquefy and transport the captured CO_2 to its disposal sites. Costs are highly uncertain but are likely to be large, increasing the cost of new coal or natural gas powerplants by 50–100 per cent.

Potential sites for storage of CO_2 include depleted oil and gas fields, deep saline aquifers, coal beds, sediments in the seabed and deep ocean water itself. The amount of CO_2 that can be securely stored in terrestrial sediments or coal beds is highly uncertain. Among the environmental issues associated with sequestration on land are potential leakage through the thousands of drill wells that are found in most populated regions, potential displacement of saline groundwater into freshwater groundwater supplies and mobilization of toxic elements in saline aquifers due to the increase in groundwater pH associated with CO_2 injection. Other issues are the preclusion of the future use of saline groundwater through desalination, preclusion of future mining of saline groundwater for trace elements and interference with compressed air energy storage or geothermal energy. Disposal of CO_2 in the oceans as anything more than a supplement to a major shift from fossil fuels to renewable energy sources is not acceptable. However, burial in sub-seabed aquifers would probably pose negligible environmental risks and risks of leakage, but is also likely to be the most expensive storage option.

At least 20 years of demonstration projects involving carbon capture from powerplants using a variety of different types of coal, and carbon storage in a variety of different geological settings, would be required before large-scale deployment of carbon capture and storage (CCS) could begin. Another 20 years would be required before a significant fraction of the world's powerplants would (through normal retirement and replacement) be equipped with CCS. Thus, even if it proves to be viable, CCS could not make a significant difference before mid-century. Carbon sequestration could nevertheless be used as an emergency measure to accelerate the later stages in the phase-out of fossil fuel CO_2 emissions. In conjunction with the capture of CO_2 released from the use of biomass, it could create negative CO_2 for many decades, if this is needed in order to reduce atmospheric CO_2 concentration.

Chapter 10 The hydrogen economy

Hydrogen has the potential to serve as an energy currency, replacing fossil fuels in all the ways in which they are used for energy today and, in combination with biomass, replacing them as a chemical feedstock. Hydrogen could be produced by electrolysis of water when solar- and wind-generated electricity are in excess, stored and later used in a fuel cell to produce electricity when there is a shortage of wind or solar power. It could also be transported from distant sunny or windy regions to demand centres. In this way, it could serve to close the spatial and temporal mismatch between intermittent renewable sources of electricity and the demand for electricity. However, much of this gap could also be closed through alternative strategies, such as long-distance high-voltage DC power transmission to link regions of good wind and solar energy with dispatchable hydroelectric facilities and electricity demand centres, and by using CAES.

The use of hydrogen in automobiles presents major challenges, particularly concerning the current high cost of fuel cells and onboard storage, as well as the difficulties of creating a whole new infrastructure to supply hydrogen fuel. Nevertheless, there are indications that the total cost of driving using hydrogen in fuel cell vehicles could become competitive with gasoline or diesel alternatives. Advanced hydrogen fuel cell vehicles are projected to be almost four times as efficient (in terms of km driven per unit of energy in the fuel tank) as current vehicles and about 1.75 times as efficient as advanced gasoline vehicles, which reduces the problem of the much greater bulk of hydrogen storage systems compared to gasoline. However, the overall effectiveness in using renewably based electricity to make hydrogen (produced by electrolysis) for subsequent use in automobiles would be only about half that of the direct use of renewably based electricity to charge batteries. Thus, the optimal solution will probably be a plug-in hybrid vehicle using onboard hydrogen in a fuel cell only to give extended driving range. Hydrogen as a transportation fuel is more promising in various niche applications, such as in railway locomotives and to power auxiliary power units in trucks. There are also important industrial niche applications for hydrogen in a fossil fuel-free world, such as its use as a reducing agent in the manufacture of iron (in place of coke), in the manufacture of nitrogen fertilizer (in place of natural gas or coal) and in the manufacture of a variety of biomass-based chemicals (in place of petroleum).

Chapter 11 Community-integrated energy systems with renewable energy

Community-integrated energy systems involve centralized production of heat and possibly chilled water that are distributed to individual buildings through district heating and cooling networks. District heat networks can be coupled with large-scale underground storage of heat that is collected from solar thermal collectors during the summer and used for space heating and hot water requirements during the winter. Heat can also be supplied with biomass (as part of a biomass cogeneration system) or from geothermal heat sources. If both heat and coldness are stored, then heat pumps can be used to recharge the thermal storage reservoirs (or to directly supply heat or coldness to the district heating and cooling networks) during times of excess wind energy. This in turn permits sizing of the wind system to meet a larger fraction of total electricity demand without having to discard as much (or any) electricity generation potential during times of high wind and/or low demand. In the long run, district heating systems with cogeneration will make it easier to make the transition to a hydrogen economy, as a new infrastructure to supply hydrogen to individual buildings would not be needed.

Chapter 12 Integrated scenarios for the future

Based on the observation that almost no region in the world is more than 3000km from regions of either good winds or semi-arid or arid regions where CSTP is applicable (and most regions are no more than 2000km from such sites), supply scenarios consisting of the following elements are constructed (with the amounts depending on the demand scenario):

- 6000–12,000GW of wind energy capacity;
- 6000–12,000GW of CSTP capacity;

- 3000–4000GW of BiPV or PV capacity;
- minor amounts of geothermal, biomass and additional hydroelectric generation;
- interconnection of the major renewable energy source regions and the major demand with a high-voltage DC power grid;
- retirement of all existing nuclear reactors at the end of an assumed 40-year lifespan.

Two scenarios for the supply of fuels for transportation, heating and industry are considered: a hydrogen-intensive scenario and a biomass-intensive scenario, with hydrogen produced through some combination of low-temperature electrolysis largely from wind-generated electricity and high-temperature electrolysis using CSTP-generated electricity. Rates of increase in the supply of C-free energy and in C-free fuels are prescribed so as to completely eliminate fossil fuel emissions by around either 2080 or 2120, with emissions in 2050 ranging from a 15 per cent increase to an 85 per cent decrease compared to emissions in 2005 (depending on the demand scenario).

Annual material flows required to build up the renewable energy system in these scenarios are not excessive compared to current material flows, and all supply components of the system quickly become a significant net source of energy. However, water could be a significant limiting factor for the biomass-intensive scenarios. The amount of biomass required in the biomass-intensive supply scenario and the higher demand scenarios considered here is unlikely to be available unless there is a worldwide shift to diets with low meat consumption.

The CO_2 emissions produced from these scenarios are used as input to a simple coupled climate–carbon cycle model in order to calculate changes in atmospheric CO_2 concentration, global mean temperature and acidification of the oceans. The CO_2 concentration peaks at values of about 430–530ppmv and, for the lowest emission scenario considered here (peaking at 8.4GtC/yr in 2015), global mean warming peaks at a value of 1.2–3.7°C before slowly declining and ocean surface water pH declines by 0.13 relative to the pre-industrial value. For the highest emission scenario considered here (peaking at 11.5GtC/yr near 2030), global mean warming peaks at a value of 1.6–4.9°C before slowly declining and ocean surface water pH declines by 0.26 relative to the pre-industrial value. If CO_2 can be removed from the atmosphere at a rate of 1GtC/yr by 2050 through geological or biological sequestration, and the sequestration sustained at this rate, then, for the lowest emission scenario, the atmospheric CO_2 concentration, global mean warming and ocean surface water pH would return to close to present conditions by the year 2500 if there are no major releases of methane or other positive climate–carbon cycle feedbacks between now and then. Otherwise, yet larger rates of CO_2 sequestration would be required, but there is a risk that methane emission rates could be such that countermeasures would be ineffective, causing global warming to slip beyond human control, with globally catastrophic impacts.

Chapter 13 Policy sketch and concluding thoughts

Although great strides have been made in the development of C-free technologies for electricity generation, research and development is needed in many areas in order to improve performance and bring down costs. The development of 'clean' coal technologies and carbon capture and storage for coal, in contrast, is not recommended. Instead, the overarching policy goal should be to phase out the use of coal altogether as rapidly as possible. Similarly, research and development related to nuclear energy should be terminated, with the possible exception of research related to the use of nuclear powerplants to consume discarded plutonium and highly enriched uranium from nuclear weapons.

We need three large transformations: a transformation to vastly greater levels of energy efficiency than at present in all end-use sectors, a rapid deployment of C-free energy sources (primarily wind and solar energy), and a whole new way of thinking that places stabilization of greenhouse gas concentrations at levels below the equivalent of a CO_2 doubling ahead of promotion of economic growth.

List of Abbreviations

°C	degrees Celsius
µm	micron
3D	three dimensional
AA-CAES	advanced adiabatic compressed air energy storage
AC	alternating current
AEI	average energy intensity
AGR	advanced gas-cooled graphite reactor
ATES	aquifer thermal energy storage
BAU	business-as-usual
BEV	battery-electric vehicle
BIGCC	biomass integrated gasification combined cycle
BiPV	building-integrated photovoltaics
BLGCC	black liquor gasification combined cycle
BOM	balance of module
BOS	balance of system
Bq	becquerel
BSF	back surface field
BTES	borehole thermal energy storage
BWR	boiling-water reactor
C	carbon
CAES	compressed air energy storage
CANDU	Canadian deuterium uranium
CBM	coal bed methane
CBP	consolidated biomass processing
CCS	carbon capture and storage
CDCS	carbon dioxide capture and storage
CFC	chlorofluorocarbon
CH_4	methane
CHP	combined heat and power
Ci	curie
CIGS	copper-indium-gallium-diselenide
CIS	copper-indium-diselenide
CLC	chemical looping combustion
CO	carbon monoxide
CO_2	carbon dioxide
COP	coefficient of performance
CPC	compound parabolic collector
CRF	cost recovery factor
CSTP	concentrating solar thermal power
CTES	cavern thermal energy storage
DC	direct current
DDG	dried distillers' grains
DHW	domestic hot water
DME	dimethyl ether

DSG	direct steam generation
DSG	distiller grains and solubles
DU	depleted uranium
EER	energy efficiency ratio
EGS	enhanced geothermal system
EJ	exajoules
EOR	enhanced oil recovery
EPC	engineering, procurement and construction
EPR	European pressurized reactor
EROEI	energy return over energy invested
ESTIA	European Solar Thermal Power Industry Association
ESTIF	European Solar Thermal Industry Federation
EU	European Union
FBR	fast breeder reactor
FCV	fuel cell vehicle
FFB	fresh fruit bunch
FRT	fault ride through
FSU	former Soviet Union
FT	Fischer-Tropsch
GCM	general circulation model
GDP	gross domestic product
GEMIS	Global Emission Model for Integrated Systems
GFDL	Geophysical Fluid Dynamics Laboratory
Gg	gigagram
GHG	greenhouse gas
GIF	Generation IV International Forum
GIS	geographical information system
GJ	gigajoule
gm	gram
GRP	gross regional product
GSHP	ground-source heat pump
Gt	gigatonne
GTCC	gas turbine combined cycle
GW	gigawatt
GWP	global warming potential
HCFC	hydrochlorofluorocarbon
HEU	high-enriched uranium
HEV	hybrid electric vehicle
HF	hydrogen fluoride
HFC	hydrofluorocarbon
Hg	mercury
HHV	higher heating value
HOMER	Hybrid Optimization Model for Electric Renewables
HRT	hydraulic retention time
HTGR	high-temperature gas-cooled reactor
HTU	high thermal upgrading
HVAC	heating, ventilation and air conditioning
HVAC	high-voltage alternating current

HVDC	high-voltage direct current
HWR	heavy-water reactor
ICE	internal combustion engine
IEA	International Energy Agency
IFR	integrated fast reactor
IGBT	insulated gate bipolar transistor
IGCC	integrated gasification combined cycle
IMAGE	Integrated Model to Assess the Global Environment
IOA	input–output analysis
IPCC	Intergovernmental Panel on Climatic Change
IRR	internal rate of return
ISCC	integrated solar combined cycle
ISL	in situ leaching
ITO	indium tin oxide
J	joule
K	kelvin
kHz	kilohertz
kg	kilogram
km	kilometre
kph	kilometres per hour
kt	kilotonne
kW	kilowatt
kWh	kilowatt hour
LDPE	low-density polyethylene
LET	linear energy transfer
LEU	low-enriched uranium
LH_2	liquid hydrogen
LHV	lower heating value
LNG	liquefied natural gas
LOLP	loss-of-load-probability
LPG	liquefied petroleum gas
LWR	light-water reactor
m	metre
mb	millibar
MCFC	molten carbonate fuel cell
MCM	mixed conducting membrane
MEA	monoethylamine
mg	milligram
MHD	magneto-hydrodynamic
MIT	Massachusetts Institute of Technology
MJ	megajoule
mm	millimetre
MOX	mixed oxide
mpg	miles per gallon
MPP	maximum power point
MSF	multi-stage flash
N	nitrogen
NCAR	National Center for Atmospheric Research
NGCC	natural gas combined cycle

NIR	near-infrared
NMHC	non-methane hydrocarbon
NO_x	nitrogen oxide
N_2O	nitrous oxide
NPP	net primary production
NREL	National Renewable Energy Laboratory
O_3	ozone
O&M	operation and maintenance
OD	outside diameter
odt	oven-dry tonne
OECD	Organisation for Economic Co-operation and Development
OMR	operation, maintenance and refurbishment
OTEC	ocean thermal energy conversion
OWC	oscillating water column
OWSC	oscillating wave surge converter
PAH	polycyclic aromatic hydrocarbon
PBL	planetary boundary layer
PC	pulverized coal
PCA	process chain analysis
PCM	phase-change material
PEC	photoelectrochemical cell
PEM	proton exchange membrane
PEMFC	proton exchange membrane fuel cell
PET	polyethylene terephthalate
PFCHEV	plug-in fuel cell hybrid electric vehicle
PGF	pressure gradient force
PHA	polyhydroxyalkanate
PHEV	plug-in hybrid electric vehicle
PICHTR	Pacific International Center of High Technology Research
PLA	polylactic acid
PP	polypropylene
ppmv	parts per million by volume
PPP	purchasing power parity
PRIS	Power Reactor Information System
PRO	pressure-retarded osmosis
PSA	Plataforma Solar de Almería
psi	pounds per square inch
Pt	platinum
PTT	polytrimethylene terephthalate
PUREX	plutonium uranium recovery by extraction
PV	photovoltaic
PVPS	PV Power Systems
PV/T	photovoltaic/thermal
PWR	pressurized-water reactor
Q	solar constant
RAR	reasonably assured resources
RCG	reed canary grass
RFC	reversible fuel cell
RMBK	reaktor bolshoy moshchnosti kanalniy

RO	reverse osmosis
rpm	rotations per minute
S	sulphur
SACE	Solar Air Conditioning in Europe
SEGS	solar electricity generating system
SEIA	American Solar Energy Industries Association
SMR	steam methane reforming
SO_2	sulphur dioxide
SOFC	solid oxide fuel cell
SRCW	short-rotation coppice willow
SRES	Special Report on Emission Scenarios
SRT	solids retention time
SSF	simultaneous saccharification and fermentation
SUV	sport utility vehicle
SWU	separative work unit
TCO	transparent conductive oxide
TIM	thermal insulation material
TW	terawatt
TWh	terawatt hour
UNFCCC	United Nations Framework Convention on Climate Change
UOX	uranium oxide
USGS	United States Geological Survey
UTES	underground thermal energy storage
V	volt
V2G	vehicle-to-grid
VRB	vanadium redox battery
VSC	voltage source converter
VVER	voda-vodyanoi energetichesky reaktor
W	watt
WinDS	Wind Energy Deployment System Model
W/m^2	watts per square metre
yr	year

1

Introduction and Key Points from Volume 1

This introductory chapter briefly outlines the scientific basis for great concern over the global warming issue and explains why greenhouse gas (GHG) emissions must be reduced dramatically and with the greatest urgency. A framework for analysing the driving factors for industrial CO_2 emissions is then presented, followed by a brief summary of the key points from Volume 1.

1.1 The scientific basis for concern about global warming

Emissions of carbon dioxide (CO_2) due to industrial activities and land use changes during the past 200 years have caused the CO_2 concentration in the atmosphere to increase by 40 per cent, from about 280ppmv (parts per million by volume) in 1800 to 390ppmv by 2010. Other GHGs have increased as well – methane (CH_4) by a factor of 2.5, nitrous oxide (N_2O) by 15 per cent and ozone (O_3) in the lower atmosphere by a factor of two to five (depending on location). A number of entirely artificial GHGs have been added to the atmosphere as well. All of these gases trap radiant heat emitted from the earth's surface and so have a warming effect on the climate. The collective heat trapping of all the GHG increases so far is equivalent to that which would arise from roughly an 80 per cent increase in CO_2 alone.

The key parameter in the science of global warming is the *climate sensitivity*, defined as the global average temperature change resulting from a fixed doubling in the atmospheric CO_2 concentration, once the climate system has had a chance to fully adjust to the increase in CO_2. There are several independent lines of evidence (computer simulations with three-dimensional (3D) climate models, analyses of temperature changes during the past 150 years, analysis of times in the geological past when climates were quite different from today, and analysis of past slow natural variations in atmospheric CO_2) that all agree in indicating that the climate sensitivity very likely lies somewhere between 1.5 degrees Celsius (°C) and 4.5°C. Thus, with the prospect of GHG concentrations reaching the equivalent of four times the pre-industrial CO_2 concentration or larger by the end of the century under business-as-usual (BAU) emission scenarios, we face the prospect of a commitment to 3–9°C by the end of this century in the global average and a realized warming of 2–6°C. The average warming would be greater over land areas and greater still in polar regions.

The current atmospheric GHG concentrations, if sustained, would probably eventually warm the climate by 1.2–3.6°C, whereas only 0.8°C global average warming has been observed so far. The observed warming is smaller than the projected warming because pollution in the form of reflective aerosols (which are produced largely from the emission of sulphur and nitrogen oxides in association with the use of fossil fuels) has offset up to half of the heating effect from the concurrent GHG buildup, and because the oceans delay the surface warming due to the mixing of heat to great depths. The reflective aerosols are themselves associated with acid rain, so as acid rain emissions are cleaned up, or when fossil fuel use eventually decreases, there will be an acceleration in the warming as the climate system attempts to 'catch up' to the GHG concentrations already found in the atmosphere.

As discussed in more detail in Volume 1 (Chapter 1, subsection 1.4.1), there is a continuously increasing risk of serious impacts as the projected warming increases from 1 to 3°C and beyond. In particular:

- widespread extinction of coral reef ecosystems with 1–2°C warming above pre-industrial levels;
- possible collapse of the Greenland ice sheet with as little as 1–2°C sustained global average warming, and almost certainly with 3–4°C sustained warming;
- likely extinction of 15–30 per cent of species of life on earth due to a mere 2°C warming happening by 2050, with greater losses with greater warming;
- reduced agricultural productivity in some major food-producing regions once local warming exceeds 1–3°C (depending on the crop and location);
- severe water shortages in semi-arid regions, including the eventual loss of glacier meltwater that supplies summer water to 25 per cent of the population of China;
- acidification of the oceans as CO_2 is absorbed by the oceans, with severe and still poorly understood impacts.

Industrial CO_2 emissions reached about 8 gigatonnes of carbon per year (GtC/yr) in 2005 and land use emissions were about 1–2GtC/yr. Total human emissions of CO_2 are thus about 9–10GtC/yr, while the observed annual increase is about 4–5GtC/yr. The balance is absorbed through a combination of increased rates of photosynthesis on land and through net diffusion of CO_2 into the ocean water and its conversion to other forms of dissolved carbon (C). These are referred to as the terrestrial biosphere and oceanic *sinks*. In the absence of changes in the strength of the sinks, stabilization of the atmospheric CO_2 concentration requires reducing total emissions only to the point where they equal the total sink strength (a reduction of about 5GtC/yr, of which 1–2GtC/yr could be achieved by ending deforestation). However, once the atmospheric CO_2 concentration stops growing the sinks would themselves weaken over time even in the absence of climatic change. Warming induced by the GHG increases will further weaken the sink strength, such that stabilization of atmospheric CO_2 at a concentration of 450ppmv requires the near elimination of human CO_2 emissions by the end of this century. Inasmuch as a CO_2 concentration of 450ppmv in combination with minimal increases in other GHGs is the equivalent of a doubling in CO_2 concentration, and given that this is very likely to cause anywhere from 1.5°C to 4.5°C warming, with the consequences outlined above, it is clear that even an atmospheric CO_2 concentration of 450ppmv is too large. Thus, efforts will almost certainly be needed to draw down the concentration as rapidly as possible, but significant negative impacts, ecosystem losses and species extinctions will be unavoidable during the process. Depending on how quickly we act, how large the climate sensitivity is and how sensitive the Greenland and West Antarctic ice sheets are to warming, it may or may not be possible to avoid triggering an eventual 10 metres (m) or so sea level rise.

1.2 Kaya identity and efficiency versus C-free energy tradeoffs

Future CO_2 emissions, expressed as a mass of carbon per year, can be written as the product of the following factors:

population × GDP (gross domestic product) per year per capita × primary energy per unit GDP × carbon emission per unit of primary energy used

or, in shorthand notation:

$$\text{total emission} = P \times (\$/\text{yr})/P \times E/\$ \times C/E \qquad (1.1)$$

where $E/\$$ (primary energy use per unit of GDP) is referred to as the *energy intensity* of the economy and C/E (carbon emission per unit of energy) is referred to as the *carbon intensity* of the energy system.[1] This decomposition of CO_2 emission is referred to as the *Kaya identity*, in honour of the Japanese scientist who first proposed it.

In Volume 1 (Chapter 10, subsection 10.2.1) we applied the Kaya identity at the global scale, considering three scenarios of global population, two scenarios of growth in GDP/capita (one with continuous growth at 1.6 per cent/year, the other where the rate of growth decreased linearly from 1.6 per cent/year in 2000 to 0.8 per cent/year growth in 2100), various rates of decrease in energy intensity (1 per cent/year and 2 per cent/year), and various rates of increase in C-free power. The product of the first two terms of the Kaya identity gives gross world product – the size of the global economy – while the product of the first three terms gives the amount of primary energy (in J, joules) required per year. The primary energy requirement in joules per year can be divided by the number of seconds per year to give the primary power requirement (W, watts), then divided by 10^{12} (the number of joules in one terawatt) to give the primary power requirement in terawatts (TW).[2] Figure 1.1a shows the variation in gross world product for the low and high rates of GDP/capita growth each combined with the low and high population scenario, while Figure 1.1b shows the variation in primary power for low GDP/capita and low population growth and for high GDP/capita and population growth, each combined with slow (1 per cent/year) and fast (2 per cent/year) rates of improvement of energy intensity. Gross world product grows from $56 trillion in 2005 to anywhere from $170 trillion to $420 trillion in 2100. Global primary power demand decreases from 15.3TW in 2005 to 6.9TW in 2100 for the lowest scenario, and increases to 44.7TW for the highest scenario.

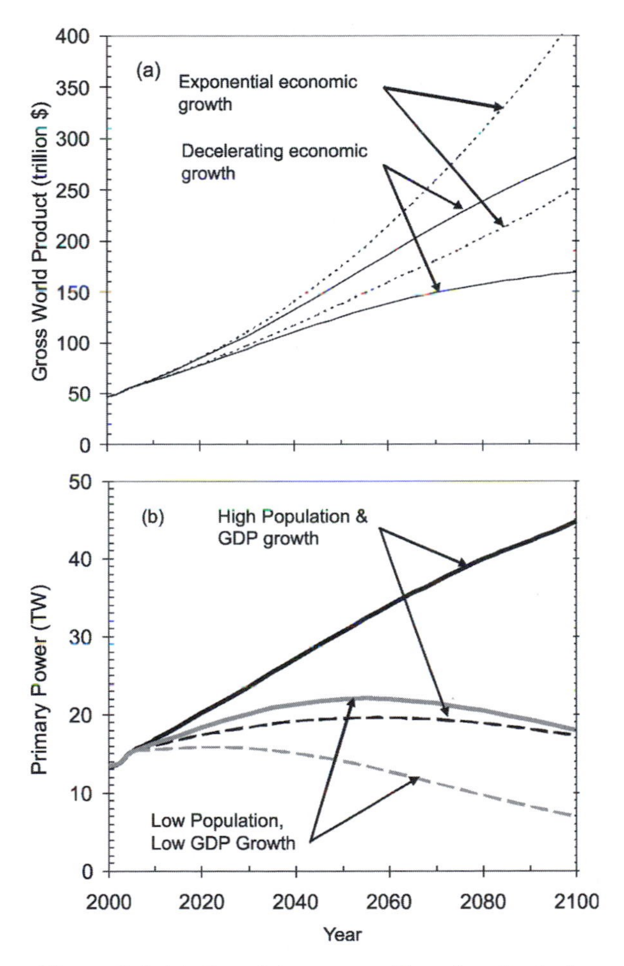

Figure 1.1 *(a) Growth in gross world product for the low and high population scenarios combined with low (solid lines) and high (dashed lines) GDP/capita scenarios; and (b) variation of world primary power demand for low population and GDP/capita growth or high population and GDP/capita growth combined with energy intensity reductions of 1 per cent/year (upper curves) or 2 per cent/year rate of reduction (lower curves)*

CO_2 emissions are shown in Figure 1.2a for four scenarios:

1 high population, high GDP/capita, low improvement in energy intensity, low C-free power;
2 high population, high GDP/capita, high improvement in energy intensity, low C-free power;
3 high population, high GDP/capita, high improvement in energy intensity, high C-free power;
4 low population, low GDP/capita, high improvement in energy intensity, high C-free power.

In Scenario 1, CO_2 emissions rise to 28GtC/yr and in Scenario 2 they peak at 9.7GtC/yr around 2050, while in Scenarios 3 and 4, emissions are eliminated by 2095 and 2070, respectively. Scenario 3 is consistent with the emission reductions needed to stabilize atmospheric CO_2 at 450ppmv, while Scenario 4 provides a margin of safety (given uncertain climate–carbon cycle feedbacks) for stabilization at 450ppmv.

From Figure 1.2a, it appears that the single most important factor in reducing CO_2 emissions is to increase the rate of improvement of energy intensity from 1 per cent/year to 2 per cent/year; the impact of quadrupling the rate of growth in C-free power, or the combined effect of lower population and GDP/capita growth, is substantially less. However, the relative impact of the different changes depends on the order in which they are implemented. This is illustrated in Figure 1.2b, which shows CO_2 emissions for Scenarios 1 and 4 as well as for the following:

2a low population, low GDP/capita, low energy intensity improvement, low C-free power;
3a low population, low GDP/capita, high energy intensity improvement, low C-free power.

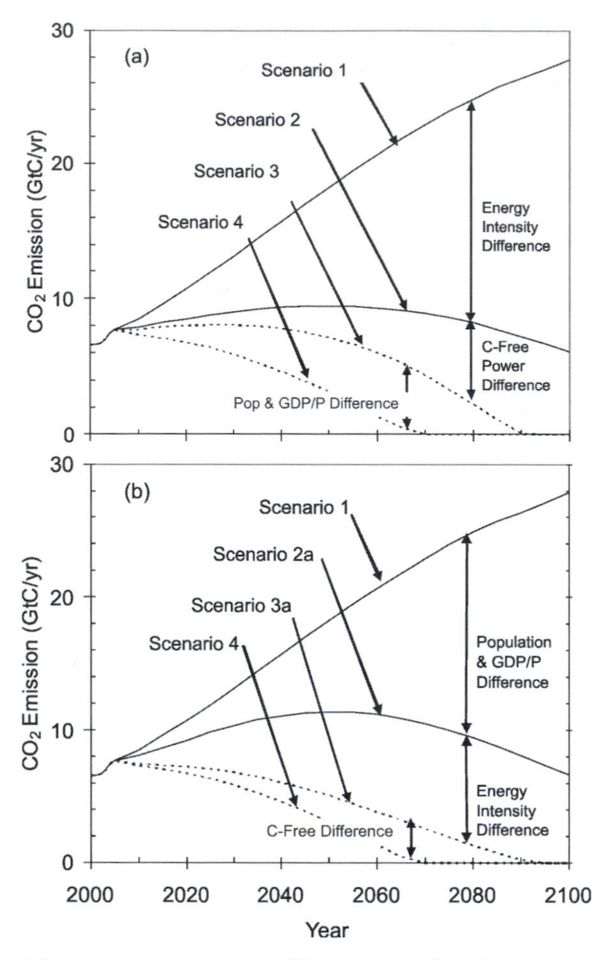

Figure 1.2 *Variation in CO_2 emissions for (a) Scenarios 1, 2, 3 and 4, and (b) Scenarios 1, 2a, 3a and 4*

Now, the combined effect of low population and low GDP/capita, when implemented first, is comparable to the effect of the higher rate of improvement in energy intensity when it is implemented first. Significant economic growth is of course needed in the developing countries in order to improve the material standard of living, but beyond some level of wealth, long since achieved in the developed countries, economic growth should no longer be the goal. Rather, the underlying goal should be human happiness that, once basic human needs are satisfied, is related to many non-material factors, including the diversity and richness of human relationships (family, friends, community), and not the accumulation of material wealth.

Figure 1.3a illustrates the tradeoffs between the product of future population and GDP/capita, the rate of reduction in energy intensity and the amount of C-free power that would need to be installed by 2050 in order for the CO_2 concentration not to exceed 450ppmv. Shown is the total primary power demand in

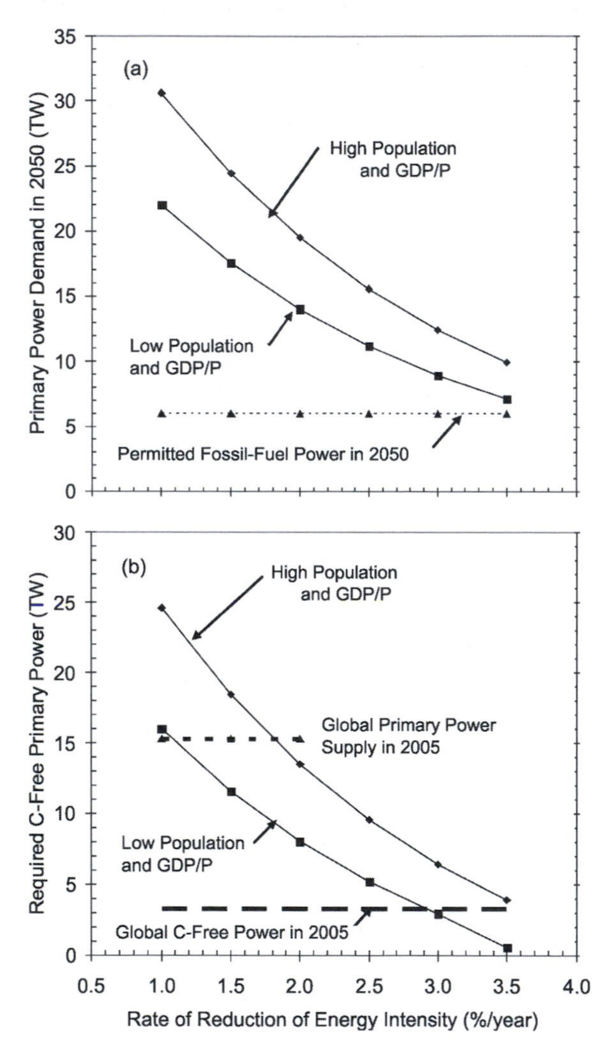

Figure 1.3 *(a) Primary power demand in 2050, as a function of the rate of decrease of global energy intensity for the high population and GDP/capita or low population and GDP/capita scenarios, and the approximate permitted fossil fuel power in 2050 if atmospheric CO_2 is to be stabilized 450ppmv; and (b) amount of C-free power required in 2050 for the same conditions as in (a), and global primary power and C-free primary power supply in 2005*

2050 as a function of the rate of reduction in energy intensity, in comparison to the fossil fuel power permitted in 2050, assuming either high population combined with high GDP/capita, or low population combined with low GDP/capita. The difference between the primary power demand and the permitted fossil fuel primary power gives the required C-free primary power, which is shown in Figure 1.3b for 2050 in comparison to the global primary power supply in 2000. Average global primary power supply in 2005 was 15.3TW, of which 12.0TW were from fossil fuels and 3.3TW from carbon-free sources. Energy intensity decreased at an average rate of 1.07 per cent/year during the period 1965–2005, and if it were to decrease at 1 per cent/year through to 2050, global primary power demand in 2050 would be 22TW and 31TW for the low and high population and GDP/capita scenarios, respectively. The permitted fossil fuel use supplies 6.0TW of primary power, so the required C-free primary power in 2050 is 16–25TW. If global average energy intensity decreases at 2.5 per cent/year, the required C-free power in 2050 is 5.2–9.6TW, or two to four times the present non-nuclear world C-free power supply of 2.45TW. The required C-free power under the low population and GDP/capita scenario is only 54 per cent that of the high population and GDP/capita scenario, which underlines the importance of low population growth and a moderation in the rate of growth of GDP/capita.

Volume 1 deals comprehensively with the prospects for accelerating the rate of decrease in energy intensity, and also touches upon the issues of growth in population and in GDP/capita. Volume 2 deals with the remaining factor in determining future CO_2 emissions: the rate at which C-free sources of energy can be deployed and ultimate limits on the deployment of C-free energy sources. The magnitude of the C-free challenge depends on how successful we are in reducing the growth in world energy demand. Under anything close to a BAU pathway with regard to energy demand, renewable energy sources simply will not be able to satisfy a large part of, much less all of, our energy needs within this century. It is therefore pertinent to briefly review the findings from Volume 1 concerning the potential to rein in our at present insatiable demand for ever more energy, before proceeding to assess the prospects for ramping up the C-free energy supply.

1.3 Potential reductions in end-use energy demand

The potential to reduce end-use energy demand in various sectors (buildings, transportation, industry and food, and municipal services), as determined in Volume 1, is reviewed here.

1.3.1 Energy use in buildings

Technologies already exist to reduce the energy use in new buildings by a factor of two to four compared to conventional practice for new buildings. This is true for buildings of all types and in all climate zones of the world. The keys to achieving such large reductions in energy use are:

- to focus on a high-performance thermal envelope;
- to maximize the use of passive solar energy for heating, ventilation and daylighting;
- to install energy-efficient equipment and especially energy-efficient systems;
- to ensure that all equipment and systems are properly commissioned and that building operators and occupants understand how they are to be used; and
- to engender enlightened occupant behaviour.

In order to design buildings with factors of two to four lower energy use, an integrated design process is required, in which the architects and various engineering specialists and contractors work together simultaneously in an iterative fashion before key design decisions are finalized. Attention to building form, orientation, thermal mass and glazing fraction is also critical. With regard to renovations of existing buildings, factors of two to four reductions for overall energy use, and up to a factor of ten reduction in heating energy use, have frequently been achieved. In many parts of the world, the cost of reductions in energy use of this magnitude is already justified at today's energy prices.

1.3.2 Transportation energy use

Urban form (in particular, residential and employment density and the intermixing of different land uses) and the kind of transportation infrastructure provided are the most important factors affecting future urban transportation energy use. Today there is almost a factor of ten difference in per capita transportation energy use between major cities with the lowest and highest transportation energy use per capita.

Existing or foreseeable technologies could reduce the fuel requirements of gasoline automobiles and light trucks (sport utility vehicles (SUVs), vans, pickup trucks) by 50–60 per cent with no reduction in vehicle size or acceleration. With a modest reduction in vehicle size and acceleration (to that of the 1980s), a factor of three reduction in fuel consumption could be achieved. Due to the inherent high efficiency of electric drivetrains compared to gasoline or diesel drivetrains, plug-in hybrid electric vehicles (PHEVs) would reduce the onsite energy requirements per kilometre (km) driven by about a factor of three to four compared to otherwise comparable gasoline vehicles. The economic viability of PHEVs depends on significant reductions in the cost of batteries and verification that they will maintain adequate long-term performance, but the prospects look good. Use of hydrogen in fuel cells could reduce onsite energy requirements by up to a factor of two compared to advanced gasoline-electric hybrid vehicles (depending on the performance of the latter) and by a factor of four compared to current gasoline vehicles, but significant problems remain concerning the cost of fuel cells and onboard storage of hydrogen. The global supply of platinum (Pt) would probably be a significant constraint on the development of a global fleet of hydrogen fuel cell automobiles.

The foreseeable feasible reductions in the energy intensity (energy use per passenger-km or tonne-km) of other modes of transportation are as follows: transport of freight by trucks, 50 per cent; transport of freight by ship, 45 per cent; diesel freight trains converted to trains using hydrogen in fuel cells, 60 per cent; air travel, 25–30 per cent; urban buses, 25–50 per cent (through use of diesel-electric hybrids); interurban buses, 50 per cent.

1.3.3 Industrial energy use

Compared to the current world average energy intensity, improved technology could reduce the primary energy requirements per unit of output by almost a factor of three for iron and steel, by a factor of two for aluminium and cement, by 25 per cent for zinc, and by 20 per cent for stainless steel. Technical improvements in the production of refined copper should roughly balance the tendency for increasing energy requirements as poorer grades of copper ore are exploited. However, much larger reductions in primary energy requirements for metals are possible through recycling combined with projected technical advances: a reduction in primary energy requirements by a factor of 7 for aluminium, a factor of 4.5 for regular steel, a factor of 2.5 for zinc, and a factor of 2 for copper and stainless steel (these savings pertain to uncontaminated materials). Recycling of glass reduces energy requirements by about 10–15 per cent compared to production of glass from virgin materials. If the world population and material stock stabilize by the end of this century, then the production of metals would be used almost exclusively for replacement of existing materials and so could be largely based on recycling, with attendant energy savings. The pulp and paper industry can become energy self-sufficient or a net energy exporter through the efficient utilization of all biomass residues. The potential energy savings in the plastics industry are unclear, but are probably at least 25 per cent through improved processes and at least 50 per cent through recycling of plastics. Potential energy reductions in the chemical industries appear to be very large but cannot be specifically identified at present. Better integration of process heat flows through pinch analysis and better organization of motor systems can save large amounts of energy in a wide variety of different industries.

1.3.4 Agricultural and food system energy use

Energy in the food system is used for the production of food, for transportation, processing, packaging, refrigerated storage and cooking. Energy use for the production of food consists of direct on-farm energy use and the energy used to produce fertilizers, pesticides and machinery used in farm operations. Fertilizers and pesticides are energy-intensive products. Fertilizer energy use can be reduced through substitution of organic fertilizers for chemical fertilizers, more efficient use of chemical fertilizers (30–50 per cent savings potential in the case of nitrogen (N) fertilizer), and more efficient production of chemical fertilizers (40 per cent savings potential in the case of N fertilizer). Pesticides are particularly energy intensive, but many jurisdictions have targets of reducing pesticide use in agriculture by 50 per cent or so through integrated pest management techniques. Organic farming systems reduce energy use per unit of farm output by 15 per cent to 70 per cent, but can also reduce yields by up to 20 per cent. However, rebreeding of crop varieties to maximize growth under organic farming systems could result in no yield reduction compared to current varieties with conventional methods. The biggest potential for energy savings in the food system is with a shift toward diets with lower meat content. Low-meat diets (and especially vegetarian and vegan diets) reduce direct and indirect fossil fuel energy inputs, and free up land that can be used to produce bioenergy crops.

1.3.5 Municipal services

Energy is used in the supply of municipal water through pumping and water treatment. Per household, this energy use is comparable to that of individual major household appliances. It can be reduced through measures to use water more efficiently (up to 50 per cent savings potential), through the reduction of leakage in water distribution systems (up to 30 per cent of input water is lost), and through the optimization of distribution system pressures and flow rates (10–20 per cent savings potential). The biggest opportunity to reduce net energy requirements at sewage treatment plants is through the recovery and use of methane from anaerobic digestion of sludge. Installation of systems (toilet, plumbing, storage tanks) to separately collect minimally diluted urine in new housing developments would facilitate energy-efficient recycling of nutrients from human wastes, something that will eventually be necessary. With regard to solid wastes, recycling of metals, plastics, paper and paper products is preferred to other management options. Dedicated anaerobic digestion with recovery and use of methane is the preferred option for organic wastes. Incineration with energy recovery is not particularly efficient but is preferred to land filling for wastes that cannot be recycled further.

1.3.6 Net result

Sector-by-sector scenarios of energy use as fuels and as electricity to the year 2100 were constructed for ten different world regions in Volume 1, taking into account differences in per capita income, floor area and travel today, and then summed to give a scenario of global demand for fuels and electricity. A low population scenario (global population peaking at 7.6 billion around 2035) combined with modest but regionally differentiated growth in GDP/person was considered along with a high population scenario (global population reaching 10.3 billion by 2100) and high growth in GDP/person. Slow (by 2050) and fast (by 2020) implementation of stricter standards for new and renovated buildings were considered along with the assumption that all existing buildings are either replaced or undergo a major renovation between 2005 and 2050 and that all buildings existing in 2050 undergo a major renovation between 2050 and 2095. Relatively slow and fast rates of improvement in automobile and industrial efficiencies to the potentials identified here were considered, but replacement of existing fossil fuel powerplants with the current state-of-the-art powerplants was not assumed to be completed before 2050.

The net result of these assumptions is that the global demand for fuels rises only modestly before peaking around 2030, then drops back to a level at or below the current global demand (depending on the population and GDP/capita scenario). Electricity demand rises more strongly (to 1.7–2.6 times the 2005 demand) and then stabilizes or decreases slightly, but remains well above the 2005 demand. However, these are quite modest increases compared to most BAU scenarios. When additional structural shifts in the economy are considered (50 per cent of baseline economic output from industry and freight transport shifted to commercial services by 2100), the resulting annual average compounded rate of decrease in the primary energy intensity of the global economy is 2.7 per cent/year between 2005 and 2050 and 1.8 per cent/year between 2005 and 2100.

This is a significant result. As seen from Figure 1.3b, a compounded annual rate of reduction in the global average primary energy intensity of 2.7 per cent/year between 2005 and 2050 implies that only 1–5TW of additional C-free primary power will be needed by 2050 (depending on the future human population and rate of growth in GDP per person), plus additional new C-free power to replace whatever portion of the existing nuclear power supply (equivalent to 0.8TW of continuous primary power) is retired without replacement. The total C-free primary power requirement is 4.3–8.3TW. To the extent that a 2.7 per cent/year average annual improvement in energy intensity is not achieved between now and 2050, greater amounts of C-free power will be needed.

The rest of the book examines the prospects for providing sufficient C-free energy to eliminate fossil fuel CO_2 emissions by about the end of this century, given the separate global demand for fuels and electricity generated by the demand scenarios produced in Volume 1. The consequences for atmospheric CO_2 concentration, climate and ocean acidification are then explored using a coupled climate–carbon cycle model, taking into account potential positive climate–carbon cycle feedbacks but also the possibility of creating negative CO_2 emissions by the end of this century.

Notes

1 As discussed in Volume 1 (Chapter 2, section 2.1), primary energy refers to energy as it is found in nature, prior to conversion to other energy forms (such as electricity or gasoline) by humans. It consists of unprocessed coal, petroleum and natural gas, plus electricity from hydropower, nuclear powerplants, wind and solar multiplied by a factor (typically 2.5 or 3) to convert the electricity production to the amount of fossil fuels that would generate the same amount of electricity as from these powerplants.

2 For the definition of tera and other prefixes used in relation to energy and power, see Appendix A.

2

Solar Energy

The flux of solar radiation on a plane per unit area (W/m^2) is referred to as the *solar irradiance*. The solar irradiance on a plane outside the earth's atmosphere, perpendicular to the sun's rays and at the mean (average) earth–sun distance is referred to as the *solar constant* (Q_s). It has a value of about $1370W/m^2$. As noted in Volume 1, Chapter 2, the global primary power demand in 2005 was about 15.3TW. Given a radius for the earth of 6.36×10^6m, the amount of solar radiation intercepted by the earth is over 11,000 times present global primary power demand. By this measure, the amount of solar energy available is enormous. The total area of deserts in the world is about 12 million km^2 (Barry and Chorley, 1982) and the annual average solar irradiance in desert areas is about $250W/m^2$. If solar radiation is converted to electricity at an efficiency of 10 per cent (already exceeded today in commercial modules), then a mere 0.7 per cent of the world's desert area ($84,300km^2$) would be required to generate the total 2005 world electricity consumption of about 18,500TW-hours (TWh) (given in Volume 1, Table 2.4). Allowing that some of the electricity would be generated when it is not needed and so would need to be converted to some other energy carrier (such as hydrogen), stored and then converted back to electricity when needed, and given energy losses in transporting the alternative energy carrier from desert areas to where it is needed, one might have to double the module area (the role of hydrogen as a complement to electricity in an entirely renewable energy-based system is discussed in Chapter 10). As a fraction of desert area alone, the required area is still very small, but compared to the cumulative production of photovoltaic (PV) modules up to present (about $25km^2$), the required area is enormous.

The solution to this challenge is to use solar energy as directly as possible, without converting it into electricity as an intermediate step. In buildings, which account for about a third of primary energy use in countries in the Organisation for Economic Co-operation and Development (OECD), energy is used for lighting, heating, ventilation, cooling and dehumidification. We have discussed (in Volume 1, Chapter 4) the opportunities to displace electric lighting with daylighting in commercial buildings, and to use solar energy for passive heating and ventilation. Solar thermal energy can also be used directly for cooling, for dehumidification and for the production of hot water. These energy uses account for a large fraction of total energy use in buildings. Those uses that can be satisfied *only* with electricity represent a small enough fraction of current electricity use in buildings that much of the required electricity could be provided with building-integrated PV (BiPV) panels that displace conventional façade materials. That is, instead of generating the needed electricity at remote (perhaps desert) locations and transporting it to demand centres, much of the needed electricity could be generated on the surfaces of the very buildings that use it. This would greatly reduce the material and land requirements.

In industry, which accounts for another third of primary energy use, much of the required energy is in the form of heat – either low-temperature (80–250°C) thermal energy in industries such as food processing and chemical manufacturing, or high-temperature (800–2000°C) thermal energy for reduction of metal ores. Here too, solar thermal energy can be used directly in place of fuels or electricity in many cases. For this to happen, those energy-intensive industries that can most readily make direct use of solar thermal energy may need to shift to desert regions.

Personal long-distance motorized transportation and air travel in a C-free energy system may require hydrogen as a fuel. However, there are many ways to produce hydrogen from renewable energy, and even using solar energy, one is not restricted to first producing electricity as an intermediate step. More will be said about this in Chapter 10.

In short, solar energy can be used in many different ways, differing greatly in complexity and cost. The simplest and least costly technology to utilize solar energy is the clothes line. Some of the next simplest ways to use solar energy – daylighting, passive heating and passive ventilation in buildings – were discussed in Volume 1, Chapter 4. The high-performance window is a critical technology for many of the passive uses of solar energy. Here, we discuss active uses of solar energy: the generation of electricity, both centrally and through building-integrated PV modules; solar thermal energy for heating and the production of domestic hot water; solar thermal energy for air conditioning of buildings; and solar thermal energy for industrial processes. In Chapter 11 (Community-Integrated Energy Systems with Renewable Energy) we go one step further, discussing how solar heat from the summer can be stored for heating buildings in winter, and how ambient coldness in the winter can be stored for air conditioning of buildings in the summer.

2.1 Seasonal, latitudinal and diurnal distribution of solar energy

As noted above, the solar constant – which pertains to the solar irradiance on a plane perpendicular to the sun's rays at the *mean* earth–sun distance – has a value of about 1370W/m^2.[1] Due to the eccentricity of the earth's orbit, the distance of the earth from the sun varies during the course of a year (the earth being closest to the sun on 3 January and furthest from the sun on 5 July). The solar irradiance on a plane perpendicular to the sun's rays varies inversely with the square of distance from the sun (because the area of the imaginary sphere over which the fixed output from the sun is distributed increases with the square of the radius of the sphere). Elementary trigonometry indicates that the flux density on a horizontal plane varies with the cosine of the solar *zenith angle*, where the zenith angle is defined as the angle between a point and the vertical. Thus, the solar irradiance on a horizontal plane at the top of the atmosphere is given by:

$$Q(\theta,t)=\left(\frac{Q_s}{d(t)^2}\right)\cos\theta \tag{2.1}$$

where $d(t)$ is the earth–sun distance at time t as a fraction of the mean distance.

As solar radiation enters the earth's atmosphere it is attenuated by absorption at ultraviolet wavelengths by ozone in the stratosphere and at near infrared wavelengths by water vapour in the lower troposphere, and by scattering at all wavelengths by air molecules and aerosol particles (sea salts, dust, sulphate and various organic compounds). Scattering is the process by which a photon changes direction. Clouds, when present, also reflect, scatter and absorb solar radiation. Solar radiation that reaches the surface without having been scattered arrives from a single direction and is referred to as *direct-beam radiation*. Some of the radiation that is scattered is scattered in the forward direction and reaches the earth's surface, but arrives from all angles and so is referred to as *diffuse-beam radiation*. On an overcast day, all of the solar radiation reaching the surface is diffuse. The diffuse-beam radiation can be divided into two parts: one that arrives at the surface after passing once through the atmosphere (this is referred to as *single scattering*), and one that arrives after at least one initial reflection from the surface and subsequent scattering by the atmosphere back to the surface (this is referred to as *multiple scattering*). On a clear day, the fraction of radiation reaching the surface that is direct exceeds 90 per cent with the sun directly overhead, but decreases as the zenith angle increases.

For a solar panel inclined to the horizontal, there is an additional contribution from radiation reflected directly off of the surrounding landscape. The fraction of incident radiation reflected by a surface is referred to as the *albedo*. The albedo depends on the type of surface and increases with increasing solar zenith angle; albedo values for a zenith angle of 60° range from 0.15 for a dense vegetation canopy, to 0.25–0.35 for a sandy surface, to 0.8 for fresh snow. As the albedo of the surrounding landscape increases, the contribution of reflected radiation (and of multiply scattered diffuse radiation) increases.

Box 2.1 Intensity of solar radiation on a flat surface

The intensity of incident solar radiation at the top of the atmosphere is given by:

$$Q(\theta,t)=\left(\frac{Q_s}{d(t)^2}\right)\cos\theta \tag{2.2}$$

where Q_s is the solar constant, $d(t)$ is the earth–sun distance at time t as a fraction of the mean distance and θ is the zenith angle (the angle of the sun from the vertical). $d(t)$ is given by (Berger, 1978):

$$d=\frac{1-\varepsilon^2}{1+\varepsilon\cos\left((\gamma-P)\frac{\pi}{180}\right)} \tag{2.3}$$

where ε is the orbital eccentricity (0.01672), P is the longitude of perihelion (282.04°) and γ is the orbital angle (in degrees) measured from the vernal equinox (20 or 21 March). $\gamma = P$ on 3 January, when the earth is closest to the sun. The solar zenith angle is given by (Sellers, 1965):

$$\cos\theta=\sin\phi\sin D+\cos\phi\cos D\cos h \tag{2.4}$$

where ϕ is latitude, D is declination (the angular distance of the sun north (positive) or south (negative) of the equator) and h is the solar angle (the angle through which the earth must turn to bring the longitude in question directly under the sun). The declination can be computed as:

$$D = -\arcsin(0.397901\cos(2\pi(DN + 11)/365.25)) \tag{2.5}$$

where DN is the day number within the year (DN = 1 on 1 January) and D will be in radians (this equation has been slightly modified from Kreider and Kreith (1981), so as to give a maximum declination of ±23.447°, equal to the obliquity of the earth's axis at present). If $\cos\theta<0$ the solar irradiance is zero.

In the case of a surface inclined at an angle δ from the horizontal with an azimuth of a', the angle θ^* between the surface normal and the sun is given by:

$$\cos\theta^* = \cos\delta\cos\theta+\sin\delta\sin\theta\cos(a-a') \tag{2.6}$$

where a is the solar azimuth and is such that:

$$\cos a=\frac{\sin\phi\cos\theta-\sin D}{\cos\phi\sin\theta} \tag{2.7}$$

For the sun, the azimuth is the horizontal direction of the sun at a given point in time, while for a module, the azimuth is the direction that it is facing if it is tilted. The azimuth is zero for directions due south, is positive when directed west of south up to a maximum value of π when due north, and negative when directed east of south up to a maximum value of $-\pi$ when due north ($\cos\pi = \cos(-\pi)$) (for example, an azimuth of 90° or $\pi/2$ radians corresponds to due west). Equations (2.6) and (2.7) are derived in Sellers (1965). For a sloping surface directed due south, $a' = 0$, so:

$$\cos\theta^* = \cos\delta\cos\theta+\sin\delta\left[\frac{\sin\phi\cos\theta-\sin D}{\cos\phi}\right] \tag{2.8}$$

The solar irradiance I_T incident on an inclined plane at the earth's surface is given by:

$$I_T = I_{Dr} + I_{Df} + I_R \tag{2.9}$$

where I_{Dr}, I_{Df} and I_R are the directly transmitted, diffuse and reflected components, respectively. Directly transmitted radiation is radiation that arrives within a 2.5° angle from the sun, whereas diffuse radiation is radiation that has been scattered in all directions and arrives from all parts of the sky. The direct, diffuse and reflected components are given by:

$$I_{DR} = Q^* D_R \cos\theta^* \tag{2.10}$$

$$I_{Df} = Q^* D\downarrow \left\{ D_R \frac{\cos\theta^*}{\cos\theta} + (1 - D_R)\frac{1+\cos\delta}{2} \right\} \tag{2.11}$$

and:

$$I_R = Q^*(D_R \cos\theta \bullet \alpha_s + D\downarrow \overline{\alpha_s})\left(\frac{1-\cos\delta}{2}\right) \tag{2.12}$$

where

$$Q^* = \left(\frac{Q_s}{d^2}\right)(1 - O_3)(1 - A_w) \tag{2.13}$$

$$D\downarrow = D_1 + D_2 \tag{2.14}$$

$$D_1 = S_F \cos\theta \tag{2.15}$$

and:

$$D_2 = \frac{(D_R \alpha_s \overline{S_B} + S_F \overline{\alpha_s} \overline{S_B})\cos\theta}{1 - \overline{\alpha_s} \overline{S_B}} \tag{2.16}$$

In the above, Q^* is the extraterrestrial irradiance perpendicular to the sun's rays and attenuated by absorption by ozone and water vapour only (which are assumed to occur entirely above multiple scattering between the surface and atmosphere), O_3 and A_w are the fractional absorption of solar radiation by ozone and water vapour, D_R is the zenith angle-dependent fraction of radiation transmitted as direct beam, α_s is the zenith angle-dependent surface albedo for direct-beam radiation, $\overline{\alpha}_s$ is the surface albedo for diffuse radiation (which is assumed to come equally from all directions, so the appropriate $\overline{\alpha}_s$ is the surface albedo averaged over all zenith angles, which in turn is roughly equal to α_s evaluated at a solar zenith angle of 60°), $D\downarrow$ is the diffuse-beam irradiance on a horizontal surface, as a fraction of Q^*, consisting of singly and multiply scattered components D_1 and D_2, S_F is a zenith angle-dependent forward-scattering fraction, and $\overline{S}_B$ is the back-scattering fraction for upward-reflected diffuse radiation, evaluated at $\cos\theta = 0.5$.

Equation (2.11) is taken from Hay (1986). In this equation, the term involving D_R is the circumsolar component of diffuse radiation, while the term involving $(1-D_R)$ is the isotropic component. D_R is used here as an index that defines what portions of the diffusely transmitted radiation should be treated as circumsolar and isotropic. On an overcast day, $D_R = 0$ and all of the radiation would be isotropic. The amount of diffuse radiation incident on a plane decreases as the plane is tilted from the horizontal because it 'sees' less of the sky and more of the surrounding terrain, whereas the reflected component is zero for a horizontal plane and increases as the tilt increases.

D_R, S_F and $\bar{S}_B$ are computed using polynomial functions of the form:

$$x = a_o + m(a_1 + m(a_2 + m(a_3 + m(a_4 + a_5 m)))) \quad (2.17)$$

where x is D_R, S_F or $\bar{S}_B$, and m is the optical air mass and is given by:

$$m = \frac{1}{\cos\theta + 0.15/(57.3(1.6386 - \theta))^{1.253}} \quad (2.18)$$

where θ is in radians. Equations (2.17) and (2.18) are from Thompson and Barron (1981). Values of the a_i coefficients for evaluation of D_R, S_F and $\bar{S}_B$ are given in Table 2.1.

Table 2.1 *Values of the coefficients used in Equation (2.17) for the computation of D_R, SF and $\bar{S}_B$*

	D_R	S_F	$\bar{S}_B$
a_0	8.9788×10^{-1}	2.519×10^{-2}	$1.73548478 \times 10^{-2}$
a_1	-1.1908×10^{-1}	6.582×10^{-2}	4.8979660×10^{-2}
a_2	7.3649×10^{-3}	-4.3399×10^{-3}	$-3.02510749 \times 10^{-3}$
a_3	-2.4966×10^{-4}	1.2697×10^{-4}	$8.87911913 \times 10^{-5}$
a_4	4.4260×10^{-6}	-1.3608×10^{-6}	$-9.65852864 \times 10^{-7}$
a_5	-3.1921×10^{-8}		

The fractional absorption of solar radiation by water vapour can be computed as:

$$A_W = \frac{2.9W_a}{(1.0 + 141.5W_a)^{0.635} + 5.925W_a} \quad (2.19)$$

where $W_a = e_a m$ and e_a is the surface water vapour pressure in millibars (mb).

Finally, an adequate formulation for the zenith angle dependence of the surface albedo for direct-beam radiation is:

$$\alpha_s = \bar{\alpha}_s \frac{1.7 - 0.751\cos\theta}{1.0 + 0.649\cos\theta} \quad (2.20)$$

The set of equations needed to compute the position of the sun in the sky at any latitude at any time and total flux of solar radiation on a solar panel of any orientation under clear skies is given in Box 2.1. The path of the sun through the sky on various days of the year at a given latitude can be plotted on a *stereographic sun path diagram*. Figure 2.1 contains sun path diagrams for latitudes 0°N, 30°N, 45°N and 60°N. The outermost circle represents the horizon, while the centre represents a point directly overhead. On the spring and autumn equinoxes (20 or 21 March and 22 or 23 September) the sun rises and sets exactly due east and west. During the summer half-year, the sun rises to the north of east and sets to the north of west in the northern hemisphere, with the amount north of east or west increasing up to the summer solstice (21 June) and increasing with latitude. Conversely, during the winter half-year, the sun rises and sets to the south of east and west, respectively, and does not rise as high in the sky at solar noon. The sun path diagram is helpful in visualizing which surfaces will be directly illuminated by the sun at any given time. For example, during the time just after sunrise or just before sunset in summer, when the sun is to the north, it will directly

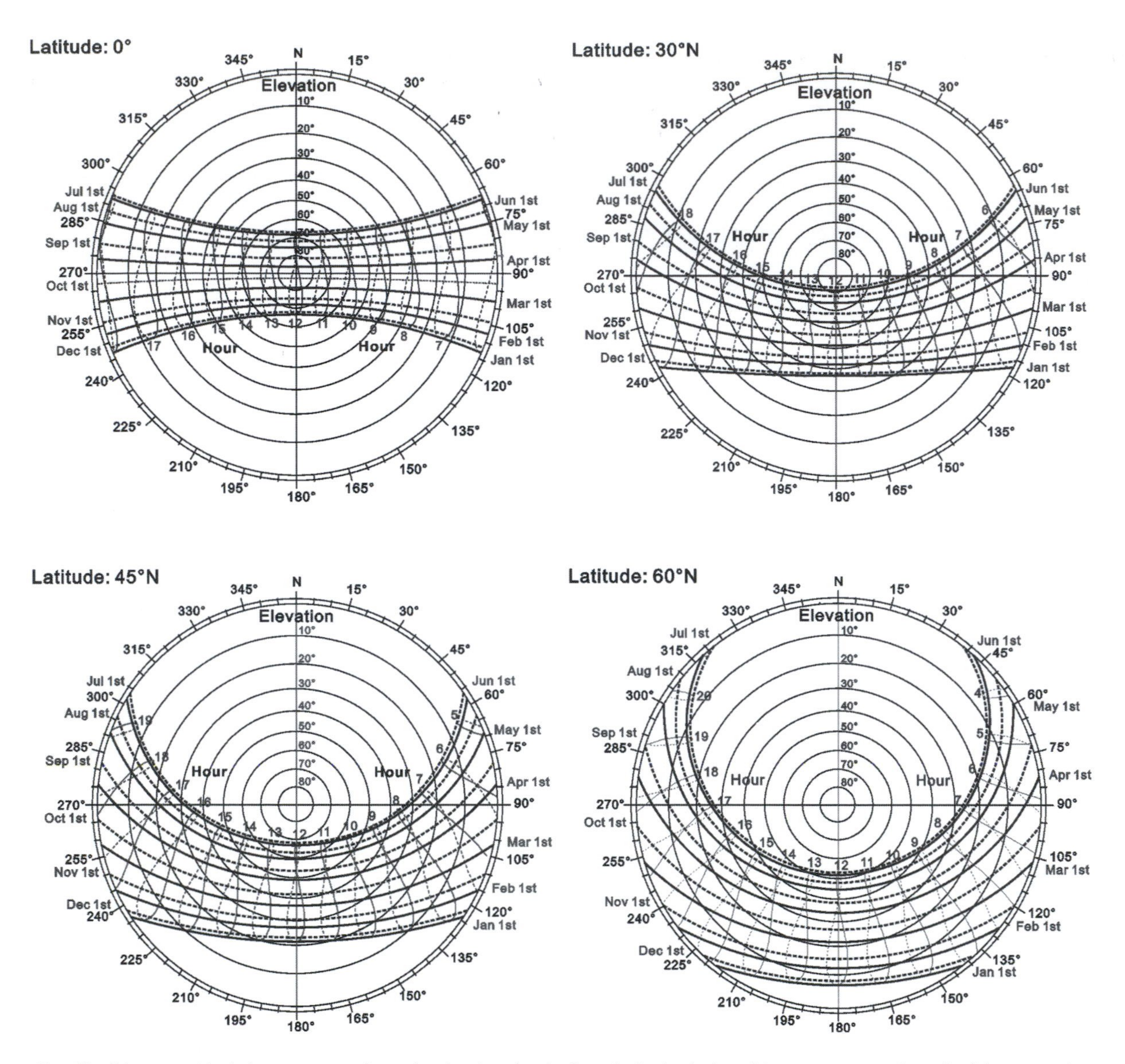

Note: The light concentric circles represent various solar elevations (angles from the horizon), the solid curves represent the path of the sun on the first day of each month from January to June, the dashed curves represent the path of the sun on the first of each month from January to June, and the helical curves give the position at various times relative to solar noon (the time when the sun is due south, and set to 12:00).
Source: Computed using *The Solar Tool* developed by Square One Research and available through Ecotech (ecotech.com)

Figure 2.1 *Stereoscopic sun paths for latitudes 0°N, 30°N, 45°N and 60°N*

illuminate north-facing walls but will not directly illuminate a panel tilted to the south (unless the sun's angle above the horizon is greater than the tilt of the panel).

Figure 2.2 shows the diurnal variation of solar irradiance at 0°N, 30°N and at 60°N on 21 June, 21 September and 21 December for cloud-free conditions. The upper panels are for a horizontal surface, the middle panels for a surface tilted to the south at an angle equal to the latitude and the lower panels for a sun-tracking panel that is always oriented perpendicular to the sun's rays. In the last case, the peak solar irradiance

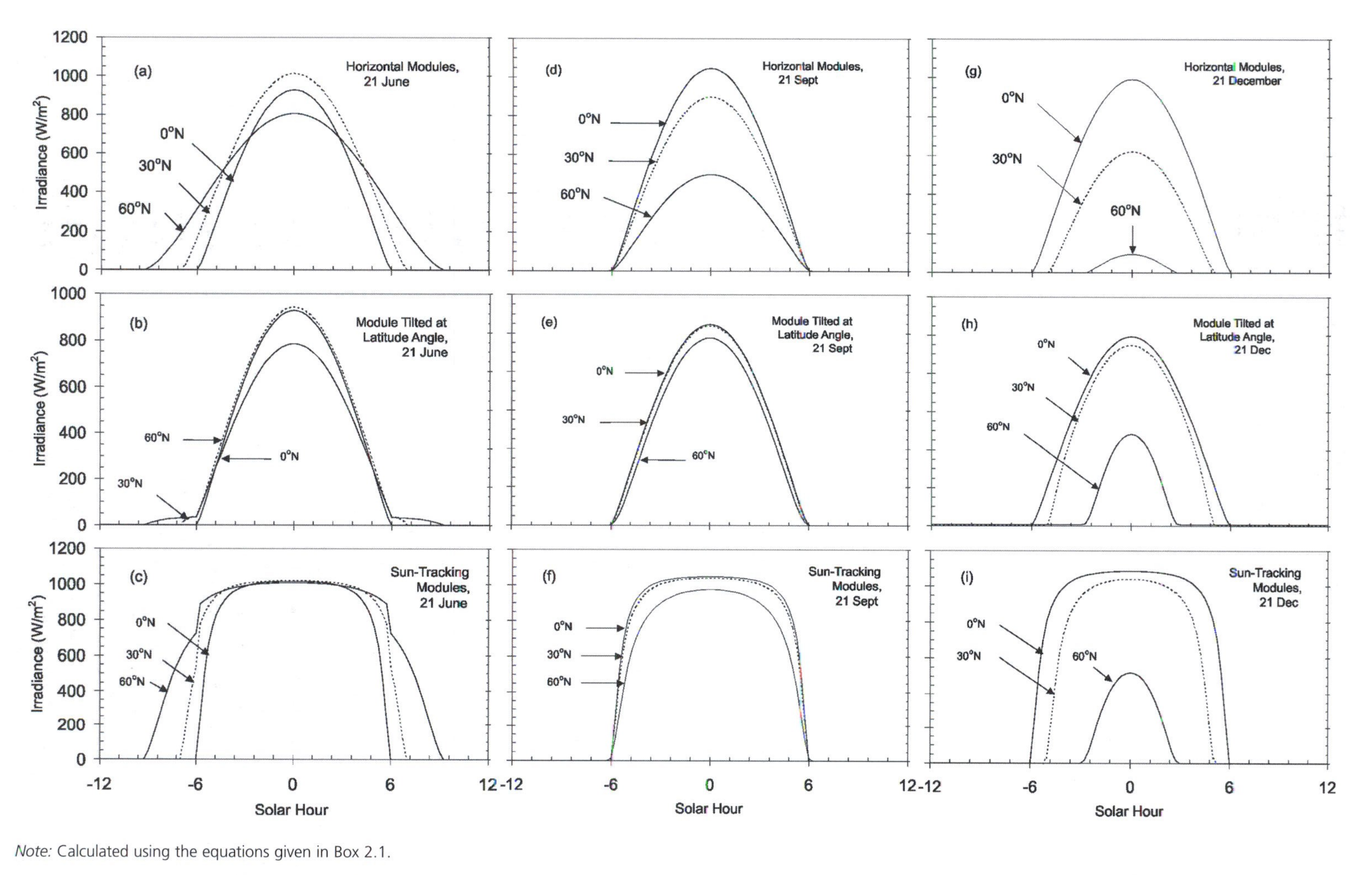

Note: Calculated using the equations given in Box 2.1.

Figure 2.2 *Diurnal variation of solar irradiance on horizontal, tilted and sun-tracking modules at 0°N, 30°N and 60°N for cloudless skies on 21 June, 21 September and 21 December*

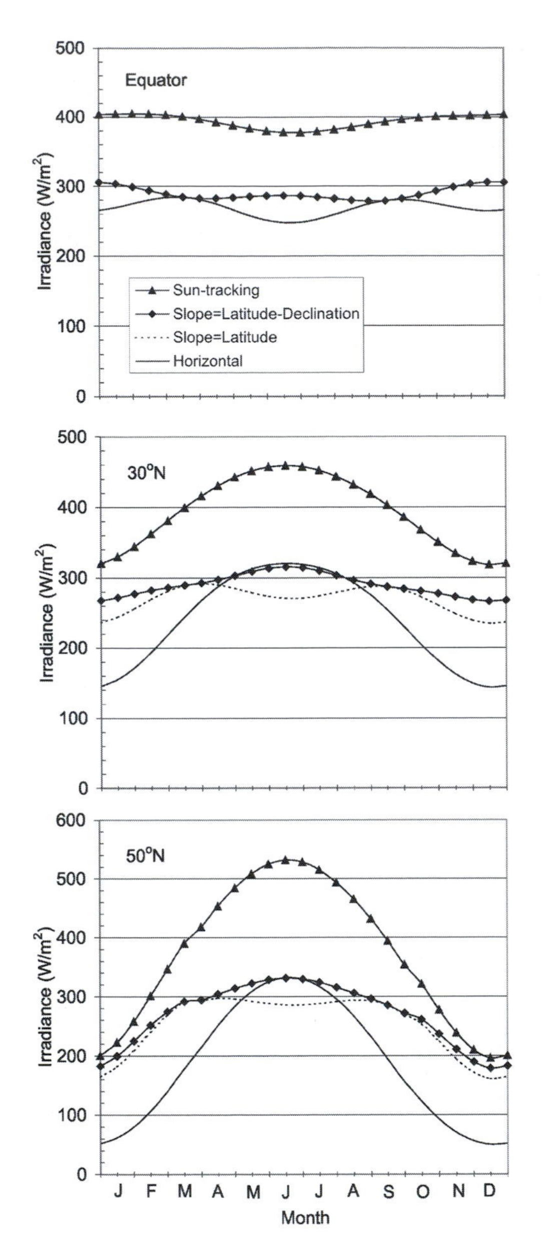

Note: Results are given for latitudes 0°N, 30°N and 50°N for cloudless skies. Calculated using the equations given in Box 2.1, assuming a surface albedo of 0.25 at all latitudes and times except for 15 October–31 March at 50°N, for which a surface albedo of 0.45 is assumed. Surface atmospheric water vapour pressure is assumed to be 24mb, 16mb and 8mb at latitudes of 0°N, 30°N and 50°N, respectively.

Figure 2.3 *Annual variation in diurnally average solar irradiance on a horizontal flat solar collector and on solar collectors inclined at the latitude angle, inclined at the latitude angle minus the seasonally varying declination, or that track the sun in two dimensions*

is about 1000W/m^2 at all latitudes from 0 to 60°. Figure 2.3 shows the annual variation in daily average solar irradiance for cloudless skies on a horizontal panel, on a panel inclined at the local latitude angle, on a solar panel inclined at the latitude angle minus the seasonally varying declination (see Box 2.1) and on a sun-tracking solar panel at latitudes of 0°, 30°N and 50°N. For a solar panel with a fixed orientation under clear skies, the mean annual solar irradiance is maximized if the panel is inclined equatorward at an angle equal to the local latitude. The solar irradiance can be increased further if the panel tilt can be adjusted each day to equal the latitude angle minus the solar declination (at 45°N and on 21 June, when the sun is 23° north of the equator, the panel would be tilted 22° to the south, while on 21 December, when the sun is 23° south of the equator, the panel would be tilted 68° to the south). The irradiance can be increased further if the panel can tilt east–west as well as north–south, so as to track the sun in its course through the sky. As seen from Figure 2.3, there is a big increase in irradiance during the winter, compared to a horizontal panel, if the panel is tilted at the latitude angle, but there is a modest decrease in summer irradiance. Thus, the irradiance is more uniform with season, which is clearly of value for some applications of solar energy (such as meeting domestic hot water loads). There is very little difference in irradiance in going from a fixed panel to one with variable tilt, the benefit of variable tilt being to eliminate the above-mentioned reduction of summer irradiance. There is a large increase in irradiance at all seasons if the panel can tilt along two axes, so as to follow the sun, but this entails substantially greater cost and complexity. The biggest benefit of sun tracking is in summer, which could be useful if solar energy is used to power air conditioning.

Under cloudy skies or if the sun is shaded by surrounding terrain, the incident radiation is entirely diffuse. Diffuse irradiance on a solar collector is maximized if the collector is horizontal, as this increases the amount of sky 'seen' by the collector. When clouds are taken into account, the optimal tilt of a fixed collector will be decreased compared to the case with continuous clear skies (where a tilt equal to the latitude maximizes annual irradiance on the collector). If winters are particularly cloudy (as is the case in many regions), the irradiance is low no matter what the tilt, so the optimal tilt is shifted toward summer conditions (thus, the optimal tilt is smaller than with seasonally uniform cloudiness).

Maximizing passive solar heat gain through windows in winter and minimizing solar heat gain in summer are important strategies in reducing both heating and air conditioning energy use. The equations presented in Box 2.1 can be readily used for calculating the irradiance incident on a wall of any orientation under clear skies (by setting $\delta = 90°$ and choosing the appropriate azimuth angle). Results are shown in Figure 2.4 for mid-June and mid-December at 45°N. Interestingly, the largest irradiance in June occurs on east- and west-facing walls rather than on south-facing walls, due to the fact that the sun is closer to the horizon early and late in the day and thus closer to perpendicular to a vertical surface. Vertical panels facing south (in the northern hemisphere) have maximum solar irradiance in winter, in contrast to any orientation considered in Figure 2.3. Thermal collectors on a south-facing wall would therefore be ideal for meeting space-heating loads.

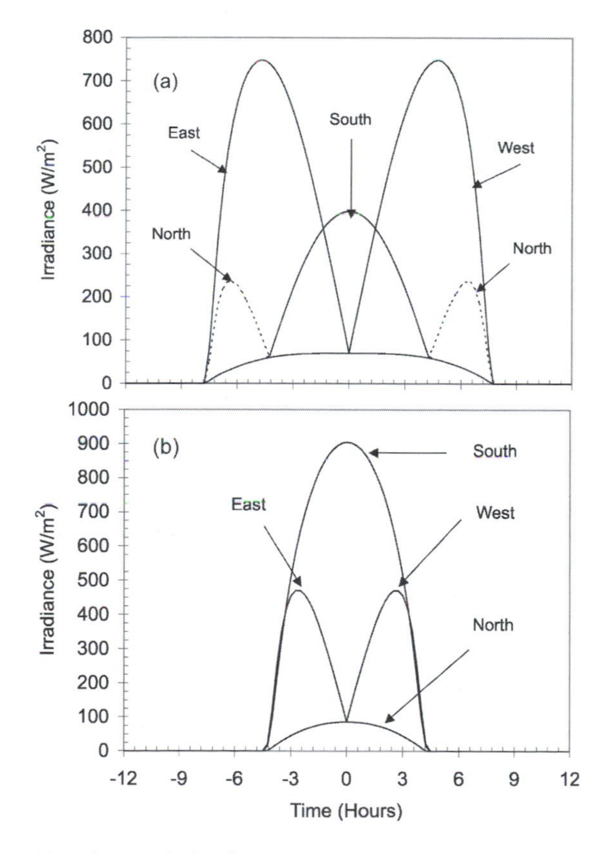

Note: Computed using the equations given in Box 2.1.

Figure 2.4 *Diurnal variation in irradiance on vertical walls of various orientations for cloudless skies at 45°N in (a) mid-June and (b) mid-December*

Figure 2.5 shows the geographical distribution of measured or estimated mean annual solar radiation on a horizontal surface, taking into account observed cloudiness.[2] The best locations have a mean annual solar irradiance of 250–275W/m^2. This is a low power density compared to the concentrated power demands in cities of the industrialized world. For example, a typical thermal powerplant (coal-fired or nuclear) would produce 2GW of electric power. For a solar power density of 200W/m^2, an area of 10km^2 would be required to intercept this much power. The area required to generate this much electric power would be several times larger, given current and near-term efficiencies for economically competitive modules of 10–20 per cent. However, and as previously noted, the total amount of solar energy intercepted by the earth is vast – about 11,000 times the total world primary power demand of 15.3TW in 2005. The difficulties arise not from a shortage of solar energy, but from its diffuse nature and from the fact that it is often most plentiful at the wrong times or wrong places compared to energy demand.

Sometimes solar irradiance is given in terms of the kWh of solar energy per m^2 per year. This is just the annual mean irradiance (in kW) times the number of hours in a year (8760). This is a convenient measure because if multiplied by the efficiency of a PV module, it immediately gives the kWh of electricity generated per year per m^2 of module. The maximum annual solar irradiance anywhere is around 2200kWh/m^2/yr, while that in central Europe or the northern US/southern Canada is around 1600kWh/m^2/yr.

2.2 Photovoltaic electricity

Solar energy can be converted into electricity using *PV panels*, into useful heat using a variety of *solar collectors*, or into both electricity and heat with integrated PV/thermal collectors. All three are discussed in this chapter, beginning with PV panels.

2.2.1 Quantum-mechanical principles

The PV conversion of sunlight to electricity ultimately depends on quantum mechanics, one of the most

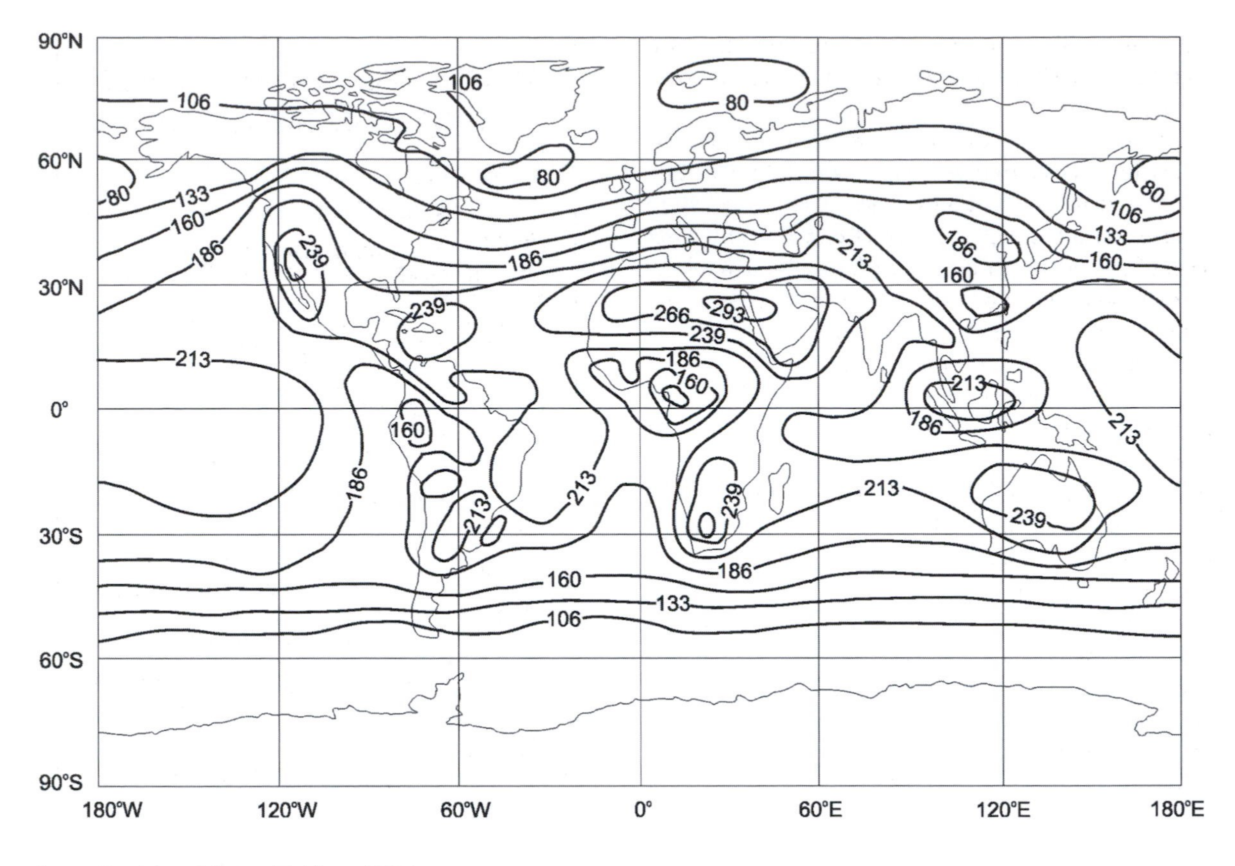

Source: Henderson-Sellers and Robinson (1986)

Figure 2.5 *Geographical variation in observed mean annual solar irradiance (W/m²) on a horizontal surface*

successful theoretical frameworks in all of physics. Under the rules of quantum mechanics, the permissible energies of electrons consist of a series of discrete levels. Energies between these levels are not permitted, so the energy of an electron can change only as a series of steps (from one level to the next). Similarly, the energy of electromagnetic radiation comes in discrete packages called *photons*. The energy of a photon (in joules) is given by its frequency, ν, times Planck's constant, h:

$$E(\text{joules}) = \nu h, \tag{2.21}$$

where $h = 6.6262 \times 10^{-34}$ joule-sec. The speed of an electromagnetic wave, c, is given by the product of its frequency and wavelength, λ. Since the wave speed in a given medium is fixed (at the speed of light, 2.99793×10^8m/s in a vacuum), it follows that the frequency of a photon – and hence its energy – varies inversely with its wavelength (in particular, $\nu = c/\lambda$). It is common (in the semi-conductor industry) to express photon and electron energies in terms of *electron* volts, which are given by dividing the energy in joules by the charge q of one electron ($q = 1.6 \times 10^{-19}$ coulombs). That is:

$$E(\text{eV}) = \frac{hc}{q\lambda} = \frac{1.242}{\lambda} \tag{2.22}$$

where λ is now in microns (μm). A photon is able to bump an electron to a higher energy level if its energy exceeds the size of the energy difference from one level to the next. Shorter wavelength light, having more energetic photons, will be more capable of doing this.

Electrons are often thought of as orbiting around the nucleus of an atom. It is more accurate to think of an electron as having a particular probability distribution around the nucleus, with regions of high probability of finding the electron called 'orbitals'. The orbitals are arranged in layers or 'shells' around the nucleus, with the innermost shell containing two orbitals (two electrons) and the next shell accommodating and therefore desiring eight electrons.

Atoms form bonds with each other through the interaction of electrons in the outer shell of each atom. In a covalent bond, atoms share electrons so as to acquire the full complement of electrons in the outer shell of each atom.[3] If the molecules are brought closer to each other, their outer shells begin to overlap and exhibit new energy levels or bands. As a solid form, two outer energy bands are formed, often separated by a gap. The lower energy band is called the *valence band* and the upper energy band is called the *conduction band*. In materials that conduct electricity, electrons occur in both the valence and conduction bands, and these bands overlap, allowing the movement of electrons. In an insulator (non-conductor), the valence band is completely filled and the conduction band is completely empty and the bands do not overlap. In a *semi-conductor*, electrons occur in both the valence and conduction bands (as in a conductor), but the valence and conduction bands do not overlap. However, the gap between the two bands is narrower than in an insulator. Silicon is a common and inexpensive semi-conductor.

The principles underlying the generation of electricity in a photovoltaic cell are illustrated in Figure 2.6. Conductors and semi-conductors have empty electron sites, called *holes*, within the valence and conduction bands. A semi-conductor that will be useful for converting sunlight to electricity should have many electrons and few holes in one band, and vice versa in the other band. Then, when an electron absorbs a photon, it will jump from one band to another with relatively little flow in the reverse direction. To enhance the net flow, impurities are added to the semi-conductor so as to create a valence band with excess electrons – a process called *doping*. Silicon has a valence of four (it has four electrons in its outer shell), so substances with a valence of five (phosphorus, arsenic, antimony) are added to a layer of silicon, forming an *n-type* layer. The extra electron from each phosphorus atom, for example, is not taken up in the crystal lattice. To create a conduction band with an excess of holes, a dopant with a valence of three is added (typically boron, aluminium, gallium or indium). This forms a *p-type* layer. The juxtaposition of the two layers creates a *p–n junction*. The energy difference between the valence and conduction bands is called the *band gap*.

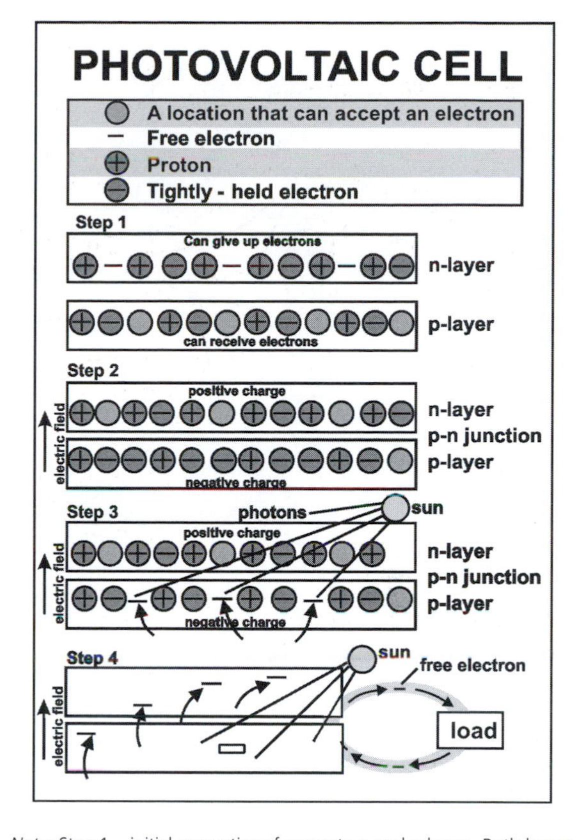

Note: Step 1 – initial properties of separate n and p layers. Both layers have overall charge neutrality, but some of the electrons in the n-layer are not taken up by the crystal lattice and so are free to move, whereas the p-layer has room for extra electrons. Step 2 – when the two layers are brought together, mobile electrons move from the n-layer to available sites in the p-layer. Thus, the n-layer has a positive charge, and vice versa, creating an electric field near the p–n junction. Step 3 – absorption of photons by electrons elevates some from the valence band energy state to the conduction band energy state. Step 4 – because of the electric field created by the p–n junction, electrons flow toward the n-layer and through the external circuit.
Source: US EIA (2007)

Figure 2.6 *Steps in the generation of electricity in a PV cell*

The layout of a solar cell is shown in Figure 2.7. When the p–n junction is first formed, some electrons in the immediate vicinity of the junction will move from the n-side to the p-side. This sets up a *reverse* electric field that is negative on the p-side and positive on the n-side. When an electron near the p-n junction absorbs a photon and jumps from the valence band to the conduction band, it leaves a hole behind and, under the influence of the reverse electric field, flows into the

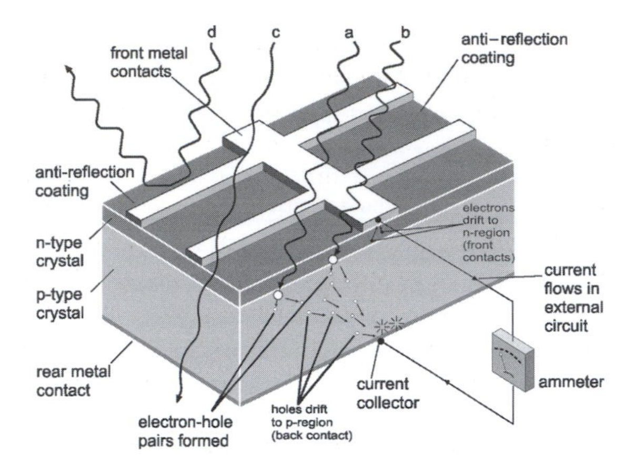

Note: Photon 'a' has an amount of energy just sufficient to dislodge an electron from its orbital next to a silicon atom near the p–n junction, causing it to enter the conduction band energy state. It then migrates into the n-type layer under the influence of the electric field straddling and set up by the p–n junction. A hole is left behind, which is filled by an electron from the p-layer, creating a hole in that layer, which progressively 'migrates' to the bottom electrical contact. The electron in the n-layer finds it way to the top electrical contact, travels through the external circuit to the bottom contact, and fills a waiting hole. Photon 'b' has more energy than needed to dislodge an electron, and the excess is dissipated as heat. Photon 'c' has less energy than the band gap and passes right through the material as if nothing was there. Other photons are reflected by the surface or absorbed by the current collectors on the cell surface.
Source: Boyle (1996)

Figure 2.7 *Layout of a silicon solar cell*

n-region. If there is an external circuit, the moving electron will eventually flow out of the semi-conductor through one of the metallic contacts. The hole that is left behind migrates[4] in the opposite direction until it reaches the other metallic contact and is filled by an electron coming in from the external circuit. The power produced by a solar cell is given by the product of voltage and current, the voltage being provided by the internal electric field at the p–n junction.

Photons with energy equal to or greater than the band gap will be able to bump electrons from the valence to the conduction band. However, for photons with energy greater than the band gap, the excess energy is converted to heat. Photons at longer wavelengths (less energy) will not be able to bump electrons, but instead pass through the semi-conductor. For a silicon semi-conductor at 20°C, the band-gap energy is 1.12 electron volts (eV), while the photon energy at the wavelength of peak emission from the sun (0.55µm) has an energy of 2.5eV. Because most photons either do not have enough energy to bump an electron, or have excess energy that is wasted, the maximum possible efficiency for converting sunlight to electricity using a single band gap is limited. The limitation on the efficiency depends on the distribution of solar radiation with respect to wavelength and the wavelength of radiation where the band gap occurs. Table 2.2 lists the band gap and corresponding wavelength, maximum theoretical efficiency and minimum required thickness in order to achieve the maximum efficiency for the semi-conductor materials of greatest interest. Many materials have theoretical efficiencies of 33–38 per cent. However, if cells with different band gaps are stacked on top of each other, with the cell having the largest band-gap energy

Table 2.2 *Semi-conductor materials of greatest efficiency for the PV generation of electricity, the band gaps and corresponding wavelength of radiation, the maximum theoretical efficiency, and the minimum thickness required for the maximum theoretical efficiency*

Material	Band gap (eV)	Wavelength (µm)	Maximum theoretical efficiency	Minimum required thickness (µm)
x-Si	1.1	1.129	33	32.4
a-Si	1.75	0.710	38	0.127
CIGS	1.21	1.026	33	0.05
CdTe	1.44	0.863	33	0.436
CdSe	1.73	0.718	29	0.211
CdS	2.42	0.513	14	0.161
GaAs	1.38	0.900	24	0.659
InP	1.35	0.920	34	0.236

Note: x-Si = crystalline silicon, a-Si = amorphous silicon
Source: Wadia et al (2009, Supplementary information)

on top so that photons with smaller energies pass through to the next cell, then the maximum theoretical efficiency increases: to 43 per cent for two cells, 49 per cent for three cells and 66 per cent for an infinite number of stacked cells. Actual efficiencies at present are far below these limits.

All of the materials listed in Table 2.2 have a minimum required thickness well below 1μm, except crystalline silicon (x-Si), for which the minimum required thickness is about 32μm. All other materials are thus referred to as 'thin-film' materials. The actual thickness of solar cells at present is generally much greater than the minimum required thickness.

2.2.2 Kinds of PV cells and modules and their efficiencies

The major different kinds of solar cells and the processes by which they are produced are described here. Green (2001) provides a detailed review of the physics, structure and manufacture of the major types of solar cells. Solar cells can be made from a single crystal of silicon, from multiple crystals, from amorphous (non-crystalline) silicon, or from other materials. A solar cell has an area of 100–200cm^2, and 50–60 cells are placed next to each other and wired together in series to form a module with a typical area of 0.4–1.0m^2. The module is covered with a weather-proof glass coating. In wireless modules, the modules are connected in parallel to their metal frame, which serves to transport the current to the inverter.

The best cell and module efficiencies (ratio of electrical energy produced to incident solar energy) achieved as of May 2009 under laboratory conditions with normal and concentrated sunlight for different technologies are given in Table 2.3. Module efficiencies are less than cell efficiencies because slight differences between adjacent cells cause some of their

Table 2.3 *Best laboratory efficiencies of PV cells and modules achieved as of May 2009 for various technologies at a temperature of 25°C*

Technology	Unconcentrated sunlight		Concentrated sunlight	
	Cell	Module	Cell	Concentration factor
		Single-junction silicon semi-conductor		
c-Si	25.0	22.9	27.6	92
m-Si	20.4	15.5		
a-Si	9.5			
thin-film m-Si	16.7	8.2		
		Single-junction compound semi-conductors		
CdTe	16.7	10.9		
CIGS	19.4	13.5[a]	21.8	14
InP	22.1			
thin-film GaAs	26.1		28.8	232
m-GaAs	18.4			
		Multi-junction semi-conductors		
a-Si/μc-Si	11.7[b]			
a-Si/a-SiGe/a-SiGe		10.4		
thin-film GaAs/CIS	25.8			
GaInP/GaAs	30.3			
GaInP/GaAs/Ge	32.0		40.7	240
GaInP/GaInAs/Ge			41.1	454
		Photochemical and organic		
Dye-sensitized	10.4			
Organic	5.2			

Note: c-Si = single-crystalline silicon, m-Si = multi-crystalline silicon, a-Si = amorphous silicon, and μc-Si = micro-crystalline silicon. The efficiency decreases with warming for all cell types except a-Si, where the temperature effect is close to zero, and also varies during the day and with atmospheric conditions due to changes in the relative amounts of solar radiation at different wavelengths. [a] Pertains to CIGSS (CuInGaSSe); [b] submodule.
Source: Green et al (2009)

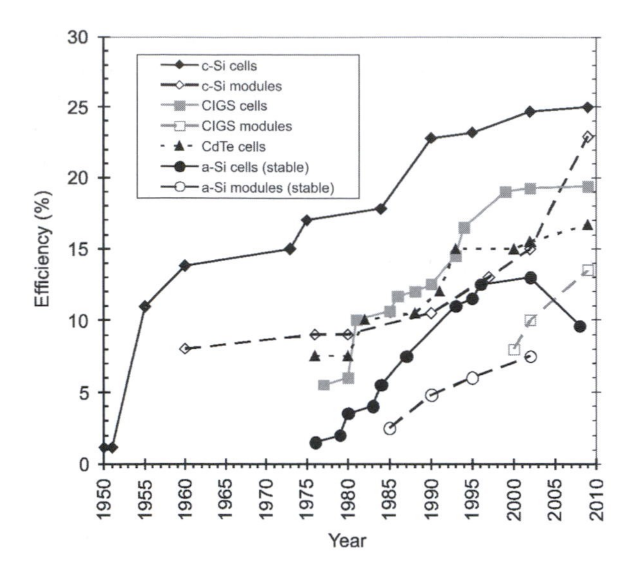

Source: IEA (2003) for data up to 2002, with 2009 data from Table 2.3

Figure 2.8 *Progression in the efficiency of crystalline silicon (c-Si), amorphous silicon (a-Si) and other types of PV cells and modules*

electrical power output to cancel out, because there are inactive regions between cells, and because of resistance losses in the wiring between cells and in the diodes used to protect cells from short-circuiting (Boyle, 1996). When sunlight is concentrated onto a cell, only small active areas are required and efficiencies tend to be larger. Efficiencies achieved in commercial modules (when available) tend to be 80 per cent of the efficiency achieved by laboratory modules, due to further imperfections introduced through mass production. The increase in the best laboratory cell and module efficiencies from 1976 to mid-2009 is shown in Figure 2.8 for selected cell and module types.[5]

Single-crystalline silicon

To produce single-crystalline silicon, silica (SiO_2) is first extracted from sand, refined and melted. As the molten silicon cools, a seed crystal is added, slowly rotated and pulled away from the melt, pulling out a large tubular ingot of crystalline silicon. A p-type dopant is usually added during this slow crystallization phase. Once the ingot has solidified, it is removed from the mould and sliced into wafers about 300μm thick (the 2004 average). The wafers are then etched in various ways in order to remove surface damage and to create square-based pyramidal structures on the surface that minimize the reflection of sunlight (leaving more sunlight to produce electricity). An n-type dopant is made to seep into the wafers through a heating process, and then top and bottom contacts in the form of a paste with silver particles and other additives are added to the wafers. Alternatively, the silver-based contacts may be embedded into the silicon material itself, improving output efficiency by up to 30 per cent over paste contacts (Green, 2000). Because each wafer is cut from a single crystal of silicon, the cells are called single- or mono-crystalline cells (c-Si).

As of May 2009, the best cell efficiency obtained in the laboratory was 25.0 per cent and the best module efficiency was 22.9 per cent (Table 2.3). Given that the efficiency of commercially available PV modules is typically 80 per cent of the efficiency measured under laboratory conditions, we are close to commercial single-crystal silicon modules with an efficiency of 20 per cent.

Multi-crystalline silicon

Multi-crystalline silicon (m-Si) consists of numerous individual grains made from lower-grade silicon and under less stringent conditions than for single-crystal silicon. A seed crystal is not used, so ingots are formed by cooling molten silicon directly in cubical moulds. Due to the presence of crystal variations and transfer boundaries, electron conduction is more frequently impeded than in a c-Si cell. The best laboratory cell and module efficiencies are a little lower than for c-Si cells and modules, but production costs are also lower. m-Si cells can be directly produced in the form of a ribbon or sheet, which saves the cost of wafering as well as wasting less material (Green, 2003). In one manufacturing process, a film of molten silicon is trapped between two wires drawn through a molten bath, with the trapped film then solidifying as a ribbon. Module efficiencies are a few per cent lower than for the usual wafer-based manufacturing process.

Both single-crystalline and multi-crystalline cells take a lot of time to produce (about two days per ingot), require a lot of energy and must be sliced into thin wafers. These limitations are overcome by thin-film cells, described next.

Thin-film amorphous silicon cells

PV cells can be made through direct deposition of a semi-conductor onto a substrate (usually glass), rather than by making wafers. Typically, less than 1μm of semi-conductor material is used, reducing the material requirements by a factor of more than 300 compared to crystalline silicon cells with the 2004 average thickness of 300μm. This in turn allows almost any semi-conductor to be economically competitive with silicon. Of course, use of such thin layers requires materials that are better than silicon in absorbing sunlight, and there are semi-conductors that are up to 100 times more effective.

Thin-film modules can also be made from amorphous silicon, but special techniques are required to give it a reasonable efficiency. The first amorphous silicon (a-Si) cells, made in 1973, had an efficiency of only 1 per cent, due in part to their irregular molecular structure. The efficiency has been dramatically improved by adding up to 10 per cent hydrogen by volume (giving a-Si:H) to make the structure more regular. However, exposure to strong sunlight undoes some of the benefit of adding hydrogen, leading to a decrease in cell efficiency during the first thousand hours of use. The norm now is to report the lower stabilized efficiencies for a-Si cells, rather than initial efficiencies.

In crystalline cells, the most active part of the solar cell is at the p–n junction, due to the electric field in this region. In a-Si cells, the active region is made to extend over almost the entire cell by making the p- and n-type regions very thin, with an undoped region between them. A thinner cell creates a stronger electric field, improving the cell efficiency. It turns out that the required cell thickness is less than that needed to absorb all of the usable incident sunlight, so several cells are stacked on top of one another so that light not absorbed by the upper cell can be absorbed by a lower cell. This works better if the cells have different band gaps, with a cell good at absorbing blue light on top and a cell good at absorbing red light at the bottom. The different band gaps are made from different alloys of a-Si. An example of a three-cell (or triple-junction) arrangement is (beginning from the top) a-Si/a-SiGe/a-SiGe, with the lower a-SiGe layer having a smaller band gap than the upper a-SiGe layer. Stabilized multi-junction a-Si module efficiencies have reached 10.4 per cent under laboratory conditions (see Table 2.3). The best stabilized efficiency in commercial modules in 2008 was 8.5 per cent for an a-Si/microcrystalline-Si module by Sharp (Hegedus, personal communication, 2008). Amorphous silicon and other thin-film cells are suited to continuous production, with continuous deposition on a moving substrate (such as flexible plastics) involving layers as thin as 10–30nm (nanometres), 1m wide and 2km long (Hegedus, 2006).

Thin-film, multi-crystalline silicon cells

Crystalline silicon is not a good absorber of light compared to some compound semi-conductors or a-Si, so early attempts to produce thin-film multi-crystalline silicon cells were not successful. However, if the light direction is randomized by striking a rough surface once inside the cell, the effective optical thickness of the cell is increased by a factor of 50 (Green, 2000). Thin-film multi-crystalline cells can be produced through high-temperature deposition of silicon onto a substrate, melting silicon after deposition, or by depositing silicon in amorphous form and then crystallizing it by heating it for prolonged periods at intermediate temperatures. A cell efficiency of 16.7 per cent has been achieved (see Table 2.3).

Thin-film compound semi-conductors

In silicon cells, the p–n junction is formed by doping two layers of the same material (silicon) whereas in compound cells, the p–n junction is formed by doping two different semi-conductors. The common compound semi-conductors are cadmium telluride (CdTe), copper-indium-diselenide ($CuInSe_2$, abbreviated as CIS) and copper-indium-gallium-diselenide (CIGS). A CIGS laboratory cell achieved an efficiency of 19.4 per cent, and a module efficiency of 13.5 per cent has been achieved (see Table 2.3). Cadmium telluride cells also give good efficiency (16.7 per cent maximum so far). Some CIGS and CdTe modules have shown long-term stable performance, while others have had deteriorating performance. It is not clear why some modules degrade and others do not, indicating a need for further fundamental research concerning these technologies (Noufi and Zweibel, 2006). Research is continuing in other thin-film technologies, primarily for use in spacecraft (where the absolute amounts used will be small and cost is not an important consideration). Among these are various alloys of gallium arsenide (GaAs), which have achieved cell efficiencies of over 30 per cent in multi-junction arrangements. The highest

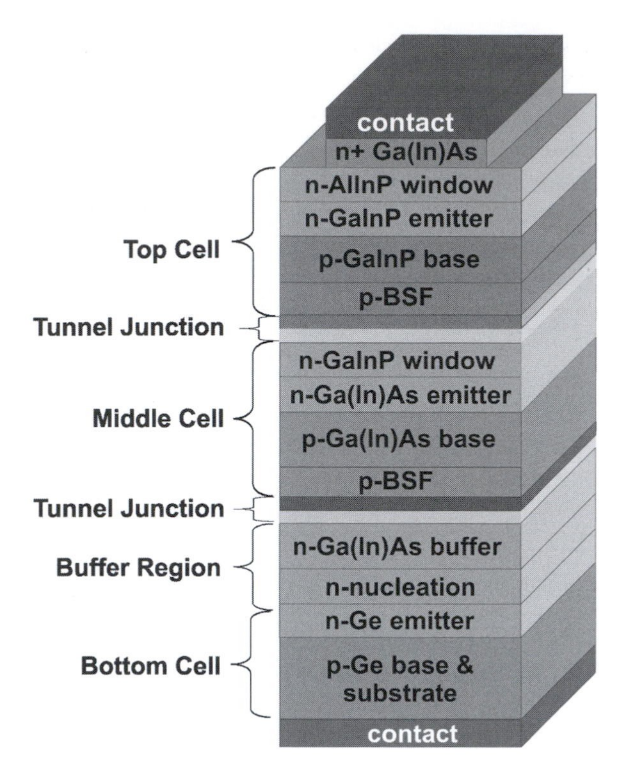

Note: As conveyed to the author by G. Kinsey, the 'emitter' layer is the n-doped side of a p–n junction. Together with the p-doped 'base', the emitter helps create the built-in electric field that allows for collection of electrons. The 'window' layer is a transparent layer that serves to passivate the front surface of the emitter. Without the window, the front emitter surface would become the termination point for the semi-conductor stack. The terminating surface of a semi-conductor has dangling bonds that lead to carrier recombination that, in turn, reduces cell voltage. The window provides an inactive stand-off layer so that surface recombination does not affect the emitter. BSF stands for 'back surface field'. It is a highly doped layer that likewise prevents carrier recombination at the back of the sub-cells.
Source: Kinsey et al (2008)

Figure 2.9 *Structure of the triple-junction GaInP/GaInAs/Ge cell*

efficiency achieved so far is 41.1 per cent for a GaInP/GaInAs/Ge cell under sunlight concentrated by a factor of 454. The structure of this kind of cell is illustrated in Figure 2.9.

Compound semi-conductors with spectral splitting

In the thin-film compound semi-conductors described above, different wavelengths of light are absorbed sequentially as the light passes through successive semi-conductors. The arrangement of different semi-conductors on top of each other constrains the performance of each individual semi-conductor because the system as a whole needs to be optimized. However, it is possible to optically concentrate sunlight, split the light into different spectral (wavelength) regions, and direct the different wavelengths onto different semi-conductors within a single optical/semi-conductor device. A research programme in the US is targeting an efficiency of 53 per cent with a six-junction cell design using sunlight concentrated by a factor of 20 (Barnett et al, 2009). The theoretical maximum efficiency for a six-junction cell with unconcentrated sunlight and a cell temperature of 25°C is 60 per cent (Kurtz et al, 2008a).

Nanocrystalline dye cells and other organic cells

A variety of organic molecules, illustrated in Figure 2.10, are semi-conductors and so can generate electricity through the formation of electron-hole pairs. Organic

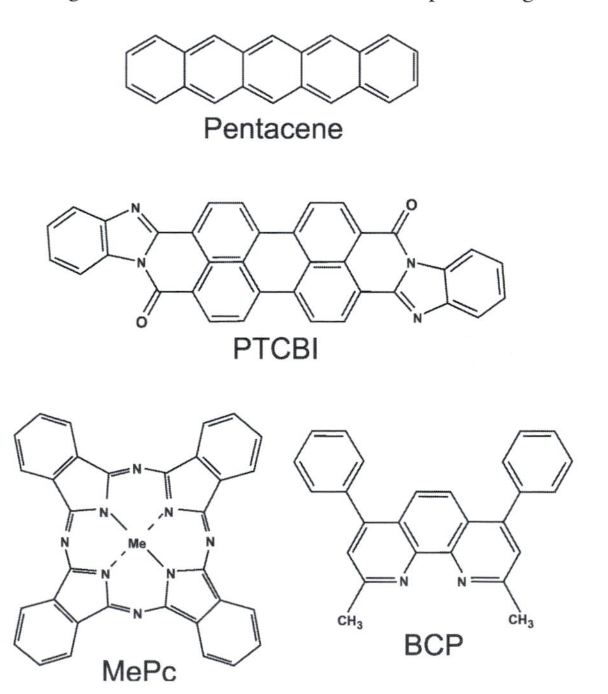

Source: Rand et al (2007)

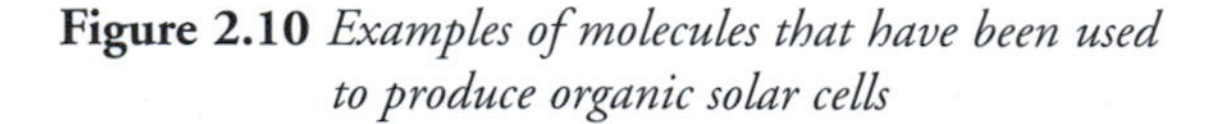

Figure 2.10 *Examples of molecules that have been used to produce organic solar cells*

semi-conductors have the potential of low cost, ease of processing and compatibility with flexible substrates. They have been used in organic light-emitting diodes, organic transistors and PV cells. The most developed organic cells are *nanocrystalline dye-sensitized cells*, which consist of a porous matrix of titanium oxide crystals that is coated with a one-molecule-thick layer of an organometallic dye. The crystals are in the order of 10nm in size (1nm = 10^{-9}m). The dye absorbs sunlight, creating an excited electron that is transferred to the titanium oxide matrix, and through the matrix to the cell contacts and an external circuit. Electrons transferred from the dye to the matrix are replaced with electrons transferred from an electrolyte to the dye. The matrix structure and the dye cell are illustrated in Figure 2.11. The dye cell differs from a semi-conductor cell in that (1) absorption of solar radiation and electron transport occur in separate media, and (2) there is no internal electric field inducing electrons to flow to a cell contact. The dye cell mimics photosynthesis, which also uses an organometallic compound (chlorophyll) to absorb sunlight, creating excited electrons that are then transferred to a transport medium. The dye cell absorbs across the visible part of the spectrum and slightly into the near infrared wavelengths. The organic dye contains ruthenium, a rare element, but significant amounts could become available in the future as a byproduct of mining platinum for automotive fuel cells (Andersson, 2000).

The current (2009) record dye-cell efficiency is 10.4 per cent, obtained under standard text conditions – 25°C temperature, AM1.5 (air mass path length of 1.5) solar radiation. At present, the cells have an operating lifespan of only five to ten years (Rand et al, 2007). Unlike other semi-conductor-based solar cells, the dye-cell efficiency increases with increasing temperature. In one experiment reported by McConnell (2002), power output increased by 25 per cent in going from 30°C to 65°C, and by another 15 per cent in going from 65°C to 80°C. The technology is still relatively new (the initial breakthrough report was published in 1991). Dye-sensitized cells can be manufactured to be transparent in the visible part of the spectrum, which means that they can be used on windows (Phani et al, 2001). They are commercially available from several manufacturers.

Plastic (polymer) cells

Plastics can be made that conduct electricity, and placement of a titanium oxide layer within such a

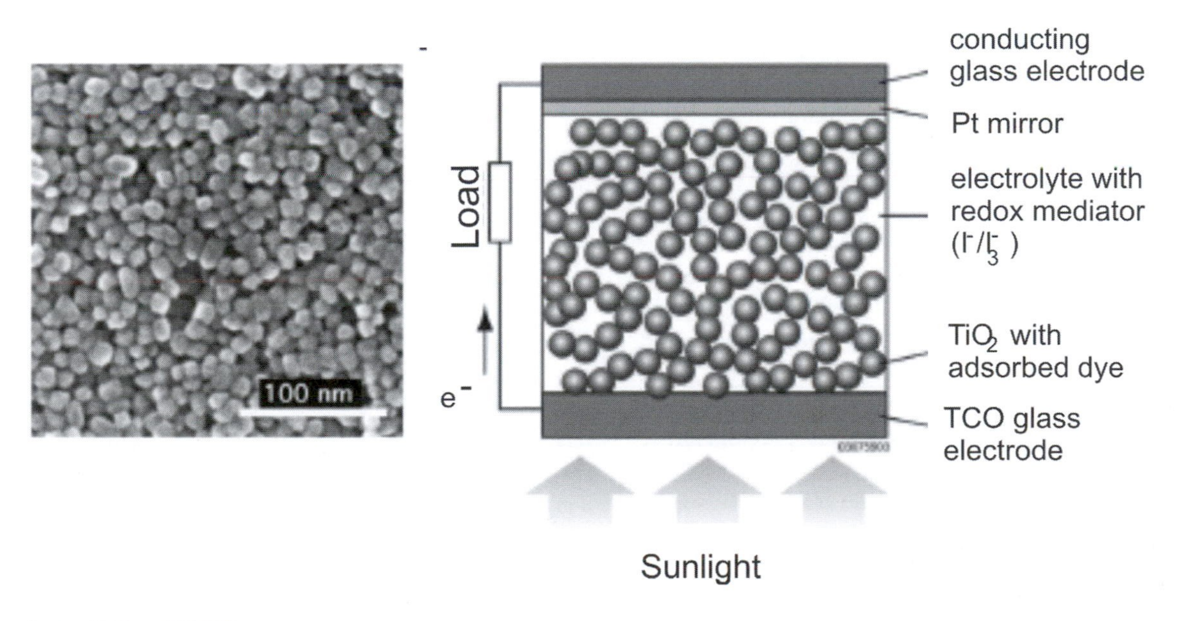

Source: McConnell (2002)

Figure 2.11 *Titanium oxide matrix of a dye-sensitized solar cell (left) and structure of a dye-sensitized solar cell (right)*

plastic makes it more sensitive to radiation near the peak solar irradiance. As of mid-2006, an efficiency of 5.6 per cent had been obtained, which is substantially below that of crystalline silicon or even dye-sensitized solar cells. At the moment, the cells last only one to two years and no commercial products are available yet. The appeal of plastic solar cells is the potential of very low cost.

2.2.3 Effect of solar irradiance, temperature and dust on module efficiency

Module efficiencies depend on the module temperature, the total solar irradiance and the proportion of the irradiance in different wavelength intervals. As the solar irradiance increases, the current generated by a solar cell at the maximum power point (MPP, defined later) and at a fixed temperature increases in direct proportion to the irradiance. If cell voltage were fixed, the output would vary in direct proportion to irradiance and the efficiency would be independent of irradiance. However, the voltage decreases slightly with decreasing irradiance (as seen later in Figure 2.14), so efficiencies would be lowest at very small irradiance and increase continuously with increasing irradiance. However, the module temperature, effective series resistance and effective leakage conductance also increase with irradiance. The first two effects reduce the efficiency by an amount that increases with irradiance while the third effect reduces the efficiency by an amount that varies inversely with irradiance (Topic et al, 2007). The net result is that cell efficiencies are largest at a solar irradiance of about 500W/m^2, dropping off sharply for an irradiance less than 200W/m^2 and slowly for an irradiance greater than 500W/m^2. The calculated variation of efficiency with irradiance for different PV technologies, and measured efficiencies for a crystalline silicon (c-Si) module, are shown in Figure 2.12. The calculated variation assumes that the module temperature increases from 20°C to 50°C at an irradiance of 1000W/m^2, which is similar to the variation associated with the measured efficiencies. However, unventilated PV panels can reach temperatures of up to 80°C, so the potential decrease in efficiency at high irradiance is

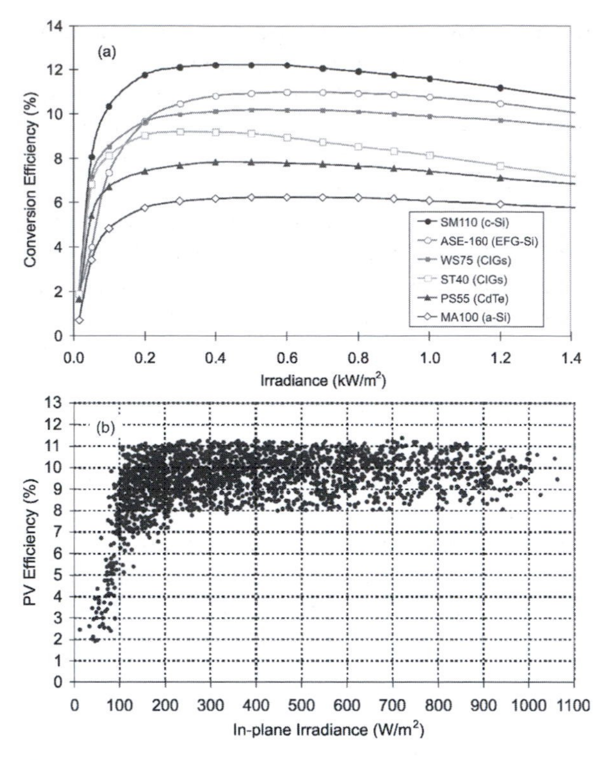

Source: (a) Topic et al (2007); (b) Mondol et al (2007)

Figure 2.12 *(a) Calculated variation of module efficiency with solar irradiance, and (b) measured variation of the efficiency of a c-Si module with irradiance*

larger than shown in Figure 2.12. Each Celsius degree increase in module temperature decreases the electricity output by 0.4–0.6 per cent for c-Si and CIGS, and by 0.2–0.3 per cent for a-Si and CdTe (Hegedus, personal communication, 2008).

Standard test conditions for solar modules are 1000W/m^2, 25°C and an air mass pathlength of 1.5 – conditions that are encountered only a few hours per year. It is therefore necessary to develop equations to relate the efficiency under standard test conditions to the conditions as they vary over the course of a year at a given location. Huld et al (2008a) have done this for crystalline silicon modules in Europe. They find that annual average electricity output in Southern Europe is up to 12 per cent less than expected based on the module rating, but only 3–5 per cent less in Northern Europe. The difference is due mainly to the effect of temperature. As noted above, the effect of temperature on efficiency is smaller for a-Si cells than for c-Si cells, so output of a-Si cells should be closer to expectations and with smaller regional variations.

Dust deposition in desert areas can significantly decrease the performance of PV modules after as little as one week, particularly in regions of high winds and for modules with low tilt, but not all desert areas are subject to significant dust, and module arrays can be located relative to hills so as to minimize dust deposition (Goossens and Kerschaever, 1999). Nevertheless, weekly cleaning could be required in some areas (Hegazy, 2001). Large utilities in the southwest US report a 6 per cent loss due to dust, which can be reduced to 3 per cent with a single washing during mid-summer. In non-desert areas, periodic rainfall is usually sufficient to keep the modules clean.

2.2.4 Balance-of-system components for PV power

In addition to the module, the other components of a PV power system are (1) the support structure, and (2) the power electronics, which is used to determine and produce the voltage–current combination in the module at each instant that maximizes the power output for the given solar irradiance and to produce alternating current (AC) electricity from the direct current (DC) electricity that the modules produce.

Support structures

Solar modules can be placed horizontally, inclined at the latitude angle with zero azimuth (which maximizes the mean annual solar irradiance for fixed modules), or mounted on one- or two-axis tracking systems. By tracking the sun, the modules can be maintained in an orientation always perpendicular to the sun's rays, which maximizes the energy output but also increases costs. Solar modules can also be built as part of structural materials, such as roof shingles, wall siding or windows – referred to as *building-integrated* PV (BiPV). This reduces the module cost further (once credit is given for the structural material that it replaces), although energy output will be smaller since the orientation is constrained by the building. BiPV modules are discussed later.

Inverter

An inverter converts the DC power output of the module to high-quality AC power at the right voltage for transfer to the grid. Some modules have an inverter mounted on the back of the module, and produce AC power; these are called AC modules. Inverters can have a low-frequency transformer, a high-frequency transformer, or no transformer at all. Figure 2.13 shows how the inverter efficiency (ratio of AC power to DC power) varies with the load (the top panel shows theoretical calculations, while the bottom panel gives measured data, the latter showing substantial scatter).

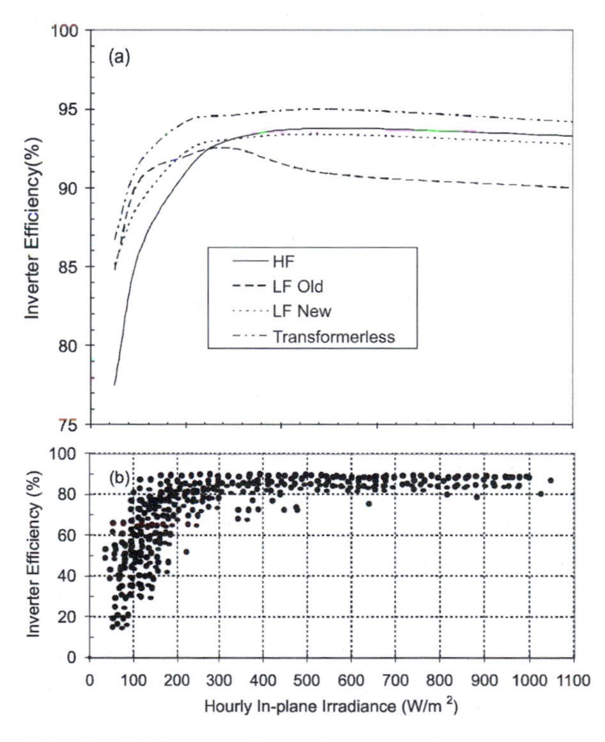

Note: HF = high frequency; LF = low frequency.
Source: (a) Data in Abella and Chenlo (2004), (b) Mondol et al (2007)

Figure 2.13 *(a) Calculated and (b) measured variation of inverter efficiency with load*

The efficiency drops significantly below 20 per cent of full load. This efficiency loss can be reduced if the PV array is configured to use several inverters at full load under conditions of maximum solar irradiance, but to use progressively fewer inverters (operating at close to full load) as the solar irradiance decreases (Abella and Chenlo, 2004). Inverter efficiency decreases with increasing module temperature.

Consider a test situation, with a solar flux of 1000W/m^2 incident on a solar cell. If the external circuit is open, there will be no current flow and the voltage is at a maximum of 0.6V (this is the *open-circuit voltage*). If the external circuit is closed with zero resistance (i.e. the cell is short-circuited), the current will be at maximum of about 2.8 amps and the voltage will be zero. As the resistance in the external circuit varies between these extremes, the voltage and current will co-vary as shown in Figure 2.14. Since power (P) is equal to voltage times current, there will be a single point called the maximum power point (MPP) at which P is maximized. The short-circuit current and the current at the MPP vary in direct proportion to the intensity of the available sunlight, while the voltage at the MPP decreases slightly with increasing irradiance. In PV systems, an electronic circuit is used to automatically vary the resistance of the external circuit so as to maintain the operation of the cell at the maximum power point as the intensity of sunlight varies. A typical cell will operate at a voltage of 0.5V. If 40 cells are connected in series (end-to-end) in a module, the module voltage will be 20V.

MPP tracking is usually done through a 'perturb-and-observe' operation (Abella and Chenlo, 2004). After a given time interval, the operating voltage is increased by a small amount and the change in power (ΔP) is measured. If ΔP is positive, the voltage is increased further; otherwise, it is decreased. When the incident solar irradiance is low, the power curve is very flat, so it is difficult to determine the true position of the MPP. In any case, the inverter voltage oscillates around the true MPP. Finally, the tracking algorithm can exhibit erratic behaviour under rapidly changing irradiance levels. The MPP efficiency is the ratio of the power output to the power output with perfect MPP tracking. For the reasons explained above, the efficiency is close to but less than 100 per cent at full load, then decreases

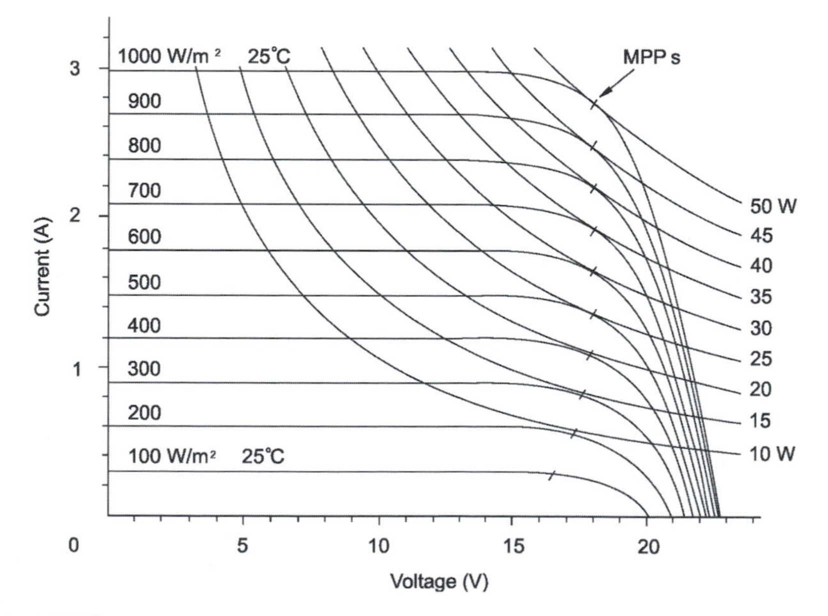

Source: Hastings and Mørck (2000)

Figure 2.14 *Variation of current with voltage in a given PV module for various solar irradiances, and the maximum power point for each irradiance*

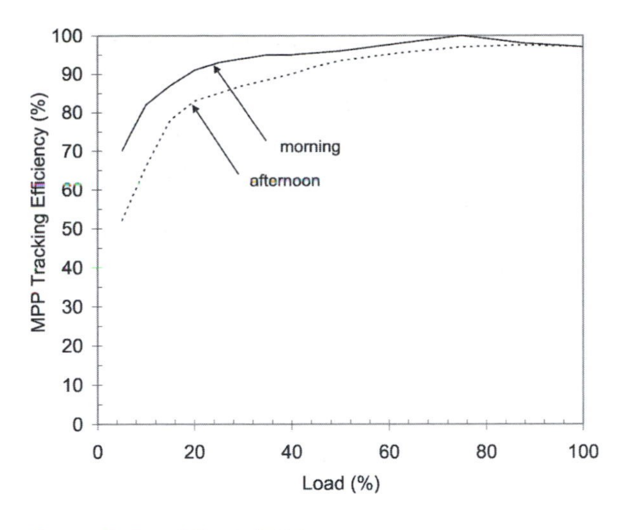

Source: Abella and Chenlo (2004)

Figure 2.15 *MPP tracking efficiencies versus load*

first gradually and then sharply with decreasing load, as shown in Figure 2.15. Measured daily average MPP efficiencies for some recent inverters are 86–96 per cent on sunny days and 42–94 per cent on cloudy days. Inasmuch as quoted module efficiencies assume that the module operates at the MPP, the efficiency in producing AC power is equal to the product of the module, inverter and MPP efficiencies.

Concentrating mirrors or lenses

Some PV systems can physically track the position of the sun so that the module is always oriented perpendicularly to the sun's rays (this is distinct from the MPP tracking discussed above). Other systems concentrate the sunlight optically onto a cell that has been reduced in size (and cost). Concentrating systems can be either tracking or non-tracking. Sun tracking and concentrating systems are discussed below in subsections 2.2.7 and 2.2.8, respectively.

2.2.5 System efficiency

The efficiency of a PV system is equal to the module efficiency times the inverter efficiency times a factor to account for further losses. These additional losses include dust on the modules, temperatures higher than the conditions for the specified module efficiency, wiring losses and module mismatches. The ratio of AC energy output (kWh) to module DC energy output for complete systems is referred to as the *performance ratio.* The average performance ratio of 19 residential PV systems monitored in California by Scheuermann et al (2002) was only 62 per cent, with a range from 46 per cent to 86 per cent. A similar range of performance ratios was found in a survey of 18 systems in Europe (Nordmann and Clavadetscher, 2003). For 333 residential systems in Japan, Nakagami et al (2003) report a range from less than 0.5 to better than 0.85, with an average of 0.71. For systems installed in Germany between 1991 and 1994, Jahn and Nasse (2004) report performance ratios of 0.30 to 0.87, with an average of 0.64 but falling over time for any given building, whereas for systems installed after 1996, the performance ratio ranges from 0.52 to 0.91, averages 0.74 and is constant over time. For a $3.5MW_{DE}$, utility-scale PV array in Arizona, Moore and Post (2008) report a performance ratio averaged over five years of 79 per cent.

2.2.6 Building-integrated PV modules

PV modules can be attached to conventional roof and façade materials, or they can replace conventional materials. In the latter case, they serve as an integral part of the building façade or roof, and so are called BiPV. PV modules can also serve as skylights in atria or as fixed or adjustable external shading devices. The major ways of adding or integrated PV into buildings are:

- Mounting onto a sloped roof – this can often be the cheapest method of mounting PV modules, due to its simplicity (there is no need for a water-tight structure). Modules can either be attached to clamps that can be attached to roofing tiles with screws, or attached to special tiles having built-in plates. An example is shown in Figure 2.16.
- Integration into a sloped roof – the modules displace conventional roofing material, must form a water-tight layer and tend to be more aesthetically pleasing than modules mounted onto the roof. They can take the form of overlapping tiles or panels, or non-overlapping units, as illustrated in Figure 2.17. Bahaj (2003) reviews the state-of-the-art concerning PV roofing tiles and shingles. Examples of houses with BiPV roofs are given in Figure 2.18. Membrane and drain trays provide weather barriers as part of the PV system.

Source: Klöber GmbH & Co. KG, Germany (www.kloeber.de)

Figure 2.16 *PV panels mounted onto a sloping roof*

Source: Top, Omer et al (2003); bottom, Peter Toggweiler, Enecolo AG, Switzerland (www.solarstrom.ch)

Figure 2.17 *PV panels integrated into a sloping roof, either as shingles on the Eco-Energy House, University of Nottingham (top) or as flush panels (bottom)*

- Tilted modules on a flat roof – this requires a support structure, but provides for cooling of the modules by airflow underneath. The building cooling load will also be reduced due to shading of the roof. Gravel on roofs can be used as ballast to hold the modules in place, along with tension cables.
- Horizontal modules on a flat roof – modules can be used in place of conventional roofing material, perhaps with attached interlocking insulation, as illustrated in Figure 2.19. Alternatively, modules can be elevated above the roof for ventilative cooling.
- Modules on façades – modules can be mounted on vertical façades in a variety of ways. They may be physically integrated, as in the Kyocera building in Austria (see Figure 2.20, left) or visually but not physically integrated, as in the façade of the Brundtland Centre in Denmark (see Figure 2.20, right). When used in curtain walls, PV modules can alternate with window glazing, as in the Condé Nast Building, New York (see Figure 2.21). A Japanese firm produces a window with laser-etched a-Si laminated between two layers of glass, with 4.1 per cent efficiency (www.msk. ne.jp). It has 10 per cent light transmission, which optimizes the balance between heat gain and daylighting in Japan. PV modules are used as fixed sunscreens in a building of The Netherlands Energy Research Foundation, as illustrated in Figure 2.22.
- Modules as skylights in atria – BiPV is now being used as a design element in its own right as a part of skylights (see Figure 2.23). When used as skylights, the overall transmittance and appearance of PV modules can be altered by varying the proportion of the skylight with and without opaque crystalline cells. In thin-film modules, a wide variety of patterns can be laser-etched.

Source: Top, Hestnes (1999)

Figure 2.18 *BiPV in the residential sector: single-family house in Pietarsaari, Finland (top); single-family house in coastal Maine (bottom)*

Source: www.powerlight.com

Figure 2.19 *Illustration of PV modules separated by an air gap from interlocking insulation*

One of the largest examples of BiPV is the Nieuwland development in Amersfoort (Netherlands), which involves about 500 houses and 1MW peak power (Schoen, 2001). Other examples, as well as a discussion of the technical characteristics, design considerations, costs and benefits of BiPV can be found in Boyle (1996), Sick and Erge (1996), Thomas and Fordham (2001), Schoen (2001), Prasad and Snow (2005) and at the International Demonstration Centre and website of the International Energy Agency's (IEA) Implementation Agreement concerning PV power systems (www.iea-pvps.org). Van Mierlo and Oudshoff (1999) discuss the solutions pertaining to a large number of non-technical problems associated with the introduction of BiPV: financing, administration, architectural integration, communication, marketing and environmental considerations. Eiffert and Kiss (2000) provide design details and examples of particular interest to architects.

Gutschner and Task-7 Members (2001) have estimated the potential power production from BiPV in member countries of the IEA. They first estimated building ground floor areas from population data, from which roof and façade areas were estimated. They then took into account the architectural suitability (based on limitations due to construction, historical considerations, shading effects and the use of the available surfaces for other purposes) and solar suitability (taking into account the relative amounts of solar radiation on surfaces of different orientation). Based on architectural considerations, they find that the suitable roof area on average is 60 per cent of the ground floor area and the suitable façade area is 20 per cent of the ground floor area. Restricting the usable roof and façade area to those elements where the solar irradiance is at least 80 per cent of that on the best elements in a given location, defined separately for roof and façade elements, they find that the suitable roof area is 40 per cent of the ground floor area and the suitable façade area is 15 per cent of the ground floor area.

Table 2.4 gives the resulting estimates of the roof and façade areas available for BiPV for residential, commercial, industrial, agricultural and other buildings, assuming a 10 per cent sunlight-to-AC conversion efficiency. Also given, in the last column, is the percentage of present total electricity demand that could

Source: Left, Stromaufwärts, Austria (www.stromaufwaerts.at); right, BEAR Architecten, The Netherlands (www.bear.nl)

Figure 2.20 *PV modules physically integrated into the façade (Kyocera commercial office building, Austria, left) or visually integrated (Brundtland Centre, Denmark, right)*

Note: The BiPV panels (dark) are mounted in exactly the same way as the glazing (light).
Source: Eiffert and Kiss (2000)

Figure 2.21 *BiPV panels integrated into the curtain wall instead of conventional spandrel panels in the Condé Nast Building, New York*

be provided by BiPV. This ranges from about 15 per cent (Japan) to almost 60 per cent (US). Although one may question the specific assumptions used in this study, there is in any case a significant potential for electricity production from BiPV. Thus, systematic incorporation of BiPV into new buildings, and into old buildings when they are renovated and whenever this is feasible, can make an important contribution to electricity supply.

The European Commission has established a free design guide and software tool entitled PVCoolBuild, which can be downloaded from www.pvcoolbuild.com. Its goal is to facilitate better integration of PV into buildings, with particular attention to use of passive or active cooling to minimize the temperature of the PV module and the associated reduction in electricity output (see subsection 2.2.3).

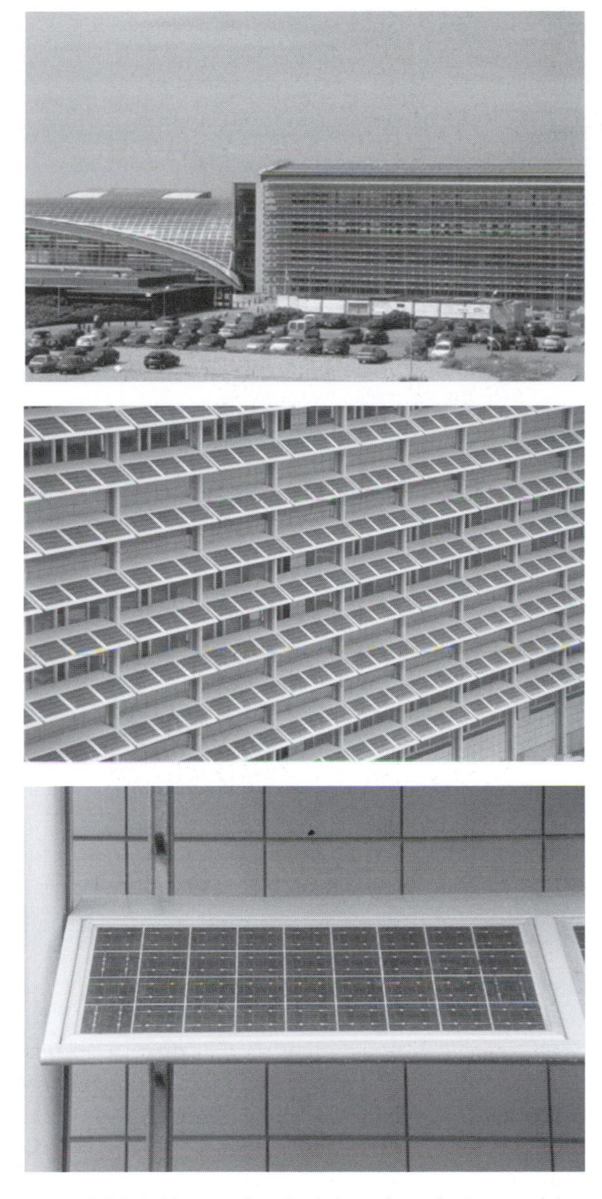

Source: BEAR Architecten, Case Study (www.bear.nl); photograph by Marcel von Kerckhoven

Figure 2.22 *PV panels as fixed sunscreens on Building No. 31 of The Netherlands Energy Research Foundation*

Source: Top, Henrik Sørensen, Esbensen Consulting; bottom, www.nsk.ne.jp

Figure 2.23 *PV providing partial shading in an atrium in (top) the Brundtland Centre, Denmark (KHR A/S, architect) and (bottom) Kowa Elementary School, Japan*

Parking lots represent another large area, close to electricity demand centres, that could be covered with PV modules while providing a significant co-benefit, namely, shading of parked cars. The Earth Policy Institute estimates the area of parking lots in the US to be 1.9 million hectares (see www.earth-policy.org/Alerts/Alert12_data2.htm). If this area were covered with PV modules at 15 per cent efficiency, and assuming an average solar irradiance of 180W/m^2, 4490TWh of electricity could be generated per year, which exceeds the 2005 US electricity demand of 4300kWh/year (see Volume 1, Table 2.4). As noted in Volume 1 (Chapter 5, subsection 5.4.5), parked bidirectional plug-in hybrid vehicles could be recharged with PV electricity at times of excess and could return some of it to the grid when there is a deficit. This system is referred to as a V2G (vehicle-to-grid) system.

Table 2.4 *Potential area and electricity generation from BiPV in selected IEA countries, and potential BiPV-generated electricity as a percentage of total domestic electricity demand*

Country	Potential BiPV area (km²)		Potential solar electricity (TWh/yr)			National electricity consumption (TWh/yr)	Potential BiPV portion (%)
	Roofs	Façades	Roofs	Façades	Total		
Australia	422.3	158.3	68.2	15.9	84.1	182.2	46
Austria	139.6	52.4	15.2	3.5	18.7	53.9	35
Canada	963.5	361.3	118.7	33.1	151.8	495.3	31
Denmark	88.0	33.0	8.7	2.2	10.9	34.4	32
Finland	127.3	33.0	11.8	3.1	14.6	76.5	19
Germany	1295.9	486.0	128.3	31.8	160.0	531.6	30
Italy	763.5	286.3	103.1	23.8	126.9	262.0	45
Japan	966.4	362.4	117.4	29.5	146.9	1012.9	15
Netherlands	259.4	97.3	26.7	6.2	31.9	99.1	32
Spain	448.8	168.3	70.7	15.8	86.5	180.2	48
Sweden	218.8	82.0	21.2	5.5	26.7	137.1	20
Switzerland	138.2	51.8	15.0	3.4	18.4	53.2	35
UK	914.7	343.0	83.2	22.2	105.4	343.6	31
US	10,096.3	3786.1	1662.4	418.3	2080.7	3602.6	58
Total/average	16,842.7	6301.2	2450.6	614.3	3063.5	7064.6	43

Source: Gutschner and Task-7 Members (2001)

2.2.7 Sun-tracking systems

A PV array can be mounted on a mechanical system that changes the orientation of the module during the course of the day. The simplest approach would be to tilt the array toward the south (in the northern hemisphere) at a fixed angle and have the module face east as the sun rises and rotate to the west as the sun sets. This is one-axis tracking, and causes the sun's rays to be closer to perpendicular to the module at most times than for a fixed module. With two-axis tracking the module can be oriented perpendicularly to the sun's rays at all times. Tracking the sun increases the intensity of direct-beam radiation on the module but has very little effect on the intensity of diffuse radiation (diffuse radiation comes from all directions, and is maximized on a horizontal module because tilted modules 'see' less of the sky). The benefits of tracking thus depend on the degree of cloudiness (and its seasonal variation), as this affects the proportions of direct- and diffuse-beam radiation. Compared to a module fixed at the optimal angle in a given location, Huld et al (2008b) finds that two-axis tracking increases the annual solar irradiance on a module by 20–30 per cent in the British Isles and other cloudy parts of Europe, by 30–40 per cent in Southern and Eastern Europe and by 40–60 per cent north of 60°N. Note that, by increasing the average irradiance on the module, the average inverter efficiency will also be larger, thereby compounding the benefit of tracking. However, the module temperature will be warmer on average, which will increase temperature-related efficiency losses (by an extra 1 per cent according to Huld et al, 2008b).

For Spain, Narvarte and Lorenzo (2008) compute that one-axis tracking increases the electricity output by 18–25 per cent while two-axis tracking increases the output by 37–45 per cent. As the two-axis system costs only 20 per cent more than fixed arrays in the Spanish market (€6 per peak watt of output (W_p) versus €5/W_p), there has been an explosive growth in the market for tracking systems in Spain, with about 70 per cent of the 106MW that were installed in 2006 being tracking systems.

If there is an array of sun-tracking (or tilted) modules, then the effects of partial shading of one module on another need to be taken into account. The degree of shading will be less the further apart the modules are spaced, thereby increasing the electricity output per m^2 of module area, but the average output per m^2 of land area will be less. The economically optimal spacing will depend on the relative costs of modules and of land. For one-axis sun-tracking modules in Spain, tilted at 45° with a ratio of module area to ground area of 0.18, García et al (2009) estimate that self-shading reduces the output by 5.4 per cent compared to the hypothetical case of no shading.

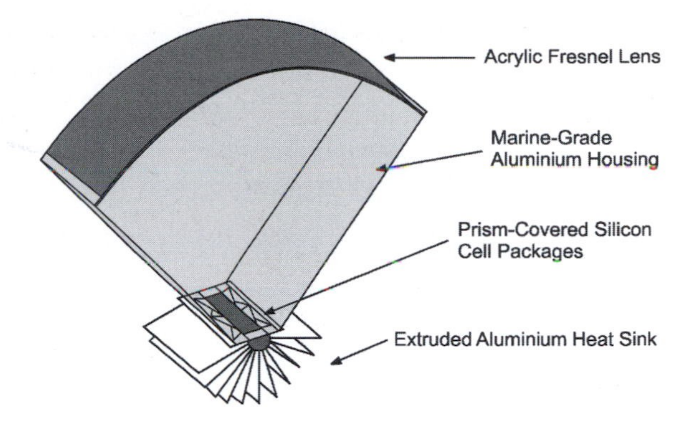

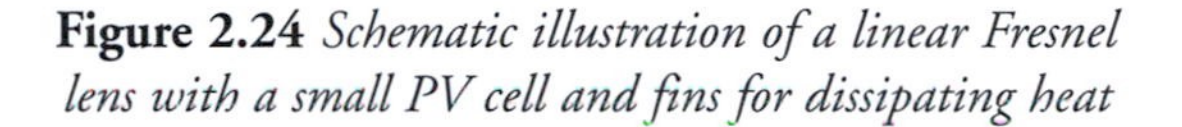

Source: www.ENTECHSolar.com

Figure 2.24 *Schematic illustration of a linear Fresnel lens with a small PV cell and fins for dissipating heat*

2.2.8 Concentrating PV systems

Solar cells, particularly the more efficient solar cells, are expensive, whereas mirrors and lenses are inexpensive. Thus, the cost of generating electricity can be reduced if lenses or mirrors are used to concentrate sunlight onto the expensive solar cell. Not only is more solar energy made available per unit of cell area, but efficiencies in converting the sunlight to electricity are usually greater (see Table 2.3).

Kurtz (2008b) identifies over 30 companies worldwide that manufacture concentrating PV systems. The *Fresnel lens* is a commonly used optical concentrator. This is a type of lens consisting of a series of concentric annular zones rather than forming a continuous curved surface. It requires significantly less material than a conventional lens, and was first used in lighthouses in 1822 in France and is now also used in traffic lights and automobile headlights. A Fresnel lens developed by Entech Corporation that concentrates sunlight by a factor of 20 is illustrated schematically in Figure 2.24. Also shown are radiator fins behind the solar cell that are used to dissipate heat to the air; these are said to maintain a PV cell temperature equal to that of a conventional cell at normal solar irradiance (1 sun). Conventional concentrating PV systems require direct-beam sunlight and must track the sun. The upper part of Figure 2.25 shows an Entech two-axis sun-tracking unit, while the lower panel shows a 100kW system with one-axis tracking units (the tilt of the modules is fixed).

A second example is the point-focus plastic lens developed by Amonix Corporation. This lens concentrates the sunlight by a factor of 250–500 (250–500 suns) onto a single silicon cell. A 25kW unit that tracks the sun along two axes is shown in Figure 2.26. It consists of five 44' × 11' (15m × 3.6m) modules with single-junction crystalline silicon cells that each produce 5kW of DC power with a solar irradiance of 850W/m^2 perpendicular to the module and a temperature of 25°C (Stone et al, 2006). The implied system efficiency is 10.9 per cent. Electricity production begins when the irradiance reaches 400W/m^2. A total of 570kW of Amonix units were installed from mid-2000 to mid-2006, all in Arizona or Nevada, of which a 300kW field was completed in Arizona at a cost of \$6000/kW_p. Tests with a GaInP multi-junction cell instead of a silicon cell indicate a cell efficiency of 27 per cent at 100 suns, 26 per cent at 250 suns and 25 per cent at 400 suns (Slade and Garboushian, 2005).

A third example is the point-focus concentrating system called FLATCON that was developed in Germany and is illustrated in Figure 2.27. A 2 millimetre (mm) diameter cell lies beneath each lens and receives 500 suns irradiance. A two-junction cell yields a module efficiency under operating conditions of 23 per cent, while a three-junction cell (36 per cent cell efficiency at 25°C) yields an operating module efficiency of 26 per cent (Peharz and Dimroth, 2005). Similar results have been achieved with a concentrating unit built at the Institute of Nuclear Energy Research

Note: Each row is 100m long, consists of 72 units and is rated at 25kW.
Source: O'Neill et al (2002) and www.ENTECHSolar.com

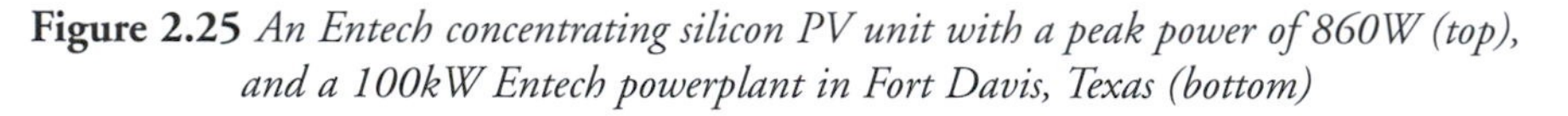

Figure 2.25 *An Entech concentrating silicon PV unit with a peak power of 860W (top), and a 100kW Entech powerplant in Fort Davis, Texas (bottom)*

Source: Amonix Corporation, www.amonix.com

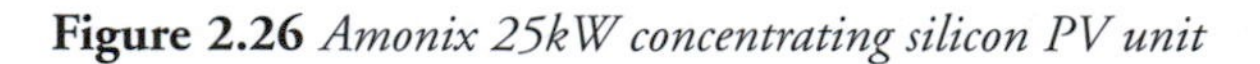

Figure 2.26 *Amonix 25kW concentrating silicon PV unit*

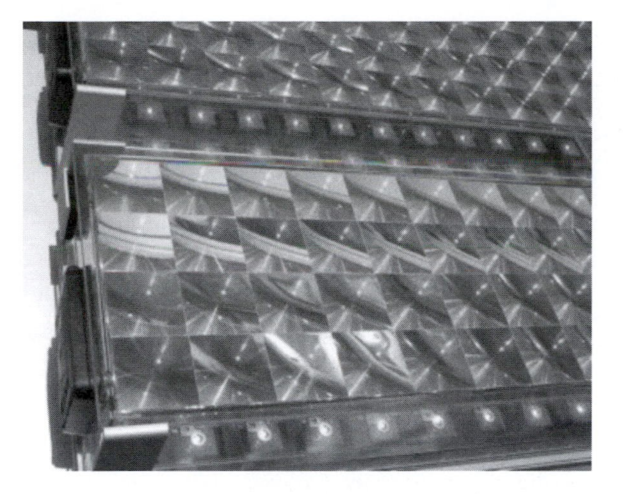

Source: Peharz and Dimroth (2005)

Figure 2.27 *The FLATCON prototype point-focus concentrating PV system*

in Taiwan (26.1 per cent system efficiency using cells with 35 per cent efficiency at 476 suns) (Kuo et al, 2009), while the Australian company Solar Systems has produced a concentrating module with a system efficiency of 24.2 per cent at 551 suns irradiance (Verlinden et al, 2006).

As noted earlier (subsection 2.2.3), module efficiency for non-concentrating systems tends to peak at an irradiance of about 500W/m^2. This represents a tendency for efficiency to increase with increasing irradiance due to the increase in cell voltage with irradiance at constant temperature but to decrease with increasing irradiance due increasing series resistance. To minimize the loss, concentrator cells are constructed with very small series resistance. The result is that peak efficiency with fixed temperature occurs at an irradiance of several hundred suns (Luque et al, 2006), with the point where the efficiency begins to decrease depending on the cell thickness and type. As with non-concentrating cells, the efficiency decreases with increasing cell temperature due largely to a decrease in voltage, but for a given temperature the relative decrease in voltage and hence in efficiency is smaller under concentrated sunlight. Some form of cooling, either passive or active (using air or water) is nevertheless essential.

Efficiencies for one particular silicon cell are presented in Table 2.5, where it is seen that the efficiency is substantially better at 250 suns than at 1 sun at a fixed temperature, while the efficiency at 250 suns and 65°C is slightly better than at 1 sun and 25°C. However, there are optical losses of up to 20 per cent associated with the concentrating lenses.

Table 2.5 *Comparison of silicon cell efficiency for different temperatures and intensity of solar radiation*

Solar intensity	Temperature	Efficiency (%)
1 sun	25°C	21.2
	65°C	18.6
250 suns	25°C	24.1
	65°C	21.5

Source: Yoon and Garboushian, www.amonix.com

Luque et al (2006) discuss a number of other issues related to concentrating PV. A high current must be extracted from the cell, which requires that thick wires or ribbons be soldered to the cell. Attention to differences in expansion coefficients is required. Differential expansion is minimized if the cells are small, but then more cells are needed for a given output, increasing the assembly costs. High precision tracking and optics are required in order to achieve the expected concentration factors. The whole structure must be quite rigid yet able to withstand commonly encountered winds, and must be able to be stowed in a safe position during periods of high winds.

Concentrating PV systems do not collect diffuse radiation, which typically accounts for 15 per cent of total solar radiation over the course of a day, so they are usually rated at an irradiance of 850W/m^2 rather than at 1000W/m^2 (as in non-concentrating systems).

Because they require direct-beam radiation, conventional concentrating PV systems can be used only in desert or semi-desert areas, where conditions are almost always cloud-free. A new concept, still under development, is the fluorescent-dye or *quantum-dot* concentrator, which can concentrate both direct and diffuse beam radiation without tracking (Gallagher et al, 2007; Currie et al, 2008).

Because of the extra complexity of concentrating systems compared to flat-plate systems, they have been slow to develop. As well, solar cells optimized for non-concentrating use need to be re-optimized for use in concentrating systems, because concentrating lenses do not reflect and concentrate all wavelengths equally. However, by concentrating the sunlight by up to a factor of 250, the solar cell area required for a given electricity production can be reduced by a comparable factor. This in turn can circumvent the resource constraints (discussed in subsection 2.2.13) for cells based on rare elements. Concentrator systems require lenses and trackers that are not needed for flat-plate systems, but the extra cost of these components can be offset by the fact that a much smaller area of expensive solar cell is needed for a given electricity output. As well, the annual solar irradiance on a sun-tracking module is about 30 per cent greater than on a fixed module. Such systems would then be in direct competition with concentrating solar thermal electricity generation (discussed in section 2.3), as both would be largely restricted to arid and semi-arid regions. Some people believe that capital costs could drop to \$1500/kW_p with a GW/year scale of production.

2.2.9 Energy payback of PV electricity and GHG emissions

The energy payback time of a PV power system is the length of time required for the energy output from the system (AC electricity) to save an amount of primary energy equal to the amount of primary energy required to manufacture the system. It is only after this payback time that there is a net energy gain from the system. The energy payback time will depend on the amount of energy used to make the PV components, the amount of solar energy incident on the modules, the module efficiency and performance ratio, and the efficiency with which electricity would otherwise be generated. If the electrical generation efficiency by fossil fuel powerplants is 40 per cent, for example, then each kWh of solar electricity saves 3.6MJ/0.4 = 9MJ. If solar electricity is produced at the point of demand (through BiPV) and displaces distant fossil fuel electricity, then savings in transmission losses at the time when solar electricity is generated should also be accounted for but usually are not (transmission losses at times of peak demand can be as large as 20 per cent).

Silicon is produced from silica (SiO_2), which is ubiquitous as quartz sand. Silica is reduced (the oxygen is removed) in electric arc furnaces by reacting it with carbon (from wood, charcoal or coal) to produce *metallurgical-grade silicon*, which is purified to *solar-grade* or *electronic-grade silicon* (the latter requiring much higher purity). Purification is done by reacting silicon with trichlorosilane at 1150°C. Trichlorosilane itself is produced by reacting pulverized metallurgical-grade silicon with hydrogen fluoride, a highly energy-intensive process that uses about 80 per cent of the initial silicon feedstock. The reduction and refining of silica require an extraordinarily large amount of primary energy: 330MJ/kg, compared to about 24MJ/kg for steel and 160MJ/kg for aluminium (see Volume 1, Chapter 6).

The purified silicon is melted and recrystallized under carefully controlled conditions to form large blocks or *ingots* of single- or multi-crystalline material. In the case of ingots produced by the electronics industry from electronic-grade silicon, the outer 40 per cent is removed (being of lower quality) and the remainder is cut into thin slabs or wafers but with 50 per cent loss by the sawing process. Altogether, only 30 per cent of the electronic grade silicon ends up in wafers. Some of these are rejected by the electronics industry due to insufficient quality, and these rejects are what the PV industry had used exclusively until recently. Alternatively, silicon ribbons can be produced directly at the thickness required for solar cells, thereby eliminating sawing losses, and thin-film a-Si cells can be produced by a vapour deposition process.

Table 2.6 compares recent estimates of the payback times for different kinds of PV systems, accounting for the embodied energy in balance of system (BOS) components (inverter, wiring and support structures). The payback times given in Table 2.6 have been adjusted to an annual solar irradiance of 1700kWh/m^2/yr, corresponding to conditions in Southern Europe or the northern US. Note that the payback time is progressively shorter as one moves from fixed to one-axis tracking to two-axis tracking systems. The energy used to make PV modules has fallen substantially over time, due in large part (in the case of silicon modules) to a reduction in the amount of silicon used per watt of peak output. According to Alsema and de Wild-Scholten (2007), the silicon loading has fallen from 32gm/W_p in 1976 to 18gm/W_p by 1992 and 10gm/W_p by 2007. Table 2.7 gives the embodied energy per m^2 of module area for the latest production facilities and as projected for the near future. A 70 per cent reduction in the energy required to produce silicon feedstock is

Table 2.6 *Length of time required for the primary energy saved by PV electricity to offset the primary energy required to make PV modules and the balance of system components, for different kinds of PV cells*

Cell type	Efficiency, performance ratio (%)	Location	Payback	Equivalent CO_2 emission (gC/kWh)	Reference
c-Si	14.0, 75	RT	2.0 years		Alsema and de Wild-Scholten (2007)
c-Si	14.0, –	RT		28–45	Fthenakis and Kim (2007)
m-Si	13.0	M	3.8 years		Stoppato (2008)
m-Si	14.8, 75	RT	2.5 years		Mohr et al (2007)
m-Si	14.0, 75	RT	2.5 years	15.5	Raugei et al (2007)
m-Si	13.2, 75	RT	1.7 years	8.4	Alsema and de Wild-Scholten (2007)
m-Si	12.8, 78	GM	2.3 years	14.3	Ito et al (2008)
m-Si	13.2, –	RT		22–37	Fthenakis and Kim (2007)
m-Si, fixed	12.4, –	GM	3.1 years		Perpiñan et al (2009)
m-Si, 1-axis tracking	12.4, –	GM	2.8 years		Perpiñan et al (2009)
m-Si, 2-axis tracking	12.4, –	GM	2.5 years		Perpiñan et al (2009)
ribbon-Si	11.5, 75	RT	1.5 years		Alsema and de Wild-Scholten (2007)
ribbon-Si	11.5, –	RT		19–30	Fthenakis and Kim (2007)
a-Si	6.9, 77	GM	3.0 years	18.5	Ito et al (2008)
a-Si	5.5, 75	RT	1.2 years		Alsema et al (2006)
CdTe	9, 75	RT	2.8 years	13.1	Raugei et al (2007)
CdTe	9, 77	GM	2.3 years	15.2	Ito et al (2008)
CdTe	9, 80	GM	1.1 years	6.8	Fthenakis and Alsema (2006)
CdTe	8, 75	RT	1.0 year	5.7	Fthenakis and Alsema (2006)
CdTe	9.5, –	RT		13–16	Fthenakis and Kim (2007)
CIS	11, 75	RT	1.5 years	25.9	Raugei et al (2007)
CIS	11, 78	GM	1.9 years	12.4	Ito et al (2008)
CIGS	11.5, 75	RT	1.3 years		Alsema et al (2006)
GaAs	23.3, 75	RT	2.9 years		Mohr et al (2007)
GaInP/GaAs	28.5, 75	RT	2.7 years		Mohr et al (2007)
GaInP/GaInAs/Ge	26, –	TR	7–10 months		Peharz and Dimroth (2005)
Dye-sensitized	9, 75		≤1 year		Greijer et al (2001)
BOS alone	–, 83.5	GM	4 months		Mason et al (2006)

Note: All payback times have been adjusted to an annual solar irradiance on the module of 1700kWh/m²/yr (corresponding to horizontal modules in Southern Europe or the northern US). BOS = balance of system, RT = rooftop, M = modules only, GM = ground-mounted, TR = tracking unit with 333 Fresnel lens concentration to 500 suns.

expected, thereby reducing the overall embodied energy in silicon modules by about 50 per cent. Energy inputs can also be reduced by further reducing the silicon loading per watt (by further reducing wafer thickness), through improvements in module efficiency and system performance ratio, and through general improvements in the efficiency of lighting and air conditioning of the building in which modules are manufactured. As seen from Table 2.6, most estimates of energy payback times of PV modules fall in the 1–2 year range for Southern European/northern US conditions, and this should decrease to less than one year in a few years.

The energy payback times given above are based on energy used at the production facility (zero-order inputs) and direct energy inputs used in making the materials used in the modules (first-order inputs). This is a process-based calculation. However, there

Table 2.7 *Embodied energy per m² of module area for multi-crystalline silicon modules using present best-practice technology and as projected with near-future technologies*

Process	Embodied primary energy (MJ/m²)	
	Present	Future
Production of silicon feedstock	1730	500
Production of ingots and wafers	570	400
Production of cells	530	500
Assembly of modules	400	400
Total	3250	1800

Note: Assumes the average European electricity supply mix.
Source: Alsema et al (2006)

are additional second-, third- and higher-order energy inputs that are not accounted for in process-based analyses. To capture higher-order inputs requires an input–output matrix representing the entire economy, but this approach is subject to its own limitations, as discussed more fully in Appendix B. Nevertheless, Crawford et al (2006) estimate that the correct energy payback time for PV modules could be 40 per cent greater than that deduced from process-based studies.

Equivalent lifecycle CO_2 emission per kWh of generated electricity are given in Table 2.6 for those studies where estimates have been included along with energy payback times. Various fluorinated gases with a high global warming potential are used in the manufacture of PV modules. According to Alsema and de Wild-Scholten (2007), emissions of these gases amortized over the lifetime energy production by PV modules amounts to the equivalent of 10gmC/kWh for multi-crystalline silicon modules at 13.2 per cent efficiency and a performance ratio of 0.75. It is not clear which studies from Table 2.6, if any, included these emissions in the estimated lifecycle emissions. By comparison, the emission for a natural gas combined cycle powerplant at 60 per cent efficiency is 90gmC/kWh. Efforts should certainly be made to reduce the use and emissions of fluorinated gases.

2.2.10 Economics of PV electricity and present-day costs

Basic relationships

The cost of a PV module can be given per m² of module area (e.g. \$/m²) or per peak watt of power output (\$/W_p), where W_p denotes watts of output for an assumed peak irradiance I_p of 1000W/m² on the module. The two measures of cost are related according to:

$$\frac{\$}{W_p} = \frac{\$/m^2}{W_p(\text{output})/m^2} = \frac{\$/m^2}{\eta_{mod} I_p} \tag{2.23}$$

where η_{mod} is the module efficiency (sunlight to DC electricity). c-Si modules cost about \$400/m² today. For a module efficiency of 10 per cent, this translates into a module cost of \$4/W_p or \$4000/$(kW_p)_{DC}$. To this must be added BOS costs (the inverter, wiring and support structures) and installation costs. The conversion from DC to AC electricity by the inverter entails energy losses, and there will be other losses due to dust and because module temperatures are often higher than assumed for quoted efficiencies. These losses are represented by the BOS efficiency η_{BOS}. Thus, the peak output is $\eta_{BOS}\eta_{mod}I_p$ and not $\eta_{mod}I_p$, so the final cost has to be divided by the additional factor η_{BOS}.

Details on how to calculate the cost of electricity from solar energy are given in Box 2.2. Table 2.8 gives the investment cost per watt of peak output and the cost per kWh of generated electricity as calculated using the equations given in Box 2.2 for a variety of illustrative module and BOS cost assumptions for poor, average and high annual solar irradiance. For module, BOS and power conditioning costs of \$400/m², \$100/m² and \$400/$(kW_p)_{DC}$, respectively, and 25 per cent indirect costs, the investment cost is \$9000/kW_p. With 5 per cent financing over 20 years electricity costs about 41 cents/kWh at 2200kWh/m²/yr irradiance and 83 cents/kWh at 1100kWh/m²/yr irradiance for 10 per cent efficient modules. If module, BOS and power conditioning costs can be reduced to \$100/m², \$50/m² and \$250/$(kW_p)_{DC}$, respectively, and the module efficiency increased to 15 per cent, electricity cost ranges from 10–19 cents/kWh. With a credit for savings in

Box 2.2 Economics of PV electricity

The cost of solar electricity depends on the initial investment, referred to as the *capital cost*, the interest rate on money used to finance the initial investment, the length of time over which the investment is to be paid back and annual operation and maintenance cost. The length of time used to pay back the initial investment (with interest) is set equal to the expected lifetime of the powerplant, so that the initial investment is exactly paid back with interest when the powerplant ceases operation. The fixed cost of electricity that allows one to exactly pay back the initial investment (along with fixed operation and maintenance costs) is referred to as the *levelized* cost of electricity. The levelized cost does not include the effects of inflation, but this is appropriate as long as the interest rate used in the calculation is the nominal interest rate minus the rate of inflation.

For a fossil fuel powerplant, capital costs are expressed in terms of $ per kW of capacity. We can express capital costs per kW of peak DC power output assuming the peak solar irradiance to be 1000W/m^2 (whether or not this assumption is true does not matter, because in the final calculation the assumed irradiance appears twice and cancels out). Given module costs initially specified in terms of $ per m^2, the cost in terms of $ per peak watt (kW$_p$) of DC output is given by:

$$C_m(\$/\mathrm{kW_p}) = \frac{C_m(\$/\mathrm{m}^2)}{\eta_m I_p} \tag{2.24}$$

where η_m is the module efficiency (solar energy to DC electricity) and I_p is the assumed peak solar irradiance of 1kW/m^2. This (and other) capital costs are then divided by the number of kWh of electricity generated per year, which is equal to the number of hours in a year times the capacity factor. Here, the capacity factor would be given by the average irradiance on the panel, I_a, divided by the assumed peak irradiance (I_p). To convert from a cost per DC kWh to AC kWh, one must divide by the efficiency (η_{BOS}) in converting from DC to AC power. The fraction of the initial investment that needs to be paid back each year is called the *cost recovery factor* (CRF), and is computed as explained in Appendix C. The contribution of the module cost to the levelized cost of electricity from solar power would be:

$$CRF\left(\frac{C_m(\$/\mathrm{kW_pDC})}{8760(\overline{I}_a/I_p)\,\eta_{BOS}}\right) = CRF\left(\frac{C_m(\$/\mathrm{m}^2)}{8760\eta_m\eta_{BOS}\overline{I}_a}\right) \tag{2.25}$$

where the denominator on the right-hand side is the number of kWh of AC electricity produced per year per m^2 of module area and $\overline{I}_a$ is in kW. The term ($\overline{I}_a/I_p$) is the ratio of average irradiance to peak irradiance, and assuming that power output varies exactly in proportion to irradiance, ($\overline{I}_a/I_p$) is equal to the module *capacity factor* – the ratio of mean annual output to peak output.

Altogether, the levelized cost of producing PV AC electricity is given by:

$$C(\$/\mathrm{kWh}) = \frac{(CRF + INS)(1 + ID)(C_m + C_b + C_p + C_{inst}) + OM}{8760(\overline{I}_a/I_p)\eta_{BOS}} \tag{2.26}$$

where C_m, C_b, C_p and C_{inst} are the module, BOS, power conditioning and installation costs, respectively (all as $/kW$_p$ DC), *OM* is the operation and maintenance cost (as $/kW$_p$(DC)-yr), *INS* is the annual insurance as a fraction of the capital costs, and *ID* is an indirect cost factor to account for site engineering, maintenance of inventories, and contingencies. In some cases, C_b and C_{inst} might be lumped together as C_b.

Table 2.8 *Illustrative costs of PV electricity (cents/kWh(AC)) for 2.5 per cent, 5 per cent and 10 per cent/year real financing costs, a 20-year lifespan, 1 per cent/year insurance, 25 per cent indirect costs, 75 per cent BOS efficiency, annual operation and maintenance costs equal to 2 per cent of the gross investment cost and other assumptions*

Module cost ($/m^2)	BOS cost ($/m^2)	Power conditioning ($/kW$_p$(DC))	Module efficiency	Investment cost ($/(W$_p$)$_{AC}$)	Solar irradiance and real interest rate								
					1100kWh/m^2/yr			1700kWh/m^2/yr			2200kWh/m^2/yr		
					2.5%	5%	10%	2.5%	5%	10%	2.5%	5%	10%
400	100	400	0.10	$9.00	69.0	82.6	114.7	44.6	53.4	74.2	34.5	41.3	57.3
400	100	400	0.15	$6.22	47.7	57.1	79.3	30.8	36.9	51.3	23.8	28.6	39.6
100	50	250	0.10	$2.92	22.3	26.8	37.2	14.5	17.3	24.0	11.2	13.4	18.6
100	50	250	0.15	$2.08	16.0	19.1	26.5	10.3	12.4	17.2	8.0	9.6	13.3
0[a]	50	250	0.10	$1.58	15.8	18.2	23.8	10.2	11.8	15.4	7.9	9.1	11.9
0[a]	50	250	0.15	$1.19	11.6	13.4	17.6	7.5	8.7	11.4	5.8	6.7	8.8

Note: a This is the net cost, assuming a $100/m^2 gross module cost and a $100/m^2 credit for savings in conventional building materials in a BiPV application. Insurance and indirect costs are computed here based on gross module costs.

conventional building cladding materials in BiPV applications of $100/m^2, the cost of electricity is 7–13 cents/kWh. By comparison, the cost of producing electricity from coal in new powerplants is 4–7 cents/kWh (see Volume 1, Table 3.11).

Present-day costs

IEA (2008) presents data on the cost of PV modules in IEA countries. Typical module costs in 2007 ranged from a low of about US$3.7/$W_p$ in Japan and the US to almost $10/$W_p$ in some parts of the world, with 11 out of 16 reporting countries having typical costs of $4–8/$W_p$. For some countries, the best reported module price is significantly lower than the typical price.[6]

With regard to invertors, costs were about $400/$kW_{DC}$ for large-scale solar installations built in the early 2000s in Arizona, but are projected to fall to $250/$kW_{DC}$ by 2015 in the US (Moore and Post, 2008). In Europe, de Wild-Scholten et al (2006) indicate inverter costs of €1000/W_{AC} for 1kW_{AC} residential systems and €350/kW_{AC} for 8kW_{AC} residential systems, and costs of €400/kW_{AC} at the 100kW scale and €220/kW_{AC} at the 500kW scale.

Module, inverter, wiring and installation costs can easily bring the total cost to $6000–9000/$(kW_p)_{AC}$ at present. IEA (2008) reports that installed prices in 2007 for grid-connected systems smaller than 10kW peak power ranged from $6/$W_p$ to $20/$W_p$, with most countries having typical prices of $8–10/$W_p$. In most countries, the unit cost of larger systems is slightly less. By comparison, large advanced natural gas combined cycle powerplants (described in Volume 1, Chapter 3, section 3.3) have an installed cost of about $400–600/kW in mature markets. Furthermore, the average electricity output as a fraction of peak output – the so-called capacity factor – will typically be 15–25 per cent of the peak output since typical annual average solar irradiance is only 150–250W/m^2, whereas the capacity factor of typical load-following fossil fuel systems is 60–70 per cent.

Tucson Electric Power (in Arizona) installed 26 c-Si PV systems between July 2001 and July 2004, each rated at 135kW_{DC}, for a total of 3.51MW_{DC} (Moore and Post, 2008). These systems were installed in a standardized approach, which brought the cost down to $5.40/$(W_p)_{DC}$ ($6.84/$(W_p)_{AC}$). At 5 per cent financing cost over 20 years, this corresponds to about 26 cents/kWh. The average performance ratio (AC output/rated DC power) was 0.79. The goal of the US Solar Energy Industries Association is to reduce the system cost to $3.68/$(W_p)_{AC}$ by 2015 (giving an electricity cost of about 14 cents/kWh in sunny locations).

IEA (2003) indicates that annual operation and maintenance (O&M) costs for modules are about 1–3 per cent of the investment cost, but that inverters need to be replaced every 5–10 years. However, inverters account for only 10 per cent or less of the total capital cost at present. At $9000/$kW_p$ capital cost, O&M at 2 per cent per year adds 8.2 cents/kWh to the cost of electricity in sunny (2200kWh/m^2/yr) locations.

Because peak module output occurs only a few times per year, the overall cost of electricity can be reduced if the inverter is undersized relative to the module capacity (Mondol et al, 2009). This means that some electricity production at times of peak solar irradiance will be discarded, but the discarded electricity would be a small fraction of annual electricity production and the economic penalty would be offset by the reduction in capital cost. The optimal inverter undersizing depends on the relative cost of the inverter and other system components.

Capacity credit

In jurisdictions where the peak in electricity demand is driven by air conditioning loads, which are largest during hot summer afternoons, the peak in solar electricity can be reliably expected to occur at times of peak electricity demand. Thus, the addition of PV to a grid allows a reduction in the required fossil fuel capacity. This reduction, as a fraction of the installed peak PV power, is called the *capacity credit*. The proper calculation of the capacity credit requires taking into account the variability of both electricity demand and PV power, as well as the reliability of the fossil fuel components of the system. There is a much larger literature concerning the calculation of capacity credits for wind than for solar power, no doubt because the amount of wind capacity installed in some jurisdictions has become large enough (10–20 per cent of total system capacity) that the proper calculation of the capacity credit matters. Concepts related to the calculation of capacity credits for wind are discussed in Chapter 3 (subsection 3.13.3) but are applicable to solar energy too. Using a concept equivalent to the loss-of-load probability discussed in Chapter 3, Pelland and Abboud (2008) calculated a capacity credit for PV power in Toronto of about 30 per cent. Thus, if 1MW of PV power is installed in Toronto, the peaking fossil fuel capacity can be reduced by 0.3MW. Where this avoids construction of new powerplants, the capital cost of the saved peaking capacity can be subtracted from the capital cost of the added PV. In warmer climates the capacity credit will normally be larger. Thus, for warm and sunny locations in the US, Perez et al (2006) calculate a capacity credit of up to 0.65 for PV systems tilted to the south at the optimal angle.

The partial supply of peak loads with PV means that the utilization of those peaking fossil power units that remain on the grid will be less, thereby increasing the cost of electricity from these units. This is an indirect cost of PV power that is insignificant at present but could become important as the share of electricity demand supplied by PV increases. A non-zero capacity credit for PV will at least partially offset this indirect cost (the analysis presented in Chapter 3, subsection 3.13.3 with regard to wind is applicable here).

Additional benefits of BiPV

BiPV – by providing electricity where it is needed – reduces peak transmission requirements as well as reducing the need for peaking fossil fuel powerplants. As transmission bottlenecks have become a significant problem in many urban centres, this is can be a significant benefit. Other benefits of BiPV include its role as a façade element, replacing conventional materials and providing protection from UV radiation; providing thermal benefits such as shading or heating; augmenting power quality by serving a dedicated load; and serving as backup to an isolated load that would automatically separate from the utility grid in the event of a line outage or disturbance. These benefits are discussed in some detail by Eiffert and Task-7 Members (2002). The grid and load-shaving benefits to the power utility are worth \$107–180 per year per kW of peak power produced in the US. Inasmuch as BiPV is used as part of an aesthetic design element in a building, it can replace rather expensive building materials. The cost of PV modules today is about \$400/m^2. By comparison, the costs of envelope materials in the US that PV modules can replace range from \$250–350/m^2 for stainless steel, \$500–750/m^2 for glass-wall systems, to at least \$750/m^2 for rough stone and \$2000–2500/m^2 for polished stone (AEC, 2002).

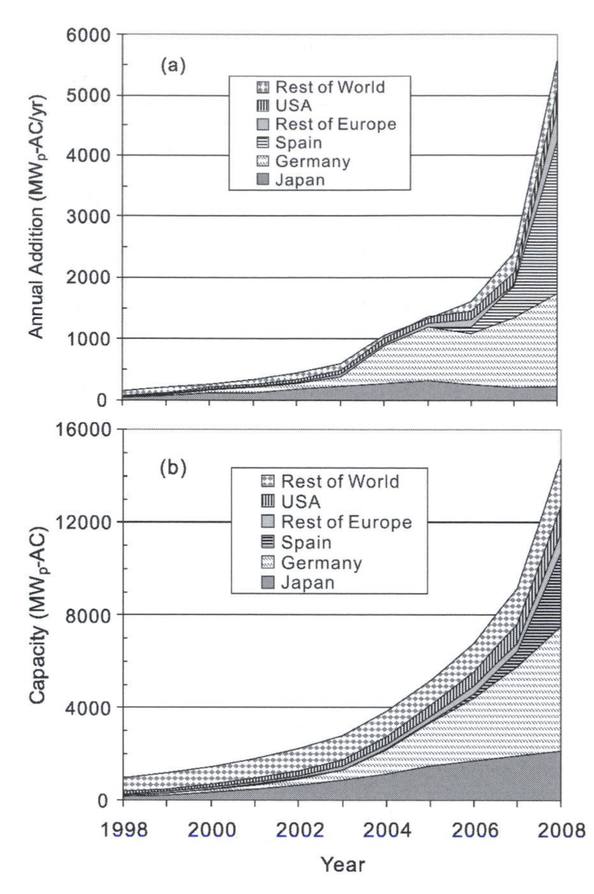

Source: Data from EPIA (2009)

Figure 2.28 *(a) Growth in annual installation of PV power, 1998–2008; and (b) growth in installed PV power capacity, 1998–2008*

2.2.11 Present-day global PV market

Figure 2.28a shows the growth in the amount of PV power capacity installed per year from 1998 to 2008. World installation of PV power amounted to 5.6GW$_p$ in 2008, having more than doubled from 2007 and grown by an average compounded rate of 35.9 per cent since 1998. Until 2007, Germany was the largest market, accounting for 47 per cent of the total installation of PV systems in 2007. However, the Spanish market grew enormously in 2008 in response to government incentives, such that more PV was installed in Spain in 2008 than in any other country. Table 2.9 gives the breakdown of world production of

Table 2.9 *Annual production of PV modules in terms of $MW_p(DC)$ or $MW_p(AC)$ of peak power capacity*

		2000	2002	2004	2006
	Production by company, $MW_p(DC)$				
Sharpe	Japan	50	123.1	324	434
Q-Cells	Germany			75	253
Kyocera	Japan	42	60	105	180
Sanyo	Japan	17	35	65	155
Mitsubishi	Japan	12	24	75	111
Motech	Taiwan		8.0	35	110
Schott Solar	Germany	14	29.5	63	96
BP Solar	UK	42	73.8	85	86
Shell Solar	UK	28	57.5	72	77
Isofotón	Spain	10	27.4	53.3	61
Total top 10			438	952	1563
World total		288	650	1195	2586
	Production by country or region, $MW_p(DC)$				
Japan		128.6	251.1	601.5	926.9
Europe		49.8	122.1	311.8	678.3
US		75	120.6	138.7	201.6
Rest of world		23.4	53.3	141.5	714
Total		277	547	1194	2521
	Production by technology, $MW_p(DC)$				
Single crystal silicon			370.9	957.8	
Multi-crystalline silicon			641.6	1174.4	
Ribbon silicon			41	68	
Other crystalline			80	150	
Crystalline total			1134	2350	
Amorphous silicon			47.1	97.7	
Cadmium telluride			13	68	
CIS/CIGS			3	4.9	
Thin-film total			63	171	
Total			1197	2521	
	Production by application, $MW_p(AC)$				
Consumer products		60	75	90	
Off-grid residential		80	110	140	
Grid-connected residential and commercial		270	700	1600	
Communications and signal		60	80	100	
PV-diesel hybrid, commercial		45	65		
Central (>100kW)		5	20	100	
Total		520	955	2200	

Note: Global sums when categorized in different ways do not exactly match.
Source: Maycock and Bradford (2006); Bradford and Maycock (2007)

PV modules from 2000 to 2006 by company, country, technology and application. Four of the top five PV manufacturing companies in 2006 were Japanese and 38 per cent of the world's production was in Japan. Single-crystal, multi-crystalline and ribbon silicon cells together accounted for 87 per cent of total production and various thin-film technologies accounted for 13 per cent, with grid-connected residential systems the single largest application. The website www.pvresources.com contains an up-to-date listing of the largest 600 PV plants in the world, along with photos and other information. Eighteen

Table 2.10 *Cumulative installed PV peak power capacity (MW) in reporting IEA countries as of the end of 2007*

Country	Off-grid		Grid-connected		Total	Per capita (W)	Total installed in 2007
	Domestic	Non-domestic	Distributed	Centralized			
Australia	27.7	38.7	15.0	1.0	82.5	4.1	12.2
Austria	3.2		22.7	1.8	27.7	3.4	2.1
Canada	8.1	14.7	2.8	0.07	25.8	0.8	5.3
Czech Republic	3.2	0.40	30.0	2.6	36.2	4.9	6.5
Denmark	0.10	0.29	2.7	0.0	3.1	0.6	0.18
France	15.9	6.7	52.7	0.0	75.2	1.2	31.3
Germany	35		3827		3862	46.8	1135
Israel	1.6	0.21	0.01	0.01	1.8	0.3	0.5
Italy	5.4	7.7	83.9	23.2	120.2	2.1	70.2
Japan	1.9	88.3	1823	5.5	1919	15.0	210
Korea	0.98	5.0	32.6	39.1	77.6	1.6	42.9
Mexico	15.5	5.0	0.30	0.0	20.8	0.2	1.0
Netherlands	5.3		44.5	3.5	53.3	3.3	1.6
Norway	7.5	0.41	0.13	0.0	8.0	1.7	0.32
Portugal	2.8		0.68	14.4	17.9	1.7	14.5
Spain	29.8		625.2		655.0	15.1	512
Sweden	3.9	0.69	1.7	0.0	6.2	0.7	1.4
UK	0.42	1.1	16.6	0.0	18.1	0.3	3.8
US	134	191	465	40.5	830.5	2.8	207
Total	265	397	6020	1159	7841		2258

Source: IEA (2008)

centralized PV systems of 20MW or greater were constructed worldwide in 2008, the majority in Spain and the largest at 60MW. Some modest sized CdTe and CIGS arrays have been constructed too: 0.5MW and 1.04MW CdTe arrays in Arizona and Germany, respectively, and a 0.25MW CIGS array in California (Hegedus, 2006).

Figure 2.28b shows the growth in installed PV power capacity from 1998–2008, while Table 2.10 gives the breakdown by country for 2006 in terms of off-grid and grid-connected PV capacity, the latter further broken into distributed and centralized. The total installed PV capacity had reached 14.8GW by the end of 2008, having grown at an average compounded rate of 26.4 per cent/year since 1997.

As noted above, crystalline silicon modules account for the overwhelming majority of present-day module production. However, current shortages of silicon in the face of strong demand in Germany, Japan and now California due to government incentive programmes, combined with bottlenecks in the supply of silicon to manufacturing plants, have led to a recent strong investment in thin-film manufacturing capability (growing to a production capacity of at least 300MW/yr by 2008 – about twice the capacity in 2006).

2.2.12 Projections of future costs and market

The future cost of PV modules or systems can be estimated based on a projection of past trends, or based on a bottom-up engineering analysis of potential changes in specific technical factors that, together, determine the costs. These two approaches are discussed below.

Progress ratio and extrapolation of recent module cost trends

An important feature of technology is that the cost of the technology decreases with the cumulative, industry-wide experience. This is represented by the *progress ratio*, which is the factor by which the cost is multiplied for each doubling in cumulative production. A progress ratio of 0.8, for example, means that the price decreases by 20 per cent (it is multiplied by 0.8) for each doubling in cumulative production. This cost function can be represented by the relation:

$$C(t) = C_o PR^{\left(\frac{\ln R(t)}{\ln 2}\right)} \quad (2.27)$$

where C_o is the initial cost, PR is the progress ratio and $R(t)$ is the ratio of cumulative production at time t to the cumulative production at the time when the cost was C_o. Figure 2.29 shows the price of PV modules versus cumulative global PV shipments up to 2007. According to van Sark et al (2008), the progress ratio for PV modules (after prices are adjusted for inflation) was 0.79 from 1976 up to 2003. The prices used for this calculation are based on deals closed and not on catalogue prices, and reflect a mixture of crystalline and a-Si products. Thus, from 1976 to 2003, the sales-weighted price of PV modules fell by about 21 per cent for each doubling in cumulative production, which has been occurring in a little more than every two years. As noted above, prices have remained constant or even increased recently in some markets, as demand has grown faster than supply.

Caution is required in extrapolating past trends or progress ratios into the future. As noted by Nemet (2006), the experience curve switched to a slower rate of cost reduction in 1980, so projections based on trends up to 1980 would have predicted a faster decrease in prices than actually occurred. Ideally, it is desirable to identify technical improvements that could give the cost reductions implied by a given extrapolation of past trends. This requires understanding the factors leading to previous costs decreases. From the 1950s to the present, the cost of crystalline silicon modules fell by a factor of 100, and from 1979 to 2002 it fell by a factor of ten, from about \$32/W to \$3.1/W in constant 2002 US\$. Swanson (2006) has assessed the main reasons for this remarkable reduction in cost. These were: a decrease in the cost of crystalline silicon, from \$300/kg to \$30/kg by 2002 (with a subsequent increase to \$40–50/kg due to temporary shortages), a byproduct of the rapid growth in the integrated-circuit industry; more productive saws with less loss of material; the advent of screen printing; a decrease in wafer thickness from 500 to 300μm; an increase in cell size from 75 to 150mm (reducing the amount of handling required for a given power); an increase in cell efficiency from 10 per cent to 15–16 per cent; and an increase in factory size from a typical capacity of 1MW/yr in 1980 to a typical size of 100MW/yr today (with the largest plant having a capacity of 400MW/yr). According to Nemet (2006), plant size, module efficiency and the cost of silicon were the most important factors in the cost reduction from 1980 to 2001 (accounting for 43 per cent, 30 per cent and

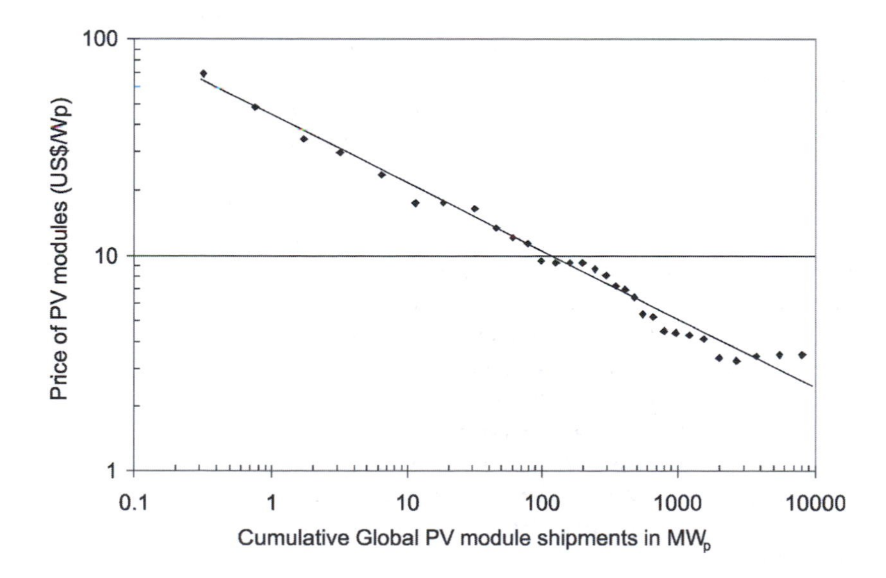

Source: van Sark et al (2008)

Figure 2.29 *Price of PV modules versus cumulative global PV shipments, 1968–2002*

12 per cent, respectively). Thus, economy of scale (and not experience) is the dominant factor. As for the future, silicon costs could drop to \$20/kg (although \$30–40/kg is more likely) and usage could drop in half (through a combination of thinner wafers and greater cell efficiency). Of the 16 advances in efficiency since 1980, most were accomplished by university teams, none of which would have learned from experience with large-scale production (Nemet, 2006). Learning-by-doing is most important for the following factors, which, however, only account for 10 per cent of the overall reduction in cost: lower defect rates, wafer size and reduction in sawing waste. Also, if profit margins have been falling recently but further decreases are restrained, then the experience curve will overestimate future decreases in price. A PR of 0.79 with 11 per cent/year growth gives \$1/W_p modules by 2027. As discussed below, however, several analysts foresee costs of \$1/W_p or less based on a bottom-up analysis.

Engineering-based estimates of future module costs

The plausibility of future costs based on extrapolation of past trends using the progress ratio approach can be assessed through a bottom-up, engineering analysis that examines the various technical factors that contribute to the cost of PV modules.

Swanson (2006) presents a pathway by which crystalline silicon module costs could reach \$1/W_p well before 2020, with installed costs of \$3/W_p. The main factors contributing to the projected cost decrease are cell efficiencies of 20 per cent, wafers decreasing from 280 to 120μm in thickness, a reduction in sawing losses from 200 to 130μm, and improved productivity in the production of crystals. Recent advances in micromachining techniques have made it possible to cut silicon wafers to a thickness of less than 50μm. Origin Energy and the Australian National University are preparing to commercialize a monocrystalline silicon cell made of wafers less than 70μm in thickness and with a demonstrated efficiency of 19.5 per cent, while the Fraunhofer Institute for Solar Energy in Germany has produced 37μm thick cells with an efficiency of 20.2 per cent (Anonymous, 2004). Although substantially thicker than amorphous silicon cells (described below), these cells are three times as efficient while using a quarter to an eighth the silicon of recent conventional crystalline cells.

More recently, del Cañizo et al (2009) foresee a cost of about €0.9–1.3/W_p for the next generation of crystalline modules (about \$1.3–1.7/W_p at the mid-2009 exchange rate), compared to a manufacturing cost of €2.0–2.3/W_p in 2005. Their projection is based in part on data provided by major European manufacturers on the current cost structure of PV modules. About half of the expected cost reduction is due to expected improvements in technology and half is due to economies of scale in manufacturing.

Thin-film solar modules may also reach costs close to \$1/W_p. Hegedus (2006) believes that a-Si modules could reach \$1/W_p if frameless modules with spray-on encapsulants are manufactured in facilities with an output of 100MW/year. There are already several a-Si plants with capacities of 10–25MW/yr under construction or being planned, so one more step would bring the industry to 100MW/yr plants. Green (2006) and Zweibel (2005) argue that \$30–45/m^2 thin-film modules at 15 per cent efficiency should be possible. This would result in a cost of \$0.20–0.30/W_p. Thin-film PV has the potential to achieve lower costs than crystalline PV due to the ability to produce PV material on large areas through a continuous deposition process, as illustrated in Figure 2.30, which shows the production of laminate roofing rolls consisting of triple junction a-Si on stainless steel.

Zweibel (2005) breaks the cost of PV modules into active-material costs (the various semi-conductor layers) and the balance-of-module (BOM) costs – items such as the substrate (which, for thin films, can be glass, stainless steel or polyimide), encapsulant and bonding materials. The largest thin-film manufacturing facilities have an output of about 25MW_p/year, so Zweibel (2005) has estimated a cost breakdown for plants of this size (actual costs are confidential). For the simplest design (involving a glass substrate), BOM costs are estimated at \$54–62/m^2. For an a-Si double-junction module, the non-BOM cost is estimated to be \$37/m^2. Combined with \$3/m^2 ES&H[7] cost, this gives a total cost of \$93/m^2, which translates into a cost of \$1.56/W_p for 6 per cent efficient modules. By comparison, real sales prices (which include marketing, administration and other costs) have been as low as \$2.25/W_p. Table 2.11 summarizes Zweibel's projections for possible cost reductions in the total module costs (\$/m^2), along with efficiencies assumed to be achieved by the time that plants of the indicated size are built, for various thin-film technologies. Module costs in the range of \$0.47–0.73/W_p (\$50–100/m^2) are projected for 1GW_p/yr plants.

Source: Hegedus (2006); www.nrel.gov/ncpv

Figure 2.30 *Production of a roll of laminate roofing with triple junction a-Si on stainless steel at the Solar Integration Technology plant in California*

A number of companies have announced intentions to produce PV modules near \$1/W in the near term, and one company – First Solar – announced that is has achieved a manufacturing cost of \$0.98/W_p (Anonymous, 2009a).

With regard to concentrating PV, Faiman et al (2007) project a cost of \$1.1/W_p for 32 per cent efficient triple junction (GaAs-based) solar cells with sunlight concentrated by a factor of 625 using a Fresnel lens, as in the Amonix system of Figure 2.26.

Future balance of system costs

Shum and Watanabe (2008) have studied the BOS costs for small grid-connected PV systems within the experience curve framework. BOS costs largely involve the inverter, support structures and labour for installation. Unlike module learning, which is global in nature, BOS learning is mostly local in nature. BOS learning involves experience with system design, integration and installation. A significant opportunity for reducing BOS costs is through standardizing the BOS to the greatest possible extent so as to minimize onsite customization.

Engineering-based estimates of future installed costs

Returning to Zweibel's (2005) scenario with module costs of \$0.47–0.73/W_p, Table 2.11 gives the assumed BOS costs and the resulting system prices (\$/W_p), which include administration, marketing, installation and profit. BOS costs are expected to be larger for rooftop glass modules than for ground-mounted modules because these modules are assumed not to be integrated into the building envelope. Even the flexible modules are assumed to have greater BOS costs than ground-mounted modules. Costs are less certain for CdTe and CIS modules, which are manufactured at present in plants of only a few

Table 2.11 *Hypothetical evolution of PV module, BOS and system costs*

Type of module	Plant manufacturing capacity (MW_p/yr)			
	25	50	200	1000
Total module cost (\$/m²)				
a-Si/a-Si, batch on glass	94	80	70	57
a-Si/a-Si, in-line on flexible substrate	130	110	99	77
CdTe on glass	110	93	82	65
CIS on glass	210	160	130	91
CIS, in-line on flexible substrate	230	170	140	100
Module efficiency (%)				
a-Si/a-Si, batch on glass	6	7	7.5	8.5
a-Si/a-Si, in-line on flexible substrate	7	8	9	10.5
CdTe on glass	8.5	10	11.5	14
CIS on glass	11	12	14	16
CIS, in-line on flexible substrate	8.5	10	11.3	14
Thin-film c-Si on glass		6.5	8	11
Module cost (\$/W_p-DC)				
a-Si/a-Si, batch on glass	1.56	1.21	0.97	0.67
a-Si/a-Si, in-line on flexible substrate	1.88	1.44	1.13	0.73
CdTe on glass	1.28	0.94	0.73	0.47
CIS on glass	1.87	1.37	0.91	0.57
CIS, in-line on flexible substrate	2.67	1.68	1.27	0.71
Thin-film c-Si on glass		1.62	1.06	0.59
BOS costs[a]				
Ground mounted				
– area-related (\$/m²)	90	70	55	40
– power-related (\$/W_p)	0.50	0.44	0.38	0.27
Commercial rooftop, glass modules				
– area-related (\$/m²)	135	110	93	75
– power-related (\$/W_p)	0.77	0.66	0.56	0.39
Commercial rooftop, flexible modules				
– area-related (\$/m²)	99	80	67	54
– power-related (\$/W_p)	0.75	0.66	0.56	0.39
Markup (%) for profit and marketing				
Ground-mounted	25	20	15	10
Commercial rooftop	40	32	24	16
Ground-mounted system prices (\$/W_p-DC)				
a-Si/a-Si, batch on glass	4.45	3.24	2.42	1.55
a-Si/a-Si, in-line on flexible substrate	4.58	3.32	2.46	1.52
CdTe	3.55	2.50	1.83	1.12
CIS on glass	4.00	2.87	1.93	1.20
CIS, in-line on flexible substrate				
Thin-film c-Si on glass		3.77	2.45	1.34
Commercial-rooftop system prices (\$/W_p-DC)				
a-Si/a-Si, batch on glass	6.38	4.65	3.48	2.50
a-Si/a-Si, in-line on flexible substrate	5.66	4.10	3.04	1.90
CdTe	5.07	3.58	2.61	1.61
CIS on glass	5.39	3.88	2.64	1.66
CIS, in-line on flexible substrate				
Thin-film c-Si on glass		5.25	3.44	1.92

Note: [a]includes hardware costs as well as non-hardware costs (design, site preparation, shipping and installation

Source: Zweibel (2005)

MW_p/yr capacity. The price of ground-mounted systems could drop to the \$1.12–1.55/W_p range, while the price of roof-mounted systems could drop to the \$1.6–2.5/W_p range. For sunny locations (2000kWh/m^2/yr, as in southern California), this translates to a cost of 7–10 cents/kWh for ground-mounted systems and 10–16 cents/kWh for roof-mounted systems, assuming 5 per cent financing over 20 years, 0.85 BoS efficiency, 1 per cent/year insurance and annual O&M equal to 2 per cent of investment cost. For plants of a given capacity, CdTe modules are potentially the lowest-cost modules. However, a-Si modules have a larger market at present and so are manufactured in plants of larger capacity, which gives them a cost advantage. Flexible thin-film a-Si is competitive with c-Si on glass for large metal roofs (an important market), whereas a-Si on glass is not competitive in any market. c-Si might be competitive for residential rooftop systems.

Keshner and Arya (2004) have presented perhaps the most aggressive scenario for future cost reductions, with an *installed* cost of \$1/W_p or less for thin-film a-Si, CdTe and CIGS modules. This requires very large production facilities (2.1–3.6GW_p/yr) with 100 identical production lines (with a learning curve/progress ratio process applying within the plant). This drives the cost down by spreading out the engineering of the production plant over many production units. Other cost savings arise from fully integrated production; a dedicated glass plant; reduced maintenance costs and disruption; elimination of intermediate shipping, packaging and breakage; and 75 per cent internal recycling of materials that would otherwise be wasted. Their calculations assume an inverter cost of \$120/kW_{AC}, compared to \$400/kW_{DC} achieved in the early 2000s by Tuscon Electric Power company and a projected 2015 cost of \$250/kW_{AC} (Moore and Post, 2008). A 3.6GW_p/yr manufacturing facility is estimated to require an investment of \$600 million, which would be out of reach for all except the largest firms, so government would need to be involved as a partner.

Table 2.12 gives the breakdown of manufacturing and installation costs and of manufacturing and retailer's profits for the \$1/kW_p installed cost scenarios. At 5 per cent financing over 20 years, a \$1/W_p installed cost translates into an electricity cost of about 4 cents/kWh for 12 per cent efficient modules in sunny locations (2000kWh/m^2/yr irradiance) and 6.7 cents/kWh for 7 per cent efficient modules.

Table 2.12 *Breakdown of costs leading to an installed cost of \$1/W_p or less for very large (2.1–3.6GW_p/yr) integrated manufacturing facilities*

	Technology		
	a-Si	CdTe	CIGS
Efficiency	7%	11%	12%
	Panel costs		
Substrate	\$0.071	\$0.045	\$0.041
Active layer	\$0.065	\$0.034	\$0.038
Packaging	\$0.161	\$0.103	\$0.094
Manufacturer's gross margin (50%)	\$0.300	\$0.210	\$0.260
Total panel cost per Wp	\$0.60	\$0.42	\$0.52
	Installation costs		
Power converter	\$0.120	\$0.120	\$0.120
AC wiring	\$0.027	\$0.027	\$0.027
Installation (4 hours @ \$75/hour)	\$0.050	\$0.033	\$0.033
Retailer's gross margin (20%)	\$0.200	\$0.150	\$0.180
Total installation cost per W_p	\$0.40	\$0.33	\$0.36
Mounting pedestals (flat roofs only)	\$0.05	\$0.033	\$0.030
	Total costs		
Sloped roofs	\$1.00	\$0.75	\$0.88
Flat roofs	\$1.05	\$0.78	\$0.91

Source: Keshner and Arya (2004)

Illustrative scenarios

By the end of 2006, the cumulative worldwide PV production was about 8.6GW, the installed capacity was 5.9GW, and the best installed capital cost was around \$4500/kW_p. Given an interest rate of 10 per cent, a lifespan of 20 years, annual insurance and O&M costs each of 1 per cent of the capital cost, an average solar irradiance of 2000kWh/m^2/yr and a balance of system efficiency of 0.85, this translates into an electricity cost of 36 cents/kWh. Figure 2.31 shows how the installed capacity and price of electricity would vary for rates of growth in annual production of 20 per cent/year and 30 per cent/year and a progress ratio of 0.8. At 20 per cent/year growth, total capacity reaches 185GW by 2020 and the cost of solar electricity – in the sunniest locations – has dropped to 13.4 cents/kWh. At 30 per cent/year growth, total capacity reaches 425GW by 2020 and the cost of solar electricity reaches 10.3 cents/kWh. Both of these costs are consistent with the engineering-based estimates (reviewed above) of what is feasible. Total worldwide electric generating capacity in 2005 was about 4100GW, so at 30 per cent/year growth in annual production, PV capacity would still constitute only 10 per cent of present-day capacity by 2020.

The market for PV today is largely dependent on government or utility subsidies. Figure 2.32 illustrates how slight differences in the progress ratio have a very large impact on the cumulative production of PV and hence in the cumulative subsidy before the price drops to a particular threshold. Note that the differences in cumulative production and subsidy seen in Figure 2.32 are much larger than might appear, given that the axes are on logarithmic scales. If the goal is to accelerate the decline in the cost of PV electricity and accelerate the deployment of PV, a subsidy equal to the difference between the cost of PV electricity and that of the cheapest alternative could be applied. A simplified calculation of the required subsidy is outlined in Box 2.3, and depends on (among other things) the rate of increase in the cost of the fossil fuel used to generate electricity and the rate at which the annual rate of installation of PV grows. Some results are given in Figure 2.33. For coal with the assumptions given in Box 2.3, a progress ratio of 0.78, annual growth in the rate of installation of solar capacity of 30 per cent/year, 4 per cent/year increase in the real cost of coal and 10 per cent real interest rate, the cumulative subsidy is $513 billion and cost parity occurs in 2021 (at 8.1 cents/kWh) with a cumulative production of 557GW. For a progress ratio only 0.04 greater (0.82), the cumulative subsidy is almost twice as large ($961 billion) and cost parity occurs in 2025 (at 8.3 cents/kWh and a cumulative production of 1594GW). If the PV manufacturing capacity grows at 10 per cent/year rather than 30 per cent/year, the cumulative

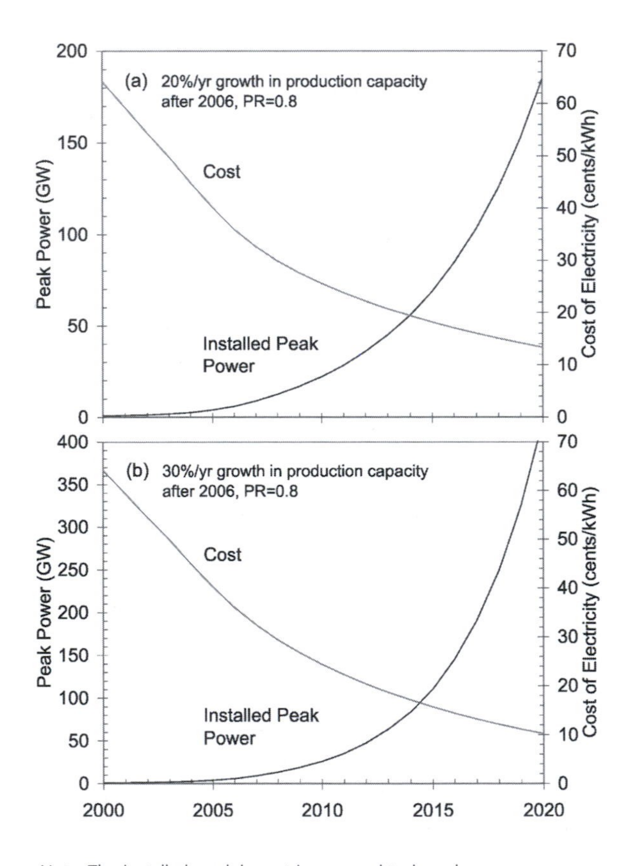

Note: The installed module cost is assumed to have been $12,000/kW$_p$ in 1997, dropping to $4680/kW$_p$ in 2006 as a result of the observed growth in capacity and assumed progress ratio of 0.8. This is consistent with the lowest observed costs for large systems.

Figure 2.31 *Growth in PV peak power and cost of PV electricity in sunny locations (250W/m² mean annual solar irradiance) for a progress ratio of 0.8 and a growth rate of (a) 20 per cent/year or (b) 30 per cent/year starting in 2006*

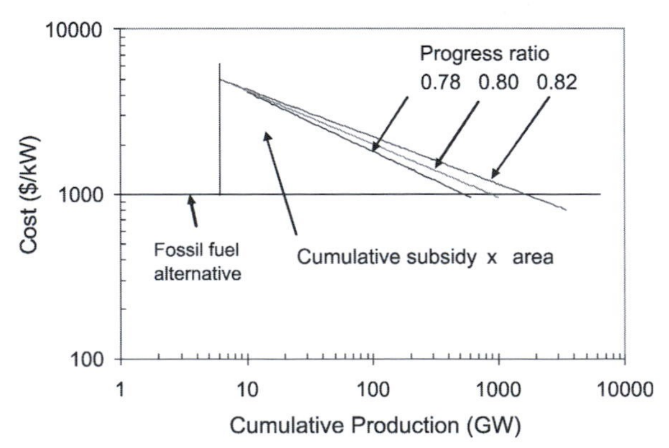

Source: Computed from Equation (2.25)

Figure 2.32 *Impact of alternative progress ratios on the cumulative PV production required for the capital cost of PV power to drop below a particular threshold*

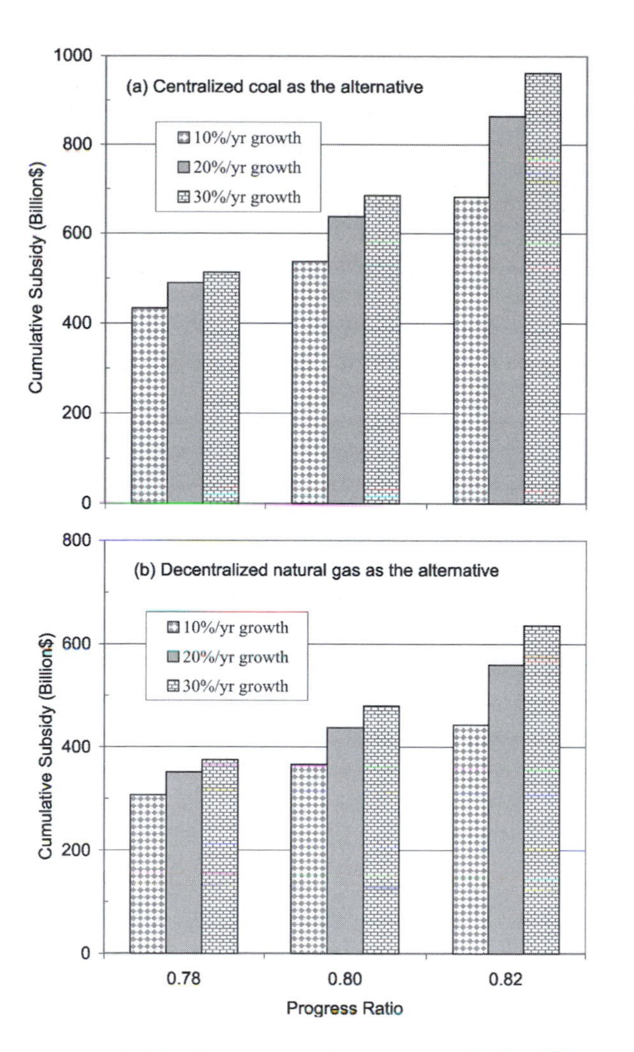

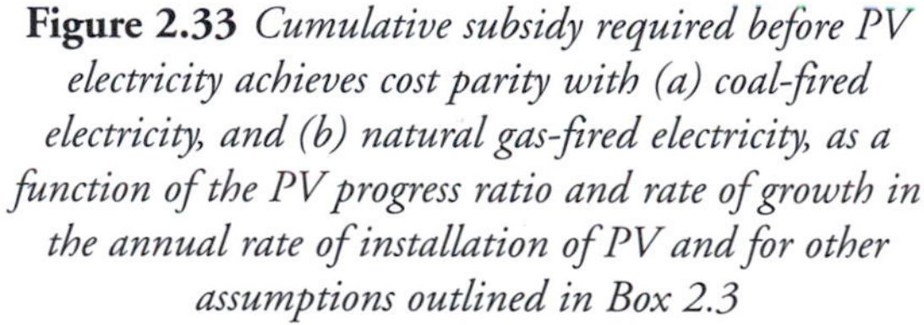

Figure 2.33 *Cumulative subsidy required before PV electricity achieves cost parity with (a) coal-fired electricity, and (b) natural gas-fired electricity, as a function of the PV progress ratio and rate of growth in the annual rate of installation of PV and for other assumptions outlined in Box 2.3*

subsidy is 15–30 per cent less because, with a smaller growth rate, a given cumulative production is reached later and hence in association with higher fossil fuel prices. If natural gas is the alternative, the cumulative subsidy is about 40 per cent less for the assumptions given in Box 2.3.

These results should be regarded as only indicative of the magnitude of the subsidies and timespan required if the costs of solar energy were to follow the historical progress ratio indefinitely. Given the importance of the progress ratio, government support of PV research and development would probably be quite

Box 2.3 Illustrative calculation of the cumulative required subsidy of solar PV power

In this box the assumptions and procedures used to calculate the cumulative subsidies shown in Figure 2.33 are explained. We use as our starting point the conditions for 2006: an installation of PV that year of 2.5GW$_p$, a cumulative installation by the end of the year of 8.61GW$_p$, and an installed cost of \$4500/kW$_p$ (DC) under near-best conditions. We assume that a subsidy equal to the difference between the cost of solar electricity and the cost of fossil fuel electricity must be paid for each kWh of solar electricity generated. The levelized cost of solar electricity is computed for the PV capacity installed in a given year and held fixed for the duration of that year's powerplant (20 years), while the cost of fossil fuel electricity increases steadily due to rising fuel costs, so the required subsidy for a given year's PV decreases during the life of the PV. The subsidy paid in a given year is the sum of the subsidies paid for PV capacity of all ages that are still operating. The cost of PV electricity is computed assuming a capacity factor of 0.2, DC–AC conversion efficiency of 85 per cent, 10 per cent interest, a 20-year lifespan and annual O&M cost equal to 1 per cent of the capital cost.

Two fossil fuel alternatives are considered. One is integrated gasification combined cycle (IGCC) coal with oxyfuel capture of CO_2 (see Chapter 9, Table 9.5) with a base cost of \$1500/kW, an efficiency of 45 per cent and transmission costs of \$1000/kW (this is the US average, but does not apply to PV as it is assumed to be BiPV and so entails no transmission cost). The second is natural gas combined cycle with 100 per cent capture of CO_2, a base cost of \$1000/kW, an efficiency of 55 per cent and transmission costs of only \$500/kW due to more modular plants and closer siting to demand. A 40-year lifespan, 10 per cent interest and annual O&M cost equal to 1 per cent of the capital cost are assumed.

effective in reducing the required cumulative subsidy between now and the time when cost parity is achieved.

2.2.13 Resource constraints on thin-film PV

The extent to which each of the PV technologies can be deployed may ultimately be constrained by the availability of limiting resources. Single-junction compound semi-conductors, multi-junction thin-film technologies and dye-sensitized cells contain various rare metals (such as Cd, Ga, Ge, In, Se and Te) that are produced only as co-products from the mining of other minerals (as described in considerable detail by Fthenakis et al, 2009a). In particular, zinc ores are a source of Cd, In, Ga and Ge, copper ores are a source of Se and Te, lead ores are a source of Cd, bauxite is a source of Ga, and platinum is a source of Ru.

As discussed by Andersson (2000), there are two kinds of material constraints on the production of PV modules: a constraint on the total stock of PV modules that is related to the ultimate availability of the most limiting resource, and a material constraint on the rate of growth of PV power due to limits on the rate of production of refined metals for PV modules. Both constraints depend on the efficiency of the modules, the loading (gm/m^2) of the limiting element, and the proportion of the refined metal that can be diverted from other potential uses to PV modules. A given module technology will be constrained by the metal most limited in supply; for CIGS, this is indium (In), while for CdTe, it is tellurium. The rare metals used in PV modules are not economically valuable enough at present to justify mining activities targeted at these metals; rather, they become available as a byproduct of the mining and extraction of the primary minerals listed above. A reduction in the rate of mining of copper and zinc in response to greater rates of recycling and movement toward a steady global stock of copper and zinc products would lead to reduced availability of many rare metals needed for solar technologies. However, demand for platinum for automotive fuel cells could generate byproduct flows of ruthenium that could be used in dye-sensitized cells.

Even crystalline silicon modules could be constrained by limited resources. Although sand (the source of silicon) is abundant, silver is needed for the electrical contact at the top of the n-type layer. Current practice is to use 20–80μm thick silver ribbons on a grid covering 5 per cent of the module area (Feltrin and Freundlich, 2008). The United States Geological Survey (USGS) estimates a global silver reserve base of 570,000 tonnes (see 'Silver' at http://minerals.er.usgs.gov/minerals/upbs/commodity). Given a silver density of 10.5t/m^3 and assuming a ribbon thickness of 50μm on 20 per cent efficient modules and that 25 per cent of the silver reserve base can be used in PV modules, the potential global power capacity from crystalline PV modules is only 1.1TW. However, with advanced evaporated/photolithographic electrodes having a potential silver thickness of only 2μm and 25 per cent module efficiency, the capacity limit would be increased to about 35TW.

Table 2.13 summarizes the resource constraints for a variety PV technologies, given current material

Table 2.13 *Present and possible future constraints on the deployment of PV power*

Technology	Assumed efficiency (%)	Limiting element and assumptions	Peak power limit
c-Si	20	silver in 50μm ribbon grid covering 5% of module area	1TW
	25	silver in 2μm evaporated electrode	35TW
a-Si	9.5	indium in indium tin oxide (ITO) electrodes	0.1TW
	9.5	Zn in ZnO electrodes	50TW
Dye-sensitized	10	indium in ITO as a transparent conductive oxide (TCO)	0.1TW
	10	tin in tin oxide (SnO_2) TCO	25TW
CdTe	16.5	Te	0.1TW
CIGS	19	In	0.1TW
GaInP/GaAs/Ge	30 and 200 suns concentration	Ge if Ge is used as a substrate (as in Figure 2.9)	0.5TW
		Ga if GaAs is used as a substrate	10TW

Source: Feltrin and Freundlich (2008)

loadings, and the higher limits that are possible with technological advances or material substitutions. The limits of potential PV power given in Table 2.13 are based on the assumption that 25 per cent of the available limiting materials are used for PV modules, with the balance used for other purposes. In combination with other C-free energy sources (such as wind and concentrating solar thermal power), and assuming a strong emphasis on efficient use of electricity, a few TW to at most 10TW of PV capacity would be needed in a future fossil fuel-free energy system (see Chapter 12). With only partial achievement of the technical advances assumed in Table 2.13, this amount of PV power could be readily provided. However, the rate at which required rare minerals become available from the mining of copper, lead, zinc and other primary minerals could limit the rate at which PV modules are deployed. Andersson (2000) estimated maximum rates of deployment of 14GW/yr for CIGS based on the In constraint, 6.7GW/yr for CdTe based on the Te constraint, 14GW/yr for a-SiGe based on the Ge constraint and 11GW/yr for dye-sensitized cells based on the Ru constraint. This gives an estimated maximum rate of deployment for these four technologies of about 46GW/yr. With the technical advances and substitutions postulated in Table 2.13, and with consideration of additional PV technologies, total PV deployment rates of several times this might be achievable. An increase in the market price of critical elements in PV modules due to the increase in demand associated with rapid deployment of PV power would stimulate efforts to increase the fraction of these elements that are recovered during the processing of zinc, copper and other ores, thereby further increasing the potential rate of deployment of PV power. An up-to-date assessment on achievable deployment rates is needed in order to confirm this and to identify other possible bottlenecks.

2.2.14 Toxicity issues for thin-film PV

Toxicity is another issue for thin-film technologies involving cadmium, arsenic and selenium. Cadmium is produced as a byproduct of zinc mining, so most of the environmental hazards associated with Cd at the mine site can be attributed to the extraction of zinc rather than to PV power. Indeed, the use of Cd in CdTe modules reduces environmental hazards compared to leaving it in the mining debris. CdTe is not water soluble and the Cd–Te bond is very strong, not amenable to breakdown by acid in landfill or by fires (Hegedus, personal communication, 2008). In the module itself, the CdTe layer is trapped between two glass sheets that have been observed to fuse together rather than crack in the event of a fire (Green, 2006). The amount of Cd that can be sequestered in CdTd modules is 5–9gm/m^2 but will probably fall over time (Fthenakis and Zweibel, 2003).

Account should also be taken of the avoided Cd emissions to the atmosphere when PV power replaces coal-fired electricity. For 10 per cent modules in a climate with 2000kWh/m^2/yr solar irradiance and a 25-year lifespan, and given an emission factor for uncontrolled emissions from coal of 0.08–0.32mg/kWh (see Accurex et al 1993, Table 4–14), the avoided emission is 0.4–1.6gm/m^2. However, scrubbers can remove up to 94 per cent of the emitted Cd.

2.2.15 End-of-life recycling

Appleyard (2009) reports on techniques that are now or can be used to recycle the materials in PV modules. Only two PV companies so far have active recycling systems, and these are for crystalline silicon modules. In many cases, intact cells can be recovered from discarded modules and reprocessed in an etching line for use in new module assemblies with no apparent loss of performance. For thin-film CdTe modules, a mixture of mineral acids and hydrogen peroxide can be used to remove the semi-conductor layer. After a series of steps, both the Cd and Te can be separately recovered. Similarly, indium and selenium can be recovered from CIGS modules. Unlike in the electronics industry, thin-film PV panels will have sufficient amounts of precious metals to pay the costs of collecting and processing the panels. The major failure mechanism of PV modules is the external wiring and power connection, in which case the modules can normally be economically refurbished in the field without the need for recycling (Fthenakis, 2000).

2.3 Solar thermal generation of electricity

Direct-beam sunlight can be concentrated by a sufficiently high factor to generate high enough temperatures to drive a steam turbine that in turn generates electricity (the operation of a steam turbine is explained in Volume 1, Chapter 3, subsection 3.2.1). This is referred to as *concentrating solar thermal power* (CSTP). Three kinds of collection systems will be described here: a line-focus system using parabolic troughs, the point-focus system based on parabolic dishes that track the sun along two axes, and the point-focus system involving a field of sun-tracking mirrors that concentrate sunlight onto a central tower. These are compared schematically in Figure 2.34. These systems are viable only in desert or semi-arid regions, where cloud cover is rare and the solar radiation will be mostly direct-beam. Diffuse radiation, as found under cloudy skies, cannot be sufficiently concentrated. PV panels, in contrast, can produce electricity under cloudy conditions (although the amount will be reduced due to the reduction in solar irradiance by clouds). However, steam turbines use water for evaporative cooling of the turbine condenser, which can be a limiting factor in desert or semi-arid regions, whereas PV power does not require water (except minor amounts for cleaning, which are also necessary with solar thermal power generation). If water is too scarce, dry cooling can be used but with about a 10 per cent reduction in electricity production.

The IEA established the SolarPACES programme to stimulate and coordinate the development of technology for the solar thermal generation of electricity. The participants in the project are Australia, Brazil, Egypt, the European Commission, France, Israel, Germany, Mexico, Russia, Spain, South Africa, Switzerland, the UK and the US (see www.solarpaces.org for information about ongoing and planned projects). A long list of demonstration projects were carried out in these are other countries (notably India, Iran, Morocco and Algeria), as summarized in Greenpeace (2005). The *Concentrating Solar Power Global Market Initiative* was launched in October 2003 by the German Federal Ministry of the Environment, the European Solar Thermal Power Industry Association (ESTIA) and the American Solar Energy Industries Association (SEIA). The objective of this initiative was to facilitate the building of 5000MW of solar thermal electrical capacity during the following ten years. As will be seen shortly, about 15GW of solar thermal electricity power supply is in the pipeline and scheduled for completion by 2014, thereby greatly surpassing this target.

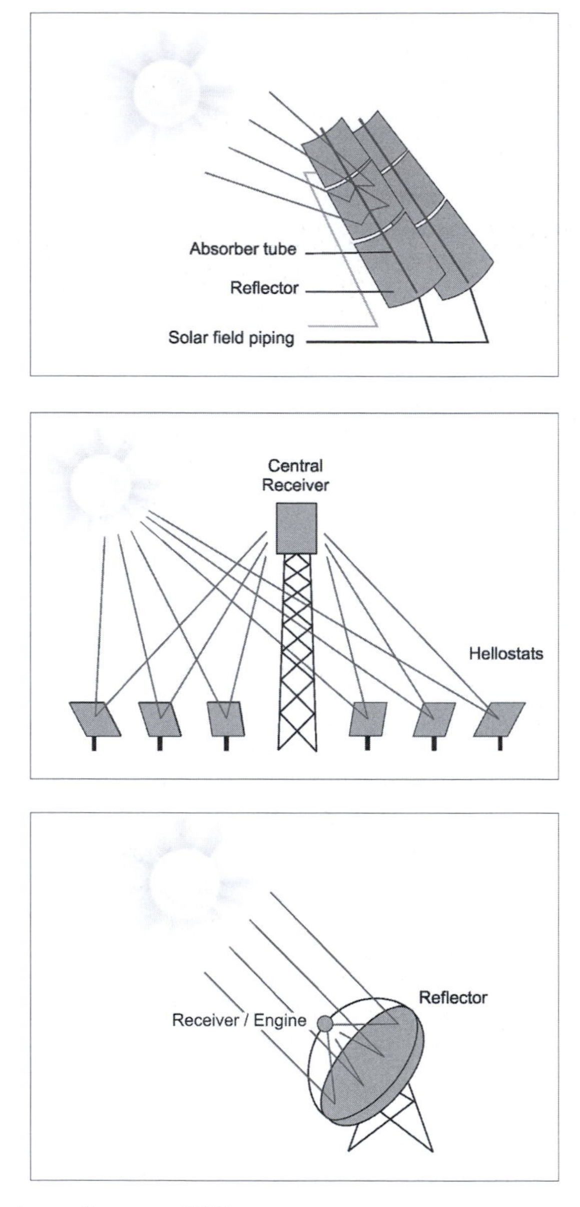

Source: Greenpeace (2005)

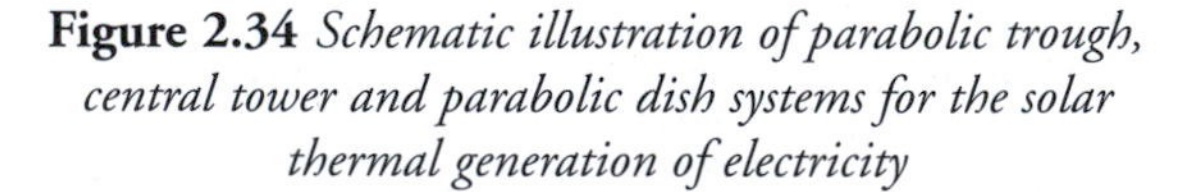

Figure 2.34 *Schematic illustration of parabolic trough, central tower and parabolic dish systems for the solar thermal generation of electricity*

2.3.1 Parabolic trough system

The first large-scale production of electricity by solar thermal means was through the parabolic trough systems developed by Luz International in the Mojave desert, California (annual solar irradiance, 2700kWh/m^2 on a horizontal surface). The 5 × 30MW system at Kramer Junction, California, is illustrated in Figure 2.35. Between 1984 and 1990,

Figure 2.35 *The 5 × 30MW SEGS parabolic trough system at Kramer Junction, California*

nine systems ranging from 13 to 80MW were built. The collectors consists of horizontal parabolic trough mirrors oriented north–south, an arrangement that maximizes summer power production (when the air conditioning load peaks in California). Solar radiation is reflected onto a linear receiver consisting of a 70mm outside-diameter (OD) steel absorber tube, surrounded by an anti-reflective evacuated glass tube with a 115mm OD. The vacuum is typically maintained at about 1.3 × 10^{-7}atm and serves to inhibit heat loss from the inner absorber tube. The collectors heat synthetic oil to 390°C, which is then used to produce steam with a heat exchanger. Natural gas burners are used to heat the oil during times of insufficient sunshine. The scale of the system is such that multi-MW steam turbines used in small fossil fuel powerplants can be used. The present total capacity in California is 354MW, with more than 2km^2 of collector area. The plants are collectively referred to as *solar electricity generating systems* (SEGS). The cost of electricity fell from 28 cents/kWh for the first plant to 9 cents/kWh (14.4 cents/kWh in 2006$) for the eighth plant (Mills and Keepin, 1994).

Heat collected by the solar collectors will either be lost through emission of infrared radiation and through convective heat transfer, or will be transferred to the working fluid. To maximize the heat transfer to the working fluid, the heat loss must be minimized. The Luz collectors have surfaces with an emissivity of 0.15–0.19 (see Volume 1, Chapter 4, subsection 4.1.1 for background to the concept of emissivity). This, combined with the relatively small area that is heated by concentrated sunlight, reduces the infrared energy loss to a few per cent. The efficiency in converting solar energy to electricity is equal to the thermal collection efficiency times the efficiency of the steam turbine to which steam from the solar collector is supplied. As explained in Volume 1, Chapter 3, the efficiency of a steam turbine is larger the greater the inlet temperature. In the Luz systems, turbine efficiencies ranged from about 30 per cent in the earliest units to 38 per cent in the latest units (Price et al, 2002). Overall solar-to-electricity conversion efficiencies reach a daily peak of 22 per cent and have an average of 14 per cent.

A second-generation 2MW parabolic trough power system was constructed at the Plataforma Solar de Almería (PSA) test facility in Spain in 1996. This is a *direct steam generation* (DSG) system, in which water rather than synthetic oil is circulated through the absorption tubes and heated under pressure to produce steam that is used directly in a steam turbine. Eck et al (2003) present initial performance results from this pilot project, while Odeh et al (2003) present computer models to simulate the behaviour of DSG and oil-based systems, and Zarza et al (2006) present the design of a follow-up DSG project. Steam conditions of 100 times atmospheric pressure and 400°C have been demonstrated. A higher steam temperature can be produced with a DSG system than with an oil-based system because the temperature drop across the heat exchanger in the latter is avoided. This, combined with reduced investment cost, is expected to reduce the cost of electricity by 10 per cent compared to an oil-based system.

Recent advances in the support structures for parabolic trough collectors result in less weight, greater resilience against high winds, better corrosion resistance, simpler manufacturing and an ability to be mounted on sites inclined up to 3 per cent (Price et al, 2002). Various polymeric materials are being investigated as alternatives to glass mirrors. These could potentially reduce the mirror weight and overall system cost, but so far no alternative to glass mirrors has been developed with the required cost, performance and lifetime characteristics. An analysis by the US solar technologies programme (US DOE, 2006) indicates that, with continued research and development, scale-up and volume production, concentrating parabolic trough power could drop in cost from about 9.5 cents/kWh in 2004 to 4 cents/kWh by 2015 (electricity from SEGS VI in 1989 is given as 17 cents/kWh).

Parabolic trough systems can store solar thermal energy for use when the sun is not shining. Storage systems are reviewed by Price et al (2002). The first SEGS plant had two tanks, one for storing cool oil and the other for storing hot oil, but was destroyed by a fire originating in one of the storage tanks. Later systems, operating at higher temperatures so as to improve the turbine efficiency, did not have storage.

The most recent parabolic trough power systems are the 50MW AndaSol projects in Spain. AndaSol-1 began commercial operation in March 2009, commissioning of AndaSol-2 began in March 2009, and construction

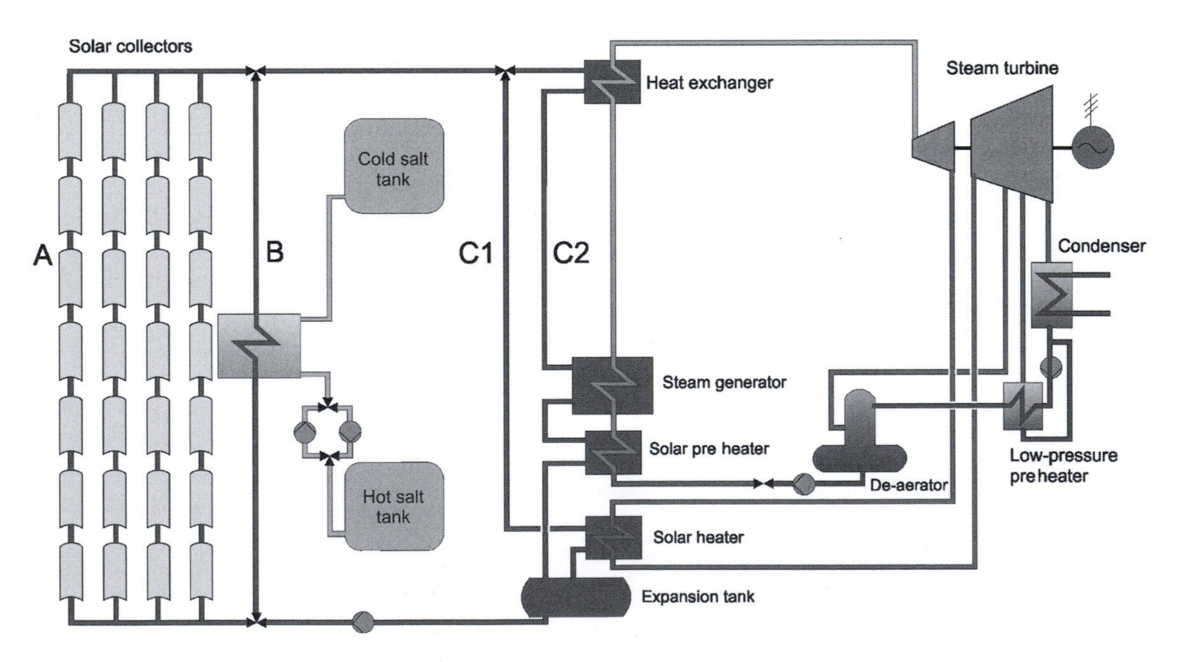

Source: Aringhoff (2002)

Figure 2.36 *Schematic layout of the 50MW AndaSol-1 parabolic trough powerplant in Andalusia, Spain*

of AndaSol-3 was scheduled to begin in October 2009. These plants use molten salts – 60 per cent sodium nitrate ($NaNO_3$), 40 per cent potassium nitrate (KNO_3) – for heat storage, as shown schematically in Figure 2.36. When the available solar energy exceeds that required to produce power, a portion of the collector working fluid follows the loop A–B while the balance follows the loop A–C1 or A–C2. Salt is pumped from the cool tank (at 240°C) to the hot tank (at 560°C), picking up heat from the collector working fluid via the heat exchanger on segment B. When there is inadequate heat from the solar collectors, the working fluid will follow the loop B–C (bypassing the collectors altogether) and salt is pumped from the hot tank to the cool tank, transferring heat to the working fluid via the heat exchanger.

The use of molten salt directly as the heat transfer fluid is another option that is being pursued. This eliminates the need for heat exchangers and allows the solar field to operate at a higher temperature (450–500°C), thereby increasing the turbine efficiency (to 40 per cent or so) and reducing the cost of thermal storage. Organic salts are another possibility. Another alternative is a thermocline system for storage of heat, consisting of a single tank with the hot fluid in the upper portion and the cooler fluid in the lower portion. This approach is potentially 35 per cent cheaper than a two-tank system. Current thermal storage concepts will not work for DSG systems, but phase-change materials might be viable.

Storage of solar thermal energy increases the capacity factor. With six hours of storage, the annual capacity factor increases from 0.22 with no concentration of sunlight, to almost 0.6 with a factor of five concentration, as shown in Figure 2.37. With 18 hours of storage and a factor of five concentration, the capacity factor reaches 0.85.

Integrated solar combined cycle powerplants

Electricity from parabolic trough plants can be combined with natural gas combined cycle plants to form *integrated solar combined cycle* (ISCC) systems. A schematic layout of an ISCC plant is shown in

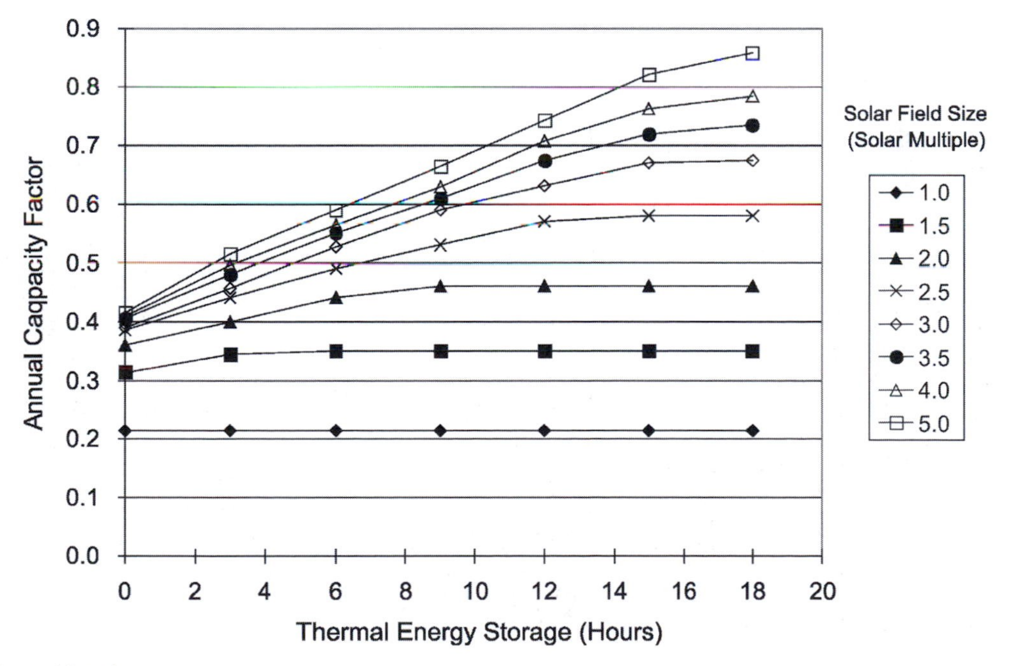

Source: Price et al (2007)

Figure 2.37 *Annual capacity factor for parabolic trough systems as a function of the number of hours of thermal storage and of the solar multiple (the factor by which sunlight is concentrated)*

Figure 2.38. The gas turbine is the same as that in a conventional natural gas combined cycle (NGCC) plant, while the heat required to produce steam for the steam turbine can be supplied from the gas turbine exhaust or the solar collectors. Solar heat at 395°C combines with gas turbine exhaust at 600°C to make steam at 540°C and 100 bar pressure (Greenpeace, 2005), both of which are higher than in SEGS or conventional NGCC plants. In a conventional NGCC, the steam turbine electricity output is about half the gas turbine output, while in an ISCC plant the steam turbine output is about 75 per cent that of the gas turbine. Unlike conventional plants, electricity production does not drop in summer (due to warmer temperatures) because the solar field delivers more energy. Because of the larger steam turbine capacity, the powerplant operates at less than peak capacity when solar energy is not available. Supplemental boilers can be used to maintain full capacity, but the efficiency for the incremental power production would be much less than the NGCC efficiency (about 37 per cent instead of 50 per cent) (Hosseini et al, 2005).

As shown by Dersch et al (2004), the net incremental (marginal) solar efficiency of a solar powerplant with supplemental use of natural gas (whether SEGS or ISCC) is given by:

$$\eta_{net\_solar} = \frac{P_{el,net} - \eta_{ref} m_{fuel} LHV}{P_{th,solar}} \tag{2.28}$$

where $P_{el,net}$ and m_{fuel} are the net electrical output and fuel use, respectively, of the ISCC plant, $P_{th,solar}$ is the thermal energy collected by the collector array and η_{ref} is the efficiency of the NGCC plant that would otherwise be built. The efficiency of an optimized NGCC plant will be greater than that of an ISCC plant without solar input, and this reduces the net solar efficiency. The solar fraction is given by:

$$f_{solar} = \frac{\eta_{net\_solar} P_{th,solar}}{P_{el,net}} \tag{2.29}$$

Calculations by Dersch et al (2004) indicate solar efficiencies of 36–38 per cent and solar fractions of

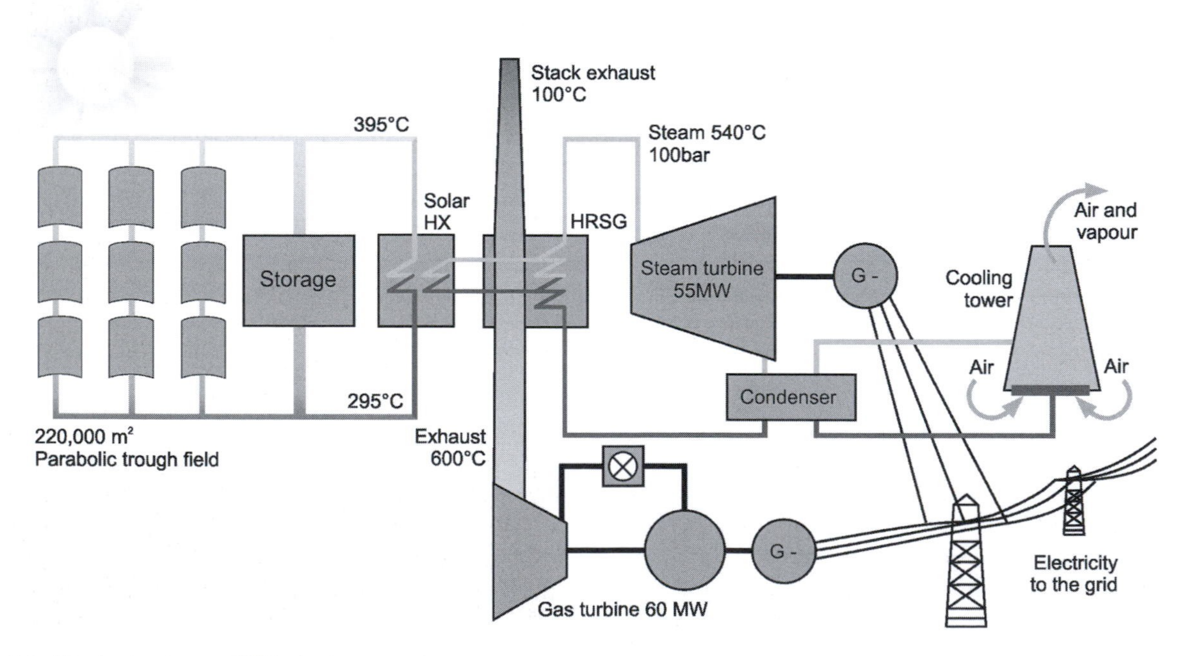

Note: HX = heat exchanger, HRSG = heat recovery steam generator.

Source: Greenpeace (2005)

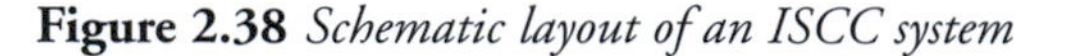

Figure 2.38 *Schematic layout of an ISCC system*

4–18 per cent. These efficiencies would need to be multiplied by the thermal collector efficiency to get the overall sunlight-to-electricity efficiency. For an ISCC plant proposed for Egypt, the solar faction is 8–9 per cent and the cost of the solar portion is 9.5–10.2 cents/kWh (depending on the specific configuration) (Horn et al, 2004).

Compared to SEGS (solar-only) plants, ISCC plants offer the following advantages: (1) lower cost because the incremental cost of a larger steam turbine in a NGCC plant is less than a dedicated steam turbine for a SEGS plant; and (2) avoidance of the efficiency loss associated with cold starts. ISCC systems are expected to eventually produce electricity at 5–6 eurocents/kWh according to Teske (2004). This is consistent with projections reported in Greenpeace (2005): mid-term costs of $2100/kW and 8 cents/kWh; long-term costs of $1800/kW and 6 cents/kWh.

2.3.2 Parabolic dish systems

Parabolic dish systems consist of a 5–15m diameter dish that tracks the sun and concentrates the sunlight by a factor of 600 to 2000 at a single point, where a small (2–50kW) thermal engine is located. Figure 2.39 shows the overall arrangement of reflectors and receiver, while Figure 2.40 shows the receiver of a 10kW system; the circular aperture (where concentrated sunlight enters) is evident at the left. The concentrating reflectors may consist of (1) a single surface in the shape of a parabola, (2) a faceted concentrator, consisting of individual spherically curved glass mirrors aligned in the shape of a parabola, or (3) a stretched membrane concentrator, in which thin membranes are stretched over both sides of metal rings. In the last of these, the membranes can be made of plastic or metal with a reflective coating. A Stirling engine is used to generate electricity. Individual parabolic dish/Stirling engine units are small, nothing larger than 52.5kW having been built. The first unit was built in California in 1984, and a total of eight different parabolic dish/Stirling systems have been built by companies in the US, Germany, Japan and Russia, many of which are described by Mancini et al (2003). The largest array is a 150kW test facility in Arizona run by Stirling Energy Systems. A number of prototypes are being tested at Sandia National Laboratory in Arizona and at the Plataforma Solar de Almería in Spain.

As discussed in Volume 1, Chapter 3, steam turbines are constrained by the rapid rise in water vapour pressure with temperature to operate at

Source: US CSP (2002)

Figure 2.39 *Parabolic dish/Stirling system*

Source: Mancini et al (2003)

Figure 2.40 *Detail of the receiver of the 10kW Stirling engine originally developed by USAB, Sweden, and used in a number of different dish-Stirling systems*

temperatures not much greater than 700°C. A Stirling engine is a piston engine in which combustion takes place outside the cylinder (rather than inside). When combined with solar thermal energy, however, combustion does not occur – heat is provided directly from the sun. A working fluid (typically hydrogen or helium) is alternately heated and cooled. The engine compresses the gas when it is cooled, but the gas expands and drives the engine when it is heated. More power is produced from the expanding hot gas then is needed to compress it when it is cooled, so there is net energy production. Figure 2.41 shows the flow of energy from incident solar radiation to net electrical production for four different dish-Stirling systems. Losses occur during reflection of solar radiation by the reflectors, interception and absorption by the receiver, in the conversion of thermal energy to mechanical energy by the Stirling engine (this being the predominant loss) and in the conversion of mechanical to electrical energy by the electrical generator. Overall annual efficiencies range from about 20–28 per cent.

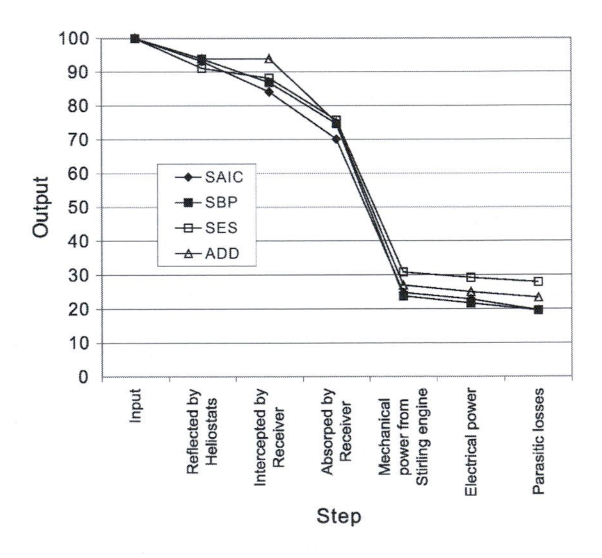

Source: Based on data in Mancini et al (2003)

Figure 2.41 *Energy flow from incident solar radiation to net electrical output for four different dish-Stirling systems*

Parabolic dish systems have low thermal inertia, so they can start up earlier in the day than parabolic trough or central tower systems, or resume full power production sooner after a cloudy period. They can also operate on cloudy days according to Tsoutsos et al (2003). They are highly modular and come in small unit sizes (25–50kW). At present they are the most costly of the solar thermal power systems, with current costs in the order of €10,000–14,000/kW. However, because of their small, modular nature, they lend themselves to mass production. It is felt that the cost could fall to €1600–2400/kW with production of 10,000–3000 units (respectively) per year, but are unlikely to fall much below that (Aringhoff and Brakmann, 2003). Similar costs are foreseen by Mancini et al (2003): $3000–5000/kW at an annual production of 5MW/yr (2000 25kW units/year) and $2000–3000/kW at an annual production of 50MW/yr (with 1–2 cents/kWh O&M and 15–20 cents/kWh total electricity cost). The US solar technologies programme (US DOE, 2006) has set a target of 24 cents/kWh by 2011 and 8 cents/kWh by 2020 (compared to a 2006 benchmark of 50 cents/kWh). This is consistent with the capital cost projections given above.

Source: US CSP (2002)

Figure 2.42 *Central tower solar thermal electricity generation*

2.3.3 Central tower systems

A central tower system consists of an array of sun-tracking mirrors called *heliostats* that reflect sunlight onto a central tower, where a working fluid (either synthetic oils or molten salt) is heated to in excess of 500°C and used to power a thermal engine. One small (0.5–1.0MW) system was built in each of Sicily, Spain and France in 1981, followed by the 10MW 'Solar One' plant at Barstow, California, depicted in Figure 2.42. It was completed in 1982 and produced electricity until 1988 using water as the working fluid (that is, water was directly heated in the central receiver). A redesigned 10MW system, 'Solar Two', was connected to the power grid in 1996. The capital cost of 'Solar Two' was $4850/W$_p$ ($6280 in 2006$, about ten times the cost of large-scale gas turbines). Solar Two used molten salt as the working fluid, which is stored in large tanks at the base of the tower for later use. This levels out the daily variation in electricity production, giving an average capacity factor of 0.38 (instead of the 0.26 that would apply to PV electricity in this location) and resulting in an electricity price of 18 cents/kWh (in 2006$). A 10MW (net) central tower powerplant (PS10), containing 624 120m^2 heliostats and a 115m high central tower, was completed in southern Spain in 2006 at a cost of 3500€/kW (Solúcar, 2006). The annual capacity factor will be about 23–29 per cent. A second central tower powerplant (PS20, with 20MW capacity) began operation in 2009.

Central tower powerplants can store heat and thereby produce electricity 24 hours per day. An inexpensive and simple storage medium is a bed of hot rocks; a more expensive system uses molten salt (as in Solar Two). Hybrid hot-rock/molten-salt storage systems are also possible. A system with storage would be designed for following the electric load rather than meeting only peak demand. This allows a smaller turbine relative to the size of the collector array (since, during times of peak solar irradiance, some of the heat goes into storage), thereby offsetting some (or all) of the cost of a hot-rock storage system on a per kWh basis.

An alternative central tower system uses air as the working fluid. A 1.5MW solar tower using air as the working fluid, the first in the world, was recently built in Jülich, Germany (Schmitz, 2009). Ambient air is sucked through a blackened porous structure on which solar radiation is focused, then goes through a heat exchanger to make steam in the same way that the hot exhaust from a gas turbine would in a combined cycle powerplant. In fact, a gas turbine can be used in place of solar heat when the sun is not shining. Thermal storage is also possible with this system, as discussed by Fricker (2004).

The land area required to collect enough solar radiation for a given powerplant depends on the extent to which empty space between the heliostats is needed in order to prevent the heliostats from shading or blocking each other.[8] If more than one tower is inserted in an array of heliostats, then not all adjacent heliostats need point in the same direction, and the fraction of solar radiation falling on a given heliostat field that can be used increases. Schramek and Mills (2003) find that up to 85 per cent of the solar radiation falling on a heliostat field, including the area between heliostats, can be directed to central receivers using this strategy. By altering the shape of the heliostats, a heliostat area greater than the ground area is possible (Schramek and Mills, 2004).

2.3.4 Solar convective towers

Another concept is referred to as a *solar chimney* powerplant. Solar radiation is trapped beneath a transparent canopy measuring several kilometres in diameter, with a tower in the centre. Air is heated to 35°C warmer than the ambient air, and the resulting buoyancy creates an airflow of $15 m/s^2$ toward the top of the tower. The resulting updraught passes through turbines at the base of the tower. The first solar chimney was tested at Manzanares, Spain, between 1982 and 1988. It achieved a solar-to-electricity conversion efficiency of 0.53 per cent (Mills, 2001). The plant had a collector radius of 122m and a chimney height of 195m.

Dai et al (2003) present a theoretical model of a solar chimney powerplant and apply it to analyse electricity production at remote locations in the semi-arid north-western region of China. They decompose the overall efficiency in the generation of electricity, η_e, as follows:

$$\eta_e = \eta_{\text{collector}} \eta_{\text{chimney}} \eta_{\text{turbine}} \tag{2.30}$$

where $\eta_{collector}$ is the efficiency in converting from solar radiation to heat, $\eta_{chimney}$ is the efficiency in converting from heat to kinetic energy and $\eta_{turbine}$ is the efficiency in converting from kinetic energy to electricity. Typical values are $\eta_{collector} = 0.7$ and $\eta_{turbne} = 0.8$, while $\eta_{chimney}$ is given by:

$$\eta_{chimney} \approx \frac{gH}{c_p T} \tag{2.31}$$

where H is the height of the chimney, T is the ambient air temperature (K), c_p is the specific heat of air and g is the acceleration due to gravity. For a chimney height of 200m (as considered by Dai et al, 2003), $\eta_{chimney} \approx 0.7$ per cent and $\eta_e \approx 0.4$ per cent. For a 500m diameter collector area, the power production is 110–190kW. Tingzhen et al (2008) have carried out computer simulations of a 10MW powerplant.

Given the extremely low efficiency in converting solar energy into electricity, and the resulting large land area requirement, the viability of solar chimney powerplants may depend on using the area underneath the transparent collector as a greenhouse for growing food, as suggested by Dai et al (2003). The ability to grow food under the collector would be a distinct advantage over other collectors. A rough analysis by Kashiwa and Kashiwa (2008) indicates that it should be possible to use a solar chimney to simultaneously generate electricity and extract water from the atmosphere, which would complement the needs of food production in arid regions. Other advantages of solar chimney powerplants are that no cooling water is needed, both direct and diffuse-beam solar radiation can be used, some electricity can be produced during the night (when heat absorbed by the ground under the collector is released), and simple and locally available materials can be used without the need to invest in high-tech manufacturing facilities (although the high towers required may pose a challenge).

2.3.5 Energy payback of solar thermal electricity

Lenzen (1999) has estimated the energy payback time of the 'Solar Two' powerplant to be about 2.5 years and that of a parabolic dish powerplant to be about 1.5 years. GAC (2006) estimated the energy payback time for parabolic trough solar thermal power imported from North Africa to Europe using 800kV DC power lines to be 4–6 months, based on material-embodied energies expected for the year 2030. Included in this estimate are materials for the powerplant including 59 hours thermal storage capacity as full load, and for the lines, assuming powerplant and transmission line lifetimes of 20 and 50 years, respectively, and material recycling at end of life.

2.3.6 Comparative performance

Some technical and economic characteristics of the AndaSol-1 plant are compared with two recent central tower powerplants, PS10 and Solar Tres in Table 2.14. The capital cost of AndaSol-1 was about €6000/kW but the cost of future plants is expected to eventually drop to €2500/kW. Combined with a 40 per cent capacity factor, this results in electricity costs of about 5 cents/kWh. Solar Tres has the largest storage capacity of the three projects, and therefore has the largest reflector field in relation to its power capacity and the smallest power capacity per hectare. However, it has the largest capacity factor (69 per cent) of the three. Electricity generation ranges from 0.4 to 0.9GWh/ha/yr, compared to about 0.2GWh/ha/yr for wind farms (see Chapter 3, Table 3.5) (however, 95 per cent of the land spanned by wind farms can be used for other purposes).

2.3.7 Seasonal variation in the supply of solar thermal electricity

Although solar thermal energy provides the option of storing heat between night and day and thereby generating electricity during the night, there can be strong seasonal variations in the ability to generate electricity from solar thermal energy. This is illustrated in Figure 2.43, which shows monthly electricity yield from solar thermal energy systems with 24-hour storage in El Kharga (Egypt), Madrid (Spain) and Freiburg (Germany). Average monthly outputs as a fraction of the peak monthly output at El Kharga are 97 per cent, 59 per cent and 26 per cent at the sites in Egypt, Spain and Germany, respectively. Assuming peak hourly output to be the same at all three locations, these numbers correspond to the annual average capacity factors. The

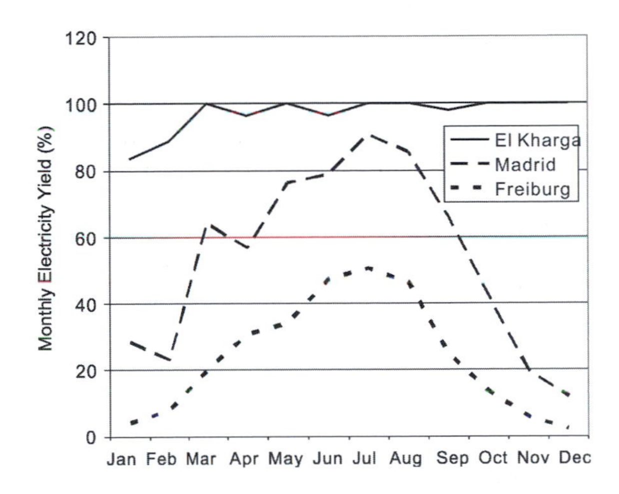

Source: GAC (2006)

Figure 2.43 *Seasonal variation in monthly electricity production from solar thermal electricity generation systems with 24-hour storage at sites in El Kharga (Egypt), Madrid (Spain) and Freiburg (Germany)*

Table 2.14 *Characteristics of recently constructed solar thermal powerplants*

	Powerplant		
	AndaSol-1	**PS-10**	**Solar Tres**
Type	Parabolic trough	Central tower	Central tower
Year operation began	2009	2006	2008
Capacity	50MW	11MW	16MW
Cost	€6000/kW	€3500[a]	€3530/kW
Storage	Molten salt, 7.5 hours	Steam, 50 minutes @ 50% of full power	Molten salt, 15 hours
Capacity factor	41%	23–29%	69%
Average efficiency	15%		
MW/ha	0.25	0.20	0.11
GWh/ha/yr	0.90	0.42	0.68

Note: [a]Solúcar (2006)
Source: EC (2007) or derived therefrom unless indicated otherwise

low annual capacity factor at Freiburg is substantially greater than for PV electricity at this location (16 per cent) in spite of the dependence on direct-beam radiation and hence on the occurrence of clear skies for the operation of solar thermal systems.

2.3.8 Projects under development

A large number of pilot projects during the period 1995–2005 demonstrated the technical feasibility of the different kinds of solar thermal power discussed above, as well as providing a basis for projecting future reductions in cost. This in turn laid the groundwork for the establishment of a fixed feed-in tariff for solar thermal electricity in Spain (originally set at 12 eurocents/kWh, in 2002, then increased to 18 eurocents/kWh in 2004 and 26.9 eurocents/kWh in 2007) and for the establishment of Renewable Portfolio Standards in the US in a number of sunbelt states.[9] The California Solar Initiative, enacted in 2006, provides $3 billion of state support to solar energy over ten years, some of which will be available for solar thermal projects. Portugal, France, Greece and Israel began offering fixed feed-in tariffs of about 16–30 cents/kWh in 2006 or 2007.

Following the increase of Spain's feed-in tariff to 18 cents/kWh in 2004, over a dozen 50MW solar thermal power projects were announced (Greenpeace, 2005). This includes the AndaSol and solar tower projects mentioned earlier. As of mid-2006, another 800MW of parabolic trough projects in Spain had been announced and were in various stages of planning. In the US, California approved plans for a 500MW dish/Stirling plant consisting of 20,000 25kW units in October 2005. The first phase consists of a 1MW test facility, followed by construction in 100MW increments. A $6 million, 1MW parabolic trough facility was built in Arizona in 2006, heats mineral oil and was expected to produce electricity at 12 cents/kWh. In Israel, a 500MW parabolic trough system is planned for the Negev desert by 2010.

The net result of this and other activity is that, by the start of 2009, 480MW of solar thermal power generation had been constructed worldwide, 800MW were under construction in Spain alone and a total of 130 projects with a total capacity of 5700MW were under development in Spain (EER, 2009). In the US, 8500GW of projects are under development and scheduled for completion by 2014. With projects expected in other countries, the worldwide solar thermal power capacity should reach 15GW by 2014 (world PV capacity had reached 14.8GW by the end of 2008). The projects in Spain are overwhelmingly parabolic trough powerplants, while those in the US are a mixture (in decreasing order of importance) of parabolic trough, central tower, parabolic dish and linear Fresnel powerplants.

2.3.9 Summary of solar thermal technologies and prospects for the future

The primary advantages of solar thermal electricity generation compared to PV electricity generation are:

- lower costs, both now and as the costs of both types of system decrease;
- the ability to generate electricity when the sun is not shining; and
- the ability to smooth out short-term variations in output, which can be significant for PV even under clear-sky conditions (see section 2.9).

Table 2.15 compares the present cost and performance of the three major solar thermal technologies for generating electricity that were discussed here. Significant reductions in the cost of solar thermal power systems are expected, especially for thermal tower systems (through mass production of heliostats) and dish-Stirling systems (through mass production of modular units). Figure 2.44 shows the projected cost of heliostats (which account for half the cost of thermal tower systems) as a function of production volume; cost reductions by a factor of four are foreseen. The EuroDish project at the Plataforma Solar Almería aims to reduce the cost of dish systems from $11,000/kW to $5000–6000/kW in the near term and by more in the long term. Current and future electricity costs are given in Table 2.15 using the capital costs and capacity factors given in Table 2.15, assuming financing at 5 per cent/year interest over 20 years, annual insurance equal to 1 per cent of the investment cost, and annual O&M equal to 5 per cent of the investment cost (based on IEA, 2003).

2.4 Solar thermal energy for heating and hot water

Our discussion so far has dealt with the use of solar energy to generate electricity, either with PV modules or through use of concentrated solar heat to drive some form of steam turbine. The latter entails temperatures of 400–600°C. Solar heat can also be used at much lower temperatures, for space heating and the production of domestic hot water. These applications are discussed next.

Table 2.15 *Comparison of the current performance and current and projected (2020–2025) cost of different solar thermal technologies for generating electricity*

Attribute	Technology		
	Parabolic trough	Parabolic dish	Central tower
Powerplant characteristics			
Peak efficiency (%)	21	29	23
Net annual efficiency (%)	13	15	13
Capacity factor without storage (%)	24	25	24
Capacity factor with 6 hours storage (%)	42–48	35–60	35–60
Current investment cost (€/kW)	3500–6000	10,000–12,000	3500–4500
Future investment cost ($/kW)	2000–3000	2000–3000	2000–3000
Current electricity cost (€/kWh)	0.13–0.23	0.27–0.32	0.17–0.22
Future electricity cost ($/kWh)	0.05–0.08	0.05–0.08	0.05–0.08
Storage system characteristics			
Medium	Synthetic oil	Battery	Molten salt
Cost ($/kW heat)	200	30	500–800
Lifetime (years)	30	5–10	30
Round-trip efficiency	95	76	99

Source: this chapter and IEA (2003)

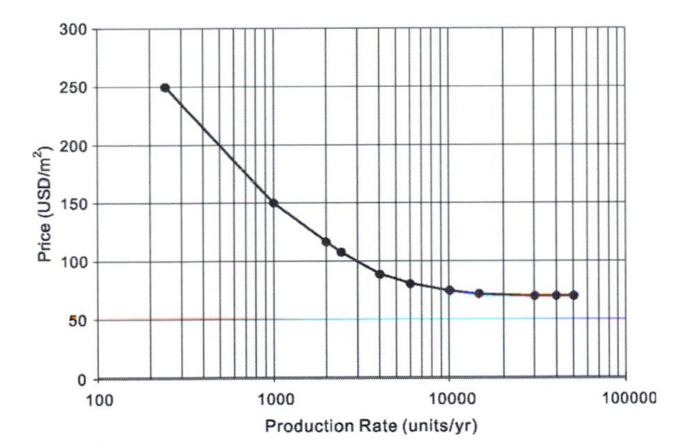

Source: IEA (2003)

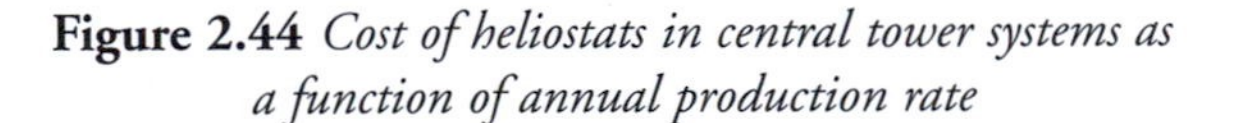

Figure 2.44 *Cost of heliostats in central tower systems as a function of annual production rate*

2.4.1 Types of solar collectors for space heating and domestic hot water

The main types of solar collectors suitable for space heating and hot water requirements are illustrated in Figure 2.45. Thermal solar energy can be collected from flat-plate or evacuated-tube collectors that are mounted on the roof or that form part of the building fabric, stored in hot water tanks and used for domestic hot water or to provide space heating. Here, we discuss the key efficiency and operating characteristics of the kinds of solar collector systems that can be used in buildings. More detailed technical information can be found in Preuser et al (2002) and Andén (2003). Preuser et al (2002) present a synthesis of 30 years of research and operating experience with solar-thermal systems for domestic hot water and space heating in Germany, including a discussion of issues pertaining to the interconnection of solar panels, solar thermal storage tanks, auxiliary water heaters and possible auxiliary storage tanks; the sizing of various components; and control sequences. Andén (2003) provides a nuts-and-bolts treatment of solar thermal collectors and related components for space and water heating, with an emphasis on examples and performance in Sweden, one of the world's northernmost countries.

Flat-plate collectors

Flat-plate collectors can be used to heat air or to heat water. Unglazed collectors contain pipes, through which water flows, overlain by an absorbing plate and

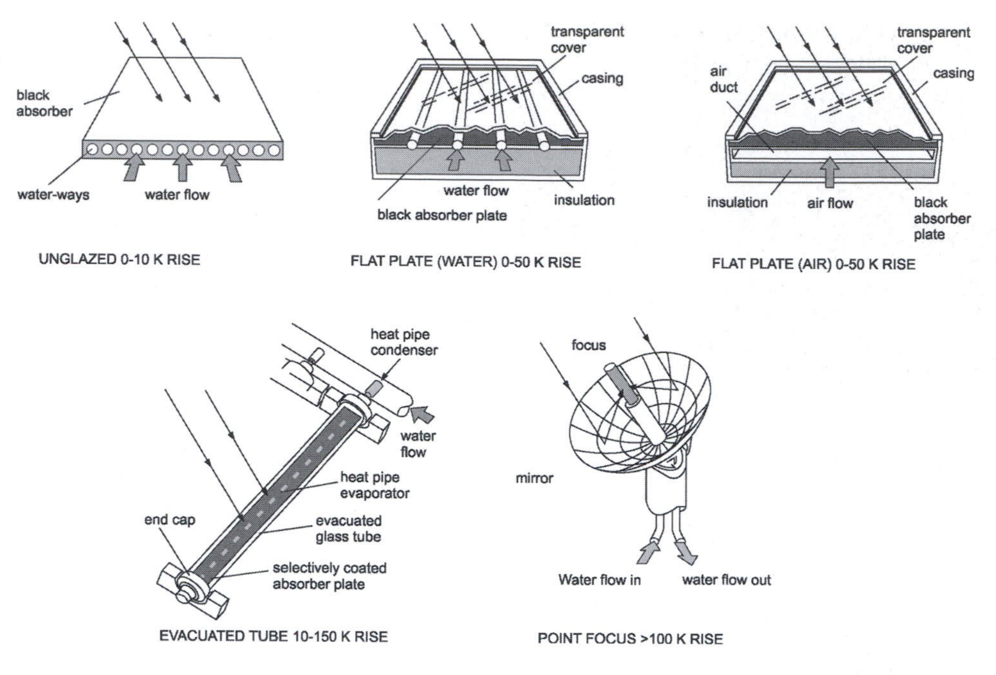

Source: Everett (1996)

Figure 2.45 *Types of solar thermal collectors to produce hot water and associated temperature rises*

give a temperature rise of 0–10K. Transparent flat-plate collectors, with one or two layers of glass, give a temperature rise of 0–50K and can be used to heat water or air. The absorbing plate has a very high absorptivity (greater than normal black paint, which still reflects 10 per cent of the incident radiation). It should have a high heat conductivity to effectively transfer heat to the water and a low infrared emissivity (see Volume 1, Chapter 4, subsection 4.1.1) to minimize radiant energy losses. Figure 2.46 illustrates the installation of large roof thermal collectors, while Figure 2.47 illustrates the artistic integration of thermal collectors into building roofs and façades.

Source: www.socool-inc.com

Figure 2.46 *Installation of solar thermal collectors*

Evacuated-tube collectors

An evacuated-tube collector consists of a series of tubes with a vacuum inside in order to eliminate convective heat losses. There are several kinds of evacuated-tube collector, but the only two with demonstrated long life under fluctuating outdoor conditions are the *heat-pipe collector* and the *all-glass collector*. The heat-pipe collector has a black absorber plate inside each vacuum tube, and within each absorber plate, a pipe that carries

Source: Top, Sonnenkraft, Austria; Bottom, AEE INTEC, Austria

Figure 2.47 *Thermal solar collectors integrated into a building façade (top) and building roof (bottom)*

a liquid at a pressure such that it will boil where it absorbs solar heat and condense onto a pipe carrying the water that is to be heated. In this way, very effective heat transfer occurs, with an effective thermal conductivity much greater than can be achieved with a solid. A temperature rise of up to 150K can be achieved, but the tubes must be inclined at a slope of at least 25°. The all-glass collector is illustrated in Figure 2.48. The solar absorbing surface is placed on the vacuum side of the inner glass tube. Heat is conducted through the glass and heats water, which rises into a cylindrical storage container, to which the evacuated tubes are attached. China is the largest manufacturer and user of all-glass evacuated-tube collectors, with an estimated production of 20 million tubes in 2001 (Morrison et al, 2004).

A disadvantage of evacuated-tube collectors is that they cannot be integrated into the roof of a building. Flat-plate evacuated collectors are available that can be integrated into the roof (Quaschning, 2004). However, they require supports to prevent collapse and have to be re-evacuated from time to time.

Compound parabolic trough collectors

In a parabolic collector, a parabola-shaped reflecting surface concentrates sunlight onto a linear absorber tube. In a *compound parabolic collector* (CPC), an additional concentrator is integrated into a glass envelope that surrounds the absorber tube. The gap

Source: Morrison et al (2004)

Figure 2.48 *All-glass evacuated-tube thermal collectors from China, integrated with storage tanks*

Source: Gee et al (2003); American Society of Mechanical Engineers

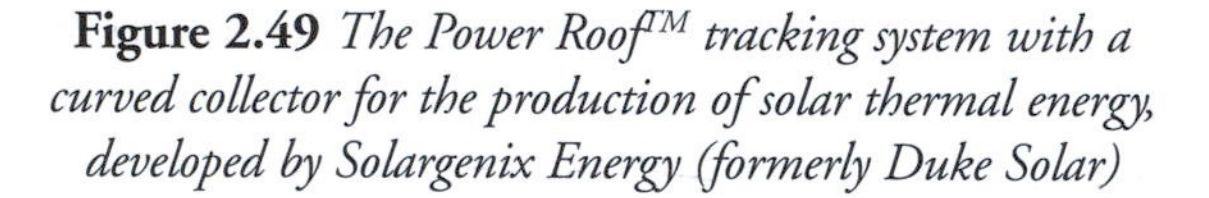

Figure 2.49 *The Power Roof™ tracking system with a curved collector for the production of solar thermal energy, developed by Solargenix Energy (formerly Duke Solar)*

between the glass envelope and the absorber tube may or may not be evacuated. The US firm Solargenix (formerly Duke Solar) has developed a system referred to as Power Roof™ that is mounted on the roofs of buildings. It consists of a fixed curved (non-parabolic) reflector that reflects sunlight onto an evacuated tube that moves during the day as the sun's position changes. The reflector, receiver and curved tracking guides are illustrated in Figure 2.49. The receiver consists of a glass cylinder with an anti-reflective coating to improve transmission and an inner absorber tube with a solar absorptance greater than 0.96 and an emissivity less than 0.10. The space between the glass tube and receiver is a vacuum and contains a silver-coated secondary reflector. The primary reflectors are arranged in rows adjacent to one another in such a way that some of the light intercepted by one reflector is reflected onto the backside of an adjacent reflector and can be used for daylighting.

2.4.2 Efficiency of solar collectors

As a solar collector absorbs solar energy and heats up, heat losses through emission of infrared radiation and through convection will increase. This increase offsets some of the absorption of solar energy, thereby limiting the net efficiency of the collector in converting solar energy into heat. A lower collector temperature, and hence a lower inlet temperature, will lead to greater efficiency for a given collector, but the heat is less useful due to its lower temperature. The efficiency of a solar collector, η_T, is represented as:

$$\eta_T = F_R \tau \alpha - F_R U_L \frac{T_i - T_a}{I} \tag{2.32}$$

where τ is the transmissivity of the cover plate, α is the absorptivity of the absorber plate, U_L is a heat exchange coefficient that depends on the collector, T_i is the temperature of the fluid entering the collector, T_a is the ambient air temperature, I is the incident solar irradiance and F_R is an empirical coefficient that differs from collector to collector. The first term represents the fraction of incident solar radiation that is absorbed, while the second term represents the effect of radiative and convective heat losses. Physically, this term should depend on the average temperature of the absorber plate, but since this is not easy to measure in practice, the inlet temperature is used instead. The effect of this difference is incorporated in the value of U_L.

There are two general ways to increase the efficiency of a solar collector: (1) through changes in the design of the collector itself, and (2) through changes in the system of which the collector is a part, so as to reduce the temperature (T_i in Equation (2.32)) of the water returning to and entering the collector.

As examples of design changes in flat-plate collectors, an additional glazing layer can be added, an anti-reflective coating can be applied to the glazing so that more solar radiation enters the collector, transparent insulation can be placed over the glazing, or the rate of heat exchange between the absorber and circulating water can be enhanced by creating an absorber having channels filled with small porous spheres. In systems to create domestic hot water (DHW), the collector efficiency is improved if the storage tank is thermally stratified, with hot water (from the collector) on top and cooler water (to be reheated by the collector) at the bottom. In systems for space heating, the collector efficiency will be increased if heat is used as it is generated rather than accumulating in a storage tank (through a continuously increasing storage temperature). This, however, requires buildings with adequate thermal mass and low rates of heat loss, so that making full use of sunshine to heat the building when the sun is shining avoids overheating during the day and maintaining comfortable temperatures during the night.

Figure 2.50 shows the efficiency of flat-plate solar collectors with one and two cover glazings, an evacuated-tube collector and a flat-plate collector with 5cm of transparent insulation. Adding an additional glazing reduces the efficiency when $(T_i-T_a)/I$ is small, due to partial reflection of sunlight by the additional glazing, but increases it when $(T_i-T_a)/I$ is large due to the reduced heat loss. Evacuated-tube collectors have a smaller efficiency when $(T_i-T_a)/I$ is close to zero, but also have a very small decrease in efficiency with increasing $(T_i-T_a)/I$, such that they have a much higher efficiency at high temperatures. The efficiency of a flat-plate collector with transparent insulation is similar to that of an evacuated-tube collector with water as working fluid. The efficiency of an evacuated-tube collector can be increased at all temperatures if supercritical CO_2 rather than water is used as the working fluid inside the collector; the result is an annually averaged efficiency in producing domestic hot water of 60 per cent (Zhang and Yamaguchi, 2008). By comparison, for 700W/m^2 solar irradiance and a 25K temperature difference (giving $(T_i-T_a)/I$ = 0.036), a typical flat-plate collector will have an efficiency of 0.50–0.55, dropping to 0.2–0.3 for a 50K temperature difference.

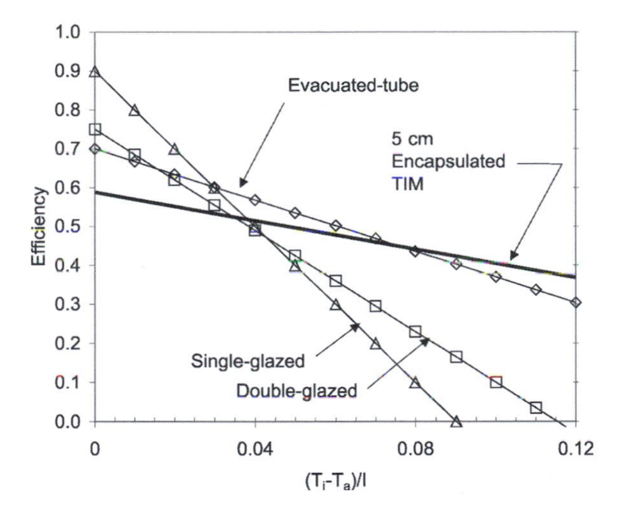

Note: TIM = thermal insulation material. The coefficient values from Atmaca and Yigit (2003) are: single-glazed flat-plate collector, $F_R\tau\alpha$ = 0.9 and F_RU_L = 10.0W/m^2/K; double-glazed flat-plate collector, $F_R\tau\alpha$ = 0.75 and F_RU_L = 6.5W/m^2/K; evacuated-tube collector, $F_R\tau\alpha$ = 0.70 and F_RU_L = 3.3W/m^2/K.

Source: For the TIM collector, Kaushika and Sumathy (2003); all others, computed from Equation (2.31) using the coefficient values from Atmaca and Yigit (2003)

Figure 2.50 *Variation of solar collector thermal efficiency with $(T_i-T_a)/I$ for flat-plate collectors with one or two glazings, for an evacuated-tube collector, and for a flat-plate collector with a 5cm encapsulated thermal insulation material*

2.4.3 Solar water heating

Active solar heat is a sensible option for providing 30–90 per cent of DHW requirements, depending on climate. Solar water heaters can be classified as *open loop* (or direct) and *closed loop* (or indirect). In an open-loop system, the potable water circulates through the collector from a hot water tank. In a closed-loop system, a fluid containing anti-freeze circulates between the collector and a heat exchanger inside the hot water storage tank, but does not mix with the potable water. In a passive or *thermosyphon* system, the hot water tank is placed above the solar collector. As water is heated in the collector, it rises naturally into the storage tank as colder and heavier water enters the bottom of the collector. A thermosyphon system is viable only where there is high solar irradiance throughout the year. Active systems use pumps to transfer the fluid between the collector and storage tank, so the tank can be located anywhere that is convenient. Morrison (2001) provides a detailed discussion of solar water heating systems.

Phase-change materials (PCMs) – materials that change from solid to liquid near the temperature at which one wishes to store heat – can be placed inside the storage tank. When they warm to their melting point, absorbed heat is used to melt the material rather then increasing the temperature further. If all the PCM melts and further heat is added, the water temperature in the tank will continue to increase. The heat used to melt the PCM is released when the PCM refreezes, as heat is withdrawn from the storage tank. PCMs thus permit a greater density of heat storage for the same tank temperature, thereby reducing heat losses from the tank for a given storage of heat (Mehling et al, 2003). As well, the collector efficiency will not fall as rapidly during the course of a day as the hot water tank heats up, since the water returning to the collector from the storage tank will be kept lower. If hot water at a temperature of 60°C is needed, then a PCM with a melting point near 60°C would be used. Hasnain (1998) provides information on many PCMs suitable for DHW (and other) applications.

Existing solar water heating systems are not optimized as integrated systems. The solar collector is heated by solar radiation only on one side (the side facing the sun). Mills and Morrison (2003) analysed a design whereby reflectors at the top and bottom of a solar collector reflect sunlight onto a portion of the underside of the collector. One reflector is oriented such that it operates only during months on the winter side of the spring and autumn equinoxes, thereby reducing the seasonal asymmetry in collected energy. This and other simple design changes increase the fraction of annual hot water demand that can be met with solar energy from 50 per cent to 80–90 per cent in Sydney, Australia. The solar fraction is limited by occasional occurrences of several days of continuously cloudy weather; increasing the collector area does not increase the solar fraction but rather increases the amount of collected energy that has to be dumped.

Diab and Achard (1999) assessed the potential to meet hot water demand in the French city of Chambéry using rooftop solar collectors. Based on data on total floor area and the ratio of roof area to floor area for different building types, they estimated the total roof area available for each building type. These numbers are then multiplied by three factors: an orientation factor (1.0 for flat roofs, 0.25 for tilted roofs) to account for some tilted roofs not having the correct orientation; an availability factor, to account for some of the roof area being taken up by equipment; and an arrangement factor (0.4 for flat roofs), to account for the spacing needed to avoid self-shading by tilted collector modules on flat roofs. They found that the available roof area greatly exceeds that needed to meet an assumed practical upper limit of 30 per cent of the city's hot water requirements.

Given that some backup energy source will be needed along with solar energy, the calculation of the energy savings using solar energy must take into account possible changes in the efficiency of the backup system. Condensing boilers used for space heating and DHW may drop in efficiency from 95 per cent in the winter to 85 per cent in the summer without solar energy, but can drop to 45 per cent or lower efficiency when solar energy is added, due to greater on–off cycling (Thür et al, 2006). As outlined in Box 2.4, the net result is that the savings when solar energy supplies 80 per cent of the hot water load is only 78 per cent of the expected savings. The reduction in boiler efficiency can, however, be reduced or eliminated through use of modulating boilers, which have a slightly higher efficiency at part load rather than lower efficiency (see Volume 1, Chapter 4, subsections 4.3.3 and 4.7.2).

Box 2.4 Energy savings through solar hot water systems

Suppose that 100 units of hot water are supplied through a boiler with an efficiency of 85 per cent. Then 100/0.85 = 118 units of fuel are required. If solar energy supplies 80 per cent of the hot water requirements, and the remaining 20 per cent is supplied by the boiler but now at an efficiency of 45 per cent (as found by Thür et al, 2006), then 20/0.45 = 44 units of energy are used. The energy savings is thus 118–44 = 74 units, rather than 0.8 × 118 = 94.4 units.

If, as suggested by Thür et al (2006), the boiler is completely shut off during the summer and electricity is used for supplemental heating of water, then the reduction in onsite energy use is 118–20 = 98 units. This would be an ideal solution if the electricity is from wind, solar or hydroenergy source. However, if the electricity is supplied by fossil fuels or biomass, the primary energy requirement is $20/\eta$, where η is the efficiency of the powerplant supplying the electricity. If $\eta = 0.4$, then the savings in primary energy is only 68 units.

2.4.4 Solar combisystems

'Combisystems' are solar systems that provide space and water heating. They are inherently more complex than systems for producing DHW, since they involve an additional heat distribution system (normally some low-temperature system, such as radiant floor heating). As well, the heating load undergoes a marked seasonal fluctuation, in contrast to the hot water load, and is largest when there is the least amount of solar energy available. During the period 1975–1985, many complex and non-standard combisystems were designed by engineers. After 1990, simple and inexpensive systems were designed by solar collector companies. Current designs are based on field experience but have not yet been carefully optimized, so it is thought that there is a great potential for cost reduction, improved performance and improved reliability. Modular collector panels for combisystems are available that replace conventional roofing or wall materials. Even when the collectors do not deliver usable heat to the storage tank, they reduce heat loss through the wall or roof element. Incorporation of solar collectors into equatorward-facing walls increases the heat collection in winter compared to roof collectors, due to the high solar zenith angle, while decreasing it in summer (see Figure 2.4).

In 2001, the total collector area installed for solar combisystems in eight European countries was 340,000m^2, two thirds of which was in Germany and most of the rest in Austria (Weiss, 2003). Assuming an average collector area of 15m^2, this translates into 22,600 solar combisystems installed in 2001 alone. In Sweden, 65 per cent of the total collector area installed in 2007 was for residential combisystems (Weiss et al, 2009). Problems to date have been mainly due to an improper control strategy when combining the solar and auxiliary parts of the system. Through design improvements, the number of pipe connections required in an optimized residential system has fallen from 17 to 8, the required collector area has fallen in half and the total mass of the system has fallen from 250kg to 160kg. In The Netherlands, a typical solar combisystem consists of 4–6m^2 of solar collector and a 300 litre storage tank, so solar energy meets a small (10 per cent) share of the heating demand. In Switzerland, Austria and Sweden, common systems for a single-family house consist of 15–30m^2 of collector and a 1–3m^3 storage tank, and 20–60 per cent of the heating demand is met by solar energy. In Gleisdorf, Austria, a system was installed in 1998 for an office building and six terraced houses. The collectors have been integrated into the roofs of the winter gardens and cover 80 per cent of the annual hot water demand and 60 per cent of the heating demand. In Austria, individual modules of 15m^2 for large systems have reduced overall costs by 30 per cent.

It is not economically attractive to build poorly insulated houses and then try to heat them through active solar heating. It is less expensive to build a very well-insulated house and to rely on passive solar heating, rather than active solar heating. To collect enough heat in winter using collector panels, the panels would have to be very large. In the summer, much of the collected heat would be wasted. In super-insulated houses built in cold climates, it is common for a third of the heating requirement to be supplied through internal heat gains, another third to be supplied through passive solar heat and the remaining third to be supplied through active heating systems, whether fossil fuel, electric or a hybrid involving solar thermal energy with storage in hot-water tanks.

2.4.5 Energy payback times

Lifecycle assessments of solar thermal collectors by Crawford and Treloar (2004) and Ardente et al (2005) indicate that the primary energy required to produce the collectors and associated pipes and storage tanks is paid back through savings in primary energy after about 0.5–2 years of use.

2.4.6 Present-day global market and costs

Figure 2.51 shows the trend in the annual rate of installation of solar thermal collectors and the growth in total installed collector area from 1999 to 2007. The cumulative area grew at an annual averaged compounded rate of 17.3 per cent/year, while the annual rate of installation grew at an average rate of 18.1 per cent/year. Table 2.16 gives the solar thermal collector area in operation by the end of 2007 in countries thought to represent 85–90 per cent of the world market. Total installed collector area reached about 210 million m^2, corresponding to a peak heat

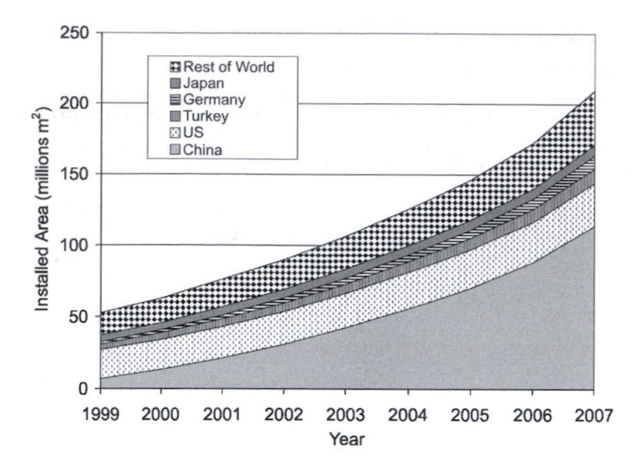

Source: Various editions of *Solar Heat Worldwide*, the latest at the time of writing being Weiss et al (2009)

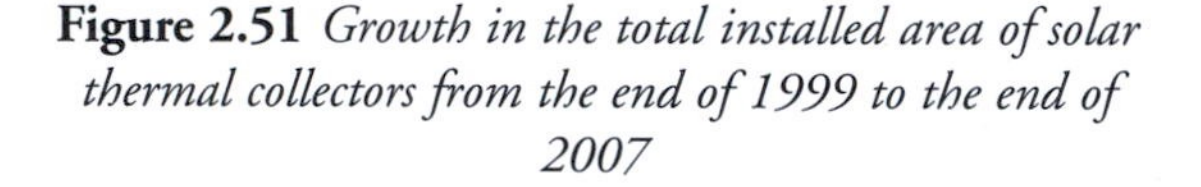

Figure 2.51 *Growth in the total installed area of solar thermal collectors from the end of 1999 to the end of 2007*

production capacity of about 147GW. This is greatly in excess of the total installed PV electrical capacity of 9.2GW and the global wind power capacity of 94GW at the end of 2007 (see Figures 2.29 and 3.1, respectively). The collectors are overwhelmingly (99.2 per cent) water collectors rather than air collectors. Of these, 32 per cent are glazed, 51 per cent are evacuated-tube and 17 per cent are unglazed collectors (the last of these are used predominantly in the US and Australia for heating swimming pools). Compound parabolic collectors have seen essentially no use to date. Also given in Table 2.16 are the collector areas installed in 2007.

Figure 2.52 compares the total collector area and collector area per 1000 inhabitants for the ten leading countries in each category. The leading countries in terms of total area are China, US, Turkey and Germany, while the leading countries in per capita collector area are Cyprus, Israel, Austria and Greece. The rate of installation has been growing by almost 30 per cent/year in China, and constitutes 12 per cent of the national water heater market according to Jones (2005). Zhai and Wang (2008) provide an interesting review of the use of solar thermal energy in China and of the integration of solar thermal collectors into building façades and roofs in China. In 1999, Barcelona City Council unanimously passed a 'solar ordinance' that requires the installation of solar thermal collectors sufficient to meet 60 per cent of the hot water needs on all new buildings and on larger buildings during renovation (ESTIF, 2007). For outdoor swimming pools, only solar heating is permitted. The Barcelona ordinance was followed by a similar requirement in Madrid in 2003 and a national-level requirement that 30–70 per cent of the hot water in new buildings or in buildings undergoing a major renovation be provided by solar energy.

The European Solar Thermal Industry Federation (ESTIF) estimates the technical potential for solar thermal energy in the 15-nation European Union (EU) for heating, hot water, air conditioning and industrial processes to be a collector area of 1.41 billion m^2 (3.74 m^2 per capita) and an energy output of 682 TWh/yr (ESTIF, 2003). Its 2007 *Solar Thermal Action Plan for Europe* calls for a target of 1m^2 of collector area per person by 2020, which would be equivalent to a thermal capacity of about 320GW and would require growth in the EU solar collector market by 31 per cent/year between 2007 and 2020.

In the US, Denholm (2007) estimates that 50 per cent of the residential roof area and 67 per cent of the commercial roof area is suitable for solar hot water collectors. For buildings with suitable roofs, an estimated average of 62 per cent of hot water needs could be met with solar energy (the solar potential ranges from 50 per cent in New England to 75 per cent in California).

Henning (2004) indicates the following costs for solar collectors, support structures and piping (but excluding storage systems, heat exchangers and pumps):

- solar air collectors, €200–400/m^2;
- flat-plate or stationary compound parabolic collectors, €200–500/m^2;
- evacuated-tube collectors, €450–1200/m^2.

Table 2.17 gives illustrative costs of solar thermal energy for total system costs of €400–1200/m^2, solar irradiance of 1100–2200kWh/m^2/yr, system efficiency of 20–60 per cent (which roughly spans the range encountered in practice), and real interest rates of 4 per cent, 6 per cent and 8 per cent. In all cases, a 25-year system lifespan and annual insurance and maintenance costs of 1 per cent and 2 per cent of the investment cost, respectively, are assumed. The system efficiency accounts for the efficiency of the solar collectors and any losses from the piping and hot water storage tank;

Table 2.16 *Total solar thermal collector area (1000s of m²) in operation by the end of 2007 in reporting countries thought to represent 85–90 per cent of the world market, and the collector area added in 2007*

Country	Water collectors			Air collectors		Total	Total added in 2007
	Unglazed	Glazed	Evacuated	Unglazed	Glazed		
Australia	4070	1660	23			5753	782
Austria	608	2950	43			3601	290
Barbados		83				83	2.7
Belgium	49	134	12			195	51
Brazil	97	3587	0.35			3685	573
Canada	666	82	4.7	130	0.19	883	61
China		10,400	103,740			114,140	21,140
Cyprus		795	0.96			796	16
Czech Rep.	15	97	15			128	31
Denmark	21	394	3.4	3.4	18	441	31
Finland	0.5	16	1.3			17	2.7
France + Terr	105	1417	33			1554	323
Germany	750	7784	864			9398	970
Greece		3566	6.8			3573	283
Hungary	2.8	41	2.6			47	8
India		2150			17	2167	257
Ireland		28	7.9			36	20
Israel	24	4936				4961	72
Italy	26	874	103			1002	249
Japan		6825	127	434	13	7399	183
Jordan		840	7.2			848	11
Mexico	468	444				911	154
Netherlands	344	330				673	48
New Zealand	6.2	103	10			119	18
Norway	1.6	11.2	0.15		1.2	14	1.0
Poland	1.3	198	37	3	2.5	241	68
Portugal	0.59	276	5.4			282	51
Slovak Republic		88	9.9			98	25
Slovenia		116	1.2			117	7.7
South Africa	627	248				876	81
Spain	3	1164	46			1213	265
Sweden	80	223	29			332	45
Switzerland	212	433	25	838		1509	78
Taiwan		1137	118			1255	135
Turkey		10,150				10,150	700
UK		278	27			305	54
US	27,639	1898	578	0.09	230	30,346	1276
Total	35,820	66,272	105,885	1409	282	209,669	28,440
Total capacity (GW_{th})	**25.1**	**46.4**	**74.1**	**0.99**	**0.12**	**146.8**	**19.8**

Note: Also given is the total world heat production capacity corresponding to the reported collector areas. The thermal capacity is computed from total area using an agreed conversion factor of 0.7kW/m².
Source: Weiss et al (2009)

in other words, it is the ratio of heat delivered to the final end use to the solar energy incident on the collectors. In moderate climates or highly insulated buildings, not all of the collected heat can be used due to overheating, but a larger storage tank will reduce the amount of collected heat that has to be discarded. The system efficiency will tend to be larger in regions with smaller solar irradiance, thereby partly offsetting the impact on cost of lower irradiance. Thus, solar thermal energy costs the same at 1100kWh/m²/yr irradiance and 40 per cent system utilization as at 2200kWh/m²/yr irradiance and 20 per cent system

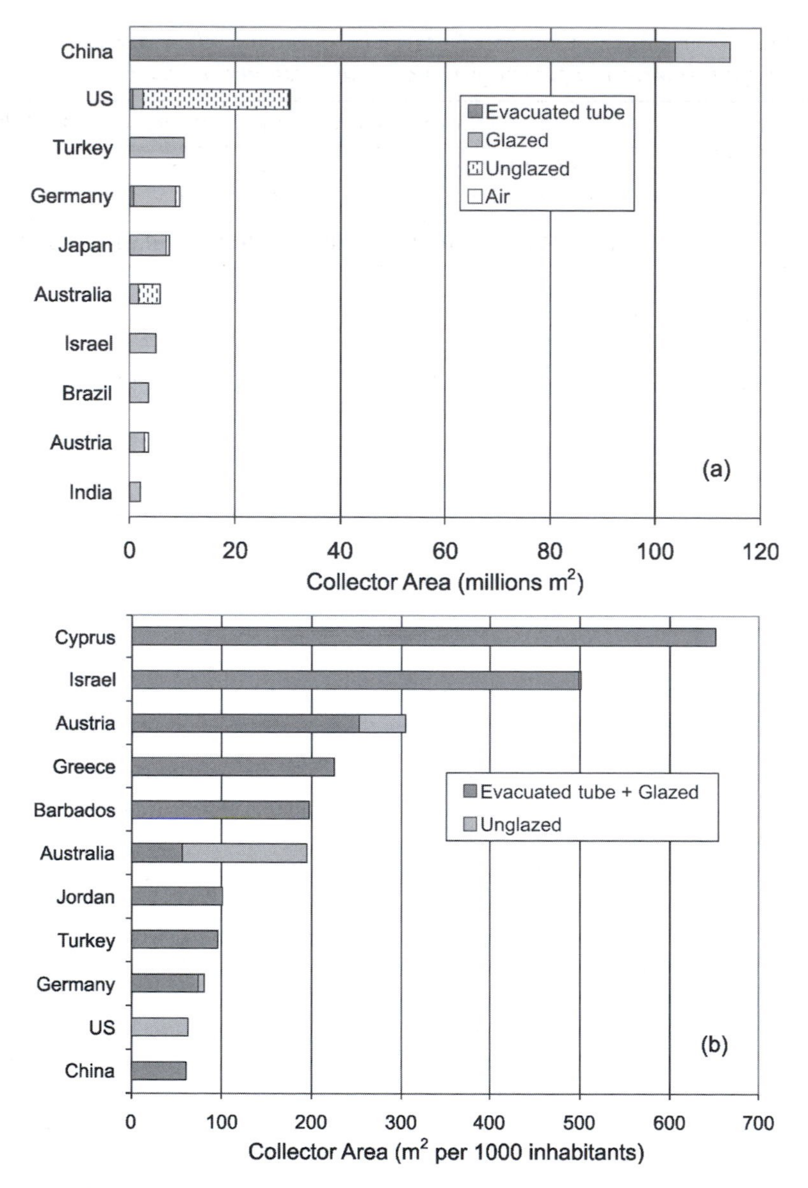

Source: Data from Weiss et al (2009)

Figure 2.52 *(a) Total solar collector area in the ten countries with the largest area, and (b) collector area per 1000 inhabitants for the ten countries with the greatest per capita collector area*

utilization. For mid-European conditions (1650kWh/m^2/yr irradiance), 4 per cent financing cost and total system costs of €800 per m^2 of collector, the cost of thermal energy ranges from about 8 eurocents/kWh at 60 per cent solar utilization (not yet achieved in practice) to 23 eurocents/kWh at 20 per cent solar utilization. By comparison, the typical cost of heat in various countries (including fuel and amortization of the heating equipment) from conventional (fossil fuel) systems is 5–12 cents/kWh.

Table 2.17 *Illustrative costs of solar thermal energy for the indicated solar irradiance and real interest rate, a 25-year lifespan, 1 per cent/year insurance, 2 per cent/year maintenance cost and the given system cost and efficiency*

System cost ($/m² or €/m²)	System efficiency	Cost of thermal energy (cents or eurocents/kWh)								
		1100kWh/m²/yr interest rate			1650kWh/m²/yr interest rate			2200kWh/m²/yr interest rate		
		0.04	0.06	0.08	0.04	0.06	0.08	0.04	0.06	0.08
400	0.2	17.1	19.7	22.5	11.4	13.1	15.0	8.5	9.8	11.2
	0.4	8.5	9.8	11.2	5.7	6.6	7.5	4.3	4.9	5.6
	0.6	5.7	6.6	7.5	3.8	4.4	5.0	2.8	3.3	3.7
800	0.2	34.2	39.4	45.0	22.8	26.2	30.0	17.1	19.7	22.5
	0.4	17.1	19.7	22.5	11.4	13.1	15.0	8.5	9.8	11.2
	0.6	11.4	13.1	15.0	7.6	8.7	10.0	5.7	6.6	7.5
1200	0.2	51.3	59.0	67.5	34.2	39.4	45.0	25.6	29.5	33.7
	0.4	25.6	29.5	33.7	17.1	19.7	22.5	12.8	14.8	16.9
	0.6	17.1	19.7	22.5	11.4	13.1	15.0	8.5	9.8	11.2

Price is thus a barrier to the widespread adoption of solar thermal water heating systems in many parts of the world. However, possible development of integrated collector-storage systems using polymer materials could break the current price barrier (Lutz et al, 2002). If a single storage tank is used for a hybrid solar/natural gas water heating system, the incremental cost of the solar component is about half of the cost when bought as a separate add-on system (see www.iea-shc.org/task24, Case Study 1). In this case, the heating coil from the auxiliary heater would be placed in the upper portion of the storage tank, so as to avoid heating all of the water in the tank when the sun is not shining or is inadequate. Care is needed to maintain a temperature stratification within the storage tank.

2.5 Solar thermal energy for air conditioning

Thermal energy can be directly used for cooling and dehumidification. Absorption chillers are widely used for chilling of commercial buildings, while desiccant systems have been used in supermarkets. Both of these options are discussed as Volume 1 (Chapter 4, subsection 4.5.4). Adsorption chillers provide a third alternative. The required heat can either be directly supplied through the combustion of natural gas or some other fuel, or can be supplied as steam produced through the production of electricity (cogeneration). As shown in Volume 1, Chapter 4, the dedicated use of natural gas to provide the heat input to absorption chillers always increases the use of primary energy compared to electric chillers, while the dedicated use of natural gas in desiccant chillers can either increase or decrease primary energy use compared to using electric chillers. When heat to absorption chillers or desiccant systems is supplied by microturbines, primary energy requirements can either increase or decrease (depending on the efficiency of the powerplant that would supply electricity to electric chillers), while supplying heat to absorption chillers from combined cycle cogeneration reduces the primary energy requirement by about 50 per cent. However, combined cycle cogeneration is not an option for desiccant systems because the scale of this type of cogeneration is too large for desiccant systems, which must operate at individual buildings because they produce cool and dehumidified air, rather than chilled water that can be produced centrally in large systems and distributed through a district cooling grid. Thus, the greatest potential for reducing energy use with desiccant systems (and, to a lesser extent, with absorption chillers) is if the heat can be supplied from solar thermal collectors.

There was a substantial effort to develop solar cooling systems in the late 1970s and early 1980s in the US, but this effort subsided and the centre of activity shifted to Japan and Europe. Task 25 (Solar-Assisted Air Conditioning of Buildings) of the IEA's Solar Heating and Cooling Programme ran from 1999 to 2004 and was directed toward improving conditions for the market introduction of solar-assisted cooling systems by defining performance criteria, identifying and further developing promising technologies, optimizing the integration of solar-cooling systems into building heating, ventilation and air conditioning (HVAC) systems, and creating design tools and concepts for architects, planners and civil engineers. Eleven demonstration projects were carried out and the book *Solar-Assisted Air Conditioning in Buildings – A Handbook for Planners* (Henning, 2004) was produced. The Chinese first tested a solar cooling system in 1987, then followed up with a further ten demonstration projects, five of which are reviewed by Zhai and Wang (2008).

Task 25 of the IEA Solar Heating and Cooling Programme was followed by Task 38 (Solar Air Conditioning and Refrigeration), which runs from 2006 until December 2010. Its goal is to accelerate the market introduction of solar cooling systems through the development and testing of pre-engineered system concepts for residential and small commercial buildings, through the development of systems for large non-residential and industrial applications, and through workshops, training sessions and publication of a second edition of the solar cooling handbook. Australia, Canada, Mexico and several European countries are participating in Task 38. The January 2009 newsletter of Task 38 lists 33 solar cooling projects that will be monitored, all of them in Europe, but activity is also continuing in China.

Many projects have also been carried out in Europe as part of the Solar Air Conditioning in Europe (SACE) programme. Balaras et al (2003) provide summary information about 21 solar air conditioned buildings in Germany alone, while Balaras et al (2007) and Henning (2007) present more recent reviews of this work. Key results are summarized below. Solar air conditioning is expensive, which provides a strong incentive to first substantially reduce cooling loads through the array of measures outlined in Volume 1 (Chapter 4, subsection 4.5.1).

2.5.1 Performance of solar cooling systems

The performance of a chiller or air conditioner is characterized by its coefficient of performance (COP) – the ratio of heat removed to energy input. Residential air conditioners have COPs of 2.5–3.5, while large commercial chillers have COPs of 5–7. Figure 2.53 plots the COP of heat-driven chillers against the driving temperatures. Liquid desiccant chillers (which are not yet widely available commercially) provide close to the highest COP (about 0.75) with close to the lowest driving temperature (65°C), whereas solid desiccant systems have among the lowest COPs (0.5) but with relatively low driving temperature (80°C).

For solar-driven chillers, the appropriate COP to use for comparative purposes is the chiller COP times the efficiency of the solar collector. This is referred to as the *solar COP*. Thus, for an absorption chiller with a COP of 0.7, supplied with heat from flat-plate solar collectors with an efficiency of 0.5, the solar COP is 0.35. This means that one unit of solar energy incident on the solar collector produces only 0.35 units of cooling. The efficiency of a solar collector tends to be higher if it operates at a lower temperature, which is possible when lower driving temperatures are required by the chiller.

In reality, the COP of different kinds of chillers depends on the size of the chiller and the desired degree of cooling. The COP of all systems is lower the deeper the desired cooling, but the dependence on cooling temperature is generally weaker for absorption systems. Syed et al (2002) found that a double-effect LiBr chiller gives the largest solar COP for chilling to about 7–15°C, the single-effect ammonia chiller gives the largest solar COP for chilling to about −12°C to −17°C, and a vapour compression chiller coupled to PV modules with an efficiency (sunlight-to-AC) of 15 per cent gives the largest solar COP for all other temperature ranges.

In addition to heat, thermally driven chillers require electricity for pumps and fans. For the solar projects surveyed by Balaras et al (2007), this amounted on average to 22.5 per cent of the cooling capacity (i.e. 225W of electricity demand per kW of cooling power). Water consumption averaged 5.3 litres/kWh of cooling.

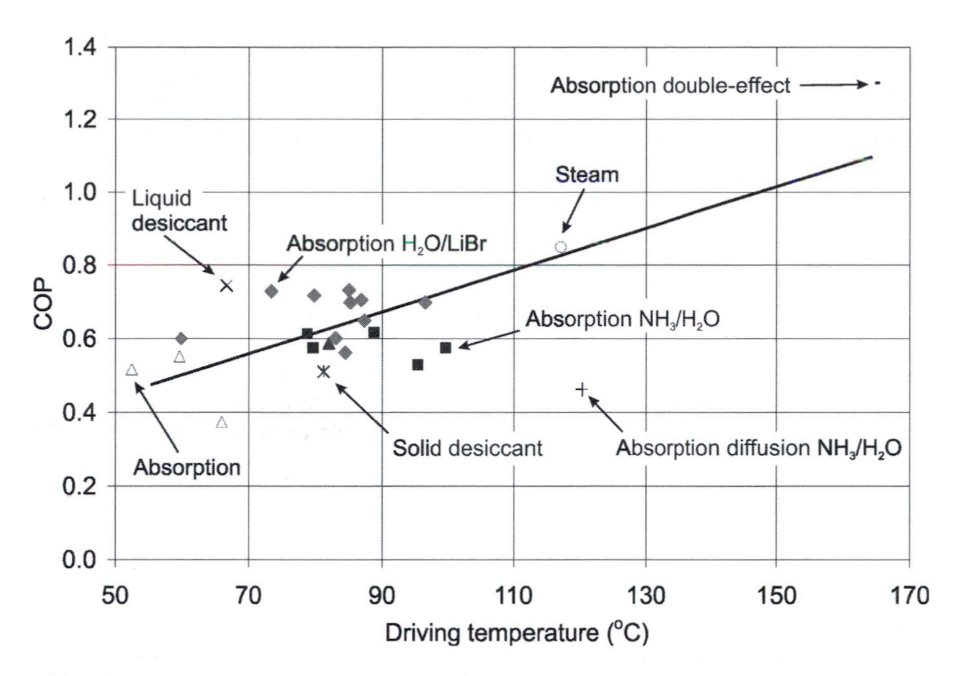

Source: Balaras et al (2007)

Figure 2.53 *Chiller COP versus driving temperature for solar thermal cooling systems in Europe*

2.5.2 Impact of the backup component on overall energy use

If a backup fossil fuel heat source is used to power an absorption chiller when solar heat is inadequate, the benefit of solar heat could be offset by the low COP of the absorption chiller (Florides et al, 2002). Assuming an electric air conditioning COP of only 4.0, an electricity generation efficiency of 55 per cent (the current state-of-the-art), a boiler efficiency of 85 per cent and an absorption chiller COP of 0.74, the benefit of obtaining even two thirds of the heat input from solar energy is largely nullified by the low efficiency of the absorption chiller when the backup boiler is needed. If solar heat provides only a third of the cooling and natural gas the rest, the primary energy use more than doubles compared to using an electric chiller.

Ways to largely or entirely eliminate the efficiency penalty of using a fossil fuel heat source as backup for a single-effect absorption chiller are:

- to use an electric chiller as backup when solar heat is not adequate;
- to use a double-effect absorption chiller that can operate as a double-effect chiller (at high temperature and with a relatively high COP) when natural gas is used, and as a single-effect chiller when solar heat (at lower temperature) is available (Lamp et al (1998) worked out the design of such a chiller, which could operate with simultaneous solar and natural gas heat as well as with either heat source alone);
- to use a dedicated double-effect absorption chiller powered by evacuated-tube collectors, CPCs or integrated evacuated-tube CPCs when the sun is shining;
- to accept that temperature will drift higher at times when solar heat alone cannot provide all of the cooling required to maintain a fixed temperature. Such floating temperatures could, in some cases, be consistent with the adaptive temperature standard shown in Volume 1, Figure 4.8, whereby warmer temperatures are acceptable when the outdoor temperature is warmer, due to psychological adaptation.

A double-effect absorption chiller with 70kW of cooling capacity (a $70kW_c$ chiller) powered by non-tracking solar collectors was installed in Sacramento, California, in 1997, and a 176kW unit powered by a tracking CPC was installed in 2002 in Raleigh, North Carolina (Guiney and Henkel, 2003). When using natural gas as a backup, the chiller COP is 1.2 (including the burner efficiency) and the solar fraction is at least 60 per cent. This gives an average COP based on fuel use alone of 3.0, compared to a primary energy COP of 1–2 for electric chiller COPs of 3–6, 35 per cent electricity generation efficiency and 5 per cent losses during electricity transmission. This is a reduction in primary use by 33–67 per cent (up to a factor of three reduction). However, if we make the comparison against a system involving new, state-of-the-art electricity generation (55 per cent efficiency) and large electric chillers (COP = 6–7), then there can be a slight increase in primary energy use even with double-effect absorption chillers and a 60 per cent solar fraction.

2.5.3 Cost of solar cooling systems

Typical costs of the components of a solar-driven chilling system in Europe according to Balaras et al (2007) are:

- flat-plate collectors, €280/m^2;
- stationary compound parabolic collectors, €420/m^2;
- evacuated-tube collectors, €620/m^2;
- absorption and adsorption chillers, €400/m^2;
- backup heater, €120/m^2;
- electric vapour-compression chiller, €620/m^2.

The cost of solar-powered absorption chilling systems is dominated by the cost of the solar collectors, which provide the heat that drives the absorption chiller. The lower the COP of the chiller, the greater the collector area needed (a lower COP means that more heat is needed to produce a given amount of cooling).

Figure 2.54 compares the initial investment cost for various projects surveyed by Balaras et al (2007) with the solar collector area required per kW of cooling capacity. Absorption chillers with higher COPs (namely, double- and triple-effect machines) require a greater input temperature, as noted above. The required temperatures can be achieved using evacuated-tube or compound parabolic collectors. These are more costly, but the required collector area is smaller due to the higher COP. Costs for absorption systems range from €1300 to 3200/kW_c and those of solid-desiccant systems range from €3200 to 4000/kW_c By comparison, large conventional cooling systems cost in the order of \$100/kW_c and medium-sized systems cost \$300–500/kW_c. Figure 2.55 plots the cost of saved primary energy against primary energy savings for

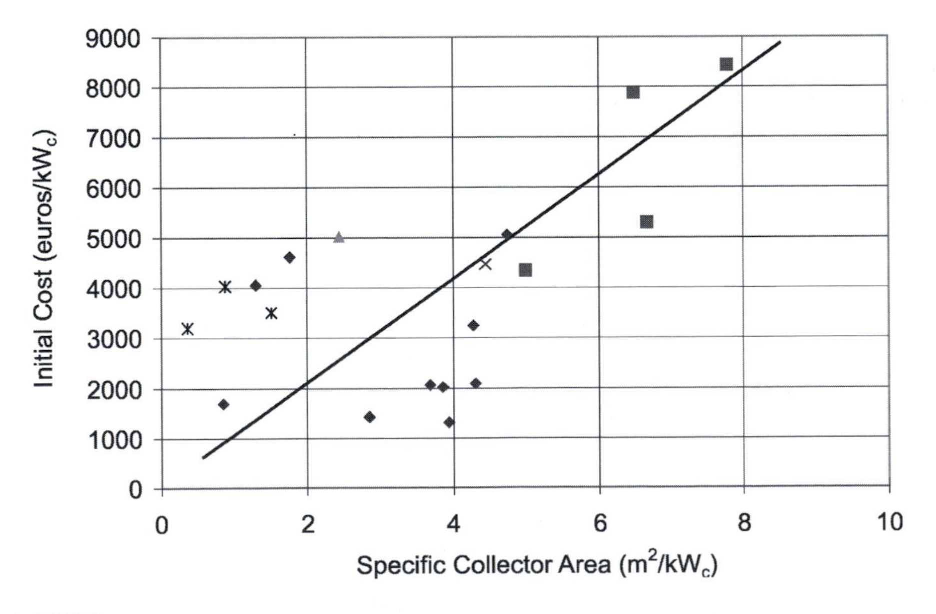

Source: Balaras et al (2007)

Figure 2.54 *Initial cost versus specific collector area for solar thermal cooling systems in Europe*

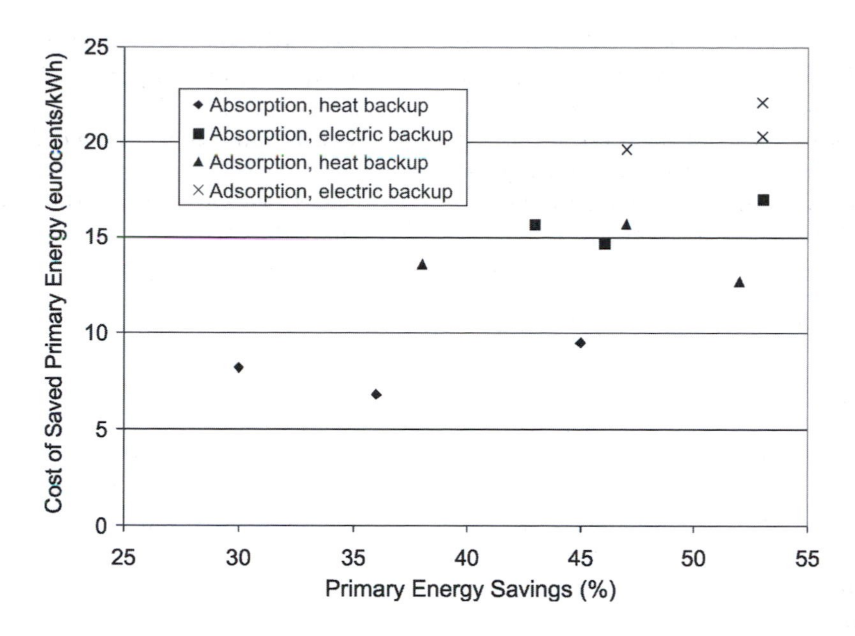

Source: Data from Balaras et al (2007)

Figure 2.55 *Cost of saved primary energy versus percentage savings in primary energy for solar thermal cooling systems in Europe*

various absorption and adsorption chiller projects surveyed by Balaras et al (2007). The solar-assisted cooling systems achieve primary energy savings of 40–50 per cent at a cost of about 10–20 eurocents/kWh. The reason the primary energy savings are not larger is because of electricity use by auxiliaries.

The economics of solar air conditioning are much more favourable in China than in Europe due to the availability of lower-cost flat-plate and evacuated-tube collectors in China – in the range of 300–350 Yuan/m^2 (€30–35/m^2) according to Yattara et al (2003), compared to €200–500/m^2 for flat-plate collectors in Europe (a factor of seven less expensive in China). Thus, Zhai and Wang (2008) report that the solar cooling system installed in a green building of the Shanghai Research Institute of Building Science (and consisting of evacuated-tube solar collectors and both liquid desiccant and adsorption chillers) had a payback period (compared to conventional cooling systems) of only two to three years when the concurrent provision of DHW by the solar collectors is accounted for (otherwise, the payback is a still-respectable seven to eight years). Thus, solar air conditioning is already economically attractive in China and probably in other large developing countries with minimal building air conditioning at present.

2.5.4 Hybrid electric thermal cooling using parabolic solar collectors

Gordon and Ng (2000) proposed a hybrid electric-thermal system for chilling using solar energy that should be able to achieve a substantially higher solar COP than is possible using only PV-electric chiller or thermal collector-absorption chiller combinations alone. The key is to begin with high-temperature (1000°C) thermal energy and to cascade the energy flow to the maximum possible extent. The proposal makes use of small (0.2m diameter), roof-mounted, mass-produced (and therefore inexpensive) parabolic dishes, each of which would concentrate sunlight onto a fibre-optic cable. A high solar collection efficiency (0.8) would be achieved as the mini-dishes themselves would remain at a low temperature. The fibre-optic cables would direct energy to a central receiver facility, where a hot gas at a temperature of 1000°C would be created. After heat exchange, hot gas at a temperature of 800°C would be supplied to a microturbine. Such turbines have an efficiency (at present) of about 0.23 if ambient air is drawn through the combustion chamber and there is no heat recovery from the exhaust gas. However, by using waste heat to preheat air to 185°C, the thermal-to-electricity efficiency would be increased to 0.35 for present turbines and to 0.42 for future, more efficient turbines. Electricity from the turbine drives a vapour-compression chiller with a COP of 3.0–4.0, while waste heat from the turbine at 360°C (minus 10 per cent losses) drives a double-effect absorption chiller with a COP of 1.35. Both the electric and absorption chillers would produce chilled water at 5°C. The net solar COP would be 1.4–1.7 with present turbines and 1.5–1.8 with future turbines. The energy flow in the latter case is shown in Figure 2.56. By comparison, the solar COP with individual solar-driven absorption or electric chillers would be 0.4–0.8.

2.6 Solar cogeneration: Integrated PV modules and thermal collectors

PV panels and solar thermal collectors can be combined to form an integrated PV/thermal (PV/T) collector. This can be done in a number of different ways, including by using a PV module as the outer cladding of an airflow thermal solar collector (as

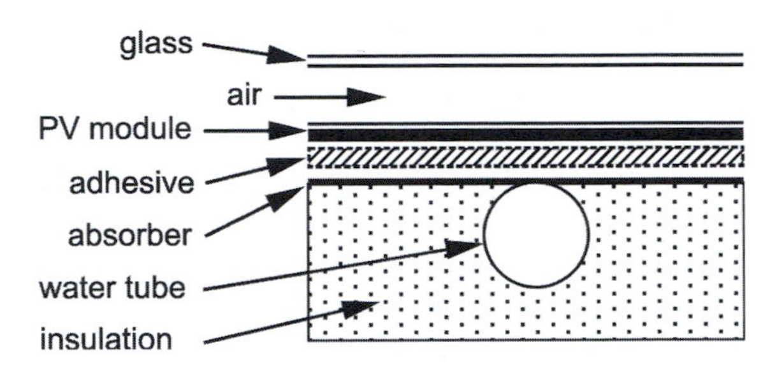

Source: Charalambous et al (2007)

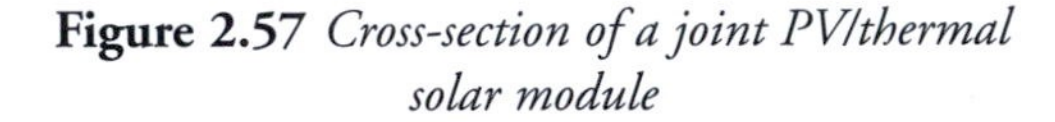

Figure 2.57 *Cross-section of a joint PV/thermal solar module*

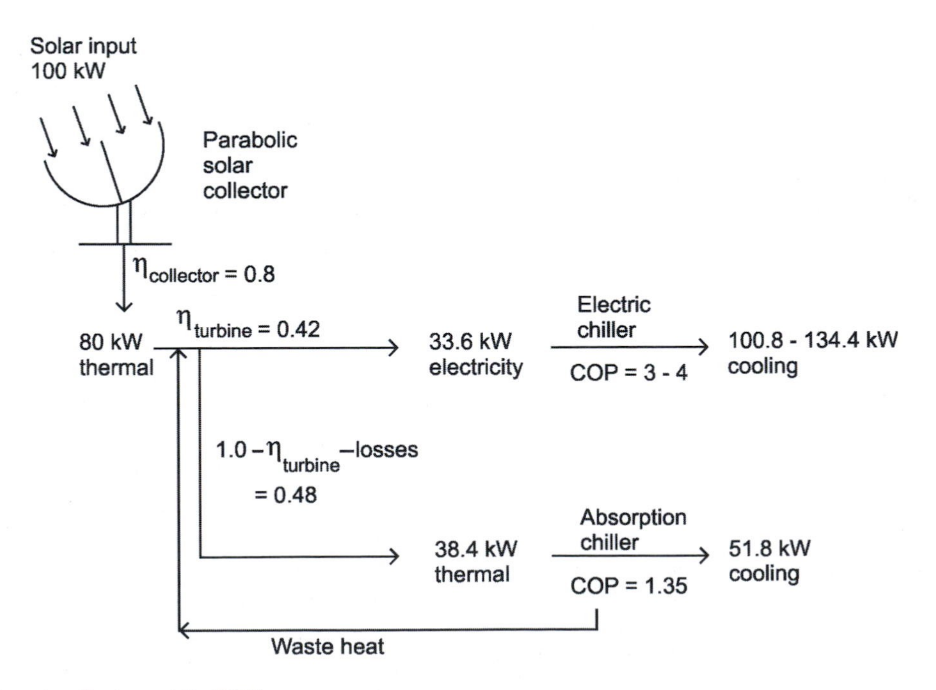

Source: Based in part on Gordon and Ng (2000)

Figure 2.56 *Energy flow in a hybrid electric thermal solar cooling system*

discussed in Volume 1, Chapter 4, subsection 4.3.1), or by pasting solar cells directly onto the black absorbing surface of a solar water heater.

In the first case, heat is carried away by air flowing underneath the PV modules, while in the second case, heat is carried by flowing water. As noted in Volume 1 (Chapter 4, subsection 4.1.2), water is much more effective than air in removing heat. A common arrangement is shown in Figure 2.57, in which water flows through tubes attached to the underside of the absorbing surface and there is an extra glazing layer in order to increase the effectiveness of the module in collecting thermal energy. Charalambous et al (2007) and Zondag (2008) provide recent reviews of integrated PV/T collectors.

The electricity production by an integrated PV/T unit may or may not be less than that of a standalone PV module, depending on the relative impact of:

- the additional reflection and absorption of solar radiation by any glazing layers;
- the effect of the glazing layers in making the PV module warmer; and
- the effect of heat removal by air or water flow on the temperature of the PV modules (the efficiency of a PV module decreases as its temperature increases, with the possible exception of a-Si modules).

The thermal efficiency of a PV/T collector is generally less than that of a standalone thermal collector because:

- PV modules have a lower absorptance than the usual collector plate;
- there is poorer thermal contact with the heat-carrying fluid; and
- some energy is extracted to generate electricity.

Saitoh et al (2003) compared the energy production of PV, thermal and integrated PV/T collectors in an experimental setup at Hokkaido University, Japan. The PV/T collector has a single glazing layer and a tube-and-sheet collector below the solar cells. Results are summarized in Table 2.18. In this case, the loss in electricity output in a PV/T unit compared to a PV unit is negligible, while the drop in thermal output for a PV/T unit compared to a standalone thermal collector is close to the electricity output. Thus, total energy output for the PV/T and thermal units is almost the same, but with about one quarter occurring as electricity for the PV/T units. If necessary, standalone PV units could be installed along with PV/T units to match the electricity to heat demand ratio of the building in question.

Table 2.18 *Comparison of annual efficiency of a standalone thermal collector, standalone PV unit and integrated PV unit*

	Thermal collector	PV unit	PV/T unit
Heat	0.462	–	0.320
Electricity	–	0.107	0.106
Total	0.462	0.107	0.426

Source: Saitoh et al (2003)

Chow et al (2003) compared the electricity output and building heat gain for BiPV, PV/T and PV/C units on a 30-storey hotel at 22°N latitude in south China. The PV/T collects heat through a warm air stream behind the solar cells that enters at the bottom and exits at the top, and so functions as a solar chimney. The PV/C unit is a PV/T unit except that the sides are open so as to permit free exchange of ambient air by wind. The purpose in this case is not to collect heat, but to keep the PV panel as cool as possible. The annual electricity output is almost identical for the PV/T and PV/C options, and is always slightly larger than for the BiPV unit. More significantly, the BiPV unit reduces the heat transfer through the wall into the building by about 30 per cent compared to a conventional building envelope, with almost another 20 per cent savings for either the PV/T or PV/C units. The difference between PV/T and PV/C in heat transfer through the wall is negligible, so a PV/T unit can be used to collect heat for preheating of water for consumptive use without noticeably increasing the cooling load (compared to PV/C) in hot climates.

Sandnes and Rekstad (2002) simulated the use of a PV/T collector as part of a floor radiant heating system. The great advantage of floor radiant heating is that a very low distribution temperature is adequate, particularly if the building is well insulated (see Volume 1, Chapter 4, subsection 4.6.2). As well, if the floor is made of concrete, it can store heat overnight, so the floor can be heated during the day when solar energy is available, while releasing heat during the night. By using solar heat in this way as it becomes available, rather than storing it in a hot water tank (as in the usual DHW or solar combisystems), the PV/T collector temperature is kept lower, since the water returning to the collector is cooler. This will increase efficiencies in collecting heat and in generating electricity. Integration of a PV/T unit with floor radiant heating can increase the daily electricity output by up to 10 per cent compared to the case where the heat accumulates in a storage tank.

Kalogirou and Tripanagnostopoulos (2006, 2007) have assessed the economics of using PV/T for production of domestic hot water and electricity, and for industrial applications. They adopted costs of €400/m² and €700/m² for a-Si and m-Si PV, respectively, compared to €650/m² and €950/m² for the corresponding PV/T units. Electricity production with PV/T modules is 27–38 per cent less than with PV, but the economics are more favourable than for PV alone. For residential applications in Cyprus, the time required to pay back the initial investment is about 28 years for m-Si PV and 18 years for a-Si PV/T, given assumed market prices for electricity of €0.1/kWh and heating fuel €0.62/litre. As the assumed system lifespan is 20 years, the PV system does not pay for itself over its lifespan. For industrial applications with a 10m³ hot water storage tank at €17,000 and requiring heat at 60°C or 80°C, PV/T is again more favourable than PV, with economic payback times of less than 20 years for PV/T and greater than 25 years for PV. For both residential and industrial applications, m-Si has better economics for PV and a-Si has better economics for PV/T. The economics would improve as the price of electricity and or heating fuel increase.

Combining solar collectors with PV modules in a single unit reduces the time required for the savings in primary energy to equal the primary energy required to make the collectors according to a comparative analysis by Tripanagnostopoulos et al (2006). For PV systems having an energy payback time of 2.7–3.2 years, the energy payback time is reduced to 1–2 years when thermal energy collection is added. Assuming a 20-year lifespan for the equipment, this gives an energy return over energy invested (EROEI) of 10–20.

2.7 Industrial uses of solar thermal energy

Solar thermal energy can be used to supply the medium-temperature heat (60–260°C) needed for food processing and in some chemical industries, as well as for desalination of water and for reduction of various metal ores. According to EcoHeatCool (2006), about 30 per cent of total industrial heat demand (13EJ/yr) in the EU-25 is required at temperatures below 100°C. If only 10 per cent of this heat load were supplied by solar energy, it would require about 143–180 million m² of collector with a thermal capacity of 100–125GW (Vannoni et al, 2008). This is comparable to total world solar thermal capacity at present (see Table 2.16). Weiss and Rommel (2008) review medium-temperature solar collectors and their use for industrial processes. A brief overview of industrial uses of solar heat is provided here.

2.7.1 Medium-temperature industrial processes

Table 2.19 gives the temperatures required for a variety of common industrial processes. In most cases, this

Table 2.19 *Temperature ranges for some medium-temperature industrial processes*

Industry	Process	Temperature (°C)
Dairy	Pressurization	60–80
	Sterilization	100–120
	Drying	120–180
	Concentrates	60–80
	Boiler feed water	60–90
Tinned food	Sterilization	110–120
	Pressurization	60–80
	Cooking	60–90
	Bleaching	60–90
Textiles	Bleaching, dyeing	60–90
	Drying, degreasing	100–130
	Dyeing	70–90
	Fixing	160–180
	Pressing	80–100
Paper	Cooking, drying	60–80
	Boiler feed water	60–90
	Bleaching	130–150
Chemical	Soaps	200–260
	Synthetic rubber	150–200
	Processing heat	120–180
	Preheating water	60–90
Meat	Washing, sterilization	60–90
	Cooking	90–100
Beverages	Washing, sterilization	60–80
	Pasteurization	60–70
Flours and byproducts	Sterilization	60–80
Timber byproducts	Thermo-diffusion beams	80–100
	Drying	60–100
	Preheating water	60–90
	Preparation pulp	120–170
Bricks and blocks	Curing	60–140
Plastics	Preparation	120–140
	Distillation	140–150
	Separation	200–220
	Extension	140–160
	Drying	180–200
	Blending	120–140

Source: Kalogirou (2003)

Table 2.20 *Solar energy collectors, concentration ratios and temperature ranges*

Motion	Collector type	Absorber type	Concentration ratio	Indicative temperature range (°C)
Stationary	Flat plate	Flat	1	30–80
	Evacuated tube	Flat[a]	1	50–200
	Compound parabolic	Tubular	1–5	60–240
Single-axis tracking	Fresnel lens	Tubular	10–40	60–250
	Parabolic trough	Tubular	15–45	60–300
	Cylindrical trough	Tubular	10–50	60–300
Two-axes tracking	Parabolic dish	Point	100–1000	100–500
	Heliostat field	Point	100–1500	150–2000

Note: [a] This pertains to absorbing flanges between the inner and outer tubes (see Kalogirou, 2009, Chapter 3).
Source: Kalogirou (2003)

heat is supplied as hot water or steam at the pressure corresponding to the highest temperature needed. However, these and higher temperatures can be achieved with a variety of solar collectors, as listed in Table 2.20. Flat-plate collectors, which are also used for DHW or space heating, produce temperatures of 30–80°C. Evacuated-tube collectors, which are also used for DHW, space heat and solar air conditioning, produce temperatures of 50–200°C. To achieve higher temperatures, solar energy must be concentrated on the absorber using mirrors, as in the parabolic trough and parabolic dish collectors that are also used to generate electricity. Concentrating collectors require some form of tracking system, so that the collector always points toward the sun, they are ineffective with diffuse radiation (i.e. under cloudy skies), and the reflecting surfaces require periodic cleaning and refurbishing so as not to lose some of their reflectivity over time. However, the efficiency in converting solar radiation into usable heat is high. This is because only a small area is heated relative to the collector area, so the heat losses are relatively small. Thus, a parabolic trough collector can maintain a collection efficiency of 0.6–0.7 throughout the entire $(T_i - T_f)/I$ range shown in Figure 2.50 (Kalogirou, 1998).

Kalogirou (2003) developed a computer simulation model to evaluate the use of solar heat for medium-temperature industrial processes. For conditions in Cyprus, the calculated cost of solar heat is 0.015–0.028 Cypriot pounds/kWh (approximately \$0.07–0.13/kWh). Karagiorgas et al (2001) discuss the experience and lessons learned from ten projects involving the use of solar heat in a variety of industries in Greece, including a winery, tannery, dairy factory and textile and cosmetics factories, and a greenhouse. Bokhoven et al (2001) discuss two systems in The Netherlands, one at an agricultural drying and conditioning facility and consisting of a 1200m^2 flat-plate collector array and a 1000m^3 underground hot water storage tank, and the other at a sweet factory and consisting of a 2400m^2 flat-plate array without storage. In all the Greek and Dutch projects, fossil fuels are used to provide heat when solar heat is inadequate. Abdel-Dayem and Mohamad (2001) present an analysis of the use of solar thermal energy in the textile industry, either to preheat water fed into the hot water boiler, or to feed the process of textile drying with the low-temperature (up to 85°C) hot water that is needed. With existing collectors on a textile factory in Egypt, about 20 per cent of the boiler energy could be saved with preheating and 80 per cent of the drying energy supplied utilizing the roof area only.

2.7.2 Reduction of metal ores and other material processing

Metals generally occur in nature as oxides. For example, iron occurs primarily as haematite (Fe_2O_3), while aluminium occurs as at least 18 different types of bauxite. As discussed in Volume 1, Chapter 6, the first step in the processing of iron ore or bauxite is to remove the oxygen atoms – a process referred to as *reduction* (the opposite is oxidation – adding oxygen atoms). The amount of energy required for this reaction is equal to the increase in the chemical energy content or *enthalpy* (H) of the final materials compared to the initial materials. Of the total enthalpy change (ΔH), a certain portion, known as the Gibbs free energy (ΔG) must be supplied either as electricity or as a reducing agent (a substance that will readily accept oxygen atoms, which is coke in the case of iron ore reduction). As the temperature of the reaction increases, the Gibbs free energy – the amount of energy that needs to be supplied as electricity or as a reducing agent, decreases. At a high enough temperature, the Gibbs free energy is zero – heat alone is sufficient for separating the metal from oxygen.

Table 2.21 lists the temperatures at which $\Delta G = 0$ for various metal oxides, as well as the temperatures at which no electricity is required if the metals are reacted with carbon or methane. These temperatures are very difficult to achieve through combustion of fossil fuels, so thermal decomposition of metal ores is not practised today. However, the required temperatures can be readily achieved by concentrating sunlight in a *solar furnace*. This has been done in a number of pilot projects, as reviewed by Bjorndalen et al (2003). Werder and Steinfeld (2000) propose the co-production of zinc and synthetic gas (for ammonia fertilizer production) in a single, integrated process using solar thermal energy. Solar heat would provide about half of the overall energy requirements for the production of these two commodities. Steinfeld and Thompson (1994) analyse the use of solar energy in iron and cement production and Murray (1999) discusses the use of solar heat in aluminium production, while Murray and Steinfeld (1999) provide an informative background discussion on the potential use of solar heat in the metals industry more generally. Flamant et al (1999) discuss experimental data in the use of solar heat for the surface hardening of steel, the production of nano-structured materials and the synthesis of fullerenes and carbon nanotubes.

The melting point of silicon dioxide (which occurs as sand in nature) is 1710°C, and it is the large amount of energy required for this process, which causes silicon PV modules to be rather energy intensive (see subsection 2.2.9). Of particular interest is the possibility of using solar thermal energy for the melting step in the production of PV modules.

Table 2.21 *Approximate temperature (K) at which the Gibbs free energy of various reactions is equal to zero*

	Thermal dissociation	Reduction with C	Reduction with CH_4
Fe_2O_3	3700	920	890
Al_2O_3	>4000	2320	1770
MgO	3700	2130	1770
ZnO	2335	1220	1110
TiO_2	>4000	2040	1570
SiO_2	4500	1950	1520
CaO	4400	2440	1970

Source: Bjorndalen et al (2003)

2.8 Direct use of solar energy for desalination, in agriculture and for cooking

Solar energy can be used in a variety ways to meet fertilizer and water needs of agriculture, in cooling greenhouses in order to enhance food production, for drying of agricultural crops and for cooking.

2.8.1 Desalination of salt water or extraction of water from air

The major methods of desalinating seawater are briefly discussed in Volume 1 (Chapter 6, section 6.13). These can be classified as thermal or mechanical processes. Reverse osmosis (RO) with energy recovery requires the least amount of energy per mass of water desalinated, 1.58kWh/m^3 for the best plant in the world and typically 3–4kWh/m^3. An alternative is multi-stage flash (MSF) desalination, which requires about 300MJ of thermal energy per m^3 of desalinated seawater. If MSF is driven by steam that is produced by solar collectors with a 50 per cent collection efficiency, the solar energy requirement for desalination is 300/0.5 = 600MJ/m^3. If, instead, a RO system with energy recovery (2kWh/m^3) is powered with PV electricity generated at 10 per cent efficiency and a pump at 90 per cent efficiency, the solar energy requirement is (2kWh/m^3) × (3.6MJ/kWh)/(0.1 × 0.9) ~ 80MJ/m^3. Thus, PV to power RO would make by far the most effective use of solar energy. However, RO requires skilled workers and a steady power supply so as not to damage the membrane. It also requires chemical pretreatment of the seawater so as to avoid scaling, foaming, corrosion and fouling. Any serious mistake during operation can ruin the membranes. Nevertheless, RO is more suitable for intermittent operation than other desalination processes according to Kalogirou (2005). Paulsen and Hensel (2007) report on the development of an RO desalination unit consisting of four modules, each of which can operate between 50 per cent and 100 per cent of peak output, so that the overall unit can operate between 12.5 per cent and 100 per cent of peak output. García-Rodríquez (2003) lists 19 desalination plants worldwide powered at least in part by PV electricity, and 14 plants powered by solar thermal energy.

Mink et al (1998) and Kalogirou (1998) analysed the use of solar thermal energy for desalination using flat-plate and parabolic trough collectors, respectively. The effectiveness of solar desalination can be expressed in terms of the rate of production of fresh water times the latent heat of vaporization (the product being equal to the rate at which heat must be supplied to first evaporate the salt water), divided by the solar radiation flux incident on the solar collector. For water at 73°C, the latent heat of vaporization is 2326kJ/kg. If some of the latent heat that is released when the water vapour condenses is used to preheat the salt water, then an effectiveness greater than 1.0 can be achieved according to the analysis by Mink et al (1998).

Another way to supply water is to extract it from air. This can be done using a desiccant that absorbs water from the air during the night and is regenerated with solar heat during the day, releasing liquid water in the process. In Egypt, such a system produced 1.5 litres of water per day per square metre of solar collector (Gad et al, 2001).

Finally, solar radiation can also be used to detoxify and disinfect water, as discussed by Blanco et al (2009).

2.8.2 Solar fixation of nitrogen

The production of nitrogen fertilizer is an essential component of modern agriculture systems, and some artificial nitrogen fertilizer will almost certainly be needed under even the most innovative alternative agriculture systems if the projected future human population (peaking at 8–11 billion) is to be fed. Today, almost all nitrogen fertilizer is produced from or consists of ammonia (see Volume 1, Chapter 7, subsection 7.4.2). Nitrogen can also be fixed through direct reaction between atmospheric N_2 and O_2 – at a temperature of over 2000°C followed by fast cooling so as to prevent the back reaction. This in turn could be carried out using recently developed high-temperature solar receivers. Epstein et al (2004) provide a preliminary analysis of this concept. The process would be coupled with the thermal generation of electricity. A $400MW_{th}$ solar plant ($60–64MW_e$ peak output) would produce 25,000 tonnes of HNO_3 and 17,000MWh of electricity per year.

2.8.3 Solar cooling of greenhouses

Evaporative cooling is widely used in greenhouses in hot countries, such as the Persian Gulf region. However, the temperatures achieved permit cultivation of lettuce (a temperate crop) only in winter, while year-round cultivation of tomatoes and cucumber (tropical crops) is just possible. Davies (2005) proposes solar-regenerated desiccants to achieve cooler temperatures. By drying the air prior to evaporative cooling, a cooler final temperature can be achieved. Conventional evaporative cooling would lower the temperature by 10K in July in Abu Dhabi, while desiccants add another 5K cooling. If desiccants are developed that absorb only in the near-infrared (NIR) part of the solar spectrum (which is of no use to plants) but are transparent in the visible part of the spectrum, and if the desiccant container is transparent, then the desiccants can be place above the greenhouse, thereby minimizing the total land area required for the greenhouse plus desiccant regeneration while providing shading in the NIR and hence further cooling.

The use of seawater to recover water from the air has been successfully demonstrated (Davies and Paton, 2005). The basic idea is as follows: if the air is cooled to close to the wetbulb temperature (see Volume 1, Chapter 4, subsection 4.1.3), it will also be saturated. Circulation of this air through the greenhouse will reduce evapotranspiration, such that the irrigation requirement can be reduced by about a factor of five compared to outside crops. After the humid air passes through the greenhouse, it is cooled with seawater using a heat exchanger. This condenses water, which is used for the evaporative cooling. There is an additional makeup water requirement that could be met through solar desalination of seawater.

2.8.4 Crop drying

Bena and Fuller (2002), Pangavhane et al (2002), Kumar and Kandpal (2005) and Sharma et al (2009) report on the use of solar energy for drying of crops and production of dried fruits, while Enibe (2002) reports on use of solar energy for incubating eggs. Bena and Fuller (2002) consider the use of biomass as a backup energy source, in conjunction with solar energy. Janjai and Tung (2005) developed a roof-integrated solar collector that is used for drying herbs and spices. The dryer also provides protection from rain and insects. A solar dryer can be integrated with desiccants to provide faster drying: during the day, a desiccant would be recharged with the same solar-heated airflow that is used to dry crops. At night, air would be circulated through the desiccant so as to continue the drying process (Shanmugam and Natarajan, 2006). Othman et al (2006) report on the performance of a solar dryer that can produce a drying temperature of 50°C with

700W/m^2 incident radiation and 27–30°C ambient conditions. The thermal collector efficiency for this application is 35–40 per cent. A PV/T system with just enough PV could be used to operate the fans, thereby creating a self-sufficient system.

2.8.5 Cooking

Solar thermal energy can be directly used for cooking food (Patel and Philip, 2000; Kumar et al, 2001; Buddhi et al, 2003). Sharma et al (2000) and Buddhi et al (2003) discuss a phase-change material (acetanilide) with a melting point of 119°C that can store heat collected before 3pm and be used to cook food at a temperature ranging from 95°C at 4pm to 82°C at 6pm for winter conditions in Dehli, India. Solar cookers could supply a third of cooking energy requirements in South Africa, saving poor rural people time (due to a reduced need to collect fuelwood) and money (Wentzel and Pouris, 2007). Solar cookers are especially beneficial for slow cooking food (such as maize porridge, soup and beans) that otherwise requires a lot of fuel. A hybrid device that can be used for cooking, pasteurizing milk, distilling small quantities of water and drying foods has been developed for use in Costa Rica (Nandwani, 2007).

2.9 Dealing with the intermittent nature of solar energy

The intermittent nature of solar energy is most critical for electricity production with PV modules, as the output responds instantly to changing irradiance. Even on sunny days in Arizona, solar irradiance can vary by up to 20 per cent from one ten-minute period to the next due to the passage of high-level, low-optical-depth clouds over the site (Curtright and Apt, 2008). The output from solar thermal electricity generation systems will be smoothed to some extent by the capacity to store heat and draw it down gradually. Similarly, the short-term fluctuations in heating, hot water, air conditioning or industrial uses of solar thermal energy will be buffered by the thermal storage tanks that are part of these systems.

As more and more PV power sources that are distributed over a wide area (on building rooftops and façades) are added to a grid, power systems will have to be modified in order to accommodate the intermittent nature of PV power. Techniques for doing so are briefly discussed here. However, it should be noted that a distributed PV system can also provide many benefits to a power grid (apart from shaving peak loads). These benefits arise from the many functions that modern inverters can perform, such as acting as a filter to reduce grid voltage harmonics, power factor regulation, reactive power control and phase symmetry control (Caamaño-Martín et al, 2008). Similar benefits are provided by modern wind turbines, as discussed in Chapter 3 (subsection 3.11.3).

2.9.1 Use of rapidly variable fossil fuel powerplants

Fluctuations in solar electricity generation can be met by compensating variations in the output of other generation units on the power grid. There are two issues when fossil fuel powerplants are the only units available for offsetting variations in PV power output: (1) loss of fossil fuel powerplant efficiency, and (2) wasted PV electricity production due to limits on the minimum fossil fuel output.

Loss of fossil fuel powerplant efficiency

When rapidly variable power units (such as simple cycle or combined cycle gas turbines) compensate for variations in PV output, they start and stop more frequently and operate at a smaller fraction of peak output on average, which reduces their average efficiency. Adding PV power to the grid has no effect on the already high fuel use per kWh of electricity generated by simple cycle turbines, but increases the fuel use by combined cycle powerplants by 6 per cent when PV supplies 10 per cent of total electricity demand and combined cycle powerplants supply the rest (Denholm et al, 2008). Thus, the energy and CO_2 emission savings are only 40 per cent as large as expected based on the fraction of electricity supplied by solar energy. The incremental benefit of adding more PV decreases because progressively more efficient generators will be displaced, and because of increased cycling.

Wasted PV output

Fossil fuel plants cannot reduce their output below some minimum level, which is typically 25–40 per cent of peak output. If the PV output is a large enough fraction of the system peak demand, there will be times when the minimum fossil powerplant output plus PV output exceeds demand. In this case, some of the PV output will be wasted, and the wasted fraction increases with increasing penetration (deployment) of PV power. Suppose that average electricity demand is 60 per cent of the peak demand. Consider a peak demand of 1kW and a 1kW array. This array on average produces 0.2kW if the mean annual solar irradiance is 200W/m^2, and this corresponds to 33 per cent of the average demand. Thus, a PV array sized to provide only 33 per cent of the total electricity demand will frequently exceed the electricity demand, and potential PV electricity will be wasted even if other powerplants can decrease their output to zero. With less than 100 per cent flexibility, the wasted PV electricity is larger still. Calculations by Denholm and Margolis (2007a) for PV systems in Texas indicate that, with 100 per cent system flexibility, about 95 per cent of each additional unit of PV electricity produced is wasted on average by the time the PV penetration reaches 50 per cent. With 60 per cent system flexibility, about 100 per cent of each additional unit of PV electricity is wasted by the time the PV penetration reaches 20 per cent. This, however, assumes no storage of excess PV electricity.

2.9.2 Aggregating geographically dispersed PV arrays

One way to reduce fluctuations in power output is to aggregate the power from many PV systems spread over a large geographical area, such that the fluctuations in the power from individual systems are largely uncorrelated with one another. Wiemken et al (2001) have compared the fluctuations in output from a single PV system with the output from 100 PV systems scattered over Germany. The variability of a typical PV system and of the total output from 100 systems (divided by the installed capacity) over three summer days is shown in Figure 2.58a. The average normalized output during one June is shown in Figure 2.58b, along with the standard deviation of five-minute average output from one system and from the ensemble of 100 systems. The standard deviation of the ensemble is about half that of any one system. The peak output of the ensemble never exceeds 65 per cent of the installed capacity and the aggregate power output averaged over five minutes never differs by more than 5 per cent from one five-minute time period to the next. The

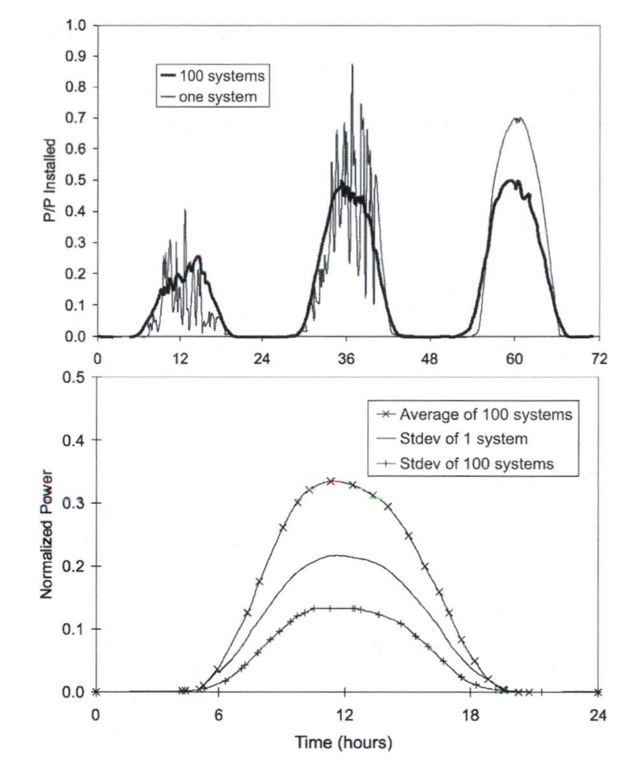

Source: Wiemken et al (2001)

Figure 2.58 *(a) Variability in the output of a typical PV system and of the total output from 100 systems (divided by the installed capacity) over three summer days in Germany; and (b) variation in average output of 100 PV systems in Germany during one June, along with the standard deviation of five-minute average output from one system (middle curve) and from the ensemble of 100 systems (bottom curve)*

smoothing effect is due to the large spatial extent of the arrays rather than the large number of arrays. This is due to the large decrease in the correlation of the output between two arrays with increasing distance between them, shown in Figure 2.59.

2.9.3 Introduction of energy storage and dispatchable loads

Greater fractions of electricity demand can be met with PV if some electric loads (such as space or water heating, air conditioning, refrigeration, irrigation pumps, or desalination equipment) can be shifted from times of insufficient PV output to times of excess PV output. Another option is to introduce storage systems, which are discussed in Chapter 3, section 3.11). Those appropriate for PV include batteries (for small-scale systems), flywheels, supercapacitors and (in the future) parked bidirectional plug-in hybrid vehicles (discussed in Volume 1, Chapter 5, subsection 5.4.6). PV electricity would be used to recharge the batteries in plug-in hybrids at times of excess, and some would be returned to the grid when there is a deficit. This system is referred to as a V2G system.

Calculations by Denholm and Margolis (2007b) indicate that, for 70 per cent system flexibility (meaning that the pre-existing power units can decrease to only 30 per cent of peak output), no more than 20 per cent of total electricity demand can be met with no storage, whereas almost 40 per cent of total demand can be met with 12 hours storage.

Compressed air energy storage (CAES) is another option. As discussed more fully in Chapter 3 (subsection 3.11.2), this involves using excess electricity to compress air and injecting it into porous underground rock formations or in underground caverns, then using the compressed air in a gas turbine at times when extra electricity is needed. A few CAES systems have already been built in the world, and another ten are being planned in the US. Current systems reduce fuel use by 60 per cent, but advanced systems that capture and make use of the heat produced during compression can almost completely eliminate fuel requirements (see Chapter 3, subsection 3.11.2). Wind energy lends itself more easily to CAES in the US, as the best wind sites (the US plains) also coincide with suitable geology for CAES, so the two together could provide near-steady electricity to a high-voltage grid for transmission to demand centres. The best solar region (the south-west of the US) is not suitable for CAES, so large-scale solar electricity from this region would be spatially separated from CAES. Mason et al (2008) and Fthenakis et al (2009b) present a preliminary analysis of how a PV-CAES system in the US might work. In spite of the spatial separation of good PV and CAES sites in the US, the conclusion is that PV-CAES could be

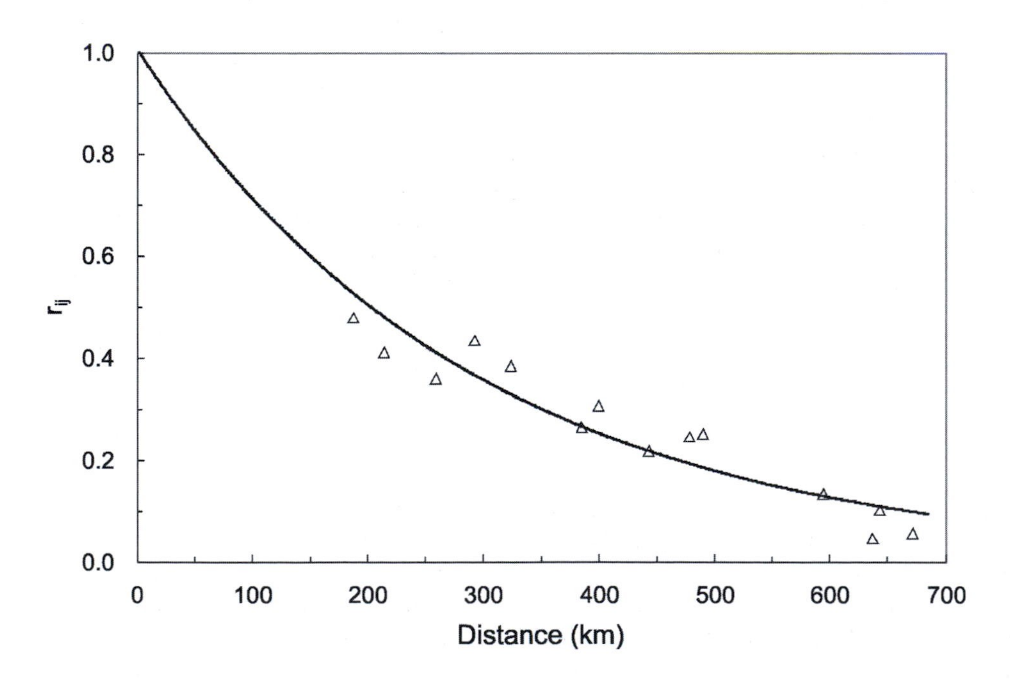

Source: Wiemken et al (2001)

Figure 2.59 *Temporal correlation between the hourly output of PV arrays in Germany as a function of the distance between the arrays, based on ten years of data from six different sites*

competitive with natural gas power generation by 2020, and that about 70 per cent of total US electricity demand and 35 per cent of electricity plus fuel demand could be met this way by 2050.

2.9.4 Linking diverse renewable energy sources

Electricity from solar energy and other renewable energy sources can be pooled over distances of 3000–4000km with minimal energy losses using high-voltage DC power generation, as discussed in Chapter 3 (section 3.12). A comprehensive scenario for supplying 80 per cent of European electricity demand in 2050 with a mixture of solar, wind, biomass, hydro and geothermal energy sources, interconnected by a grid of high-voltage DC power lines, is presented in Chapter 12 (subsection 12.2.2).

2.10 Synthesis and global potential

The solar energy resource is enormous, the amount of solar radiation intercepted by the earth being over 11,000 times present world primary energy demand. As stressed in Volume 1, Chapter 4, much of the energy used in buildings and currently provided by electricity (lighting, ventilation, cooling) or fuel (space heating) can be provided passively with solar energy (daylighting, passive ventilation arising from indoor–outdoor temperature differences, passive cooling through use of cool outside air or earth pipes combined with thermal mass, and passive heating using high-performance windows and thermal mass). Other energy needs in buildings, and energy needs for industry and transportation, could be met through direct use of solar thermal energy, through hydrogen fuel produced from solar energy (as discussed in Chapter 10), and through solar-generated electricity. Solar thermal energy can be used for space heating (with seasonal storage, as discussed in Chapter 11), for hot water needs (with diurnal to weekly storage), for air conditioning and dehumidification (as discussed in this chapter), and as process heat in a variety of industries.

The options available for generating electricity from solar energy are:

- to generate electricity from large, centralized PV arrays in sunny areas, using either flat collectors or concentrating collectors;
- to generate electricity from solar thermal energy, which requires direct-beam radiation and would therefore be restricted to desert or semi-desert areas;
- to use BiPV to generate electricity where it is needed.

Table 2.22 compares PV and solar thermal methods of generating electricity in terms of advantages and disadvantages, cost and efficiency. PV electricity is currently expensive (20–40 cents/kWh in good locations with favourable financing) but should decrease in cost by at least a factor of two. Concentrating solar thermal power is less expensive than PV power (as little as 12 cents/kWh) and should also decrease in cost by a factor of two or more.

As noted earlier (subsection 2.2.6 and Table 2.4), BiPV alone could provide 15–60 per cent of total present-day electricity demand in individual OECD countries according to a study commissioned by the IEA PV Power Systems (PVPS) Implementing Agreement. A complementary project, IEA PVPS Task VI/Subtask 50, 'A Preliminary Analysis of Very Large-Scale Photovoltaic Power Generation (VLS-PV) Systems', has assessed the potential for large-scale PV power generation in desert regions. Key results are summarized in Table 2.23, where it is seen that the total power capacity is estimated to be 1235TW (compared to 15.3TW world primary energy demand and 4.3TW electrical capacity in 2005). This estimate assumes a module efficiency of 14 per cent, something that has already been achieved by multi- and single-crystalline silicon modules and should soon be achieved by various compound single-junction and multi-junction semi-conductors. The factor limiting the amount of energy that could be collected from desert arrays is not the availability of land or sunlight, but the availability of rare elements that are used either in the semi-conductor materials or in the electrodes attached to the semi-conductors. However, as discussed in subsection 2.2.13, the limit on future deployment of PV power posed by resource constraints is likely to be several tens of TWs if concentrating PV is used (whereby sunlight is concentrated by up to a factor of 500). This is several

Table 2.22 *Summary of PV and solar thermal methods of generating electricity*

		Costs	Efficiency	Advantage	Disadvantage
PV	Crystalline silicon (single, poly)	30–40 cents/kWh in best locations, 9–13 cents/kWh for best projection	10–15% modules, 20% eventually	Silica is abundant, higher efficiency than amorphous	Material inputs, greater embodied energy
	Thin film amorphous silicon		5–6% modules, 10% hoped for	Can go on anything, suited to continuous production	Low efficiency
	Multi-junction thin films using various rare substances (e.g. CIS, CdTe)	Highest cost	25–32% cells, 41% under concentrated sunlight	High efficiency	Uses toxic or rare elements, high cost
	Nanocrystalline dye cells		10% cell at 25°C	Efficiency increases with temperature, can be transparent to visible radiation	Requires rare ruthenium
Concentrating PV	crystalline Si CIGS thin-film GaAs GaInP/GaAs/Ge		28% @ 92 suns 22% @ 14 suns 29% @ 232 suns 41% @ 454 suns	Potentially lower cost, stretches rare elements	More complicated, requires direct-beam solar radiation
	quantum dot			Concentrates direct and diffuse radiation without tracking	Still under development
Thermal	Parabolic trough	12–20 cents/kWh now, 5–10 cents kWh future	15–20%, 42–48% capacity factor with six-hour storage	Large scale, lots of demo projects, some storage	Thermal storage more difficult than for other thermal methods
	Parabolic dish	€10,000–14,000/kW, eventually $2000–3000/kW (8–24 cents/kWh)	20–28%	Suitable for isolated villages, low infrastructure costs, quick start	Expensive at present, limited heat storage ability
	Central receiver	18–32 cents/kWh today	10–15%	Most amenable to 24-hour electricity	Each mirror must individually track the sun

times current world primary power demand. The potential identified in Table 2.23 also applies to concentrating solar thermal power (CSTP), given that sunlight-to-electricity conversion efficiencies for CSTP (10–15 per cent for central tower, 10–20 per cent for parabolic trough and 20–28 per cent for parabolic dish systems) equal or exceed the 14 per cent efficiency assumed in Table 2.23. Furthermore, electricity can be generated continuously over a 24-hour period with central tower and parabolic trough CSTP systems due to the thermal mass of the heat storage medium.

As discussed in subsequent chapters, solar energy is only one of several forms of renewable energy that could supply a significant fraction of world energy needs. Hydrogen produced from renewable energy or biofuels produced from biomass will be needed where renewable energy cannot be directly used. The critical questions involve the best mix of renewable energy sources, the cost of energy and the overall efficiency of the future energy system.

Table 2.23 *Potential electricity production from PV arrays in desert areas*

Region	Desert	Area (10^4 km^2)	Average irradiance		Possible PV array capacity (TW)	Annual electricity generation (10^3 TWh)
			W/m^2	kWh/m^2/yr		
North America	Great Basin	49	235	2060	34.3	49.5
	Chichuahuan	45	228	1995	31.5	44.0
	Sonoron	31	199	1745	21.7	26.5
	Majave	7	245	2145	4.9	7.4
	Subtotal	*132*			*92.4*	*127.3*
South America	Patagonian	67	148	1299	46.9	42.6
	Atacama	14	256	2239	9.8	15.4
	Subtotal	*81*			*56.7*	*58.0*
Australia	Great Victoria	65	250	2187	45.5	69.7
	Great Sandy	40	267	2343	28	45.9
	Simpson	15	250	2187	10.5	16.1
	Subtotal	*120*			*84*	*131.7*
Asia	Arabia	233	257	2255	163.1	257.4
	Gobi	130	191	1676	91	106.8
	Kara kum	35	189	1657	24.5	28.4
	Kyzly kum	30	189	1657	21	24.4
	Takla Makan	27	187	1641	18.9	21.7
	Kavir	26	212	1858	18.2	23.7
	Syrian	26	209	1835	18.2	23.4
	Thar	20	248	2174	14	21.3
	Lut	5	244	2138	3.5	5.2
	Subtotal	*532*			*372.4*	*512.3*
Africa	Sahara	860	272	2385	602	1004.9
	Kalahari	26	261	2285	18.2	29.1
	Namib	14	261	2285	9.8	15.7
	Subtotal	*900*			*630*	*1049.7*
Grand total		**1765**			**1235.5**	**1878.9**

Source: Kurokawa (2003)

Notes

1 Despite the name, Q_s is not constant. Rather, it varies in value by about ±1.0W/m^2 during the course of the 11-year sunspot cycle, and perhaps by a few times this over a timescale of a century or longer.

2 Regional colour maps of the irradiance on a horizontal surface, as well as maps of the irradiance on a fixed surface tilted toward the equator at the latitude angle and maps of the ratio of the two can be found in the online PowerPoint presentation for this chapter.

3 For example, carbon has four outer-shell electrons and oxygen has six, but oxygen has a stronger affinity for electrons than carbon. The four outer carbon electrons are shared with four oxygen electrons, two from each of two oxygen atoms, so that – through sharing – the carbon atoms and both oxygen atoms have a full complement of eight outer-shell electrons. The result is a molecule consisting of a central carbon atom double-bonded to an oxygen atom on each side – the CO_2 molecule.

4 The hole 'migrates' in the sense that if an electron to the right of the hole fills in the hole, a new hole has appeared to the right. If this process continues, new holes appear successively to the right, as if the original hole were moving through the substance.

5 a-Si shows a downward trend from 2002 to 2009 because the efficiency of an a-Si cell tends to decrease during the first few hundred hours of exposure to sunlight before stabilizing. The efficiency given for 2009 is a stabilized efficiency, while some earlier results were not stabilized efficiencies.

6 For up-to-date retail costs of solar modules, invertors and electronic controls in the US and Europe, for lists of suppliers in the US and Europe, and for information on funding programmes in the US and Europe, consult www.solarbuzz.com.

7 Environment, Safety and Health.

8 Shading occurs when the passage of sunlight from the sun to one heliostat is blocked by another heliostat, while blocking occurs when the passage of sunlight from one heliostat to the central tower is blocked by another heliostat.

9 These standards require that a minimum amount of electricity be produced from renewable energy. Some specific examples are: California, 20 per cent by 2010; New Mexico, 10 per cent by 2011; Nevada, 20 per cent by 2015; Arizona, 15 per cent by 2055.

3

Wind Energy

3.1 Introduction

Wind energy is one of the fastest-growing sources of energy in the world, and is competitive or almost competitive with the market price of fossil fuel-generated electricity. Significant development of wind power capacity has occurred in those jurisdictions where government policy has been to support wind energy. Figure 3.1 shows the growth in the annual rate of installation of new wind capacity (the wind energy market) and in the cumulative installed capacity from 1995 to 2009. The wind energy market grew at an average compounded rate of 23.9 per cent/year over the period 1995–2009, while the installed capacity grew at an average rate of 24.9 per cent/yr. Figure 3.2 gives the installed capacity at the end of 2009 for the ten countries with the largest capacity, as well as the expansion of wind capacity in 2009 for the ten countries with the largest expansion. World wind capacity reached 157.9GW by the end of 2009, equal to 3.7 per cent of the worldwide electrical capacity of 4267GW in 2005. Germany and the US each accounted for about 20 per cent of the world wind capacity at the end of 2009, while the US, Spain and China were the three largest markets for wind energy in 2009. India is also experiencing a rapid growth of wind energy, and several other countries are planning rapid growth in wind capacity (several thousand MW each) during the coming years. Table 3.1 lists the world's ten largest wind turbine manufacturers in terms of market share in 2006.

In terms of electricity generation, the installed wind capacity was projected to supply 20 per cent of Denmark's electricity demand in 2007, 13 per cent of Spain's, 12 per cent of Portugal's and Ireland's, and 7 per cent of Germany's. Wind supplied only 1.3 per cent of US electricity consumption in 2008, but it supplied 7.5 per cent in Minnesota and 7.1 per cent in Iowa (AWEA, 2009). Texas had the most wind capacity of any state at the end of 2009, 9410MW out of 35,159MW for the entire US, and this provided about 4.3 per cent of the state's electricity demand.

3.2 Components and characteristics of wind turbines

A typical wind turbine consists of the following components:

- foundation;
- tower;
- rotor, consisting of three rotor blades on a horizontal axis, and a hub;
- nacelle, which sits atop the tower and houses the equipment listed below;
- gearbox, which makes the shaft rotate about 50–90 times faster than the rotor;
- high-speed shaft (about 1500rpm), which drives the electrical generator, and has an emergency mechanical brake to be used in case of failure of the aerodynamic brake or when the turbine is being serviced;
- electrical generator, usually an induction (asynchronous) generator;
- electronic systems to monitor the condition of the wind turbine and to control the yaw mechanism;
- cooling unit, consisting of an electric fan for the generator and oil for the gearbox;
- yaw mechanism, which turns the rotor to face into the wind at all times;
- anemometer and wind vane, to measure the speed and direction of the wind.

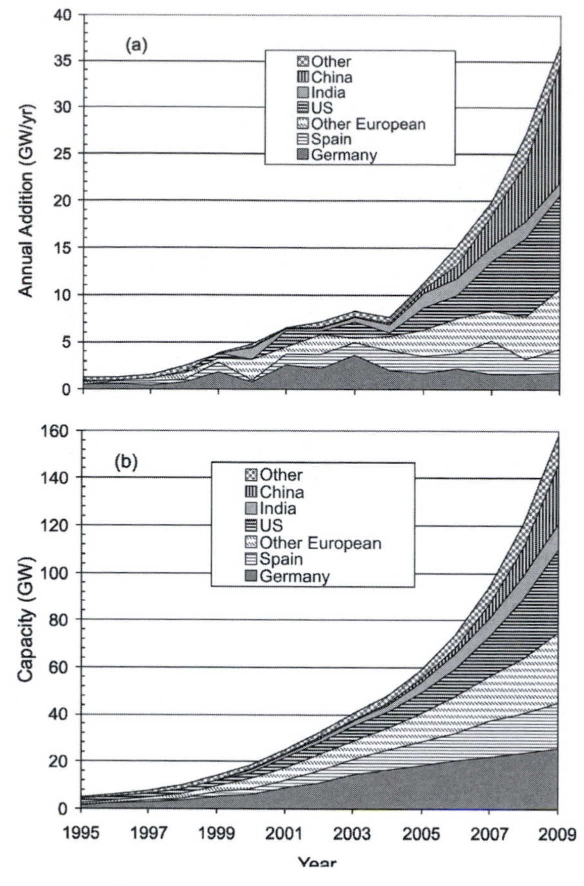

Source: Data from Greenpeace (2005), WWEA (2009) and GWEC (2010)

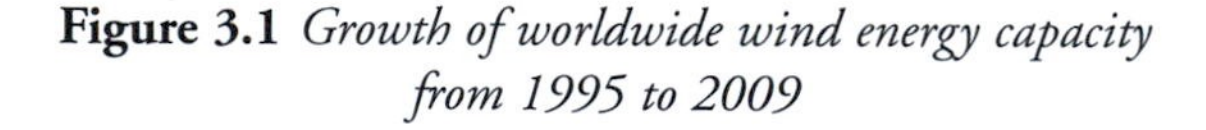

Figure 3.1 *Growth of worldwide wind energy capacity from 1995 to 2009*

Table 3.1 *Market shares of the world's leading wind turbine manufacturers in 2006*

Vendor	Market share	Country
Vestas	26.5	Denmark
Gamesa	14.6	Spain
GE Wind	14.6	US
Enercon	14.5	Germany
Suzlon	7.2	India
Siemans	6.9	Germany
Nordex	3.2	Germany
REpower	3.0	Germany
Acciona	2.6	Spain
Goldwind	2.6	China
Others	4.3	

Source: BTM Consult Press Release (March 2007)

Figure 3.3 is a photograph of a collection of wind turbines on a wind farm in Alberta, Canada.

There are many choices to be made when selecting a wind turbine. Among these are:

- whether to use stall, pitch or active-stall mechanisms to shut down the rotor during very high winds;
- whether to use a synchronous generator (with a fixed rotor speed), an asynchronous generator (having a slightly variable rotor speed), or a variable-speed generator;
- the rotor size to use in relation to the peak generator power (or generator size);
- the tower height to use in relation to the rotor size.

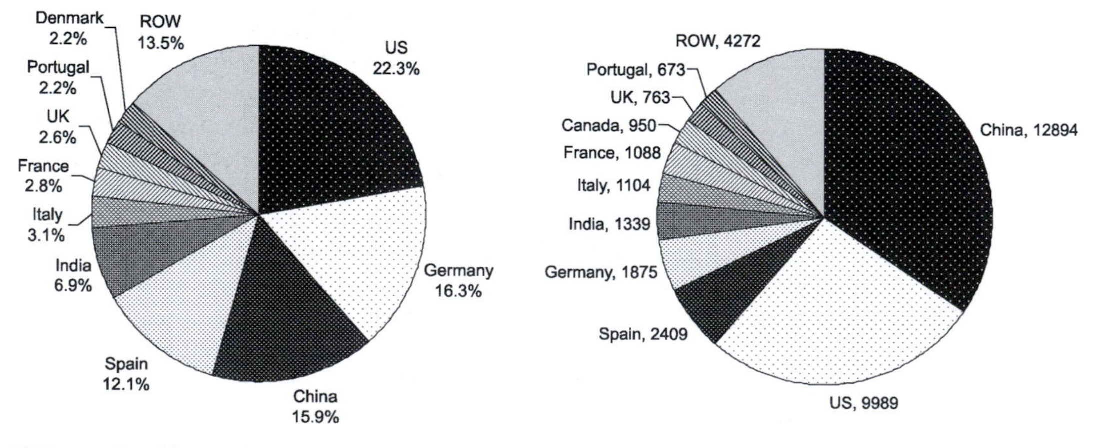

Note: ROW = rest of world.
Source: Data from WWEA (2009) and GWEC (2010)

Figure 3.2 *Distribution of installed wind capacity at the end of 2009 for the top ten countries (left) and wind capacity (MW) added in 2009 for the ten largest national markets (right)*

Source: Garry Sowerby

Figure 3.3 *A wind farm at Pincher Creek, Alberta, Canada, with the Rocky Mountains as the backdrop and illustrating the use of the majority of the land for other purposes*

These choices are explained in the following subsections.

3.2.1 Trends in wind turbine size and constraints on future size increases

Wind turbines are referred to in terms of the 'size' or peak power output of the generator. The peak power output is really the *rated power* of the generator, not a firm limit, as it is usually possible to exceed the rated power for short periods of time.

Turbine capacities and sizes have grown significantly during the past 25 years. In 1985, the 55kW turbine was the dominant unit. The 600kW turbine was introduced in 1995, the 2.5MW turbine in 2000 and the 5.0MW turbine in 2006. Figure 3.4 shows the progression in maximum turbine size and capacity over

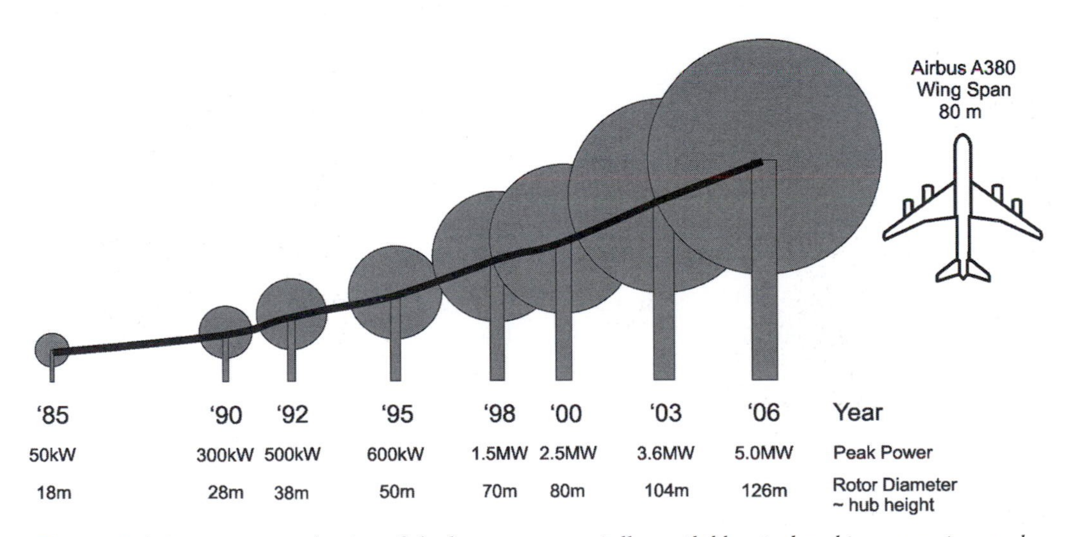

Figure 3.4 *Progression in the size of the largest commercially available wind turbines over time and, for comparison, the outline of an Airbus A380 aircraft*

time. During the period 2003–2005, the bulk of installed wind turbines were in the 1.5–2.0MW range, although by the end of 2005, 101 Vestas 3MW turbines were operational and 96 offshore units were installed in 2006, while 14 of Siemen's 3.6MW turbines were operational by the end of 2006 and 2 of REpower's 5MW turbines were erected in 44m-deep water at the Beatrice wind farm off the Scottish coast in 2007.

Logistical difficulties in transporting rotor blades of turbines much larger than 3.6MW on land are likely to restrict 5MW turbines largely to offshore applications. In the US, loads larger than the standard dimensions for trailer transport of 4.1m high and 2.6m wide require expensive rerouting in order to avoid obstructions and often require law enforcement or utility assistance along roadways. Lindenberg et al (2008) suggest that transportation costs could be reduced through onsite manufacturing. This would require segmented moulds that would be moved into temporary buildings close to the site of a major wind farm. Cranes could still be a factor limiting turbine size, as the larger the turbine, the larger the required crane. Large cranes are difficult to transport and require large crews, and may require repeated disassembly for travel between turbine sites in rough or hilly terrain. Tower diameters greater than about 4m would also incur severe transportation cost penalties. Large diameters are preferred because this spreads out the load and requires less material because the walls can be made thinner. Reducing the mass of the nacelle and related materials reduces the required tower diameter for a given height.

The constraints on the size of offshore wind turbines are much less severe than for onshore turbines. Bard Engineering of Germany is planning on scaling up its 5MW offshore turbine to 6.5MW, and a half dozen other companies are developing turbines in the 5–6MW range (de Vries, 2009a). American Superconductor Corporation and the US Department of Energy are assessing the economic feasibility of a 10MW direct-drive turbine that would use a high-temperature superconductor, making it smaller, lighter and more efficient and reliable than with conventional wires (Anonymous, 2009b).

3.2.2 Characteristics of existing turbines

Rotor blades are made of fibreglass, but the fibreglass has been partially replaced with carbon fibres in the latest wind turbines, resulting in stronger and lighter rotors. As a result, the Vestas 3MW turbine is no heavier than the older 2MW turbine (Cavallo, 2007). The 5MW turbine involves new aerodynamic shapes and pre-bent blades with new materials, rather than a simple scaling up of smaller machines (which would not have been feasible).

Early wind turbines used lattice towers, which are less expensive than tubular towers (requiring only half the material for the same stiffness). Modern turbines use tubular towers, which are visually more pleasing and do not attract birds. Fortunately, the required tower mass (and associated cost) per kW of turbine capacity decreased by almost 50 per cent during a recent five-year period as turbines became larger.

Table 3.2 lists some of the characteristics of the turbines currently produced by the ten turbine manufacturers listed in Table 3.1, along with data for the Bard 5MW offshore wind turbine. The rotor diameter increases with increasing generator size, but the rotor rotation rate decreases. Thus, rotors range in diameter from 40–50m for 600–850kW turbines, to 100–120 m for 3–5MW turbines, while rotation rates range from 15–30rpm (depending on wind speed) for 850kW turbines to about 8–15rpm in the 3MW and larger turbines. In the majority of turbines available today, gears are used to step up from the rotor rotation rate to a generator shaft rotation rate that is within a factor of two to three of the required AC frequency (see Volume 1, Chapter 3, section 3.1, on the relationship between rotor rotation rate and electrical frequency). Power electronics are used to convert the AC electricity generated by the generator to the exact grid frequency (60Hz in North America, 50Hz elsewhere). As the rotor rotation rate decreases (i.e. as turbine generator size increases), the gearbox ratio is increased to compensate. The Enercon turbines use a *direct-drive* system, meaning that the rotor shaft and the generator shaft are the same, and a gearbox is avoided altogether. The generator frequency is thus the same as the rotor frequency, so a much larger jump in frequency is required by the power electronics to bring the output up to the grid frequency. Recent advances in power electronics have made this economical. The elimination of the gearbox reduces the mechanical resistance in the turbine and, all else being equal, allows power generation to begin at lower wind speeds.

Three wind speed parameters are given for the turbines in Table 3.2: the *cut-in wind speed*, at which

Table 3.2 *Technical characteristics of wind turbines that are available from the vendors listed in Table 3.1*

Model	Power (MW)	Rotor diameter (m)	Hub height (m)	Rotor rpm	Wind speeds (m/s)			Mass (tonnes)		
					Cut-in	Rated	Cut-out	Rotor	Nacelle	Tower
Vestas turbines, www.vestas.dk										
V52-850	0.85	52	44–74	14–31.4	4	16	25	10	22	45
V82-1.65	1.65	82	70–80	14.4	3.5	13	20	43	52	105
V90-1.8	1.8	90	80–105	9–14.4	3.5	12	25	38	68	150
V90–2.0	2.0	90	80–105	9–14.4	2.5	13	25	38	68	150
V90–3	3.0	90	80–115	8.6–18.4	4	15	25	41	70	160
V112–3.0	3.0	112	85–119		3	12	25			
Gamesa turbines, www.gamesacorp.com/en										
G52–850	0.85	52	44–65	14.6–30.8	4	15	25	10	23	45
G58–850	0.85	58	44–65	14.6–30.8	3	15	21	12	23	45
G80–2	2.0	80	60–100	9.0–19	4	16	25	38	70	127
G87–2	2.0	87	67–100	9.0–19	4	15	25	37	70	153
G90–2	2.0	90	67–100	9.0–19	3	15	21	36	70	153
GE Wind turbines, www.gepower.com										
GE–1.5sle	1.5	77	65–80		3.5	14	25			
GE–1.5xle	1.5	82.5	80		3.5	12.5	20			
GE–2.5xl	2.5	100	75–100		3.5	12.5	25			
GE–3.6s	3.6	111	111	8.5–15.3	3.5	14	27			
Enercon turbines, www.enercon.de										
E33	0.33	33.4	44–50	18–45	3	13	28–34			
E44	0.9	44	55	12.0–34	2	16	28–34			
E48	0.8	48	50–76	16–30	3	14	28–34			
E53	0.8	52.9	73	12.0–29	3	13	28–34			
E70	2.3	71	58–113	6–21.5	3	15	28–34			
E82	2.0	82	78–138	6–19.5	3	13	28–34			
Suzlon turbines, www.suzlon.com										
S52	0.6	52	75		4	13	25			
S64	1.25	64	57–75		3	14	22			
S66	1.25	66	57–75		3.5	14	25			
S82	1.5	82	79		4	14	20			
S88	2.1	88	79		4	14	25			
Siemens, www.powergeneration.siemens.com										
SWT–2.3–82VS	2.3	82.4	80	6.0–18				82	54	158
SWT–2.3–93	2.3	93	70–80	6.0–16				82	60	134
SWT–3.6–107	3.6	107	80	5.0–13				125	95	
Nordex, www.nordex-online.com										
S70	1.5	70	65–85	10.6–19	3		25			
S77	1.5	77	60–100	9.9–17.3	3.5	13	25			
N90–2.3	2.3	90	70–100	9.6–16.8	3	13	25			
N80–2.3	2.5	80	60–80	10.8–18.9	3	15	25			
N90–2.5	2.5	90	60–120	9.6–16.8	3	13	25			
N100–2.5	2.5	100	100	9.6–14.9	3	13	20			
REpower, www.repower.de										
MD77	1.5	76.5	62–100	9.6–17.3						
MM70	2.0	70	55–80	10.0–20						
MM82	2.0	82	59–100	8.5–17.1	3.5	13	25			
MM92	2.0	92.5	69–100		3	11.2	24			
3XM	3.3	104	78–100	7.1–13.8	3.5	12.5	25			
5M	5.0	126	117	6.9–12.1	3.5	13	25			
Acciona, www.acciona-energia.com										
AW–70/1500	1.5	70	60–80	20.2	4	11.6	25	31.6	52.5	95
AW–77/1500	1.5	77	60–80	18.3	3.5	11.1	25	31.6	52.5	95
AW–82/1500	1.5	82	80	16.7	3	10.5	20	32.3	52.5	95

Table 3.2 *Technical characteristics of wind turbines that are available from the vendors listed in Table 3.1* (Cont'd)

Model	Power (MW)	Rotor diameter (m)	Hub height (m)	Rotor rpm	Wind speeds (m/s)			Mass (tonnes)		
					Cut-in	Rated	Cut-out	Rotor	Nacelle	Tower
AW–100/3000	3	100	100–120	14.2	4	11.7	25	66	118	
AW–109/3000	3	109	100–120	13.2	3.5	11.1	25	66	118	
AW–116/3000	3	116	120	12.3	3	10.6	20	66	118	
Goldwind, www.goldwind.cn/en										
GW43	0.6	43	40–50	17.9–26.8				13	23	
GW48	0.75	48	50–60	22				14	22	
GE62	1.2	62	70	11.0–20				30	50	
GW70	1.5	70	65–85	9.0–19				28	50	
Bard offshore turbine, www.bard-offshore.de										
Bard 5	5.0	122	90		3	12.5	25			

Source: Technical data from the company websites (given in the table), obtained in March 2009

the turbine begins to generate electricity, *the rated wind speed*, at which the turbine output reaches its rated power output, and the *cut-out wind speed*, at which the turbine shuts down for safety reasons. The cut-in wind speed ranges from 2.5m/s to 4.0m/s, the rated wind speed from 11m/s to 16m/s, and the cut-out wind speed is usually 25m/s. Turbine manufacturers also give information on the maximum wind gust of various durations that the turbine can survive, typically in the range of 50–60m/s (180–220km/hr).

It can be seen from Table 3.2 that many companies offer a variety of rotor sizes with a turbine having a given generator rating. Placing a larger rotor on a given generator results in a lower cut-in speed (as there is more torque from the larger rotor to overcome the given generator inertia) but also, in some cases, requires a lower cut-out speed (as the larger rotor is more susceptible to wind damage). Combining a larger rotor with a given generator would thus be better for locations with lower wind speeds, whereas combining a smaller rotor with a given generator would be more appropriate at windier sites.

It can also be seen from Table 3.2 that a range of tower heights is possible for each turbine generator and rotor size. Higher towers have three advantages. First, higher towers expose the turbine blade to a larger average wind speed. Second, the rate of increase of wind speed with height decreases with increasing height, so the difference in wind speed between the top and bottom blades (and associated fatigue loads) will be smaller with a higher tower. Third, turbulence decreases with increasing height, resulting in more uniform power output and an additional reason for less wear on the turbine with higher towers. On aesthetic grounds, most people prefer a hub height comparable to the rotor diameter.

Figure 3.5 plots rotor size and hub height versus turbine power rating for all the turbines listed in Table 3.2. As discussed later, the available wind power is proportional to the area swept by the rotor, which increases with the square of the rotor diameter, so the rotor diameter increases roughly with the square root of the turbine rating. Both minimum and maximum hub heights available for a given turbine also increase with turbine rating, but the relationship shows much more scatter than the relationship between rotor diameter and turbine rating. Figure 3.6 plots hub height against rotor diameter. A 10m increase in rotor diameter is associated on average with a 7m increase in hub height.

Larger turbines could be chosen for the following reasons:

- to obtain economies of scale, especially concerning the foundation, road, grid connection and construction mobilization (e.g. crane) costs per kW;
- to have taller towers (to obtain the advantages given above);
- to have smaller rotational speeds (an aesthetic consideration).

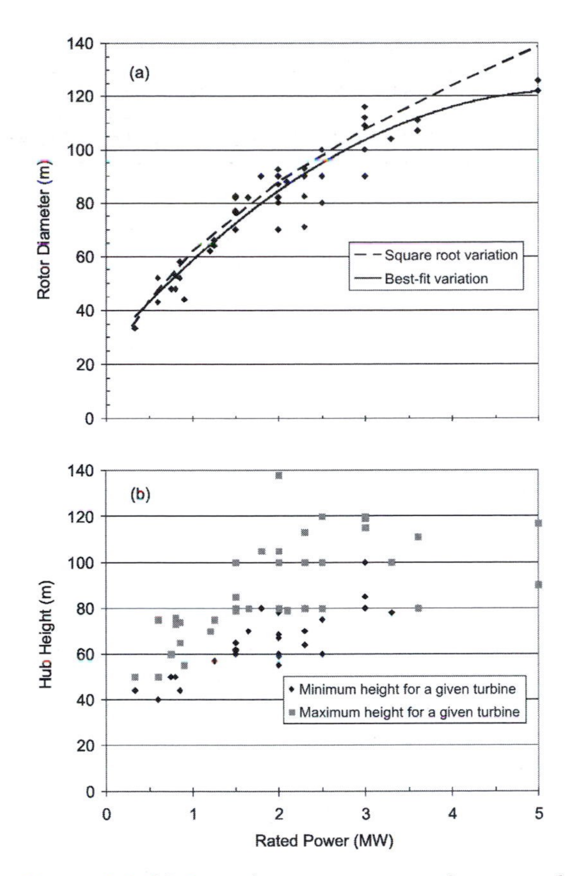

Figure 3.5 *(a) Rotor diameter versus rated power and (b) hub height versus rated power for the turbines listed in Table 3.2*

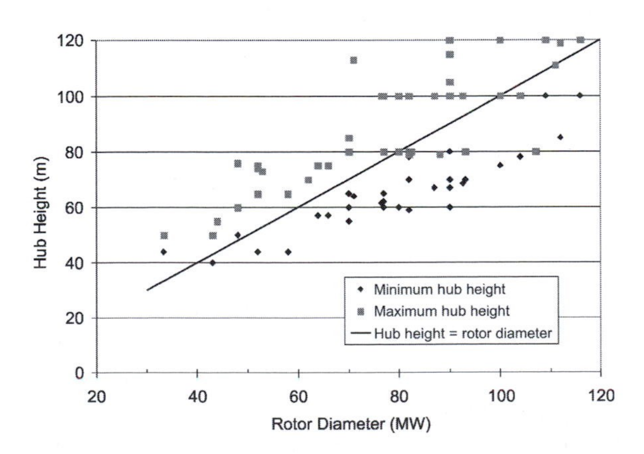

Figure 3.6 *Hub height versus rotor diameter for the turbines listed in Table 3.2*

Smaller turbines could be chosen:

- if the electrical grid is weak (so less power is fed into the grid at any one point); or
- so that there will be less fluctuation in the total power output from an array of turbines, as there will be greater opportunity for cancellation of fluctuations in the power from individual turbines if there is a larger number of smaller turbines; or
- to spread the risk in case of temporary machine failure; or
- because of easier transportation of smaller turbine components.

An advantage of wind turbines over thermal power generation (fossil fuels, biomass, nuclear) in arid regions is that water is not needed for cooling purposes (see Volume 1, Chapter 3, section 3.5).

Figure 3.7 plots the variation in the turbine output with wind speed for selected turbines from Table 3.2. This variation is referred to as the wind turbine *power curve*. Figure 3.7a shows the effect of combining different sized rotors with the same generator, using the Gamesa 2MW turbine as an example. Using a larger rotor results in greater electricity production at all wind speeds up to the rated wind speed than using a smaller rotor, but can result in a smaller cut-out speed. Figure 3.7b compares the output of the Nordex N90-2.3 turbine (90m rotor with 2.3MW generator) with that of the N80-2.5 and N100-2.5 turbines (80m and 100m rotors with a 2.5MW generator). The output of the 2.3MW turbine is greater than that of the N80-2.5 turbine for wind speeds of up to 13m/s. Thus, depending on the distribution of wind speeds over time, the smaller turbine (that is, the turbine with the lower generator capacity) could have a greater average power output.

3.3 Wind turbine aerodynamics

The movement of the rotor of a wind turbine is created through lift forces rather than through drag. Consider the wing of an aeroplane, which has a cross-sectional shape as shown in Figure 3.8. As air flows around the wing, the air flowing across the top of the wing is forced to travel faster than the air flowing underneath the wing. The faster-moving air on top exerts a smaller pressure on the wing than the air below, giving a net upward force (lift) (this is known as the *Bernoulli effect*). The *angle of attack* is the angle between the wing

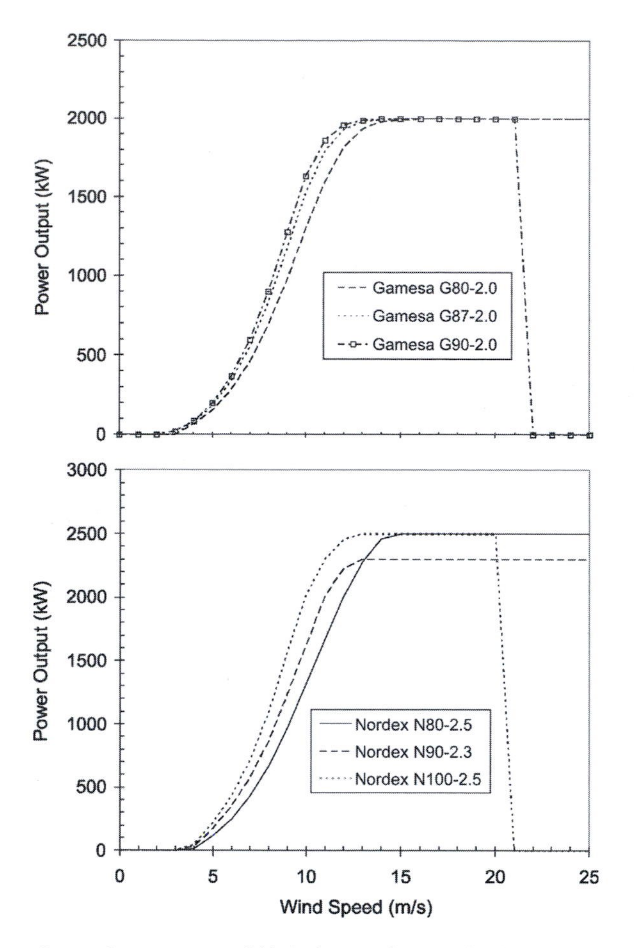

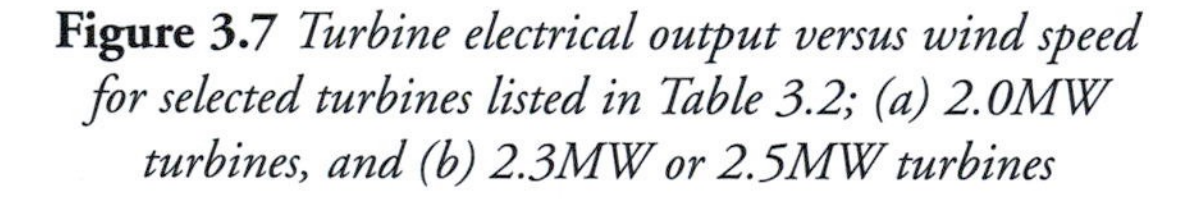

Source: Power curves available in the manufacturers' brochures

Figure 3.7 *Turbine electrical output versus wind speed for selected turbines listed in Table 3.2; (a) 2.0MW turbines, and (b) 2.3MW or 2.5MW turbines*

and the wind direction. If we tilt the wing up, we have increased the angle of attack. As the tilt of the wing increases, a point is reached at which turbulence suddenly appears on the upper back side of the wing. The lift suddenly disappears; *stall* has occurred. The blade of a turbine has a cross-section similar to that of an aeroplane wing, with the blunt edge facing the direction that the blade rotates.

Now consider a bicycle moving at 7m/s with a cross-wind at 7m/s to the left. A banner attached to the bicycle will be directed to the left and behind bicycle at an angle of 45°. To a person on the bicycle, this *is* the direction of the wind. If the bicycle travels faster, the banner will be directed more to the rear, while if the bicycle travels more slowly, the banner will be directed more across the path of the bicycle. As we move from the hub of a turbine blade to the tip, the speed of the blade increases. Thus, analogously to the moving bicycle, the air speed seen from the turbine blade is coming from a steeper angle closer to the centre of the blade than at the tip of blade. As a result, the rotor blade has to be twisted so as to maintain an optimal angle of attack throughout the length of the blade.

3.3.1 Relationships among velocity and force vectors

The relationships between the velocity and force vectors are illustrated in Figure 3.9. The rotor is constrained to rotate in a plane perpendicular to the wind direction. In Figure 3.9, V is the wind direction, U is the rotor direction and W is the wind speed as seen from the rotor. The lift force L acts perpendicularly to W and the drag force D acts

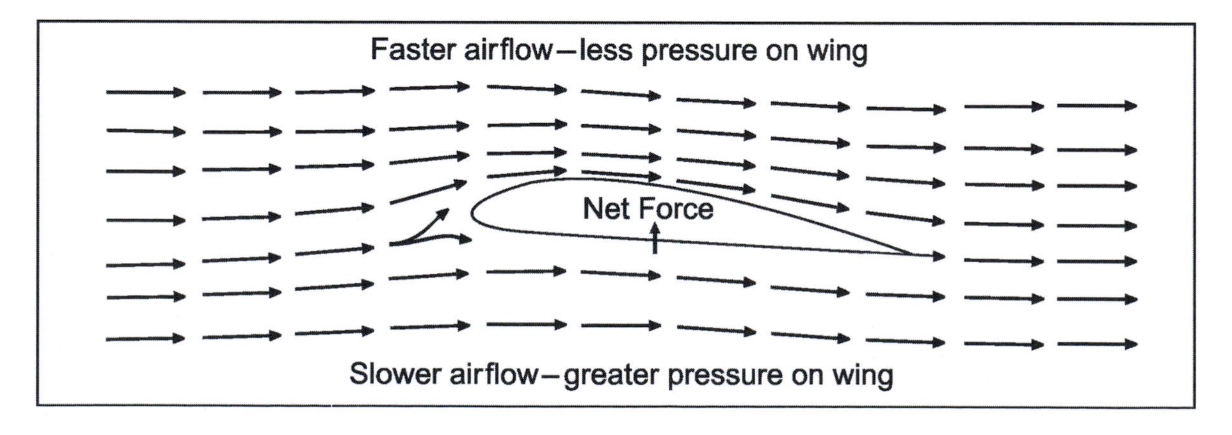

Figure 3.8 *Airflow past an aeroplane wing*

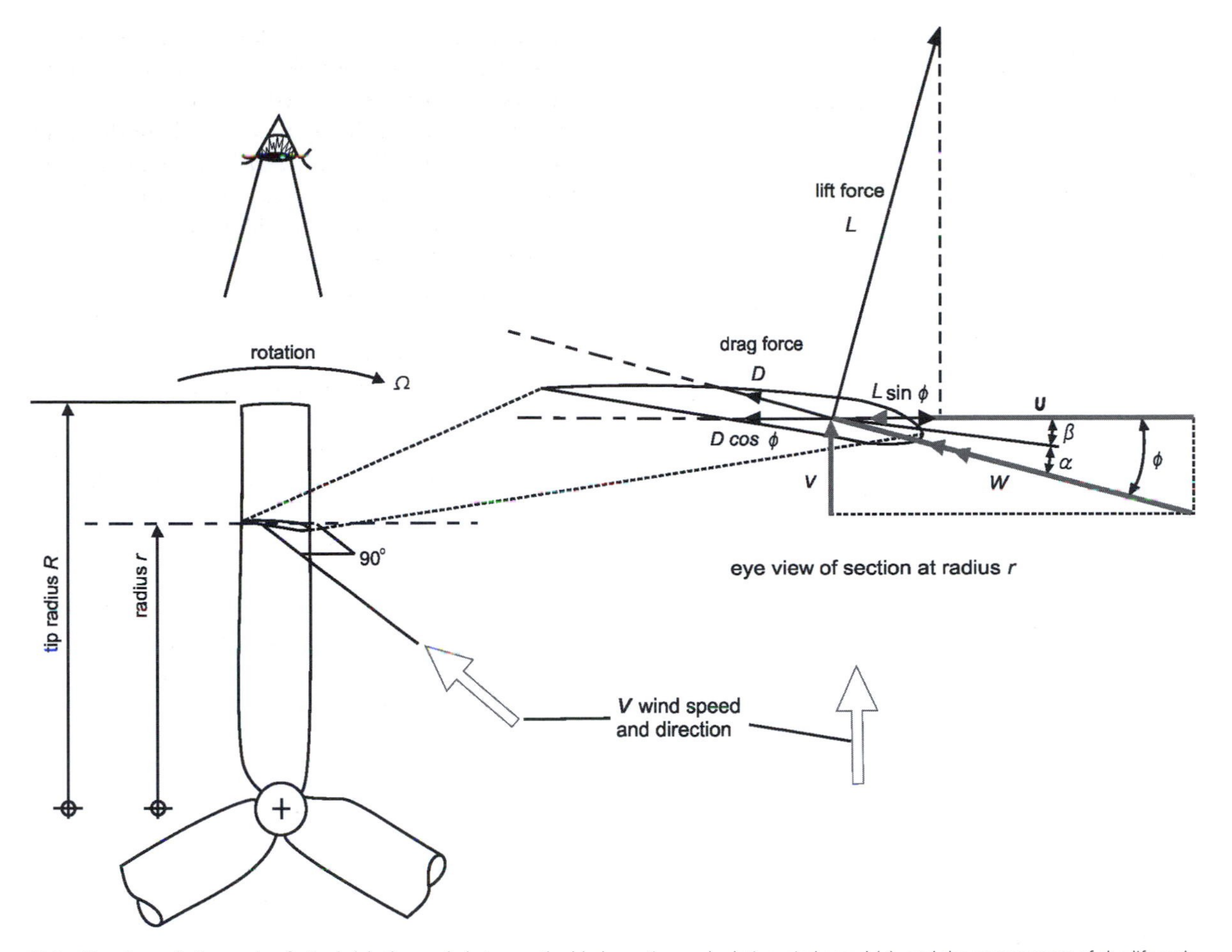

Note: Also shown is the angle of attack (α), the angle between the blade motion and relative wind speed (ϕ), and the components of the lift and drag forces along the axis of motion.
Source: Danish Wind Turbine Manufacturers' Association

Figure 3.9 *Vector diagram showing the movement of a horizontal axis turbine blade* (U), *the absolute wind velocity* (V), *the relative wind velocity* (W), *the lift force* (L) *and the drag force* (D)

parallel to W. The net force acting on the rotor at a distance r from the hub is given by the difference between the component of L in the direction U and the component of D in the direction U. This difference times r gives the torque exerted by the rotor at that point.

The angle between the blade motion and relative wind speed, ϕ, is such that cotan $\phi = U/V$. This angle varies with distance along the blade, since V is constant but the blade velocity varies distance from the hub. If R is the distance from the hub to the tip of the blade and ω is the rate of rotation of the blade, then $R\omega$ is the velocity of the tip of the blade. Then the *tip-speed ratio* γ is given by:

$$\gamma = \frac{R\varpi}{V} = \frac{U}{V} = \text{cotan } \varphi \qquad (3.1)$$

where U and ϕ now pertain to the tip of the blade. Maximum aerodynamic efficiency requires that γ be fixed at 7–10, depending on the shape of the blade (Twidell, 2003). This implies that (1) ω should vary with $1/R$; that is, larger rotors should rotate more slowly at a given wind speed; and (2) the rotation rate should vary in direct proportion to the wind speed. However, as discussed in the next section (on turbine generators), this requires more complex electronics which, until recently, was uneconomic. Instead, turbines were designed to rotate at a fixed speed, with the design chosen so as to have the optimal tip-speed ratio at the most frequent wind speeds encountered at a given site.

3.3.2 Shutting down the rotor

As can be seen from Figure 3.9, if the blade speed is fixed and the wind speed increases, the angle of attack will increase. Beyond some angle of attack, the blade will stall (there is no further lift force, as with an aeroplane wing). In *stall-controlled* turbines, the blade is built so that it stalls gradually from the centre of the blade outward as the wind speed increases above some critical value. This avoids the need for moving parts and complex controls systems, but creates a complex design problem, as it is necessary to avoid stall-induced vibrations. As well, the rated power of the turbine is exceeded at the beginning of a gust of wind, and the turbine runs at less than the rated power at high wind speeds. About two thirds of the turbines being installed today are stall-controlled. An alternative is *pitch-control*, in which an electronic controller checks the power output several times per second. When the power is too high, the blades are pitched (turned) slightly out of the wind, a fraction of a degree at a time. A third possibility is *active stall-control*, in which normal pitch control occurs up to the rated power output. At this point, the angle of attack of the blade (α in Figure 3.9) is increased to make the blades stall (this is opposite to what happens in pitch-controlled turbines). This gives more accurate control of the power output (no overshooting in gusts). An increasing number of larger (>1MW) turbines use this mechanism.

3.3.3 Orientation of the rotor

Almost all wind turbines have the rotor on the upwind side of the tower, facing the wind. This minimizes wind shadow effects (but does not completely eliminate them, as the wind is affected by the presence of the tower downwind from the rotor). The disadvantage is that the blades have to be rather inflexible and placed some distance from the tower (to avoid striking the tower when they are bent by strong winds).

It is important that the plane of rotation of the turbine blades be oriented perpendicularly to the wind direction. Otherwise, the blades will be bent forward and backward with each rotation, thereby increasing the fatigue loads on the turbine. There is a tendency for the blades to automatically orient themselves perpendicular to the wind, since the blade closest to the incoming wind will be subject to the largest forces. However, wind turbines have an active yaw mechanism consisting of electric motors and gearboxes to keep the turbines yawed against the wind. This minimizes the above-mentioned fatigue loads, since the direction of the wind is checked several times per second. As well, there is a cable twist counter to avoid accidentally yawing the turbine continuously in the same direction for several revolutions. When necessary, the active yaw mechanism will untwist the cable running through the tower to the turbine.

3.3.4 Efficiency

The efficiency of a wind turbine can be defined as the ratio of electrical energy generated to the kinetic energy of the wind passing through the area swept by the rotor blades. This efficiency is the product of three component efficiencies:

- the aerodynamic efficiency (ratio of the mechanical power of the rotor to the wind power);
- the mechanical efficiency (ratio of mechanical power in the generator axis to mechanical power in the rotor axis);
- the electrical efficiency (ratio of electric power fed into the grid to generator mechanical power).

As kinetic energy is extracted from the wind, the wind slows down. If a wind turbine were 100 per cent efficient, the wind speed would drop to zero, meaning that no wind would pass the turbine and there would be no power output, which contradicts the assumption of 100 per cent efficiency. According to *Betz' law*, the maximum theoretical aerodynamic efficiency of any wind turbine with a disc-like rotor is 16/27 = 59.3 per cent. This would occur if the wind turbine slows down the wind to two thirds of its original speed. The aerodynamic efficiency of a real turbine varies with the wind speed, typically peaking at around 44 per cent for some intermittent wind speed and producing an efficiency averaged over a typical distribution of wind speeds of about 25 per cent. The aerodynamic efficiency tends to be larger for larger rotors because the drag is smaller relative to the lift with larger rotors. For very large rotors, the aerodynamic efficiency at the optimal wind speed can approach the Betz limit. The speed at which the aerodynamic efficiency peaks is a deliberate design choice. At low wind speeds, the efficiency is not very important to the annual electricity production, since there is little wind energy to begin with. At high wind speeds, the turbine must waste any excess energy above what the generator was designed for, so there is a sharp dropoff in efficiency. The goal, however, is not to design a turbine with the highest possible efficiency. Rather, the goal is to design a turbine with the lowest *cost* in extracting kWh from the wind, efficiency being only one parameter that determines the cost.

The mechanical efficiency accounts for losses in the bearings and gearbox. It is in the 96–99 per cent range. The electrical efficiency accounts for the combined electric power losses in the generator, inverter, switches, controls and cables. For small (0.5–10kW) wind turbines, the electrical efficiency is only 60–70 per cent, whereas for large (2.5–3.0MW) asynchronous generators (described below), the electrical efficiency is about 96–97 per cent (Mertens and de Vries, 2008).

Figure 3.10 shows the variation in the power output and the overall efficiency with wind speed for the Nordex N90-2.3 turbine. The efficiency rises from zero at wind speeds below the cut-in speed, peaks at 43.6 per cent at a wind speed of 8m/s and drops to only 5.2 per cent at a wind speed of 25m/s, before cut-out.

3.4 Wind turbine generators

Wind turbine generators produce electricity at 690 volts. This is increased by a transformer to 10–36kV for transmission to the grid.

3.4.1 Types of generators

Wind turbine generators can be either synchronous or asynchronous. The latter can be further divided into fixed-speed and variable-speed generators.

Synchronous generators

The operation of a synchronous electric generator was explained in Volume 1, Chapter 3, subsection 3.1.3, and will be briefly summarized here. A generator consists of a stator with a minimum of three electromagnets, each connected to a different phase of the 3-phase power grid, and a rotor. The rotor has either a permanent magnet or, more often, an electromagnet. The electromagnet requires DC electricity to be magnetized, which is obtained from the grid (with AC to DC conversion) and transferred to the rotor with brushes and slip rings. If there is one rotor magnet, there will be two poles (one north and one south). As the

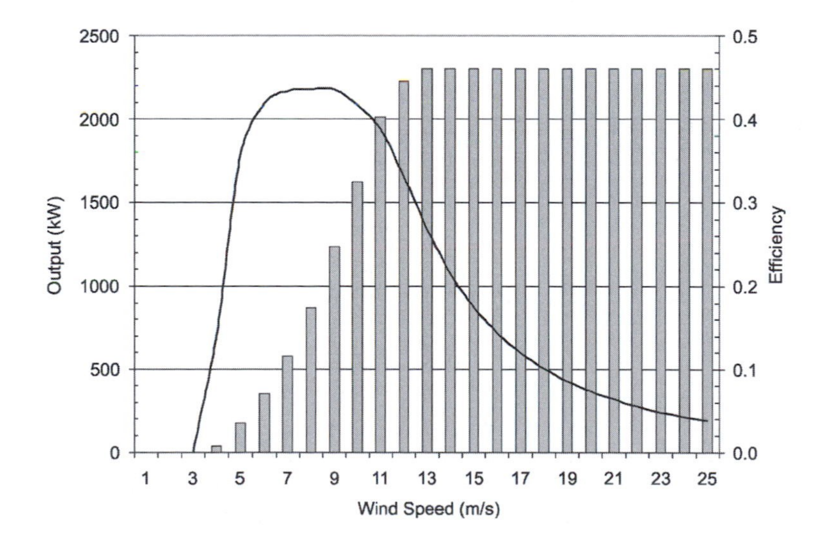

Source: Output data and rotor swept area (used to compute efficiency) from manufacturer's brochure

Figure 3.10 *Variation of power output (bars) and efficiency (line) with wind speed for the Nordex N90-2.3 turbine for an air density of 1.225kg/m*^{3}

rotor is forced around, a rotating magnetic field is created. This interacts with the stator magnets, and current is sent into the grid. The more force (torque) applied to the rotor, the more electricity that is generated, but the rotation speed does not change in a synchronous generator, it being dictated by the frequency of the electrical grid. In a 2-pole generator, the rotor speed will be the same as the grid frequency, while in a 4-pole generator it would have half the frequency (since a complete cycle of one north pole and one south pole occurs with every half rotation).

Asynchronous generators

Most wind turbines use a 3-phase *asynchronous* or *induction* generator. If the rotor is made to rotate at the synchronous speed of the generator (i.e. at 1500rpm for a 4-pole generator with 50Hz current), the magnetic field from the rotor rotates at the same speed as the stator magnetic field associated with the AC electric grid. As a result, nothing happens. However, if the rotor is made to rotate faster than the rotating magnetic field in the stator, a strong current is induced in the rotor. The harder one cranks the rotor, the more power that is transferred as an electromagnetic force to the stator, converted to electricity and fed into the grid.

A major advantage of asynchronous generators for wind turbines is that the speed of the generator varies with the applied torque (rather than being fixed, as in synchronous generators). The difference in speed between peak power and idling is only about 1–3 per cent, and is called the generator *slip*. Allowing the rotation rate to vary slightly with wind speed (and especially in response to sudden gusts) results in less wear on the gearbox and smaller fatigue loads on the tower and blades. This is the main reason for using an asynchronous generator. Energy from the wind gust is stored as additional rotational energy, to be released later when the rotor slows down after the wind gust, giving better power quality.

Variable-speed generators

With a variable-speed generator, the rotor speed can vary from, say, 8rpm to 16rpm, which in turn allows a better matching of the rotor aerodynamics to the wind speed. The result is that average power output increases by 10–15 per cent (Enslin et al, 2004). As well, there is less mechanical stress on the rotor and tower, it is easier to control the natural vibration frequencies because the structure is lighter and less rigid, power output is more constant because wind speed variations go into changing the rotor speed before changing the power output, there is less noise at low wind speeds and the foundations are cheaper. However, a more complex power electronics and gearbox are required, so as to always produce electricity at 50 or 60Hz in spite of a variable rotor rotation rate.

Variable-speed turbine operation can be achieved only if the electrical grid frequency can be decoupled from the mechanical motor frequency. This can be done either with:

- a *doubly-fed induction generator*, in which 25–30 per cent of the power generated is fed into the grid through an AC–DC–AC frequency inverter, with the balance bypassing the inverter altogether;
- a *direct-drive synchronous generator*, where 100 per cent of the power generated is passed through an inverter that converts the AC power to DC and then to the AC power at the desired frequency (Slootweg and de Vries, 2003); or
- *hydrodynamic coupling* between the rotor and generator, which allows a variable rotor to drive a synchronous generator at a fixed speed with no need for an inverter, as in the DeWind D8.2 2MW turbine.

The first turbines with hydrodynamically adjustable drives were delivered in 2007, but such drives have been used extensively in the gas turbine industry for many years. Variable-speed turbines are normally offered with pitch rather than stall regulation.

3.4.2 Inverters

All modern wind turbines are variable speed (with the exception of turbines that are equipped with a hydrodynamic torque converter to regulate generator speed), and so must be connected to the grid through an inverter. Inverters produce a high-quality sine wave that is synchronized with the grid, but also supply reactive power (see subsection 3.11.3). Inverters are also used in PV systems, and more information on inverters is provided in that context (Chapter 2, subsection 2.2.5).

3.4.3 Starting a wind turbine

Generators typically do not have a permanent magnet, but rather, use electromagnets that require a DC current to be magnetized (although there has been a recent move toward permanent-magnet generators). The power needed to magnetize the electromagnet comes from the grid itself. If a turbine were switched on with a normal switch, there would be a brownout in the neighbourhood as current was drawn to magnetize the generator, followed by a power surge. There would also be extra wear on the gearbox. Instead, a wind turbine has to be started gradually with a switch using a *thyristor*. As the generator starts, it must be connected to the grid exactly synchronized with the alternating current of the grid. The phase-locking between the generator rotor magnetic field and the rotating magnetic field in the stator provides resistance to the rotor. In the absence of this phase-locking to the electrical grid, there would be only the mechanical resistance of the gearbox and generator to prevent the rotor from accelerating and overspeeding.

3.5 Variation of wind speed and turbulence with height near the earth's surface

The lower few hundred metres of the atmosphere, called the *planetary boundary layer* (PBL), are subject to the influence of friction from the earth's surface. Above the PBL, the winds blow parallel to the contours of atmospheric pressure, with a speed that is greater the greater the horizontal pressure gradient. These winds are referred to as *geostrophic winds*, as they arise from a balance between the horizontal pressure gradient force and the horizontal component of the Coriolis force, the latter arising from the rotation of the earth. As the large-scale pressure patterns change over the course of a few days, the geostrophic wind will change too, but on a short timescale, we can regard the geostrophic wind as given. At the surface the wind speed is zero and at the top of the planetary boundary layer the wind is geostrophic, with a variation of wind speed in between that depends on the surface roughness and vertical stability of the atmosphere.

Within the PBL layer, the wind speed normally varies with the natural logarithm of height, according to:

$$U(z) = \frac{u_*}{\kappa} \ln\left(\frac{z}{z_o}\right) \tag{3.2}$$

where u_* is called the friction velocity, κ is a constant (von Karmen's constant) and z_o is the surface roughness (it is comparable to the height of roughness elements protruding from the ground surface, but it is really an effective roughness that takes into account the arrangement of individual roughness elements, and depends on the scale at which the surface is examined). An example of the logarithmic velocity profile is shown in Figure 3.11. If wind speed is plotted versus the logarithm of height, the result will be a straight line with slope u_*/κ. The height at which the velocity extrapolates to zero is z_o. A rougher surface will be characterized by a larger z_o. Figure 3.12 shows the wind speed profile in the lower 100m of the atmosphere for three different surface roughnesses. A smooth surface will generate fewer turbulent eddies than a rough surface, so the frictional influence of the surface is not as pronounced. As a result, the geostrophic wind speed penetrates closer to the surface and there is a sharp velocity gradient close to the surface. A rougher surface generates more turbulence, which tends to smooth out the variation in average velocity near the surface. Thus, the sharp velocity gradient seen in Figure 3.12 for the smoother surface cannot be sustained, and there is a more gradual increase of wind speed with increasing height.

Atmospheric stability refers to the ease with which air parcels can move up and down. In an unstable atmosphere, an air parcel, if lifted slightly, will tend to keep rising. In a stable atmosphere, vertical motions are suppressed. If vertical motions are suppressed, the frictional influence of the surface is also limited in vertical extent, so the wind speed can increase more rapidly with increasing height. Thus, a smooth surface

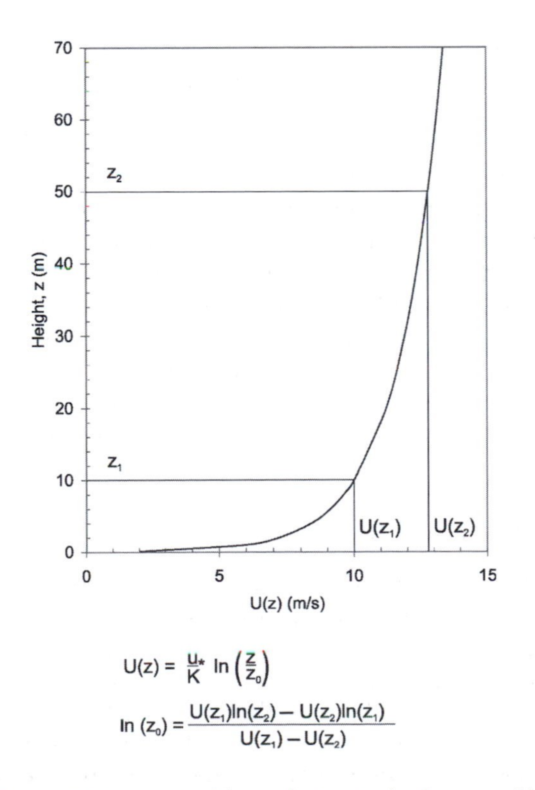

Figure 3.11 *A typical logarithmic wind velocity profile*

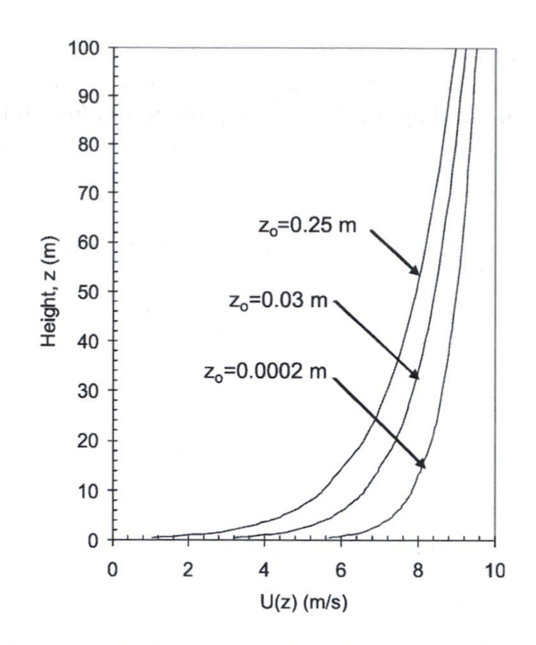

Figure 3.12 *Comparison of wind speed in the lowest 100m for three different surface roughnesses, with a wind speed of 10m/s at a height of 200m in all cases*

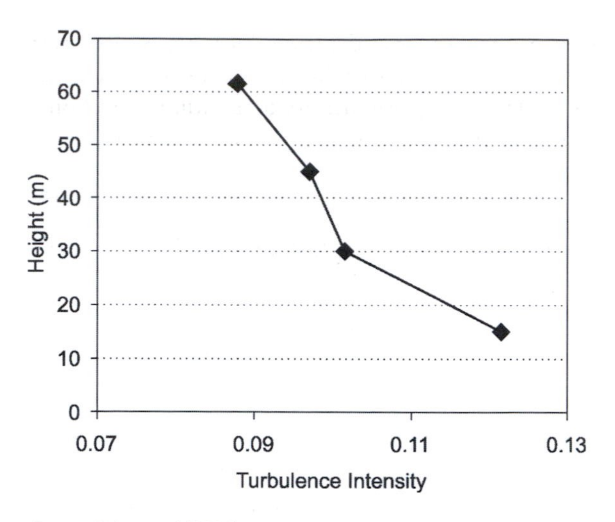

Source: Söker et al (2000)

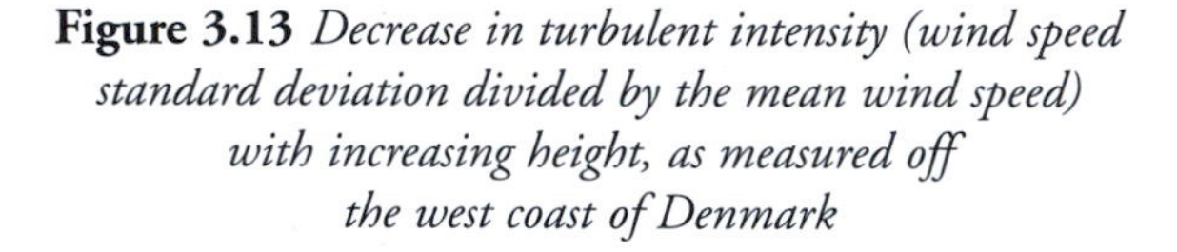

Figure 3.13 *Decrease in turbulent intensity (wind speed standard deviation divided by the mean wind speed) with increasing height, as measured off the west coast of Denmark*

or a stable atmosphere, by minimizing vertical motions, both allow a greater increase in wind speed with height (for a given geostrophic wind speed above the PBL) than a rough surface or an unstable atmosphere.

When wind speeds need to be estimated at a height greater than the highest point with data (usually data from no higher than 50m are available), a common procedure is to estimate the wind speed at greater heights assuming a power relationship. That is:

$$\frac{U_H}{U_{ref}} = \left(\frac{H}{h_{ref}}\right)^n \quad (3.3)$$

where U_H is the unknown velocity at height H, U_{ref} is the measured wind speed at the reference height h_{ref} and n is called the *shear factor* and can be estimated from the observed variation of wind speed with height up to h_{ref}. Typically, $n = 0.15–0.16$, so that wind speeds at an 80m hub height would be 7–8 per cent greater than at 50m, and the power output (which varies with wind speed to the third power) about 25 per cent greater. Alternatively, the logarithmic law (Equation 3.2) can be used. The logarithmic law is theoretically valid only in a neutral atmosphere (neither stable nor unstable). The power law (Equation 3.3) has no theoretical basis but provides a good fit to observed wind profiles. Archer and Jacobson (2003) used both approaches in fitting curves to observed wind profiles.

At the same time as average wind speed increases with increasing height above the surface, the turbulence decreases. This is illustrated in Figure 3.13, which shows the standard deviation of the wind speed divided by mean wind speed at various heights, as measured off the west coast of Denmark. Turbulent intensity (normalized by average wind speed) drops by about 15 per cent as height increases from 30m to 60m.

3.6 Power output from a wind turbine

In this section we discuss the relationship between wind speed, wind power and turbine output; the probability distribution of wind speeds; and how the probability distribution can be combined with the curve of turbine power versus wind speed to compute the relationship between mean wind speed and mean turbine power output.

3.6.1 Basic relationships

The kinetic energy of an object of mass m, velocity U, is equal to $\frac{1}{2}mU^2$. The mass of wind passing through a plane perpendicular to the wind direction, per unit area and per unit time, is given by the air density times the air speed. Thus, the power density of the wind (W/m^2 on a plane perpendicular to the air flow), P_W, is given by:

$$P_W = 0.5\rho U^3 \tag{3.4}$$

where ρ is air density (kg/m^3) and is a function of air temperature T (°C) and pressure P (kPa), given by:

$$\rho = 1.225\left(\frac{288.15}{T+273.15}\right)\left(\frac{P}{101.325}\right) \tag{3.5}$$

The wind power density times the area swept by the rotor blades gives the wind power intercepted by the rotor. This times the efficiency of the turbine (η) gives the turbine power output, P_T. That is, at any given wind speed:

$$P_T(U) = \eta\pi R^2 P_W(U) = \frac{1}{2}\eta_a\eta_m\eta_e\pi R^2\rho U^3 \tag{3.6}$$

where R is the rotor radius and η_a, η_m and η_e are the aerodynamic, mechanical and electrical efficiencies, respectively, all of which tend to be larger for larger turbines and are multiplied together to give the overall efficiency η. As previously noted (subsection 3.3.4), the maximum possible aerodynamic efficiency is 0.593 (the Betz limit) and is approached by the most efficient turbines when the wind speed is optimal, while the mechanical and electrical efficiencies are in the 0.96–0.99 and 0.96–0.97 ranges, respectively.

The average power output can be computed from the power output for the wind speed at the centres of a number of small wind speed intervals times the probability of each interval, summed over all intervals and divided by the sum of the probabilities (which is exactly 1.0 if all wind speeds are considered). That is:

$$\overline{P_T} = \frac{\int_0^\infty P_T(U)f(U)dU}{\int_0^\infty f(U)dU} = \frac{\sum_{i=1}^{N} P_T(U_i)f(U_i)\Delta U_i}{\sum_{i=1}^{N} f(U_i)\Delta U_i} \tag{3.7}$$

where f(U_i) is the probability of the wind speed occurring over an interval of width 1m/s centred at wind speed U_i, ΔU_i is the width of the interval and N is the number of intervals.

The average power output from a wind turbine divided by the rated power output is referred to as the *capacity factor*. The capacity factor is thus the average output as a fraction of the peak output. Since a wind turbine generates electricity at its rated capacity only for wind speeds of 11–16m/s or greater (depending on the turbine), and since wind speeds are well below this wind speed most of the time in most locations, typical capacity factors are quite small – as little as 0.15 in a poor wind regime, to 0.4 or more in very good regimes.

3.6.2 Weibull wind speed probability distribution function

The probability distribution of wind speeds can be described by the *Weibull distribution*, which is

Box 3.1 Wind velocity distribution

The wind speed fluctuates continuously, with some wind speeds occurring frequently and other winds speeds rarely occurring. The probability of observing a wind speed within a given wind speed interval is given by the *Weibull distribution*. This is an empirical function with two adjustable constants, the *scale factor*, *c* (m/s), and the *shape factor*, *k* (dimensionless), but it describes the wind distribution quite well in most locations. The Weibull distribution is given by:

$$f(u) = \frac{k}{c}\left(\frac{u}{c}\right)^{k-1}\exp\left[-\left(\frac{u}{c}\right)^k\right] \tag{3.8}$$

Note that f(u) has dimensions of (m/s)$^{-1}$, since it is the probability *per* unit distance on the velocity axis. If f(u) is evaluated at 1m/s wind speed intervals, multiplied by the corresponding wind speed and summed over all possible wind speeds (that is, from zero up to some arbitrarily large wind speed), the result will be the average or *mean* wind speed for that distribution. This is the point in the distribution at which a pile of bricks representing the various wind speeds would balance. The mean wind speed depends almost entirely on c, while large k causes the distribution to be more peaked with almost no effect on the mean. The *modal wind speed* is the wind speed at which f(u) is largest (i.e. the most likely wind speed).

The cumulative probability distribution is given by:

$$F(u) = 1 - \exp\left[-\left(\frac{u}{c}\right)^k\right] \quad (3.9)$$

where F(u) is the probability of a wind speed up to or equal to u. Unlike f(u), F(u) is dimensionless. The *median* wind speed (the wind speed that is exceeded 50 per cent of the time) is the wind speed at which F(u) = 0.5.

Figure 3.14 shows combinations of scale factors and shape parameters fitted to wind data from 31 different sites; the maximum observed scale parameter tends to increase with increasing scale factor. Figure 3.15 gives a typical Weibull distribution, using k = 1.6 and c = 5.0m/s. The Weibull distribution has a long tail to the right

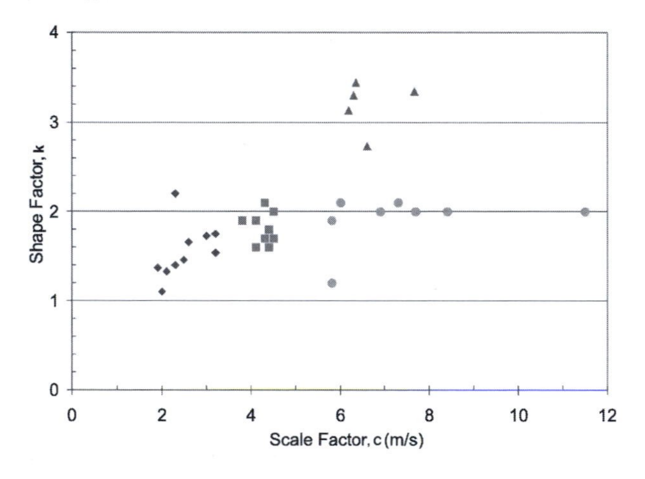

Source: Data from Hu and Cheng (2007)

Figure 3.14 *Observed pairings of the Weibull scale parameter and shape factor from 31 different sites*

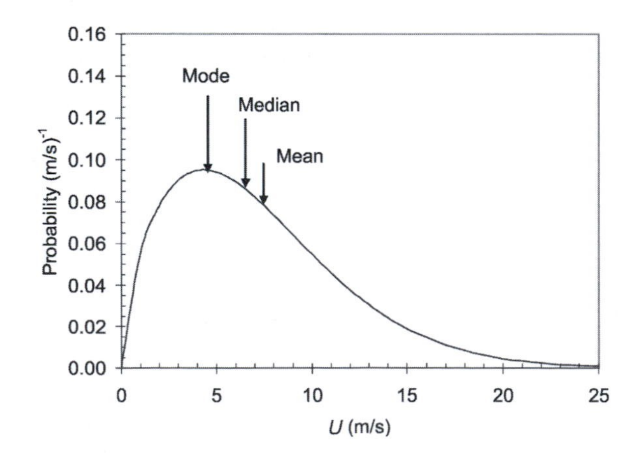

Figure 3.15 *A typical Weibull velocity distribution (obtained using c = 5m/s and k = 1.6), showing the relationship between modal wind speed, median wind speed and mean wind speed*

(toward high wind speeds), so the mean wind speed is greater than the median wind speed, which in turn is greater than the modal wind speed. Sometimes the wind data are better represented by the superposition of two Weibull distributions; techniques for estimating the resulting five parameters in that case are given in Carta and Ramírez (2007).

explained in Box 3.1. The Weibull distribution is governed by two parameters: a scale parameter c and a shape parameter k. Figure 3.16 shows the distribution of wind speeds calculated using the Weibull distribution for three different values of the Weibull scale factor and two different representative values of the Weibull shape parameter. The mean (or average) wind speed increases linearly with c and depends only slightly on k. For a given k, greater mean wind speeds (larger c) are associated with a wider and flatter distribution of wind speeds, while for a given mean wind speed, a larger k produces a narrower and more peaked distribution of wind speeds. The more peaked the distribution, the less frequent the occurrence of high-speed winds. As high winds contribute disproportionately to the mean power due to the cubic dependence between wind speed and power seen in Equation (3.4), a more peaked distribution (large k) results in a smaller average power for a given average wind speed. This is illustrated in Figure 3.17, which shows the variation of average wind power with average wind speed for three different values of the Weibull shape parameter. The mean power output from a turbine cannot be determined from mean wind speed alone, as the relationship between mean wind speed and mean power output depends on how the wind speed probabilities are distributed on either side of the mean wind speed, due to the cubic dependence of wind power on wind speed.

The cubic dependence of wind power on wind speed also means that the wind power computed at the average wind speed will be less than the average wind power computed using the wind speed probability distribution. Table 3.3 compares the wind power computed at wind speeds of 3–9m/s with the average

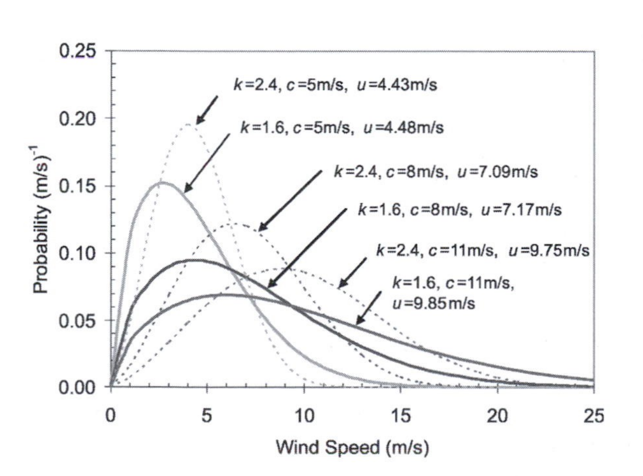

Figure 3.16 *Comparison of Weibull wind speed probability distributions for k = 1.6 and 2.4, and for c = 5, 8 and 11m/s*

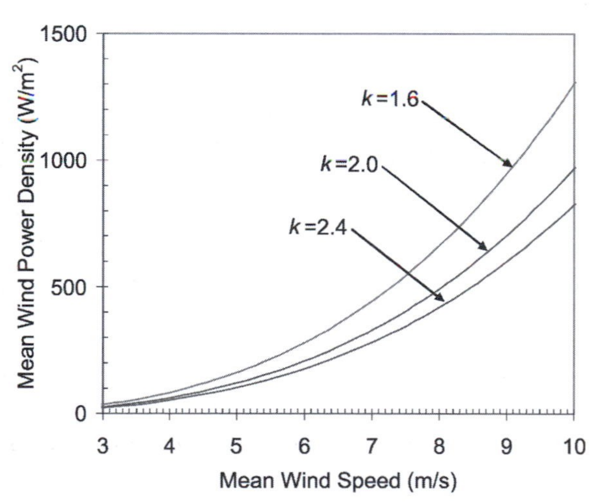

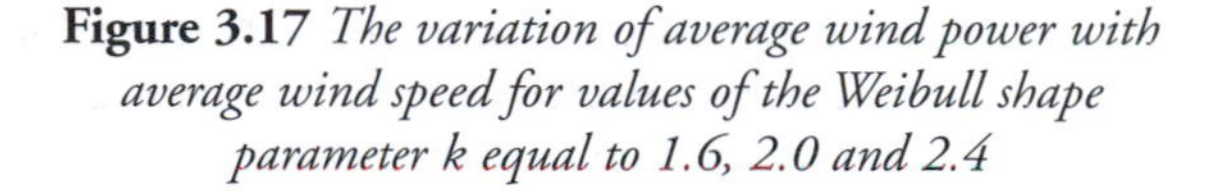
Note: A larger shape parameter (k) corresponds to a more peaked wind speed distribution.

Figure 3.17 *The variation of average wind power with average wind speed for values of the Weibull shape parameter k equal to 1.6, 2.0 and 2.4*

Table 3.3 *Comparison of wind power computed at the average wind speed with the average wind power computed over a distribution of wind speeds giving the same average wind speed*

Average wind speed (m/s)	Wind power (W/m^2) computed at average wind speed	Wind power (W/m^2) computed over a distribution of wind speeds having the same average wind speed as used in column 2		
		k = 1.6	k = 2.0	k = 2.4
3	16	40	31	26
5	75	187	144	122
7	207	513	394	336
9	439	1088	838	715

wind power computed over Weibull distributions having the same average wind speed, for different values of the shape parameter. Average wind power computed using realistic distributions of wind speeds is about twice that computed using the average wind speed. For the same average wind speed, changing the shape parameter from 2.4 to 1.6 results in a 50 per cent larger average wind power, a factor of enormous importance to the economic viability of wind energy. For this reason, several years of wind observations are needed to accurately characterize the wind speed distribution before building turbines. Variations in air density between summer and winter in temperate climates also have a non-negligible effect on wind turbine output, as air at −20°C (for example) is 16 per cent denser than air at 20°C.

3.6.3 Computing the mean wind turbine efficiency

As noted above (Equation 3.7), the mean power output of a turbine is computed using the power output centred at various wind speed intervals multipled by (that is, weighted by) the probabilities of each interval. However, the efficiencies must be weighted by the product of probability and wind power, rather than by probability alone, in computing the mean efficiency. The mean efficiency computed in this way, times the mean wind power and swept area, will give the mean wind turbine power output. That is, if:

$$\bar{\eta} = \frac{\int_0^\infty \eta(U) P_W(U) f(U) dU}{\int_0^\infty P_w(U) f(U) dU} \quad (3.10)$$

and:

$$\overline{P_W} = \frac{\int_0^\infty P_W(U) f(U) dU}{\int_0^\infty f(U) dU} \quad (3.11)$$

then:

$$\pi R^2 \bar{\eta} \overline{P_W} = \frac{\int_0^\infty \pi R^2 \eta(U) P_W(U) f(U) dU}{\int_0^\infty f(U) dU} = \overline{P_T} \quad (3.12)$$

For a given turbine, the mean efficiency decreases with increasing mean wind speed beyond some minimal mean wind speed because efficiencies decrease rapidly as the wind speed increases beyond the rated wind speed (13m/s in Figure 3.10) and these low efficiencies will occur more frequently as the mean wind speed increases. This is illustrated in Figure 3.18, which shows the variation of mean efficiency with mean wind speed for the Nordex N90-2.3 turbine (the same one illustrated in Figure 3.10). The mean efficiency computed using the Weibull distribution with k = 2.0 when $\overline{U}$ = 4.4m/s, for example, is 0.36 but only 0.21 when $\overline{U}$ = 9.3m/s. Mean power output thus increases by only a factor of 6.8 rather than a factor of 9.4 as the mean wind speed increases from 4.4 to 9.3m/s. Of course, different turbines would be chosen for different wind regimes, each tending to have the highest efficiency when placed in the chosen wind regime. This is illustrated in Figure 3.18b, which compares the variation of mean efficiency with mean wind speed for three different turbines with k = 2.0. For the chosen shape parameter, the mean efficiency of one turbine peaks at a mean wind speed of 6m/s, while the mean efficiency of the other two peaks near a mean wind speed of 5m/s.

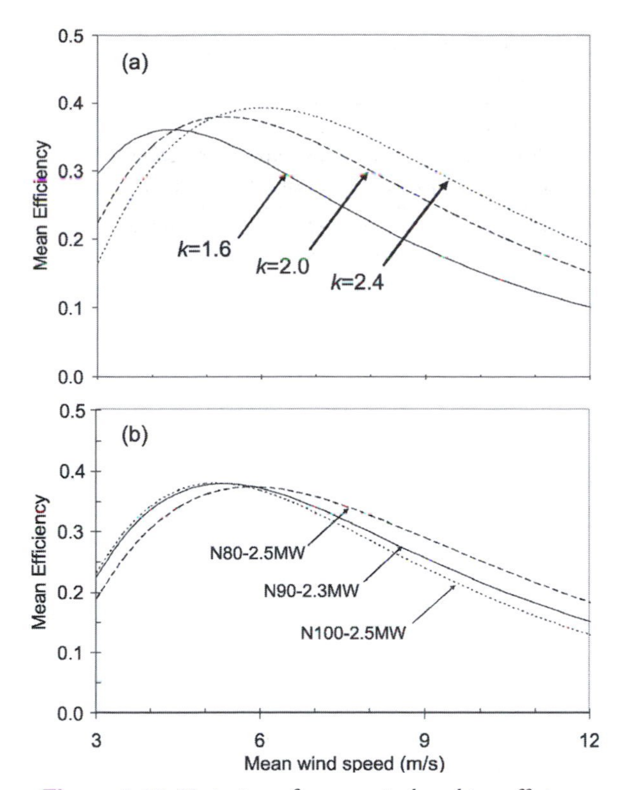

Figure 3.18 *Variation of mean wind turbine efficiency with mean wind speed for selected turbines or for different Weibull shape parameters: (a) results for the Nordex N90-2.3 turbine using a Weibull wind speed distribution with shape parameters of 1.6, 2.0 and 2.4; and (b) results for the Nordex N80-2.5, N90-23 and N100-2.5 turbines with k = 2.0*

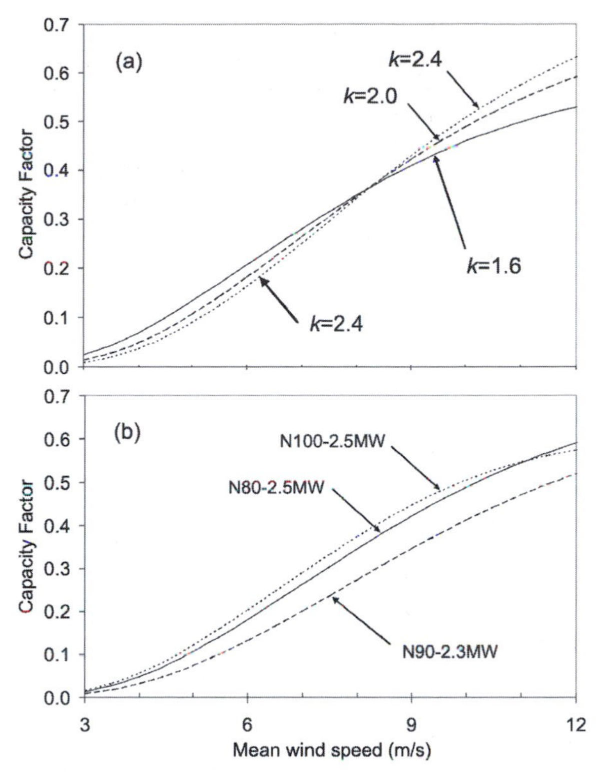

Figure 3.19 *Variation of wind turbine capacity factor with mean wind speed for selected turbines or for different Weibull shape parameters: (a) results for the Nordex N90-2.3 turbine using a Weibull wind speed distribution with shape parameter k having values of 1.6, 2.0 and 2.4; and (b) results for the Nordex N80-2.5, N90-23 and N100-2.5 turbines with k = 2.0*

3.6.4 Wind turbine capacity factor and its dependence on mean wind speed and its variability

The capacity factor of a given turbine in a given wind regime can be easily computed as the average power output (given by Equation 3.7) divided by the rated power. Figure 3.19 shows the variation of wind turbine capacity factor with mean wind speed for selected turbines. The relationship between capacity factor and mean wind speed depends on the turbine power curve (illustrated for several turbines in Figure 3.7) and the shape of the wind speed probability distribution (illustrated in Figure 3.16 for different shape parameters). Figure 3.19a shows the relationship for the Nordex N90-2.3 turbine for Weibull shape parameters of 1.6, 2.0 and 2.4 (which span most of the observed values given in Figure 3.14). For an average wind speed of 5.0m/s (typical of many inland locations), the capacity factor is only 0.10–0.15 for this turbine (depending on the shape parameter), while for an average wind speed of 10.0m/s (characteristic of the very best wind regimes), the capacity factor is around 0.5–0.6.[1] Figure 3.19b compares the curves of capacity factor versus mean wind speed for the Nordex N80-2.5, N90-2.3 and N100-2.5 turbines. The N100-2.5 has a greater capacity factor than the N80-2.5 at all mean wind speeds shown in spite of its lower cut-out speed, as the greater output at lower windspeeds (shown in Figure 3.7b) more than compensates for the more frequent shutdown due to high wind speeds.

Table 3.4 *Average wind turbine capacity factors in 2001*

Country	Capacity factor
UK	0.32
Greece	0.29
Denmark	0.26
Spain	0.24
Netherlands	0.24
China	0.24
Sweden	0.24
Italy	0.23
Germany	0.21
India	0.20

Source: BTM Consult (2002)

Table 3.4 gives the average capacity factor for various countries in 2001. The average ranged from lows of 0.20 in India and 0.21 in Germany, to highs of 0.29 in Greece and 0.34 in the UK. For offshore wind farms in the North Sea, where mean wind speeds at hub height are generally in excess of 9m/s, the capacity factor will be 0.4–0.5.

3.7 Available wind resources

Atlases of mean wind speed and, in some cases, mean wind power have been prepared for a number of countries or regions. Regional maps for North America, Europe and China are shown in Figures 3.20 to 3.22 (and are available in colour in the online material). Good wind resources (mean wind speeds >8m/s) occur in Europe along the coast of England, northern France to Denmark, in the North Sea, along the French Mediterranean coast, in isolated parts of Spain and in Crete. Good US wind resources occur in the northern central plains and in parts of a band running from Colorado to Washington state, while good Canadian wind resources are found in Labrador and parts of the central prairies. Both countries have excellent offshore wind potential in the Great Lakes and along parts of both the Pacific and Atlantic coast. Lake Erie is the most promising of the Great Lakes because it is shallow (much of it is less then 25m deep). The best Chinese wind resource occurs in Inner Mongolia (and Mongolia itself also has an excellent wind resource).

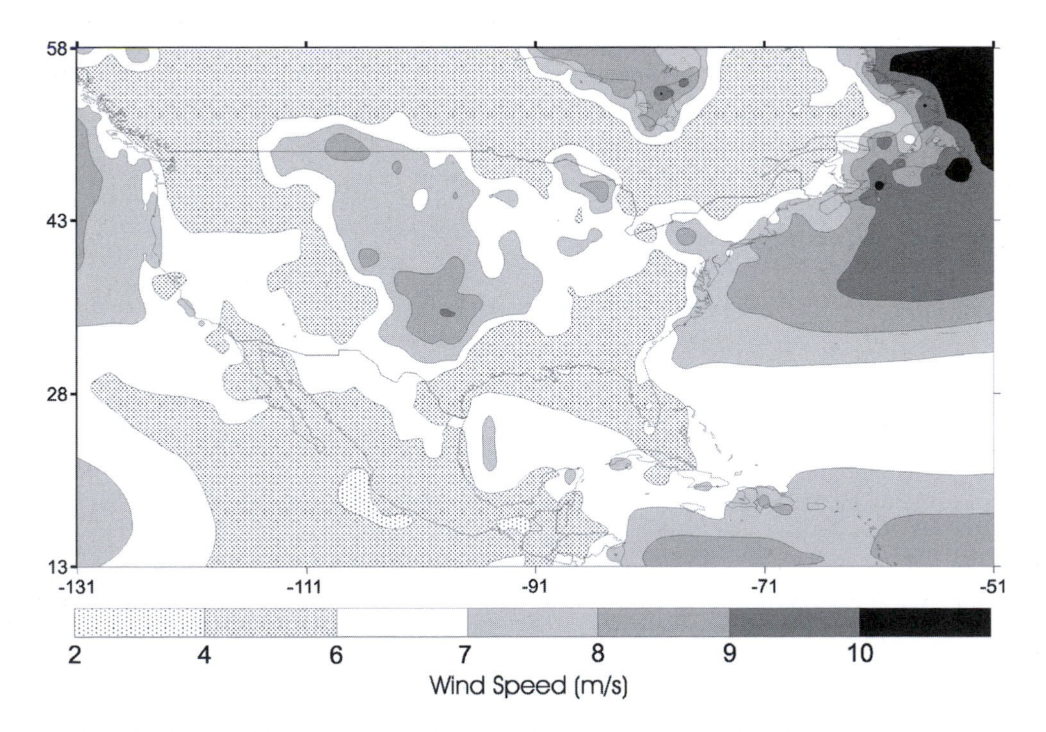

Source: Produced from digital data available at the NASA Surface Meteorology and Solar Energy website, http://power.larc.nasa.gov

Figure 3.20 *Mean wind speed at a height of 100m over North America*

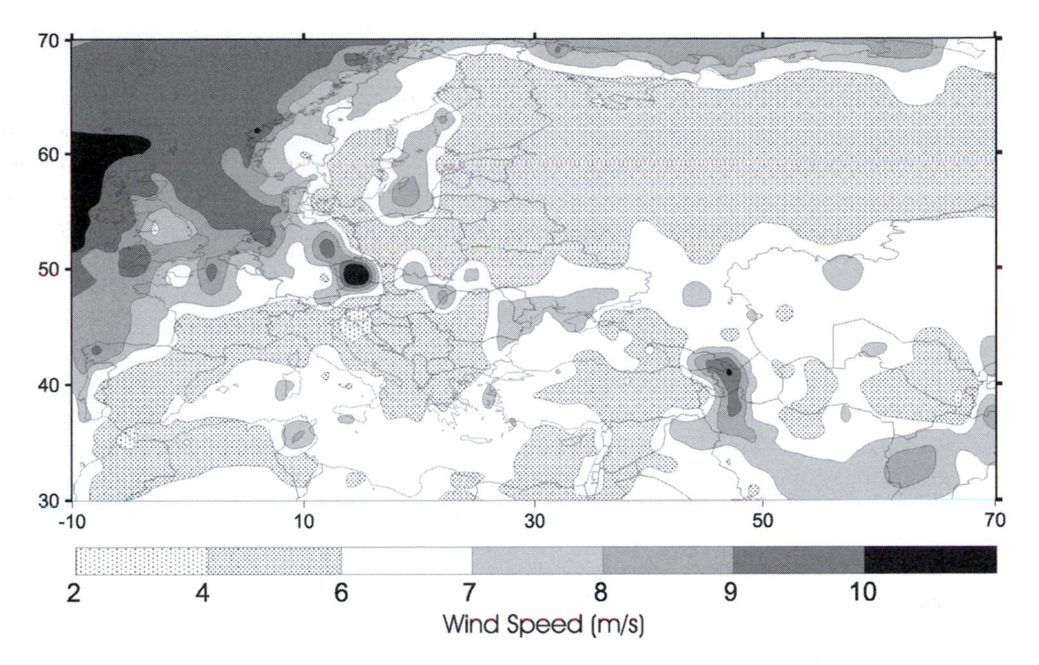

Source: Produced from digital data available at the NASA Surface Meteorology and Solar Energy website, http://power.larc.nasa.gov

Figure 3.21 *Mean wind speed at a height of 100m over Europe*

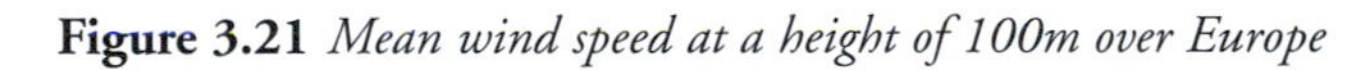

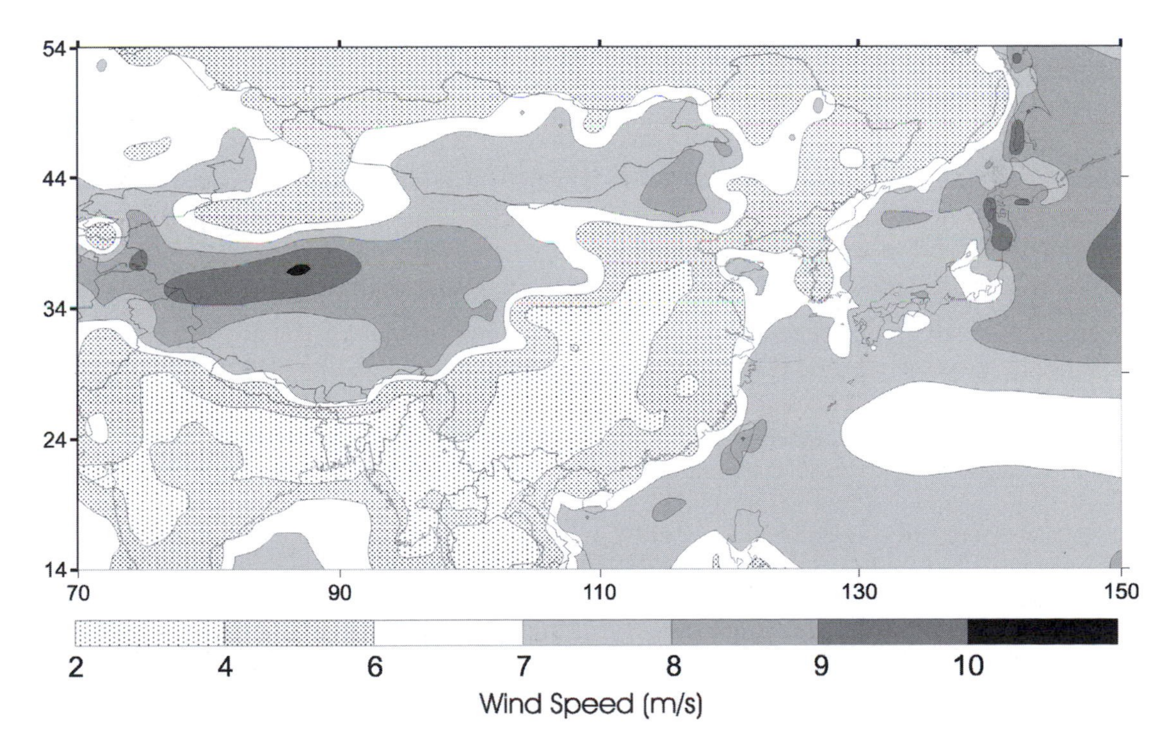

Source: Produced from digital data available at the NASA Surface Meteorology and Solar Energy website, http://power.larc.nasa.gov

Figure 3.22 *Mean wind speed at a height of 100m over China and surrounding regions*

3.8 Wind farms

Wind turbines can be grouped together in a regular array to form a wind farm. To minimize aerodynamic interference of one turbine in the wind farm with another, the turbines are spaced 5–9 rotor diameters apart in the along-wind direction and 3–5 rotor diameters apart in the cross-wind direction. With these spacings, power output per turbine is reduced by about 5 per cent compared to what would be expected from isolated turbines. However, this can be more than offset by the fact that wind turbines in wind farms are concentrated at the best wind sites, and by the reduction in construction mobilization costs and grid connection costs when turbines are grouped together in a wind farm. The economic benefit of clustering wind turbines in wind farms is particularly strong in offshore applications; indeed, clustering of turbines as wind farms is mandatory in offshore settings on economic grounds.

3.8.1 Impact on wind speed, turbulence and surface temperature

Atmospheric kinetic energy is generated by the differential heating of the earth's surface and atmosphere by solar energy, creating horizontal pressure gradients within the atmosphere. Due to the rotation of the earth, atmospheric winds not subject to friction tend to blow parallel to the contours of atmospheric pressure that would be seen on a weather map, but the winds are not exactly parallel to the contours, so there is a component of the winds across the contours in the direction of the pressure gradient force (PGF) (which is perpendicular to the contours). The component of the PGF acting in the wind direction leads to an acceleration of wind speed – the generation of kinetic energy. Kinetic energy is largely generated in the upper troposphere and at large (10^3 to 10^4km) spatial scales, then cascades to successively smaller scales (as large-scale motions break up into successively smaller motions). There is a downward flux of kinetic energy to the earth's surface that averages about 1.5–2.0W/m^2 over land. This sets an ultimate constraint on how much energy can be extracted from the wind. Over the entire land surface area, this ultimate constraint is about 220–300TW (current global primary power demand is about 15TW). This kinetic energy flux is ultimately dissipated as heat, as friction generates turbulent eddies that eventually cascade down to molecular-scale random motion.

An isolated wind turbine extracts kinetic energy from the wind, slowing the airflow downwind from the turbine. This is compensated by transfer of kinetic energy into the affected layer from the airflow above and to the side, such that the original wind speed will be restored some distance downwind from the turbine. However, for a large array of wind turbines in a wind farm, the overall wind speed will be reduced due to the extraction of kinetic energy from the wind. Additional kinetic energy is converted into turbulence, further reducing the large-scale wind flow and hence the power output from the wind farm. Thus, the reduced wind speed reflects both energy extracted from the mean airflow and converted to electricity (about 50 per cent of the extracted energy), and the generation of turbulence by the turbines. The reduction in wind speed extends beyond the wind farm and reduces the dissipation of kinetic energy by surface friction beyond the wind farm, such that the global total kinetic energy dissipation is largely unaltered, which must be the case if it is to equal the rate of kinetic energy generation by large-scale pressure gradients (which may change only slightly). The increased turbulence and associated vertical mixing will alter the temperature and humidity of the air next to the land surface.

Roy et al (2004) have investigated these effects using a mesoscale atmospheric model with three nested grids: a 250km × 250km domain with 2km resolution inside a 616km × 616km domain with 8km resolution, which in turn is inside a 1568km × 1568km domain with 32km resolution. A wind farm consisting of a 100 × 100 array of turbines at a 1km spacing is placed inside the innermost grid. The innermost grid cannot of course resolve the slower airstreams and turbulence created by individual turbines, so the impact of turbines is emulated by reducing the kinetic energy of the grid-scale (i.e. the resolved) airflow by an amount equal to the area-average rate of extraction of kinetic energy by the wind farm (for conversion to electricity), and by converting some of the grid-scale kinetic energy to sub-grid-scale turbulent kinetic energy in the model. The simulations were carried out for one particular 15-day period in summer in Oklahoma with wet and cool soil conditions. The results indicate a 30 per cent reduction in electricity production compared to the case with no impacts on the wind flow and a slight warming and drying of the surface during the early morning, but a less than 10 per cent increase in evapotranspiration. Two thirds of the reduction in electricity production is due to the effects of turbulence on the airflow, implying that designing rotors to reduce turbulence will increase power output not only by increasing the turbine efficiency but also by reducing the impact on the large-scale airflow. Further work is required using a wide variety of surface and atmospheric conditions in order to get a clearer picture of the impacts of wind farms on the wind field itself and on surface climate, preferably using models with higher horizontal resolution so that three-dimensional mixing and the benefits of, for example, staggered turbine arrays can be investigated.

Keith et al (2004) have investigated global-scale effects using two different global atmospheric general circulation models (GCMs), one developed at the National Center for Atmospheric Research (NCAR) in Boulder, Colorado, and the other developed at the Geophysical Fluid Dynamics Laboratory (GFDL) in Princeton, New Jersey. These models have rather course resolution (about 200–250km). The effects of wind farms over broad areas were investigated by imposing an increase in the surface roughness or a reduction of wind speed in the lowest two atmospheric layers in the model. The effect of increased turbulence on the vertical fluxes of heat and moisture was not simulated.

The exact relationship between increased surface roughness as seen by the atmosphere and wind farm characteristics is not known, and in any case, will depend on atmospheric conditions. What Keith et al (2004) did was impose hypothetical changes in surface roughness or drag over the wind farm area; the model computes the increase in dissipation over that area, as well as changes in surface temperature and winds on a worldwide basis (the change in kinetic energy dissipation on a worldwide basis is negligible, as explained above). Given that about half of the increase in dissipation over the wind farm area is due to the generation of electricity (and the other half due to increased turbulence), the modelled changes in surface temperature can be related to the average wind farm power production. For an increase in surface drag in North America, Western Europe and China such that kinetic energy dissipation over these areas increases by 21TW, which would correspond to an average electric power generation by wind turbines of 10–11TW, temperature increases by up to 1K in the annual mean and by up to 2K in the North American prairies, but there are comparable decreases in temperature in the European part of Russia. Both the increases and decreases are due to shifts in global-scale wind patterns related to the surface friction caused by the wind farms. For smaller rates of electricity generation, the temperature effects would be proportionately reduced.

Over a period of 100 years, the reduced regional climate warming due to avoided CO_2 emissions would be several times larger than the warming effect in regions where wind farms induce warming, but initially, the direct warming effect of the wind farms would dominate. This is because the effect on temperature of a given installation of wind farms is a one-time effect, whereas the CO_2 buildup associated with sustained electricity production from fossil fuels is cumulative.

To sum up, the reduction in wind speeds in the wind farm region has a small one-time effect on regional climate, and also reduces the electricity production by the wind turbines themselves. The latter effect is accounted for by an array efficiency in the calculation of the annual electricity production by a wind farm, as discussed in subsection 3.13.1.

3.9 Scaling relationships and implications for required land area

This section examines the extent to which the wind turbine capacity and electricity output per unit of land area vary with the size of the wind turbine, assuming that turbines are laid out in a wind farm with a spacing that is a fixed multiple of the rotor diameter. We begin by developing the expected relationships based on simple scaling relationships. The scaling relationships are as follows:

1 For constant turbine efficiency, the output power of a turbine at a given wind speed will vary linearly with the rotor swept area. Figure 3.23 shows the variation of swept area with rated power based on the data in Table 3.2 – swept area increases at a decreasing rate as the rated power increases, implying that the efficiency has improved slightly as the turbines have increased in size. This results in rotor diameter increasing slightly more slowly than the square root of the rated power (Figure 3.5a). Conversely, rated power output increases

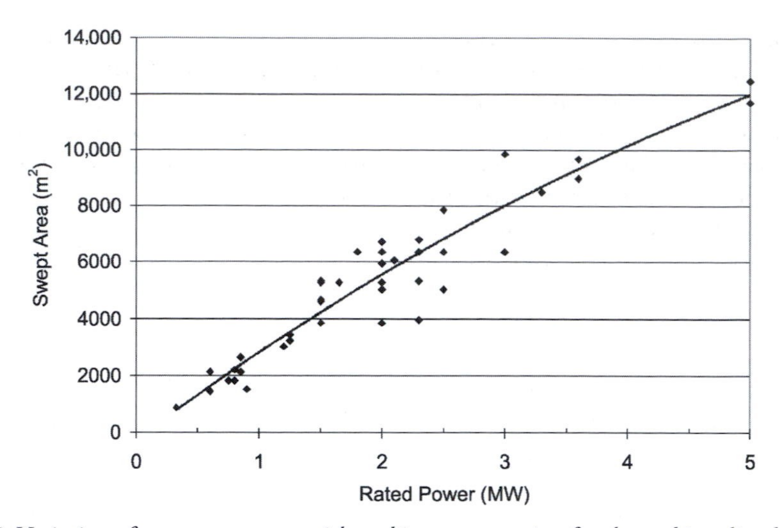

Figure 3.23 *Variation of rotor swept area with turbine power rating for the turbines listed in Table 3.2*

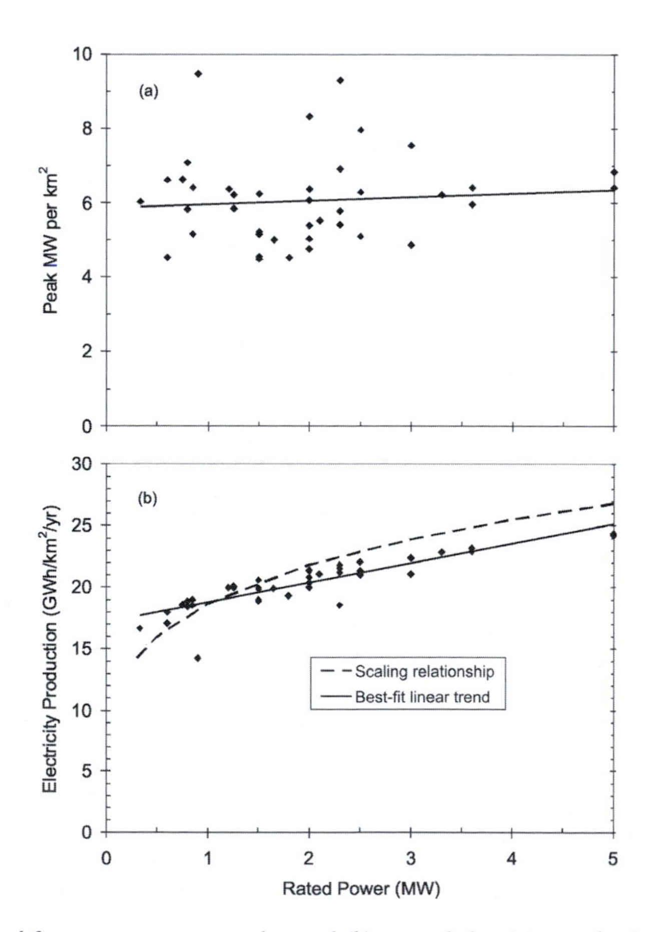

Figure 3.24 *(a) Peak wind farm power per square km and (b) annual electricity production per square km vs turbine size (rated power) for the turbines listed in Table 3.2, assuming the turbines to be arranged on a grid with a spacing of 7D × 7D (where D is the rotor diameter)*

slightly faster than the square of the rotor diameter. It will be assumed here that rated power varies exactly with the square of the rotor diameter.

2 As shown in Figure 3.6, the minimum hub height H of commercially available wind turbines is slightly less than and increases more slowly than the rotor diameter, while the maximum hub height is greater than and increases more quickly than the rotor diameter, so it will be assumed here that the hub height H is equal to the rotor diameter D.
3 As discussed in section 3.5, wind speed typically varies with $H^{0.15}$ while, as discussed in subsection 3.6.1, wind power density varies with wind speed cubed.
4 Assuming the turbine spacing in a wind farm to be a fixed multiple of the rotor diameter, the land area required per turbine varies with D^2.

Because the rated power of the turbines listed in Table 3.2 increases slightly faster than the rotor diameter squared, one can expect that peak power per hectare will increase slightly with increasing turbine size. This is confirmed by Figure 3.24a, which plots MW of peak power per square km of land taken up by a wind farm versus the rated power for all the turbines listed in Table 3.2, assuming a turbine spacing of $7D$ in both directions. MW/km^2 increases slightly with rated power, but with substantial scatter. Assuming the rated power to vary exactly with the rotor diameter squared, the peak power per unit of land area would be independent of the size of the turbine.

However, large turbines have higher hubs and so are exposed to greater average wind speeds, giving a larger capacity factor. From the scaling relationships given above, it follows that energy output per turbine will vary with $D^{2.45}$, that the energy output per unit of land area will vary with $D^{0.45}$ and that the electricity output per unit of land area increases with the turbine rating to the power 0.225. Thus, if the turbines are spaced apart at a fixed multiple of the rotor diameter, a doubling in the rating of the turbines used leads to roughly a 17 per cent increase in electricity production per unit of land area.

Archer and Jacobson (2005) find that, to within a few per cent, the capacity factor of the wind turbines is given by:

$$CF = 0.087\overline{V} - \frac{P_{rated}(kW)}{D^2} \tag{3.13}$$

where $\overline{V}$ is the mean wind speed, P_{rated} is the turbine capacity and D is the rotor diameter. This relationship, which was derived from wind data in the US, can be expected to depend on the wind speed probability distribution and the shape of the wind turbine power curve. The average power output $\overline{P}$ per km^2 of land area is given by:

$$\overline{P} = \delta P_{rated}\, CF \tag{3.14}$$

where δ is the number of turbines per km^2.

The above relationship is used here to compute the annual electricity production per square km for all of the turbines listed in Table 3.2, assuming that $H = D$, that the mean wind speed is 8.0m/s at a height of 80m and varies with height to the power of 0.15, and that the turbines are placed on a $7D \times 7D$ grid. Results are shown in Figure 3.24b along with the best-fit linear trend and the variation expected from the scaling relationship derived above (whereby annual electricity production per unit of land area varies with turbine rating to the power of 0.225). The slope of the scaling relationship and the best-fit linear trend are almost the same for turbines larger than about 1.0MW. Table 3.5 shows the mean wind speed at the hub height, the number of turbines per km^2, the capacity factor computed from Equation (3.13), the peak and average power output, the annual electricity production per km^2

Table 3.5 *Average power output per unit land area for turbines of various sizes and rotor diameters D, assuming a hub height equal to the rotor diameter and turbine spacing of 7D in both the cross-wind and along-wind directions and a mean wind speed of 8m/s at a height of 80m*

P_{rated} (kW)	D (m)	Turbines per km^2	Mean wind speed at hub (m/s)	Capacity Factor	Peak power (MW/km^2)	Average power (MW/km^2)	GWh/km^2/yr	Hectares per GWh/yr
600	43	11.0	7.3	0.31	6.62	2.05	18.0	5.57
1500	77	3.4	8.0	0.44	5.16	2.27	19.9	5.04
3600	107	1.8	8.4	0.41	6.42	2.65	23.2	4.31
5000	126	1.3	8.6	0.43	6.43	2.76	24.2	4.13

Table 3.6 *Rotor diameter, tower height, foundation diameter, annual electricity production per m² of land area taken up by the foundation and access road, and land area required to supply the global 2005 electricity demand of 18,454TWh using the indicated turbines*

Power rating (kW)	Rotor diameter (m)	Tower height (m)	Foundation diameter (m)	Mean wind speed at hub height (m/s)	Capacity factor	kWh/m²/ year	Global area (1000s km²)
900	44	45	12.5	7.3	0.23	1695	10.9
		55	13.9	7.6	0.24	1759	10.5
2300	71	64	15.4	7.7	0.26	3123	5.9
		113	16.4	8.4	0.31	3669	5.0

Note: Capacity factors were computed from the power curve for each turbine using a Weibull parameter chosen to given the indicated mean wind speeds, which in turn are based on an assumed mean wind speed of 8m/s at a height of 80m and varying with height according to Equation (3.3) with $n = 0.15$. The area of the access road is assumed to by 21Dm² (where D is the rotor diameter), which in turn is based on the assumption of a 3m-wide road right of way and turbines spaced 7D apart.
Source: Tower and foundation data from Pascal Pettinicchio, Enercon

and the number of hectares required to generate a GWh of electricity for selected turbines ranging in size from 0.6MW to 5.0MW. Annual average electricity output per square km increases by 29 per cent as the turbine size increases from 0.6MW to 3.6MW, but by only 4 per cent in going from 3.5MW to 5.0MW.

The electricity production shown in Figure 3.24b is (with one exception) 17–24GWh/km²/yr, which is equivalent to 17–24kWh/m²/yr. This is an 18th to a 12th of what would be produced in sunny locations using PV modules with an efficiency of 20 per cent (about 300kWh/m²/yr). However, in the case of wind energy, almost all of the land area can be used for other purposes (only the area taken up by the turbine foundation and any access road could not be used for agricultural purposes). Table 3.6 gives information on the foundation area required for various Enercon wind turbines where soil conditions are firm. This information is used to calculate the annual electricity output per m² of land area taken up by the turbine foundation and an access road that is assumed to run directly from one turbine to another in a rectangular wind farm array. The electricity output is in the order of 3500kWh/m²/yr (an average power production of 400 W/m²) for the 2.3MW turbine. This is about 30 times the output from a tilted PV array with an efficiency of 20 per cent and a module/land area ratio of 0.25 (which would be typical). The last column of Table 3.6 gives the global land area that would be taken up by turbine foundations and access roads if the entire 2005 electricity demand were supplied with onshore wind

Source: Author

Figure 3.25 *The Middelgrunden wind farm, next to Copenhagen*

turbines. The total area is in the order of 5000km^2 for the 2.3MW turbine, and an area equal to 0.0034 per cent of the world's land area.

3.10 Offshore wind power

Increasingly, offshore wind farms are being built. Examples include the Middelgrunden wind farm (20 × 2MW), built next to Copenhagen in 2001 in 3–5m deep water and illustrated in Figure 3.25, and the Horns Rev wind farm (80 × 2MW), built during a three-month window of calm weather in 2002, 15km off the west coast of Denmark in water ranging from 7 to 13m deep. A new generation of wind turbines is being developed that will be optimized for offshore applications, and the next few years will see several thousand MW of wind farms constructed in the North Sea and in the Baltic Sea.

3.10.1 Characteristics, advantages and disadvantages

Wind turbines that are optimized for offshore conditions will differ in several important respects from wind turbines designed for onshore applications (CA-OWEE, 2001; Breton and Moe, 2009). First, as noise will be less of a constraint, faster rotor speeds are possible. This in turn results in lower torque and permits a lighter blade, both of which reduce the cost of the tower. As well, faster rotation speeds require a smaller gearbox ratio. Second, lack of noise constraints will permit turbines where the rotor is located downwind of the tower rather than upwind. This will permit less expensive and more efficient blade designs. Third, lack of aesthetic constraints will permit use of truss rather than solid tubular towers, cutting material requirements for the tower in half. These cost reductions could offset some of the additional cost required to protect the tower and turbine against the salty marine environment and associated with the greater cost of foundations, installations and connection to the grid. For economic reasons, offshore wind turbines will be clustered in wind farms and larger turbines are preferred. However, as discussed by Breton and Moe (2009), offshore wind projects entail more risk than onshore projects, and it is very important that new turbine concepts be thoroughly tested in onshore settings before being applied at sea due to the 5–10 times greater cost of making repairs or fixing problems offshore. Some of the ways in which offshore turbines are being designed specifically to reduce maintenance requirements are outlined in Box 3.2.

Figure 3.26 shows average offshore wind speeds in Europe, which can be compared with the onshore

Box 3.2 Reducing the maintenance requirements of offshore wind farms

Minimizing the frequency of scheduled and unscheduled maintenance will be critical to the economically competitive production of electricity from offshore wind farms. Specific features of offshore wind turbines that are designed to minimize the maintenance requirements include:

- elimination of the gearbox (the number one cause of mechanical failures);
- use of an induction generator rather than a synchronous generator, as the former requires less maintenance and does not require a DC power source;
- use of closed-system water cooling or air-to-air heat exchangers to cool the generator, rather than open-air cooling;
- replacement of hydraulic systems in the yaw, blade pitch and braking systems with electronic controls;
- improved electronic control systems, including automatic and remote reset systems (which are becoming common in new onshore turbines);
- possible elimination of the transformer, although alternative systems are heavier, which might be a problem for 4–5MW concepts;
- use of onsite diesel power backup for maintenance work when the grid connection is lost.

If realized, the elimination of the gearbox and transformer, along with improvements in the generator efficiency, could increase turbine output by 10 per cent.

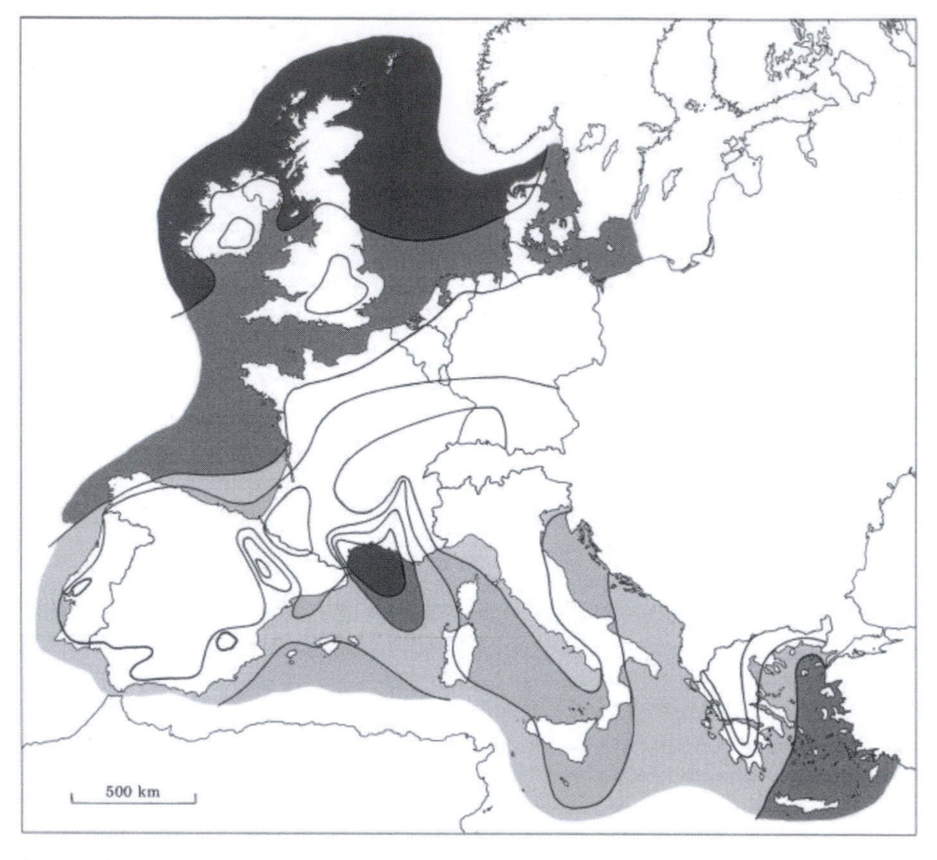

50 m		100 m		200 m	
m/s	W/m²	m/s	W/m²	m/s	W/m²
> 9.0	> 800	> 10.0	> 1100	> 11.0	> 1500
8.0-9.0	600-800	8.5-10.0	650-1100	9.5-11.0	900-1500
7.0-8.0	400-600	7.5-8.5	450-650	8.0-9.5	600-900
5.5-7.0	200-400	6.0-7.5	250-450	6.5-8.0	300-600
< 5.5	< 200	< 6.0	< 250	< 6.5	< 300

Source: European Wind Energy Association

Figure 3.26 *Mean wind speed and mean wind power density at heights of 50m, 100m and 200m in the open seas (>10km from shore) around Europe*

wind speeds shown in Figure 3.21. The best winds are in the North Sea off the coast of England and Scotland, where wind speeds at a 100m height are in excess of 10m/s and wind power densities are in excess of 1100W/m². As can be seen from Figure 3.19, the capacity factor (and hence, average power output) at this wind speed is about twice that for an average wind speed of 7m/s, which would be considered to be a very good wind speed in onshore settings. However, capital costs are also up to twice as large as for onshore locations, depending on the water depth and distance to shore (see subsection 3.13.1).

Offshore turbines have other advantages besides greater average wind speeds. Winds are less turbulent in offshore locations, which reduces fluctuations in power output and should result in a longer lifetime – 25 to 30 years instead of 20 years (as for onshore turbines). The foundations should last 50 years, allowing for two generations of turbines on the same foundation. The cost-effective tower height is a bit lower than for onshore turbines due to the smoother surface, such that wind speed initially increases more rapidly with increasing height than for a rough surface, and then more slowly (see Figure 3.12).

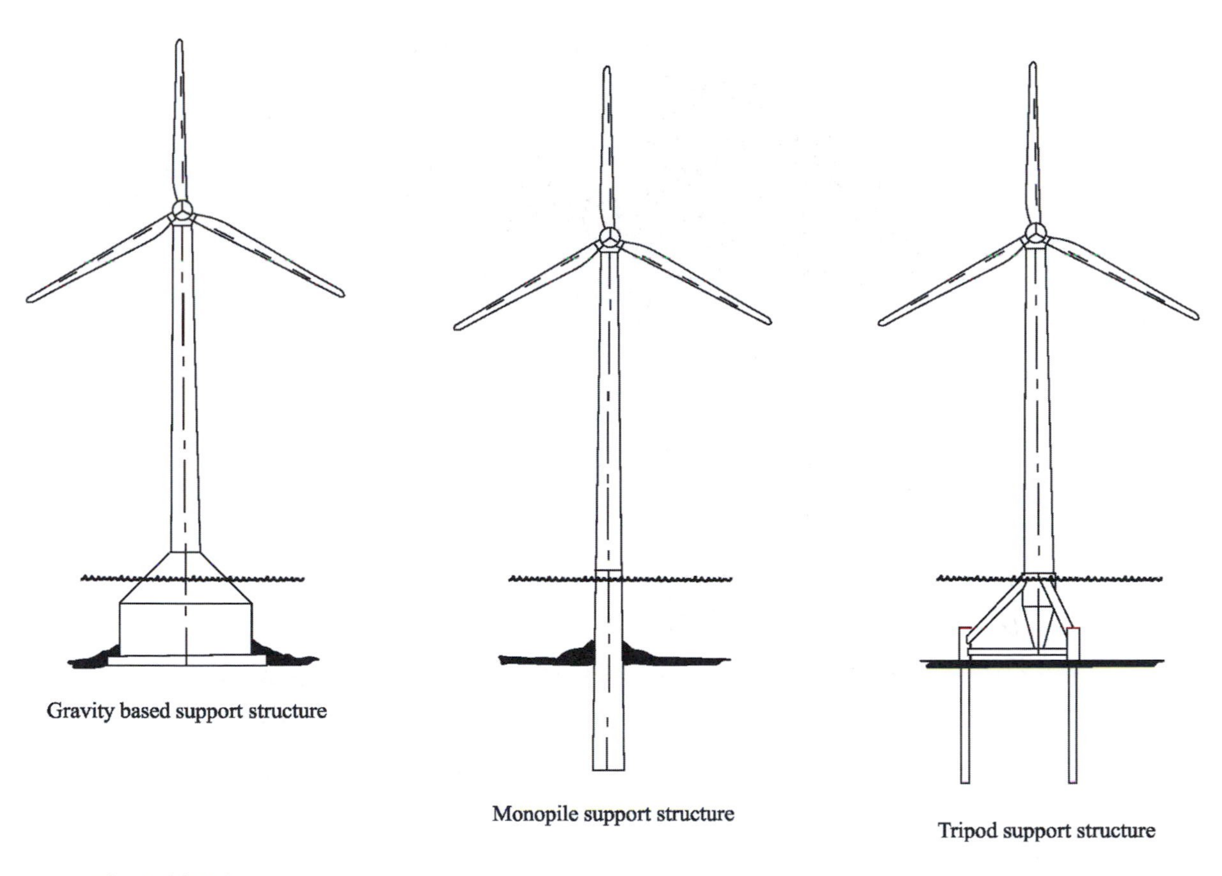

Source: Söker et al (2000)

Figure 3.27 *Support structures that can be used for offshore wind turbines*

3.10.2 Foundations

Söker et al (2000) provide a thorough discussion of the engineering and economics of offshore wind turbines. A variety of support structures is possible, as illustrated in Figure 3.27. The simplest is a gravity-based support structure, in which the weight of a concrete caisson is used to keep the complete structure in an upright position while being exposed to overturning forces from wind and waves on the turbine and support structure itself. The heave loads from waves depend on wave heights, which in turn depend in part on the water depth. The weight of the foundation has to be greatly increased with increasing water depth, so economic considerations limit this type of foundation to water depths of 10m or less. The gravity-based foundation also requires extensive seabed preparation, which is more costly at greater depth.

A second choice is the monopile foundation. The 'pile' is a steel tube that is driven into the seabed to a depth of 18–25m using a vibrating or piling hammer. The maximum water depth for a monopile support system is about 25m. A tripod structure can be used with greater water depths.

3.10.3 Floating turbines

A number of companies and consortia are working on the development of floating wind turbines, or of wind turbines that can be floated while towed to sea and then anchored to the seabed. The development of floating platforms draws heavily on the experience of the offshore oil and gas industry, which makes use of floating oil and gas rigs. In June 2009, StatoilHydro of Norway and Siemens installed a floating wind turbine approximately 12km southeast of Karmøy, Norway, in about 220m water (de Vries, 2009b). A 2.3MW wind turbine with an 82.4m rotor diameter and 65m hub height was attached to a ballasted steel float that extends 100m beneath the surface and is fastened to the seabed by three anchor wires.

Two concepts of particular interest are the *Floating-to-Fixed Wind Energy Concept* (F2F) and the self-orienting *WindSea* concept (de Vries, 2008). The F2F concept is under development by the Dutch firm Sea of Solutions BV. It consists of a long central steel column below and flanged to the turbine tower, forming a single vertical assembly. Three hollow vertical pipes are attached to the central structure and act as floatation devices. They also serve as legs once the turbine is in place; once the structure has been towed into place, valves are opened and the legs fill with water. Each leg would be equipped with a suction bucket to firmly anchor the structure to the seabed. Seabed preparation would not be necessary because the structure would be levelled by adjusting the vacuum in each individual leg. Other cost savings would arise from the use of conventional, non-dedicated tugboats for towing the structure into place, rather than requiring specialized barges (as are needed for turbines that are fixed to or placed on the seabed). The structure could be refloated and towed to a sheltered harbour for major overhauls.

The WindSea concept is under development by the Norwegian firm Force Technology, which has worked on the design and maintenance of offshore structures for three decades. The concept consists of a triangle-shaped floating platform that would have a 3.2MW turbine at each corner on an outward-inclined tower. Each turbine would be 100m from the other. Two turbines would be located on the upwind side of the platform and oriented perpendicular to the wind with a hub height of 60–80m, while the third turbine would be located downwind with a hub height of about 100m so as to minimize the wake from the two upwind turbines. The rear rotor would be located on an aerofoil-shaped tower that would act like the tail of an aeroplane, serving to continuously position all three rotors perpendicular to the wind. In this way, individual yaw mechanisms would not be needed for each turbine, thereby providing some cost savings.

Another consortium with previous experience in the offshore oil and gas industry, the Dutch Blue H Group, installed a prototype floating wind turbine in the summer of 2008 off the coast of Tricase, southern Italy. A 90MW wind farm is planned for the site, with the first operational unit under construction as of mid-2009 (see www.bluehgroup.com).

3.10.4 Grid connection

Currently available wind turbines generate electricity at 690 volts. The electricity is transformed to higher voltages (typically 20–36kV) for the connection between turbines in a wind farm and for transmission to shore, as the transformer energy loss is more than compensated by the reduced transmission loss at higher voltage. At the Horns Rev wind farm, output from individual wind turbines is relayed to an offshore transformer platform at 36kV and raised to 150kV for transmission 15km to shore. Cables are buried 1–2m below the sea floor to provide at least partial protection from ships' anchors (which can apparently dig themselves up to 13m into the seabed) and from strong currents. At the moment, no redundancy is built into the connection to the shore.

All currently operating offshore wind farms use an AC connection, although DC connections might be used in later offshore wind farms. Two DC options are available: high-voltage direct current (HVDC) and voltage source converter-based (VSC) HVDC technology (Ackermann, 2002). Either HVDC option is likely to be economically attractive, if at all, only for distances of 30–50km or for very large (⩾300MW) wind farms. If a large number of large wind farms is deployed offshore, than a submarine HVDC backbone grid would be an attractive option, as discussed in section 3.12.

3.10.5 Existing and potential offshore wind farms and their cost

Figure 3.28 shows the location of existing and planned offshore wind farms in Europe as of the end of 2009, while Table 3.7 gives cost and other information for each of the offshore wind energy projects shown in Figure 3.28. The total of existing and planned capacity is almost 45 GW. More detailed information for selected projects is given in Table 3.8. The projects in the 1990s were largely demonstration projects, with costs around €2000/kw. Costs have been as low as €1500/kW, but costs for many of the large planned projects are about €3500/kW ($4900/kW). This is more than twice the typical cost of onshore projects, but is largely compensated by twice the electricity production or more per unit of capacity. Many offshore wind projects have been proposed for Canada and the US but have yet to be approved. The Great Lakes represent a large potential offshore wind resource, particularly western Lake Erie, which is no deeper than 25m.

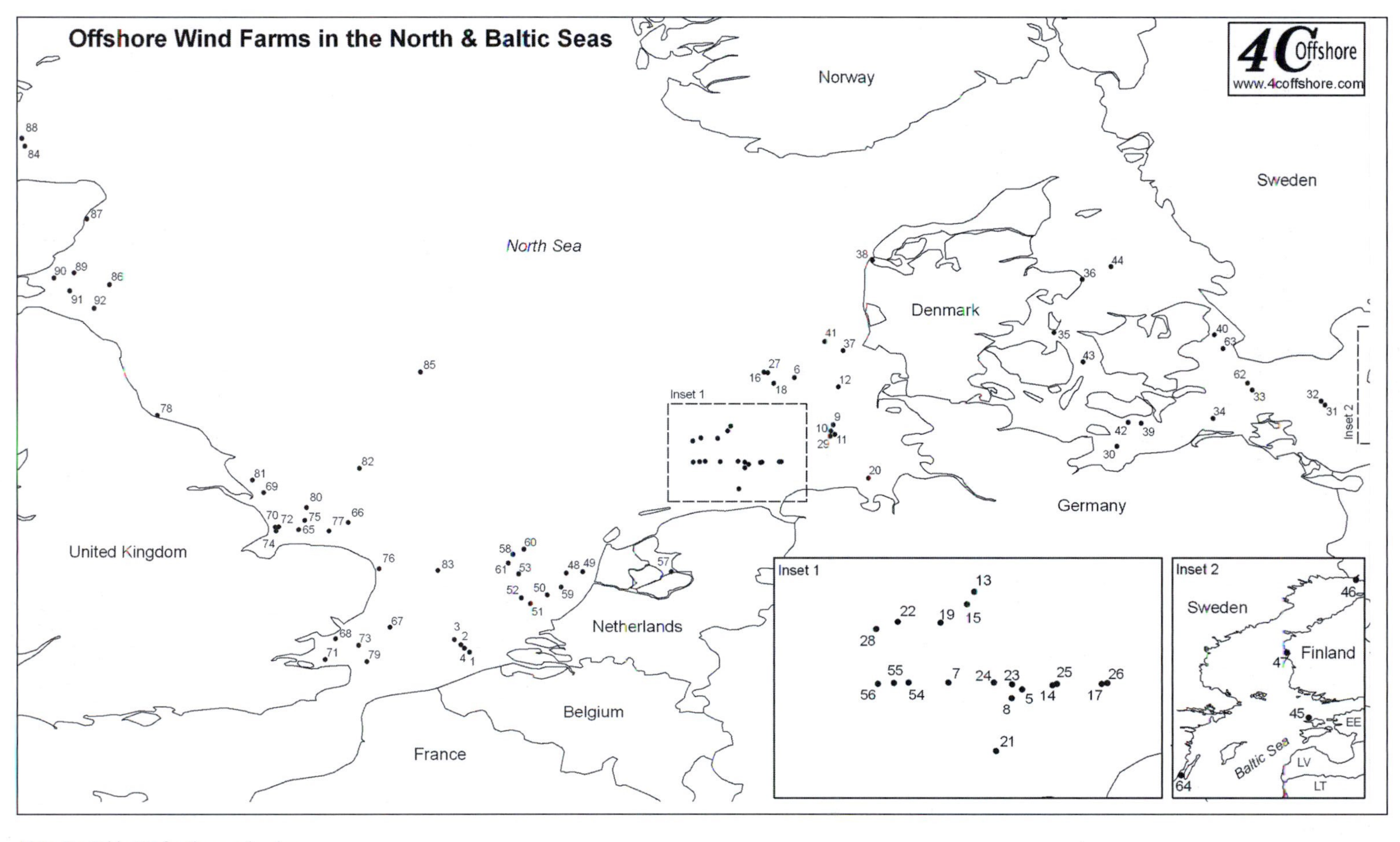

Note: See Table 3.7 for the number key.

Source: Richard Harrod, 4COffshore Ltd, UK

Figure 3.28 *Offshore wind farms developed in or planned for the North Sea and Baltic Sea by 2009.*

Table 3.7 *Key to the wind farm numbers shown in Figure 3.26, and characteristics of the wind farms*

Wind farm no	Wind farm name	Country	Capacity (MW)	Turbine manufacturer & model	Year online	Estimated cost	
						(million€)	(1000€/kW)
1	Thornton Bank	BE	300	REpower 5M	2009	672	2.24
2	Eldepasco	BE	216		2012	749	3.47
3	Belwind	BE	330	Vestas V90-3.0 MW	2010	1145	3.47
4	RENTEL	BE	288		2015	999	3.47
5	Alpha Ventus	DE	60	Multibrid M5000 & REpower 5M	2009	134	2.23
6	DanTysk	DE	400		2013	1388	3.47
7	Borkum Riffgrund West Phase 1	DE	280		2015	972	3.47
8	Borkum Riffgrund	DE	231		2015	802	3.47
9	Amrumbank West	DE	400		2011	1388	3.47
10	Nordsee Ost	DE	400	Siemens SWT-3.6	2012	1388	3.47
11	Meerwind Ost	DE	200		2013	694	3.47
12	Offshore-Bürgerpark Butendiek	DE	288	Siemens SWT-3.6-107	2012	999	3.47
13	Global Tech I	DE	400	Multibrid M5000	2013	1388	3.47
14	OWP Delta Nordsee 1	DE	240			833	3.47
15	Hochsee Windpark 'Nordsee'	DE	360		2011	1249	3.47
16	Sandbank 24	DE	480		2012	1666	3.47
17	Gode Wind	DE	400	Vestas unknown	2010	1388	3.47
18	Nördlicher Grund	DE	360		2012	1249	3.47
19	Hochsee Windpark He dreiht	DE	400		2015	1388	3.47
20	OWP Nordergründe	DE	90	REpower 5M	2010	312	3.47
21	OWP Riffgat	DE	220			763	3.47
22	BARD Offshore 1	DE	400	BARD 5.0	2010	1388	3.47
23	MEG Offshore I	DE	400	Multibrid M5000		1388	3.47
24	Borkum West II	DE	400	Multibrid M5000	2012	1388	3.47
25	OWP Delta Nordsee 2	DE	160			555	3.47
26	Gode Wind II	DE	400	Vestas unknown	2012	1388	3.47
27	Sandbank 24 extension	DE	200			694	3.47
28	Veja Mate	DE	400	BARD 5.0	2012	1388	3.47
29	Meerwind Süd	DE	200		2013	694	3.47
30	GEOFReE (GErman Offshore Field for Renewable Energy)	DE	25			86	3.44
31	Arkona Becken Sudost	DE	400	REpower 5M		1388	3.47
32	Ventotec Ost 2	DE	400	Vestas unknown	2010	1388	3.47
33	Kriegers Flak	DE	288	SWT 3.6-107	2013	999	3.47
34	Baltic I	DE	48.3	Siemens SWT-2.3-93	2010	167	3.46
35	Samso	DK	23	Siemens SWT-2.3-82	2003	34	1.48
36	Grenaa Havn	DK	18		2015	62	3.44
37	Horns Rev	DK	160	Vestas V80-2.0 MW	2002	241	1.51
38	Roenland	DK	17	Siemens SWT-2.3-93	2002	25	1.47
39	Nysted	DK	166	Siemens SWT-2.3-82	2003	250	1.51
40	Middelgrunden	DK	40	Bonus 2 MW	2000	60	1.50
41	Horns Rev 2	DK	209	Siemens SWT-2.3-93	2009	468	2.24
42	Rodsand II	DK	207	Siemens SWT-2.3-93	2010	463	2.24
43	Sprogo	DK	21	Vestas V90-3.0 MW	2009	72	3.43
44	Djursland Anholt	DK	400		2012	1388	3.47
45	Hiiumaa	EE	1000		2018	3472	3.47
46	Kemi Ajos	FI	30	WinWinD 3MW	2008	45	1.50
47	Kristiina	FI	400		2011	1388	3.47
48	Princess Amalia Wind Farm	NL	120	Vestas V80-2.0 MW	2008	268	2.23
49	Offshore Windpark Egmond aan Zee	NL	108	Vestas V90-3.0 MW	2006	163	1.51

Table 3.7 *Key to the wind farm numbers shown in Fig. 3.26, and characteristics of the wind farms* (Cont'd)

Wind farm no	Wind farm name	Country	Capacity (MW)	Turbine manufacturer & model	Year online	Estimated cost	
						(million€)	(1000€/kW)
50	Beaufort (formerly Katwijk)	NL	279			968	3.47
51	Scheveningen Buiten	NL	212.4			737	3.47
52	West-Rijn	NL	284	GE 3.6 MW	2012	986	3.47
53	Breevertien II	NL	349	GE 3.6 MW	2013	1211	3.47
54	BARD Offshore NL1	NL	300	BARD 5.0		1041	3.47
55	EP Offshore NL1	NL	275	BARD 5.0		954	3.47
56	GWS Offshore NL1	NL	300	BARD 5.0		1041	3.47
57	Irene Vorrink	NL	17	Nordtank NTK600/43	1996	32	1.88
58	Tromp Binnen	NL	295	REpower 5M	2015	1024	3.47
59	Q10	NL	165		2015	572	3.47
60	Den Helder I	NL	468	REpower 6M		1624	3.47
61	Brown Ridge Oost	NL	282	Vestas V90-3.0 MW		979	3.47
62	Kriegers Flak II	SE	640		2015	2222	3.47
63	Lillgrund	SE	110	Siemens SWT-2.3-93	2007	166	1.51
64	Utgrunden II	SE	86			298	3.47
65	Docking Shoal	GB	500			1736	3.47
66	Dudgeon	GB	560		2013	1944	3.47
67	Greater Gabbard	GB	504	Siemens SWT-3.6-107	2010	1749	3.47
68	Gunfleet Sands	GB	173	Siemens SWT-3.6-107	2009	387	2.24
69	Humber Gateway	GB	298.8			1037	3.47
70	Inner Dowsing	GB	97	Siemens SWT-3.6-107	2009	146	1.51
71	Kentish Flats	GB	90	Vestas V90-3.0 MW	2005	136	1.51
72	Lincs	GB	270	Siemens SWT-3.6-107	2012	937	3.47
73	London Array	GB	630	Siemens SWT-3.6-120	2012	2187	3.47
74	Lynn	GB	97	Siemens SWT-3.6-107	2009	146	1.51
75	Race Bank	GB	620			2152	3.47
76	Scroby Sands	GB	60	Vestas V80-2.0 MW	2004	90	1.50
77	Sheringham Shoal	GB	316.8	Siemens SWT-3.6-107	2011	1099	3.47
78	Teesside	GB	90		2010	312	3.47
79	Thanet	GB	300	Vestas V90-3.0 MW	2010	672	2.24
80	Triton Knoll	GB	1200		2017	4166	3.47
81	Westernmost Rough	GB	240		2012	833	3.47
82	Hornsea	GB	3000		2018	10416	3.47
83	Norfolk	GB	5000		2018	17360	3.47
84	Moray Firth	GB	500		2018	1736	3.47
85	Dogger Bank	GB	9000		2018	31248	3.47
86	Firth of Forth	GB	500		2018	1736	3.47
87	Aberdeen	GB	115			399	3.47
88	Beatrice	GB	920			3194	3.47
89	Inch Cape	GB	905			3142	3.47
90	Bell Rock	GB	700			2430	3.47
91	Neart na Gaoithe	GB	450			1562	3.47
92	Forth Array	GB	415			1440	3.47

Source: Richard Harrod, 4COffshore Ltd, UK

Table 3.9 gives the technical potential to generate wind from offshore locations in Europe, subdivided according to distance from shore and water depth. The technical potential from all locations within 40km of the coast and water not exceeding 40m in depth is estimated to be about 3000TWh/yr, which is the same as the total electricity demand of about 3000TWh in 2005 for all of Western and Central Europe.

Kempton et al (2007) estimate the potential power generation in waters off the states of Massachusetts through to North Carolina to be an average power output of 50GW using 46 per cent of the area with water depths

Table 3.8 *Characteristics of selected completed and planned offshore wind farms in Europe*

Name	Country	Year built	Number × size (MW)	Total capacity (MW)	Make	Water depth (m)	Minimum distance to shore (km)	Capacity factor	Cost from original source	Cost in 2008$/kW
Nogersund	SE	1990	1 × 0.22	0.22	Wind World	6	0.25			
Vindeby	DK	1991	5 × 0.45	5	Bonus	2–5	1.5	0.26	€2100/kW	2240
Tunø Knob	DK	1995	10 × 0.5	5	Vestas	3–4.7	6	0.29	€2100/kW	2240
Middelgrunden	DK	2001	20 × 2	40	Bonus	3–6	2	0.25	€1300/kW	1325
Horns Rev	DK	2002	80 × 2	160	Vestas	6–12	14		€1900/kW	3125
North Hoyle	UK	2003	30 × 2	60	Vestas	10–20	6	0.46	£1200/kW	2470
Kentish Flats	UK	2005	30 × 3	90	Vestas	5	5	0.36	£1167/kW	2410
Beatrice	UK	2007	2 × 5	10	REPower	44	5.5		£4100/kW[a]	7000
Borkum West	DE	2007	12 × 5	60	REPower	30	43			
Arklow Bank	IR	2007	200 × 3.6	500	GE Wind	2–5	10		€1260/kW	
Solway Firth	UK	2008	60 × 3	180		3–21	8		£1610/kW	
Q7-WP	NL	2008	60 × 2	120	Vestas	20–24	23	0.41	€3190/kW	4920
Robin Rigg	UK	2009	60 × 3	180	Vestas	3–21	9		£2100/kW	4030
Borkum-2	DE	2009		400			100[b]			
Greater Gabbard	UK	2009–2010	140 × 3.6	504	Siemens	24–34	25		£2580/kW + connection cost	
Rødsand II	DK	2010	90 × 2.3	207	Siemens					
Nordsee I	NL	2011–2015	150–180 5 & 6MW	1000	REPower	26–34	40			
Butendiek	DE	2006	80 × 3	240	Vestas	20	34	0.38		
London Array	UK	planned		1000						

Note: [a] Cost includes research. [b] To be connected by ABB using its HVDC Light technology. This involves relatively small AC–DC and DC–AC converters (compared to the truly massive ones used in full HVDC) and can be applied where the DC connection is over a relatively short distance. There are even cases where the DC link is trivially short, and HVDC Light is applied specifically to isolate two AC control regions from each other, allowing the two regions to transfer power between themselves without needing to keep their AC sine waves synchronized (both in frequency and in phase) (Jim Prall, University of Toronto, personal communication, 2009).

Source: CA-OWEE (2001), Wood (2003), Zaaijer and Henderson (2003), IEA (2006b), Aubrey (2008), Krohn et al (2009), Breton and Moe (2009), Snyder and Kaiser (2009) and www.offshorewindenergy.org. Costs in the last column are from Snyder and Kaiser (2009) and are based on exchange rates at the time of construction

Table 3.9 *Technical potential to generate electricity (TWh/yr) with offshore wind in Europe according to distance from shore and water depth*

	Distance from shore		
Water depth	**Up to 10km**	**Up to 20km**	**Up to 30km**
Up to 10m	551	587	596
Up to 20m	1121	1402	1523
Up to 30m	1597	2192	2463
Up to 40m	1852	2615	3028

Source: BTM Consult (2001)

of 0–20m and a further 117GW using 40 per cent of the area with water depths of 20–50m. This gives an average power output of 167GW, compared to average power demands in the region of 73GW as electricity, 29GW for gasoline and 83GW for building fuels.

3.11 Mitigating the adverse effect of fluctuations in available wind energy

The fluctuating and unpredictable nature of wind energy might be thought of as an impediment to its widespread use. Such fluctuations might require substantial reserves of rapid-response fossil fuel capacity, so-called 'spinning' reserves that consist of power units running at less than full capacity. However, a small proportion of fossil powerplant units operate as spinning reserve in any case, so the critical question is whether the addition of wind energy requires an increase in the spinning reserve. In general, the addition of wind energy will increase the variability of demand minus wind power beyond that from fluctuating demand alone, but the increase is less than the variability of the wind alone at all timescales of interest. The magnitude of the increase depends on how large the wind energy component is and the extent to which the variability of wind energy output can be dampened or short-term storage capacity used.

The problem with an increase in the variability of net electricity demand is that, if it is met through additional spinning reserve or through greater reliance on units that can start up and shut down rapidly, it will result in reduced average efficiency of the fossil fuel portion of the electricity supply. With the exception of fuel cells (which are not yet economically competitive), all fossil fuel power units operate at lower efficiency at part load than at full load (see Volume 1, Figure 3.11), so increasing the spinning reserve by running more units at part load will reduce average efficiency. If greater reliance is placed on simple natural gas turbines, this will also be at the cost of reduced efficiency as gas turbines have efficiencies of 30–35 per cent rather than 40–45 per cent (for state-of-the-art steam turbines) or 60 per cent (for state-of-the-art natural gas combined cycle powerplants). Both combined cycle and pure steam turbine systems require too long a time (one to three hours) to reach full load, and both are adversely affected by stops and restarts (as explained by Starr, 2007). Natural gas reciprocating engines would seem to be the best choice for fossil fuel backup, as they have relatively high full-load efficiency (up to 45 per cent), little loss in efficiency at part load and rapid start-up capability. In a worst-case scenario, only half of the emissions savings that would otherwise be expected can occur with high rates of wind penetration in a predominantly coal-based power system (Rosen et al, 2007).

In the remainder of this section, techniques for reducing or coping with the variability in the power output of wind turbines without recourse to additional fast-response fossil fuel units are outlined.

3.11.1 Rapid fluctuations (seconds to minutes)

There are several ways to reduce rapid fluctuations in wind energy output, as outlined below. Issues related to the integration of variable wind output into wholesale competitive markets are discussed by Hirst (2002).

Use asynchronous or variable-speed generators rather than synchronous generators

With synchronous generators, brief gusts of wind lead to a slight (1–3 per cent) increase in the rotor speed rather than an increase in power output (see subsection 3.4.1). Asynchronous generators are already widely used in the wind industry for this reason. With variable-speed generators, much larger (factor of two) variations in rotor speed are possible.

Build a wind farm with many wind turbines

The relative variation in total output from a large number of wind turbines is less than the variation in output from any one turbine. This is due to partial cancellation of the effects of turbulent fluctuations at individual turbines. Turbulent eddies have a spatial scale of a few tens of metres and last several seconds in duration. Given that the turbine spacing is larger than the spatial scale of the turbulent eddies, wind speed fluctuations of only a few seconds duration at individual turbines will be largely independent of one another, and so will cancel out to some extent. The larger the contribution that wind makes to an electric grid, the larger the number of wind turbines and the larger the area over which they will be spread (of necessity), so the greater this cancellation effect. For example, Archer and Jacobson (2003) find that, for one meteorological station in Kansas (a state with good wind), the frequency of 80m wind speeds <3m/s varied from 3.9 per cent at 0800–1100 hours, to 7.6 per cent at 1200–1500 hours. When three stations were considered, the frequencies of area-average winds <3m/s dropped to 0.4 per cent and 2.6 per cent, respectively, while for eight stations, there were no area-average winds <3m/s.

Store excess energy in short-term storage

Wind energy could be stored for short periods of time (minutes) and later released using flywheel systems, super-capacitors and superconducting magnetic storage. The first two of these options are also under consideration for fuel cell-powered automobiles. Flywheel systems are used today in the London Underground to handle peak power demands. Energy losses occur due to friction and so are greater the longer that energy is stored; the instantaneous efficiency is 85 per cent, decreasing to 75 per cent after five hours and 45 per cent after one day (Ibrahim et al, 2008). Capacitors store electricity by storing an electric charge on electrodes separated by a non-conducting material, whereas batteries store electricity chemically. They can be quickly charged and discharged with only 5 per cent energy loss, but their energy density is only 15–20Wh/kg, compared to 20–40Wh/kg for lead acid batteries and up to 200Wh/kg for lithium-ion batteries (see Volume 1, Chapter 5, subsection 5.4.6). However, some people have speculated that developments in nanotechnology may eventually permit capacitors (referred to as *supercapacitors*) with energy densities of 30–60Wh/kg, such that they could play a role in grid stabilization in small-scale renewable energy systems or in V2G systems (see subsection 3.11.2).

Make use of dynamic demand

When a power supply is lost from the grid, the grid frequency decreases until new supply can be brought on or the load is reduced. Conversely, the grid frequency increases when new supply becomes available without a compensating increase in load. Devices such as refrigerators or air conditioners, which do not need to run continuously and which contain compressors, can be designed so as to automatically switch off and on when the grid frequency drops below or rises above certain thresholds. In this way, instant compensation for changes in wind output (or in the use of other power units) can occur. This is referred to as *dynamic demand* (Infield and Watson, 2007).

3.11.2 Fluctuations over periods of minutes to days and longer

Fluctuations in wind power output at timescales of minutes to days can be reduced by linking together geographically dispersed wind farms, or they can be levelled out through intermediate timescale storage systems. Storage systems can be characterized by the round-trip efficiency: the ratio of electricity generated from the stored energy to the amount of electricity required to put the energy into storage. Fluctuations on timescales of hours to days can be anticipated through wind forecasting and taken into account in planning the operation of fossil fuel components of the electricity system. Fluctuations on a timescale of minutes to up to a few hours can be handled through adjustable end loads (particularly irrigation, desalination and district heating and cooling systems with heat pumps). These options are briefly discussed below.

Building and linking together many wind farms over a broad region

Fluctuations in wind velocity of increasing duration are related to variations in wind flow with an increasingly larger spatial scale. The largest spatial scale involves synoptic-scale disturbances, spanning several 100 to 1000 kilometres and typically requiring several days to pass a given location. Thus, the output from two wind farms that are far enough apart that they are not encompassed by a single synoptic-scale system will be at least partly uncorrelated even at a timescale of a few days. In general, the larger the spatial scale of the wind power plant, the longer the temporal scale over which some cancellation in wind variability will occur. Thus, the variability in the relative power output of two wind farms will be less than that of either wind farm, and smaller the further apart that the wind farms are located. At very large distances in the north–south direction, even the seasonal variation of wind can be different.

Wan and Bucaneg (2002) compared wind speed variability with power output for two central US wind farms, one consisting of 138 turbines in Minnesota, the other consisting of 262 turbines in Iowa. The variability in wind farm output is only about half the variability of wind speed cubed. When the two wind farms are combined, the variability in total power output is reduced by a further 5–20 per cent.[2] About 95 per cent of the minute-by-minute changes in wind farm output are within ±2.8 per cent of the total capacity. When four wind farms are spread across the UK, the variability in total hourly power output decreases by 36 per cent compared to locating all the wind turbines in one region, although average power output decreases by 9 per cent because not all turbines are in the best region (Drake and Hubacek, 2007).

Czisch and Giebel (2000) have assessed the variability of wind farms in high-wind areas within and around Europe. The regions considered and the potential wind capacity and energy production were northern Russia (350GW, 1100TWh/yr), Kazakhstan next to the Caspian Sea (210GW, 550TWh/yr), southern Morocco (120GW, 400TWh/yr), Mauritania (105GW, 320TWh/yr) and prime sites within the EU (primarily Ireland, UK and Norway) (150GW and 400TWh/yr). These sites add up to 960GW and 2800TWh/yr electricity production, compared to a total demand in 2005 for Europe (including Russia) of about 4800TWh (see Volume 1, Table 2.4). Table 3.10 compares the standard deviation at different averaging timescales and the frequencies of extremes for wind produced within Denmark plus Germany alone, within the EU-15 and spread over all of the sites listed above. The standard deviation drops dramatically as the wind powerplant area increases. Over Denmark plus Germany alone, there are times when the total wind output is zero, but over all regions, the lowest wind output is 4 per cent of peak. Over Denmark plus Germany, output less than 20 per cent of peak output occurs 46 per cent of the time, whereas over all regions, output less than 20 per cent of peak occurs only 10 per cent of the time. Figure 3.29 shows the variation in monthly wind electricity production for individual wind farm sites and for all sites combined. Significant cancellation in the seasonal variation in wind energy output occurs when all sites are combined.

To conclude this section, Table 3.11 shows the progressive decrease in maximum hourly change in wind power as larger areas are considered, based on a variety of studies in Europe. At the scale of several countries, the largest hourly variation is about 10 per cent.

Table 3.10 *Impact on the statistical properties of wind energy of spreading wind farms over increasingly larger areas in and around Europe, or over all the regions discussed in the text*

	Region		
	Denmark + Germany	**Europe**	**All regions**
Standard deviation divided by mean power production (%)			
Six-hour means	88	59	33
Weekly means	64	49	22
Monthly means	46	41	16
Extremes of wind power (%)			
Maximum	100	80	67
Minimum	0	3	4
Frequency of occurrence of extremes of wind power (%)			
Over 60 per cent	18	8	1
Under 20 per cent	46	37	10

Source: Czisch and Giebel (2000)

Making use of short-term wind forecasts

The impact on system operation of hourly to daily fluctuations in wind output can be mitigated through forecasts (one to several hours in advance) of wind velocity variations using computer models. This will allow better use of slowly responding power units (such as steam turbines). However, for predictions of up to six hours, assuming the wind conditions during the previous hour to persist gives a better prediction than computer model forecasts (Redlinger et al, 2002, Chapter 3). For computer models, the error in predicted hourly wind turbine output is fairly constant at about 15 per cent of actual output for predictions ranging from 1 to 36 hours. Forecast errors decrease as larger regions are considered, due to partial cancellation of local errors.

Hourly electricity demand must also be forecast in advance in order to permit use of slowly responding

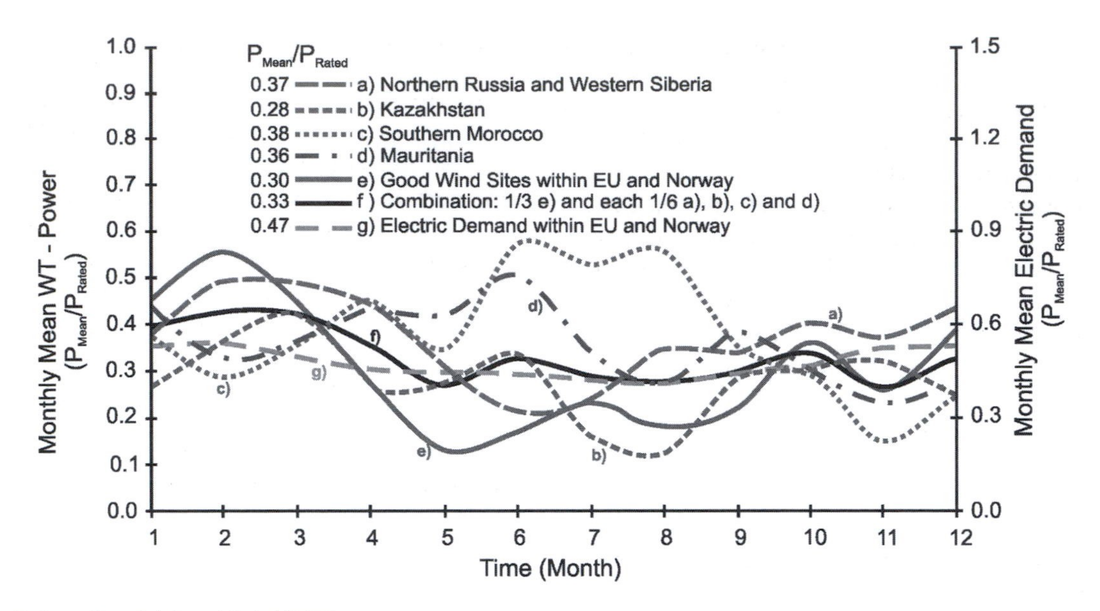

Source: Redrawn from Czisch and Giebel (2000)

Figure 3.29 *Variation in monthly mean electricity production from wind power within selected high-wind areas in and around Europe and for the combination of all regions (one sixth each from regions (a) to (d) and a third from region (e))*

Table 3.11 *Largest variation in wind power over different time periods and averaged over differently sized regions*

Areal dimensions	Largest variation up or down (%)	Location
Hourly variations		
100 × 100 km	50	UK
200 × 200 km	30	Denmark
400 × 400 km	20	Germany, Denmark, Finland
Group of countries	10	
4–12 hour variations		
One country	40–60	Denmark
	80	Germany
Larger area	35	Nordic area
400 × 400 km	4 hours: 80%	
	6 hours: 80%	UK
	12 hours: 90%	

Source: EWEA (2005)

units. Demand and wind forecast uncertainties are not added to give the uncertainty in the demand minus wind supply. Rather, this uncertainty is given by the square root of the sum of the squares of each uncertainty. Thus, if the uncertainty in demand is 20MW and the uncertainty in the supply is 20MW, the uncertainty in the net demand is 28MW (not 40MW).

Use of electrolysers and fuel cells

Excess wind-generated electricity can be used to generate hydrogen with an electrolyser. The hydrogen would be compressed, stored and used to generate electricity with a fuel cell when there is a wind deficit. The round-trip efficiency is given by the product of the AC/DC inverter, electrolyser, fuel cell and DC/AC inverter efficiencies. Inverter efficiencies are 90 per cent at present but could be increased to 95 per cent, while alkaline electrolysers have an efficiency of 63–73 per cent (see Table 10.3) and alkaline fuel cells have an efficiency of 50–70 per cent (see Table 10.7). The round-trip efficiency is thus only 30–50 per cent (assuming 95 per cent inverter efficiencies). A further problem is that the electrolyser can operate at no less than 20 per cent of its peak rate, and interruptions lower the efficiency, reduce the operating lifetime and reduce the purity of the hydrogen produced. A solution is to add a battery to smooth the power input to the electrolyser (Samaniego et al, 2008) but at the expense of a further 15–25 per cent energy loss. Partially offsetting the energy losses associated with using hydrogen to store wind energy is the fact that there will be less reduction in the efficiency of any fossil fuel powerplants in the system, due to smoother electricity output from the wind fuel cell system (Schenk et al, 2007).

Troncoso and Newborough (2007) carried out a detailed analysis of the use of electrolysers to absorb excess wind in a system also containing fossil fuel powerplants. The hydrogen in their system is used in other sectors, rather than to generate electricity at times of weak wind. Instead, the fossil powerplant provides this function, but a goal in their analysis is to avoid any curtailment of wind output at times of low demand and to minimize the installed fossil fuel capacity but with maximal and steady utilization of the installed fossil fuel powerplant. If an average CO_2 emission of 6.8kgC/GJ H_2 is permitted,[3] then the fossil powerplant can be used at times to power the electrolyser, resulting in a fossil powerplant capacity factor of 0.9 and steadier electrolyser operation. For very large wind penetrations, 1.1MW of electrolyser capacity is required per MW of wind power capacity. However, at the moment fuel cells are very expensive – about $3000–5000/kW.

Flow batteries (regenerative fuel cells)

Flow batteries, also known as regenerative fuel cells, could be used to store excess wind energy. Unlike other fuel cells, regenerative fuel cells do not use stored hydrogen to generate electricity, but rather, involve storage of two different electrolytes that react at electrodes. The electrodes are immersed in a single container separated by a membrane that allows passage of only certain ions, as illustrated in Figure 3.30. There are two major kinds of flow batteries: the polysulphide battery, developed by the British firm Regenesys, and the vanadium redox battery (VRB). The composition of the two electrolytes, the reactions during charging and discharging, costs and other characteristics are given in Table 3.12. For both flow batteries, deep discharging and rapid recharging and discharging are possible, and the power and energy ratings are independent of one another (power depends on cell area, energy storage depends on the volume of electrolyte). VRBs are more durable than conventional lead-acid batteries as there are no liquid–solid phase changes to damage the electrodes, and fluctuating loads and overcharging do not damage the battery. VRBs can switch from charging

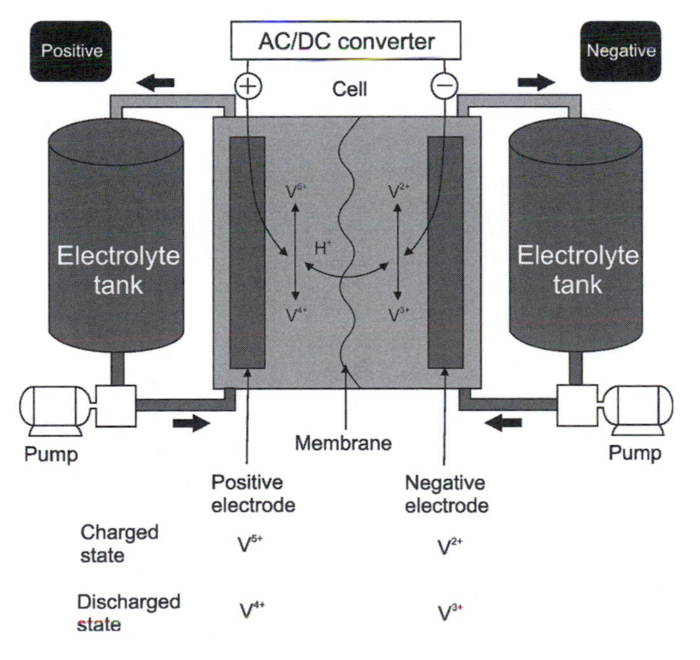

Source: Modified from Lotspeich and van Holde (2002)

Figure 3.30 *Schematic representation of a flow battery*

Table 3.12 *Characteristics of the Regenesys and vanadium redox flow batteries*

Characteristic	Regenesys	Vanadium redox
Electrolyte at positive electrode	Sodium bromide	V^{5+} / V^{4+}
Electrolyte at negative electrode	Sodium polysulphide	V^{2+} / V^{3+}
Charging reaction, positive electrode		$V^{4+} \rightarrow V^{5+} + e^-$
Charging reaction, negative electrode		$e^- + V^{3+} \rightarrow V^{2+}$
Discharging reactions	Reverse of above	Reverse of above
Membrane type		PEM
Ion crossing membrane	Na^+	H^+
Waste products	$NaSO_4$	None
Volume	0.32m^3/kWh	
Power capacity cost, current	$1500/kW	$1500–5500/kW
Power capacity cost, projected	$750/kW	$1000/kW
Energy storage capacity cost	$160–180/kWh	$300–1000/kWh
System cost	$1300–2500/kW[a]	$4400/kW[b]
Round-trip efficiency[c]	62%	71%

PEM = proton exchange membrane

Note: [a] For a 10–15MW system with 100–150MWh (i.e. ten hours at peak power) storage capacity.

[b] For a 2.5MW system with 10MWh storage capacity. [c] From Table 3.22.

Source: Lotspeich and van Holde (2002)

to discharging in a thousandth of a second and can sustain up to twice their rated power output for up to several minutes. The electrolytes have an indefinite lifetime, while the membrane is expected to need replacing every eight to ten years (but no flow batteries have yet been operated this long). A site capable of storing 100MWh of energy would occupy less than one hectare. The round-trip efficiency of the Regenesys flow battery is about 75 per cent according to Ibrahim et al (2008) but only 62 per cent according to calculations presented later (in Table 3.22). The VRB round-trip efficiency is likely to be around 70 per cent.

Both flow batteries are expensive, especially the VRB, but would be used to absorb only small fluctuations in wind turbine output and for short periods of time (a few hours). Thus, for a system costing $4000 per kW of charging or discharging power but designed to accept no more than 10 per cent of the peak turbine output, the cost would be $400 per kW of wind turbine capacity.

Underground compressed air energy storage

Another option is to store power that is temporarily in excess as underground compressed air in underground aquifers, hard rock caverns, solution-mined salt caverns, depleted gas reservoirs and abandoned mines (Cavallo, 1995). In a simple cycle gas turbine, half to two thirds of the power produced by the turbine is used to compress the air that is supplied to the combustor, with the balance used to drive an electric generator (see Volume 1, Chapter 3, subsection 3.2.3). In compressed air energy storage (CAES), the air fed to the gas turbine has already been compressed, so the output of the turbine can be largely directed to the electric generator. The turbine efficiency (based on natural gas input only) increases from about 37 per cent (in a modern turbine) to 84 per cent (DeCarolis and Keith, 2006). Natural gas (or some other fuel) is needed only to offset the adiabatic cooling that otherwise occurs as the gas expands. Part-load efficiency is also good, decreasing from 84 per cent at full load to 76 per cent at 20 per cent of full load (Succar and Williams, 2008).

The best geological formations for CAES are salt domes, salt beds and porous sedimentary rocks. These types of rocks underlie about 75 per cent of the US, including the regions with the best wind resources. Salt domes closely coincide with the best wind resource regions in Europe (see Succar and Williams, 2008, Figure 6). However, detailed site-specific geological assessments are required in order to determine the true resource base for CAES. Excavation of caverns in hard rock is possible but would be considerably more expensive. CAES would not significantly compete with sites for potential geological sequestration of CO_2, as the latter requires a minimum depth of 700m for CO_2 to become supercritical (see Chapter 9, subsection 9.6.1), which is near the upper limit of the acceptable depth for CAES.

CAES was first developed in the 1970s as a method to match daily variations in electricity demand, the idea being to use inflexible nuclear electricity when supply exceeds demand to compress air and to use the compressed air to generate additional electricity at times of peak demand. When nuclear energy did not expand as projected, and with falling oil prices in the early 1980s, interest in CAES declined, but two plants were completed, one in Germany in 1978 with a capacity of 290MW and 4 hours of storage, and the other in Alabama in 1991 with a capacity of 110MW and 26 hours of storage. Both plants store compressed air in salt caverns, the first operating between pressures of 48atm and 66atm and the second between pressures 45atm and 74atm. A third CAES facility is planned for Ohio, to be built in 300MW increments, beginning with an idle limestone mine and operating between pressures of 55atm and 100atm. Finally, a 280MW aquifer CAES project is planned for Iowa and plans for several projects in Texas have been announced, including a 540MW system utilizing a previously developed brine cavern (Succar and Williams, 2008).

Greenblatt et al (2007) indicate a storage volume of $0.14m^3/kWh_{output}$ for salt caverns with pressure varying between 45 and 72atm, or a layer 10m thick over an area of $12km^2$ for storage of 1GW-week of power in an aquifer with an effective porosity (that is, actual porosity times the pore fraction that can be filled with air) of 0.2. This is equal to a rock volume of $0.81m^3/kWh_{output}$. By comparison, the volume of electrolytes in the Regenesys flow battery is $0.032m^3/kWh_{output}$.

The round-trip efficiency computed for CAES should take into account the opportunity cost of the natural gas that is used to produce electricity from the stored compressional energy. That is, it should be based on the electricity that could have been produced by the natural gas if it were instead used in a dedicated powerplant. This depends on the efficiency η of dedicated electricity production using natural gas. Thus, the round-trip storage efficiency η_s is given by:

$$\eta_s = \frac{e_{out}}{e_{in} + F\eta} \quad (3.15)$$

where e_{out} is the electrical energy in 1kWh (3.6MJ), F is MJ of natural gas energy used per kWh of output electricity and e_{in} is the input electrical energy (in MJ) per kWh of electricity generated from storage. According to Denholm (2006), a modern CAES system uses 0.7kWh of electricity (for the initial compression of air and its injection into storage) and 4.5MJ of fuel to produce 1kWh of electricity. This gives a round-trip efficiency of 70 per cent (and an efficiency of 80 per cent based on fuel use alone) if η = 0.6. Greenblatt et al (2007) give slightly better numbers: inputs of 0.67kWh electricity and 4.22GJ natural gas, which yield a round-trip efficiency of 73 per cent.

Figure 3.31 gives an example of the energy flows for a wind/CAES powerplant supplying constant power, as computed by Denholm (2006). Some wind energy directly supplies electricity demand, some goes into storage (when supply exceeds demand) and some is 'spilled' (wasted) when supply exceeds demand plus the capability of air compressors to absorb excess wind energy. Fuel consumption per unit of electricity produced is a factor of six less than that of a state-of-the-art natural gas combined cycle powerplant at 60 per cent efficiency.

Denholm et al (2005) have assessed the reduction in CO_2 (and pollutant) emissions for a wind farm/CAES system in the US plains. Calculations were performed for wind farms with existing turbines (having a 33–37 per cent capacity factor at midwestern US sites) and for future turbines with a capacity factor of 46.3 per cent. As the system capacity factor increases

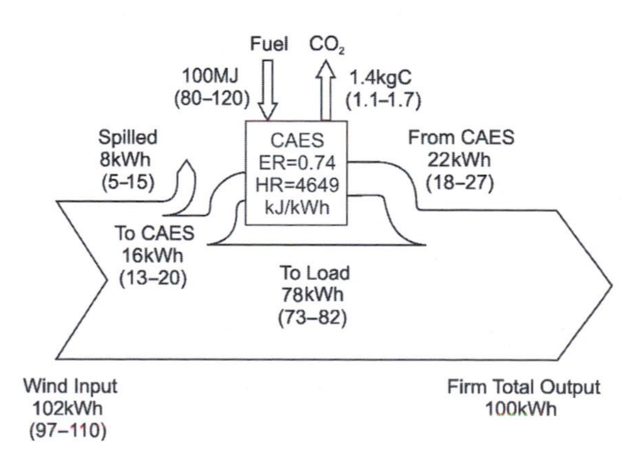

Source: Denholm (2006)

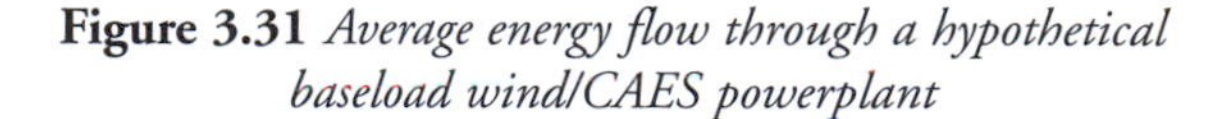

Figure 3.31 *Average energy flow through a hypothetical baseload wind/CAES powerplant*

by adding more storage and gas turbine capacity, the average CO_2 emission per kWh of generated electricity increases, due to the relatively greater use of the turbine. However, even at 90 per cent capacity factor, the average CO_2 emission is about a tenth that of a coal-fired powerplant.

Gasified biomass can be used instead of natural gas, and Table 3.13 presents the steps and inputs needed in order to compute that land area needed to provide the amount of biomass that would be used in conjunction with the generation of 1GWh/yr of electricity. This amounts to 4.7–11.1ha for the Midwestern US, with a best-guess value of 6.9ha/GWh/yr. As seen from Figure 3.31, 102 units of electricity produced by a fluctuating wind powerplant would, with storage, produce 100 units with steady output. Assuming 3.6MW turbines and an average wind speed of 10m/s at the 104m hub height, 3.4ha of farmland in the Midwest would be needed to produce 1GWh/yr of electricity, which is about half the cropland area needed to supply sufficient biomass to the wind/CAES system. Thus, some cropland outside the wind farm area would be needed to support the wind/CAES system. Denholm (2006) suggests that this cropland could come from a 1km wide swath on either side of transmission corridors, thereby reducing opposition to new transmission corridors (as the farmers would benefit now from the transmission corridor).

In summary, CAES is a promising technology for storing wind energy. It is the only technology with storage capabilities comparable to that of hydroelectric reservoirs, but at lower cost. As discussed later (subsection 3.14.2), wind CAES is close to being economically competitive with conventional natural gas electricity generation without carbon capture and storage, and is highly competitive with coal-based electricity generation with carbon capture and storage.

Advanced adiabatic compressed air energy storage

When air is compressed it warms because work is done on it, and when it expands it cools. In advanced adiabatic compressed air energy storage (AA-CAES), currently under development, heat released during the compression of the gas will be stored separately from the air at a temperature of about 650°C and used later when the compressed air expands, thereby potentially reducing the fuel use to zero (Bullough et al, 2004). The round-trip efficiency in this case is the ratio of electricity input to electricity output, which is about 70 per cent at present (Succar and Williams, 2008) but likely to improve over time.

Use of hydroelectric reservoirs

In regions where there is adequate hydroelectric capacity, the fluctuations in wind energy output (having been smoothed by one or more of the above mechanisms) could be matched by inverse variations in hydroelectric output. This would allow a larger fraction of total demand to be met by wind without requiring additional fossil fuel-powered spinning reserve. Hydroelectric reservoirs could be used for matching wind energy output on a seasonal basis as well. If seasons with high winds coincide with lower-than-annual-average reservoir water levels, then hydroelectric power production will be shifted toward times of higher-than-average water level, thereby increasing the annual hydroelectric power output

Table 3.13 *Derivation of cropland area requirement for a baseload wind/CAES plant using biomass as the fuel source*

Component	Base case	Estimated range
Average syngas fuel requirement (MJ/kWh), from Figure 3.31	1.0	0.8–1.2
Gasifier efficiency (%)	70	65–75
Switchgrass feedstock requirement (MJ/kWh)	1.43	1.07–1.85
Switchgrass yield (tonne/ha/yr)	11.3	9.1–12.5
Switchgrass yield (GJ/ha/yr assuming 18.3 GJ/tonne)	207	166–228
Switchgrass cropland yield (MWh/ha/yr)	145.2	89.9–214.0
Switchgrass cropland intensity (ha/GWh/yr)	6.9	4.7–11.1

Source: Denholm (2006)

because, as explained in Chapter 6, hydroelectric power for a given rate flow is larger when the water level is higher. That is, the addition of wind energy *increases* the efficiency of hydroelectric power generation.

In mountainous regions lacking reservoirs that can be filled with the available runoff, water can be pumped into relatively small but high reservoirs at times when there is excess wind energy, and allowed to flow out of the reservoir through hydroelectric turbines when there is a wind energy deficit. This system is referred to as *pumped hydro storage*. The round-trip efficiency ranges from 65 per cent to 80 per cent (Ibrahim et al, 2008).

Use of heat pumps and thermal energy storage in district energy systems

Excess electricity production potential can also be utilized if a district heating system with cogeneration of heat and electricity is available (see Chapter 11). When total wind power output exceeds total electricity demand, there is of course no need for electricity production through cogeneration. In this case, the heat requirement of the district heating system can be met by operating heat pumps, in effect increasing the electricity demand (and saving heating fuel) to utilize the available wind turbine output that would otherwise be in excess of demand. For a scenario in which 50 per cent of Danish electricity demand in 2030 is met by wind, integrated operation of the electricity and heating systems cuts the wasted electricity production potential in half (Redlinger et al, 2002, Chapter 2). Much of the heating in Denmark today is provided through cogeneration, with the amount of electricity so produced tied to the concurrent heat load rather than being freely dispatchable.

Lund and Münster (2006) have assessed the economics of using heat pumps in Denmark in combination with wind power, cogeneration and the export and import of electricity to and from the Nordic power grid. In a system where heat is provided by cogeneration, the use of heat pumps can absorb excess wind energy in two ways: by increasing electricity demand and by reducing the non-wind electricity supply by reducing the amount of heat that must be provided through cogeneration (which also reduces the amount of electricity produced through cogeneration). Taking into account the variation in wind energy supply, electricity demand and the price of electricity sold to or purchased from the Nordic grid, they find that the incorporation of wind energy and heat pumps is profitable up to the point where wind supplies 76 per cent of the total demand if there is a carbon tax of €48/tC (€13/tCO_2), but that the profit is maximized when wind supplies only 43 per cent of total demand. At a larger carbon tax, a greater amount of wind energy would be optimal.

Use of other flexible AC end-use loads

If wind energy is used in part to supply electric loads where only the cumulative energy supply rather that than the detailed timing matters, the wind farm can be sized larger than the peak of the remaining loads and the excess used when available. Example loads include irrigation pumps and saltwater desalinization using electric methods (see Chapter 2, subsection 2.8.1).

Direct coupling to DC loads

Zhao et al (2009) propose that wind farms in Jiangsu Province in China (on the east coast) be directly coupled via a 500kV DC line to industries that require DC power (and which otherwise have to convert AC power to DC power): the chlor-alkali industry, aluminium smelting, silicon production and large-scale electrolytic desalination and hydrogen production. As discussed later (section 3.12), transmission by DC normally requires AC–DC transformers at both ends of the transmission line, but beyond a distance of a few hundred kilometres, energy losses and costs are smaller for high-voltage DC than for high-voltage AC transmission. The economics and losses would be more favourable with this proposal because transformation from DC back to AC at the end of the line would not be needed, while the initial transformation from AC to DC would simply displace transformers that would otherwise be needed in the above-mentioned industries. Furthermore, by not having to connect wind turbines to an AC grid, turbine electronics would be reduced in cost and there would be greater flexibility in the choice of rotor speeds. Materials such as potassium (an important ingredient of fertilizer), bromine and magnesium would be separated from seawater as a byproduct of electrolytic desalination. According to Zhao et al (2009), fluctuations in voltage related to variations in wind power output would not pose a problem for electrolysis. Thus, for one significant electricity load, variability in the smoothed output produced by large wind farms would not be a problem.

Use of parked plug-in hybrid vehicles

If plug-in hybrid vehicles (described in Volume 1, Chapter 5, subsection 5.4.6) become a large part of the automobile fleet, they could be used as short-term storage for fluctuating wind power. Inasmuch as most of the vehicles would be parked most of the time, they could – if plugged in with appropriate electronic controls – absorb excess wind energy by using it to recharge the onboard battery, and deliver energy from the batteries (up to some limit) when there is a deficit of wind energy. This is referred to as *vehicle to grid* or V2G. The round-trip efficiency would be comparable to that of other storage technologies but would not require the construction of additional storage devices. Total societal cost would therefore be reduced. Kempton and Tomić (2005) calculate that, with wind supplying one half of total US electricity demand, only 3 per cent of the US automobile fleet would be needed for 'regulation' (keeping the frequency and voltage on the grid steady) while only 8–38 per cent would be required to provide 'spinning reserves' (providing storage and backup on a timescale of hours to a day).

3.11.3 Improving overall system stability with the addition of advanced turbines

The preceding discussion has treated wind turbines as a liability as far as system stability is concerned, due to the fluctuating and largely unpredictable nature of wind speed. However, modern wind turbines can actively contribute to the stabilization of the electrical grid.

As discussed in Volume 1 (Chapter 3, subsection 3.1.2), a phase shift between the voltage and current variation in an AC system gives rise to *reactive power*. Reactive power does not carry energy but subtracts from real power, so a goal in electric power systems is to have zero reactive power, that is, to have zero phase difference between voltage and current variation. Modern wind turbines can compensate for shifts in the system reactive power. They can also maintain the voltage at the point of connection to the grid to within a preset range. Some can also remain connected to the grid when faults (such as lightning strikes, equipment failure or third-party damage) cause large transient voltage drops over a large part of the network (EWEA, 2005). This so-called *fault ride through (FRT) capability* adds to the cost of the turbine, and so may be justified only where large numbers of turbines are to be connected to the grid. FRT capability is still under development.

Previously, grid operators required wind turbines to disconnect from the network in the event of a drop in frequency (which would occur in response to a loss of power by other units or in response to an increase in demand that has not yet been matched by an increase in power from other units). Now, wind turbines are required to stay connected and operate over a wide frequency band (from 47Hz to 53Hz in the case of Denmark, but over a slightly smaller range for other countries) (EWEA, 2005). A wind turbine can increase its output in response to a drop in system frequency. However, this requires limiting wind power output at other times, so there is a cost in terms of foregone electricity generation. This cost has to be compared with the cost of alternative ways of providing the same service. Control systems permit the wind powerplant voltage to change so as to offset changes in the system voltage at the point of interconnection. For more information on this topic, see Lindenberg et al (2008).

3.12 Long-distance transmission

Some of the best wind resources (the North American prairies, Labrador, Inner Mongolia) are 1000–2000km from major demand centres, so long-distance transmission will be required. Technical considerations in long-distance transmission of electricity will be discussed in this section, while costs are discussed in subsection 3.13.5.

Transmitted power is equal to voltage times current, but resistance energy losses vary with current squared. Thus, energy losses during transmission can be reduced by first transforming up to high voltage; doubling the voltages reduces the resistance losses by a factor of four. High-voltage transmission can occur as three-phase AC or as DC. Three-phase AC transmission requires three conductors, whereas DC transmission requires only two conductors (bipolar case) or one conductor (monopolar case), with the current flowing back via the earth. This leads to lower line costs and line losses. However, transformer costs and losses are larger, so it is not worthwhile to use DC transmission over short distances. Worldwide, there are over 75GW of HVDC transmission capacity in over 90 projects. Transmission voltages of up to 1200kV have been used in Russia to span long distances. Voltages greater than 800kV are referred to as ultra-high voltage.

Some disadvantages of DC transmission are that DC power is not directly transformable to another voltage, it is not easy to switch off the current, and branching of the power is difficult. This would make it difficult to have many entry points of wind power into the DC grid. Instead, the HVDC grid could be easily used to transport power from large and remote wind farms to major demand centres. However, recent advances in HVDC technology using insulated gate bipolar transistor (IGBT) converters offer the possibility of constructing HVDC lines with multiple terminals (EWEA, 2005).

The advantages of HVDC systems include reactive power control and full controllability of power flows. A problem with AC mesh networks is that the flow in any one segment cannot be controlled; as a result, there is sometimes congestion on one transmission line while

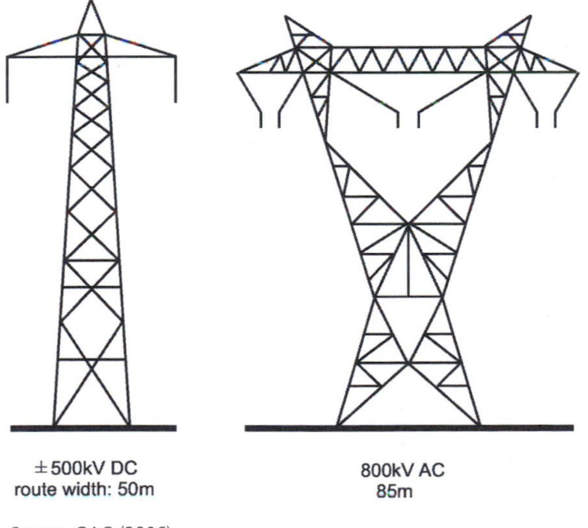

Source: GAC (2006)

Figure 3.32 *Comparison of typical high-voltage DC and AC pylons for overhead transmission of power*

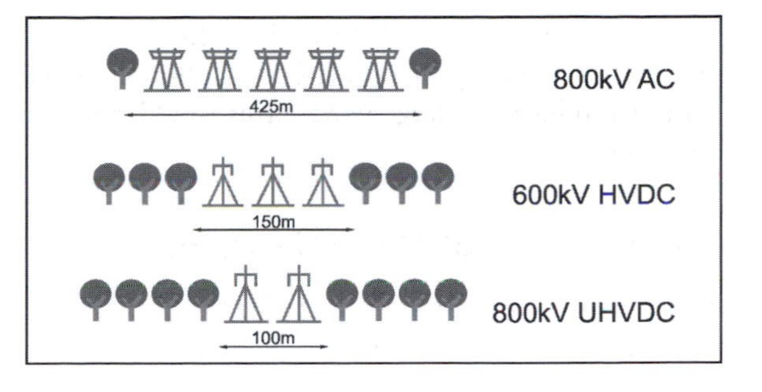

Source: GAC (2006)

Figure 3.33 *Comparison of the required number of parallel pylons and the right of way required to transmit 10GW of power using various AC and DC schemes*

there is spare capacity on alternative lines (van Hulle et al, 2009). With the construction of an offshore HVDC supergrid for Europe, the offshore wind farms could collectively operate at a variable speed and frequency, independently of the land grid, thereby permitting more efficient turbine and generator operation and offsetting at least some of the increased transmission loss due to greater transmission distance (EWEA, 2005). Finally, DC transmission generates negligible magnetic fields, whereas magnetic fields associated with AC transmission lines have been one source of public opposition.

It is interesting to compare the pylons and rights of way required for AC and DC transmission. These are shown in Figures 3.32 and 3.33, respectively. A 500kV DC pylon requires two wires, a route width of 50m and a pedestal area of 50m^2/km, whereas an 800kV AC pylon requires six wires, a route width of 85m and a pedestal area of 100m^2/km. The transmission of 10GW would require five 800kV lines of pylons (30 wires) using AC and a 425m right of way, or two 800kV lines of pylons (four wires) using DC and a 100m right of way. Thus, impacts on the landscape are substantially less with high or ultra-high voltage DC transmission than with high-voltage AC transmission. HVAC in turn is advantageous compared to lower voltage AC. According to AWEA/SEIA (2009), a single 765kV AC line (with a 65m-wide right of way) can carry as much power as three double-circuit 345kV lines (150m total right of way) or six single-circuit 345KV lines (300m right of way). Combining all this information, the transmission of 10GW of power requires:

- two 800kV DC lines and a 100m right of way, or
- five 800kV AC lines and a 325–425m right of way, or
- ~ 15 double-circuit 345kV lines and a 750m right of way, or
- ~ 30 single-circuit 345kV lines and a 1500m right of way.

The balance of advantages and disadvantages of HVAC and HVDC is such that HVDC tends to be preferred in Europe (as in the proposals of the German Aerospace Center (GAC, 2006)) while HVAC tends to be preferred in the US (as in the proposals of the National Renewable Energy Laboratory and American Electric Power (AEP, 2007)). In either case, construction of a new high-voltage backbone grid would provide the means of linking distant high-wind regions with the major demand centres using state-of-the-art technology, as well as relieving bottlenecks in the existing lower-voltage grid due to reduced use of existing powerplants, which would in turn allow greater flexibility in the operation and maintenance of the existing grid. According to a 2006 press release by the Swedish–Swiss power equipment manufacturer Asea Brown Boveri, India plans to build five 800kV, 6GW DC lines over the then next ten years, while China is planning to build about one 5–6.4GW line per year for ten years (ABB, 2006). Limited transmission capability is currently a serious impediment to further development of wind energy in both China and the US (AWEA/SEIA, 2009; Cyranoski, 2009).

It may not always be necessary to build new transmission corridors to handle large wind power inputs to the grid, as upgrading of old transmission corridors using modern technology could lead to a significant increase in transmission capacity. In particular, new conductors consist of a core of aluminium reinforced with high-strength aluminium oxide fibres surrounded by aluminium alloy wire, whereas conventional wires consist of a steel core surrounded by aluminium alloy wire. The new wires have the same strength as steel but four times the conductivity and significantly less weight, and can carry 1.5–3.0 times the current and hence power of conventional wires (Cavallo, 2007).

3.13 Economics

To assess the cost of wind energy, we must first consider the direct cost of electricity generated by a wind turbine, independent of other electricity sources in the electric grid, indirect costs arising from the impact of wind energy on the operation of other electricity sources, and the cost of long-distance transmission of electricity (where applicable).

3.13.1 Direct costs

The direct cost of electricity from wind turbines can be computed as the fixed annual revenue requirements (per kW of capacity) divided by the number of kWh of electricity generated per year (per kW of capacity), plus costs that vary in direct proportion to the amount of electricity generated. The fixed revenue requirement consists of the revenue required to pay back the initial investment and the fixed O&M costs. This is equal to:

$$(CRF_{WT} + OM_{WT\text{-}f}) \times CC_{WT} \tag{3.16}$$

where CC_{WT} is the capital cost ($/kW), $OM_{WT\text{-}f}$ is the annual O&M cost as a fraction of the capital cost and CRF_{WT} is a cost recovery factor that depends on the real interest rate (after adjusting for inflation) and the number of years over which the investment is financed (see Appendix C).

The number of kWh generated per year per kW is given by the number of hours in a year (8760) times the wind turbine capacity factor CF_{WT} (the average power output as a fraction of the turbine peak power) times the fraction of the time (f_a) that the turbine is operational (available) times an additional factor (η_s) to take into account various energy losses that reduce the power output below that expected based on the turbine power curve. The availability factor is typically 0.95–0.98 and allows for downtime for maintenance and repairs as well as unplanned shutdowns. Energy losses and their typical magnitudes are as follows (Krohn et al, 2009):

- rotor blade soiling losses, about 1–2 per cent;
- grid losses within the collection grid inside a wind farm, 1–3 per cent;
- losses due to imperfect tracking of the wind direction by the yaw mechanism, 1 per cent;
- array losses in wind farms due to wake effects of the turbines on each other, 5–10 per cent.

Array losses are larger the larger the wind farm and the more closely spaced the wind turbines. If the turbines are arranged in a straight line, the array losses would depend strongly on the wind direction. Conversely, if winds are consistently in one direction, the cross-wind spacing can be quite small (as little as 1.5D) without adversely affecting η_s (this would permit two to four times as much power generation per unit of land area).

Given a wind turbine capital cost CC_{WT} ($/kW), a cost recovery factor CRF_{WT}, fixed annual O&M costs $OM_{WT\text{-}f}$ (as a fraction of the initial capital cost) and variable O&M cost $OM_{WT\text{-}v}$, the cost of wind-generated electricity ($/kWh) is given by:

$$C_{e\text{-}WT} = \frac{(CRF_{WT} + OM_{WT-f})CC_{WT}}{\eta_s f_a 8760 CF_{WT}} + OM_{WT\text{-}v} \quad (3.17)$$

Table 3.14 gives data on recent capital, O&M and final electricity costs for onshore and offshore turbines. The cost structure of a 2MW onshore turbine costing €1227/kW (about $1700/kW, assuming an exchange rate of $1.4 per euro) is shown in Table 3.15. About 75 per cent of the cost is the turbine itself (including transport to the site), with the remaining 25 per cent taken up by the foundation, electric systems, grid connection and other miscellaneous expenses. Figure 3.34 compares total turbine costs in different countries in 2006, while Figure 3.35 shows the variation over time in the turbine capital cost and in the cost of electricity from new turbines in Denmark. The installed cost of onshore turbines ranged from €950/kW to €1350/kW in 2006 (~ $1300–1900/kW at 1.4$/€). Offshore turbines are about 40 per cent more expensive (€1300–1900/kW or $1800–2600/kW) than onshore turbines, due primarily to greater foundation and grid connection costs, but electricity costs are only slightly more expensive than for coastal onshore sites and tend to be less expensive than for low-wind onshore sites due to the greater capacity factor offshore. In spite of the recent upturn in the cost of wind turbines (which is related in part to an excess of demand over supply), Krohn et al (2009) expect the total costs of onshore turbines in Europe to fall to about €800/kW and that of offshore turbines to fall to about €1200/kW by 2030. Figure 3.36 gives illustrative costs of wind energy for capital costs ranging from $800 to 2000/kW (or €800 to 2000/kW) and interest rates (or return on investment, ROI) ranging from 3 per cent to 12 per cent, all for a 20-year lifespan, a capacity factor of 0.25 and a fixed annual O&M cost of 2 per cent of the capital cost. For a middle capital cost of $1200/kW, the cost of electricity

Table 3.15 *Cost structure for a typical 2MW onshore turbine installed in Europe*

Component	Cost (€/kW)	Share of total cost%	
		Here	Range
Turbine	928	76	68–84
Grid connection	109	8.9	2–10
Foundation	80	6.5	1–9
Land purchase	48	3.9	1–5
Electric installation	18	1.5	1–9
Consultancy	15	1.2	1–3
Financial costs	15	1.2	1–5
Road construction	11	0.9	1–5
Control systems	4	0.3	
Total	1228	100	

Source: Krohn et al (2009)

Table 3.14 *Summary of recent (2006) costs of onshore and offshore wind turbines and of the final cost of wind-generated electricity*

	Onshore	Offshore	Reference
Capital cost	€950–1350/kW ($1300–1900/kW)	€1300–1900/kW ($1800–2600/kW)	Krohn et al (2009)
Fixed O&M	$11.5/kW/yr (0.007 CC_{WT}/yr)	$15/kW/yr (0.00625 CC_{WT}/yr)	Lindenberg et al (2008)
Variable O&M	$0.007/kWh	$0.021/kWh	Lindenberg et al (2008)
Electricity cost in Europe	€0.09–0.11/kWh, low-wind areas €0.05–0.07/kWh, coastal areas	€0.06–0.09/kWh, projects built since 2000	Krohn et al (2009)

Note: CC_{WT} = capital cost of the wind turbine.

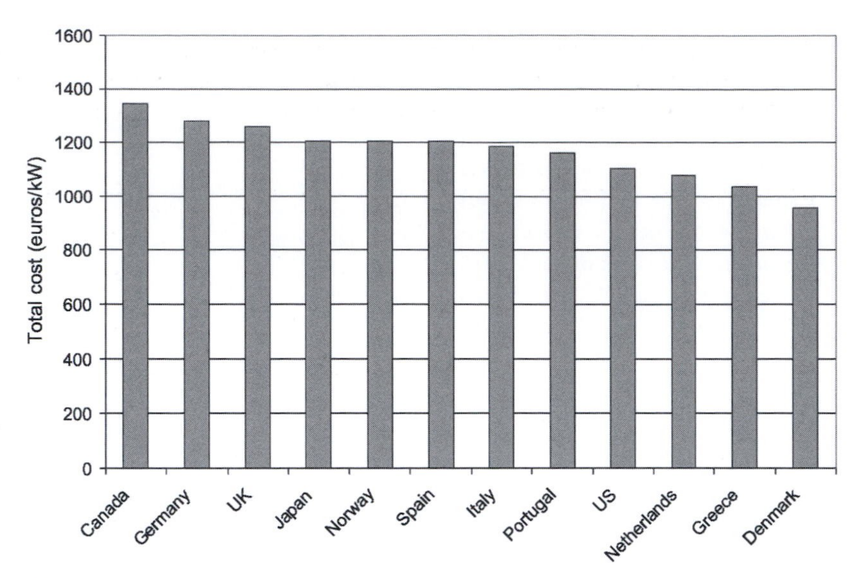

Source: Krohn et al (2009)

Figure 3.34 *Total wind turbine costs (including foundation and grid connection) in various countries in 2006*

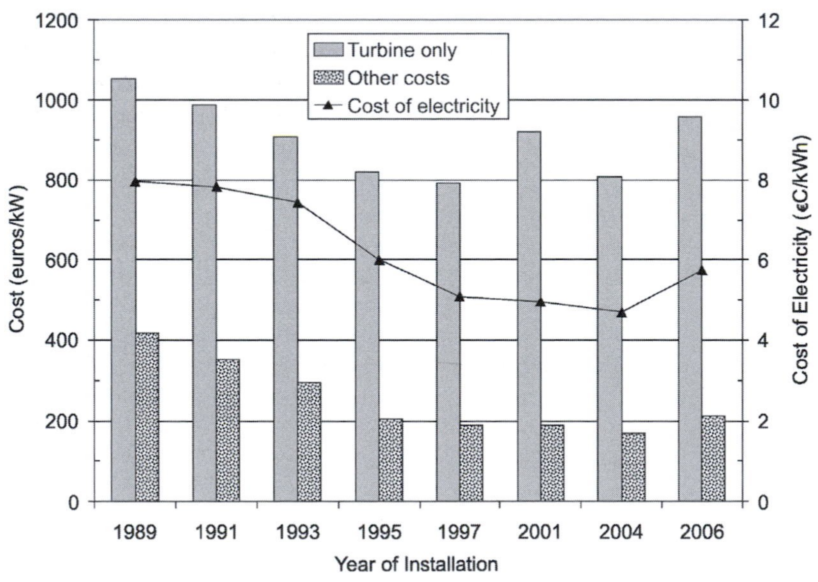

Source: Kohn et al (2009)

Figure 3.35 *Variation from 1989 to 2006 in wind turbine costs and electricity costs for wind energy in Denmark*

ranges from 4.8 cents/kWh (at 3 per cent interest) to 8.4 cents/kWh (at 12 per cent interest). These costs of course account for the relatively low power output from a turbine through the capacity factor, which occurs in the denominator of Equation (3.17).

O&M costs in reality will not be fixed over time. However, since the largest and most advanced turbines have not been used for more than a few years, there are no data on the variation of O&M costs over long periods of time. Data compiled in 2002 indicate that

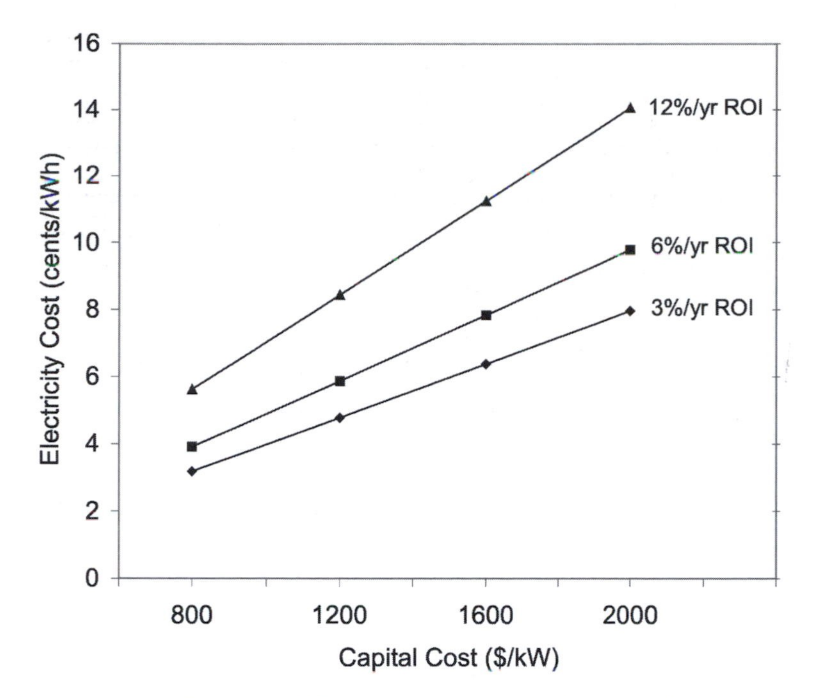

Note: In all cases, a 20-year operating life, 2 per cent/yr O&M and a capacity factor of 0.25 are assumed.

Figure 3.36 *Illustrative cost of wind energy (excluding transmission and administration) as a function of the capital cost, for different rates of return on investment*

O&M costs per kWh of electricity generated fall sharply with increasing turbine size and, for a given turbine size, increase slightly over time (Krohn et al, 2009). Thus, average O&M costs for three-year-old turbines decrease from about €0.035/kWh for 55kW units to €0.014/kWh for 150kW units and €0.007/kWh for 600kW units, while the cost for 55kW units increases from €0.035/kWh at three years to €0.045/kWh at ten years and that of 150kW units increases from €0.014/kWh at three years to €0.015/kWh at ten years (600kW units were introduced only in 1995, so ten-year data were not available). To account for O&M costs varying over time, a discounted cash flow analysis is required, as explained in Box 3.2 of Volume 1. For an installed cost of US$1000/kW, 10 per cent real interest rate, a 20-year lifetime, annual O&M costs increasing smoothly from 1.0 per cent to 4.0 per cent of the capital cost and a capacity factor of 0.3 (characteristic of sites with a mean wind speed of about 7.5m/s according to Figure 3.19), the cost of wind-generated electricity is 5.16 cents/kWh based on a discounted cash flow analysis and 5.42 cents/kWh if Equation (3.17) is used (incorrectly) with the average O&M cost.

3.13.2 Dependence of cost on wind turbine size and wind farm scale

With fixed blade proportions, the volume of material and therefore the rotor mass and cost should increase with the cube of the rotor diameter, whereas energy output varies with the rotor diameter squared. Thus, the cost per unit of energy output should increase as the rotor size increases. In reality, blade mass has increased with the rotor diameter to the power of 2.3 (Ashwell, 2004). This smaller increase in rotor mass has been accomplished through better design and through partial substitution of fibreglass with carbon fibre. As a result, the cost per kW is largely independent of turbine size, as shown by Junginger et al (2005, Figure 4). However, other costs (foundations, grid connection and project planning) may decrease with larger turbines. A more significant factor is the number of turbines ordered at once. As discussed by Junginger et al (2005), the purchase price has been reduced by up to 45 per cent for orders of 500–1600 turbines. This cost reduction is possible because the production plant can operate continuously and because the manufacturer can bargain for lower prices from suppliers.

Figure 3.37 shows the total unit cost (including grid connection) of completed offshore wind farms in Europe as a function of the size of the turbines used and of the total wind farm capacity. There is no particular relationship between unit cost and either of these size parameters. Rather, depth and distance to

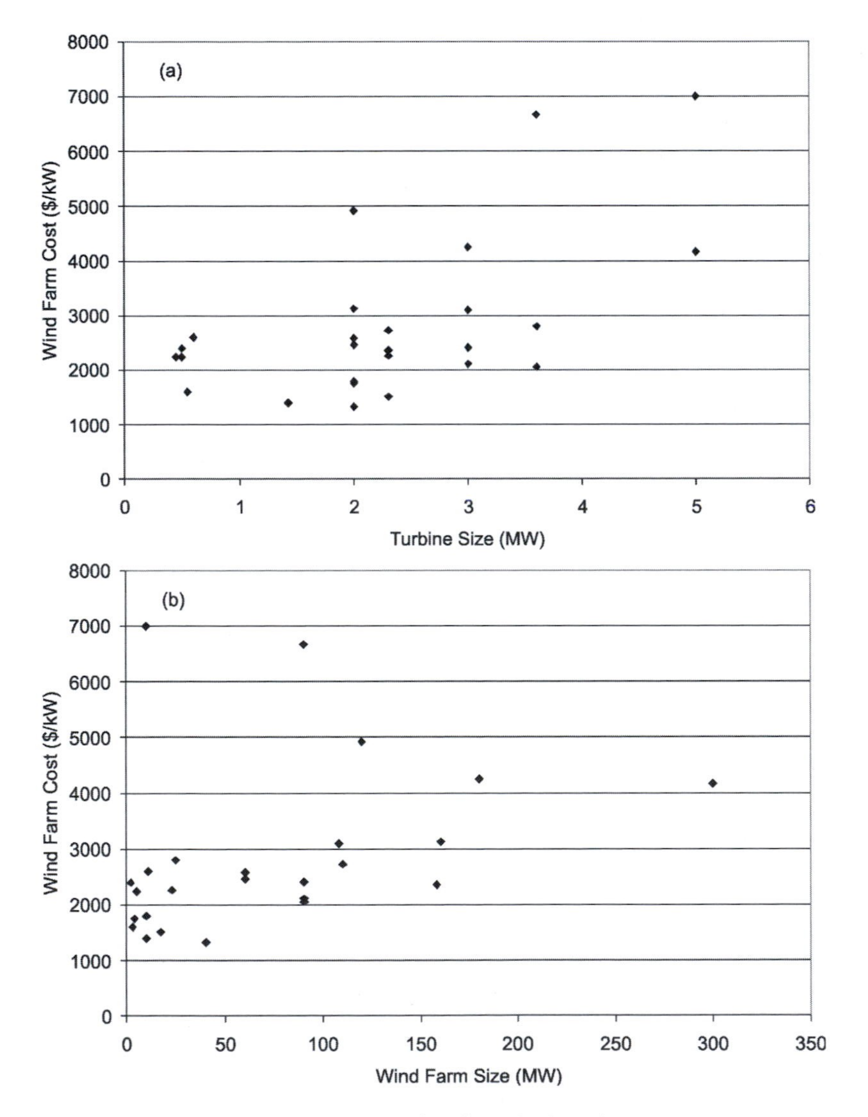

Note: Costs of wind farm constructed at different times were adjusted for inflation by the authors.
Source: Snyder and Kaiser (2009)

Figure 3.37 *Variation in the total capital cost of completed offshore wind farms as a function of (a) the size of the turbines used, and (b) the total wind farm size*

shore are more significant factors for the cost of offshore wind farms.

3.13.3 Capacity credit of wind turbines

The capacity credit of a wind turbine is the amount by which the power capacity of non-wind generating plants in an electricity supply system can be reduced due to the addition of the wind capacity. It is given as a fraction of the added wind capacity. Because periods of zero power output from the wind turbines might coincide with periods of maximum energy demand, one might expect that full backup of the wind turbine power output is required, that is, that the capacity credit is zero because no reduction in non-wind capacity is permitted. However, this is not correct.

The appropriate capacity credit for wind power is determined by use of the *loss-of-load-probability* (LOLP) approach (Redlinger et al, 2002, Chapter 3). No power source is ever available 100 per cent of the time, due to occasional breakdowns and scheduled maintenance. Indeed, most non-wind generators are out of service 10–15 per cent of the hours in a year through a combination of planned outage (for maintenance) and forced outage (due to malfunctions), as indicated in Table 3.16. For this reason, it is common to have 15–20 per cent backup capacity. The capacity credit of wind is computed as the difference in the amount of non-wind generation needed, with and without turbines present, in order to have the same LOLP. Thus, suppose that adding 4MW of wind power capacity to an existing system is needed to give the same LOLP as adding 1MW of nonwind capacity. In that case, the capacity credit would be 25 per cent. Note that, if the backup powerplant were 100 per cent reliable, the capacity credit for wind would indeed be zero. The capacity credit for wind is non-zero only because the non-wind powerplants are not completely reliable.

Table 3.16 *Outage rates for various electricity generators*

Generator type	Outrage rate (%)	
	Forced	Planned
Hydro	2.0	5.0
Gas turbine	10.7	6.4
Gas combined cycle	5.0	7.0
Existing coal	7.9	9.8
New coal	7.9	9.8
IGCC	7.9	9.8
Nuclear	5.0	5.0

Source: Denholm and Short (2006)

An exact computation of the capacity credit for a given turbine or wind farm requires hour-by-hour calculations, something that requires considerable modelling effort and data collection. Voorspools and D'haeseleer (2006) have developed a simple formula that provides a good fit to the results of such calculations. The capacity credit as a fraction of the capacity factor decreases with increasing wind penetration (that is, with increasing wind turbine capacity as a fraction of the system capacity). It also depends on how dispersed the wind turbines are. The capacity credit for wind farms is given by:

$$CCr = \frac{32.8}{0.306 + d}\frac{CF_{wind}}{R_{system}} \times \left(1 + 3.26d e^{-0.1077(0.306+d)(x-1)}\right) \quad (3.18)$$

where CF_{wind} is the capacity factor of a turbine or wind farm (and should include the η_s and f_a factors found in Equation 3.17), R_{system} is the system reliability, x is the percentage wind turbine penetration as a percentage of peak power and δ is a dispersion coefficient equal to 1.0 when the turbines are concentrated in one location and equal to 0.0 when they are fully dispersed. The forced outage rate of North American fossilfuel powerplants is rather large – about 8 per cent – compared to a forced plus unforced outage rate in modern wind turbines of about 2 per cent (Archer and Jacobson, 2002). Figure 3.38 shows the variation of capacity credit with wind penetration for capacity factors of 0.1, 0.2, 0.3 and 0.4 and δ = 0.57 (appropriate for turbines dispersed throughout The Netherlands). The capacity credit is slightly larger than the capacity factor for a wind penetration of 1 per cent (due to R_{system} being < 1.0) and drops to about half the capacity factor for a penetration of 30 per cent, with little further decrease with larger penetrations. The capacity credit as a function of x and δ for a capacity factor of 0.4 is shown in Figure 3.39.

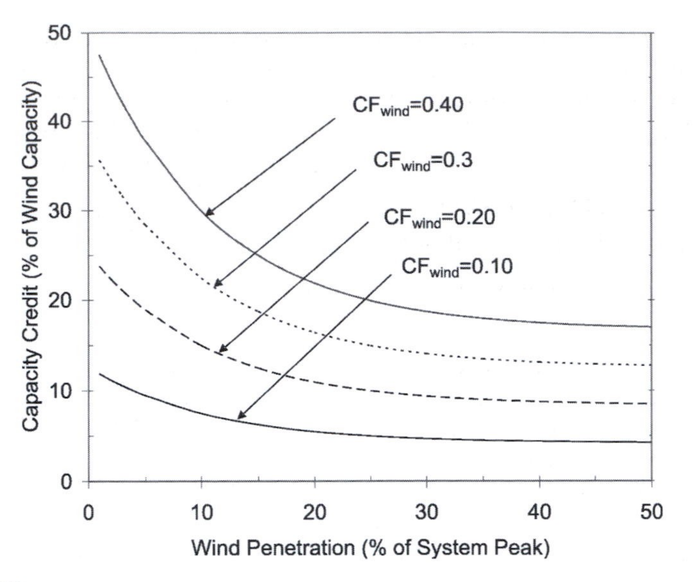

Source: Based on Equation (3.18)

Figure 3.38 *Variation of capacity credit with wind penetration for a single turbine with various capacity factors with R_{system} = 0.9 and δ = 0.57*

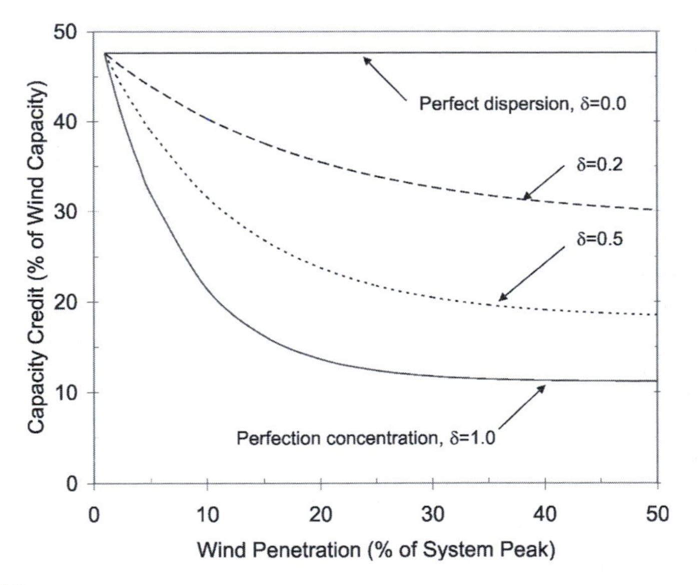

Source: Based on Equation (3.18)

Figure 3.39 *Variation of capacity credit with wind penetration for a wind farm with various dispersion factors, a capacity factor of 0.40 and R_{system} = 0.9*

3.13.4 Indirect costs

There are three ways in which indirect costs of wind energy can arise: by reducing the average use (capacity factor) of existing power supplies, through restraints on electricity production when wind energy supply would otherwise exceed total demand, and by possibly increasing the required spinning reserve as the fraction of total electricity generated by wind increases (which reduces the efficiency of the fossil fuel plant). These issues are discussed below. Measures to reduce fluctuations in wind output and the associated need for spinning reserve were discussed in section 3.11.

Reduced energy production by non-wind generators partly offset by less required non-wind capacity due to a wind capacity credit

If one adds wind turbines to an existing power system (and neglecting possible growth in energy demand), the average power output of the existing powerplants will be reduced. This will increase the cost of the non-wind-generated electricity because the fixed annual capital cost repayment is amortized over fewer kWh of electricity. However, in the case of new additions of both wind and non-wind energy sources to a power system, this cost increase can be mitigated if the peak power capacity of the non-wind components can be reduced due to the installation of the wind turbines, that is, if wind can be credited with reducing the required capacity of other components in the system.

Thus, if a peak demand of 1kW requires a fossil fuel capacity of (1+*R*) kW (where *R* is the backup as a fraction of peak demand), then when 1kW of wind capacity with a capacity credit of *CCr* is added, the required fossil fuel capacity is (1+*R*–*CCr*) kW.

The cost of electricity generated in a hybrid wind-fossil fuel system can be computed as the total annual revenue requirement divided by the total number of kWh generated per year per kW of wind capacity. The number of kWh generated is given by the number of hours in a year times a system capacity factor CF_{system}, which is the average electricity demand divided by the peak demand. Thus, for a system where the wind turbine capacity matches the peak demand and is accompanied by sufficient fossil fuel capacity to give the same overall reliability as a fossil fuel-only system, the cost of electricity is given by:

$$C_{e\text{-}hybrid} = \frac{(CRF_{WT} + OM_{WT\text{-}f})CC_{WT} + (CRF_f + OM_{f\text{-}f})CC_f(1+R-CCr)}{8760CF_{system}} + (1-f_f)OM_{WT\text{-}v} + f_f OM_{f\text{-}v} + f_f \frac{0.0036C_{fuel}}{\eta_f} \quad (3.19)$$

where CRF_{WT} and CRF_f are the cost recovery factors for the wind and fossil fuel powerplants, CC_{WT} and CC_f are the corresponding capital costs, $OM_{WT\text{-}f}$ and $OM_{f\text{-}f}$ are the fixed O&M costs for wind and fossil fuel powerplants, $OM_{WT\text{-}v}$ and $OM_{f\text{-}v}$ are the corresponding variable O&M costs, C_{fuel} is the cost of fuel ($/GJ), 0.0036 is the number of GJ per kWh, η_f is the efficiency of electricity generation with fossil fuel and f_f is the fraction of electricity supplied by the fossil fuel plant. The latter can be computed as:

$$f_f = \frac{CF_{system} - \eta_s f_a CF_{WT}}{CF_{system}} \quad (3.20)$$

assuming that no wind energy is wasted, an assumption that is relaxed below.

By comparison, the cost of electricity from a fossil fuel plant alone is given by:

$$C_{e\text{-}fossil} = \frac{(CRF_f + OM_{f\text{-}f})CC_f(1+R)}{8760CF_{system}} + OM_{f\text{-}v} + \frac{0.0036C_{fuel}}{\eta_f} \quad (3.21)$$

Wasted electricity production

As discussed above, the calculation of the appropriate capacity credit for wind is related to the possibility of low wind power production at times of high electricity demand. At the other extreme is the possibility of very high wind production at times of low demand. Fossil fuel units cannot operate below some minimum load when running, and some units will need to be kept running in order to provide a rapid-response capacity. This minimum load is typically 25 per cent and is referred to as the 'must-run' capacity. Wind power output will need to be constrained so as not to exceed demand minus the must-run fossil capacity, so that fewer

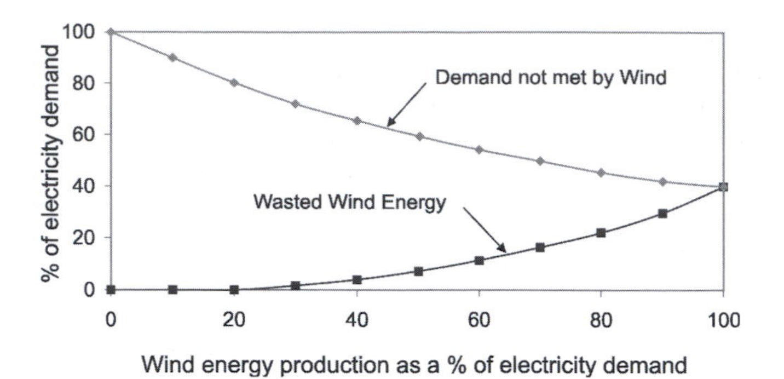

Source: Redlinger et al (2002)

Figure 3.40 *Excess electricity production potential and residual demand as the potential wind energy production increases from 0 to 100 per cent of annual demand for Danish conditions*

kWh are generated than expected based on the wind speed probability distribution. This wasted electricity production potential will lead to a higher kWh cost by reducing the capacity factor in Equation (3.17) or by increasing f_f by reducing CF_{WT} in Equation (3.20).

Figure 3.40 shows how the fraction of electricity generation potential that is wasted varies as the generation potential increases from zero to 100 per cent of annual electricity demand for Danish conditions. For potential wind energy production of up to 20 per cent of annual demand, there is no waste, while at potential generation equal to 100 per cent of annual demand, about 40 per cent is wasted. Thus, in a system scaled to produce an amount of electricity equal to annual demand in Denmark (and without storage), only 60 per cent of the annual electricity demand would be supplied in practice by wind, and the cost of the wind-generated electricity would be multiplied by a factor of 1.67 (1/0.6). Similar curves have been computed for OECD Europe and the US by Hoogwijk et al (2007) based on detailed data for electricity demand and wind energy potential on a 0.5° × 0.5° latitude–longitude grid. They find that the wasted fraction increases sharply as the potential wind generation increases beyond 30 per cent of total electricity demand if there is no storage.

Lund and Kempton (2008) have assessed the extent to which having a fleet of battery-electric vehicles (BEVs) that can be charged with wind electricity when available (and otherwise charged with fossil fuel electricity if necessary), utilizing PHEVs with V2G capability and utilizing V2G PHEVs along with heat pumps can reduce the fraction of wasted wind potential (see subsection 3.11.2 with regard to the benefits of both V2G and heat pumps in combination with wind power). For their analysis they used a model called EnergyPLAN that integrates energy for electricity, transport and heat, and includes hourly fluctuations in human needs and environmental conditions (wind resource and the weather-driven heating demand). They consider two situations: one, where a large fraction of the heating load is met by combined heat and power (CHP), and one where there is no CHP. When CHP is present, heating demand has priority, which means that during times of cold weather, a considerable amount of electricity will obligatorily be produced. The potential of having excess wind capacity, which would be wasted, is therefore greater than without CHP. Table 3.17 gives

Table 3.17 *Fraction of wind energy that is wasted in a mid-latitude country with and without CHP when the potential annual electricity production from wind equals total annual electricity demand (including the demand for transportation when electric vehicles are used)*

Case	Wasted fraction	
	With CHP	Without CHP
No BEV	0.67	0.38
With BEV	0.58	0.33
With intelligent BEV	0.53	0.27
With V2G	0.51	0.22
With V2G and heat pumps	0.21	0.16

Source: Estimated from the graphical results presented in Lund and Kempton (2008)

the fraction of potential electricity production that is wasted for the two situations in a mid-latitude country (such as Denmark), in both cases assuming a wind powerplant capacity sufficient to generate an amount of electricity over the course of a year equal to annual electricity demand. The combination of V2G with heat pumps reduces the wasted wind fraction to 21 per cent for the case with CHP, and to 16 per cent for the case without CHP.

Denholm et al (2005) have calculated the fraction of wasted wind energy as a function of the capacity factor for wind systems in the midwestern US with the ability to store 24 hours of peak output power. For a 90 per cent capacity factor, the spill rate is 25–43 per cent for existing systems (having turbine capacity factors of 33–37 per cent) but 17 per cent for a simulated future system (having a turbine capacity factor of 46 per cent). Existing wind farms could achieve a capacity factor >70 per cent with <10 per cent spill rate.

Increased system balancing and load-following costs

Gross et al (2007) review over 200 reports and papers related to the costs of intermittent generation of electricity by wind. Compensation for fluctuations in either supply or demand at timescales of seconds to hours is call *system balancing* or *regulation*, and *reserve* is defined as the power capacity required for system balancing. This is separate from the backup capacity, needed to supply unexpected peak demand or to compensate for failure of generating units (it is typically 15–20 per cent of total system capacity). Ignoring a small number of outliers, where conditions were not favourable for wind, balancing costs are generally £0.5–3/MWh (about 0.08–0.5 cents/kWh) for wind penetrations of 5–45 per cent, with only a weak increase in balancing costs with penetration. The cost of maintaining system reliability, after taking into account the capacity credit of wind turbines applicable to the British electricity system and the cost of thermal powerplants, is estimated to be £3–5/MWh (0.5–0.8 cents/kWh). In Denmark, where about 20 per cent of electricity demand is supplied through wind, the cost of integrating variable wind power into the electrical system is estimated to be €0.03–0.04/kWh (0.5–0.7 cents/kWh) (Krohn et al, 2009). As the nominal wind penetration in the UK increases from 20 per cent to 100 per cent, Milborrow (2007) indicates a cost of variability (through increased balancing requirements and reduced utilization of thermal plants) increasing from about £8/MWh to about £14/MWh (a variation from about 1.2 cents/kWh to 2.2 cents/kWh). A US Department of Energy study concluded that integration costs associated with wind supplying 20 per cent of US electricity demand would amount to 0.5 cents/kWh (Lindenburg et al, 2008).

3.13.5 Cost of transmission

As many of the best wind resource regions are far removed from the major electricity demand centres, transmission costs could be significant. Transmission costs involve two components: the capital cost and the cost of electricity that is lost during transmission (some electrical energy is dissipated as heat during transmission due to resistance to the movement of electrons). The capital cost involves the cost of transformers at each end of the transmission line and the line cost. Transmission losses occur at transformers and along the line, and are proportional to the transmission capacity factor (average power transmitted divided by peak transmission).

Table 3.18 compares the costs and energy losses associated with AC and DC transmission at various

Table 3.18 *Cost and performance of high-voltage AC and DC transmission for a transmission capacity of 5GW*

Parameter	Unit	HVAC		HVDC	
Operating voltage	kV	750	1150	±600	±800
Overhead line loss	%/1000km	8	6	5	2.5
Sea cable line loss	%/100km	60	50	0.33	0.25
Terminal loss	%/station	0.2	0.2	0.7	0.6
Overhead line cost	€ millions/1000km	400–750	1000	400–450	250–300
Sea cable cost	€ milllions/1000km	3200	5900	2500	1800
Terminal cost	€ millions/station	80	80	250–350	250–350

Source: GAC (2006)

voltages for overhead lines and undersea cables as given by a report of the German Aerospace Center (GAC, 2006). Losses are particularly severe for AC transmission by undersea cable, due to the inability to effectively dissipate the generated heat. Table 3.19 compares these costs with other estimates of costs, all of which are quite uncertain. Costs are lower the greater the transmission voltage, and the low costs for 800kV DC transmission given by the German Aerospace Center (€0.05–0.06/kW/km) seem to be roughly consistent with the estimates given by most other sources for lower transmission voltages. Costs at all voltages can be expected to be greater through rugged wilderness areas than through inhabited plains, due to the greater cost of access. Figure 3.41 compares the variation in investment cost and energy loss with distance for 1150kV AC and 800kV DC transmission using the unit costs given in GAC (2006). The break-even distance between AC and DC transmission is around 250km in terms of loss and 750km in terms of cost, DC transmission being less costly and entailing smaller losses than AC transmission at greater distances.

Table 3.19 *Comparison of line costs and station costs for high-voltage AC and DC transmission, as given by different sources*

Voltage	Cost	Comment	Reference
		Line costs	
230kV AC	$1.00/kW/km	Baseline US, up to 40 per cent greater in specific regions	Lindenberg et al (2008)
750kV AC	$0.23–0.45/kW/km	Average over 30,000km proposed network in the US, each line carrying 3.6–7.2GW, depending on the technology	AEP (2007)
345kV DC	$0.21/kW/km	Hypothetical 3GW lines in Texas	ERCOT (2008)
408kV DC	$0.17/kW/km		DeCarolis and Keith (2006)
450kV DC	$0.28/kW/km (€0.2/kW/km)		Hurley et al (2007)
750kV AC	$0.11–0.21/kW/km) (€0.08–0.15/kW/km)		GAC (2006)
600kV DC	$0.11–0.13/kW/km (€0.08–0.09/kW/km)		GAC (2006)
800kV DC	$0.07–0.084/kW/km (€0.05–0.06/kW/km)		GAC (2006)
765kV AC	$0.076/kW/km	Applies to a 6000MW line	ABB (2006)
500kV DC	$0.064/kW/km	Applies to a 6000MW line	ABB (2006)
800kV DC	$0.046/kW/km	Applies to a 6000MW line	ABB (2006)
HVDC	$0.27/kW/km	Cost of lines through wilderness areas in Quebec during the 1980s (derived from line cost of $680/kV/km and 800kV for 2GW)	Lund and Ostergard (2000)
HVDC	$0.73/kW/km (€0.52/kW/km)	Cable costs for an offshore HVDC grid in the North Sea.	EWEA (2005)
HVDC	$0.28/kW/km (€0.2/kW/km)	Onshore costs in Europe	Hurley et al (2007)
		Station costs	
345kV DC	2 × $87/kW	Inverter between 345kV DC line and 345kV AC grid in Texas	ERCOT (2008)
450kV DC	2 × $265/kW (€189/kW)		Hurley et al (2007)
750kV AC	2 × $22/kW (€16/kW)		GAC (2006)
800kV DC	2 × $70–98/kW (€50–70/kW)		GAC (2006)
765kV AC	2 × $54/kW	Applies to a 6000MW line	ABB (2006)
500kV DC	2 × $70/kW	Applies to a 6000MW line	ABB (2006)
800kV DC	2 × $74/kW	Applies to a 6000MW line	ABB (2006)

Note: Costs originally in euros are in brackets and have been converted to dollar costs assuming that 1€=$1.4.

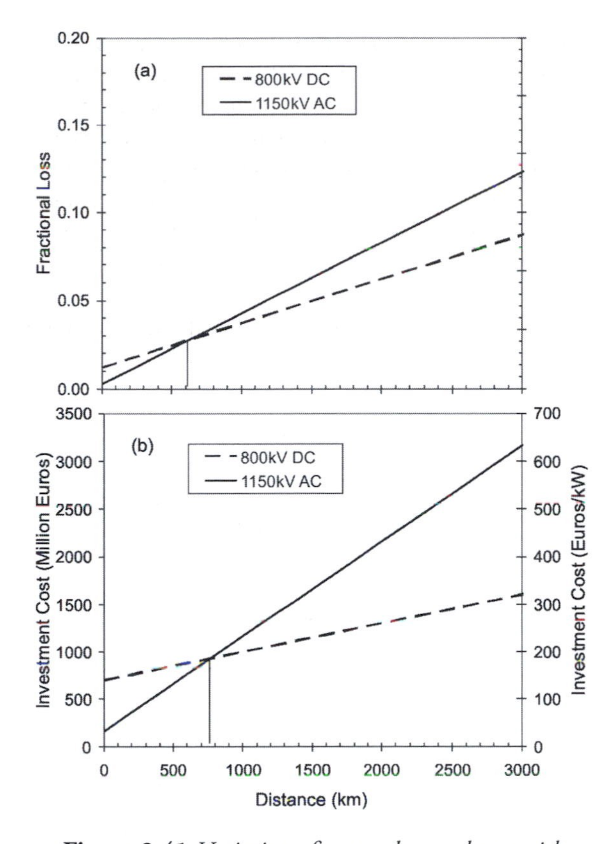

Figure 3.41 *Variation of energy loss and cost with distance transmitted for 1150kV AC and 800kV DC transmission, based on energy loss and cost data given in Table 3.18 (with twice the high-cost estimates used for DC transmission)*

The contribution of the transmission investment to the cost of electricity is given by:

$$C_{e\text{-}TL} = \frac{(CRF_{TL} + OM_{TL})CC_{TL}}{8760CF_{TL}} \tag{3.22}$$

where CRF_{TL}, OM_{TL}, CC_{TL}, and CF_{TL} are the transmission line cost recovery factor, operation and maintenance factor, capital cost and capacity factor, respectively (CRF_{TL} will be different from CRF_{WF} if the transmission line is financed over a longer period of time or at a different interest rate, while CF_{TL} will differ from CF_{WF} if the wind farm is oversized relative to the transmission link, as discussed later). O&M is given by GAC (2006) as 1 per cent of the investment cost per year.

For illustrative purposes, assume a line cost 50 per cent greater than the upper value given by GAC (2006) for 800kV HVDC, that is €0.09/kW/km or \$0.13/kW/km, and terminal costs of \$100/kW at each end of the line. Capital costs are then \$460/kW and \$590/kW for transmission distances of 2000km and 3000km, respectively. For 6 per cent interest and a 40-year lifespan, the cost of 2000km transmission is 0.8 cents/kWh with a transmission line capacity factor of 0.5 and 0.54 cents/kWh with a capacity factor of 0.75. An extra 1000km transmission adds 0.23 cents/kWh with a capacity factor of 0.5 and 0.15 cents/kWh with a capacity factor of 0.75.

The effect of transmission loss is to increase the amount of electricity that must be generated in order to supply a given demand. This component of the transmission cost is given by the required extra electricity times the cost of electricity, which has components $C_{e\text{-}WT}$ (given by Equation 3.17) and $C_{e\text{-}TL}$ (given by Equation 3.22). The total cost of electricity, including transmission investment and energy losses, is thus given by:

$$\begin{aligned} C &= \frac{C_{e\text{-}WT} + C_{e\text{-}TL}}{1 - T_{loss}CF_{TL}} \\ &= (C_{e\text{-}WT} + C_{e\text{-}TL})(1 + \varepsilon + \varepsilon^2 + ...) \\ &= C_{e\text{-}WT} + C_{e\text{-}TL} + \delta C_{e\text{-}WT} + \delta C_{e\text{-}TL} \end{aligned} \tag{3.23}$$

where T_{loss} is the fractional loss at full capacity, $\varepsilon = T_{loss}CF_{TL}$ and $\delta = \varepsilon/(1 - \varepsilon)$.[4] The cost of lost electricity is thus given by:

$$C_{e\text{-}loss} = \delta C_{e\text{-}WT} + \delta C_{e\text{-}TL} \tag{3.24}$$

Transmission losses vary in proportion to the transmitted power, and so will vary in proportion to the average transmission line capacity factor. Increasing the average power transmitted through the transmission line reduces $C_{e\text{-}TL}$ by increasing the transmission line capacity factor, but tends to increase the final cost of electricity by increasing the transmission losses.

The relative magnitudes of the three terms related to transmission cost ($C_{e\text{-}TL}$, $\delta C_{e\text{-}WT}$ and $\delta C_{e\text{-}TL}$) and how they and their sum changes with transmission capacity factor are shown in Figure 3.42 for the particular case of $C_{e\text{-}WT}$ = 8 cents/kWh and CC_{TL} = \$590/kW with

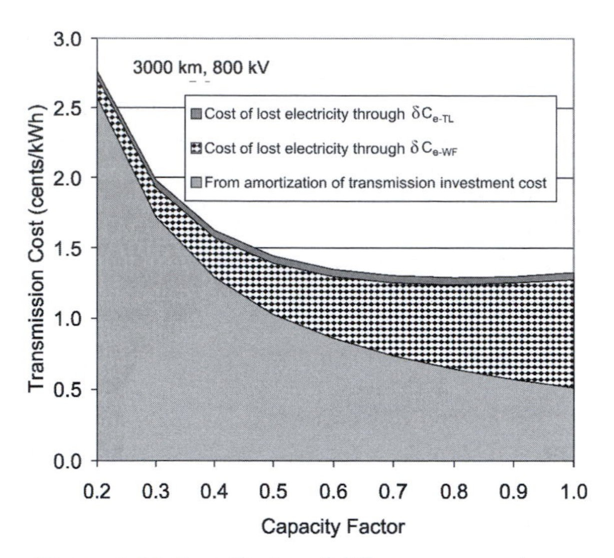

Figure 3.42 *Contribution of different terms to the cost of transmission, appropriate for 800kV DC transmission over a distance of 3000km and a cost of electricity of 5 cents/kWh*

6 per cent financing over 40 years and a full-load loss of 8.7 per cent (the cost and loss correspond to a transmission distance of 3000km using the adjusted GAC (2006) unit costs). With increasing transmission capacity factor, $C_{e\text{-}TL}$ decreases at an ever-decreasing rate while $\delta C_{e\text{-}WT}$ increases due to the increasing energy loss, such that there is little change in transmission cost as CF_{TL} increases above 0.5. $\delta C_{e\text{-}TL}$ also increases with capacity factor but is small.

Table 3.20 gives distances between potential high-wind source areas and large electricity markets. Also given are typical wind speeds in the source regions at a 110m hub height, the approximate capacity factor (computed from Equation (3.13)), and the transmission cost and loss based on the information given above. The greatest distance considered is 3000km (Central Mongolia to Hong Kong), with a full-load loss of 8.7 per cent.

3.13.6 Progress ratios and long-term cost projections

As discussed in Chapter 2 (subsection 2.2.12), historical data for many technologies indicate that the cost of a technology is multiplied by a roughly fixed fraction, called the *progress ratio*, for each doubling of cumulative production. A progress ratio of 0.8, for example, implies that the cost decreases by 20 per cent for each doubling of cumulative production. Junginger et al (2005) discuss some of the difficulties in computing progress ratios for wind farms (which take into account not only reductions in wind turbine costs but also in design, construction and integration costs). Progress ratios computed for individual countries are often not meaningful, because the costs reflect global cumulative experience but only the national cumulative capacity is counted. Junginger et al (2005) estimated a global progress ratio for wind farms based on wind farm prices in the UK and Spain, correlated against cumulative global installed capacity. These prices are assumed to

Table 3.20 *Long-distance transmission of wind energy. Costs are based on the higher estimates in Table 3.18 for 800kV HVDC (with line costs × 1.5) and are indicative only, but would tend to be more over mountainous terrain but less in China due to lower labour costs*

Transmission link	Distance (km)	Mean wind speed (m/s): At given height	Mean wind speed (m/s): At 110m height	Capacity factor	Cost at \$0.09/km/kW and \$100/kW per terminal	Full-load loss @ 2.5%/1000km and 0.6% per terminal
North Dakota to Washington DC	2000	7–8 @ 50m	8–9	0.44	$460/kW	6.2%
Labrador to Washington DC	2000	8–10 @ 80m	9–11	0.58	$460/kW	6.2%
Idaho to Los Angeles	1200	8–9 @ 50m	9–10	0.53	$356/kW	4.2%
Western Oklahoma to Miami	2300	6.4–7 @ 50m	7–8	0.36	$500/kW	7.0%
Southern James Bay to Toronto	750	7.5–8.5 @ 80m	8.4–9.6	0.49	$300/kW	3.1%
Central Mongolia to Shanghai	2000				$460/kW	6.2%
Central Mongolia to Hong Kong	3000				$560/kW	8.7%

Note: Mean wind speeds are taken from Figures 3.20–3.22.

reflect competitive world prices, with no trend due to diminishing quality of sites. They obtained a progress ratio of 0.77–0.85 (with a best-guess value of 0.81).

Although extrapolation using the progress ratio approach is highly speculative, Figure 3.43 shows the variation in the installed turbine capacity and the price of wind-generated electricity from new turbines, assuming a starting capital cost of $1500/kW in 2005 (the approximate global average), a capacity of 59.1GW at the end of 2005, a progress ratio of 0.9 (rather than the recent observed progress ratio of 0.81), a fixed capacity factor and a rate of growth in the global market (that is, in the rate of installation of new turbines) of 20 per cent/year until 2020. The result is the cost of wind energy drops from 9.9 cents/kWh in 2005 to 6.3 cents/kWh by 2020.[5] If, as the best wind turbine sites are taken, the average capacity factor for new turbines decreases, then the price of wind-generated electricity would not fall as sharply. As well, achieving the high installed capacity in this scenario implies substantial development of offshore wind energy, which tends at present to have modestly higher costs per kWh when the greater energy production is combined with greater capital costs. However, the wind resource is up to twice as large for offshore sites as for onshore sites, so declining capacity factor is unlikely to be an issue. As well, innovative and radical changes in turbine design (rather than incremental improvements with existing concepts) could see the average capacity factor held constant even as the best wind turbine sites are taken, and there would be a gradual increase in the average capacity factor later as a second-generation of turbines is placed on early sites. Thus, there are several factors that could continue to drive costs lower.

Not included in these cost projections is the impact of increases in the cost of steel and copper. Prices of these commodities spiked in 2008 and then subsequently collapsed, although the prices will rise in the long run for commodities (such as copper) that are likely to be scarce (see Volume 1, Chapter 6, subsection 6.5.5). Due to the recent spike in material costs and an excess of demand over supply, the turnkey price of onshore wind turbines in Europe rose 74 per cent from 2005 to 2008, to €1380/kW, while the price of offshore wind turbines rose 48 per cent to €2230/kW (Jackson, 2008). In the US, the price of wind turbines increased from a low of about $700/kW during the 2000–2002 period to an average price of $1240/kW in 2007. Turbines constitute about 75–80 per cent of an overall project cost, so average project costs increased from $1340/kW in 2001–2003 to $1710/kW in 2007 (Bollinger and Wiser, 2008). Early data indicate that an index of world wind turbine costs

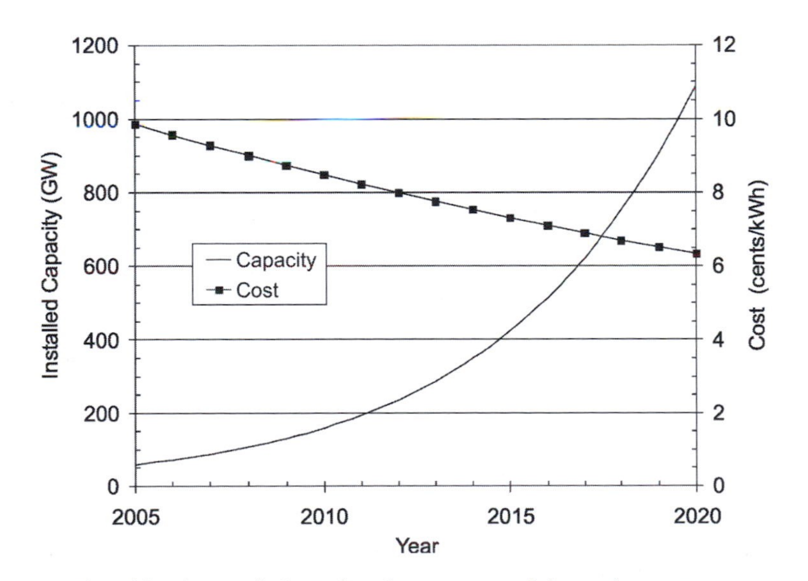

Figure 3.43 *Variation of worldwide installed wind turbine capacity if the total capacity grows at a rate that declines linearly from 20 per cent/year in 2008 to 0 per cent/year by 2050, and the corresponding cost of electricity in good wind sites, assuming a progress ratio of 0.81*

fell by about 10 per cent from the second to the third quarter of 2008 (Beck and Harrmeyer, 2009) – a trend that is likely to continue for some time.

3.14 Baseload wind-derived electricity: Strategies and costs

The average power output from a wind turbine as a fraction of its peak power output – the capacity factor – varies from about 0.2 at wind speeds of 6m/s to about 0.4 at wind speeds of 9m/s (see Figure 3.19). This is substantially less than the world average electric powerplant capacity factor of 0.49 (given in Volume 1, Chapter 2, Table 2.5) or the capacity factor of 0.8–0.9 expected from baseload powerplants. However, there are several ways in which the average output from wind farms can be increased relative to the peak power supply. These are:

- to oversize the wind farm relative to the transmission link;
- to add some form of energy storage combined with an ability to generate electricity from stored energy; and
- a combination of these two.

3.14.1 Oversizing wind farms

The idea of deliberately oversizing wind farms relative to the transmission link was first proposed by Cavallo (1995). The idea is based on the observation that wind speed is rarely strong enough to produce even half the rated power output of a turbine, much less the full output. Thus, a wind farm could be doubled in size (for example) with no increase in the transmission link, and very little potential electricity production would be wasted. Lew et al (1998) applied this concept to the delivery of electricity from wind farms in Inner Mongolia (which has a good wind resource) to distant demand centres in China, while DeCarolis and Keith (2006) have applied it to power production in the central plains of the US.

Oversizing by a factor of two is illustrated in Figure 3.44 using the output from a 1.8MW turbine and the wind speed probability distribution using a Weibull shape parameter of 2.0 and a scale factor of 6m/s (these give a mean wind speed of 5.3m/s). The electricity production is limited by the transmission capacity, so the wind power that is potentially wasted as a result of oversizing is the area between the scaled system output and the transmission capacity (the area indicated as

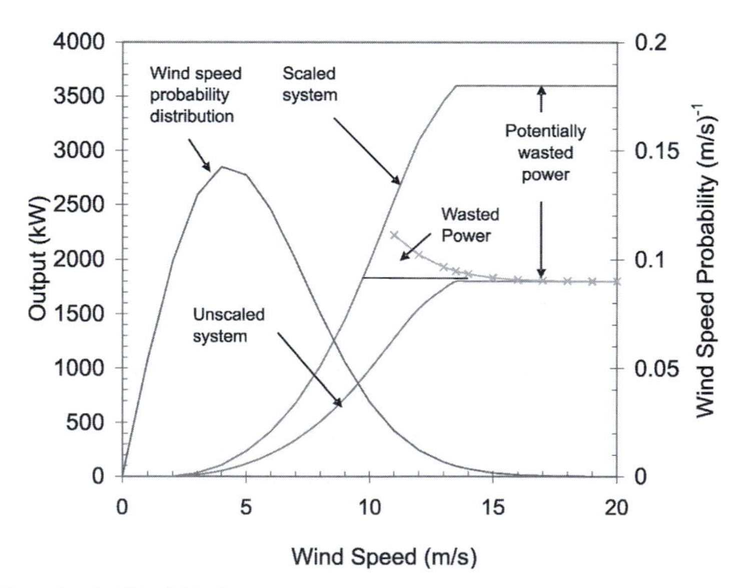

Note: Also given is the wind speed probability distribution.

Figure 3.44 *Variation of power output with wind speed for a single 1.8MW turbine and for two 1.8MW turbines with total output capped at 1.8MW*

'wasted power' in Figure 3.44). The wastage is small because wind speeds high enough to require this wastage (greater than about 10m/s) do not occur very often (as indicated by the probability distribution). Average power output and hence the capacity factor would be almost doubled. The wasted electricity generation potential will increase the cost per kWh of electricity generated, but partly (at least) offsetting this will be a reduction in the per kWh transmission cost, due to the fact that more kWh will be transmitted through the transmission link. If the transmission cost is a significant fraction of the wind farm cost (as would be the case for wind farms located in good wind areas but removed from demand centres), the savings in transmission cost can largely or entirely offset the increased generation cost, resulting in little or no increase in total cost per kWh while substantially increasing the ratio of average to peak transmitted power – what I shall call the *demand-side capacity factor* (it is the same as the transmission line capacity factor).

Figure 3.45a shows the capacity factor of a wind farm when scaled to have a capacity one to four times the transmission link capacity, computed using the power curve for the N100-2.5 turbine. If there were no loss of electricity generation, the capacity factor would be the same for all scalings. Thus, the fraction of wasted electricity generation potential is $(CF_1 - CF_n)/CF_1$, where CF_n is the capacity factor when the wind farm is scaled by a factor of n and CF_1 is the capacity factor with no scaling. The wasted fraction is shown in Figure 3.45b as a function of the mean wind speed. Greater waste occurs at higher mean wind speeds. For a mean wind speed of 7m/s (an excellent onshore regime), 24 per cent of the potential electricity generation is wasted when the wind farm is oversized by a factor of two and 50 per cent is wasted with a factor of four oversizing. However, the transmission capacity factor (shown in Figure 3.45c) ranges from 0.28 with no oversizing to 0.57 with a factor of four oversizing. Thus, the wind farm capital cost per kWh of generated electricity will be doubled but the transmission cost will be halved. If the two are the same without oversizing, then the total costs increase by only 25 per cent when the wind farm is oversized by a factor of four.

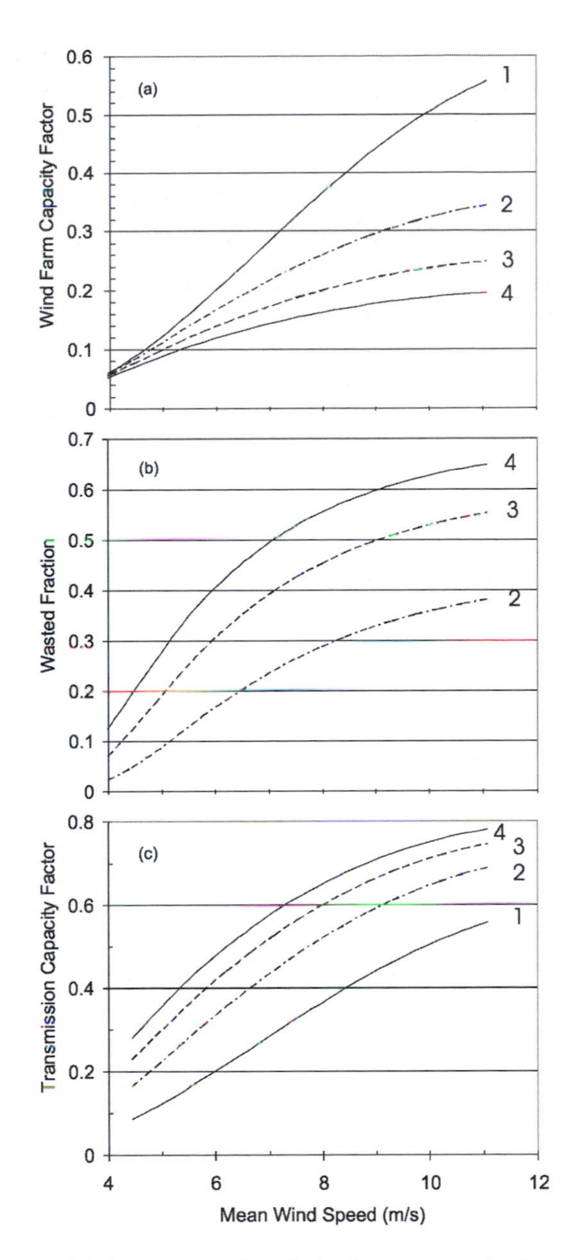

Note: Calculations were performed using the power curve for the Nordex N100-2.5 turbine.

Figure 3.45 *(a) Variation of wind farm capacity factor with wind speed for various wind farm scaling factors, computed from the Weibull distribution with* k = *2.0 and using the turbine output data of Figure 3.19; (b) the fraction of potential electricity generation that is wasted when the wind farm is scaled by various factors; (c) the transmission link capacity factor for different wind farm scaling factors*

Figure 3.46 gives the cost of electricity as computed from Equation (3.23), as a function of the mean wind speed and wind farm scaling factor, assuming an interest rate of 6 per cent, lifespans of 20 years for the wind farm and 40 years for the transmission link, and annual operation and maintenance costs of 2 per cent and 1 per cent of the

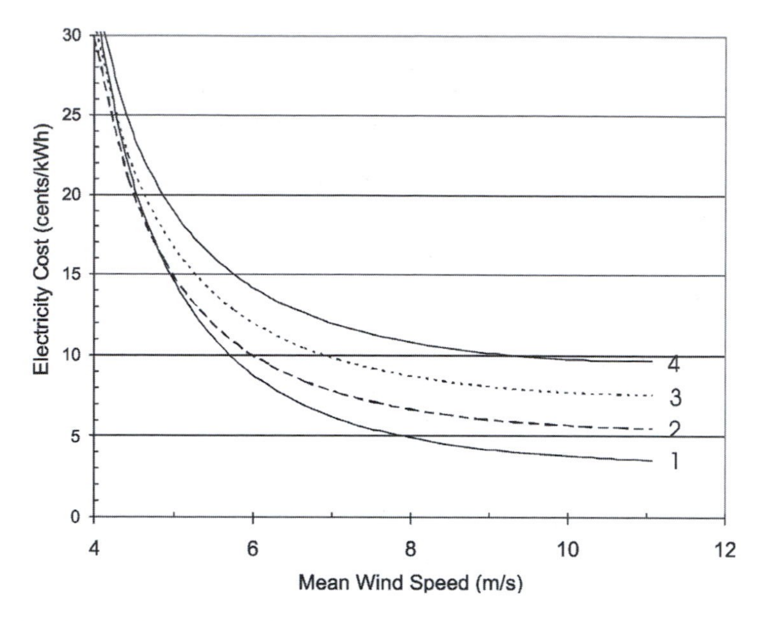

Note: Calculations were performed using the power curve for the Nordex N100-2.5 turbine.

Figure 3.46 *The cost of wind energy (including transmission costs) as a function of mean wind speed, for a wind farm capital cost of $1500/kW with various wind farm scaling factors (indicated by the numbers to the right of each line) and a transmission cost of $460/kW (corresponding to a distance of 2000km)*

initial investment for the wind farm and transmission link, respectively. Results are given assuming a wind turbine cost of $1500/kW, a transmission cost of $460/kW (derived as 2000km × $0.13/kW/km plus two transformer stations at $100/kW each) and a 5.0 per cent full-load loss. Electricity from wind farms sized to 2–3 times the transmission capacity is less expensive than from the unscaled wind farm for low mean wind speeds and more expensive at high wind speeds (at low wind speeds, relatively little potential electricity generation is wasted with oversizing while better use is made of the transmission link). Figure 3.47 gives the breakdown of the total cost as the scaling factor increases from one to four for a mean wind speed of 9m/s. The decrease in the contribution of transmission investment cost to the levelized cost of electricity with increasing scaling factor (and transmission capacity factor) is partly offset by the increasing resistance loss during transmission.

It is more relevant to compare the cost of wind farms at low wind speeds, no oversizing and minimal transmission costs with the cost of wind farms at high wind speeds, substantial oversizing and high transmission costs, on the grounds that the best wind sites tends to be far from the major demand centres. It is also likely that wind farms in remote regions (where many of the best winds are found) would be more expensive than wind farms close to demand centres, due to the cost of delivering turbines and cranes for installing the turbines. Such a comparison is made in Figure 3.48, which gives the cost of electricity for:

- an unscaled wind farm with zero transmission cost and loss and $1000/kW turbine installed cost (representing a wind farm adjacent to demand centres and entailing only local distribution);
- unscaled wind farms with $460/kW transmission cost, 5.0 per cent full-load loss, and either $1000/kW or $1500/kW turbine costs; and
- a wind farm oversized by a factor of three with $460/kW transmission cost, 5.0 per cent full-load loss, and $1500/kW turbine costs.

If local winds average 5m/s, the cost of electricity is 8.0 cents/kWh and the capacity factor is 0.16. If the distant oversized and more expensive ($1500/kW) wind farm is in a region with 9m/s average wind speed, the cost of

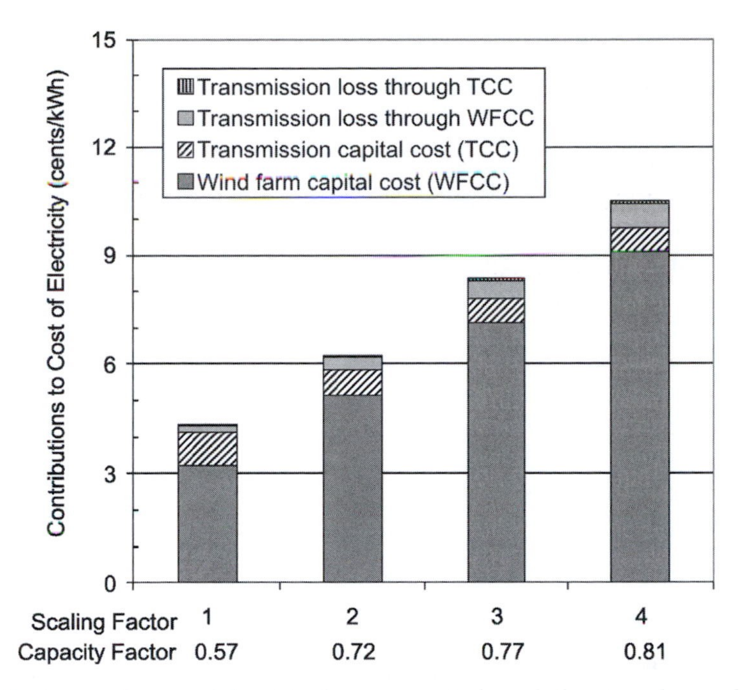

Figure 3.47 *Contributions to the cost of electricity for a mean wind speed of 9m/s as the wind farm scaling factor increases from one to four, assuming capital costs of \$1500/kW for wind turbines and \$590/kW for the transmission link (corresponding to a distance of 3000km)*

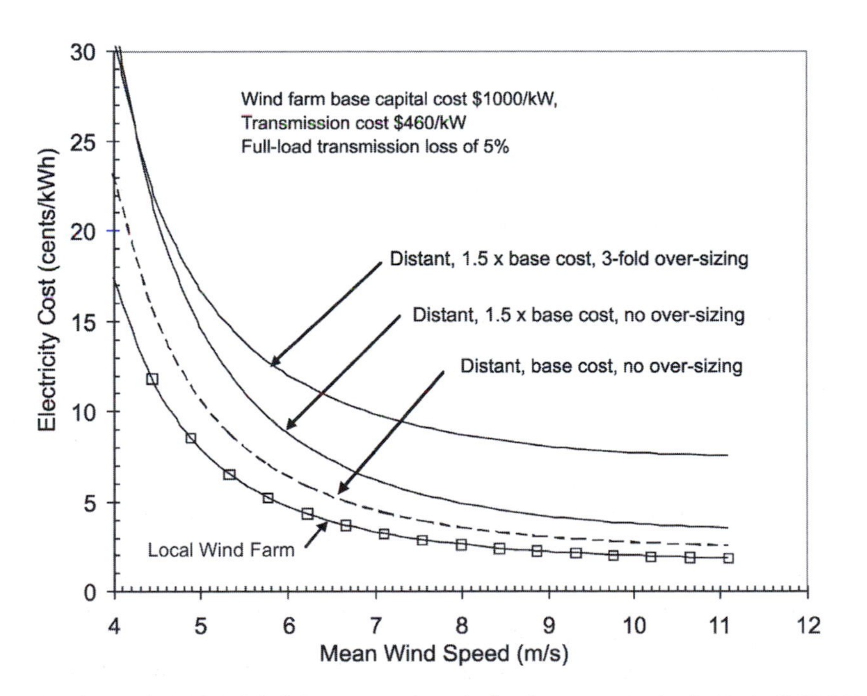

Note: The base turbine capital cost is \$1000/kW. Calculations were performed using the power curve for the Nordex N100-2.5 turbine.

Figure 3.48 *Comparison of the cost of electricity from local wind farms without oversizing, from a 2000km distant wind farm without oversizing, and from a 2000km distant wind farm oversized by a factor of three*

electricity is 8.1 cents/kWh. However, the capacity factor seen on the customer side of the transmission link (the demand-side capacity factor) is over four times greater for the distant oversized wind farm – 0.66 instead of 0.16.

If the output of a wind farm that is oversized by a factor of n is going to be restricted to $1/n$ of the peak output, then some cost savings could be obtained by installing generators and power conditioning in each of the turbines that are smaller by a factor of n as well. The cut-in wind speed might be smaller and the output at each wind speed slightly larger, since less torque is needed to start a smaller generator. Thus, oversizing the wind farm by a factor of n would probably increase the wind farm cost by somewhat less than a factor of n and give a slightly higher capacity factor than otherwise expected (that is, with less of a penalty due to wasted wind potential). This would tend to make the economics of oversized distant wind farms more attractive than indicated above.

Another factor that improves the economics of oversizing is the fact that the extra turbines double as backup capacity. In a non-wind power system, to supply 1GW of reliable power would typically require installing 1.2GW of capacity. Thus, if a wind farm is oversized by a factor of four (4GW of wind farm capacity to service 1GW of peak demand), the wind farm is only a factor 3.3 larger than the non-wind alternative.

In practice, some storage capacity might be added at the wind farm and the degree of oversizing reduced. The optimal amount of storage would depend on the relative costs of storage, wind turbines and transmission links, as well as the wind speed distribution and the round-trip loss in and out of storage. The purpose of the calculations presented here is not to present definitive calculations of the cost of wind energy, but rather, to demonstrate the potential to achieve high capacity factors through oversizing the wind farm and at relatively little additional cost. Studies assessing the cost of storage are discussed next.

3.14.2 Compressed air energy storage

As discussed in subsection 3.11.2, excess wind energy can be used to compress air, which would be stored underground and used with supplemental natural gas or biogas to generate electricity when wind power is unable to meet demand. The cost of electricity supplied through CAES can be computed as:

$$C_{e\text{-}storage} = \frac{CRF_{storage}(CC_{power} + CC_{storage}h_s)}{8760CF_{storage}} + 0.0036HR \cdot C_{fuel} + C_{e\text{-}WT}ER + OM_{storage} \quad (3.25)$$

where $CRF_{storage}$ is the cost recovery factor for the storage system, CC_{power} is the power-related capital cost ($ per kW of output capacity), $CC_{storage}$ is the energy-related capital cost ($ per kWh of storage capacity), h_s is the number of kWh of storage per kW of output power capacity, HR is the heat rate (GJ of fuel used per GJ of electricity supplied), ER is the ratio of electricity input to electricity output and $OM_{storage}$ is the O&M cost per kWh of electricity supplied from storage. As noted earlier, advanced adiabatic CAES systems provide the prospect of reducing the fuel requirements to zero. A typical ER value for CAES is 0.735. CC_{power} can be computed as:

$$CC_{power} = CC_{compressor}R + CC_{expander} \quad (3.26)$$

where $CC_{compressor}$ is the compressor cost ($/kW of input power), $CC_{expander}$ is the expander cost ($/kW of output power) and R is the ratio of peak input power to peak output power. Representative costs are: $170/kW for compressors, $185/kW for expanders, $0.1/kWh for storage in aquifers, $1/kWh for storage in excavated salt caverns and $30/kWh for caverns excavated in hard rock (Greenblatt et al, 2007). The average cost of electricity from distant wind farms with CAES is given by:

$$C_e = C_{e\text{-}WT}\beta + C_{e\text{-}storage}(1-\beta) + C_{e\text{-}TL} + C_{t\text{-}loss} \quad (3.27)$$

where β is the fraction of electricity directly supplied by wind turbines.

Greenblatt et al (2007) assess the competition between baseload electricity supply using natural gas combined cycle powerplants next to demand centres, distant wind farms with natural gas backup and distant wind farms plus CAES (using natural gas). To calculate costs, they first generated a time series of hourly wind speeds using the Weibull probability distribution function. Electricity output from the wind farm is split between directly transmitted power, power to CAES and power that is spilled (curtailed). The wind farm is assumed to be large enough (2GW) and hence spread over a large enough area that wind speed variations at individual sites that are shorter than one hour cancel out. Selected input cost assumptions and the resulting total capital costs for the three systems are given in Table 3.21. System capital costs are \$660/kW for natural gas, \$1640/kW for wind plus natural gas and \$2270/kW for wind plus natural gas plus CAES. Electricity costs as a function of the cost of natural gas are given in Figure 3.49. In spite of the much greater capital costs for the two wind systems, wind plus natural gas is the least-cost system for natural gas costs of \$7–9/GJ and wind plus CAES is the least-cost system for a natural gas cost of \$9/GJ or greater. CO_2-equivalent emissions for the wind plus CAES system (32gC_{eq}/kWh) are a factor of four less than for natural gas combined cycle and a factor of nine less than current typical coal powerplants (276gC_{eq}/kWh). Wind plus CAES would also be competitive with coal powerplants having carbon capture; for a capital cost of \$2200/kW and 40 per cent electricity generation efficiency (both being highly optimistic assumptions for coal plants equipped with carbon capture, as discussed in Chapter 9, subsection 9.3.5), annual O&M equal to 2 per cent of the capital cost, coal at \$3/GJ and 10 per cent financing over 40 years, the cost of electricity would be 6.5 cents/kWh plus additional costs for compression, transport, and sequestration of captured CO_2.

For CAES using gasified biomass rather than natural gas as the fuel, Denholm (2006) computes a levelized cost for electricity from storage followed by long-distance transport of 5–6 cents/kWh, assuming turbine

Table 3.21 *Input assumptions for and cost of electricity from natural gas combined cycle, from wind with natural gas backup, and from wind with CAES*

Item	Electricity supply system		
	Natural gas combined cycle	Wind + natural gas	Wind + CAES
	Capacities (GW)		
Gas simple cycle		0.2	
Gas combined cycle	2.0	1.8	
Wind turbines		2.16	4.61
CAES compressor			2.53
CAES expander			2.08
CAES storage			352GWh
Transmission inlet		2.16	2.08
Transmission outlet		2.0	2.0
	Overall costs and emissions		
Simple cycle turbines		\$60 m	
Combined cycle turbines	\$1.3 billion	\$1.19 b	
Wind farm		\$1.51 b	\$3.23 b
CAES			\$168 m
Transmission link		\$520 m	\$630 m
Total	\$1.3 billion	\$3.28 b	\$5.54 b
System cost (\$/kW)	\$660	\$1640	\$2270
CO_2 emission (gC_{eq}/kWh)	120	73	32
	Average outputs as a fraction of transmission capacity		
Wind farm		39%	83%
Wind direct transmission			55%
Electricity from CAES			26%
Spilled electricity			2%

Note: All systems provide 2GW of baseload power. All costs are costs projected for 2020 but in terms of year 2002 dollars. Projected unit costs are \$240/kW and \$580/kW for simple and combined cycle turbines, \$700/kW for wind turbines, \$170/kW and \$185/kW for compressors and expanders, and \$1/kWh for storage in salt caverns.
Source: Greenblatt et al (2007)

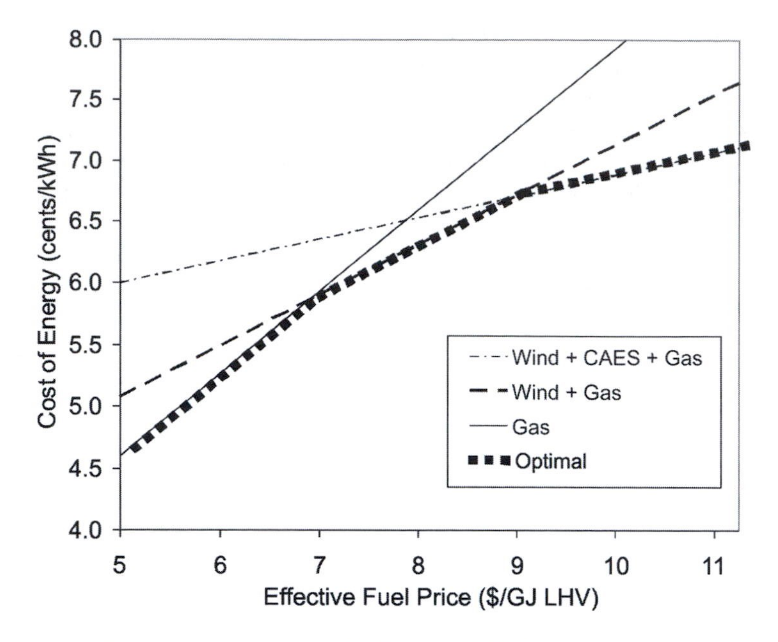

Note: The two wind systems assume transmission over a distance of 750km using HVDC from a region with a mean wind speed of 8.2m/s and 650W/m^2 wind power density, whereas the natural gas system generates electricity on site.
Source: Greenblatt et al (2007)

Figure 3.49 *Cost of electricity versus the cost of natural gas for baseload electricity generated using natural gas combined cycle, wind and natural gas or wind with CAES (using natural gas as a supplemental fuel)*

and CAES costs of $700/kW and $560–890/kW, respectively, a biomass syngas cost of $6.9/GJ and rather expensive transmission costs of $0.86/kW/km (for a 200MW, 240km link) or $0.42/kW/km (for a 2000MW, 750km link).

DeCarolis and Keith (2006) assessed the cost of large-scale baseload wind power in the US for a system consisting of the following components: one or five (geographically dispersed) wind farms in the central plains of the US; some combination of simple cycle gas turbines, combined cycle gas turbines and CAES combined with a gas turbine as backup for when the wind power alone is unable to meet demand; and HVDC transmission corridors sized to meet some fraction of the wind farm(s) capacity (the optimization model allows the wind farms to be oversized relative to the transmission capacity). For current component capital costs, the direct cost of wind energy produced in western Iowa and transmitted to Chicago is 4.1 cents/kWh – about the same as for an all-gas baseline with natural gas at $4/GJ. This cost includes transmission cost and line losses but ignores intermittency. If wind is scaled up to supply 50 per cent of the electricity demand, with some natural gas backup and some oversizing of the wind plant (and hence some wasted wind power), the average cost of electricity increases by 1.2 cents/kWh if there are five dispersed wind farms and by 1.6 cents/kWh if there is only one wind farm.

3.14.3 Use of dispatchable loads such as reverse osmosis desalination

As noted earlier, an increasing fraction of the potential wind-generated electricity will be wasted as the potential annual electricity production approaches the annual demand, due to supply–demand mismatches, and this in turn drives up the cost of wind-generated electricity. An alternative to suppressing electricity generation at these times would be to use the electricity for some nearby dispatchable load. In coastal semi-arid parts of the world, there is a growing need for desalination of seawater in order to provide fresh water. The most efficient method of desalination is by reverse osmosis, which is driven by electricity rather than heat, and may in the future require as little as 2–3kWh/m^3 (see Volume 1, Chapter 6, section 6.13). Electricity for desalination plants could be sold at a deep discount so that at least some revenue is generated from generation potential that would otherwise be wasted. However, reliance on intermittent power would significantly increase the capital cost contribution to the total cost of desalinated water due to the low utilization of the desalination equipment. The tradeoff between these two factors is briefly examined here, and costs are compared with those using fossil fuel-derived electricity but with desalination equipment operated at close to full load.

Facilities for desalination using reverse osmosis cost a minimum of \$900 per m^3/day of desalination capacity. If *CRF* is the cost recovery factor, *CF* the capacity factor and CC_{desal} the capital cost, then the capital cost contribution to the cost of water is:

$$C_{cap} = \frac{CRF \cdot CC_{desal}}{365 \cdot CF} \quad (3.28)$$

For CC_{desal} = \$900/m^3/day, 6 per cent interest over a 20-year lifespan and CF = 1.0, C_{cap} = \$0.215/m^3, whereas for an electricity requirement of 3kWh/m^3 and for electricity at a non-discounted price of 8 cents/kWh, the electricity cost contribution to the cost of water is \$0.24/m^3. Membrane replacement every five years would add another \$0.03/m^3 and chemicals would add another \$0.02/m^3 or more to the cost (Wilf, 2004), but these would be largely independent of the capacity factor, while labour costs would be about \$0.03/m^3 and would vary inversely with capacity factor. Thus, even modest reductions in the capacity factor due to intermittent supply of electricity would overwhelm the benefit of even a 100 per cent discount in the cost of electricity. For example, with electricity at 1 cent/kWh instead of 8 cents/kWh and a capacity factor of 0.25 instead of 0.9, the cost of desalinated water is \$1.06/m^3 instead of \$0.54/m^3.

Because the cost of desalination using reverse osmosis is dominated by the capital cost component, an alternative strategy for reducing the cost of desalinated water would be to build oversized wind farms in coastal semi-arid regions, next to desalination facilities, and to accept some loss in wind generation potential in exchange for being able to operate the desalination equipment at closer to full load on average. That is, peak desalination capacity and power requirement would be less than the peak wind energy output, so much of the analysis in subsection 3.14.1, pertaining to transmission capacity undersized relative to peak wind farm capacity, can be applied here. In particular, we will use the demand-side capacity factors and electricity costs derived for wind farm scaling factors of one to four, but assuming zero transmission costs and losses because the wind farm is assumed to be located next to the desalination plant.

Figure 3.50a shows the variation in the capital cost, energy cost, membrane replacement and chemical cost, and labour cost components of the total cost of desalinated water, as a function of the mean wind speed with the Weibull shape parameter k = 2.0 for the case of no wind farm oversizing. Figure 3.50b shows the variation in total water cost for wind farm scaling factors of one to five. Oversizing the wind farm by a factor of two relative to the desalination power requirement at full load reduces the cost of water at all wind speeds considered, with the greatest benefit at small mean wind speeds. However, there is relatively little benefit in oversizing the wind farm by more than a factor of two. The quantitative results are of course subject to error, but serve to demonstrate that oversizing of wind farms to some extent when coupled to a desalination load will reduce the cost of water, due to the importance of capital costs to the total cost of water and the benefit of increasing the utilization of the desalination equipment.[6]

3.15 Energy payback time and GHG emissions

The energy payback time for a wind turbine is the length of time that the turbine must operate in order for it to save an amount of primary energy equal to that used to manufacture, transport and install the turbine. The turbine generates electricity but saves the primary energy that would otherwise be used at the fossil fuel powerplant it displaces. The saving in primary energy thus depends on the electricity output as well as the efficiency of the fossil fuel powerplant that would otherwise operate. The amount of primary energy used to produce the turbine depends on the amount of thermal and electrical energy used to produce the steel, cement and other materials (first-order inputs), the amount of energy used to produce the factories that make the steel, cement and other materials (second-order inputs), and so on. As the efficiency of the fossil fuel powerplants that the wind turbine displaces

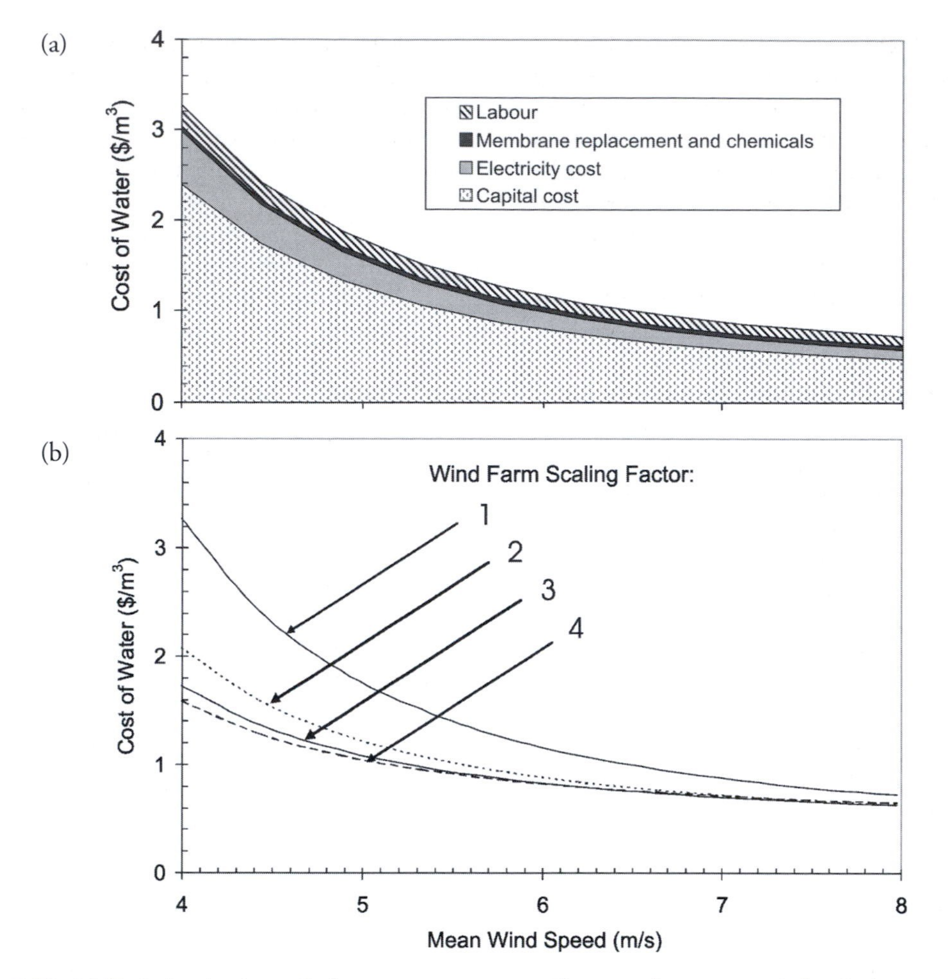

Figure 3.50 *(a) Variation in the capital cost, energy cost, membrane replacement and chemical cost, and labour cost components of the total cost of desalinated water, as a function of the mean wind speed with the Weibull shape parameter* k = *2.0 for the case of no wind farm oversizing; (b) variation in total cost of desalinated water for wind farm scaling factors of one to five*

improves (see Volume 1, Chapter 3, sections 3.2 and 3.3), the savings in primary energy will decrease. However, inasmuch as some of the energy used to manufacture the turbine is electrical, the primary energy input will also decrease. Energy used in transporting raw materials to the factory, and in transporting turbine components from the factory to the site, should be but is not always included. For transport 1500km by truck, this energy input can dominate the overall energy inputs (Tremeac and Meunier, 2009).

Lenzen and Munksgaard (2002) present a comprehensive compilation and analysis of studies up to 2000 of the energy input to energy output ratio for wind turbines. Energy intensity ratios (r) – the ratio of lifetime primary energy input (including maintenance and decommissioning, with credits for end-of-life recycling) to lifetime electrical output – are generally in the range of 0.02–0.10. The time required to pay back the energy input is equal to $r \times T \times \eta_{ff}$, where T is the turbine lifetime used in the analysis (20 years) and η_{ff} is the efficiency of the displaced fossil fuel powerplant (typically 0.33). The corresponding payback times are in the order of two to eight months. The tower tends to account for one third to one half of the energy used to produce a wind turbine. Wagner and Pick (2004) provide a recent breakdown of the energy used in making different components of a wind turbine. Computed overall payback times are three to six months for wind farms in Germany, the same range as computed by Ardente et al (2008) for wind farms in Italy. Khan et al (2005) estimated a payback time of eight months for a system in Newfoundland and Labrador consisting of wind turbines, electrolysers and fuel cells; the greater payback time in this case is due to the materials in the electrolysers and fuel cells.

Voorspools et al (2000) summarize previous studies of the amount of materials used in wind turbines. Noting that there is little relationship between material inputs and the turbine capacity, they give the following material intensities: concrete, 360t/MW; steel, 125t/MW; plastics, 9.7t/MW; copper, 4.4t/MW; and aluminium, 0.4t/MW. These material intensities can be compared with those for nuclear powerplants given in Chapter 8 (Table 8.20). For a site with a capacity factor of 0.34, the estimated primary energy use is 0.12MJ/kWh$_e$ and the estimated CO_2-equivalent GHG emission is 2.5gC/kWh. By comparison, a coal powerplant at 40 per cent efficiency would emit 225gC/kWh.

Denholm and Kulcinski (2004) estimate the energy used to store 1GWh of electricity in various storage devices, taking into account the energy used to manufacture and decommission the storage device and any additional energy inputs (apart from electricity) needed to access stored energy (such as natural gas in the case of CAES). Summary results are given in Table 3.22. The main energy input to the storage system is the electrical energy input, which is stored in one form or another. *ER* is the ratio of electricity input to electricity output; the reciprocal of *ER* would be the round-trip storage efficiency if there were no other energy inputs. The additional energy input, E_{aux}, involves the primary energy used during the operation of the storage facility, E_{op} (MJ/MWh$_{stored}$) and the primary energy used for the construction of the facility, EE_s (GJ per MWh of storage capacity) divided by the lifetime MWh of energy stored per MWh of capacity, E_L^S. E_L^S in turn is given by the lifespan of the storage facility in years times the number of equivalent complete charge/discharge cycles per year. E_{aux} is given by:

$$E_{aux}(\text{MJ/MWh}_{\text{stored}}) = EE_{op}(\text{MJ/MWh}_{\text{stored}}) + \frac{EE_s(\text{GJ/MWh}_{\text{cap}})(1000\text{MJ/GJ})}{E_L^S(\text{MWh}_{\text{stored}}/\text{MWh}_{\text{cap}})} \quad (3.29)$$

All of these inputs, as given by or deduced from Denholm and Kulcinski (2004), are given in Table 3.22. The system efficiency should take into account both *ER* and the amount of electricity that could have been generated with E_{aux}, if it had not been used for the construction and operation of the storage facility. This depends on the efficiency η of electricity generation. The system efficiency is then given by:

$$\eta_s = \frac{1}{ER + (E_{aux}/3600)\eta} \quad (3.30)$$

where the factor 1/3600 converts MJ to MWh (so E_{aux}/3600 is dimensionless). System efficiencies are given in Table 3.22 for η = 0.4 and 0.6 (the latter representing state-of-the-art gas combined cycle power plants) and range from about 0.62 to 0.76.

Pehnt et al (2008) have assessed the net effect on CO_2 emissions of offshore wind farms in Germany. The

Table 3.22 *Lifecycle energy use and system efficiencies for various storage systems*

Storage system	E_{es} (GJ/MWh$_{cap}$)	E_{op} (MJ/MWh$_{stored}$)	Lifespan (years)	Cycles per year	E_L^S (MWh$_{stored}$/MWh$_{cap}$)	E_{aux} (MJ/MWh$_{stored}$)	ER	η_s	
								η = 0.4	η = 0.6
PHS	373	25.8	60	155	9300	66	1.35	0.737	0.735
PSB	1755	54	20	220	4400	453	1.54	0.629	0.619
VRB	2253	45	20	220	4400	557	1.33	0.718	0.703
CAES	265.7	5210[a]	20	220	4400	5270	0.735	0.757	0.620

Note: PHS = pumped hydro storage, PSB = polysulphide battery, VRB = vanadium redox battery and CAES = compressed air energy storage. [a] Of which 4649MJ are fuel energy, 518MJ are energy used to deliver the fuel, and 42MJ are for O&M and materials consumed by pollution controls.
Source: Based on Denholm and Kulcinski (2004)

avoided emission is 249gC/kWh, assuming that 90 per cent of the generated electricity displaces coal that generates electricity at an efficiency of 43.2 per cent. Emissions from the construction of the wind farm and from grid extension amount to 6gC/kWh, while an extra 19gC/kWh are emitted due to a 0.5 per cent drop in the efficiency of the remaining coal generation. The net reduction in CO_2 emission is thus 224gC/kWh.

3.16 Noise and impact on birds

It is sometimes claimed that wind turbines are noisy or pose a large hazard to birds. Neither claim is true, as can be seen from the indicative noise levels for wind turbines and various activities given in Table 3.23, and from the estimated mortality rates for birds in the US due to various human-related causes given in Table 3.24. The noise from a wind turbine at a distance of 350m is similar to the background noise in a typical home. Estimated mortality rates due to modern wind turbines are well below one bird per turbine per year (modern turbines have a solid rather than lattice support structure, and so are unattractive as nesting sites, while the rotation rate is smaller the larger the turbine). By contrast, domesticated cats, road vehicles and telecommunication towers are each estimated to kill several tens of millions of birds per year in the US, and estimates of the bird mortality due to collisions with buildings in the US are as high as 1 billion per year.

It also should not be overlooked that the alternatives to wind farms, coal powerplants in particular, are responsible for some bird deaths too, through loss of breeding grounds from mining operations. More importantly, global mean climatic warming by as little as 2°C by 2050 compared to pre-industrial conditions is likely to provoke extinction of 15–33 per cent of species worldwide, with even greater losses by the end of the century with unrestrained emissions (see Volume 1, Chapter 1, subsection 1.4.1). Nevertheless, wind farms should not be located in the midst of bird migration routes or next to important breeding areas. Some bat mortality occurs due to wind turbines, but further study is required in order to determine the magnitude of this problem.

Table 3.23 *Noise level of wind turbines compared to noise levels from various other sources*

Source	Indicative noise level (decibels)
Threshold of pain	140
Jet aircraft at 250m	105
Pneumatic drill at 7m	96
Truck at 48kph at 100m	65
Busy general office	60
Car at 64kph at 100m	55
Wind development at 350m	35–40
Quiet bedroom	35
Rural night-time background	20–40

Source: GWEC (2006)

Table 3.24 *Main human-related causes of bird deaths in the US*

Cause	Estimated number of deaths per year
Utility transmission and distribution lines	130–174 million
Collisions with road vehicles	60–80 million
Collisions with buildings	100–1000 million
Telecommunications towers	40–50 million
Agricultural pesticides	67 million
Cats	39 million

Source: GWEC (2006)

3.17 Benefits to farmers

It is of interest to compare the revenue that farmers could earn by allowing wind turbines on their land, with typical revenue from food crops. Average power production in good wind sites (8–9m/s at hub height) would range from about 2–4MW/km^2. Assuming a wholesale electricity price of 6 cents/kWh and a farmer's income equal to 2 per cent of the value of the electricity generated on his land, the revenue is \$200–400/ha per year. By comparison, the net revenue (income minus costs) from a variety of different crops ranges from \$100–400/ha/yr in North America but can sometimes be negative. Thus, placement of large wind farms on agricultural land in windy areas would provide a large source of rural income that is relatively stable (±10–15 per cent variation and always positive!).

3.18 Overcoming local opposition

There is a widespread, general acceptance of wind energy – referred to by Wüstenhagen et al (2007) as 'socio-political' acceptance – but wind energy has frequently encountered a high level of local opposition to specific development sites. A number of studies have examined the reasons for local opposition to wind energy projects. These include a perceived negative impact of wind turbines on the landscape (Johansson and Laike, 2007; Wolsink, 2007a) or perceived environmental impacts, including direct bird mortality and disruption of wetland habitats (Wolsink, 2000). As discussed by Wolsink (2000), the characterization of the opposition as NIMBYism (not in my backyard) is too simplistic.

Opposition is greater when wind projects are promoted by outside developers who are perceived as exploiting public spaces solely for private profit (Toke, 2005; Firestone et al, 2009). Public perceptions of the fairness and openness of the decision-making process also exert a considerable influence over the degree of opposition to or acceptance of wind energy (Gross et al, 2007). The 'decide-announce-defend' approach to siting of wind turbines, with minimal public involvement, has been repeatedly shown to antagonize the public. Wolsink (2007b) describes a case where a 178MW project in The Netherlands failed to move forward because of the government's refusal to negotiate with a local environmental group concerning the specific site chosen. Jobert et al (2007) describe a case where a private company attempted to build a wind farm without consulting the local community, leading to mistrust that played an important role in inciting opposition. In other cases, where developers contracted with private landowners to receive turbines, other members of the community felt that the local peace had been disrupted, with some landowners profiting while all members of the community had to look at the turbines. Where developers must bid for the right to develop wind energy in a competitive process in response to a government request for proposals, the necessary secrecy in such a process prevents consultation with communities at an early stage. Instead, consultation is possible only after the project has been approved (Cowan, 2008).

Experience in Europe indicates that a key to gaining widespread approval of wind projects is to involve the community from the beginning. Communities can be encouraged to bring forth their own project proposals, or outside developers can involve key members of the community at the beginning. In the case of a project in the UK, support or opposition to the project was largely based on how people learned about the project (Hinshelwood, 2000). Local people can play an important role in disseminating information about a project, and opposition will decrease as more respected local individuals get involved. Hinshelwood (2000) stresses that it is necessary to allow time for sufficient debate and for people to discuss and resolve their concerns (although there will always be some opposition to most projects). Dedication of some of the profits from a wind farm to a local charitable trust that can support local community activities and development will also probably reduce the local opposition. Local environmental groups can be invited to participate in the analysis of local environmental impacts (Jobert et al, 2007).

Of particular interest here is the study by Firestone et al (2009) of community attitudes to offshore wind farms in Massachusetts and Delaware, US. They found that there was *less* opposition to wind farm projects that would be the first of many projects that would eventually supply half of coastal states' electricity, with associated environmental benefits, than to isolated projects that would not noticeably alter the electricity supply. This is encouraging, as wind farm developments at this very scale are what are needed to address the global warming issue.

3.19 Global and regional wind energy potential and cost

In this section we review recent estimates of the amount of electricity that could be generated worldwide while incurring costs up to various cost limits. Any such estimate begins with data on the available wind resources, so we begin with some comments on how such data sets are produced.

3.19.1 Wind assessment methodology

Assessment of the wind resource in a region usually begins with the analysis of historical data collected at the nearest measurement site. The historical data have usually been affected by nearby obstacles. Computer programs exist that attempt to remove the effects of obstacles; to do so requires information on wind direction as well as wind speed. The cleaned data can then be used as one of several inputs in computer simulation models that simulate the wind field at all locations over some region. The region simulated may contain complex terrain with highly non-uniform winds, but the simulation serves as a guide as to where to expect the best winds. Computer simulations must be supplemented by at least two years of data (preferably) from the most promising sites. The relevant information is the wind speed distribution at the turbine hub height, which ranges from 80m above ground for a 2.5MW turbine to about 120m above ground for a 5MW turbine (see Figure 3.5). However, until recently the highest meteorological towers were only 50m high, so extrapolation to greater heights is required, and most meteorological stations only provide surface or near-surface (10m height) wind speed data. In a region where several wind turbines have already been constructed, the wind turbines themselves are a source of information about wind in the region. The uncertainty in long-term wind speed estimates is about ±15 per cent if there is only one year of data, ±12 per cent if there are two years of data and ±10 per cent with three years of data.

3.19.2 Regional and global assessments

Denholm and Short (2006) assessed the wind potential in various wind classes for the US. The amount of land in each wind class was based on the mean wind speeds at the 50m height, but the wind farm power density is independent of the wind class and assumed to be $5MW/km^2$. This density times the available land areas in the various wind classes (taking into account exclusions due to slope, forests, human settlements, national parks and other considerations) gives the total wind power potentials shown in Table 3.25. The potential peak power from Class 7 winds alone is 84GW, while that from Classes 6 and 7 alone is 639GW. However, the assumed $5MW/km^2$ power density is somewhat conservative as, according to Table 3.5, the power density of the 1.5MW turbine if laid out with a spacing of seven rotor diameters in both the cross-wind and along-wind directions would be $5.2MW/km^2$, while that of the 3.6MW turbine would be $6.8MW/km^2$. Thus, based on the 3.6MW turbine, the potential power capacity in the US from Class 6 and 7 winds alone is 869GW, compared to the total US powerplant capacity in 2005 of 1076GW (see Volume 1, Table 2.4). Assuming a hub height of 104m, the actual mean wind speeds at the hub height would be greater than the wind speeds used for classifying the various grid points. The adjusted wind speeds can be used with Equation (3.13) to calculate the average capacity factor for turbines falling in the various initial wind class categories, and then multiplied by the adjusted potential capacity in each category to get the potential annual electricity generation (shown as the last column of Table 3.25). Given a turbine power density of $6.8MW/km^2$, the land area required to generate an amount of electricity equal to total US electricity consumption in 2005 (about 4300TWh/yr) is 384km × 384km (this almost exactly uses up all the Class 6 and Class 7 sites; if only a portion of Class 6 and 7 sites are used, then the total required land area would be greater due to the use of some less favourable Class 5 sites). The area taken up by these wind farms would amount to 1.9 per cent of the area of the lower 48 US states, and only about 1–2 per cent of that

Table 3.25 *Data column 1: middle wind speed for different wind speed categories; data columns 2–5: potential generating capacity at sites in the US falling within different wind class categories based on mean wind speeds at a height of 50m; data columns 6–8: adjusted wind speed at these sites based on a hub height of 104m rather than 50m, capacity factor computed from Equation (3.13) using the adjusted wind speed, and potential electricity generation (given by the potential generating capacity adjusted to a power density of $6.8MW/km^2$ × capacity factor × hours in a year)*

Wind class	Mean wind speed (m/s)	Potential generating capacity (GW)				Wind speed (m/s) at 104m	Capacity factor	Potential generation (TWh/yr)
		Onshore	Shallow offshore	Deep offshore	Total			
Class 7	9.2	22.7	0.9	60.8	84.4	10.3	0.56	563
Class 6	8.4	78.5	44.2	432	555	9.4	0.48	3700
Class 5	7.8	393.4	46.2	111	551	8.6	0.43	3673
Class 4	7.3	2349.2	12.6	87	2449	8.1	0.38	16,334

Source: Data for data columns 1–5 from Denholm and Short (2006)

would be taken up by the turbine foundations, with the rest available for agricultural or other land uses.

Lindenberg et al (2008) carried out a comprehensive assessment of how wind could meet 20 per cent of a projected US electricity demand of 5800TWh in 2030. They used the *Wind Energy Deployment System Model* (WinDS). WinDS accounts for the geographical variation in wind energy potential and in electricity demand, and takes into account the existing transmission grid. Wind capacity would have to reach 300GW, and the rate of installation would have to reach 16GW/yr by 2030 in order to reach the 20 per cent target (by comparison, 10.0GW were installed in the US in 2009). The analysis indicates that as much as 600GW of wind resources could be available at costs of 6–10 cents/kWh, including the cost of connection to the existing grid.

Figure 3.51 gives an alternative estimate of the wind energy potential in the US, this time by state (for the ten windiest states) rather than by wind class category. The total for these ten states is about two and half times the total US electricity demand (about 10,000TWh/yr versus 4300TWh/yr). These states are within 2000km of the furthest load centres, a distance that can be spanned with HVDC with minimal energy losses.

With regard to the rest of the world, Archer and Jacobson (2005) used a combination of surface wind speed (typically at the 10m height) from 7753 meteorological stations and the variation of mean wind speed with height in the atmosphere as observed at 446 sounding stations to estimate the distribution of mean wind speeds at the 80m height over all land areas. The average wind speed of the Class 3 or better sites is 8.44m/s. Table 3.26 gives the total land area by continent, the per cent of land area with Class 3 or better winds, the absolute land area with Class 3 or better winds, and the resulting regional wind energy potentials for the 1.0 and 3.6MW turbines of Table 3.5, along with the current regional electricity demand. The potential average wind power from onshore sites ranges from 46TW using 1MW turbines to 65TW using 5MW turbines, compared to a global *primary* power demand of 15.3TW in 2005. The North American prairies, the region of Labrador in Canada, the North Sea, the Inner Mongolia autonomous region in China and the plains of Russia are some examples of extensive areas with good wind resources. The potential electricity generation ranges from 402,000TWh/yr to 566,000TWhy/yr, compared to a worldwide electricity demand in 2005 of about 18,500TWh. If only 10 per cent of the technical potential can be utilized in practice (given land constraints and distance from demand centres), the potential electricity production from wind still exceeds current demand in every continent except Asia.

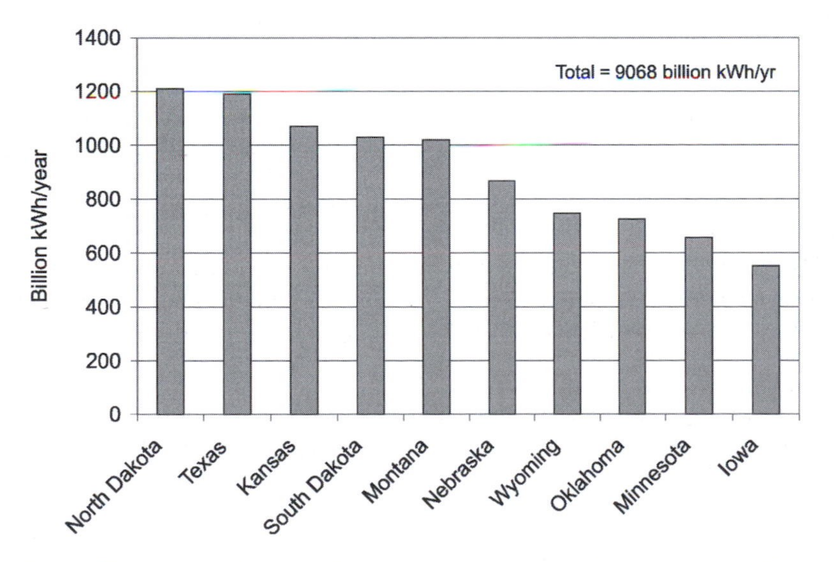

Source: Data from Makhijani (2007), from the American Wind Energy Association, in turn based on an inaccessible 1991 report

Figure 3.51 *Wind energy potential from the ten states in the US with the greatest potential, based on winds at a relatively low height (probably 30–50m)*

Table 3.26 *Land areas, percentage of area with Class 3 or better winds (≥6.49m/s) at a height of 80m, average power output, and annual electricity production using 1MW and 5MW wind turbines*

Region	Area (Mkm^2)	Percentage as Class 3 or better	Area as Class 3 or better (Mkm^2)	Average power (TW)		Electricity production (1000s TWh/yr)		Electricity demand in 2005 (TWh/yr)
				1MW	5MW	1MW	5MW	
Europe	10.5	14.2	1.5	5.6	7.9	49	69	4840
N America	24.2	19.0	4.6	17.3	24.4	152	214	5280
S America	17.8	9.7	1.7	6.5	9.2	57	80	820
Oceania	8.5	21.2	1.8	6.8	9.6	59	84	305
Africa	30.3	4.6	1.4	5.2	7.4	46	65	540
Asia	43.6	2.7	1.2	4.4	6.2	39	55	6640
Global	135.0	9.0	12.2	45.9	64.7	402	566	18,450

Source: Based on Archer and Jacobson (2005)

For China, McElroy et al (2009) estimated the wind energy potential using wind data at 0.5° longitude, 0.67° latitude and 6 hour temporal resolution from a meteorological assimilation dataset. Land use data were used to eliminate areas that are unsuitable for wind farms. They find that a suite of 1.5MW turbines in onshore regions with favourable wind resources could generate more than four times the 2005 Chinese electricity consumption (11PWh/yr vs 2.5 PWh/yr) at a cost of up to 0.55RMB/kWh (the highest contract prices currently offered for wind energy in China), and ten times the 2005 Chinese electricity consumption at a cost of up to 0.67RMB/kWh (see their Figure 4).

Hoogwijk et al (2004) assessed the global wind energy potential assuming 1MW turbines with a hub height of 69m. They extrapolated surface (10m height) wind speed data to the hub height with the following constraints: only areas with a surface wind speed of 4m/s or greater and a land surface altitude of 2000m or less were considered, urban areas were excluded, and only a portion of the remaining lands was assumed to be suitable for wind turbines (this portion ranges from 0 per cent of bio-reserves and tropical forests, to 10 per cent of boreal forest, 70 per cent of agricultural land, 80 per cent of tundra and grasslands, 90 per cent of savannah and 100 per cent of hot deserts). The result is that only 8.6 per cent of the ice-free land surface area (or 11.23 million km^2) was judged to be suitable for wind turbines. Wind turbine output was adjusted based on air density (which decreases with increasing altitude) and reduced by 10 per cent to account for interference of the turbines in a wind farm with each other. They assumed an average installed wind power capacity of $4MW/km^2$ in the suitable areas. This leads to a predicted potential electricity production from the suitable areas of 96,000TWh/yr. Hoogwijk et al (2004) calculate a global potential wind energy potential of 21,000TWh/yr up to a cost of 7 cents/kWh and 53,000TWh/yr up to a cost of 10 cents/kWh (present global electricity demand is about 18,000TWh/yr). These costs are based on an investment cost of $1000/kW, 10 per cent real interest, a 20-year lifespan and fixed O&M equal to 3 per cent of the investment cost per year, but do not include the cost of transmission, wasted electricity, or storage.

Costs arising from transmission, wasted electricity, backup capacity (after accounting for the capacity credit) and the need for extra spinning reserve were computed in a follow-up study (Hoogwijk et al, 2007), but the results depend on the extent to which interconnected wind farms are dispersed geographically, as having dispersed and well-connected wind farms simultaneously reduces the need for spinning reserve and backup capacity and the need to discard electricity. Being able to connect to hydro and quick-start natural gas powerplants, having dispatchable loads and having good wind forecasts will also reduce costs.

3.19.3 Impact of climatic change on wind resources

Future warming of the earth's climate will be spatially non-uniform, and this will lead to changes in horizontal pressure gradients that in turn will lead to changes in wind speeds. There have been several assessments (reviewed by Sailor et al, 2008) using climate simulation models of the potential impacts of projected changes on wind energy potential. Generally, increases or decreases in wind power in specific regions and seasons of up to (but usually much less than) 40 per cent have been found.

3.20 Scenario of future wind energy use

A reasonable representation of how global wind capacity could vary over time is through the logistic function, whereby:

$$C(t) = \frac{C_u}{1+(C_U - C_o)/C_o e^{-a(t-t_o)}} \quad (3.31)$$

where C_u is the ultimate wind power capacity, C_o is the capacity in year t_o, and a is the initial exponential growth rate tendency (see Volume 1, Box 2.1 for the derivation of Equation (3.31)). Equation (3.31) is applied starting from the observed global wind capacity of 59.1GW at the end of 2005 and assuming an ultimate wind capacity of 12,500GW. The solid line in Figure 3.52a shows the growth of installed wind capacity when the growth parameter a = 0.2yr^{-1} while that in Figure 3.52b shows the result using a = 0.15yr^{-1}.[7] The solid bell-shaped area in each figure shows the variation in the rate of growth in total installed capacity and the broken area shows the rate of installation of replacement turbines (which is equal to the total rate of installation of new and replacement turbines 20 years before, assuming a 20-year turbine lifespan). With the more rapid buildup in total turbine capacity, the market for new turbines enters a 20-year oscillation beginning in 2035, while with the slower buildup the oscillation is smaller in amplitude and begins later. The rate of installation of new turbines averaged over these cycles is equal to the assumed steady turbine capacity divided by the assumed turbine lifespan, namely, 625GW/yr. This is 23 times the 2008 market of 27.1GW. The global wind energy market reaches 183GW/yr and 72GW/yr for the two scenarios. Interestingly, China alone is targeting a rate of installation of new wind turbines of 30GW per year by 2020 (Yang, 2007).

As noted in subsection 3.14.1, if wind farms are located in regions with a mean wind speed at the hub height of 9m/s (which would be typical of good wind regions), and wind farms are oversized by a factor of 3 relative to the transmission link, the average transmitted power would be 0.66 times the transmission capacity but wind farm capacity factor would be only 0.16. Assuming an average capacity factor of 0.16, the 12,500GW eventual capacity produces 17,500TWh/yr of electricity – about the same as current total world electricity demand but only 40–80 per cent of the global electricity demand in 2100 under the aggressive energy efficiency scenarios presented in Volume 1 (Chapter 10, section 10.4). The total wind farm area reaches 2.1 million km^2, which is 1.4 per cent of the total global land area but located in the best wind sites and generally no more than 2000km (often much less) from the major electricity demand centres. The long-term average rate of construction of

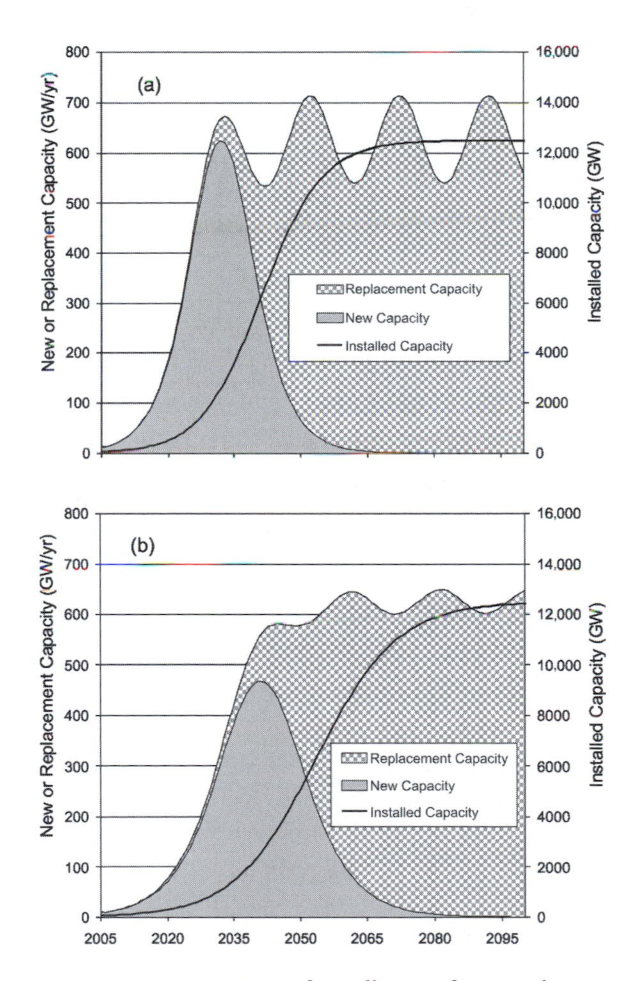

Figure 3.52 *Rate of installation of new and replacement wind turbine capacity, and growth in installed capacity, for a scenario in which the total capacity grows following a logistic curve to an ultimate capacity of 12,500GW with (a) growth parameter a = 0.2 and (b) with a = 0.15*

new turbines (625GW/yr) is about a sixth of the mechanical powerplants built in the form of cars and light trucks in 2008 (about 40 million vehicles with an average engine power of 100kW, giving a total powerplant construction of about 4000GW/yr).

3.21 Summary

Wind energy is a rapidly growing carbon-free energy source for the generation of electricity, the installed capacity having grown by an average compounded rate of 25 per cent/year since 1995 and reaching 157.9GW by the end of 2009. This rapid growth has been due to its relatively low cost among renewable-energy options, with installed costs of onshore turbines having dropped to $800–1200/kW by 2001 (but recently rising to $1300–1900/kW or €950–1350/kW due to a spike in material costs that affected all types of powerplants) and electricity costs reaching a highly competitive 5–8 cents/kWh in many jurisdictions. Offshore wind systems currently cost about 50 per cent more than onshore systems (about $1800–2600/kW or €1300–1900/kW). Until recently, the cost of wind energy had fallen by about 20 per cent for each doubling in the global cumulative production of wind turbines. The downward trend was reversed recently due to a sharp spike in commodity prices, and although commodity prices will probably be high in the future, further improvements in technology and learning-by-doing should permit wind turbine costs to return to a downward trend. This is especially the case for offshore turbines because they are a less mature technology and because larger turbines (which entail installation and grid-connection cost savings) can be installed offshore than onshore. The time required for a wind turbine to save an amount of primary energy at fossil fuel powerplants equal to the amount of primary energy required to make it (including the foundation) is less than one year, and the physical footprint (the land area taken up by the foundation and access road) is small.

The potential wind resource at a cost of up to 7 cents/kWh (taking into account expected future cost reductions) is probably greater than present world electricity, and larger still with larger cost ceilings. The difficulties with regard to wind relate to its intermittency and to the fact that some of the best wind regions are 1000–3000km from major electricity demand centres.

There are a number of strategies available for dealing with the intermittent and to some extent unpredictable nature of the wind resource. Fluctuations at a timescale of seconds can be reduced by aggregating the output of several wind turbines in a wind farm, or by using short-term storage such as flywheels, super-capacitors and superconducting magnetic storage. Fluctuations at longer timescales will be reduced when geographically dispersed wind farms are linked together. Electrolyser/fuel cell systems, flow batteries, underground compressed air, hydroelectric reservoirs and pumped hydro can store energy at a timescale of a day to months. Flexible loads, including heat pumps with thermal energy storage, can be employed to match fluctuations in wind energy. Parked plug-in hybrid vehicles could provide important regulation (voltage and frequency control) and storage functions; only 3 per cent of the US automobile fleet would be needed for voltage regulation and only 8–38 per cent would be required to provide spinning reserves if wind were to provide one half of US electricity demand.

Because the average output of a wind turbine is typically only 20–40 per cent that of the peak output, and the turbine rarely generates electricity at the rated output, it is possible to oversize a wind farm compared to the transmission capacity with relatively little wasted electricity generation potential. Although wasting some potential generation due to transmission limitations would increase the generation component of the electricity cost, it would reduce the transmission component of the cost because more energy would be transmitted through the given transmission link. If wind farms are located in the best regions, they will tend to be 1000–3000km from the major demand centres. Thus, transmission costs will be a significant fraction of total investment cost, and the savings in transmission cost per kWh of transmitted electricity can partly offset the increase in generation cost per kWh when the wind farm is oversized. However, the increased power output from the high-wind regions can also compensate for the added cost of transmission, particularly when the distant wind farm is oversized.

As an illustration, if wind turbines can be installed for $1000/kW locally and for $1500/kW at distant sites, if the mean wind speeds in the region of the demand centre and in a region 2000km away are 5m/s and 9m/s, respectively, and if the cost of transmission over a distance of 2000km is $460/kW with 5.0 per cent energy loss at times of peak transmission, then the cost of electricity produced locally is 8.0 cents/kWh and the average power output (the capacity factor) is 16 per cent of the peak output, while if the distant wind farm is oversized by a factor of three, the cost of the electricity is 8.2 cents/kWh but the capacity factor is 66 per cent.

Notes

1 One might wonder why a larger k (corresponding to a more peaked wind speed distribution) leads to a decrease in the capacity factor for mean wind speeds less than about 8m/s while having the opposite effect for mean wind speeds larger than about 8m/s. The explanation is that wind speeds in the higher ranges disproportionately contribute to the mean power output due to the cubic dependence of power on wind speed. At a low mean wind speed, the probability of wind speeds between 8–15m/s is reduced when the distribution becomes more peaked, while at a high mean wind speed, a more peaked distribution produces greater probabilities of winds throughout the 6–16m/s range. This is evident from Figure 3.16. Thus, there are opposite effects on the capacity factor.

2 For two equally sized wind farms with zero correlation in their output, the standard deviation of the joint output would be $1/\sqrt{2}$ (= 0.707) times the standard deviation of the individual turbine outputs.

3 This is one third that of petroleum fuels, but hydrogen can be used with twice the efficiency in fuel cell vehicles, so the effective emission factor for hydrogen used in the transportation sector would be as little as a sixth that of petroleum fuels.

4 The term $(1+ \varepsilon + \varepsilon^2 + \ldots..)$ arises from a Taylor Series expansion of $1/(1-T_{loss}CF_{TL})$, while the expression for δ is the expression for the sum of an infinite series involving ε. Some workers have given the cost of lost electricity as $(C_e+C_t)\varepsilon$, that is, neglecting the higher order terms in ε. The first-order term (ε) accounts for the extra generation required to account for the loss of electricity, but some of this extra generation is itself lost, which is accounted for by the second-order term (ε^2), and so on.

5 Assuming 10 per cent financing over 20 years and 1.5 per cent of the investment cost per year as O&M.

6 Note, however, that if desalination were to be applied on a very large scale, additional costs would be incurred in building pipes to disperse the salty brine that is produced over a sufficiently large area of the ocean to reduce impacts to an acceptable level.

7 In the first case, installed wind capacity reaches 107.2GW by the end of 2008 and in the second case it reaches 92.4GW, whereas actual world wind capacity reached 121.1GW by the end of 2008. Thus, the recent surge in the rate of growth capacity would be regarded as a temporary deviation from a more gradual buildup under these scenarios. At present, limited transmission capacity is becoming a real constraint on the rate at which wind farm capacity can be increased in many regions.

4

Biomass Energy

4.1 Introduction

The term 'biomass' refers to plant materials that were produced recently enough through the process of photosynthesis, using energy from the sun, that they are still present in unaltered form. Fossil fuels are biomass that has been transformed by chemical or thermo-chemical processes at elevated temperature and pressure after burial, and represent stored solar energy that accumulated over periods of millions of years.

Primary biomass energy can take a number of forms and can be transformed into secondary energy by a large number of different processes. In developing countries, biomass supplied about 16 per cent of total primary energy demand in 2005 according to data in Volume 1, Table 2.2 (40EJ out of 251EJ), largely as wood and wood waste, agricultural waste and animal dung. This energy is largely used for cooking, but at very low efficiency (10–20 per cent) and with very high pollutant emissions. In developed countries, biomass supplied about 3.5 per cent of total primary energy demand in 2005 – 8.2EJ out of 232EJ (Volume 1, Table 2.1). Globally, biomass provided 10 per cent of total primary energy demand in 2005 (48 out of 483EJ).

Figure 4.1 gives an estimate of the breakdown of biomass energy use in 2007 for traditional biomass uses (for heating and cooking) and for commercial energy uses as part of the market economy. The latter includes biomass used to generate electricity, to produce heat (in wood-burning stoves, industrial boilers or for district heating systems) and to produce biofuels (ethanol and biodiesel). Figure 4.2 gives an estimate of the amount of biomass used worldwide from 1900 to 2000, broken down by end use (domestic and industrial), source (fuelwood, crops, dung and charcoal) and by region. The uncertainty in the total biomass use indicated in Figure 4.2a is quite large. A large amount of biomass is used in the pulp and paper industry, where forestry residues and biomass wastes are used to produce some of the electricity and heat needed by the pulp and paper facilities (see Volume 1, Chapter 6, subsection 6.9.4).

As discussed in this chapter, biomass energy can be used much more efficiently and cleanly than at present, and could provide a significant fraction of future global energy needs. It provides the following advantages over other forms of renewable energy:

- It can supply energy at all times, without the need for expensive storage (such as batteries or production of hydrogen), in contrast to wind and solar energy.
- It provides rural income and opportunities for local recycling of energy expenditures, as discussed by Kartha and Leach (2001).
- It is or can be cleaner than the use of coal with regard to most pollutant emissions.
- Energy plantations can be fertilized with sewage sludge, or fertilized and irrigated with sewage water, providing a means of treating this material.
- Conversion of animal waste to high-quality, gaseous energy provides an effective method for managing agricultural and other wastes, while the production and combustion of woody biomass provide a method for removing heavy metals from soils (which become concentrated in specific small-volume ash fractions).

Among the disadvantages of biomass energy are that it is land intensive, it is more complex to initiate and manage than packaged technologies (such as wind and PV), it must be tailored to the biophysical and socio-economic circumstances of each region and it is susceptible to adverse climatic variations and adverse regional climatic changes in association with global warming. However, large streams of biomass residues from agricultural and forestry operations are already

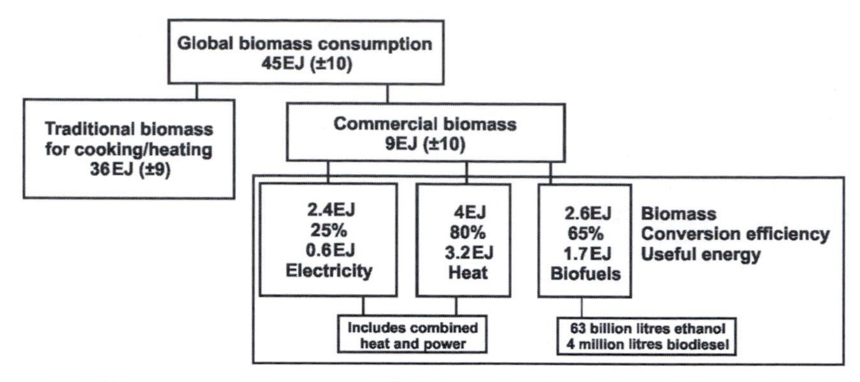

Note: Also given, for commercial biomass, are average conversion efficiencies to secondary energy (electricity, heat, biofuels) and the amount of secondary energy supplied.
Source: Sims (2010)

Figure 4.1 *Estimated breakdown of biomass energy use in 2007 into traditional and commercial uses*

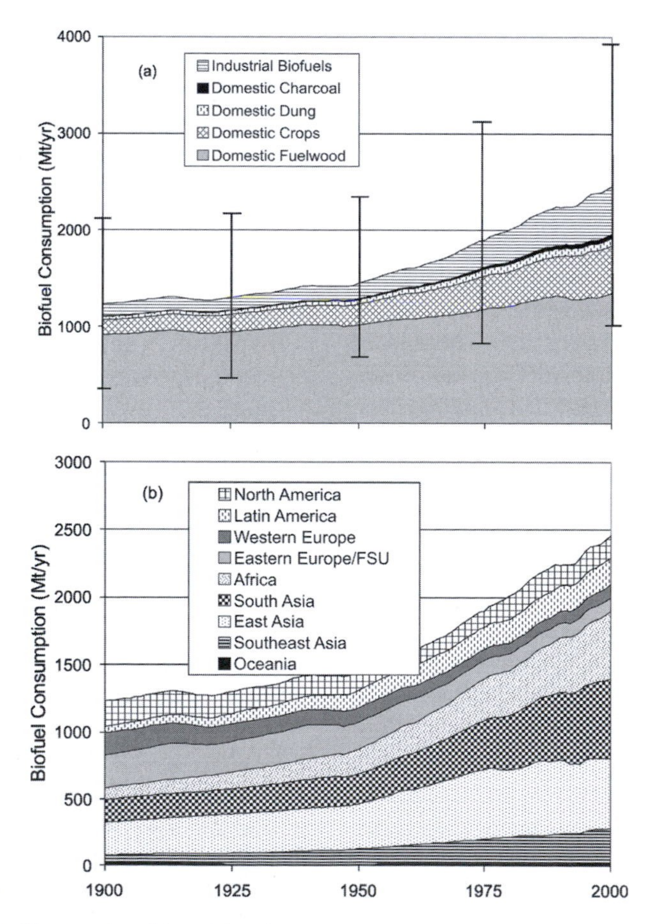

Note: The error bars represent the 95 per cent confidence intervals.
Source: Modified from Fernandes et al (2007) using data in electronic form kindly provided by D. Streets.

Figure 4.2 *Estimated consumption of biofuels from 1900 to 2000 by (a) fuel type and (b) world region*

available in many regions that can be or already are being used as an energy source. With the dramatic improvements in conversion efficiency that are possible, these existing biomass streams could provide far more energy services than they provide at present.

Energy is extracted from biomass (or from liquids or gases produced from biomass) in the form of heat when it is burned. Table 4.1 gives the heating value of typical biomass fuels and compares this with that of fossil fuels. The energy in biomass is less concentrated than in any of the fossil fuels on a weight basis, and is less concentrated than in oil or coal on a volumetric basis. This means that biomass fuel is relatively bulky, which increases the cost and energy required to transport it, and this in turn limits the distance that it is worth transporting. Table 4.2 compares the chemical composition and mass of carbon emitted during combustion per GJ of heat energy produced for biomass and fossil fuels. Biomass has a relatively low H to C ratio, and thus relatively few C–H bonds per carbon atom, which results in a relatively high carbon emission per unit of energy released. However, to the extent that biomass is grown sustainably, the carbon released during the combustion of biomass is simply the carbon that was removed from the atmosphere by photosynthesis when the biomass was growing. If fossil fuel energy is used in harvesting the biomass, in producing fertilizers that might be used to enhance growth, or in transporting the biomass to where it is used, there would still be some GHG emissions. There will also be emissions of N_2O – a powerful GHG – associated with fertilization of biomass crops. There may also be releases of CO_2 to the atmosphere from loss of soil carbon, or conversely, carbon could be built up in the soil over the course of several biomass growth and harvest cycles.

Table 4.1 *Average energy content of biomass, ethanol and fossil fuels*

Fuel	Energy content	
	GJ/tonne	GJ/m³
Wood (air dried, 20 per cent moisture)	18–20	13
Paper (stacked newspapers)	17	9
Dung (dried)	16	4
Straw (baled)	14	1.4
Sugarcane (air-dried stalks)	14	10
Domestic refuse (as collected)	9	1.5
Commercial wastes	16	
Grass (freshly cut)	4	3
Vegetable oils	37–39	
Ethanol	30	24
Oil	42	34
Coal	28–31	50
Natural gas	55	0.038

Source: Ramage and Scurlock (1996), except for wood

Table 4.2 *Chemical composition of biomass and of fossil fuels*

	Ratio of atoms			% by weight			kg C per GJ of heat output
	C	H	O	C	H	O	
Coal	1	1	<0.1	85	6	9	24–25
Oil	1	2	0	85	15	0	19
Natural gas	1	4	0	75	25	0	13.5
Wood	1	1.5	0.7	49	6	45	21

Source: Ramage and Scurlock (1996)

The remainder of this chapter discusses sources of biomass, processes for converting biomass energy to heat or electricity or to gaseous or liquid fuels, the net energy gain or loss using biomass energy, environmental considerations in the use of biomass, and global scenarios of future biomass energy use, taking into account competition between land for biomass and land for other purposes (primarily agriculture and forestry).

4.2 Sources of biomass for energy

Hoogwijk et al (2003) have identified seven categories of biomass, listed in Table 4.3. The first two categories involve dedicated biomass crops grown either on surplus agricultural land or on degraded land that is no longer suitable for agricultural crops. The third and fourth categories are biomass residues from agricultural and forestry operations. These can be divided into primary residues – parts of the plant left on the field – and secondary residues – material generated during the

Table 4.3 *Biomass energy resource categories*

Category I	Biomass grown on surplus agricultural land that becomes available as agricultural yields increase
Category II	Biomass grown on degraded or deforested land that is still suitable for reforestation
Category III	Agricultural residues (primary, taken from the field; or secondary, generated during food processing)
Category IV	Forestry residues (primary and secondary)
Category V	Animal manure
Category VI	Organic wastes (such as municipal solid waste)
Category VII	Biomaterials (used as a feedstock in place of petroleum)

Source: Hoogwijk et al 2003

processing of crops for food and wood for timber or paper (an example would be sawdust and wood scraps at a sawmill operation, or black liquor from the production of paper). The fifth and sixth categories involve animal and human waste. The seventh category is represented by used biomaterials (such as wood, paper, or (eventually) plastics made from biomass instead of from fossil fuels) that can be converted to heat and electricity through incineration or other processes.

Agricultural and forestry residues, animal manure and biosolids from sewage treatment plants are not 'wastes' as, in nature, there is no such thing as waste. Agricultural and forestry residues contain nutrients that, if left on the field, are returned to the soil, and they also contribute to the maintenance of soil carbon. Biosolids from sewage waste and the ash from incineration of municipal solid waste also contain nutrients that in principle can be returned to the land that produced human food or biomaterials. The recycling of nutrients, the implications for sustainability and the need for fertilization of croplands and forests are discussed in detail later. The use of land for dedicated bioenergy crops also has important implications for nutrient recycling, as well as for biodiversity and the availability of land for food – issues that are also discussed in some detail later. Here, we briefly elaborate upon the seven biomass energy categories identified in Table 4.3.

4.2.1 Bioenergy crops

A wide variety of crops can be grown specifically to serve as a biomass energy source. General biomass yields from annual (agricultural) crops, grasses and

Table 4.4 *General mass yields from different kinds of bioenergy crops*

	Yield (t/ha/yr)	Region	Reference
		Grasses	
Switchgrass	10–13	Eastern Canada	
	10–16	Central and Northern US	Samson et al (2005)
	15–24	Southern US	
Switchgrass	8–13	US today	
	10–26	US by 2025	Table 4.49
	13–43	US by 2050	
Miscanthus	10	Europe, today	Faaij (2006)
	20	Europe, future	
Napier grass	30	Brazil	Samson et al (2005)
Jatropha curcas	12.5	India, peak production of seeds after five years of plant growth with good soils	Hooda and Rawat (2006)
		Annual food crops	
Sunflower	1.5	US	Pimental and Patzek (2005)
Soybeans	2.9	US	Pimental et al (2008)
Wheat	2.6	World average	Sims et al (2006)
Maize	4.2	World average	Sims et al (2006)
Winter wheat	11–14	Netherlands	Table 4.37
Sugar beet	60.0	European average	Sims et al (2006)
Sugarcane	61.8	Brazilian average	Sims et al (2006)
		Trees	
Loblolly pine	4–5	Louisiana	Scott and Dean (2006)
Pine	11–12	Georgia (US)	Sartori et al (2006)
Poplar	9–13	Today, future	Table 4.6
Poplar	12–18	Northern US	Sartori et al (2006)
SRWC	10–15	Today, future	Table 4.6
Willow	15–22	New York state, with fertilization and irrigation	Sartori et al (2006)
Eucalyptus	10–20	India	Hooda and Rawat (2006)
Eucalyptus	23	1985–1991 average in Brazil	
	50	Max commercial stand in Brazil, 1986–1991	Larson (2006)
Acacia	17.8	Brazil	Patzek and Pimentel (2005)

Note: SRWC = short-rotation woody crops. See Table 4.48 for potential future yields for tropical forest plantations on degraded land.

Table 4.5 *Total land area and land areas devoted to various plantations in 1995*

Region	Land area (millions of hectares)							Rate of planting (Mha/yr)
	Total land area	Non-forest plantations				Forest plantations	Total	
		Rubber	Coconut	Oil Palm	Sugar-cane			
N America	1832				0.4	17.5	17.9	0.23
C & S America	2053	0.24	0.27	0.27	9.8	10.5	21.1	0.51
Europe	1002				<0.1	32.1	32.2	0.05
Africa	2964	0.53	0.46	0.92	1.6	8.0	11.5	0.19
Asia	4376	8.72	10.55	4.59	9.6	115.8	149.3	3.50
Oceania	811				0.5	3.2	8.7	0.05
Total	13,038	9.49	11.28	5.77	21.9	187.1[a]	235.7	4.53

Note: [a] This represents 5% of the global forest area.

Source: FAO (2002) for rubber, coconut and oil palm plantations, Fischer et al (2008) for sugarcane and FAO (2005) for forest plantations

woody crops are summarized in Table 4.4. To provide perspective, the total land areas devoted to various kinds of plantations in different continents and globally are given in Table 4.5. Hardwood species account for 57 per cent of the total forest plantation area, of which 32 per cent (10.3Mha[1]) is *Eucalyptus.* Softwood species account for 43 per cent of the total plantation area, of which 61 per cent is pine.

Annual bioenergy crops include:

- Starch-rich crops such as maize (increasingly used to produce ethanol in the US), wheat and potatoes;
- Sugar-rich crops such as sugarcane (used to produce ethanol in Brazil) and sugar beets;
- Oil-rich crops, used to produce biodiesel fuel, such as coconut oil (Philippines), palm oil (Malaysia) and sunflower oil (South Africa).

As discussed later, the fossil fuel energy inputs required to produce fuels from annual crops, and the net energy gain, are quite different for different crops, with ethanol from sugarcane being most favourable and ethanol from corn (maize) or wheat being the least favourable. Sugarcane used for the production of sugar has a yield of 30–40t/ha/yr, while varieties of sugarcane used for energy production have yields of 60–70t/ha/yr, with most of the extra production as increased fibre content. A sugarcane plantation in Brazil, the harvesting operation and cut sugarcane are illustrated in Figure 4.3, while palm oil fruit bunches and fruit are illustrated in Figure 4.4.

Source: Photography by Hannes Grobe (upper), Edrossini (centre) and Rfino Uribe (lower), www.wikipedia.org

Figure 4.3 *Sugarcane plantation, sugarcane harvesting and cut sugarcane*

Source: Left, photo by Jeff McNeely in Howarth and Bringezu (2009, Executive Summary); upper right, Stone (2007, reprinted with permission from AAAS), lower right, Koh and Wilcove (2007)

Figure 4.4 *Palm oil plant (left), fruit bunches being moved by rail (upper right) and close-up of oil palm fruit (lower right)*

The primary grasses considered for bioenergy are:

- Switchgrass (*Panicum virgatum*);
- *Miscanthus*;
- Napier grass (*Pennisetum purpureum*); and
- *Jatropha curcas.*

In addition, systems with multiple grasses growing together have been considered (as discussed later). All of these grasses are perennials.[2] Switchgrass is a hardy grass that can grow up to 2.2m in height, as illustrated in Figure 4.5. It has a C_4 carbon fixation path,[3] giving it an advantage in conditions of drought and high temperature. It was one of the dominant species of the central North American tall grass prairie and non-forested areas of the eastern US before the arrival of Europeans (Parrish and Fike, 2005). It has great value in reducing erosion and runoff losses of N and P, and enhancing water infiltration. Few other herbaceous plants are as broadly adapted to the range of growing conditions found in North America. Nevertheless, substantial research effort is required for the economical and energy-efficient use of switchgrass (Sanderson et al, 2006). *Miscanthus* is a genus of about 15 species of grasses native to subtropical and tropical Africa and southern Asia, with one species (*M. sinensis*) extending north into temperate eastern Asia. It can grow to heights of 3.5m over five to six years, with a dry-weight yield of 25t/ha. It is illustrated in Figure 4.6. Napier grass (*Pennisetum purpureum*) is a tropical grass native to the grasslands of Africa. It typically grows at a height of 2.0–4.5m, rarely up to 7.5m, and has razor-sharp leaves 30–120cm long and 1–5cm wide. Napier grass is used on a wide scale for cattle feed and is fertilized with P, K and micronutrients but not N, as it appears to support biological nitrogen fixation.[4] It is

Source: US government public domain

Figure 4.5 *Switchgrass* (Panicum virgatum)

better for controlling soil erosion than sugarcane, as it can form a full canopy three weeks after harvest, compared to six to eight weeks for sugarcane. It is also illustrated in Figure 4.6. Finally, *Jatropha curcas* is a poisonous scrub weed that originated in Central America and has been used around the world for lamp oil and soap. It grows well on marginal and semi-arid lands, as illustrated in Figure 4.7. Biodiesel can be produced from the oily seeds, and a pilot project in India involves growing *Jatropha* on 500,000 hectares of government land; if this phase goes according to plan, cultivation of *Jatropha* would be expanded to 12 million hectares and privatized (Fairless, 2007). For more information on the properties of *Jatropha* and its use in producing biodiesel, see Achten et al (2008).

There is a larger potential land base in the tropics for dryland grasses (such as switchgrass) than for sugarcane, which requires higher-quality land. Grasses can be grown on marginal lands where food crops have difficulty growing, such as denuded hillsides and lands with thin, heavy or drought-prone soils, or on soils exhausted from intensive maize cropping. Many of the grasses evolved in Africa on similar soils. Some grasses (such as *Erianthus*) tolerate alternating dry and wet seasons well by sprouting deep (up to 2.5m) roots even when there are hardpans.

Woody crops (trees) can be produced in two ways. First is short-rotation coppicing. *Coppicing* refers to the

Source: www.wikipedia.org

Figure 4.6 *Upper:* Miscanthus sinensis *in Japan; lower: Napier grass* (Pennisetum purpureum)

Source: Left, photo by Jeff McNeely in Howarth and Bringezu (2009, Executive Summary); right, Fairless (2007)

Figure 4.7 *Close-up of* Jatropha *(left), and degraded lands in India prior to (upper right) and after (lower right) cultivation of* Jatropha

periodic cutting of branches and a portion of the stem of a tree, allowing the tree to sprout again. Willow (*Salix*) and poplar (*Populus*) are commonly harvested in this way, and are planted at a density of 5000–20,000 trees/ha (0.5m × 1m to 1m × 2m spacing). After five to seven coppice rotations, the entire tree is removed and new seedlings are planted. Figure 4.8 shows a combined harvesting and chipping operation in Sweden during winter, when the soils are frozen (upper), and irrigation of coppice regrowth using log-yard runoff from an adjacent sawmill (lower). There are about 350 species of willow.

The second production method is modified conventional forestry. Trees are grown to maturity as in conventional forestry, then harvested for energy purposes. In Sweden, 5000 trees are planted per hectare and vigorously thinned later. Europe has a potential for 10EJ/yr of primary energy in this way, using 5 per cent of its land area (10EJ/yr is twice the present-day total primary energy use in the UK). Yields can be enhanced through fertilization with chemical fertilizers or with wastewater and/or biosolids from municipal sewage treatment plants. *Eucalyptus* is a fast-growing tropical tree species (10–15 years to harvesting) that is widely used as an energy crop in the tropics. *Eucalyptus* wood is converted to charcoal for the iron and steel industry, and is also a source of pulp and furniture (Wright, 2006).

Source: Photography by P. Aronsson in Dimitriou and Aronsson (2005)

Figure 4.8 *Coppice harvesting of willow in Sweden (upper) and irrigation of new growth (lower)*

Source: Upper left, Doug Maguire, Oregon State University, www.forestryimages.org; right, Oak Ridge National Laboratory on the NREL Photo Exchange (www.nrel.gov/data/pix, PIX 04645); lower left, Dennis Haugen, www.forestryimages.org

Figure 4.9 *Five-year-old* Acacia nilotica *(upper left), four-year-old* Eucalyptus *in Hawaii (right), and 14-year-old* Pinus taeda *(loblolly pine) in Georgia (US) (lower left)*

Acacia and *Pinus* are also suitable as bioenergy crops. Five-year-old *Acacia nilotica*, four-year-old *Eucalyptus* in Hawaii and 14-year-old *Pinus taeda* (loblolly pine) in Georgia (US) are illustrated in Figure 4.9.

Faaij (2006) reviews the status of bioenergy crops in Europe. Rapeseed and cereals are cultivated for energy purposes, intermixed with conventional agricultural production and used to produce transport fuels. Short-rotation coppice willow is grown in Sweden on about 14,000ha; cuttings are harvested every 2–5 years over a period of 20–25 years before the plant is replaced. Poplar, *Miscanthus* and sweet sorghum are other energy crops, but their use for energy is currently negligible. Growing perennial energy crops (such as willow and *Miscanthus*) on fallow land is not popular, because agricultural land is typically rotated in and out of fallow. In the long run, the highest energy yields in Europe are expected from sugar beets and *Miscanthus*, but sugar beets require high-quality land and *Miscanthus* is restricted to warmer regions. Woody biomass can be produced in northwest Europe for about €3–6/GJ, compared to €1–2/GJ for imported coal.

Table 4.6 compares the expected near-term (next few years) and long-term (2025 or later) performance characteristics of some current or potential bioenergy crops in Europe. The column 'energy input' refers to the energy required for fertilizers and mechanical harvesting and transport, while the energy yield is given by the product of biomass yield (t/ha/yr) times the energy content per tonne. The ratio of the energy output to energy input, the EROEI, ranges from about 10 to 60 for the crops and conditions given in Table 4.6. These crops may be transformed into gaseous and liquid fuels, with further energy inputs and losses, or used directly for heating or the generation of electricity, as discussed later.

Table 4.6 *Characteristics of current or potential future energy crops in Europe*

Crop	Time horizon	Energy input (GJ_{prim}/ ha/yr)	Typical mass yield (t/ha/yr)	Typical net crop energy yield (GJ/ha/yr)	EROEI	Cost (€/GJ)	Location and conditions
Rapeseed	Short term	11	5.5[a]	110	11	20	Germany, France. Requires
	Long term	12	8.5[b]	180	16	12	good-quality land
Sugar beet	Short term	13	14	250	21	12	Requires good-quality land and
	Long term	10	20	370	38	8	agri-chemicals
SRWC-willow	Short term	5	10	180	37	3–6	Sweden, UK. Suited for colder and
	Long term	5	15	280	57	<2	wetter climates
Poplar	Short term	4	9	150	39	3–4	Currently planted mainly for
	Long term	4	13	250	64	<2	pulpwood production
Miscanthus	Short term	13–14	10	180	14	3–6	Limited commercial experience so far.
	Long term	13–14	20	350	26	~2	Suited for warmer climates. Improvement potential hardly explored

Note: Energy yields for the various crops are computed assuming higher heating values (HHVs) per dry tonne of 19GJ/tonne for wood and straw, 28GJ/tonne for rapeseed and 16GJ/tonne for sugar beet. [a] Includes seed (2.9t/ha/yr) and straw (2.6t/ha/yr). [b] Includes seed (4.0t/ha/yr) and straw (4.5t/ha/yr).
Source: Faaij (2006)

4.2.2 Agricultural residues

Table 4.7 lists tropical crop residues, their energy content and current uses. Significant amounts of residue are generated by the sugar industry by virtue of the size of the industry. These include the sugarcane leaves and tops, which are traditionally burnt off in the field before harvest, and bagasse – the fibre from sugarcane (see Figure 4.10). Most sugar factories already use bagasse as a source of heat to make steam, but deliberately use it inefficiently in order to avoid a surplus of bagasse. Some also produce electricity, but such production is often limited by the existence of electric utility monopolies that refuse to purchase independently produced electricity (enhanced production of electricity from sugar mills is discussed later, in subsection 4.4.6).

Rice husks – which constitute a fifth of the dry weight of unmilled rice – are another important tropical crop residue, due to the large production of rice. Rice husks have a high silica (ash) content, but their uniform texture makes them well suited for gasification. It is estimated that in 1996, the crop residues left in the field or generated from agricultural processing in China amounted to 790 million tonnes, with an energy content of 11EJ (Kammen et al, 2001). If half of this were used to generate electricity at 25 per cent efficiency (appropriate for small-scale generation today), it would equal about 20 per cent of the electricity generated in China from fossil fuels in 2005.

The main agricultural residue in temperate climates is straw from crops such as barley, corn and wheat.[5] Potato and sugar beet tops are other temperate crop residues. In Denmark, straw provides 11.9PJ (1.4 per cent of national primary energy), both as a supplementary fuel (co-firing) with coal in large CHP plants and as the sole fuel in smaller CHP plants.

Corn usually produces the largest amount of residue per hectare of all the main crop types. This residue is referred to as *stover* and consists of the stalks and leaves. It could be a good choice for the cellulosic production of ethanol (see subsection 4.3.5), particularly with widespread use of no-till cultivation methods, as less of the crop residue needs to be left on the field for prevention of erosion with no-till cultivation.[6]

4.2.3 Forestry residues

Forestry residues can come from the thinning of plantations, trimming of felled trees or from timber

Table 4.7 *Crop residues, residue to crop ratios or residue yields, residue energy content and typical current uses of residues*

Crop	Residue	Residue ratio or yield	Residue energy (MJ/dry kg)	Typical current residue uses
Barley	Straw	0.8–2.5 (4.3t/ha/yr)	17.0	
Coconut	Shell	0.1kg/nut	20.6	Household fuel
Coconut	Fibre	0.2kg/nut	19.2	Mattress making, carpets, etc.
Coconut	Pith	0.2kg/nut		
Cotton	Stalks	0.95–2.0 (6.7t/ha/yr)	18.3	Household fuel
Mustard Cotton	Gin waste	0.1	16.4	Fuel in small industry
Groundnut	Shells	0.3		Fuel in industry
Groundnut	Haulms	2.0		Household fuel
Maize	Cobs	0.3	18.8	Cattle feed
Maize	Stalks	1.5	17.7	Cattle feed, household fuel
Millet seed	Straw	1.2		Household fuel
Millet seed	Stalks	1.8		Household fuel
Other seeds	Straw	2.0		Household fuel
Pulses	Straw	1.3		Household fuel
Rapeseed	Stalks	1.25–2.0		Household fuel
Rice	Straw husk	1.5–0.25 (6.7t/ha/yr)	16.3	Cattle feed, roof thatching, field burned
Sorghum	Stalks	0.85–2.0 (8.4t/ha/yr)	16.1	Fuel in small industry, ash used for cement production
Soybeans	Stalks	0.8–2.6	15.9	
Sugarcane	Bagasse	0.15	17.3	Fuel at sugar factories, feedstock for paper production
Sugarcane	Tops/leaves	0.15		Cattle feed, field burned
Tobacco	Stalks	5.0		Heat supply for tobacco processing, household fuel
Tubers	Straw	0.5	14.2	
Wheat	Straw	1.1–2.5 (5t/ha/yr)	17.5	Cattle feed

Note: Also given are absolute residue yields from WWI (2006).
Source: Kartha and Larson (2000), except for corn, sorghum and all residue ratios given as ranges, which are from WWI (2006)

Source: Left: http://energyconcepts.tripod.com/energyconcepts/bagasse.htm; right: Kartha and Larson (2000)

Figure 4.10 *Bagasse, the fibre residue from milling of sugarcane*

processing (sawdust, offcuts). Wood wastes are widely used for cogeneration of heat and electricity in the pulp and paper industry, as biomass is already present onsite as a byproduct of the industrial activity. Mechanical harvesters and chippers can be used.

4.2.4 Animal waste and sewage sludge

Processing livestock manure or sewage sludge in an anaerobic digester to produce biogas is a method for waste management, as well as to produce energy. Small biogas plants for human and animal wastes are common in China and India, while both centralized and farm-scale digesters are relatively common in Denmark.

4.2.5 Municipal solid waste

Municipal waste contains biomass materials (primary food waste, paper and paper products) that can be processed in a variety of ways so as to extract useful energy, as discussed in Volume 1, Chapter 8, section 8.3. The options are disposal in landfills with capture of the methane that is produced from the anaerobic decomposition of organic matter, incineration (preferably with prior removal and recycling of non-combustible materials) combined with the generation of electricity and the use of waste heat, anaerobic digestion of the readily decomposable organic fraction, gasification or pyrolysis followed by combustion to generate electricity and useful heat, and use of the high-energy waste components as a fuel in cement kilns. As discussed in Volume 1, neither disposal in landfills with (of necessity only) partial recovery of methane nor incineration are particularly attractive from an energy savings point of view. Issues associated with the net energy gain or loss associated with incineration of paper and paper products are discussed in Volume 1, Chapter 6, subsection 6.9.7.

4.3 Processes for extracting energy from biomass

The processes by which biomass can be used for energy other than electricity are described below, while the production of electricity from biomass is discussed in section 4.4. Environmental issues associated with all uses of biomass are discussed in section 4.5 and costs are discussed in section 4.10.

4.3.1 Direct combustion

Direct combustion can be used for cooking, for heating or for generation of electricity using a steam turbine. Coal is widely used to generate electricity and for heating in district heating plants (and also in individual households in China), but biomass can be used in its place. As seen in Table 4.2, woody biomass has a smaller carbon percentage and thus a smaller heating value per unit weight than coal. It has also a smaller nitrogen, sulphur and ash content. The use of grasses for heating, by contrast, has been limited by the inability so far to develop grasses that can be used with advanced combustion technologies, due to the higher ash content of herbaceous fuels. A reduction in the silica content of the grass would reduce the amount of ash produced, while reducing the amount of K and Ca in the fuel would reduce the tendency for agglomeration in boilers.

Cooking

Conventional cooking stoves use biomass with an efficiency of 10–20 per cent. In most sub-Saharan African countries, firewood provides 90 per cent or more of the energy used for cooking, with small amounts from gas, kerosene and/or charcoal (Karekezi and Kithyoma, 2004). Major efforts have been made to encourage the adoption of more efficient cooking stoves in developing countries, but decisions concerning the kind of stove to use involve cost and a range of social considerations (Kammen et al, 2001). Three different methods of cooking using woody biomass in Kenya are illustrated in Figure 4.11: using the traditional open fire with three stones, using a ceramic wood-burning stove and using a charcoal stove.[7] Another approach to cooking with solid biomass is the wood-burning hearth, illustrated in Figure 4.12 and compared with a stove using biogas. Figure 4.13 compares the cost and efficiency of various cooking stoves using wood and alternative fuels. Improved wood-burning stoves give an efficiency of 30 per cent. As of 2006, there were an estimated 220 million improved stoves in use worldwide, including 180 million in China and 34 million in India (REN21, 2007). Much higher efficiencies are possible with

Source: Bailis et al (2003)

Figure 4.11 *Three different methods of heating using woody biomass in Kenya: the traditional open fire with three stones (upper), the ceramic wood-burning stove (centre); and the charcoal stove (lower)*

gaseous fuels (45–60 per cent) and with electricity (62 per cent). If gaseous fuels are produced from biomass, the overall gain in cooking efficiency will

Source: Kartha and Larson (2000)

Figure 4.12 *A wood-burning hearth (top) and replacement stove using biogas (lower) in China*

depend on the efficiency with which the biomass is gasified (see below).

Pellet heating

Fiedler (2004) reviews state-of-the-art pellet-based heating systems in Sweden, Austria and Germany. By 2001, there were 30,000, 12,000 and 10,000 pellet boilers or stoves in these three countries, respectively. Pellet boilers have a maximum heat output of 10–40kW, and some can automatically modulate from 30–100 per cent of full output. Pellet stoves are used to heat single rooms, compact apartments or entire low-energy houses, have a maximum heat output of 10kW and can be regulated manually or automatically based on room temperature. Helical screws automatically deliver pellets to the combustion area as needed, and ash is automatically removed and compressed before being stored in an ash container. As discussed later, modern pellet boilers have significantly lower emissions of CO, NO_x or dust compared to log wood or wood chip boilers. Austrian pellet boilers have efficiencies of

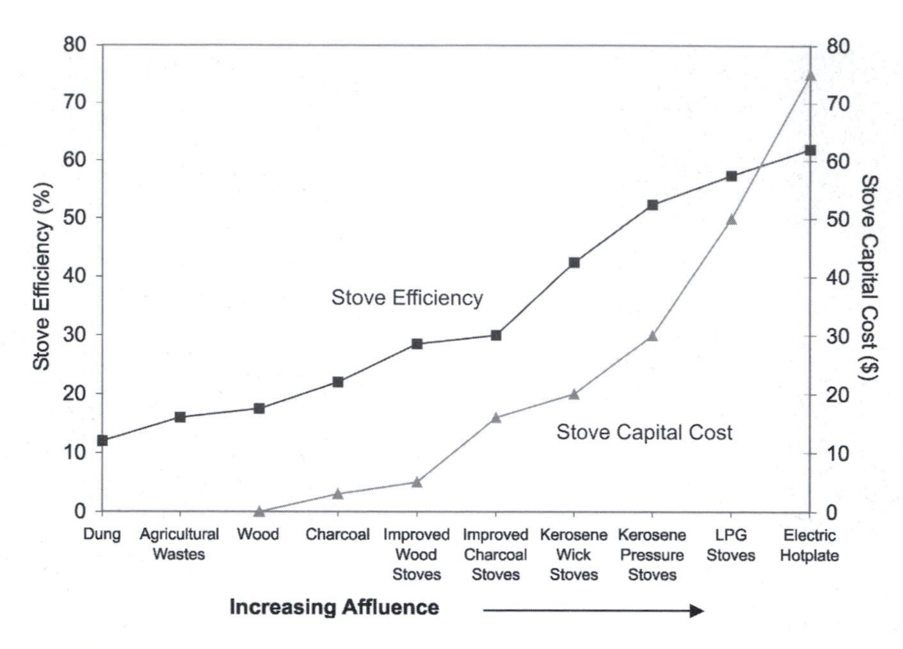

Source: Kammen et al (2001)

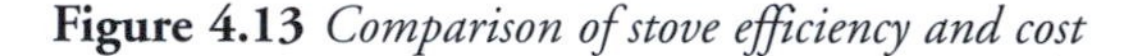

Figure 4.13 *Comparison of stove efficiency and cost*

86–94 per cent. Examples of commercially available products can be found at www.pelletstove.com.

Figure 4.14 shows 4–10mm diameter standardized biomass pellets that are being produced in increasing quantities in Europe. They are produced largely from dry sawdust and wood shavings, but also from bark, straw and crops, and have a heating value of 18GJ/tonne and a water content less than 10 per cent. They can be delivered by tanker truck to individual houses, or used at centralized district heating plants, either for dedicated production of heat or for production of heat and electricity through cogeneration. Sweden is the largest producer and consumer of biomass pellets, with production increasing from 5000 tonnes/year in 1992 to 667,000 tonnes/year in 2001 (Egger et al, 2003). Denmark, Germany and Austria are also significant markets for biomass pellets, with some pellets used in Austria supplied from Canada. Figure 4.15 gives an example of a pellet-burning stove and a pellet-burning furnace.

Source: Unknown

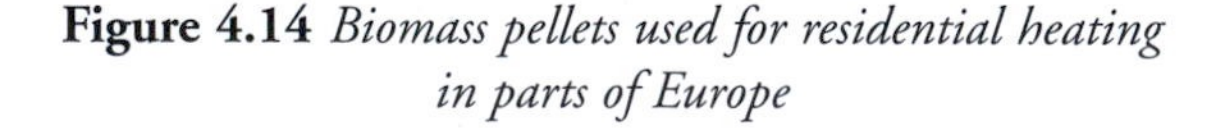

Figure 4.14 *Biomass pellets used for residential heating in parts of Europe*

District heating

Figure 4.16 shows the growth in the use of biomass in Sweden for district heating from 1980 to 2007, by which time biomass (much of it short-rotation willow) provided two thirds of the energy for district heating nationwide, which in turn provided 30 per cent of the total space heating load in Sweden (Swedish Energy Agency, 2008, Figure 14). As of 1995, Austria had 266 biomass district

Source: www.harmonstoves.com

Figure 4.15 *A pellet-burning stove (left) and furnace (right)*

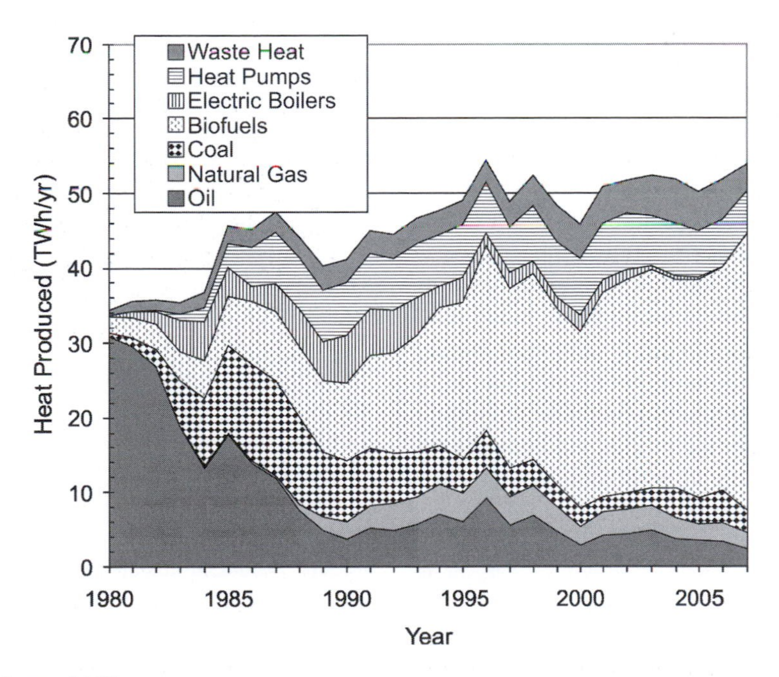

Source: Swedish Energy Agency (2008)

Figure 4.16 *Fuels used for district heating in Sweden, 1980–2007*

heating plants with capacities of 0.5–10MW$_{th}$, a total capacity of 374MW$_{th}$ and a total primary energy input of 3.0PJ/yr (Obernberger et al, 1997).

4.3.2 Thermo-chemical gasification

Thermo-chemical gasification involves two steps: heating the biomass to a temperature of 300–500°C in the near-absence of air to drive off the easily vaporized or *volatile* material – a processes called pyrolysis – and then gasifying the remaining solid material (called *char*) by heating it to a temperature of 850–900°C in the presence of steam. Two thirds of the energy content of the fuel appears in the volatile material and is lost if the gases are not captured. This is the case in the manufacture of charcoal, as widely used in developing countries.[8] Figure 4.17 compares the fractional loss of biomass and coal mass due to volatilization, as a function of temperature, and the rate of conversion of the remaining char to gases as a function of temperature. Biomass has a much higher volatile content than coal (80–90 per cent versus 35–45 per cent), and pyrolysis occurs at a lower temperature. However, coal is easily ground into small spheres, whereas fibrous wood gets stuck in the gasifier, creating uneven flow, unless special pretreatment is carried out (Kintisch, 2008).

Gasification of biomass (or coal) can produce a mixture of CO, H_2, CH_4 and CO_2 (sometimes called *producer gas*, after the first 'gas producers' in the 1800s) or just CO and H_2 (called *synthesis gas*, because it can be used to synthesize almost any hydrocarbon). Synthesis gas can be used to produce methanol through a series of chemical reactions involving high temperatures and pressures and an expensive production facility. This is of interest because methanol is a liquid fuel that can be used in place of gasoline in an internal combustion engine or in fuel cells. Synthesis gas can also be shifted to an essentially pure hydrogen output through reaction of the CO with water to produce CO_2 and additional H_2. The conversion efficiencies from biomass to methanol and hydrogen are 55–63 per cent and 57–69 per cent, respectively (Larson, 1993), depending on the conversion process used.

There are two basic kinds of gasifiers, illustrated in Figure 4.18. In an *updraught fixed-bed gasifier*, air is injected at the bottom and biomass enters at the top and is gasified as it falls down. These gasifiers are relatively efficient due to the effective heat exchange between rising air and descending biomass, but they produce significant amounts of tars that need to be removed from the gas (Kartha and Larson, 2000). This reduces the overall efficiency, as the energy content of the tar is lost. In the *downdraught fixed-bed gasifier*, the gas produced by the gasifier is drawn out from the bottom of the gasifier. This forces all the gasification products to pass through the hottest zone, where almost all of the initially produced tars are broken into gases.

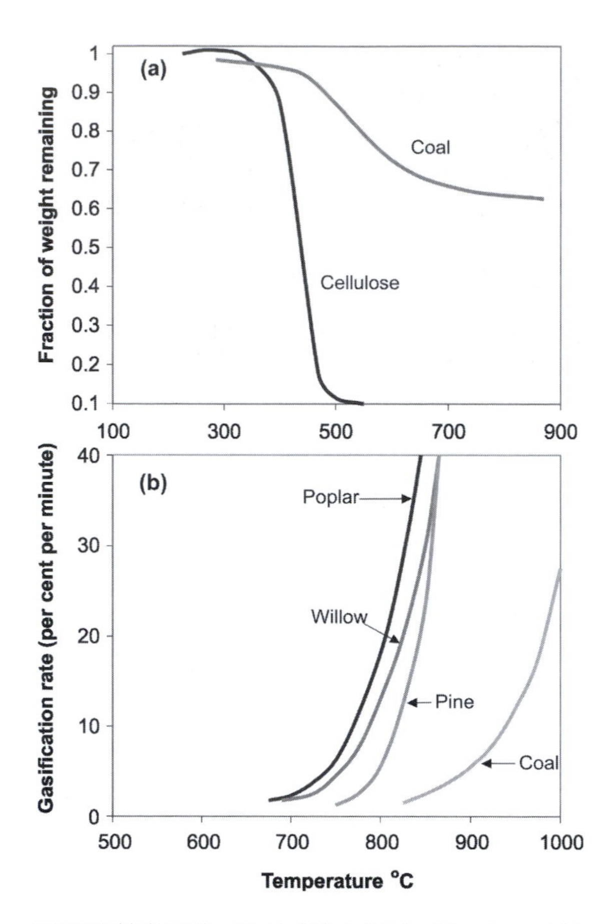

Note: Part (a) shows the rate at which coal and cellulose lose weight as they are heated in the near-absence of air (pyrolysis). Part (b) shows the rate at which the remaining solid residues (char) of biomass and coal are converted into carbonaceous gas.
Source: Larson (1993)

Figure 4.17 *Comparison of thermo-chemical reaction rates for biomass and coal*

Gasification products can be used for:

- direct heating, using an industrial boiler, furnace, or kiln that is directly coupled to the gasifier to minimize heat losses;

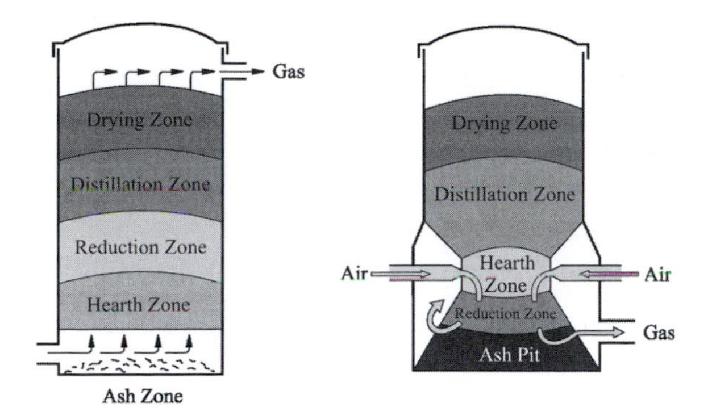

Source: Kartha and Larson (2000)

Figure 4.18 *Illustration of the updraught fixed-bed gasifier (left) and updraught fixed-bed gasifier (right)*

- cooking;
- electricity generation;
- combined heat and power generation (cogeneration); or
- synthesis of various fuels (methanol, Fischer-Tropsch liquids, dimethyl ether).

In cooking applications, the gas needs to be cleaned, stored and distributed with a piping system. The carbon monoxide component of the original gas is a health concern. For small-scale generation of electricity (5–100kW), a diesel engine is often used; the gas must be carefully cleaned to avoid fouling the engine. In the future, large-scale (5–100MW) generation of electricity using integrated biomass gasification combined cycle (steam and gas turbine) powerplants is likely, as discussed in subsection 4.4.3. Biogas can also be blended with natural gas in combined cycle powerplants, as discussed in subsection 4.4.4. Dimethyl ether is well suited for use in diesel engines and is relatively clean burning.

Gasification on dairy farms

Young and Pian (2003) assess the feasibility of using gasification of dairy farm animal wastes. Manure would be first separated into solid and liquid portions using a commercial auger press, and the liquid manure returned to the field as fertilizer. Gasification of the solid portion at a temperature of 1350–1400°C and an efficiency of 65–85 per cent is possible, depending on the gasifier operating configuration. Ash production corresponds to about 9 per cent of the dry manure by weight, and could also be returned to the field. The biogas would be used to produce heat and electricity – about twice the energy needs of a dairy operation in New York state.

4.3.3. Biological gasification (anaerobic digestion)

Anaerobic digestion, like pyrolysis, is a process that occurs in the absence of air, but using bacteria rather than heat. It can be applied to solid organic waste, sewage sludge and animal manure. In all three cases, a gas consisting largely of methane is produced.

Digestion of municipal solid waste

Uncontrolled anaerobic digestion of solid waste occurs in sanitary landfill sites (i.e. those landfills that alternate waste and impermeable clay layers). Ideally, 150–300m^3 of *landfill gas* can be produced per tonne of municipal waste. This gas is 50–60 per cent CH_4, which has a heating value of 38MJ/m^3, so the energy yield ideally is 4–7GJ per tonne of waste (the remaining gas is CO_2, which has no heating value). This implies a conversion efficiency of 45–70 per cent if the waste is 50 per cent organic matter with a heating value of 18–20GJ/tonne. The gas is collected through a series of perforated underground pipes, as illustrated in Figure 4.19. In practice, yields are around 2GJ per tonne at present (~4GJ/tonne of organic waste, an efficiency of only 20 per cent). Landfill gas is produced from organic waste over a period of five to ten years. As discussed in Volume 1 (Box 8.1) only about 15 per cent or less of the methane escaping from a landfill needs to leak to the atmosphere for this leakage to completely negate the benefit of using the remaining escaping methane to displace coal for the generation of electricity.

Organic solid waste can also be processed under carefully controlled conditions in specially designed digesters. An energy yield of 10GJ/tonne of organic waste can be obtained with full utilization of all gases and residues and recovery of low- and high-temperature heat, as illustrated in Figure 4.20. This is substantially better than the yield of about 2GJ/tonne of waste obtained through collection of landfill gas, requires much less land area than landfills and should not entail leakage to the atmosphere. Given a typical heating value of biomass of about 18–20GJ/tonne, the implied conversion efficiency is 50–55 per cent.

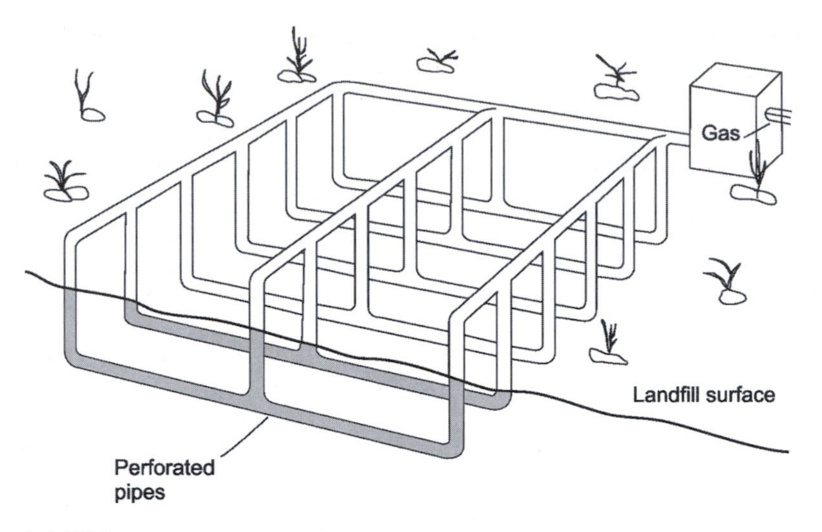

Source: Ramage and Scurlock (1996)

Figure 4.19 *Extraction of landfill gas with perforated pipes*

Digestion of sewage sludge, animal manure and other farm wastes

Anaerobic digestion of sewage sludge or animal manure produces *biogas* over a period of ten days to several weeks. Yields are 200–400m^3 of biogas per tonne of input, of which 50–75 per cent is CH_4 (Ramage and Scurlock, 1996). This gives an energy yield of 4–11GJ/tonne. The maximum efficiency in energy conversion from animal manure is around two thirds, although typical efficiencies are much lower (10–15 per cent). The production of biogas is greatly

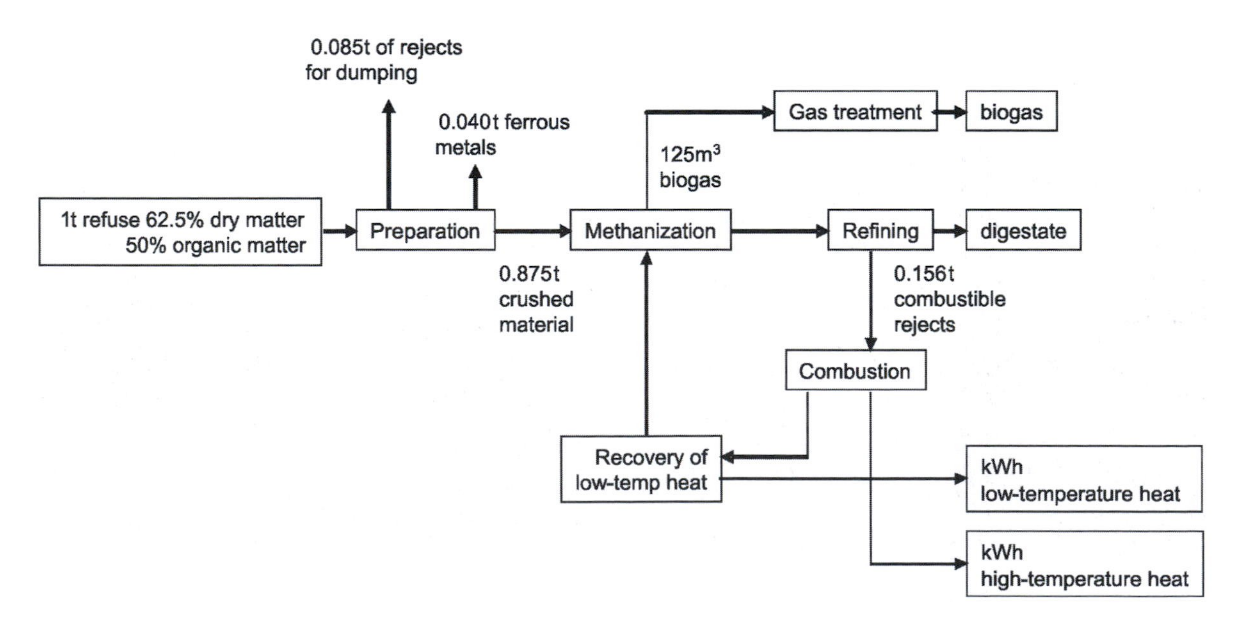

Source: Ramage and Scurlock (1996)

Figure 4.20 *Flow diagram for the anaerobic digestion of municipal solid waste*

increased with the addition of up to 20 per cent organic waste (Junginger et al, 2006). The leftover effluent is a concentrated nitrogen fertilizer with the pathogens in the original feedstock largely eliminated by the warm temperatures in the digester tank.

At present, it is estimated that there are 5 million household cattle dung digesters in China in working condition, 500 large-scale digesters operating at pig farms and other agro-industrial sites and 24,000 digesters at sewage treatment plants. As well, there are several thousand digesters in other developing countries and 5000 digesters in industrialized countries, primarily at large livestock processing facilities and municipal sewage treatment plants (Kartha and Larson, 2000). As of 2006, an estimated 25 million households worldwide used biogas for their cooking and lighting needs, including 20 million in China and 3.9 million in India (REN21, 2007). A cattle dung digester in India is illustrated in Figure 4.21 and a digester on a pig farm in England is illustrated in Figure 4.22. In Germany, 25 per cent of manure waste provides energy through anaerobic digestion, while in Denmark, about 7 per cent of manure is used in this way (Tijmensen and van den Broek, 2004).

Different digesters differ in the relationship between the average length of time that the influent feed remains in the digester (the hydraulic retention time, HRT) and the average length of time that solids remain in the reactor (the solids retention time, SRT). Short HRTs improve the economics of biomass digestion by increasing the throughput, while long SRTs are needed to maximize the conversion of biomass to biogas (i.e. to give higher efficiency). For small-scale applications in developing countries, simple *unmixed-tank digesters* have been developed with long HRTs and SRTs (several weeks). Two different designs, one common in India and the other in China, are illustrated in Figure 4.23. In the Indian floating-cover design, the input slurry displaces the output sludge. In the Chinese fixed-dome design, the buildup of gas pressure forces out the effluent sludge. The difficulty has been in constructing a leak-proof dome. In *retained-biomass digesters*, solids settle to the bottom of the digester and remain there a long time, while the HRT is short (as little as a few hours). These digesters are widely used in industrialized countries and for

Source: Kartha and Larson (2000)

Figure 4.21 *Anaerobic cattle dung digesters in the village of Pura, Karnataka, India*

Source: Unknown

Figure 4.22 *A biomass digester on a pig farm in England*

industrial and municipal applications in developing countries.

Denmark is one of the leading countries in the industrialized world in the implementation of biogas plants to turn farm wastes into energy, with 20 centralized plants (output of 23–345GJ/day) and over 35 farm-scale plants (Raven and Gregersen, 2007). The digested manure is returned to farm fields as fertilizer; the digestion process improves the fertilizer characteristics of the manure while killing pathogens and weeds. Biogas costs fell from $127/GJ in 1984 to $12/GJ in 1995 (Maeng et al, 1999). This is close to competitive with recent peaks in the cost of natural gas (which has fluctuated between $6/GJ and $12/GJ). The digesters were developed by local district heating companies. Maeng et al (1999) estimate that use of 50 per cent of the available manure in Denmark for biogas production would allow almost the complete elimination of oil for heating and of coal for the generation of electricity in Denmark.

Anaerobic digestion is also suited to the extraction of energy from semi-solid agricultural residues (having a dry matter content less than 25 per cent), which

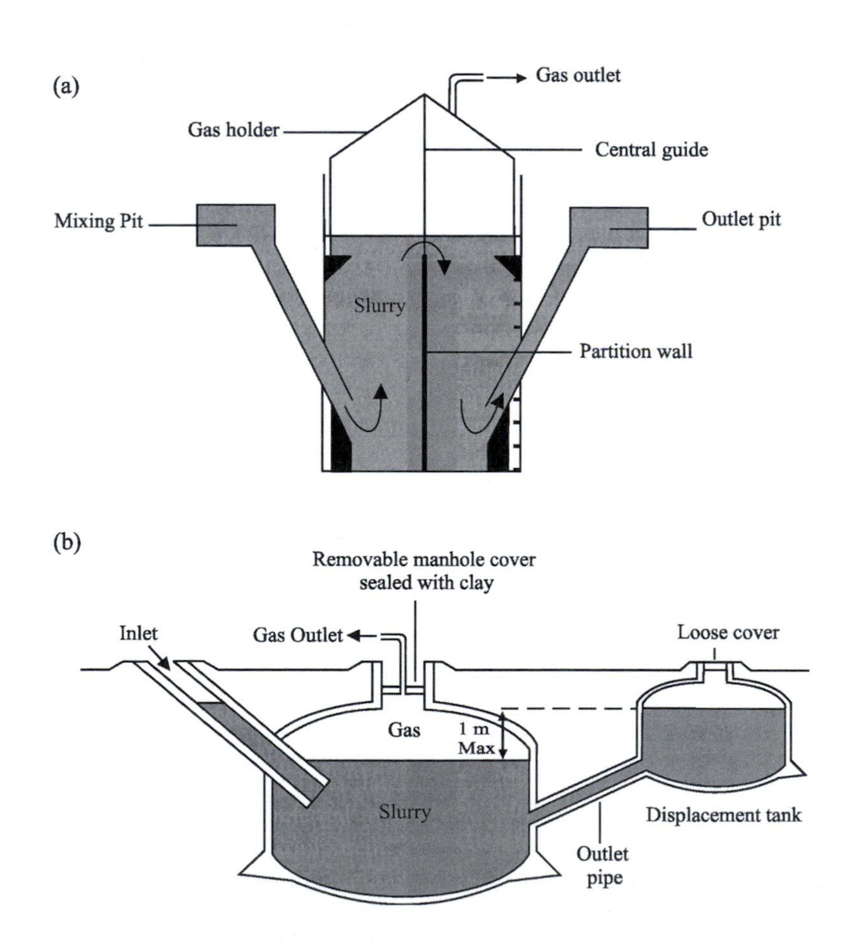

Source: Kartha and Larson (2000)

Figure 4.23 *Basic design of (a) the floating-cover biogas digester, common in India, and (b) the fixed-dome biogas digester, common in China*

cannot be combusted for energy (Parawira et al, 2008). Examples of such wastes include potato and sugar beet residues.

4.3.4 Production of ethanol by fermentation of non-woody biomass

Fermentation is another anaerobic process, but one requiring an input rich in sugars. The product is ethanol (C_2H_5OH), which is a liquid fuel that can be used in an internal combustion engine (such as used in current automobiles) or in fuel cells. In an internal combustion engine it has no clear advantage compared to gasoline in terms of air pollution (see subsection 4.5.4).

Sugarcane is the best biomass source for producing ethanol; the sugar in sugarcane occurs as sucrose. Sucrose is a combination of glucose plus fructose, both of which are sugars with six carbon atoms. Potatoes, corn and wheat contain starch – long chains of glucose molecules – that can be broken into the constituent sugars, but some initial processing is required. For grains this is done in one of two ways: wet milling, in which the grains (or starches) are soaked in water, usually with a sulphurous acid, to separate the starch-rich endosperm from the high-protein germ and the high-fibre husks; and dry milling, in which the grains are simply ground down.

Wet milling produces corn oil, gluten feed, germ meal, starches, dextrin and sweeteners such as corn syrup as co-products, whereas dry milling produces dried distillers' grains (DDG) as a co-product. DDG is high in protein (28 per cent) and fibre, and is used primarily as cattle feed. Almost all new ethanol facilities in the US are dry-milling operations, as they are less expensive and complex (but also less flexible in terms of their output). Corn and wheat can be stored for long periods of time between harvesting and processing, whereas sugarcane must be processed within 24–48 hours of harvest.

4.3.5 Production of ethanol by hydrolysis and fermentation of woody biomass

Efforts are underway to develop ethanol from lignocellulosic biomass sources (agricultural and forestry residues, herbaceous and woody crops). Lignocellulose is a mixture of cellulose (35–50 per cent), hemi-cellulose (20–35 per cent) and lignin (12–20 per cent). Cellulose consists of long chains of glucose sugars and is referred to as a *polysaccharide*. Hemi-cellulose consists of short, highly branched chains of different sugars, mainly xylose (a five-carbon sugar) but also arabinose (five-carbon), galactose, glucose and mannose (all six-carbon). Lignin is a heterogeneous substance with a phenol-propene backbone. Table 4.8 gives the relative proportions of cellulose, hemi-cellulose, lignin and other constituents of three different kinds of woody biomass and for corn stover and switchgrass, as well as the amounts of the different sugars within the hemi-cellulose fraction. Cellulose sugar and the hemi-cellulose sugars are converted into ethanol with varying effectiveness, so their relative proportions affect the overall ethanol yield. Lignin is highly resistant to breakdown and so forms a solid residue that can be dried (from 60 per cent to 15 per cent moisture) and burned to provide heat and electricity, or gasified and used in a combined cycle powerplant to generate heat and electricity. Some of the electricity and all of the heat would be used for the ethanol plant's own energy needs, with the excess electricity exported to the power grid.

The steps in the production of ethanol from lignocellulosic biomass are:

1. pretreatment to separate cellulose from lignin and hemi-cellulose;
2. breakdown of cellulose and hemi-cellulose into their constituent sugars;
3. fermentation of sugars; and
4. concentration of the ethanol produced by fermentation.

These steps are briefly outlined below based largely on information in Hamelinck et al (2005a), where more details may be found.

Pretreatment

Hemi-cellulose and lignin form a protective sheath around cellulose, which must be modified or removed in a pretreatment process in order for cellulose to be broken into its constituent sugars. A major area of research with regard to the production of ethanol from

Table 4.8 *Typical composition (per cent) and heating value of different lignocellulosic biomass materials, with the sugars composing cellulose and hemi-cellulose and their identification as five-carbon (5C) or six-carbon (6C) sugars*

	Hardwoods		Softwood	Corn stover	Grass
	Eucalyptus	Hybrid poplar	Pine		Switchgrass
Cellulose glucan (6C)	49.5	44.7	44.56	37.4	32.0
Hemi-cellulose	13.1	18.6	21.9	27.5	25.1
– xylan (5C)	10.7	14.6	6.3	21.1	21.1
– arabinan (5C)	0.3	0.8	1.6	2.9	2.8
– galactan (6C)	0.8	1.0	2.6	1.9	1.0
– mannan (6C)	1.3	2.2	11.4	1.6	0.3
Lignin	27.7	26.4	27.7	18.0	18.1
Ash	1.3	1.7	0.3	5.8	6.0
Acids	4.2	1.5	2.7	3.2	1.2
Extractives[a]	4.3	7.1	2.9	8.1	17.5
Heating value ($GJ_{HHV}/tonne_{dry}$)	19.5	19.6	19.6	14.9	18.6

Note: The heating value for corn stover given here was deduced from Sheehan et al (2004), and is smaller than the value of 16.5MJ/kg given by Lind and Wang (2004, Endnote 10). [a] Extractives are low molecular-weight materials such as aromatics and terpenes, some of which may be toxic to ethanol-fermenting organisms and may cause deposits during some pretreatment processes.

Source: Hamelinck et al (2005a) except for corn stover, which is from Sheehan et al (2004)

lignocellulosic biomass concerns the development of more effective pretreatment processes. Pretreatment involves: (1) grinding the biomass to 1–3mm size wood chips, and (2) removal of lignin and rendering all or part of the hemi-cellulose soluble (hemi-cellulose is easily broken down at this stage because of its branched structure). Mechanical grinding uses about a third of the power of the entire plant. This is followed by chemical, physical or biological pretreatment to remove lignin. The most common chemical pretreatment is to soak the ground biomass in dilute sulphuric acid for ten minutes at 100°C, then heating it to 160°C for ten minutes. Physical pretreatment methods include *steam explosion*, in which the material is heated with high pressure steam (20–50 bar, 210–290°C), then suddenly decompressed, and use of compressed hot liquid water. As noted above, chemical treatment is fast (minutes), whereas steam-based pretreatment can take up to a day. Biological treatment uses fungi to solubilize the lignin.

Breakdown of cellulose

Cellulose is broken down by adding water to it, a process called *hydrolysis* or *saccharification*. The most common method uses a dilute acid, and is essentially the same as the dilute acid pretreatment option described above. An alternative is to use concentrated acids. A key to the economic viability of this method as well as to minimizing environmental impacts is to be able to cost-effectively separate the acid for reuse. Current methods recycle 97 per cent of the acid with loss of 2 per cent of the sugar. An alternative is *enzymatic hydrolysis*, making use of the *cellulase* enzyme produced by organisms living on about 2 per cent of the cellulosic material that is sent to a separate vessel. Research focuses on enhancing the production of cellulase by both bacteria and fungi, enhancing the effectiveness of the enzyme and recovering and recycling the enzyme. Many experts believe that the development of low-cost enzymatic hydrolysis is the key to economically competitive ethanol production.

When acids rather than enzymes are used for the breakdown of cellulose, the acids and degradation products need to be removed prior to fermentation. This can be done either by adding lime ($Ca(OH)_2$) or through recovery and recycling of the acids. Recovery and recycling is expensive, whereas adding lime produces hydrated gypsum ($CaSO_4{\cdot}2H_2O$), which precipitates out as a solid and needs to be disposed of. About 0.2kg of gypsum is produced per kg of feedstock. Some can be used as agricultural soil conditioner, but with large-scale ethanol production from woody materials, the quantities of gypsum produced would exceed the amount that could be used in this way.

Fermentation

Once cellulose and hemi-cellulose have been broken into their constituent sugars, the sugars are fermented to ethanol. Fermentation can be carried out by bacteria, yeast or fungi. The reactions for five-carbon and six-carbon sugars are:

$$3C_5H_{10}O_5 \rightarrow 5C_2H_5OH + 5CO_2 \quad (4.1)$$

$$C_6H_{12}O_6 \rightarrow 2C_2H_5OH + 2CO_2 \quad (4.2)$$

In both cases, a third of the carbon is lost as CO_2 during the fermentation step (and is thus potentially amenable to capture and sequestration, as discussed in Chapter 9). Glucose (a six-carbon sugar) is readily fermented into ethanol by yeast, but there are no naturally occurring organisms that convert five-carbon (pentose) sugars into ethanol. Some organisms metabolize five-carbon sugars, but they produce a variety of acetic and lactic acids as fermentation products instead of ethanol. In 1985, genes that cause fermentation of sugar into ethanol were inserted in Escherichia coli, an organism that can metabolize five-carbon sugars. The genes caused E. coli to convert 90–95 per cent of the biomass sugars into ethanol, but it could originally tolerate only 4 per cent ethanol in the final solution. That amount has since been increased to 6.4 per cent, and work continues to increase the tolerance so that more concentrated ethanol can be produced during fermentation, thus requiring less concentration of the ethanol afterwards. Work is also underway on increasing the speed of fermentation; yeast can convert a batch of glucose to ethanol within a few hours, whereas organisms working on a mixture of sugars require one to two days (Service, 2007a). Some micro-organisms co-produce cell mass at the expense of ethanol.

When enzymes are used for the breakdown of cellulose, the breakdown and subsequent fermentation of sugars can be carried out in a single vessel called the *simultaneous saccharification and fermentation* (SSF) unit (Stephanopoulos, 2007). A single vessel is used so that the products of cellulose breakdown do not accumulate, which would otherwise inhibit the hydrolysis enzymes. This also potentially reduces costs, as fewer reactor vessels are needed. Further integration, involving enzyme production, hydrolysis and fermentation in a single vessel, might be possible. However, greater cost reductions might occur through optimizing the separate reactors, as the best conditions for growth of enzymes, breakdown of cellulose and for fermentation are different (Hamelinck et al, 2005a).

Concentration of ethanol

The product of the fermentation step is a mixture of ethanol (no more than 10 per cent by weight), cell mass and water. The ethanol is recovered in a distillation or beer column, then further concentrated in steps to a concentration of almost 95 per cent. This is an expensive and energy-intensive step.

Utilization of solid residues

The solid residues produced from the processing of lignocellulosic biomass can be combusted and used to produce heat and electricity, some of which can be used for the ethanol production process and some of which (if available in excess) can be exported from the system. Figure 4.24 gives an overview of the integrated production of ethanol, heat and electricity.

Future possibilities and costs

Genetic engineering is a key focus of research in the effort to improve the yield and reduce the cost of ethanol from lignocellulose. Efforts are underway to develop organisms that can use all of the sugars released during the hydrolysis step, and to improve the ethanol yields. Stephanopoulos (2007) and Himmel et al (2007) speculate that it might be possible to develop a single organism that produces the cellulase enzyme (for hydrolysis of cellulose) and that ferments the products of cellulose breakdown. This concept is referred to as *consolidated biomass processing* (CBP). It would involve either adding cellulase-producing genes to a fermenting organism, or adding fermentation capability to a cellulose-producing organism. Another possibility is to re-engineer plant cell walls to make it easier for microbes to break down the intertwined networks of lignin and cellulose and to reduce the amount of

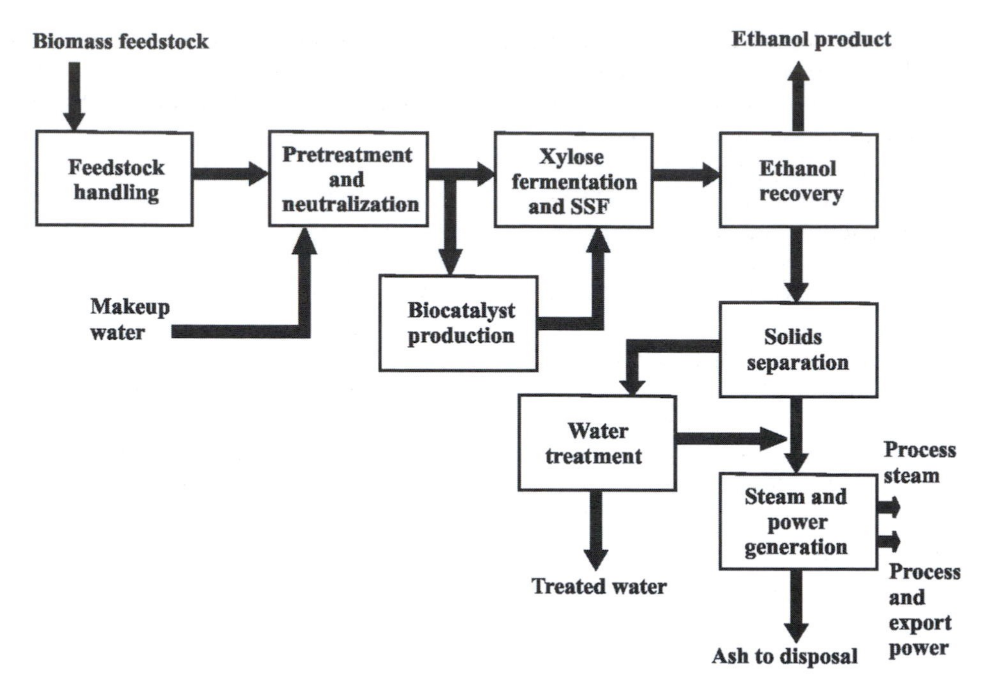

Note: SSF = simultaneous saccharification and fermentation.
Source: Wyman (1999)

Figure 4.24 *Production of ethanol from lignocellulosic biomass according to the process proposed by the National Renewable Energy Laboratory in the US*

cellulase enzyme needed, or to boost the ability of microbes to tolerate ethanol, so that higher concentrations can be produced prior to the distillation step (Service, 2007b). Such alterations may, however, increase the susceptibility of the plants to insect pests and may alter soil structure and fertility by altering the properties of plant litter (James et al, 1998).

Hamelinck et al (2005a) performed a techno-economic assessment of possible near-term, medium-term and long-term systems for ethanol production from woody biomass. The efficiency of ethanol production (ethanol energy over biomass energy) is projected to increase from 35 per cent in the near term to 48 per cent in the long term, while projected investment costs range from €2100/$kW_{ethanol}$ in the short term to about €850/$kW_{ethanol}$ in the long term. For a biomass feedstock cost fixed at €3/GJ, the ethanol cost is €22/GJ in the short term and €11/GJ in the long term (oil at $60/barrel is equivalent to $10.4/GJ and gasoline at $1/litre is equivalent to $30.9/GJ). Faaij (2006) is even more optimistic, projecting short- and long-term efficiencies of 46 per cent and 53 per cent, respectively, and a long-term cost of €4–7/GJ. This is only part of the story, however, as the residual solids and digestion gas remaining after production of ethanol can be used to generate electricity, initially using a gas or steam turbine but eventually through gasification combined cycle at an efficiency of 45 per cent. Overall utilization efficiency (ethanol plus electrical energy produced divided by biomass energy input) may therefore reach 40–50 per cent.

4.3.6 Production of biodiesel by transesterification of vegetable oils

The term 'biodiesel' refers to a fuel derived from vegetable oils or animal fats that is equivalent to diesel fuel and can be used in an unmodified diesel engine. Vegetable oils are almost always triacylglycerols, meaning a glycerol with three fatty acid esters. The fatty acids are typically 16–20 carbon atoms long, and odd-number chains are rare. Biodiesel is made from vegetable oils through the process of *transesterification*, which involves heating a mixture of 80–90 per cent oil and 10–20 per cent alcohol (such as methanol, ethanol, propanol or butanol) in the presence of a catalyst (typically NaOH or KOH) to break the molecule of the raw vegetable oil into methyl or ethyl esters, with glycerine ($C_3H_8O_3$) as a byproduct.[9] Methanol is the most commonly used alcohol for transesterification, in part due to low cost (Demirbas, 2005). It is produced from natural gas, although it could also be produced from biomass or from coal. Ethanol is preferred on environmental grounds because it can be produced from biomass and is non-toxic (unlike methanol). Methyl and ethyl esters can be burned directly in a diesel engine with very little deposit formation, unlike the direct use of vegetable oils. The major exporters of vegetable oil are Malaysia, Indonesia, Philippines, Brazil and Argentina – all except the last being countries with tropical rainforests that have been (or could be) cleared for vegetable oil production. Rapeseed oil is the most common source of biodiesel fuel in Europe, while soybean is dominant in the US.

4.3.7 Production of biodiesel from oily algae

Some researchers have suggested that oil-rich algae could be grown on a massive scale atop buildings and in deserts. The Aquatic Species Program of the US National Renewable Energy Laboratory (NREL) concluded that 28 billion litres of biodiesel (12 per cent of the current US diesel and fuel oil consumption of 240 billion litres/year) could be produced per year on 200,000 hectares (2000km^2) of desert land – a yield of 142,000 litres/ha/yr (about 4650GJ/ha/yr), compared to a maximum yield for any other oily crop of 6000 litres/ha/yr for palm oil in Malaysia.[10] Microalgae systems can use marine or fresh water. If fed with CO_2 from a coal-fired powerplant, more than 90 per cent of the injected CO_2 could be taken up by the algae (NREL, 1998). Thus, if 90 per cent of the CO_2 produced from the use of coal to generate electricity were captured (using techniques discussed in Chapter 9, subsection 9.3.3) and fed to the algal pond, 81 per cent of the coal C atoms would be used twice before being emitted to the atmosphere – once when the coal is combusted to produce electricity and once as a transportation fuel (the same double use could also be applied to biomass used for the generation of electricity, with capture of the initial biomass CO_2 as described in Chapter 9, subsection 9.3.4). Algal bioreactors, however, would capture CO_2 only during the daytime, so night-time emissions could be captured only if they were stored and passed through the bioreactor during the daytime. However, such temporary, small-scale storage could be an alternative to large-scale, permanent storage of captured CO_2 (discussed in Chapter 9).

Huntley and Redalje (2007) discuss privately funded research carried out after the termination of the US Aquatic Species Program. A fundamental limitation in using algae to produce oil is that a larger fraction of the algal mass is converted to oil when the algae are stressed, but under stress conditions the growth rate is smaller. A two-stage process has been developed that overcomes this conflict. The two stages are: (1) cultivation of algae in long, 38cm diameter plastic bioreactors to build up a cell mass under carefully controlled conditions, and (2) transfer of the cell culture to a 12cm deep, nutrient-depleted open pond to stress the cells, causing conversion of 25 per cent of the dry algal mass to oil within two days. The open pond would be emptied, sterilized and a fresh batch of cells added every two days. The photosynthetic efficiency is reported to be 3.0 per cent in the bioreactor and 4.4 per cent in the open pond. Average, maximum and projected future yields are summarized in Table 4.9. The projected yields assume conversion of 35 per cent of the algal biomass to oil and 5 per cent and 20 per cent photosynthetic efficiencies in the bioreactor and open pond, respectively, the latter already having been obtained for short periods of time. The current average oil energy yield (420GJ/ha/yr) is about twice that of biodiesel from palm oil in Malaysia (220GJ/ha/yr, the highest from any crop anywhere), while the projected future yield (3200GJ/ha/yr) is 16 times that using palm oil in Malaysia. The projected energy content of the waste products alone is about 900GJ/ha/yr. Seawater would have to be piped to the desert algal reactors and salt would accumulate.

Table 4.9 *Average and maximum measured yields from a microalgae bioreactor pond system*

	Average yield	Maximum yield	Projected yield
Dry biomass (t/ha/yr)	38	92	200
Energy in biomass (GJ/ha/yr)	763	1836	4100
Energy in oil (GJ/ha/yr)	422	1014	3200

Note: Projected yields are speculative.
Source: Huntley and Redalje (2007)

In 2009, ExxonMobil announced that it will spend up to $600 million over five to six years in an effort to develop a viable method for large-scale production of biofuels from algae (Service, 2009). Thousands of algal species will be tested in an effort to find or engineer strains of algae that can continuously produce biofuels in bioreactors. Also in 2009, a Florida-based company announced an $850 million project with Mexican partners in an effort to develop ethanol from algae.

4.3.8 Production of Fischer-Tropsch liquids by gasification and catalytic reactions

Fischer-Tropsch synthesis is a method for converting solid hydrocarbons (including cellulosic biomass) or natural gas into liquids. It can be used to produce a diesel fuel that is cleaner than regular diesel, thereby easing the resistance to diesel vehicles in some countries (see Volume 1, Chapter 2, subsection 2.7.1 for emission of air pollutants by diesel and gasoline vehicles). The first step is thermo-chemical gasification to produce a mixture of H_2 and CO under high temperature and pressure. The gases then proceed to a reactor where catalysts induce the carbon and hydrogen to reunite to form chains of varying length. Impurities such as mercury and sulphur inhibit the catalyst and so must be removed. Plant material that contains too much lignin and not enough cellulose for use in cellulosic ethanol systems could still be used in a Fischer-Tropsch system. The process is more expensive using biomass than using coal, requiring a price of oil of $70–80/barrel to be economic using biomass but only $50–55/barrel using coal (Ledford, 2006). An overall biomass-to-fuels efficiency of 46–52 per cent by 2010 is expected (Takeshita and Yamaji, 2008). Demonstration-scale biomass plants have been built in The Netherlands, and a 15,000t/yr commercial plant to produce diesel with a conversion efficiency for the gasification step greater than 80 per cent is due to be completed in Germany in 2010 (see www.choren.com/en/).

Gasification of biomass can also be used to produce electricity (as discussed in section 4.4), but gasification to produce liquids through the Fischer-Tropsch process requires more thorough gas cleaning so as not to damage the downstream catalytic gas-processing equipment. Other challenges involve scaling up from pilot to commercial scales and process integration.

4.3.9 Production of biofuels through solar-driven biomass gasification

Hertwich and Zhang (2009) propose using high-temperature heat from solar thermal towers (described in Chapter 2, subsection 2.3.3) to drive the biomass gasification reaction as the first step in producing synthetic liquid fuels. As solar thermal energy can be stored overnight, the production facility would be able to operate continuously. They calculate that only a third of the biomass and hence only a third of the land area would be required to make a given quantity of fuels as with biomass gasification and fuel synthesis without solar energy. For a biomass yield of 25t/ha/yr (a reasonable yield under favourable conditions according to Table 4.4), the fuel production would amount to about 30t/ha/yr (about 42,000 litres/ha/yr if ethanol) as essentially all of the C atoms in the biomass would become part of the fuel.[11] This is about four times the fuel production per unit of land devoted to biomass that can be expected for production of ethanol from sugarcane in Brazil in 2030 (see subsection 4.6.1). As the best regions for producing biomass and the desert regions suitable for concentrating solar thermal energy are separated by thousands of kilometres, biomass would need to be transported by ship (and train if not located on the coast) between the two regions, but – based on analysis of the energy cost of importing biomass to Europe from distant locations presented later in Table 4.35 – this would not be a large additional energy input. The land area required per tonne of fuel produced per year is estimated to be about $330m^2$ for biomass plantations and $52m^2$ for the solar thermal energy. This is instead of $1000m^2$ using biomass alone, so the total required land area is reduced by a factor of 2.5.

4.3.10 Production of hydrogen by gasification of biomass

As discussed in Chapter 10 (subsection 10.3.2), hydrogen can be produced through the gasification of biomass at an expected eventual efficiency of 50–60 per cent and a capital cost of $700 or less per kW of H_2 production rate capacity.

4.3.11 Liquid fuels summary, global production and comparative yields

The relationships between biomass feedstocks, the biomass transformation processes and the various liquid or gaseous fuels that can be produced are summarized in Figure 4.25. These fuels can be used for transportation or other applications. Gasification and anaerobic fermentation can be used to produce fuels for heating and cooking, while gasification and pyrolysis can be used as the first step in the production of a variety of chemical compounds that can substitute for petrochemicals (see section 4.7). Feedstocks and processes that are targeted at the production of transportation fuels (such as fermentation of sugar/starch crops or esterification of oily plants) produce other commodities as secondary co-products. These co-products and the associated energy credits are discussed later (subsection 4.6.1) in the context of the net energy yield in producing biofuels from biomass.

Figure 4.26 shows the distribution of bioethanol and biodiesel production in 2008, while Figure 4.27 shows the growth in world production of these two fuels from 1975 to 2008. About 50 per cent of the world's bioethanol is now produced in the US (largely from corn) and 40 per cent is produced in Brazil (largely from sugarcane). China produces small amounts of ethanol from corn, wheat and sugarcane, largely for industrial use, while Europe produces small amounts from cereals and sugar beets. The production of biodiesel is more widely distributed, with Germany (18 per cent), the US (17 per cent) and France (13 per cent) being the largest producers. Rapeseed and sunflower are the dominant sources of biodiesel.

Brazil launched an ethanol programme in 1975 in the wake of the first oil price shock and during a time of depressed world sugar prices (Kartha and Larson, 2000). By the mid-1980s, ethanol consumption exceeded gasoline consumption on a volumetric basis. In 1985, 96 per cent of new car sales were cars that ran

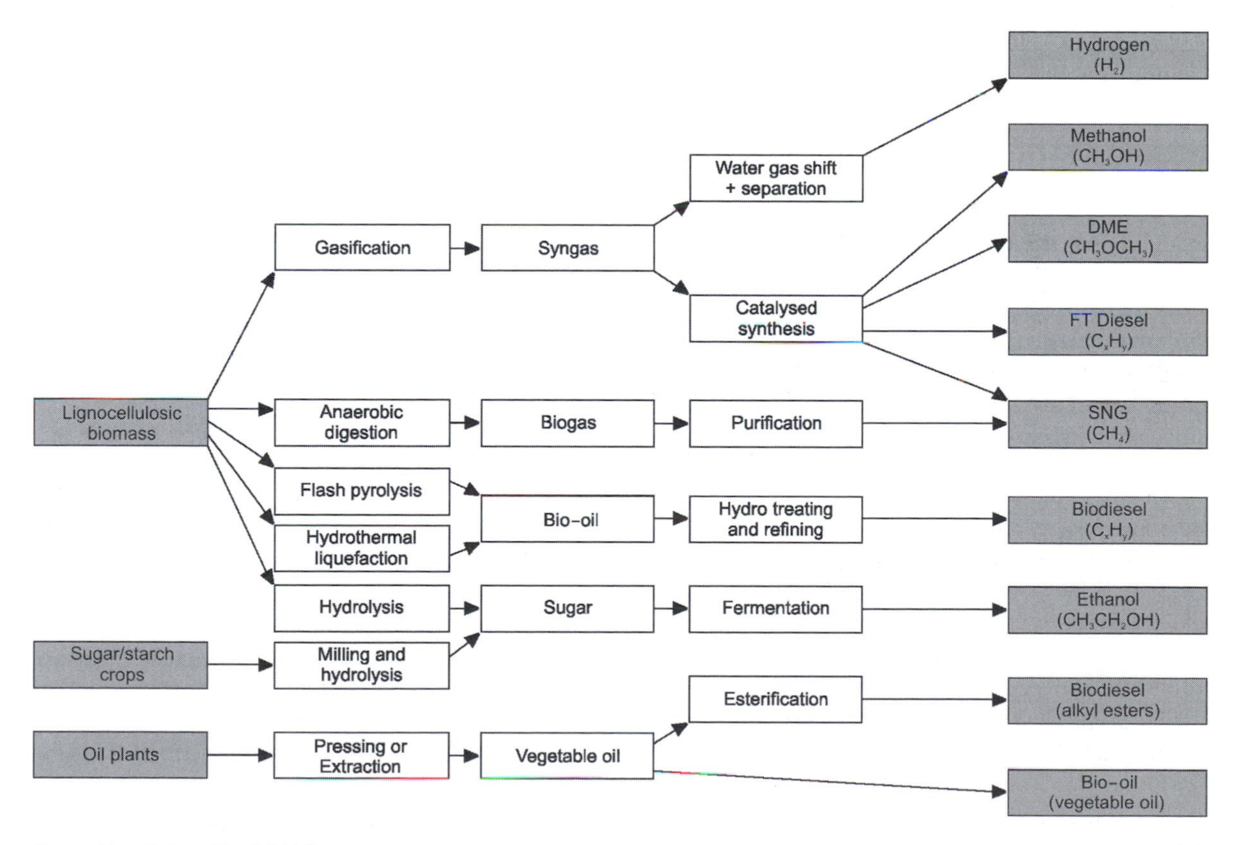

Source: Hamelinck and Faaij (2006)

Figure 4.25 *Summary of biomass feedstocks and processes used to produce various transportation fuels*

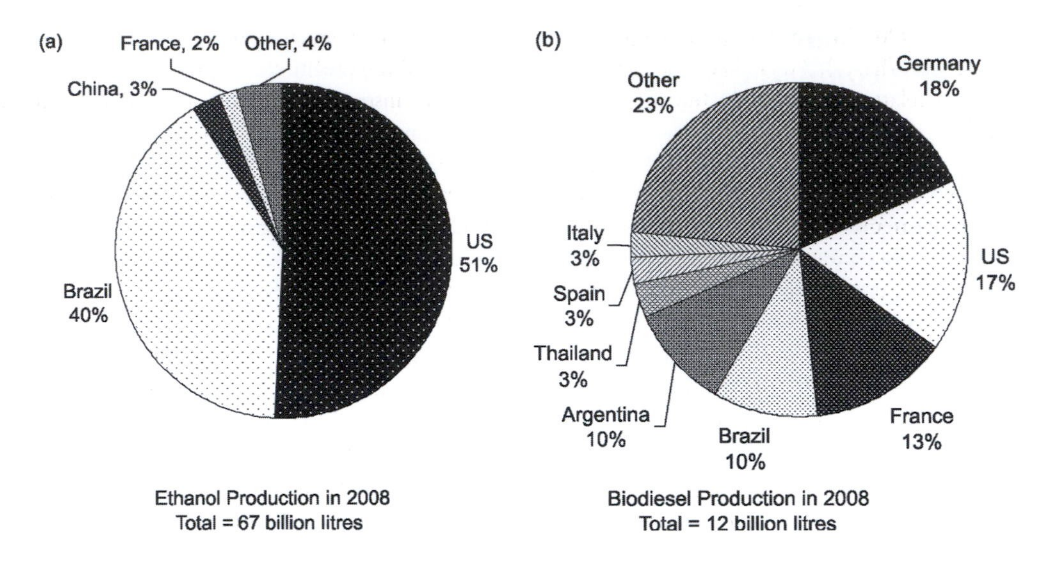

Source: REN21 (2009)

Figure 4.26 *Distribution of world ethanol and biodiesel production in 2008*

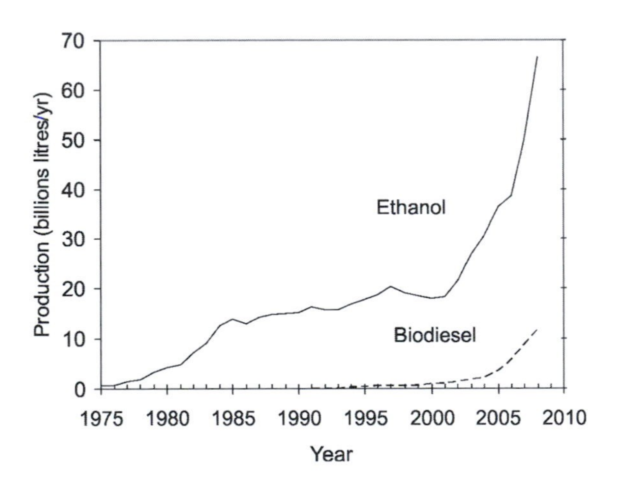

Source: REN21 (2009)

Figure 4.27 *Growth in world ethanol and biodiesel production from 1975 to 2008*

on pure ethanol, but by 1996 this had dropped to less than 1 per cent as the government-supported price difference between ethanol and gasoline decreased (Kheshgi et al, 2000). 'Flex-fuel' cars were introduced to the mass market in Brazil in 2003 and by 2006 they accounted for more than 75 per cent of new car sales; they can switch between pure ethanol (which is widely available) and gasohol (23 per cent ethanol), which constitutes most of the gasoline sold in Brazil.

Table 4.10 *Typical mid-2000s biofuel yields (litres per hectare per year) from different biofuel sources in different countries*

	US	EU	Brazil	India	Malaysia
Ethanol from agricultural crops					
Sugarcane			6000–7000	5300	
Sugar beet		5500			
Corn	3100				
Wheat		2500			
Barley		1100			
Ethanol from lignocellulosic biomass					
Corn stover	1100–2200				
Miscanthus	7300				
Switchgrass	3100–7600				
Poplar	3700–6000				
Biodiesel from oily plants					
Palm oil					6000
Rapeseed (canola)	500–700[a]	1200			
Sunflower seed		1000			
Soybean	500	700			
Jatropha				700	

Note: [a] From Table 4.26
Source: Ethanol from lignocellulosic biomass, Sanderson (2006); others, WWI (2006, Table 3-1)

Table 4.10 compares the yields (litres of fuel per hectare per year) for ethanol and biodiesel, and for different feedstocks in different regions of the world. The highest yields come from the tropics – sugarcane in Brazil (up to 7000 litres/ha/yr) and palm oil in Malaysia (6000 litres/ha/yr). Sugar beets, a temperate root crop, have a relatively high yield but are a more chemical- and energy-intensive crop than sugarcane. Other temperate crops have much lower ethanol yields than sugarcane. Similarly, biodiesel yields from temperate crops are many times lower than from palm oil. Rapeseed is further restricted by the fact that it should only be grown every fifth year in order to avoid the spread of diseases, with other crops planted in between. Within the temperate zone, biodiesel yields from oily plants tend to be less than ethanol yields from wheat or corn; however, oilseeds require less processing, so they tend to have a more favourable energy balance (as discussed in subsection 4.6.1). Soybeans are a nitrogen-fixing crop that can grow in both temperate and tropical climates. Switchgrass is the most promising of all, with ethanol yields of almost 11,000 litres/ha/yr in test plots (WWI, 2006). The energy balance of switchgrass (and of *Miscanthus* and hybrid poplar) is further improved by the fact that it needs less fertilizer and pesticides than other crops, it does not require tilling and it can sequester carbon in the soil.

It is instructive to note the dramatic increase in ethanol yield per unit land area in Brazil over the past 25 years. The average ethanol yield per tonne of sugarcane in Brazil increased by about 25 per cent from 1975 to 2002 (from 73 litres/tonne to 90 litres/tonne). The best values are a further 10–20 per cent higher and are expected to become the new norm within a few years. Combined with increases in sugarcane yield (t/ha/yr), the ethanol yield per unit land area has increased from 2000 litres/ha/yr in 1975 to about 6000 litres/ha/yr in 2005, with yields as high as 7000 litres/ha/yr under good conditions (WWI, 2006, Chapter 4). As discussed later (subsection 4.6.1), ethanol yields from Brazilian sugarcane could approach 14,000 litres/ha/yr in the future.

4.4 Electricity from biomass

An overview of the capacity, efficiency (on a lower heating value, LHV, basis; see Appendix D) and status

Table 4.11 *Capacity, efficiency and status of various technologies for using biomass for heat and/or electricity generation*

Conversion option		Scale	Net efficiency (LHV)	Comments
Biogas	Anaerobic digestion	Up to several MW$_e$	10–15% electrical	Applicable to wet organic waste streams and wastewater. Well developed in Denmark and The Netherlands
Combustion	Heat	1–5MW$_{th}$	70–90% for modern furnaces	Pellet systems replacing traditional fireplaces in Austria, Sweden, Germany
	CHP, conventional boilers	0.1–1MW$_e$ 1–10MW$_e$	60–90% overall 80–100% overall	Widely used in Scandinavian countries, Austria and Germany in district heating systems
	CHP, fluidized bed combustion	50–80MW$_e$	30–40% electrical	Finland is on the cutting edge
	Electricity	20–100MW$_e$	20–40%	
	Co-combustion	5–20MW$_e$ at existing coal plants, more at new plants	30–40% electrical	High electrical efficiency due to economies of scale of the existing plant. Investment costs low to negligible
Gasification	CHP	100skW$_{th}$	15–25% electrical, 80–90% overall	Small-scale systems never took off despite major efforts
	BIGCC	30–100MW$_e$	40–50% electrical	Demonstration projects in Brazil, Sweden and elsewhere. Currently €3500–5000/kW$_e$, needs to reach €1500–2000/kW$_e$. Many technological issues to resolve
	Solid oxide fuel cell		45–50%	Fuel cells still under development, no testing yet with biomass
	Molten carbonate fuel cell		46–54%	

Note: BIGCC = biomass integrated gasification combined cycle.
Source: Faaij (2006)

of various technologies for generating electricity or heat and electricity from biomass is given in Table 4.11. Electricity can be produced from biomass using a steam turbine, an internal combustion engine, gas turbine or fuel cell. Technical characteristics are discussed here, while costs are discussed in Section 4.10.

4.4.1 Steam turbines and co-firing with coal

The generation of electricity from coal using a steam turbine was explained in Volume 1 (Chapter 3, subsection 3.2.1). Coal steam turbines typically operate at a steam pressure of 290atm and a temperature of 580°C, which yields an efficiency of around 33–35 per cent. Biomass-based steam turbines in operation today operate with a steam pressure of around 60atm and a temperature of around 480°C, yielding an efficiency of 14–28 per cent. These low pressures and temperatures are chosen so as to minimize capital costs (since lower-quality steels can be used). The low efficiencies are geared toward the use of low-cost biomass feedstocks – primarily forestry and agricultural residues.

An example of the use of biomass for combined heat and power outside the forestry or sugarcane industries is provided by the growing use of reed canary grass (RCG) in Finland (Pahkala et al, 2008). RCG is a perennial grass that is native to Finland. It is harvested in the spring because the crop is dry and its quality as a fuel is high at this time, and is combusted together with peat and forestry residues.

4.4.2 Internal combustion engines

Either compression-ignition (diesel) or spark-ignition (gasoline) internal combustion engines can be used for small-scale (5–100kW) generation of electricity. Millions of such engines are used in developing countries for pumping irrigation water or for village lighting. Biogas can replace 75–95 per cent of the diesel fuel used in a diesel engine (some diesel fuel is needed to start the engine, since biogas will not self-ignite under compression). Tar buildup in engines using biogas has been a persistent problem.

4.4.3 Integrated biomass gasifier/gas turbine

The combination of gas and steam turbines (combined cycle) for electricity generation is described in Volume 1 (Chapter 3, subsection 3.2.3). The biomass thermo-chemical gasification process was described earlier in this chapter. The two can be integrated to give an overall biomass-to-electricity conversion efficiency of 40–45 per cent (Larson, 1993) – two to almost three times that achieved with existing biomass-fired steam turbines. Figure 4.28 illustrates the layout of an integrated biomass gasifier/gas turbine combined cycle system (BIGCC). The hot (900°C) products from the gasifier are only partially cooled (to 350–400°C) for cleaning (by condensing out alkali vapour), then burned in the combustor. Pressurized gasification avoids some of the energy losses associated with compressing the gas prior to combustion. Unlike internal combustion engines, removal of tar from the gases produced by gasification is not necessary.

Gasification/combined cycle systems using biomass are still under development. Existing gas turbines require modifications in order to work well with gases having a low heating value (biomass air gasification produces a gas with an LHV of 5–6MJ/Nm^3, while gasification in oxygen-rich air produces a gas with a LHV of 10–18MJ/Nm^3, compared to 38MJ/Nm^3 for methane).[12] Dry feedstock is required, as this increases the biogas heating value. Drying also reduces the susceptibility of the biomass to decomposition and reduces the weight and possibly transport costs. Drying below 10–15 per cent moisture is not optimal.

As for the BIGCC powerplant size that can be supplied with biomass from land within a reasonable radius, suppose that 10 per cent of the land in a region of 40km radius around the powerplant is devoted to energy crops and that the net energy yield is 100GJ/ha/yr (at the low end of what can be expected for woody crops). At 40 per cent conversion efficiency, this would supply a 250MW$_e$ powerplant.

4.4.4 Co-firing gasified biomass with natural gas

The composition and heating content of gas derived from gasification of biomass fuels and of natural gas are quite different, as illustrated in Table 4.12 for gasification of biofuel pellets. Biogas contains substantial amounts of N_2, H_2O and CO_2, and has a heating value of only 4.4MJ/kg, compared to 47.6MJ/kg for natural gas of the composition shown in Table 4.12. This would require increasing the fuel mass flow rate, which in turn requires reducing the operating

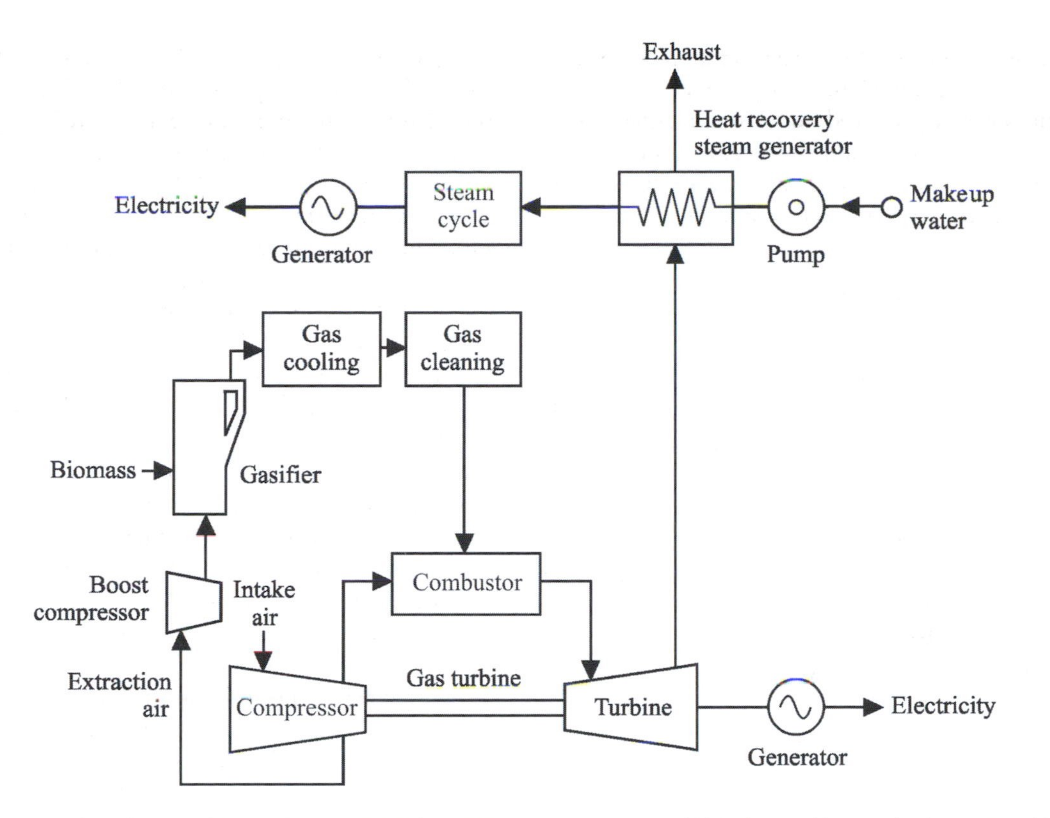

Note: The hot gas from the gas turbine produces steam in a heat recovery steam generator, which is then used in a condensing steam turbine.
Source: Larson (1993)

Figure 4.28 *A biomass gasifier/gas turbine combined cycle*

Table 4.12 *Comparison of a typical natural gas and biogas composition (mole per cent) and heating value*

	Natural gas	Biogas
CH_4	87.6	5.0
Other hydrocarbons	10.9	–
CO_2	1.2	12.0
CO	–	16.0
H_2O	–	12.0
H_2	–	11.0
N_2	0.3	44.0
LHV (MJ/kg)	47.62	4.4

Source: Marbe et al (2006)

temperature so as to avoid excessive pressures, which in turn (through the Carnot Cycle limitation discussed in Volume 1, Chapter 3, subsection 3.2.1) reduces the electricity generation efficiency and power output. Gasification requires using some of the waste heat from the steam turbine that could otherwise be supplied to a heating customer, so the overall cogeneration efficiency is also reduced. Using a powerplant simulation programme for combined cycle cogeneration, Marbe et al (2006) find that co-firing of up to 40 per cent biogas (on an energy basis) and 60 per cent natural gas in a combined cycle powerplant is possible with minimal adjustment in the turbines and with a decrease in electrical and overall efficiencies of less than 2 per cent. Rodrigues et al (2003) utilized a computer simulation model for a combined cycle powerplant and found a reduction in electrical efficiency of about 0.2 per cent with 40 per cent biogas, about 2.0 per cent with 90 per cent biogas and about 6 per cent with 100 per cent biogas. Thus, blending of some natural gas with biogas is advantageous from an efficiency viewpoint. It also permits larger turbines (for a given biomass supply), which further increases efficiency (and reduces unit cost).

4.4.5 Integration of gasification with fuel cells for cogeneration of electricity and heat

A fuel cell is an electrochemical device that uses a hydrogen-rich gas to produce electricity. Since biomass can be gasified to produce hydrogen gas, fuel cells can be used as the electricity-generation technology. The main advantage of using biomass-derived gases instead of coal-derived gases in fuel cells, apart from the greater ease of gasifying biomass, is the very low sulphur content of biomass. Sulphur contamination is a major concern in fossil fuel-based fuel cell systems. It should also be more economic to construct biomass gasifiers than coal gasifiers at the small scales that would be compatible with planned fuel cell systems (ORNL, 2001).

Both solid oxide fuel cells (SOFCs) and molten carbonate fuel cells (MCFCs) are well suited to integration with gasification of biomass, as the operating temperature of these fuel cells (600–1000°C) is comparable to that required for gasification (700–800°C). Thus, waste heat from the fuel cell can be used in part for gasification and in part to produce steam that can be used in a steam turbine to generate additional electricity. SOFCs and MCFCs are themselves still under development, so most of the work so far on integrated biomass gasification/fuel cell systems is theoretical. The overall electrical efficiency of an integrated gasification/SOFC/steam turbine system depends on the hydrogen content of the biomass fuel: as the H content by mass increases from 6 per cent to 10 per cent, the SOFC electrical efficiency (on an LHV basis; see Appendix D) increases from 28 to 42 per cent and the overall (SOFC plus steam turbine) efficiency increases from 41 per cent to 55 per cent according to calculations by Athanasiou et al (2007). With good process integration, an electrical efficiency of about 50 per cent should be possible with gasifier/SOFC systems (Karellas et al, 2008) and 46–54 per cent with gasifier MCFC systems (Sims, 2002). As the waste heat from the fuel cell that is not used for the gasification process itself will be available at a relatively high temperature (several hundred °C), it will normally be possible to get very high (>90 per cent) overall efficiencies if the waste heat is put to some use (such as process heat in industry if the biomass gasifier/fuel cell system is on an industrial site or for a district heating system).

4.4.6 Cogeneration of steam and electricity in the sugar, palm oil and pulp and paper industries

Sugar refineries require both electricity and heat for the processing of sugarcane, but have access to substantial amounts of biomass residue that are used to produce electricity and heat through cogeneration. Figure 4.29 compares the electricity output per tonne of sugarcane processed in a typical sugarcane refinery, and with

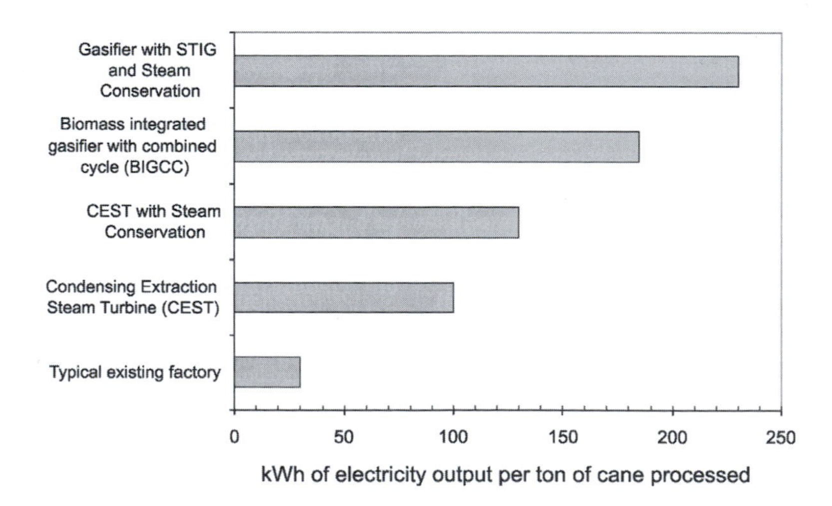

Note: STIG = steam injected gas turbine.

Source: Kammen et al (2001), except that current electricity generation is the average for Brazil as given by Moreira (2006)

Figure 4.29 *Comparison of technical options for generating electricity from sugar refinery wastes*

various improved technologies. The average refinery in Brazil produces 30kWh of electricity per tonne of sugarcane. With existing, commercially available steam turbines, this can be increased to 100–130kWh/t. With biomass-based gasification combined cycle systems still under development, 180–230kWh/t of sugarcane processed should be possible using only the bagasse. With use of 60 per cent of field residues that are currently burned (causing significant pollution problems), electricity generation could be conservatively increased to 500kWh/t according to Moreira (2006). This is an electricity production of 1.8GJ/t, compared to an average ethanol production in Brazil in 2002 of 1.9GJ/t (90 litres/t). For the 2004 sugarcane harvest of 350Mt in Brazil, the potential electricity generation is 175TWh/yr, compared to a total electricity generation in Brazil of 403TWh in 2005 (see Volume 1, Table 2.4). The low electricity output of typical current refineries is because the sole purpose of cogeneration had been simply to dispose of biomass wastes while just meeting the mill electricity requirements, combined with the refusal (until recently) of power utilities to accept independently generated electricity. Greater electricity production becomes technically feasible through reductions in the mill steam requirements, the development of advanced (gasification-based) electricity generation and the use of additional biomass residues as fuel that cannot at present be used.

Cogeneration is also used to generate heat and electricity in the palm oil industry. Conventional systems use a back-pressure turbine and generate 25–40kWh per tonne of fresh fruit bunch (FFB), whereas existing condensing extraction turbines could generate 75–160kWh/t FFB (Arrieta et al, 2007). Gasification combined cycle systems could presumably increase that to the level cited above for sugarcane mills.

As discussed in Volume 1 (Chapter 6, subsection 6.9.4), cogeneration of heat and electricity using forest residues and/or black liquor is common in the pulp and paper industry. Typical efficiencies of electricity generation are about 10 per cent in old systems and 15 per cent in new systems, but could be increased to 27–30 per cent with gasification of black liquor (an organic waste product of chemical pulping) and use of the gasified liquor in combined gas and steam turbine power generation (alternatively, as discussed in Volume 1, Chapter 6, subsection 6.9.5, black liquor can be used to produce dimethyl ether, which can be used in place of diesel fuel for transportation).

4.4.7 Comparison of combustion/steam turbines and gasification/combined cycle cogeneration

Marbe et al (2004) have compared the supply of heat and power from biomass using combustion combined with steam turbines, and using gasification combined cycle. In both cases it is assumed that the biofuel has an initial moisture content of 50 per cent and needs to be dried to 15 per cent moisture. Biofuels can be dried using either flue exhaust heat or steam. Gasification can be carried out at close to atmospheric pressure, or at high pressure (20 atmospheres); the latter gives higher electrical efficiency but is not yet commercially available. Table 4.13 compares the electrical and overall efficiencies for steam turbines and for atmospheric and pressurized BIGCC systems supplying heat to a district heating network or to an industrial load (with the latter presumably requiring higher temperature heat). The steam cycle has a low electrical efficiency (particularly when providing heat to an industrial load) but higher overall efficiency than gasification. The steam cycle can

Table 4.13 *Comparison of electrical and total efficiencies (LHV basis) for various cogeneration systems using biomass*

Technology	Power (MW_e)	Electric efficiency		Total efficiency		Power to heat ratio	
		Industry	District heat	Industry	District heat	Industry	District heat
Steam	2–26	0.17	0.27	0.87	1.10	0.24	0.33
Atm BIGCC	6–77	0.33	0.38	0.71	0.76	0.87	1.00
Press BIGCC	8–104	0.38	0.43	0.72	0.90	1.12	0.91
Biofuel boiler				0.87	0.90		

Source: Marbe et al (2004)

operate down to 30 per cent of full load, whereas gasification of biomass and use in combined cycle cogeneration powerplant can operate down to only 60 per cent of full load at present. The cogeneration plant will need to be shut down at times of smaller heat load, and boilers used to supply heat. As well, BIGCC plants must be of relatively large size (40–60MW$_e$) in order to operate profitably, but it is not always possible to find heat loads at this scale. If the heating load is district heating, a steam cycle is preferred because of its ability to cycle down to smaller loads. For an industrial heat load, which will be nearly constant year round, the gasification technology is favoured.

4.5 Environmental and social considerations in the use of biomass for energy

There are a number of ways in which the industrial use of biomass for energy could have negative impacts on the environment, but there also ways and circumstances under which industrial-scale biomass energy could have beneficial environmental impacts above and beyond reduced emissions of GHGs. Large-scale production of biomass for energy purposes may also compete for land that can be used to produce food, thereby affecting the price and availability of food and the incidence of hunger and starvation. Issues related to competition with other uses of land will be discussed later, in the context of long-term scenarios (see section 4.9).

Here, we discuss the following environmental issues pertaining to the production of biomass feedstocks and their transformation into various forms of secondary energy (fuels, electricity) or their use for heating: impacts on soil fertility and long-term productivity; water use; atmospheric and water pollution; generation of solid waste (gypsum); increased use of pesticides and herbicides; risks with genetically modified bacteria; an increase or decrease in nutrient leaching; removal of heavy metals from soils; utilization of ash; treatment of municipal wastewater, landfill leachate and sewage sludge; impact on erosion; impact on biodiversity; and indoor air pollution.

It is often reported in the literature that the environmental impacts of the production of bioenergy crops are less than the environmental impacts of producing food crops. This is especially the case for woody perennial bioenergy crops. If total production of food for human consumption is fixed, then replacing food crops with bioenergy crops on a given plot of land merely displaces the environmental impacts of food production elsewhere. However, if there were to be a shift toward diets lower in meat, substantial amounts of cropland (half of which is used to produce food for animals rather than for direct human consumption) would be freed up that could be used to produce bioenergy crops instead. Thus, there would be a net conversion from production of food crops to bioenergy crops, and depending on the types of bioenergy crop produced and how they are produced, there could be a reduction in environmental impacts.

4.5.1 Impact on soil fertility and long-term productivity

Agricultural and forestry wastes are regarded as a potential sustainable source of biomass. However, as discussed by Patzek and Pimentel (2005), true sustainability is conceivable only for a cyclic process, that is, a process in which all inputs are recycled within the system. In such a system, there is no such thing as 'waste'. Natural ecosystems almost totally recycle all nutrients; small amounts of nutrients are lost as river flows to the ocean, but over long periods of time these are balanced by the rate of release of nutrients from the weathering of rocks or by deposition from the atmosphere.

Nutrients are lost whenever biomass is removed from the field. In the case of industrial-scale biomass plantations, nutrients are lost through:

- clearing of the original natural forest (in the tropics, most nutrients are in the surface litter, not in the soil, which is generally of very low fertility, so when the litter and overlying biomass are removed, nutrients are removed too);
- nutrient export through frequent harvesting and associated activities such as slash burning, both of which remove nutrients by removing biomass;
- accelerated erosion due to human activities.

Soil fertility is the ability of soil to support plant growth. The nutrient content is one dimension of soil fertility. Other dimensions of fertility and the impact of industrial-scale plantations are:

- *soil organic matter*, which can bond to large inorganic nutrients, hold water, enhance weathering that supplies new nutrients and improve soil structure (the organization of soil into blocks), and which can be lost or built up, but is commonly lost, in biomass plantations;
- *soil structure*, which aids the infiltration of water and is damaged by excessive removal of biomass, tilling and compaction, as a result of which the soil loses its ability to absorb water and roots are less able to grow;
- *soil biodiversity* (bacteria, fungi, earthworms, insects), which plays a key role in the breakdown of organic matter and recycling nutrients and in maintaining soil structure. Soil biodiversity is reduced by frequent tillage and by frequent application of insecticides and herbicides, which are used in biomass plantations.

Sugarcane has been grown for centuries in Brazil, with indications that average productivity fell by 50 per cent over a 360-year period (Sparovek and Schnug, 2001). Conversely, it has been grown for over 300 years in infertile soil on the Nansei Islands of Japan with no decrease in yield (Ando et al, 2001). The current intensive industrial practices of sugarcane cultivation in Brazil (involving use of bagasse and leaves for energy production) began only 15–20 years ago, so their long-term impact on yield is not known.

Some strategies to maintain productivity and reduce the need for fertilizers are: (1) leaving residues in the field (although this can make replanting more difficult), (2) harvesting grasses after senescence,[13] (3) returning ashes (which contain many micronutrients) to the field (this also resolves disposal problems), (4) reducing tillage, and (5) minimizing the use of herbicides.

Nutrient recycling

In order to extend the life of biomass plantations, nutrients should be recycled by returning all residues not used for energy to the field. When ethanol is produced from sugarcane in Brazil, the final residues (stillage) are recycled back to the field at a rate of about $100m^3$/ha/yr, which provides about 125kg K_2O/ha/yr. This reduces but does not eliminate soil erosion compared to food crops; Moreira and Goldemberg (1999) report an average erosion rate of 12.4t/ha/yr for sugarcane, compared to rates of 20–40t/ha/yr for most food crops.

The nutrient concentration in biomass increases with height,[14] so the harvest of tree tops and branches for energy (in addition to stem-wood for timber) will disproportionately remove nutrients, which strengthens the need for nutrient recycling by returning the ash to the plantation (Rytter, 2002). Nutrient concentrations decrease with increasing tree age, so longer rotation periods would reduce nutrient losses. Site preparation by burning of biomass leads to loss of N through volatilization. Root systems in short-rotation plantations may be less extensive than in undisturbed forests, leading to increased leaching loss of nutrients. Woody plants generally contain a lower ratio of plant nutrients (such as nitrogen) to energy than do non-woody plants, so the loss of nutrients from the soil when woody plants are harvested will be smaller.

Required N inputs for switchgrass can be greatly reduced by harvesting only after the above-ground biomass has senesced (during the autumn) and translocated some nutrients to the roots. The harvested biomass is reduced but long-term productivity is probably enhanced (or N fertilizer needs are reduced). Letting the crop stand from November to February causes no further loss of biomass, but mineral components can leach from the plant into the soil, improving soil fertility and feedstock quality. The crop can also trap and retain snow, the melting of which enhances spring soil moisture. Mid-winter harvesting of some biomass systems occurs in Sweden due to easier access on wet fields once the soils are frozen. Long-term sustained yields of 20t/ha/yr seem to be feasible (Parrish and Fike, 2005).

Animal manure contains nutrients that should also be returned to the soil in order to maintain soil fertility, but manure can also be digested to produce biogas. However, digested manure is more useful as a fertilizer than raw manure, as organic-bound nitrogen is converted to ammonium during the digestion process. This increases the fraction of N as ammonium from about 70 per cent to 85 per cent, and also reduces nitrogen leaching by about 20 per cent (Börjesson and Berglund, 2007).

Phosphorus could be the long-term limiting nutrient, as replacement of lost P relies on mining phosphate ore deposits, and there is no known alternative natural resource for this (see Volume 1, Chapter 7, section 7.3). The other macronutrients (Mg, Ca and K) are abundant.

Changes in soil carbon

The amount of carbon in the soil can either increase or decrease with production of biomass for energy. Even where soil carbon content decreases, the change is small compared to the avoided CO_2 emission when biomass replaces fossil fuels for electricity generation (Cowie et al, 2006). However, with careful management, it is possible to build up the amount of organic matter in some agricultural and plantation soils. Rates of soil carbon sequestration under different short-rotation woody crops in different US locations are summarized by Lemus and Lal (2005, Table 8) and Sartori et al (2006, Table 3). These rates are frequently in the range 0.5–2.0tC/ha/yr. There is generally an initial decrease in soil carbon after the establishment of a plantation, followed by a recovery requiring 10–50 years. Clifton-Brown et al (2007) measured the average yield and cumulative buildup of soil carbon during 16 years of growth and harvesting of *Miscanthus* from sites in Ireland. Average yield with spring harvest (to minimize the removal of nutrients from the ecosystem) was 9 tonnes of biomass/ha/yr, while the average rate of soil C buildup was 0.6tC/ha/yr. However, the authors caution that it is not known if the buildup of soil C is permanent. The establishment of N-fixing trees can assist in the buildup of soil carbon in degraded soils. Equivalently, fertilization can sometimes increase soil carbon under perennial species by enhancing plant growth. Liebig et al (2005) measured the carbon content in soils from 42 paired switchgrass/cropland sites in Minnesota and North and South Dakota. Over a 120cm sampling depth, they found on average 15.3t/ha more soil carbon content under switchgrass than under agricultural crops.

Tilman et al (2006) examined biomass production and carbon sequestration in 152 plots of 1, 2, 4, 8 or 16 species of perennial grasses planted in 1994 on degraded soil.[15] The plots were unfertilized, and irrigated only during initial establishment. Biomass production increased with the logarithm of the number of species planted, with plots of 16 species having an average yield (65GJ/ha/yr) about 3.4 times the yield of plots with a single species.[16] Even switchgrass gave yields only a third those of the highest-diversity plots. All above-ground biomass was removed every year, yet soil carbon sequestration steadily increased from about zero with one species to about 2.7t/ha/yr with 16 species. Tilman et al (2006) believe that this rate could be maintained for a century. An additional 1.7t/ha/yr sequestration occurred in roots, but most of this occurred during the first ten years and so would not continue. The grasses in mixed plots were combined with legumes that fixed N, so after ten years the soil N content had increased by 25 per cent (in dry habitats, some N fertilizer would be useful due to the lack of efficient N-fixing species). Thus, low-input, high-diversity perennial grasses can be grown on abandoned farmland with minimal inputs and net sequestration of carbon in soils, making them a better choice than food crops for biomass energy.

In Brazil, Zinn et al (2002) report no net change in soil carbon for *Eucalyptus* on loamy soils compared to native *cerrado*-type vegetation, a 17 per cent loss (9t/ha) for *Eucalyptus* on sandy soils and a 9 per cent loss (11t/ha) for *Pinus* plantations on clayey soils. Reduced tillage during soil preparation could reduce these losses. Woody perennials in agricultural areas can recycle leached nutrients to near-surface layers, and several woody perennials stimulate N fixation.

Deep groundwater contains dissolved CO_2 at a higher partial pressure than atmospheric CO_2 (up to 0.01atm versus 0.00038atm for a CO_2 concentration of 380ppmv) and if this water is used to enhance net primary productivity through irrigation, the resultant outgassing of CO_2 may offset any buildup of soil carbon (Schlesinger, 2000). Finally, herbicides can significantly reduce the carbon content of soils (by up to 25 per cent after three years) (Sartori et al, 2006).

4.5.2 Water use

Water is used both while bioenergy crops are growing (through evapotranspiration) and for the processing of bioenergy crops into liquid biofuels. There are two issues related to evapotranspiration water losses: the absolute water flow through bioenergy crops compared to food crops or natural vegetation, and the water use per unit of biomass produced for different bioenergy crops.

Berndes (2002) has assessed the evapotranspiration water requirements per unit of biomass produced for different bioenergy crops. Results are presented in Table 4.14. Evapotranspiration water requirements are smallest for ethanol from sugarcane (~500–1000 litres/litre) and

Table 4.14 *Evapotranspiration water use associated with biomass energy*

	Water use effectiveness (kg DM/ha/mm)	Water use (kg/kg DM)	Feedstock as a fraction of above-ground DM production	Fuel yield (litres per kg of feed-stock)	Water use	
					(tonne/GJ of feedstock	(litres/litre of fuel)
			Biodiesel			
Rapeseed	9–12	833–1111	0.5	0.37	80–110	2250–3000
			Ethanol			
Corn	7–21	476–1429	0.40	0.44	60–180	1080–3250
Wheat	6–36	278–1667	0.45	0.45	30–185	620–3700
Sugar beet	9–24	417–1111	0.42	0.60	50–132	700–1850
Sugarcane	17–33	303–588	0.27	0.60	56–110	500–980
Lignocellulosic crops	10–95	105–1000	0.80	0.30	6.6–63	350–3300

Note: DM = dry matter.

Source: Data in columns 2, 4 and 5 are from Berndes (2002); data in column 3 are computed from column 2; data in column 6 are computed from columns 3 and 4; data in column 7 are computed from columns 3 and 5

largest for ethanol from wheat (620–3700 litres/litre) and corn (~1000–3000 litres/litre), but can be quite small or large for lignocellulosic crops.

Chiu et al (2009) estimate the water required per litre of ethanol produced from corn in the US. Their assessment is based only on irrigation and process water requirements, and the result is a requirement of anywhere from 5 to 2138 litres of water per litre of ethanol produced, depending on regional irrigation practices.

When grasses are grown in place of traditional agriculture crops, the water use efficiency (tonnes of dry matter production per tonne of water used) increases. Nevertheless, the absolute water requirements can increase. This is because the fractional increase in biomass yield can exceed the fractional increase in water use efficiency. If tree crops replace shallow-rooted grasses or food crops, then there would be increased withdrawal of groundwater through the tap roots. Conversely, the presence of enhanced tree cover can induce greater water supply by facilitating the interception and infiltration into the soil of rainwater that would otherwise be lost to runoff. In agro-forestry systems, rainfall that is intercepted outside the growing season could be available later for crops, increasing food crop yields. However, crop residues enhance infiltration of rainwater, so their removal for energy may increase runoff and reduce recharge of soil moisture. Although agro-forestry systems may enhance water supply, the bioenergy productivity is much less than the 10–20t/ha/yr sometimes assumed in global scenarios.

Water is also consumed in the conversion of biomass energy into fuels or electricity, such as in biomass-based steam turbine powerplants (0.5t/GJ biomass), the gasification of biomass (0.1t/GJ), or in the production of ethanol (0.5–0.7t/GJ biomass) (Berndes, 2002). The last of these corresponds to 25–35 litres water/litre ethanol in the case of ethanol from corn (assuming a yield of 20 litres ethanol/GJ dry corn, deduced from Table 4.21). Patzek and Pimentel (2005) estimate a somewhat higher use for the production of ethanol from corn, 25–175 litres/litre of ethanol for process applications (e.g. yeast and molasses preparation, steam generation) and non-process applications (e.g. cooling water, making potable liquor). In either case, the water use is small compared to the evapotranspiration water fluxes given above.

Competition between water for energy crops and for food production has occurred in parts of India and Africa (Reijnders, 2006). Hybrid poplar in the north-central US requires 1m/yr of groundwater use during mid-rotation (Tuskan, 1998). In southern China, surface runoff from *Eucalyptus* plantations in China is less than from bare land but greater than from nearby mixed forests, due to smaller interception and percolation of rainfall in *Eucalyptus* plantations (Zhou et al, 2002). As a result, the water table is 50cm lower under *Eucalyptus* than under nearby mixed forests. In some western US regions with irrigated corn, groundwater is being pumped ten times faster than the natural recharge (Pimentel and Patzek, 2005).

The water used in producing and processing bioenergy crops for transportation fuels should be compared with that used in processing crude petroleum. This amounts to about 0.25–0.35m^3 per barrel of processed oil, or about 2–3 litres water/litre gasoline (given production of about 140 litres gasoline/barrel of oil). This is at least ten times smaller than the amount of water used for processing corn into ethanol. However, wastewater discharges from petroleum refineries are about two to three times those of biofuel-processing facilities (WWI, 2006). Additional water is usually used to cool waste products distilled from crude, so as to return them to liquid form (WWI, 2006). Extraction of tar sands requires three barrels of water for every barrel of oil produced (about 3 litres/litre gasoline), and this is in addition to the water required for refining oil. The total would then be 5–6 litres/litre of gasoline, which is smaller than the processing water use of 25–35 litres/litre of ethanol produced from corn, or the evapotranspiration water losses of 1000–3250 litres/litre.

4.5.3 Use of fertilizers, herbicides and pesticides during production of feedstocks

Different bioenergy crops and different cultivation techniques (such as till versus no-till systems for annual crops) have different requirements for fertilizer, herbicides and pesticides (see Volume 1, Chapter 7, Table 7.1 for fertilizer requirements). The production of corn (used for ethanol) uses more herbicides, pesticides and N fertilizer per unit mass of crop than any other crop in the US, and so is a major contributor to groundwater and river water pollution (Pimentel and Patzek, 2005). Perennial energy crops generally require substantially less fertilizer, herbicide and pesticide impacts than annual crops. Fertilizers are of concern because they can leach into groundwater (contaminating drinking supplies) and into rivers and lakes (where they can cause eutrophication – the excess growth of algae). The amount of fertilizer that leaches depends on both rates of application and the fraction of the applied fertilizer that is taken up by the plant.

Dominguez-Faus et al (2009) calculated that the production of 1 litre of ethanol (or the energy equivalent of biodiesel) from various annual crops requires anywhere from about 20 to 60gm of N fertilizer, with sugarcane requiring the least and potatoes the most. They also question the common assertion that switchgrass does not require N fertilizer, arguing that too little is known about the plant at present, although they do acknowledge that switchgrass uses the applied N more efficiently and seems to be able to tap nitrogen from sources that other plants cannot tap. They suggest a fertilizer requirement of about 40gmN/litre ethanol.

Table 4.15 compares the rates of N fertilizer application and N leaching for various crops grown under Dutch conditions as reported by Faaij et al (1998). Leaching involves fertilizers not taken up by plants, so leaching rates will be tied to the efficiency of fertilizer uptake (the fraction of fertilizer that is absorbed by plants). Typical efficiencies of fertilizer uptake are, for N, 37–46 per cent; for P, 3–8 per cent (but most not taken up by plants is immobilized through the formation of Al- and Fe-phosphates in tropical regions); and for K, Ca and Mg, 70 per cent (Patzek and Pimentel, 2005). Perennial energy crops such as willow entail significantly less leaching compared to annual crops because they require less fertilizer input, have a longer growing season, provide year-round soil cover and have a more extensive root system. Börjesson (1999) indicates that nitrogen leaching from short-rotation forest and ley production is 30–50 per cent and 75 per cent less, respectively, than from grain production.

Conversely, perennial energy crops planted between open streams and food crop fields can absorb nutrients that would otherwise be lost from cropland. Börjesson (1999) recommends 50m wide strips, half of which would be harvested at one time. They would absorb

Table 4.15 *Comparison of rate of nitrogen fertilizer application and leaching for potential energy crops under Dutch conditions*

Crop	N fertilization rate (kg N/ha/yr)	N leaching rate (kg N/ha/yr)
Willow	76	9–16
Cereals	280	30–40
Potatoes	250	46–63
Maize	309	70–94

Note: For leaching rates, lower values pertain to clay soils and larger values pertain to sandy soils.
Source: Faaij et al (1998)

70kg N/ha/yr if nitrogen from upslope fields leaches at a rate of 15kg/ha/yr or more, and would also absorb phosphorus. For continuous absorption, the biomass would need to be regularly harvested.

Pesticides and herbicides directly kill beneficial soil organisms and wildlife in the treated areas, while infiltration into groundwater and river flows causes widespread downstream contamination. There is concern in Australia over nutrient and pesticide impacts on the Great Barrier Reef along the Queensland coast if E10 (gasoline with 10 per cent ethanol) were to be mandatory (Niven, 2005), while chemical runoff in the Mississippi River (which drains the US corn belt) is said to have killed off marine life over a 30,000km^2 area in the Gulf of Mexico. However, if weed species are harvested along with the primary energy crop, they could be used along with the primary energy crop in cellulosic, gasification, combustion or Fischer-Tropsch conversion processes (as long as having a heterogeneous biomass feedstock is not a problem). In that case, they do not need to be controlled with herbicides, although there may or may not be a decrease in overall biomass yield.

4.5.4 Atmospheric and water pollution associated with processing through to end use of bioenergy

Atmospheric and water pollution are associated with the production, transport, storage and use of biomass energy.

Production of ethanol

The current system of production of sugarcane in Brazil involves burning the fields before manual harvesting twice per year, as illustrated in Figure 4.30. This causes severe air pollution and acidification of already poor soils (Martinelli and Filoso, 2007). Although a law was passed in 2002 to reduce such burning, the deadlines for compliance have been

Source: Howarth and Bringezu (2009, Executive Summary), photo by Edmar Mazzi

Figure 4.30 *Smoke plumes from sugarcane plantations in Brazil that are burned prior to manual harvesting*

delayed several times. As discussed earlier (subsection 4.4.6), these currently burnt materials are a potentially significant source of biomass for gasification/combined cycle generation of electricity.

Vinasse is a black-reddish, viscous and acidic material that is left over after distillation (concentration by removing water) of sugarcane ethanol. In Brazil, all vinasses from sugar and ethanol mills have been monitored by official environmental bodies since the early 1980s and used as a fertilizer (Moreira, personal communication, 2009). Vinasses are transported by gravity through ducts or by trucks and then distributed in the field. The volume applied is limited to $400m^3$/ha/yr so as to avoid contamination of groundwater, but where soils are highly permeable, this approach cannot be used. The vinasses are instead treated in biodigesters or in aerobic lagoons before being discharged to rivers.

Transport and storage of ethanol and biodiesel

Ethanol–gasoline mixtures have a greater risk of corroding steel underground storage tanks than pure gasoline, increasing the risk of leakage. The likelihood of leaks from underground fuel feed lines and other components is likely to increase with mixtures of 20 per cent ethanol or greater. Once in contact with groundwater or surface water, ethanol increases the solubility of petroleum contaminants such as benzene, toluene and xylenes that may already be in the soil, and inhibits their biodegradation.

Unlike diesel fuel, biodiesel fuel is biodegradable and non-toxic. It is far more water-soluble than diesel fuel, enabling marine animals to survive higher concentrations in the event of a spill. This has been a factor motivating the use of biodiesel in China, for example, where gasoline and diesel spills have polluted groundwater. Biodiesel made from rapeseed oil can biodegrade in half the time required for petroleum diesel, and also speeds the rate of degradation of diesel itself in blends. This is in contrast to ethanol–gasoline blends, where the ethanol slows the breakdown of gasoline by depleting the available oxygen in water and soil.

Use of ethanol and biodiesel

Ethanol (C_2H_5OH) contains oxygen, unlike gasoline (a variety of molecules ranging from C_5H_{12} to $C_{12}H_{26}$). This oxygen leads to easier combustion and thus lower emissions of CO, hydrocarbons and some toxic chemicals. Ethanol has been mixed with gasoline in a 10:90 ratio (E10) and in a 85:15 ratio (E85). Compared to gasoline, E10 generally produces lower tailpipe emissions of CO, hydrocarbons, toluene and xylene, but increases emissions of acetaldehyde, ethanol and NO_x, and has mixed results for 1.3-butadiene, formaldehyde and benzene (Niven, 2005). Fuel evaporative losses are larger than for gasoline, such that E10 has greater total hydrocarbon and air toxic emissions and increases the ozone-forming potential. Compared to gasoline, E85 has substantially greater emissions of acetaldehyde (by up to a factor of 27) and formaldehyde, but substantially less emissions of benzene, volatile organic compounds (VOCs) and 1.3-butadiene. Emissions of NO_x and particulates are comparable or higher. Thus, there is no clear benefit to the use of ethanol over gasoline in terms of atmospheric pollution.

Biodiesel fuel contains oxygen, whereas diesel fuel (a variety of molecules ranging from $C_{10}H_{22}$ to $C_{15}H_{32}$) does not. This results in more complete combustion and smaller emissions of CO, particulates and visible smoke. However, higher combustion temperatures lead to an increase in NO_x emissions. When blended with diesel fuel, most of the reduction in CO, particulates and smoke occurs at 50 per cent biodiesel, with little further decrease at more than 50 per cent, whereas there is negligible increase in NO_x emissions for up to 50 per cent biodiesel according to data reported by Makareviciene and Janulis (2003). A 50:50 biodiesel to diesel blend would therefore probably be close to optimal in terms of emissions of air pollutants. Biodiesel can cause deterioration of the engine lubricant, and because it has detergent characteristics, it can bring fuel tank sludge into suspension (Carraretto et al, 2004).

Combustion for heat and/or power

Elements contained in biomass will be partly released to the atmosphere during combustion and partly retained within the ash residue. New compounds may be formed from the elements present in the biomass; examples include polycyclic aromatic hydrocarbons (PAHs), chlorinated dioxins and benzofurans (Reijnders, 2006).

The impact of substituting biomass for coal on emissions of various elements depends in part on the ratios of the elements in plant matter to coal. These are shown for 29 elements in Figure 4.31. Among the elements of environmental concern, biomass has substantially smaller concentrations of S, Cr, As, Se, Hg and Pb than coal; a comparable concentration of Cu, Zn, Mo, Ag and Cd; and a substantially greater concentration of Co. The concentrations of the macronutrients N, Mg, P, K and Ca are also quite high in biomass, implying potentially significant losses if ash is not recycled. The concentrations of trace elements in biomass depend in part on the concentrations in soils, and soils in polluted areas tend to have higher concentrations than elsewhere (due to decades of atmospheric deposition), so there could be large departures from the ratios shown in Figure 4.31 in any given region. Concentrations also vary with the type of biomass (woody or non-woody) and species.

Gustavsson and Karlsson (2002) have compared the emissions associated with different methods for heating houses (other than district heating). These emissions are related to the energy expended in producing (for biomass) or extracting (for fossil fuels) the fuel, and emissions either at an electric powerplant (in the case of electric resistance heating or heat pumps) or at the point of use (for biomass or fossil fuel boilers). The emissions depend in part of the amount of primary energy required per unit of heat delivered. Figure 4.32 compares the primary energy use and hydrocarbon, SO_x and NO_x emissions per MWh of heat delivered to the house.[17, 18] Wood and oil boilers have particularly large hydrocarbon emissions, but these are greatly reduced for pellet boilers. The oil boiler has particularly

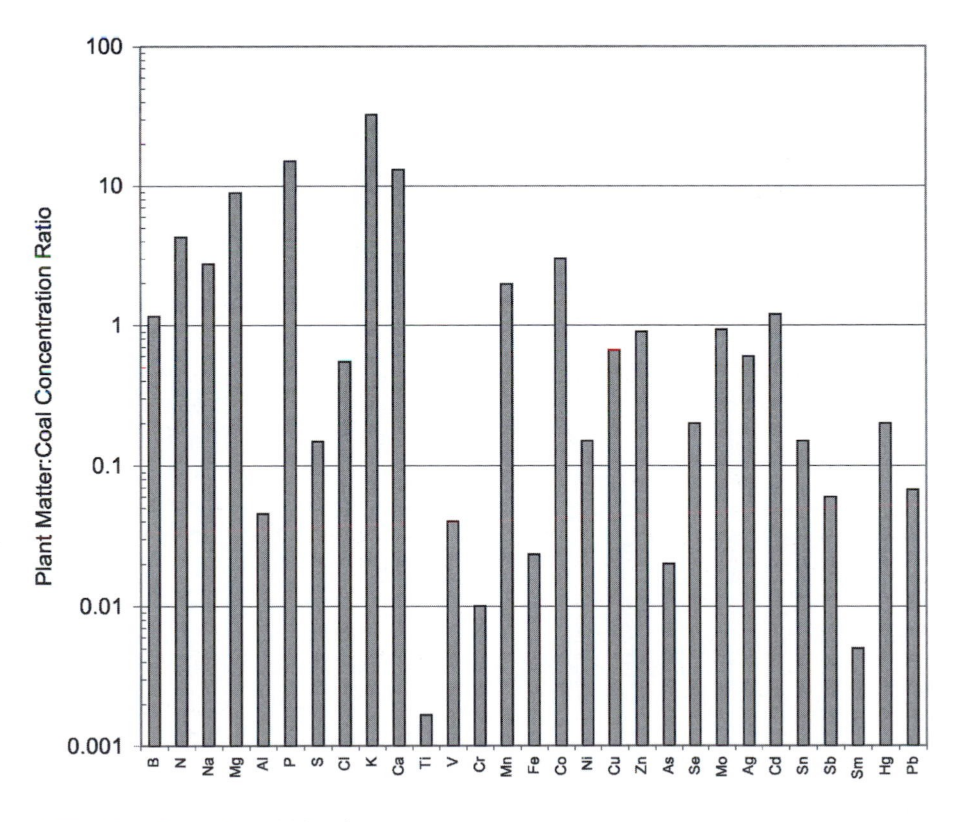

Source: Computed from data in Klee and Graedel (2004)

Figure 4.31 *Ratio of the average concentration of various elements in plant dry matter to the average concentration in coal*

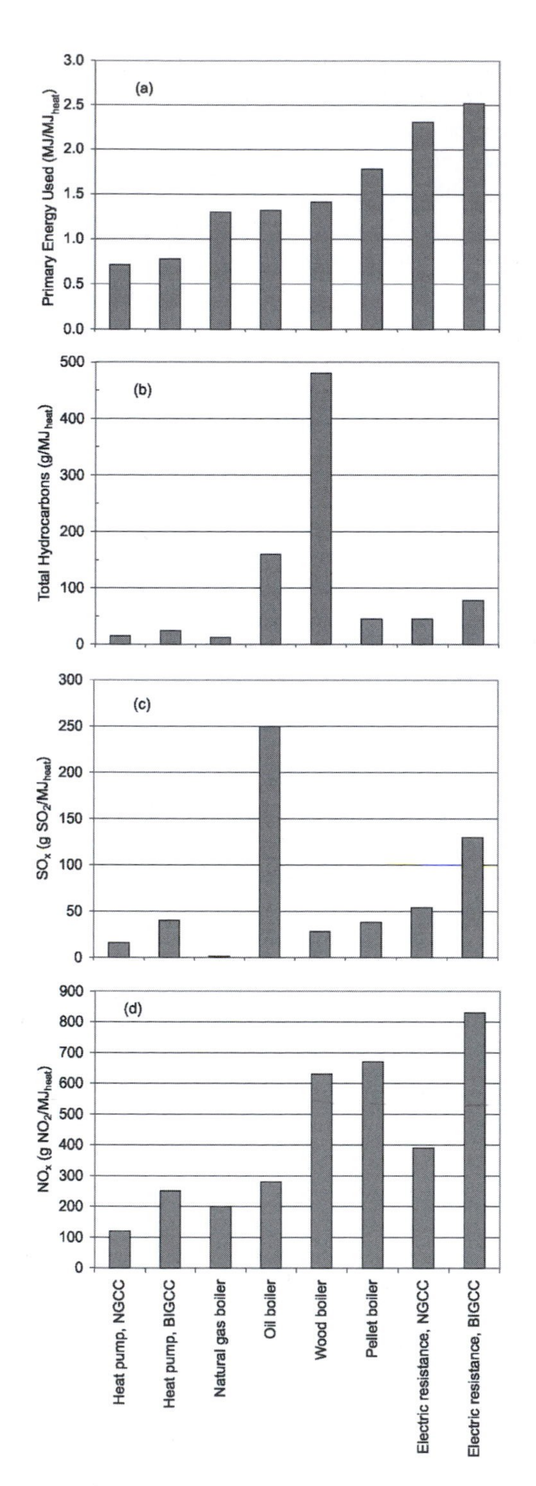

Source: Data from Gustavsson and Karlsson (2002)

Figure 4.32 *Primary energy use and emissions of hydrocarbons, SO_x and NO_x per MJ of heat delivered to detached houses in Sweden with various heating systems*

large SO_x emissions, while all forms of biomass have large NO_x emissions.

During combustion, S and Cl in biofuels produce compounds that can cause corrosion of the boiler. Woody biofuels have significantly higher concentrations of S and Cl than straw and cereals, so corrosion can be a problem in biomass combustion plants fired with straw and cereals. Straw and cereals also have greater N concentration, which contributes to greater NO_x emissions.

Gasification for heat and/or power

Carpentieri et al (2005) have compared the lifecycle atmospheric emissions of various pollutants associated with BIGCC powerplants and compared that with integrated coal gasification combined cycle (traditionally referred as IGCC) powerplants. The BIGCC system in this case involves capture of 80 per cent of the CO_2 in the flue gas, which is then available for burial in deep rock strata (as discussed in Chapter 9). The overall efficiency from biomass to electricity (including energy used for gasification and CO_2 capture) is 34 per cent and the powerplant CO_2 emission is 0.0485kgC/kWh. As this represents 20 per cent of the carbon in the biomass, the growth of biomass absorbs five times this amount, while the energy expenditure associated with harvesting and transporting the biomass amounts to only a few per cent of the biomass energy. The net result is a carbon sequestration (assuming the captured carbon to be permanently buried) of 0.194kgC/kWh, compared to an emission for a coal plant at 39 per cent efficiency with 80 per cent carbon capture of about 0.035kgC/kWh. If the harvest is 5t/ha/yr (which gives 100GJ/ha/yr at 20GJ/tonne) and the buildup of soil matter is 1t/ha/yr, the additional soil sequestration is 0.045kgC/kWh, bringing the total carbon sequestration to about 0.24kgC/kWh. With regard to pollutant emissions, the BIGCC system is slightly worse than IGCC for production of smog and for eutrophication and acidification, and is significantly worse with regard to heavy metals and carcinogenic substances (given the assumed characteristics of the soil on which the biomass is grown and the assumed emission control measures).

Faaij et al (1998) have computed atmospheric emissions associated with the use of hypothetical short-rotation willow in The Netherlands to generate

Table 4.16 *Comparison of pollutant emissions (gm/kWh) and CO_2 emissions (gmC/kWh) associated with BIGCC using short-rotation willow fuel, and associated with new coal powerplants coming into operation during 2003–2008*

	Willow BIGCC				Coal powerplant	
	Crop production	Crop transport	Powerplant	Total	Pulverized coal	IGCC
SO_2	0.024	0.001	0.071	0.096	0.38	0.2
NO_x	0.267	0.006	0.214	0.487	90.3	0.64
Dust	0.029	0.000	0.029	0.057	0.05	0.07
CO	0.267	0.006	0.290	0.563		
HC	0.057	0.002		0.059		
CO_2	5.85	0.55		6.40	222	217

Source: Faaij et al (1998)

electricity using BIGCC (at present there are no energy crops in The Netherlands). Results are given in Table 4.16, broken down into emissions associated with crop production and transport, and at the powerplant. Also given are emissions for a pulverized coal powerplant (at 44 per cent efficiency) and an IGCC powerplant. The CO_2 emissions shown for biomass pertain to fossil fuel energy inputs; direct emissions at the powerplant are offset by photosynthesis during growth. NO_x, SO_x and ash (dust) emissions are significantly lower for BIGCC compared to IGCC, and more so relative to pulverized coal.

Production of biogas from food crops and various wastes

Börjesson and Berglund (2007) have compared emissions to the atmosphere of NH_3, N_2O, NO_x, SO_2, CH_4, fossil CO_2, CO and particulates, and emission to water of nitrate (NO_3^-) for various systems for producing biogas through anaerobic digestion and for alternative biomass-based and fossil fuel-based energy systems. Replacement of fossil fuel systems and conventional systems for handling organic waste with biogas produced from organic wastes leads to reduced nitrogen leaching and reduced emissions of ammonia and methane, in addition to reduced emissions of CO_2 and air pollutants.

Rasi et al (2007) examine emissions of trace toxic materials from various biogas plants (landfills, sewage digesters, farm biogas plants). Emissions of H_2S, benzene and toluene are all potentially large from landfills but tend to be small from farm biogas plants and sewage digesters.

4.5.5 Risks with genetically modified bioenergy crops and bacteria

Genetically modified crops could be used as bioenergy sources, and genetically modified micro-organisms could be used to improve the yields of ethanol for cellulosic feedstocks. However, the use of genetically modified crops poses a number of risks and environmental hazards (summarized in Sidebar 12-1 of WWI, 2006). Fortunately, there is a vast potential to increase the yield of bioenergy crops through conventional breeding, without genetic engineering, as almost no effort has been made in this direction so far (in contrast to food crops, where factors of three to five yield increases have occurred over the past century). Most food crops used for bioenergy were bred for their starch or oil content, but not for overall biomass. However, the biomass is easier to manipulate, so an aggressive breeding programme could increase the yields of cellulosic feedstocks (such as corn stover or switchgrass). A factor of two yield increase would be a conservative projection.

The development of ethanol from cellulose is dependent on the creation of genetically modified bacteria that can digest cellulose and hemi-cellulose. However, genetically modified micro-organisms would be designed to survive and thrive only in the unusual conditions (such as 70°C temperatures) encountered during the fermentation of cellulosic materials, and so would pose no hazard in the wild.

4.5.6 Removal of heavy metals from soils

In Sweden, the cadmium content of soil has increased by 33 per cent during the past century, and the cadmium content of harvested grain sometimes exceeds the WHO/FAO limit (Börjesson, 1999). This is no doubt a problem in many other countries as well. However, *Salix* (willow) is much more effective than most other crops in absorbing cadmium and other heavy metals from soils. The cadmium and other metals can in turn be removed from the ash after the *Salix* is combusted, as explained in subsection 4.5.7. There are significant differences in the effectiveness of different species of *Salix* in removing Cd. The net removal from the surface soil layer will be reduced to some extent as deep metals are brought up by roots and deposited on the surface through litter fall. Cadmium levels could be reduced over a period of 35 years, after which the plant-available fraction remaining in the soil would be considerably reduced, but perhaps by that time the rate of atmospheric deposition and loadings from phosphorus fertilizer and sewage sludge would also be reduced through reductions in human pollution sources.

4.5.7 Utilization of ash and separation of heavy metals

Table 4.17 compares the typical ash production for various biomass fuels. The ash content ranges from essentially zero (soybeans) to as high as 13 per cent (sugarcane stems). The ash content of switchgrass harvested in the spring is lower than if harvested in the fall. Bark, straw and cereals produce 5–7 per cent ash, whereas wood chips produce around 1 per cent ash. The most abundant elements in ash are Si, Ca, Mg and K, with modest amounts of Al, Cl, Na, P, S and Fe, and small amounts of trace elements (Cu, Zn, Co, Mo, As, Ni, Cr, Pb, Cd, V and Hg) (Obernberger et al, 1997). There is little nitrogen in ash because most of it is volatilized during combustion, so returning the ash to the soil will usually still need to be accompanied by addition of nitrogen fertilizer. The concentration of heavy metals is 2–10 times higher in the ashes produced from wood chips and bark than from straw or cereals. This is due to the longer growing period of wood, higher atmospheric deposition rates in forests and the lower pH of forest soils, which increases the mobility of elements such as Cd and Zn and hence their availability for uptake by plants.

Table 4.17 *Ash content (percentage of dry biomass) of potential bioenergy crops*

	Ash content (%)
Switchgrass, spring harvest	2.0–3.2
Switchgrass, fall harvest	4.5–5.2
Wheat straw	4.5–8.3
Wood pellets	0.6
Napier grass	2.8–4.3
Alfalfa stem	7.4
Corn kernel	1
Sugarcane stems	13
Soybean	0
Poplar wood chips	4

Source: First six entries, Samson et al (2005); last four entries, Kheshgi et al (2000)

Ash can be separated into four different fractions, which are precipitated in the following order: bottom ash (70–90 per cent of total ash, produced in the combustion chamber), cyclone fly ash, boiler fly ash (the last two forming 3–25 per cent of total ash) and filter fly ash (2–10 per cent). A fifth ash component, fine dust (0.1–3.0 per cent), is too small to be precipitated or filtered, so it is emitted with the flue gas. The volatile heavy metals (Cd, Pb and Zn) are driven from the hot early-stage ashes and instead are concentrated in the cooler, later-stage ashes. Thus, although filter fly ash constitutes only 2–10 per cent of the total ash by weight, it contains 30–65 per cent of the Cd, 40–60 per cent of the Pb and 25–50 per cent of the Zn in the ash. The most effective separation of Cd and Zn occurs if the cyclone fly ash is precipitated at very high temperatures (850–900°C) followed by an effective filter fly ash operation at as low a temperature as possible, as in recent pilot plants in Sweden (Obernberger et al, 1997). Concentrations are 1.2–2.0 times higher in the fine dust, but the amount of fine dust is normally only 5 per cent the amount of filter fly ash. Conversely, 90–95 per cent of the least volatile heavy metals (Cr, Cu and Ni) are found in the bottom ash, but this is of no ecological significance as the concentrations are comparable to those found naturally in soils. The majority of plant nutrients (>90 per cent of Ca, P and Mg, >85 per cent of Na, and >80 per cent of K) occur in the bottom ash and cyclone ash. Thus, the bottom ash and cyclone ash can be returned to the soil as a fertilizer with no adverse effects, while the filter fly ash should be separately collected and either disposed of in a landfill or used in industry. However, return of the nutrients found in ash to the soil will not increase the biomass yield if other nutrients (such as nitrogen) are lacking (Park et al, 2005).

If biomass is co-fired with coal, the ash mixture is often considered to be unfit for agricultural applications – so the P and other nutrients from the biomass cannot be recycled. Even the ash from burning chicken manure cannot be returned to the field because of elevated concentrations of Cd and Zn (unless first separated out, as explained above). The high Cd and Zn concentrations arise from the presence of these elements in feed additives, illustrating another surprising linkage related to the question of long-term sustainability. As shown in Figure 4.31, the hazardous elements Hg, Pb, Cr, As and Se tend to have lower average concentration in dry biomass than in coal, while the reverse is true for Cd and Co. However, concentrations of many heavy metals have increased by factors of two to nine in European soils, and this probably leads to increased concentrations in biomass. The treatment of wood with chromated copper and arsenic salts for preservation purposes precludes their combustion for energy. Thus, sustainable use of biomass energy – which requires recovery of nutrients in biomass ash – is linked to strict control of toxic elements in the biomass waste stream (construction wood, municipal organic waste).

Another issue related to the use of biomass ash is the possible presence of polycyclic aromatic hydrocarbons (PAHs) in the ash. These are toxic compounds that form in the flue gases during the combustion process (rather than being part of the biomass) and end up in the fly ash fraction (which is captured from the flue gas). In one study of ash from a boiler using pulverized wood as the fuel, 40 per cent of the fly ash consisted of organic compounds (Sarenbo, 2009). PAH ranged in concentration from 91 to 300mg/kg, but reburning the ash reduced the concentration to 0.24mg/kg (a reduction by a factor of 360–1200).

4.5.8 Treatment of municipal wastewater, landfill leachate and sewage sludge

Energy crops are being tested in Sweden as filters of municipal wastewater and drainage water (Börjesson, 1999; Börjesson and Berndes, 2006). Wastewater is first pretreated (reducing the amount of solid particles as well as reducing the N content by 15–20 per cent), then distributed by pipe over nearby plantations, as illustrated in Figure 4.33. Two systems have been used: the first a *summer option*, whereby only wastewater produced during the summer is treated, and the second a *whole-year option*, in which wastewater produced during the winter is stored in ponds. The root systems absorb 75–95 per cent of the nitrogen and phosphorus in the applied wastewater if 500–1000mm/yr of wastewater are applied. In so doing, biomass yields are increased by 30–100 per cent compared to rainfed plantations (depending on location). The relative concentrations of nutrients in wastewater match the needs of the plant. In areas with low rainfall, the concentration of salts and heavy metals in the soil may increase. The cost of vegetation treatment of wastewater (per unit of treated wastewater) increases with increasing size of the village or community so served, due to longer wastewater distribution pipes. In Sweden, 70 per cent of the population lives in communities small enough that vegetation treatment of wastewater would be less expensive than conventional treatment plants. The required area of energy crops would be equal to 3.6 per cent of the arable land.

An example of the integrated use of sewage, biomass plantation, district heating and ash is provided by the city of Enköping, Sweden. An 80ha willow plantation was established in 2000 on farmland near

Source: P. Aronsson

Figure 4.33 *Distribution of wastewater for the irrigation and fertilization of a willow plantation in Sweden*

Source: Photography by P. Aronsson in Dimitriou and Aronsson (2005)

Figure 4.34 *Sewage treatment plant (foreground), aeration pond (middle ground) and willow plantation*

the sewage treatment plant, for the district heating system, as shown in Figure 4.34. The sewage plant operator bore the cost of establishing an irrigation system and building a 3ha storage pond to store wastewater during the winter, while the farmer bore the cost of establishing the willow plantation. The city district heat and powerplant operator is obliged to buy the harvested willow at the market price and has the right to recycle the wood ash back to the plantation. The economic value of the sewage treatment services provided by the plantation is greater than the cost of the plantation.

Sewage sludge (the solids extracted from sewage after initial treatment) is used as an agricultural fertilizer in many countries, in spite of the possibility of contamination of the sludge with heavy metals and toxic organic compounds. Sewage sludge can also be used as a fertilizer of energy crops, with less risk to humans. Soil humus levels can be increased as a result. However, sewage sludge has an excess of phosphorus relative to nitrogen (unlike wastewater), so supplemental nitrogen fertilizer would need to be applied. Heavy metal leaching and/or accumulation are insignificant in short-rotation coppice willow fertilized with biosolids according to studies from Sweden and Canada cited in Heller et al (2003). If the accumulation of heavy metals in biomass reaches high levels, they can be removed from the ash through flue gas cleaning when the biomass is combusted, as discussed in Section 4.5.7. Börjesson (1999) estimates that sewage sludge can be used on energy crops for a period of 25 years, so it will be necessary to improve the quality of sewage sludge over that time period so that it can subsequently be safely used on agricultural fields.

On experimental willow plantations in Quebec, Canada, application of granulated sludge from a sewage treatment plant resulted in biomass yields as high as 45t/ha/yr, compared to 15–20t/ha/yr without sludge fertilization (Labrecque and Teodorescu, 2001). As 80 per cent of the willow root mass occurs in the top 20cm of the soil, the willow roots form a dense network that acts as a vegetation filter, retaining heavy metals and other materials that would otherwise be leached into the groundwater.

About 10 of the 300 sanitary landfills in Sweden use *Salix* or grass for treatment of landfill leachate. The amount of nitrogen in the leachate can be reduced by 90 per cent. Vegetation filters will be needed for the entire period of anaerobic decomposition – up to 100 years. Tests with *Populus* (poplar) in the US indicate that irrigation with landfill leachate enhances growth, but that some clones respond more than others (Zalesny et al, 2009). Leachate is rich in the plant nutrients N, P, K, Na, Cl, B and Fe, although concentrations of Na, Cl, B and Fe in the leachate can sometimes be high enough to pose a problem.

4.5.9 Impact on erosion

Bioenergy crops can have either positive or negative effects on erosion. Swedish and Danish studies cited by Börjesson (1999) indicate that wind speeds are reduced by 30–90 per cent in fields with shelter belts less than 25m wide and less than 100m apart, with an average yield increase from the intermixed annual crops of 10 per cent. Energy plantations used as shelter belts would need to be 50m wide, so that half could be harvested at one time while still giving a 25m width.

Bioenergy crops can influence erosion in other ways. Perennial energy crops such as switchgrass will reduce water erosion and runoff compared to annual food or energy crops. A risk of increased erosion arises from the soil compaction and disturbance associated with harvesting. Removal of agricultural and forestry crop residues for energy rather than leaving them in the field may also increase erosion substantially.

4.5.10 Impact on biodiversity

Both short-rotation forests and grasses cultivated for energy serve as a habitat for birds and would serve to increase biodiversity if displacing a portion of land devoted to agricultural crops in areas of uninterrupted cropland (Londo et al, 2005). Intercropping of grasses and trees can further enhance the diversity of animal life. However, in a region with several different kinds of vegetation cover, inclusion of energy crops within farmland may not increase the overall regional biodiversity. Assuming a ten-year rotation, each year one tenth of the area under short-rotation woody crops would be harvested, one tenth would be planted, one tenth would be in its second year of growth, and so on. This leads to both spatial variation in stand age, as well as variation at a given site over time. Different clones may be planted in different sectors of the plantation or even intimately intermixed, leading to greater spatial variability and associated resistance to the spread of leaf diseases, although possibly at the cost of reduced average yield. Harvesting should occur during non-nesting periods.

Demand for large-scale, industrial production of biomass fuels from tropical regions is already leading to the clearing of tropical forests in order to meet export demand, with destruction of forest habitat and loss of biodiversity. Palm oil plantations in Malaysia, for example, are being established on forestland rather than on abandoned farmland, because the cleared forests need less fertilizer and the timber can be sold. According to sources cited in WWI (2006), this expansion now threatens the orang-utan with extinction within ten years. Peatlands have been drained and burned in order to establish palm oil plantations to produce biodiesel for the European market, thereby vastly overwhelming the expected GHG emission benefits (as discussed in subsection 4.6.2).

Sugarcane is one of the most effective biomass sources for the production of ethanol for transportation uses, increasing the pressure to expand sugarcane plantations that in turn would encroach upon ecologically important areas. Sugarcane requires a dry season, and so could not spread into the Amazon, but it can displace crops such as cotton and soybean, as well as livestock, that can be raised in the Amazon. The Brazilian *cerrado*, illustrated in Figure 4.35, is considered to be a natural expansion area for sugarcane and soybeans in Brazil, but it is also home to over 900 species of birds and almost 300 species of mammals, many of which are threatened with extinction (WWI, 2006).

Source: (c) Luiz Claudio Marigo/naturepl.com

Figure 4.35 *The Brazilian* cerrado*, an ideal region for bioenergy crops but also home to 935 species of birds and almost 300 mammal species*

Many lands presently used for cattle ranching are also excellent for growing sugarcane, and about two thirds of the recent expansion of sugarcane plantations in Brazil has occurred on former cattle ranches (the other one third has largely displaced other crops) (Bustamante et al, 2009). The average stocking density on Brazilian pasturelands is quite low (about 100 animals/km^2), leaving room for substantial increases, which could free up some land for increased sugarcane production.[19] However, existing pasturelands are already undergoing intensification, yet expansion of pastureland into the Amazon rainforest continues. As much of the land currently used for cattle ranching and for soybean production can also be used for sugarcane production, the amount of beef consumed in markets supplied by Brazilian beef and soybeans (much of which is fed to cattle) will have a large impact on the ecological pressures resulting from (the almost certain) increases in sugarcane production in the future.

Bustamante et al (2009) propose that one way to avoid indirect expansion of pastureland or cropland into the Amazon or other ecologically sensitive areas as a result of expansion in sugarcane cultivation is to confine its expansion to degraded pastures, and to put into a place a system to assure that for each hectare of pasture utilized, other pastures are intensified sufficiently to compensate for the lost pastureland. This would require some degree of government coordination and the use of some of the revenues from biofuel production to support intensification of pastureland.

4.5.11 Indoor air pollution and CO_2-equivalent emissions with traditional and advanced biomass for cooking, compared to natural gas

Approximately half the world's population relies on some form of biomass (such as wood, crop residues, dung or charcoal) for household energy use. Use of such biomass is a significant source of indoor air pollution, serving as a major factor in respiratory illness. Indoor air pollution is estimated to cause 400,000–550,000 excess deaths per year in India, 420,000/year in China and 390,000/year in Africa (Sagar and Kartha, 2007).

Ezzati et al (2000) and Bailis et al (2003) have compared the emissions and global warming effects of three different methods of heating using woody biomass in Kenya: using the traditional open fire with three stones, using a ceramic wood-burning stove and using a charcoal stove. These technologies were illustrated earlier, in Figure 4.11. Table 4.18 gives emissions per kg of fuel for the three technologies as measured by Bailis et al (2003). Also given in Table 4.18 are daily average fuel use (kg) and the resulting daily total CO_2-equivalent emissions. The global warming impact of non-CO_2 pollutant gases (CH_4, CO, NMHCs) is over twice that of the CO_2 emissions for charcoal, largely due to large emissions during the production of charcoal.[20] In spite of requiring only about half as much charcoal as fuelwood for the same cooking, total CO_2-equivalent emissions using charcoal are three to four times larger than with any wood-based technology. Emission ratios (ceramic–wood/three stone–wood and charcoal/three stone–wood) at the point of use for specific gases obtained in this and other studies are compared on Figure 4.36. As seen in Table 4.18 and Figure 4.36, charcoal entails substantially greater emissions of CO_2 and CO but substantially smaller emissions of particulates according to all studies, with contradictory results for CH_4 and NMHCs. In one study, ceramic stoves entail substantially greater particulate emissions than traditional three-stone stoves, in spite of their greater efficiency.

Smith et al (2000) present measurements of cooking efficiency (MJ of heat delivered to a pot divided by MJ of fuel energy used) for various fuels and technologies in India. The fuels considered were biogas (whose composition is compared with that of natural gas in Table 4.12), liquefied petroleum gas (LPG), kerosene and various biomass fuels. For biomass fuels, the technologies considered were a basic three-rock stove (3R), traditional mud stove (tm), improved metal stove (imet), improved vented mud stove (ivm), improved vented ceramic stove (ivc) and hara (a hollowed-out, mud-lined pit). Using the emission factors (gm per MJ of heat delivered to the pot) for CO_2, CO, CH_4, NMHCs and N_2O given in Smith et al (2000), and using corresponding 20-year global warming potentials[21] of 1.0, 4.5, 22.6, 4.1 and 290, respectively, the equivalent CO_2 emission (per MJ of delivered heat) can be readily computed for each fuel/technology combination. For biomass that is grown sustainably, only the non-CO_2 emissions need be considered. Figure 4.37 gives the efficiency for each case and the ratio of CO_2-equivalent emission to that

Table 4.18 *Emissions (gmC per kg of fuel) associated with different technologies for using woody biomass for heating*

	Three stones, wood		Ceramic, wood		Charcoal	
	C	C equivalent	C	C equivalent	C	C equivalent
Use phase						
CO_2	379.1	379.1	381.8	381.8	621.8	621.8
Non-CO_2		202.2		181.8		775.1
CO	33.9	135.4	31.7	126.9	111.4	445.7
CH_4	2.4	54.0	1.9	42.2	13.5	303.8
NMHCs	1.1	12.8	1.1	12.8	2.1	25.6
TSP	1.1	0.0	5.9	0.0	0.4	0.0
Total		581.3		563.7		1396.9
Production phase						
CO_2					490.9	490.9
Non-CO_2						1855.6
CO					94.3	377.1
CH_4					33.0	742.5
NMHCs					61.3	736.0
TSP					8.2	0.0
Total						2346.6
Production plus use						
CO_2		379.1		381.8		1112.7
Non-CO_2		202.2		181.8		2630.7
Total		581.3		563.7		3743.4
Non-CO_2/CO_2		0.53		0.48		2.36
Daily fuel use, kg		14.3		11.9		6.9
Daily CO_2-equivalent emission (kgC)		8.3		6.7		25.8

Note: NMHC = non-methane hydrocarbons, TSP = total suspended particulates. The C equivalent of CO, CH_4 and NMHCs have been computed here assuming molar global warming potentials of 4, 22.5 and 12, respectively.
Source: Emission data from Bailis et al (2003)

using gas for cases with and without consideration of CO_2 itself, as computed here. For unsustainable use of biomass (where CO_2 emissions must be included), the CO_2-equivalent emissions range from 3 to 14 times that using gas. The worst case (factor of 14) occurs for dung-hara and is a result of non-CO_2 emissions equivalent to that of CO_2 alone combined with a seven times smaller efficiency compared to gas. Even with sustainable use of biomass, CO_2-equivalent emissions generally range from one to four times that using natural gas.

In summary, the global warming effect of many traditional forms of cooking with biomass greatly exceeds that using natural gas, due to the heating effect of non-CO_2 pollutants. These pollutants also exert a substantial toll on human health. The efficiency at the point of use is several times greater using natural gas than using solid biomass fuels. With conversion efficiencies from solid biomass to biogas in the order of 60–70 per cent, overall efficiency could be greatly improved and pollutant emissions and global warming effects greatly reduced if biogas fuel is used in place of solid biomass for cooking.

4.5.12 Impact on employment and rural poverty

Kartha and Leach (2001) discuss the potential role of modern bioenergy systems in alleviating rural poverty. The development of modern bioenergy systems can have positive or negative impacts on rural poverty and employment. When energy crops replace food crops, employment usually falls, but when energy crops are additional to food crops (such as trees planted on field boundaries or on unused village land), employment increases. Employment will be created only if wage levels are low ($1–2/day); with increasing wage levels, which might be catalysed by the growth of local

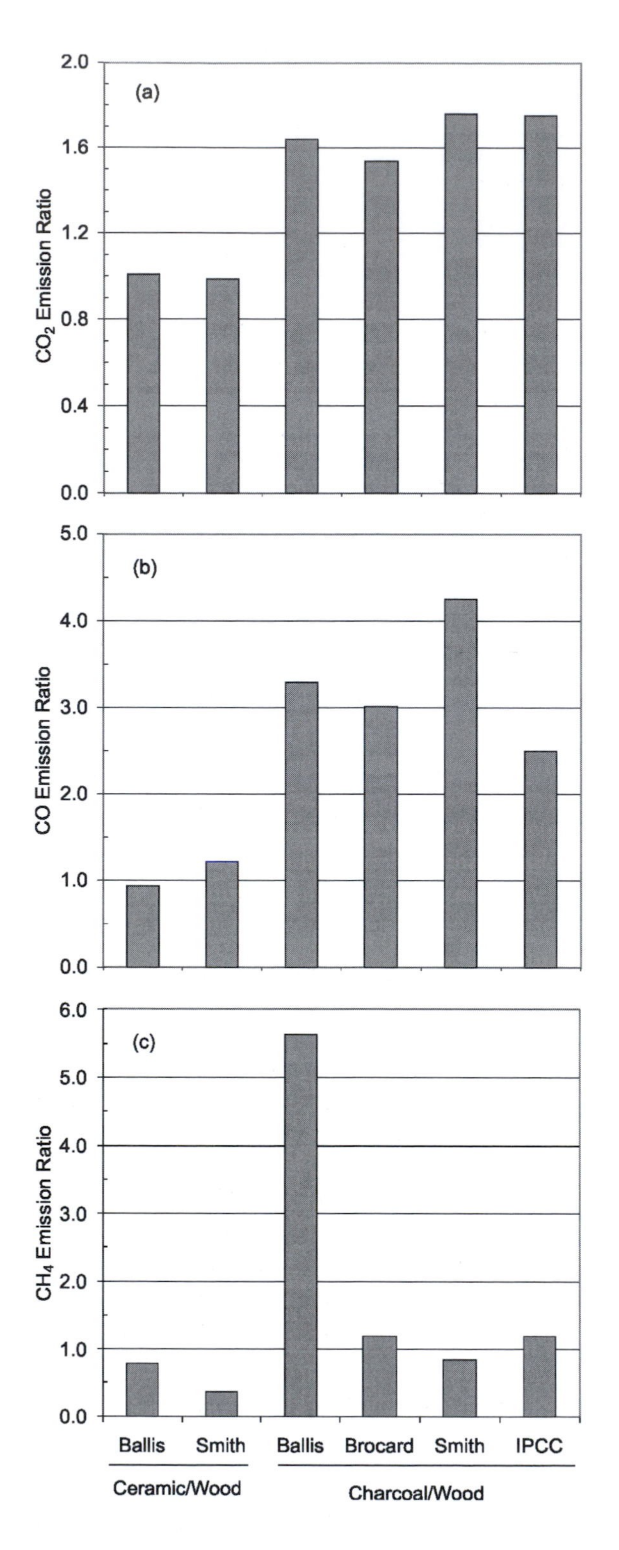

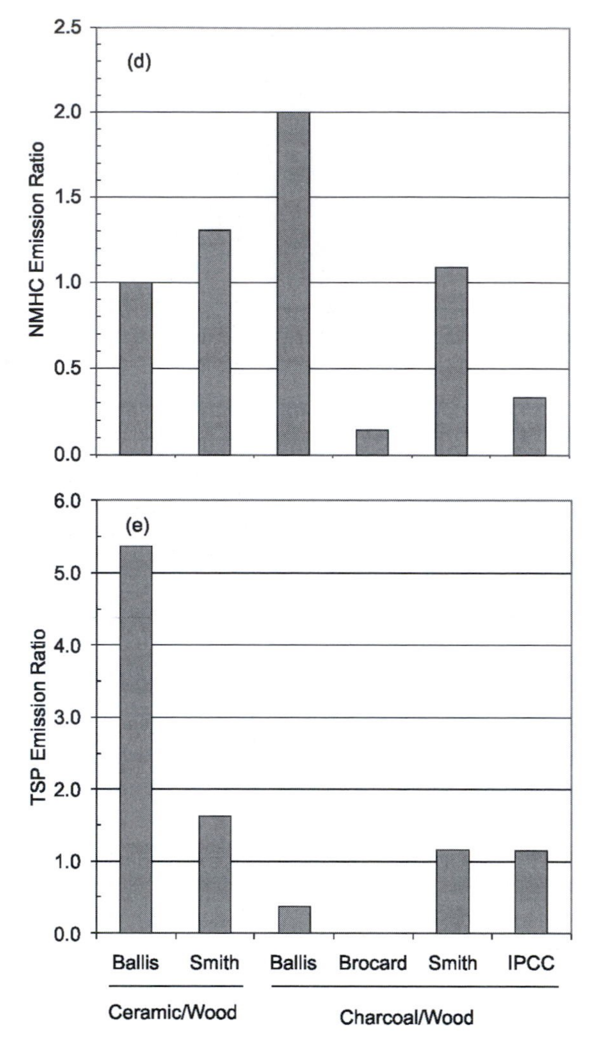

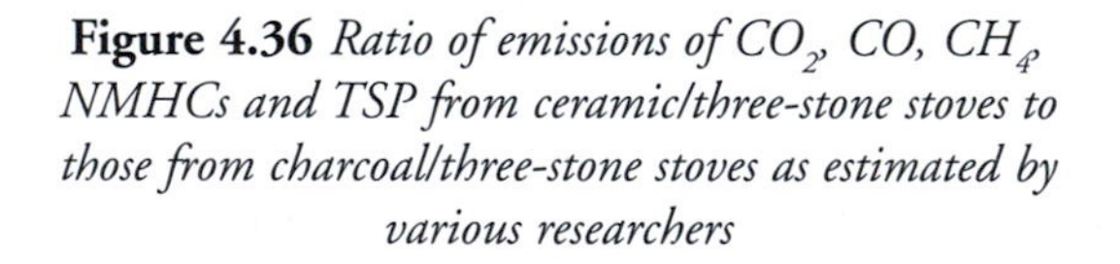

Figure 4.36 *Ratio of emissions of CO_2, CO, CH_4, NMHCs and TSP from ceramic/three-stone stoves to those from charcoal/three-stone stoves as estimated by various researchers*

enterprises using biomass energy, there will be a tendency for increased mechanization. Crops such as *Jatropha*, which can be grown along the boundaries of farmers' fields, in understocked forests and on denuded land, would increase income for farmers and/or create employment opportunities in poor areas of India according to Hooda and Rawat (2006).

Provision of more efficient and cleaner technologies for cooking will reduce the amount of time required for

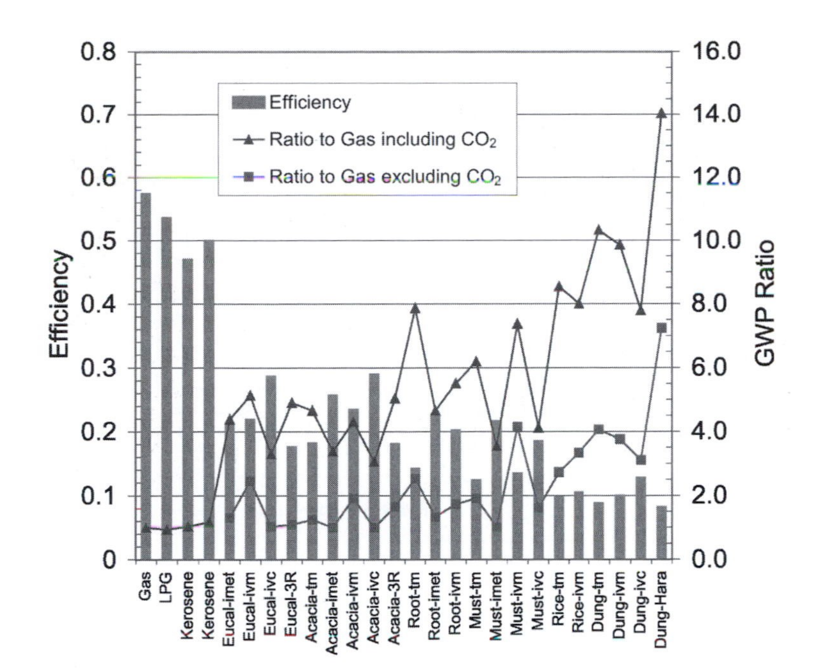

Note: Emission ratios are given including and excluding CO_2 itself, the latter being appropriate for sustainably grown biomass.
Source: Computed from data in Smith et al (2000) assuming emissions using natural gas to be the same as using biomass

Figure 4.37 *Efficiency of different fuel technology combinations for cooking, and the ratio of CO_2-equivalent emissions (accounting for CO_2, CO, CH_4, NMHCs and N_2O) for each fuel technology to those using gas*

collecting wood and reduce chronic illness associated with pollutant emissions. Development of small-scale power systems (such as microturbines using biogas) can provide electricity for basic services (such as water pumping, thereby freeing up the hours of time otherwise spent per household per day fetching water) and for small, income-generating enterprises. Efficient sources of process heat (for drying) and electricity (for grinding, milling, husking) could enable farmers to process agricultural output, increasing their revenue and also reducing food losses due to spoilage.

As noted in section 4.10, biomass-derived electricity in rural India (and elsewhere) is already less expensive than diesel-based electricity and in some cases is also less expensive than grid electricity. Development of biomass electricity in place of fossil electricity would in these cases contribute to rural development by reducing the cost of energy.

4.6 Net energy yield and GHG balance of biomass energy

The net *energy yield* of biomass energy is equal to the energy provided by the final biomass products minus the various energy inputs required to produce the biomass. The net energy yield of biofuels for transportation, solid fuels for heating and gaseous fuels for various purposes will be discussed in the following subsections. A critical parameter is the ratio of energy supplied by a given amount of biomass to the energy input (other than solar energy) required to produce, deliver and process the biomass. This is referred to as the EROEI. The EROEI was about 100:1 for the early use of petroleum in the 1930s, and is about 15:1 for petroleum today (Cleveland, 2005). That is, about 15GJ of oil energy are obtained on average today for every GJ used in the exploration, drilling and extraction of oil. As will be seen below, the EROEI for biomass ranges from less than 1.0 to as high as 70–80.

4.6.1 Net energy yield from liquid biofuels

Liquid biofuels are liquid fuels such as ethanol, diesel or various oils that are made from biomass sources. Biomass feedstocks can be used to produce several different energy and non-energy products. For example, corn grain can be used to simultaneously produce ethanol,

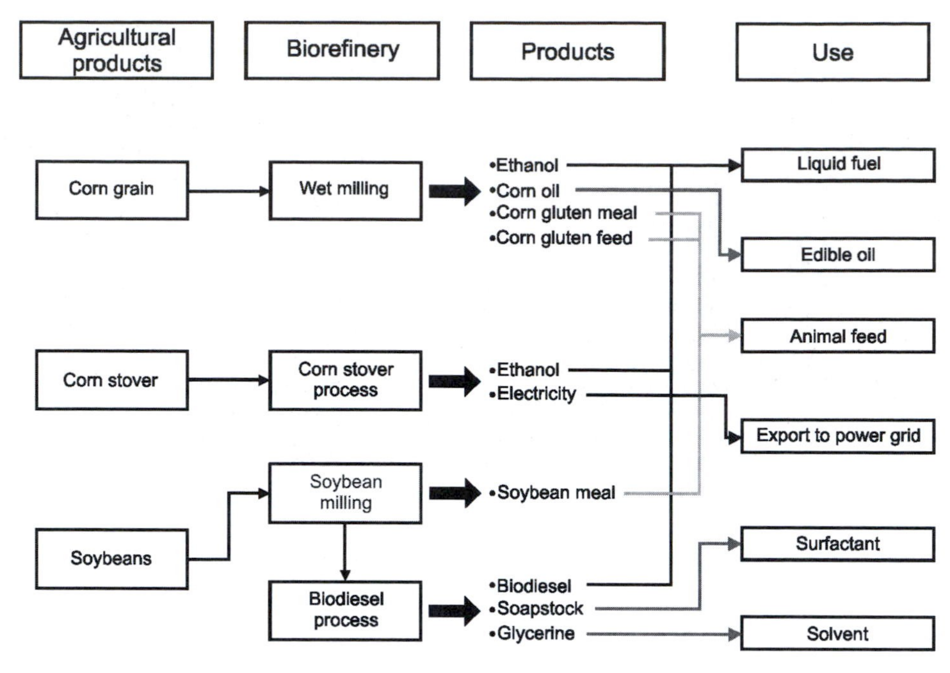

Source: Kim and Dale (2005b)

Figure 4.38 *Overview of the production of biofuels and co-products in a corn–soybean system*

corn oil (used for edible oil) and corn gluten meal and feed (used as animal feed). Corn stover (the stalks and leaves of the corn plant) can be used to produce ethanol (from the starch sugar) and electricity (from lignin residue). Soybeans can be used to produce biodiesel fuel, soybean meal, soapstock and glycerine. These relationships are summarized in Figure 4.38. Glycerine produced as a byproduct of biodiesel production can replace glycerine made from petroleum and natural gas.

The energy inputs used to make biofuels occur as: (1) the energy used during the production of the biomass feedstock; (2) the energy used during the processing of biomass feedstock into biofuels; and (3) the energy used to make all the materials and machinery used in the production of biomass and its later processing into fuels. The last category is referred to as the *embodied energy* of the machinery and equipment.

The principle energy inputs used in making biomass feedstocks are:

- the production of fertilizers, particularly nitrogen, which has an embodied energy (MJ/kg) many times that of other fertilizers and is used in greater quantity;
- the production of herbicides and pesticides;
- the production of seeds (for annual crops) or seedlings (for trees);
- the energy used to pump water used for irrigation;
- the energy used for preparing the soil and seeding (in the case of annual crops) and in harvesting;
- in some cases, the energy used in drying biomass crops prior to transportation.

The energy inputs during processing include:

- the energy used by workers in commuting to and from the work sites;
- the energy used in transporting the biomass to the processing sites, and in transporting solid wastes to the disposal or recycling sites;
- the energy used at the processing site;
- the embodied energy of the ethanol plant;
- the energy used to supply water to the processing plant and to treat wastewater (such as electricity to operate water treatment plants), which is particularly applicable to ethanol and biodiesel plants.

The embodied energy in farm equipment and in the processing plant must be divided by the total amount of biofuel produced over the lifetime of the equipment in order to determine the contribution to the total energy input per unit of fuel produced. Figure 4.39 illustrates an

Source: Tollefson (2008)

Figure 4.39 *A 450 million litre/year ethanol-from-corn plant in South Dakota, US*

ethanol-from-corn plant in South Dakota capable of producing 450 million litres of ethanol per year.

It is often assumed that nitrogen added in the form of manure subtracts from the fertilizer embodied energy that would otherwise be required if all of the nitrogen were added as chemical fertilizer. However, manure applied to the bioenergy crop is manure not returned to the soil that produced the food for the animals that produced the manure, and the associated nutrients will therefore have to be replaced with chemical fertilizer at some point. Thus, all nitrogen added to the bioenergy crop, whether as manure or as chemical fertilizer, should be assigned an embodied energy equal to that of the replacement chemical fertilizer.

Apart from non-biomass energy inputs at the processing plants, there will be losses in energy during the conversion process, as not all of the initial biomass energy ends up in the biofuel product.

Ethanol from corn

There have been numerous estimates of the energy flows in producing ethanol from corn, but most researchers have not considered all of the terms listed above, or have accounted for them in different ways. Farrell et al (2006, henceforth referred to as F2006) have presented a detailed analysis (in the online supplement to their paper) of the differences between a number of previously published assessments for the US corn belt, along with their own assessment. Their assessment is presented here in Table 4.19, along with updated data from Wang et al (2007) on energy use for dry-milling ethanol plants in the US using either coal or natural gas as the primary fuel.[22] More detailed information is given in Table S4.1 in the online supplement. Shown are the energy inputs required to produce the corn crop, the energy inputs at the ethanol plant and energy credits for co-products. In addition to energy inputs at the ethanol plant, there is a modest loss of feedstock (4–7 per cent) in going from glucose to ethanol, the mass loss instead appearing in the effluent. Both energy inputs and mass loss have been decreasing over time in successive ethanol plants. The mass loss is represented by a *mass conversion efficiency*, with values of 0.93–0.96 in the latest ethanol plants.[23]

The energy inputs at the dry-milling ethanol plants given by Wang et al (2007) are summarized in more detail in Table 4.20. Dry-milling plants produce distiller grains and solubles (DSG) as a co-product. DSG is used as animal feed and must be dried to 10 per cent moisture for transport to feedlots and for long shelf life. Fuel energy use is given in Table 4.20 in terms of total energy use for the production of ethanol and for drying DSG, and in terms of the energy use above and beyond that required for drying DSG. Plants using solid fuels (coal or wood chips) require substantially more fuel than plants using natural gas. As well, plants using solid fuels have greater electricity requirements. Table 4.19 (data columns 2 and 3) uses the incremental energy requirements from Table 4.20 rather than total energy requirements, so, to avoid double counting, no energy

Table 4.19 *Annual energy inputs used to produce corn or woody biomass (cellulose), and energy used in producing ethanol from these feedstocks as estimated by various studies*

	Corn			Cellulose
	F2006	W2007	W2007	F2006
		Production of crop		
Fertilizer embodied energy (GJ/ha/yr)	9.86			2.49
Herbicide plus insecticide energy (GJ/ha/yr)	1.07			0.14
Direct energy use (GJ/ha/yr)	6.26			4.36
Other energy inputs (GJ/ha/yr)	2.26			0.43
Total energy to produce corn (GJ/ha/yr)	18.65	18.65	18.65	7.38
Crop yield (kg moist crop/ha)	8746	8746	8746	13450
Total energy to produce crop (MJ/kg moist crop)	**2.13**	**2.13**	**2.13**	**0.55**
Crop energy yield (GJ/ha/yr)[a]	139.8	139.8	139.8	242.1
EROEI so far	7.5	7.5	7.5	32.8
		Production of ethanol		
Glucose energy yield (GJ/ha/yr)[b]	77.2	77.2	31.3	
Mass conversion efficiency[c]	0.934	0.934	0.934	
Ethanol energy yield (GJ/ha/yr)	62.5	62.5	62.5	113.2
Ethanol energy/crop energy	0.447	0.447	0.447	
Coal use at ethanol plant (MJ/L)	8.31	7.26		
Natural gas use at ethanol plant (MJ/L)	5.54		6.08	
Electricity primary energy (MJ/L)		2.2	1.8	
Energy use at ethanol plant (MJ/L)	13.9	9.4	7.9	0.0
Other energy inputs	1.57	1.57	1.57	1.65
Energy to produce crop (MJ/L ethanol)	6.34	6.34	6.34	1.39
Total fossil fuel energy inputs (MJ/L)	21.76	17.33	15.79	3.04
Gross energy yield (MJ/litre)	21.25	21.25	21.25	21.25
Net energy yield (MJ/litre)	**−0.50**	**3.93**	**5.47**	**18.22**
Credits (MJ/litre)	4.13	0.00	0.00	4.79
Net energy yield with credits (MJ/litre)	3.63	3.93	5.47	23.01
Net energy yield with credits (GJ/ha/yr)	10.67	11.55	16.08	123
EROEI without credits	0.98	1.23	1.35	7.00
EROEI with credits	1.17	1.23	1.35	8.58
EROEI based on saved PE	1.55	1.63	1.79	11.39

Note: The studies are: F2006, Farrell et al (2006); and W2007, Wang et al (2007). W2007 present updated estimates of the energy used in corn ethanol plants using coal or natural gas, which are combined here with corn energy inputs from F2006. [a] Computed here as (0.001GJ/MJ) × (kg moist crop/ha) × (kg dry crop)/(kg moist crop) × (MJ LHV)/(kg dry crop), where the crop LHV on a dry weight basis is assumed to be 188 MJ/kg in all cases (which is the correct value for corn). [b] Computed here for corn ethanol as (0.001GJ/MJ) × (kg moist corn/ha) × (0.85kg dry corn/kg moist corn) × (0.66kg glucose/kg dry corn) × (15.74MJ/kg glucose), the last three factors having been taken from Patzek (2004). [c] Sometimes inferred here so as to give the stated ethanol yield when the latter is directly given.

credits for co-products are assigned. For the other cases in Table 4.19, the energy credits were computed by the original authors based on the *system expansion method*, which is explained later in the discussion of biodiesel from oily crops.

The net result, taking into account energy credits for co-products where applicable, is a calculated EROEI for ethanol from corn ranging from 1.17 to 1.35. These EROEI values are based on a comparison of the primary energy used to make ethanol with the energy content of the ethanol. A more appropriate EROEI compares the primary energy required to make ethanol with the petroleum primary energy that is saved when ethanol is used in place of gasoline. This requires taking into account differences in fuel consumption (litres/km) and energy content of the two fuels, and the ratio of petroleum primary energy to gasoline energy. Given energy contents of 21.25MJ/litre and 31.34MJ/litre for ethanol and gasoline, respectively, the displacement of 0.75 litres of gasoline per litre of ethanol used, and a primary to secondary energy ratio of 1.2 for gasoline, the EROEI ratios given above are multiplied by a factor

Table 4.20 *Fuel and electricity requirements for new dry-milling ethanol plants*

	Energy input	
	Fuels (MJ/litre)	Electricity (kWh/litre)
Natural gas, total	9.29	0.20
Natural gas, incremental	6.08	0.20
Coal	11.22	0.24
Coal, incremental	7.26	0.24
Wood chips	11.22	0.24
Natural gas + syrup, natural gas	5.85	0.24
Biomass	3.90	0.24

Note: The separate Natural gas and biomass inputs are given for the case Natural gas + syrup.
Source: Wang et al (2007)

of 1.33. The resulting EROEI ratios are given as the final row of Table 4.19 and range from 1.5 to 1.8. As will be seen shortly, even the higher EROEI estimate of 1.8 is unimpressive.

The impact on global warming of using corn ethanol to displace gasoline could be considerably less positive than implied by the EROEI, depending on the kinds of fuels (coal, natural gas, or wood chips) used at the ethanol plant. Wang et al (2007) compute a 3 per cent GHG increase if the fuel is coal and a 52 per cent decrease if woods chips are used (for a case where EROEI ~1.25). Emissions of N_2O and CH_4 are included in this assessment.[24] Kim and Dale (2005a) also estimated the energy inputs and GHG emissions (including N_2O) associated with the production of ethanol from corn. They simulated a no-tillage case using a detailed soil carbon and nutrient model to compute rates of carbon buildup in the soil and rates of N_2O emission. Buildup of soil carbon is computed to sequester about 220gC/kg Corn/yr, while emissions of N_2O are equivalent to about only 100gC/kg Corn/yr, so the overall effect of these two terms is favourable.

Ethanol from wheat

Lechón et al (2009) estimate that the production of 1MJ of ethanol fuel from wheat in Spain requires 0.75MJ of fossil fuel energy inputs. This gives an EROEI of 1.33, which can be multiplied by the factor of 1.33 derived above to give the EROEI in terms of the primary energy that would be used to make gasoline that is displaced. The result is an EROEI of 1.7.

Table 4.21 *Derivation of average crop yields for corn in the US and sugarcane in Brazil, breakdown of 1kg of crop and derivation of ethanol yields from corn and sugarcane*

	Corn	Sugarcane
Derivation of average crop yields		
Incident solar irradiance (W/m^2)	170	220
Sunlight to biomass efficiency (per cent)	0.6	1.2
Average yield (W/m^2)	1.02	2.64
Annual yield (GJ/ha/yr)	321.7	832.6
LHV (MJ/kg dry matter)	18.0	18.0
Above-ground dry biomass yield (t/ha/yr)	17.9	46.3
Moisture fraction	0.15	0.655
Above-ground moist biomass yield (t/ha/yr)	21.0	132.2
Harvested fraction	0.35	0.53
Harvested moist crop yield (t/ha/yr)	7.4	70.0
Breakdown of 1kg of moist crop		
Water	0.15	0.655
Dry matter, of which	0.85	0.355
– bagasse		0.140
– tops and leaves		0.045
– fermentable sugars	0.561	0.160
Derivation of ethanol yield		
Moist crop yield × sugar fraction		
× stoichiometric ratio (kg ethanol/kg sugar)[a]	0.511	0.511
× mass conversion efficiency	0.9	0.9
gives kg ethanol/kg crop	0.258	0.074
Litres ethanol/kg crop	0.324	0.092
Ethanol yield (litres/ha/yr)	2385	6475
Ethanol yield (GJ/ha/yr)	50.7	137.6

Note: The breakdowns of 1kg of moist corn and sugarcane are from Patzek (2004) and Patzek and Pimentel (2005), respectively. [a] This is the ratio of mass of ethanol (C_2H_5OH) to the mass of glucose ($C_6H_{12}O_6$) in the reaction $C_6H_{12}O_6 \rightarrow 2C_2H_5OH + 2CO_2$.

For ethanol produced from wheat in Canada, (S&T)[2] Consultants Inc (2005a) compute an energy requirement of 0.73MJ/MJ of ethanol without co-product credits and 0.56MJ/MJ with credits. The latter corresponds to an EROEI of 1.78 in terms of the ethanol fuel energy and 2.4 in terms of saved primary energy.

Ethanol from sugarcane

The factors contributing to yields of corn in the US and sugarcane in Brazil, and the factors determining the ethanol yields (per hectare), are compared in Table 4.21. The ethanol yield per kg of moist crop is about 3.5 times smaller for sugarcane than for corn, but the average sugarcane yield per hectare is about ten times greater, resulting in almost a factor of three greater ethanol yield per hectare in Brazil. The ethanol yield from sugarcane derived in Table 4.21 is 92 litres/t; Moreira (2006) gives a best-practice value in Brazil of 90 litres/t and foresees an average yield of 114 litres/t by 2030.

At present, sugarcane can be harvested manually in Brazil with burning of the cane residues in the field at the time of harvesting, or can be done mechanically, generally without burning. Efforts are underway to reduce cane burning, which is a cause of severe regional air pollution. In the state of São Paulo, where most Brazilian sugarcane is produced, the harvested area subjected to burning decreased from 82 per cent in 1997 to 63 per cent in 2004 and is supposed to reach 0 per cent by 2016 (Smeets et al, 2008). This reduction will be facilitated by the transition to mechanical harvesting, but mechanical harvesting is not fully possible in 40–50 per cent of the sugarcane fields because the slope is too large, so a semi-mechanical harvesting technique will be increasingly used.

Table 4.22 *Energy balance for ethanol produced from sugarcane in Brazil*

	Average	Best	Future
Agricultural operations (GJ/ha/yr)			
– manual	1.8	1.8	1.8
– mechanical	4.1	4.1	4.1
Fertilizer, lime, chemicals (GJ/ha/yr)	5.9	4.7	4.5
Seed embodied energy (GJ/ha/yr)	0.4	0.4	0.4
Machinery embodied energy (GJ/ha/yr)	2	2	1
Transport (GJ/ha/yr)	3	2.5	1.5
Total (GJ/ha/yr), manual	13.1	11.4	9.2
– mechanical	15.4	13.7	11.5
Sugarcane yield (t/ha/yr)	69	69	140
Sugarcane yield (GJ/ha/yr)	428	428	869
Energy for ethanol production (GJ/ha/yr)	3.4[a]	2.7	4.4
Ethanol yield (litres/kg wet crop)	0.090	0.096	0.098
Ethanol yield (litres/ha/yr)	6211	6634	13720
Ethanol yield (GJ/ha/yr)	132	141	292
Surplus bagasse (GJ/ha/yr)	11.6	21.7	0
EROEI, manual	8.7	11.5	21.4
– mechanical	7.6	9.9	18.3
Potential electricity generation (kWh/t cane)	100	400	400
Energy credit (GJ/ha/yr)	49.7	198.7	403.2
EROEI, mechanical	9.7	20.7	43.7

Note: 'Average' and 'Best' in 2002 are as estimated by Macedo et al (2004) (cited in Smeets et al, 2008), while 'Future' is based on a doubling in sugarcane yield and additional, minor improvements.
[a] Already down to about 1.7GJ/ha/yr (23.6MJ/t of sugarcane) in 2005/2006 according to Macedo et al (2008).

Table 4.22 presents an energy balance for the production of ethanol from sugarcane in Brazil with manual and mechanical harvesting. Results are given for current average practice and current best practice combined with current average sugarcane yield (about 70t/ha/yr), based on work reported in Smeets et al (2008), and for a possible future sugarcane yield of 140t/ha/yr. Including a small amount of surplus bagasse, the EROEI is about 9–12 with manual harvesting and 8–10 with mechanical harvesting under current conditions. Only small amounts of electricity (up to 30kWh/t of cane) are generated at typical sugarcane mills, just sufficient to meet the mill's own electricity needs. However, as discussed earlier (subsection 4.4.6), there is the potential to vastly increase the electricity generation through the use of advanced technologies (up to 500kWh/t with biomass integrated gasification and combined cycle power generation, depending on how much residue that is currently burnt in the field is instead collected and used). In the last row of Table 4.22 the EROEI is computed assuming the production of 100kWh/t or 400kWh/t of surplus electricity (but excluding the 12–22GJ/ha/yr of surplus bagasse, which would now be used along with field residue to generate electricity) and assuming mechanical harvesting. The resulting EROEIs are 10 and 20 with the current sugarcane yield. Moreira (2006) expects the average sugarcane yield in Brazil to reach 140t/ha/yr by about 2030. If this is combined with modest reductions in fertilizer, machine and operational energy inputs (due to improved efficiency), and with an increase in ethanol yield per tonne of wet sugarcane (from 0.090 to 0.098), then the EROEI ranges from 18.3 if there are no energy credits (due to all bagasse and residues being returned to the field) to 43.7 (if 400kWh/t of electricity can be produced while sustaining sugarcane production in the long term).

Patzek and Pimentel (2005) argue that current accounting methods do not fully account for the long-term losses of soil nutrients through both erosion and removal of biomass from the field. When the energy cost of fully replacing lost nutrients is accounted for, the fertilizer energy input (about 5GJ/ha/yr in Table 4.22) would be increased, but even if it is doubled, there would still be a substantial EROEI (about 8–15 with 100–400kWh/t of electricity production and the present average sugarcane yield of 69t/ha/yr). At present, it is unclear if intensive industrial production of sugarcane is truly sustainable. Substantial field research is required to answer this question, and also – if intensive sugarcane production is sustainable – to determine the optimal use of sugarcane residues if they do not all need to be returned to the field. Among the options discussed by Smeets et al (2008) are: (1) use of bagasse and residues to generate electricity; (2) use of bagasse and residues to produce additional ethanol through cellulosic production methods (increasing the ethanol yield by 34 litres/t using 30 per cent of excess bagasse and 50 per cent of field residues, compared to a yield of 90 litres/t from sugarcane sugar at present); and (3) use of hydrolysed bagasse as an animal feed.

Ethanol from woody biomass

The production of ethanol from sugarcane or starchy crops such as corn is relatively simple, as 6C sugars

Table 4.23 *Factors determining the yield of ethanol from lignocellulosic (woody) materials*

	Transformation	Efficiency
	Cellulose, 32–50% of total biomass	
Pretreatment mass conversion efficiency	Cellulose to glucose	50–90% now 75–95% by 2020
Stoichiometric yield	Glucose to ethanol	0.511
Fermentation mass conversion efficiency	Glucose to ethanol	As low as 60% now 92.5% with process integration
Overall conversion factor (mass ethanol/mass feedstock)	Cellulose to ethanol	0.153–0.368 now 0.355–0.449 future
	Xylan, 6–21% of total biomass (from hemi-cellulose)	
Pretreatment mass conversion efficiency	Xylan to xylose	60–90% now 88–98% in 5–10 years
Stoichiometric yield	Xylose to ethanol	0.511
Fermentation mass conversion efficiency	Xylose to ethanol	80–92% now 92–95% future
Overall conversion factor (mass ethanol/mass feedstock)	Xylan to ethanol	0.35–0.42 now 0.41–0.48 in 5–10 years
	Mannan, Galactan, Arabinan, 2–16% of total biomass (from hemi-cellulose)	
Pretreatment mass conversion efficiency	Mannan to mannose Galactan to galactose Arabinan to arabinose	67.5% demonstrated at bench scale or larger 90% expected future yield
Stoichiometric yield	Mannose, galactose and arabinose to ethanol	0.511
Fermentation mass conversion efficiency	Mannose, galactose and arabinose to ethanol	0.0% now 85% expected
Overall conversion factor (mass ethanol/mass feedstock)	Mannan plus galactan plus arabinan to ethanol	0.0% now 40% future

Overall ethanol yield and energy conversion efficiency from 1t of dry biomass assuming a pretreatment mass conversion efficiency of 0.9 for all sugars and fermentation mass conversion efficiencies of 0.95 for glucose and 0.85 for all other sugars, and using the amounts of the different sugars and feedstock heating values given in Table 4.8

Eucalyptus: 267.5kg, 336 litres, 36.6% energy conversion efficiency
Switchgrass: 238.3kg, 299 litres, 34.2% energy conversion efficiency
Corn stover: 270.9kg, 340 litres, 48.6% energy conversion efficiency

Note: Future pretreatment yields and fermentation efficiencies are plausible possibilities. Overall conversion factors are given by the product of pretreatment yield, stoichiometric yield and fermentation efficiency.
Source: Pretreatment yields and fermentation efficiencies are taken from Sheehan et al (2004) and Hamelinck et al (2005a); stoichiometric yield is given by the ratio of mass of ethanol to mass of sugar on the two sides of the fermentation equations given in the text

(glucose and fructose in sugarcane, glucose in starchy crops) that are easily separated from the rest of the crop are fermented to ethanol with a mass conversion efficiency of 93–96 per cent. The process of producing ethanol from woody biomass is much more complicated than the production of ethanol from corn or sugarcane, as described in subsection 4.3.5. The major energy-consuming steps are the separation of lignin from hemi-cellulose and cellulose, and the rendering of all or part of the hemi-cellulose, soluble. The composition of different kinds of biomass was compared in Table 4.8. The fraction of biomass as cellulose and hemi-cellulose, and the relative amounts of different sugars within the hemi-cellulose fraction, are important for the final ethanol yield, as different components are converted to ethanol with different mass conversion efficiencies. Grasses have less lignin and cellulose than wood, but have a greater ash and extractive content, which makes them less attractive (see the note to Table 4.8).

The steps in calculating the yield of ethanol from wood are given in Table 4.23. The ethanol yield for unit mass of biomass is given by the cellulose and hemi-cellulose masses, times the pretreatment mass conversion efficiency times the stoichiometric ratio times the fermentation mass conversion efficiency. The stoichiometric ratio is 0.511kg ethanol/kg sugar, and is simply equal to the ratio of ethanol mass to sugar mass on the right and left sides, respectively, of the reaction $C_6H_{12}O_6 \rightarrow 2C_2H_5OH+2CO_2$(6C sugars) or $3C_5H_{10}O_5 \rightarrow 2C_2H_5OH+2CO_2$ (5C sugars). Both the pretreatment efficiency and the fermentation efficiency are subject to improvement, and the present and possible future efficiencies are indicated in Table 4.23. 1kg of wood produces as little 0.11kg of ethanol today, but should eventually be able to produce 0.34kg of ethanol. The chemical energy of the 0.11–0.34kg ethanol (at 26.7J/kg) is 2.9–9.1MJ, compared to 19.8MJ for the original wood, implying an energy conversion efficiency of 15–46 per cent. For biomass with 80 per cent cellulose plus hemi-cellulose and for 100 per cent pretreatment and fermentation mass conversion efficiencies, the energy conversion efficiency would be 54 per cent. The remaining 20 per cent of the biomass would be largely lignin, which can be burned or gasified and used to produce heat and electricity.

A projected energy balance for production of ethanol from wood is presented in abbreviated form in Table 4.19 and in more detail in Table S4.1. The expected EROEI is 7.0 without consideration of co-product credits, 8.6 with consideration of such credits, and 11.4 when based on the saved gasoline primary energy. However, these high (for ethanol) EROEIs depend on the many steps involved in making ethanol from lignocellulose improving as hoped (see subsection 4.3.5)

Ethanol from corn stover and comparison with production from corn grain

Corn stover (the stalks and leaves of the corn plant) is a lignocellulosic material that constitutes about half of the corn plant and could be used to produce ethanol, with the lignin residue used to produce heat and electricity. At present, it is largely left on the fields, where it protects the soil from erosion and returns nutrients to the soil. Harvesting of corn stover might lower soil carbon and nitrogen content and increase soil erosion, but this might be prevented through cultivation of a winter cover crop. Sheehan et al (2004; henceforth S2004) and Kim and Dale (2005b, henceforth KD2005b) assess the impact of cultivation of corn with removal of stover compared to the common system (in the US) of rotating between corn and soybeans on a two-year cycle. Both groups made use of the CENTURY soil carbon-nutrient cycle model to assess the impact on soil carbon, nutrients and erosion of alternative cropping systems. S2004 find that 70 per cent of the stover can be collected with 'tolerable' soil erosion under continuous no-till corn cultivation. The removal of stover and the cultivation of a winter cover crop would reduce N_2O emissions according to model simulations by KD2005b. There is no corn stover ethanol industry at present, so ethanol yields and process energy requirements are estimated based on the projected performance of lignocellulosic systems.

The energy balance over a 40-year period as computed by KD2005b is given in Table 4.24 for the base case corn–soybean system and for various corn stover systems. The corn stover systems have a very substantial (for biofuels) EROEI of 4.1–5.4 without energy credits from co-products, and 6.8–8.2 with such energy credits.

Ethanol from sugarcane residue

Rather than producing ethanol from the sugar in sugarcane, ethanol could be produced from the bagasse residue of sugar mills (shown in Figure 4.10) using the techniques for producing ethanol from cellulosic materials. As noted in subsection 4.4.6, however, the bagasse is a potentially large source of electricity using gasification/combined cycle power generation. Botha and

Table 4.24 *Expected energy balance over a 40-year period for the production of ethanol from corn stover and biodiesel from soybeans (where applicable) for various cropping systems in Iowa, US*

	Units	CS	CC	CC50	CwC70
		Biomass products			
Corn grain	dry tonne/ha	155	313	309	325
Corn stover	dry tonne/ha			154	228
Soybean	dry tonne/ha	52.1			
		Fuel products			
Ethanol	tonne/ha	54.8	111	155	182
Biodiesel	tonne/ha	9.3			
Fuel energy	TJ/ha	1.71	2.96	4.14	4.86
		Co-products			
Corn gluten meal	dry tonne/ha	8.7	17.5	17.3	18.2
Corn gluten feed	dry tonne/ha	37.4	75.4	74.4	78.4
Corn oil	dry tonne/ha	6.9	14.0	13.8	14.5
Soybean meal	dry tonne/ha	43.1			
Glycerine	tonne/ha	1.6			
Electricity	MWh/ha			34.8	51.3
Co-product energy Credit	TJ/ha	0.07	1.89	2.28	2.58
		Non-renewable energy inputs			
Agricultural inputs	TJ/ha	0.461	0.718	0.804	0.908
		Overall balance			
Net energy gain	TJ/ha	1.32	4.14	5.61	6.53
EROEI without credits		3.71	4.13	5.15	5.35
EROEI with credits		3.86	6.76	7.98	8.19

Note: CS = corn–soybean rotation, CC = continuous corn, CC50 = continuous corn with removal of 50 per cent of the stover, CwC70 = corn with winter crop cover and removal of 70 per cent of the stover.
Source: Kim and Dale (2005b)

von Blottnitz (2006) have compared the environmental benefits of these two options in South Africa. Assuming a yield of 65t/ha/yr of wet cane, one hectare of cultivated land yields either 7.8t of sugar, 1200 litres of ethanol, 2100kWh of electricity and 2.6t of molasses, or 7.8t of sugar, 6900kWh electricity and 2.6t of molasses.

The second option assumes production of only 105kWh per tonne of sugarcane, whereas Moriera (2006) regards 500kWh/t as a conservative upper limit. Nevertheless, Botha and von Blottnitz (2006) find that maximal production of electricity from the lignocellulosic residues of sugarcane processing for sugar generates greater GHG, acidification and eutrophication benefits than producing a mix of ethanol and electricity from the residues (assuming that the electricity so produced displaces coal-generated electricity and that 0.63kg ethanol displaces 1kg of gasoline).

Biodiesel from vegetable oils

Production of biodiesel from oilseeds involves extraction of oils by pressing, followed by esterification using methanol or ethanol, as discussed in subsection 4.3.6. A typical mass balance for the production of biodiesel and C-containing co-products from soybeans is given in Table 4.25. Only about 12 per cent of the carbon in a soybean ends up in the biodiesel fuel, but

Table 4.25 *Mass flow in the production of biodiesel fuel from soybeans*

	Primary product flow	Co-product flow
Soybean C	1.0000	
		0.8634
C in fat and oil	0.1366	
C in wastewater		0.0007
C in meal residual oil		0.0062
kg biodiesel C per kg fat and oil C		
C in soy oil	0.1297	
C in glycerine and soapstock		0.0067
C in wastewater		0.0019
C in solid wastes		0.0014
C in biodiesel	0.1197	

Source: Sheehan et al (1998)

Table 4.26 *Comparison of the energy balance for biodiesel fuel made from soybean and canola*

	Soybeans		Canola	
	P2008	**S2005**	**P2008**	**S2005**
Crop production				
Fertilizer embodied energy (GJ/ha/yr)	3.45	1.62	11.20	5.73
Herbicide plus insecticide energy (GJ/ha/yr)	0.54	0.41	1.05	0.68
Direct energy use (GJ/ha/yr)	3.20	3.11	2.87	4.45
Other energy inputs to produce crop (GJ/ha/yr)	5.18	5.19	2.96	2.96
Total energy to produce crop (GJ/ha/yr)	12.38	10.33	18.07	13.82
Crop yield (kg/ha/yr)	2890	3366	1568	1966
Total energy to produce crop (MJ/kg)	4.28	3.07	11.53	7.03
Crop energy yield (GJ/ha/yr):	43.55	50.72	23.63	29.63
EROEI	**3.52**	**4.91**	**1.31**	**2.14**
Production of Biodiesel				
kg oil/kg crop	0.18	0.18	0.3	0.35
kg biodiesel/kg oil	0.95	0.95	0.95	0.95
Biodiesel yield (kg/ha/yr)	494.2	575.5	446.9	653.8
Biodiesel yield (litres/ha/yr)	558	650	505	739
Biodiesel energy yield (GJ/ha/yr)	18.3	21.3	16.5	24.2
Biodiesel energy/crop energy	0.420	0.420	0.700	0.816
Energy to produce biodiesel (MJ/litre)	15.74	9.24	15.75	9.14
Energy to produce crop (MJ/litre)	22.16	15.88	35.79	18.71
Total fossil fuel energy inputs (MJ/litre)	**37.91**	**25.12**	**51.54**	**27.86**
Net energy yield (MJ/litre)[a]	−5.16	7.62	−18.79	4.89
Glycerine credit (MJ/litre)	11.00	11.00	11	11
Protein meal credit (MJ/litre)	−1.71	0.00	−4.07	−0.16
Net energy yield with credits (MJ/litre)	4.13	18.62	−7.79	15.89
EROEI without credits	**0.86**	**1.30**	**0.64**	**1.18**
EROEI with credits	**1.15**	**1.74**	**0.85**	**1.57**
EROEI with credits, primary energy basis	**1.40**	**2.18**	**0.98**	**1.96**

Note: The energy credits for co-products assigned here were computed as explained in the accompanying text, rather than taken from the original analyses. [a] Using an ethanol LHV of 32.74MJ/litre.
Source: Pimentel et al (2008) and (S&T)^2 Consultants Inc (2005b)

the vast majority of the rest ends up in other useful products and co-products.

Energy balances for biodiesel from soybeans and canola (a low-acid variety of rapeseed that was breed in Canada in the 1970s, and is now widely grown in the US and Canada) are presented in abbreviated form in Table 4.26 and in more detail in Table S4.2. Soybean biodiesel is estimated to have a better EROEI than canola biodiesel (about 2.0 versus 1.0–1.4 with co-product credits).

The co-products of biodiesel from soybeans are soybean meal and glycerine, while the co-products of biodiesel from canola are canola meal and glycerine. Both soybeans and canola are used to provide animal protein meal (about 70 per cent worldwide is from soybeans,11 per cent from canola). Soybean dry matter typically consists of about 17 per cent oil and 34 per cent protein, while canola dry matter typically consists of about 40 per cent oil and 20 per cent protein (the exact proportions depend on growing conditions and other factors). Thus, net production of animal protein alone or of oil alone can be accomplished by increases and/or decreases in both soybean and canola. For example, using the oil and protein proportions given above, an increase in canola seed by 6.66kg and a reduction in soybean by 3.91kg results in a net production of 2kg of oil, while a soybean increase of 5kg and a canola decrease of 2.15kg results in the net production of 1.27 kg of protein.

Box 4.1 The system expansion method for assigning credits for co-products

The accepted method for assigning energy credits to co-products is the *system expansion method*, described by Weidema (2000) and Kim and Dale (2002). Because soybean oil and canola oil have the same properties and are substitutable, and a change in the production of soybean oil would lead to a change in the production of canola oil, and vice versa, the net energy consumption assigned to the production of soybean oil ($E_{soybean\text{-}oil}$) in this method is the same as the net energy consumption assigned to the production of canola oil ($E_{canola\text{-}oil}$). However, soybean meal contains 48 per cent protein while canola meal contains 36 per cent protein, so 1.33kg of canola meal is equivalent to 1.0kg of soybean meal. Thus, the energy assigned to the production of soybean meal ($E_{soybean\text{-}meal}$) is 1.33 times that assigned to canola meal ($E_{canola\text{-}meal}$). The two known quantities are the total energy required to produce and mill soybeans and canola ($E_{soybean}$ and E_{canola}). Neglecting small mass losses, one can form four equations in four unknowns:

$$0.17E_{soybean\text{-}oil} + 0.83E_{soybean\text{-}meal} = E_{soybean} \quad (4.3)$$

$$0.40E_{canola\text{-}oil} + 0.60E_{canola\text{-}meal} = E_{canola} \quad (4.4)$$

$$E_{soybean\text{-}oil} - E_{canola\text{-}oil} = 0 \quad (4.5)$$

$$E_{soybean\text{-}meal} - 1.33\ E_{canola\text{-}meal} = 0 \quad (4.6)$$

This system can be solved to get the energy credits per kg of soybean and canola meal, namely:

$$E_{soybean\text{-}meal} = 1.62\ E_{soybean} - 0.73\ E_{canola} \quad (4.7)$$

and:

$$E_{canola\text{-}meal} = 1.21\ E_{soybean} - 0.55\ E_{canola} \quad (4.8)$$

As explained in the main text, $E_{soybean}$ = 3.7MJ/kg and E_{canola} = 8.3MJ/kg according to one set of calculations, but the resulting credits are $E_{soybean\text{-}meal}$ = −0.003MJ/kg and $E_{canola\text{-}meal}$ = −0.041MJ/kg. Not included in this analysis are the effects of changes in the prices of all four commodities. Assessment of these effects would require simulations with a general equilibrium economic model.

The procedure for the computation of credits for either soybean or canola meal depends on the energy required to produce and mill both soybeans and canola, because of market interactions between the two. The accepted procedure, called the *system expansion method*, is explained in Box 4.1. The energy credits for soybean and canola meal are given by:

$$E_{soybean\text{-}meal} = 1.62\ E_{soybean} - 0.73\ E_{canola} \quad (4.9)$$

and:

$$E_{canola\text{-}meal} = 1.21\ E_{soybean} - 0.55\ E_{canola} \quad (4.10)$$

where $E_{soybean}$ and E_{canola} are the energies required to produce and grind soybeans and canola, respectively. The calculation of $E_{soybean}$ and E_{canola} is presented in Table S4.3, and involves the energy required to produce the crop (MJ/kg, given in Table 4.26) plus the additional energy used for crushing the seeds. Canola meal is over twice as energy intensive as soybeans (8.3MJ/kg versus 3.7MJ/kg) according to the calculations in Table S4.3 based on (S&T)[2] Consultants Inc (2005c), so the resulting credits are rather small due to the negative E_{canola} term in Equations (4.9) and (4.10): −0.003MJ per kg of soybean meal produced as a co-product of soy oil, and –0.041MJ per kg of canola meal produced as a co-product of canola oil. These credits are then multiplied by the amount of soybean or canola meal produced per litre of biodiesel. This would be about 1.3kg soybean meal per litre of biodiesel produced from soybeans, and 3.9kg canola meal per litre of biodiesel produced from canola ((S&T)^2 Consultants Inc, 2005c, their Table 4-32). Thus, the overall credits are −0.004MJ/litre of biodiesel from soybeans and −0.161MJ/litre of biodiesel from canola.

It may seem counter-intuitive that the energy credit from the soybean meal that is produced as a co-product of soy oil production (–0.003MJ/kg) is so much smaller than the energy required to produce soybean meal as a primary product (3.7MJ/kg). The explanation is as follows: when we produce soy oil for biofuel purposes, we have also produced more soybean meal as a co-product, so we have to reduce the milling of soybeans that are otherwise used to produce soybean meal as the primary product. However, that milling had soy oil as a co-product that was supplying pre-existing (non-biofuel) markets, so there is now less of it for those markets. Thus, there must be an increase in canola milling directed toward the non-biofuel vegetable oil market, and this increase subtracts from the energy savings due to the reduction of soybean milling for soybean meal that would otherwise occur. This argument assumes that soybean meal is the marginal protein meal in the market (that is, that change in the demand for protein meal is met by changes in the production of soybean meal) and that canola oil is the marginal vegetable oil in the market. The net result is no change in the supply of vegetable oil for non-biofuel purposes or in the supply of protein meal.

Depending on the energy intensity of canola production (which increases) and soybean production for non-biofuel purposes (which decreases), it is possible that the energy credit for soybean and canola meal as co-products of biofuel production could be negative. This is the case for the energy credits computed from the soybean and canola production energies of Pimentel et al (2008) (see Table 4.26). Canola requires more energy to produce than soybean, thereby reducing the net energy credit, and the fundamental reason for this is that soybean is a nitrogen-fixing crop (with minimal N fertilizer requirements) and canola is not.

The other primary co-product of producing biodiesel from soybeans and canola is glycerine. Glycerine occurs as triglycerides in all animal fats and vegetable oils, constituting on average about 10 per cent of these materials. It can also be produced from petrochemical feedstocks. The amount of glycerine produced is 0.0878kg/litre of biodiesel, independently of the feedstock vegetable oil. The primary energy required to produce synthetic glycerine, and hence the energy credit, is 125MJ/kg according to one estimate cited in (S&T)^2 Consultants Inc (2005c) and 238MJ/kg according to another. This corresponds to an energy credit of 11–21MJ/litre. Given that the energy intensity of glycerine, like that of other manufactured products, is decreasing over time, the lower estimate is probably more accurate. As the LHV of biodiesel is 32.8MJ/litre, the lower estimate of the glycerine credit is a third the energy value of the biodiesel.

The EROEIs given above are computed as the ratio of the energy content of the biodiesel fuel to the non-solar energy used to make the biodiesel fuel. However, the biodiesel is being used in place of regular diesel fuel, so the energy gained is not the energy of the biodiesel fuel, but the total energy that would be required to make the amount of diesel fuel that the biodiesel replaces. As with gasoline, this depends on the relative amounts of diesel and biofuel needed to travel a given distance, the relative energy contents of the two fuels per unit volume and the ratio of primary to secondary diesel energy. Test data for buses reported by Carraretto et al (2004) indicate greater fuel consumption using biodiesel – 59 litres/100km versus 50 litres/100km, so 1 litre of biodiesel replaces 0.85 litres diesel. The avoided diesel primary energy per litre of biodiesel is the diesel energy content (36.12MJ/litre) × 0.85 × the ratio (1.2) of primary energy input (mostly crude petroleum) to secondary energy (diesel) output when making diesel fuel. This divided by the biodiesel energy content (32.75MJ/litre) gives the factor by which the EROEIs given in Table 4.26 should be multiplied. The resulting EROEIs are given as the final row of Table 4.26.

As shown in Table 4.10, the biodiesel yield (litres/hectare/year) from palm oil in Malaysia is six times that from sunflower in Europe and almost nine times that from soybeans in Europe. This should result in a considerably higher EROEI and, as long as tropical rainforests are not cleared or peat bogs drained to make way for palm plantations, should also result in a favourable GHG balance. However, these land use changes are in fact occurring and, as discussed below, the associated emissions overwhelm the savings obtained by substituting biodiesel for diesel fuel.

Biodiesel from *Jatropha*

Jatropha is a non-edible plant that produces oil seeds that can be easily converted to biodiesel, with the residues used either as a soil fertilizer or digested to produce biogas (CH_4). Achten et al (2008) indicate an energy requirement for the production of 1000MJ of methyl ester from *Jatropha* in India of 160MJ with low cultivation intensity, and 216MJ with high cultivation intensity. These correspond to EROEI values of about 5.3 and 4.6, respectively. By comparison, the EROEI for methyl ester from rapeseed (canola) is given as 2.3. In all cases these EROEIs are based on allocation of the total energy input among biodiesel and various co-products in proportion to the energy content of the various products (rather than in proportion to the energy that is displaced). Details of the calculation are not given.

Prueksakorn and Gheewala (2008) have calculated the EROEI of biodiesel produced from *Jatropha* in Thailand. The best-estimate EROEI is 6.03 when co-products are included, but the uncertainty range is 1.9–12.0. When co-products are excluded, the best-estimate EROEI and uncertainty range are 1.42 and 0.5–2.7, respectively. The main co-product is seed cake, which can be used as a fertilizer (in which case it is credited with the energy that would be used to make an equivalent fertilizer) or can be used as a fuel (in which case it is credited with its fuel energy value). A major source of uncertainty is the distance that the various inputs and outputs need to be transported.

4.6.2 GHG emission savings from liquid biofuels

Table 4.27 compares estimates of the savings of lifecycle GHG emissions per kilometre travelled when ethanol replaces gasoline and biodiesel replaces diesel fuel. The savings are related to the EROEI for both biofuels and fossil fuels, the relative fuel economies and the mix of primary energy sources used. If one unit of ethanol or biodiesel requires X units of fossil fuel energy (giving a fossil fuel EROEI of $1/X$, with X normally <1) and one unit of gasoline or diesel requires Y units of fossil fuel input energy (including the feedstock, with $Y>1$), then the fossil fuel energy used to drive a given distance using biofuel, as a fraction of the energy used with gasoline or diesel, will be:

$$\frac{X}{Y}\frac{E_{biofuel}}{E_{fossil}} \qquad (4.11)$$

where E_{fossil} and $E_{biofuel}$ are the two fuel economies (litres/km). Computation of relative GHG emissions requires taking into account differences in the primary energy sources used for the two fuels, and emissions of

Table 4.27 *Range of estimates of the reduction in CO_2-equivalent GHG emissions per km travelled when ethanol replaces gasoline and biodiesel fuel replaces diesel fuel for transportation*

Biofuel feedstock	Emission change (%)	Number of studies
	Conventional ethanol	
Corn, E90	+3.3	1
Corn, E100	−13 – −38	6
Sugar beet, E100	−35 – −56	5
Molasses, E85	−1– −51	1
Sugarcane, E100	−87– −96	2
Wheat, E100	−19– −47	5
	Cellulosic ethanol	
Wheat straw	−57	1
Corn stover	−61	1
Hay	−68	1
Grass	−37 – −73	4
Waste wood	−81	1
Crop straw	−82	1
Poplar	−51 – −107	2
	Biodiesel	
Rapeseed	−21 – −48	11
Pure plant oil	−42	1
Soybeans	−53 – −78	4
Soybeans	+0 – +100	1
Tallow	−55	1
Waste cooking oil	−92	1

Source: WWI (2006, Tables 4.1 to 4.3)

additional GHGs (primarily CH_4 and N_2O) associated with the production of the two fuels. Estimated savings range from 21–68 per cent for biodiesel from rapeseed, 53–78 per cent for biodiesel from soybeans (with one exception), 55 per cent for biodiesel from tallow and 92 per cent for biodiesel from waste cooking oil.

Ethanol production from sugarcane provides a good opportunity for the collection of CO_2 and its sequestration in deep saline aquifers, as a CO_2-rich gas stream is produced. From Equation (4.2) it can be seen that 180gm of 6C sugars would yield 92gm of ethanol and 88gm of CO_2 (24gC), all of which can be readily captured and compressed (48gC would be released when the ethanol is burned). Based on the mass balance in Möllersten et al (2003), another 75gC could be captured from the combustion of bagasse (assuming 90 per cent capture), and even more if field residues that are currently burned are instead used to generate additional electricity (as discussed in subsection 4.4.6). Capture of CO_2 from bagasse combustion requires steam whose provision reduces the electricity output, and compression of both streams of CO_2 uses some of the electricity output, the replacement of which would entail the emission of about 24gC if supplied by coal. Thus, the net CO_2 emission is $48+24-75 = -3$gC per 92gm of ethanol produced, or a net sink of 26gC/litre. This would make ethanol from sugarcane a close to carbon-neutral form of transportation energy when other energy inputs are accounted for (geological C sequestration is discussed at length in Chapter 9).

However, full accounting of the impact on GHG emissions from the production of biofuels must take into account global economic linkages. For example, higher costs of corn due to the diversion of corn to ethanol production leads to higher costs for soybeans (as demand for soybeans increases due to substitution of soybean for corn in cattle feed). This in turn leads to additional clearing of Amazonian rainforests for soy farms (Morton et al, 2006). Farmers also purchase large expanses of cattle pasture for soy production, pushing ranchers further into the Amazonian rainforest (Nepstad et al, 2006). Similarly, a large expansion in sugarcane plantations for ethanol could indirectly lead to land-use emissions that greatly exceed several decades of direct emission savings. Prevention of such indirect emissions would require a gradual expansion of sugarcane plantations in step with growth of surplus agricultural land or net reductions in beef consumption due to a shift toward diets with lower meat (which is not happening at present).[25]

Searchinger et al (2008) assess the impact of US ethanol production using a model of global agricultural economics. They accounted for the following:

- increases in the area devoted to all the major temperate and sugar crops by country or region in response to diversion of corn to the production of ethanol;
- replacement of one third of the animal feed diverted to ethanol production with dried distillers' grains (so that only two thirds of the diverted corn feed needs to be replaced);
- higher grain prices slightly increasing the cost of meat, so that a small percentage of the diverted grain is never replaced;
- lower grain productivity on lands outside the US that are brought into production to replace US exports.

They find that production of ethanol from corn in the US increases GHG emissions by 100 per cent over a 30-year period rather than decreasing them by 20 per cent, and that production of biofuels from switchgrass in the US would increase emissions by 50 per cent. Their conclusion is that biofuels should only be produced from biomass residues.

Fargione et al (2008) calculated the immediate CO_2 emissions associated with conversion of various non-agricultural land types to various crops for the production of biodiesel. The original land types, crop and fuel produced, carbon debt, percentage of the carbon debt allocated to biofuels (the balance being allocated to biofuel co-products) and the number of years estimated to be required for the annual CO_2 emission savings to repay the biofuel portion of the carbon debt are given in Table 4.28 (the payback times

Table 4.28 *Carbon debt (tC/ha) from the conversion of various non-agricultural lands to the production of various biofuels, the portion of the debt assigned to biofuels and the estimated number of years required for C emission savings to repay the biofuel portion of the debt*

Original land type	Crop and biofuel	Initial debt (tC/ha)	Portion of debt assigned to biofuels (%)	Time required to pay back initial debt (years)
Indonesian and Malaysian peatland	palm oil biodiesel	941	87	423
Indonesian and Malaysian lowland tropical forest	palm oil biodiesel	191	87	86
Brazilian Amazon	soybean biodiesel	201	39	319
Brazilian *cerrado*	soybean biodiesel	23	39	37
Brazilian *cerrado*	sugarcane ethanol	45	100	17
US central grassland	corn ethanol	37	83	93
US abandoned farmland	corn ethanol	19	83	48

Source: Fargione et al (2008)

depend on the estimated fossil fuel energy inputs required for the production of biofuels which, as shown above, are the subject of considerable disagreement). Payback times range from a low of 17 years (for ethanol from sugarcane on the Brazilian *cerrado*) to over 400 years (for biodiesel from palm oil on drained and cleared peatland). However, conversion of degraded tropical lands to palm oil plantations can result in a significant buildup of soil C, such that the overall reduction in C emission is 150 per cent that of the diesel fuel that is displaced by biodiesel produced from the palm oil (Wicke et al, 2008).

4.6.3 Solid biomass fuels

Solid biomass fuels include wood, wood pellets and compressed grasses that can be used for heating or cooking. Summary data on biomass yields were presented in Table 4.4 and data on current plantation areas were presented in Table 4.5.

Woody biomass plantations

Heller et al (2003) have assessed the external energy inputs required to produce and deliver willow biomass, including GHG emissions associated with energy use and emissions of N_2O from fertilizers. Table 4.29 gives the energy and GHG emissions for a 65ha willow plantation in New York state, based on collected field data and projections. The analysis pertains to a system with partial harvesting (coppicing) of above-ground biomass every three years after initial establishment of the willow, for a total of seven harvests over a 23-year period. After the last harvest, willow stumps are removed in preparation for a new cycle. The energy value of the biomass is 19.8GJ/tonne, while the energy inputs amount to only 0.358GJ/tonne (of which 46 per cent is energy used by farm equipment and 37 per cent is the energy used to manufacture the fertilizers), giving an EROEI of 55.3. Over the first rotation, the EROEI is only 16.6. Assuming that willow crops are grown within an 80km radius from a central facility with a road tortuosity factor of 1.8, the average transport distance is 96km. This is estimated to require an additional 0.189GJ/tonne for transportation, which reduces the EROEI to 36.2. If biosolids are used instead of chemical fertilizers, the EROEI is about 48. Not included in the energy inputs are energy used to dry the biomass or to transport it from the central plant to where it is used, or the energy embodied in the machinery and farm equipment. N_2O emissions as a result of application of fertilizer and decay of leaf litter represent a more significant emission (in CO_2-equivalent terms) than CO_2-emissions from energy use. Nevertheless, the total CO_2 equivalent emission is only 0.9kgC/GJ, compared to about 14kgC/GJ for natural gas and 24kgC/GJ for coal. There is also the possibility that carbon will be sequestered in soils, which according to the estimates given in Table 4.29, almost completely offsets the emissions.

Table 4.29 *Energy and GHG balance of a willow plantation in New York state, US, over a 23-year cycle involving seven harvestings*

Energy balance		
Yield	11.93odt/ha/yr = 236.3GJ/ha/yr @ 19.8GJ/odtr = 5435GJ/ha over 23 years	
Input, production with fertilizer		0.358GJ/odt
Input, production with biosolids		0.259GJ/odt
Input, transport 96km by truck		0.189GJ/odt
EROEI, production only with fertilizers		55.3
EROEI, production with fertilizers + transport		36.2
EROEI, production with biosolids + transport		47.8
	GHG emissions	
	tC/ha	kgC/GJ
Diesel	0.87	0.16
Material	0.92	0.17
N_2O from fertilizer	1.08	0.20
N_2O from leaf litter	1.98	0.37
Total	4.85	0.89
C sequestration in soils	−3.84	−0.71
Net	1.01	0.19

Note: CO_2-equivalent emissions include emissions of N_2O but do not include CO_2 emissions associated with burning the willow as a fuel, as these emissions are offset by the absorption of CO_2 from the atmosphere by photosynthesis as the willow grows. odt = oven-dry tonne.
Source: Heller et al (2003)

The south-eastern pine forests of the US account for 60 per cent of US industrial wood production and 16 per cent of total world production. In the early to mid-1990s, a number of experimental sites were established in the south-eastern US in order to investigate the tradeoff between removal of biomass residues from loblolly pine plantations, which would reduce long-term productivity by removing nutrients from the forest, and the use of fertilizers in order to maintain fertility, along with chemical control of non-crop vegetation. Scott and Dean (2006) report results for 13 such sites from the Gulf Coastal Plain. At ten sites, pine seedlings were planted on nine different plots after a clear-cut timber harvest, with three different degrees of soil compaction for each of three different amounts of removal of organic matter, with each plot further split into two plots with and without removal of competing vegetation manually and by spraying herbicides (on the other three sites, a less extensive array of treatments was assessed). The energy balance for a site in Louisiana ten years after a single harvest is presented in Table 4.30. As seen from Table 4.30, harvest of the whole tree (residues in addition to stem wood) reduces the extent of regrowth after ten years if no fertilizer is applied (this is a general result, although there are sites where, in the short term, harvesting of forest residues did not reduce the extent of regrowth). Application of fertilizer to replace lost nutrients, or of herbicides to inhibit competing vegetation, is decidedly beneficial from an energy point of view, in that the gain in biomass energy is several hundred times the energy required to produce and apply the fertilizer or herbicide at these sites.

Raymer (2006) estimates the energy used for harvesting, transport to the processing site, processing, and transport to the consumer for various forms of woody biomass energy in Norway. Results are given in Table 4.31. The energy used for these processes is in the order of 0.3–0.9 per cent of the energy content of the biomass, yielding an EROEI of about 8–29. These EROEIs exclude energy use by fertilizers, as it is assumed that no fertilizers are used. However, this is not a sustainable situation, so the sustainable EROEI would be smaller.

Matthews (2001) has assessed the energy balance of coppiced short-rotation willow in the UK. He computes an EROEI of 20–64, with 13 per cent of the energy input use for building fences, 30 per cent for harvesting and chipping and 40 per cent for storing and drying the wood. Again, it is assumed that no fertilization is required. CO_2-equivalent emissions are estimated to be 1–2kgC/GJ, compared to 14kgC/GJ for natural gas fuel and 25kgC/GJ for coal. The CO_2-equivalent emissions include CH_4 and N_2O but not CO_2 during combustion because the CO_2 released during combustion is offset by CO_2 absorbed from the atmosphere during photosynthesis.

For production of biomass chips from commercial forestry residues in Finland, Wihersaari (2005) estimates that collection, transportation and chipping require an energy input (diesel fuel and electricity) equivalent to 1.9–2.6 per cent of the energy of the final product. Return of the ash to the forest soil requires an energy input of 0.1 per cent of the chip energy, while production of N fertilizer to replace the N lost with removal of the residues requires an energy input equal to 1.4 per cent of the chip energy. The overall EROEI is about 25–30. Emissions of CH_4 and N_2O occur during storage and combustion, while soil carbon is lost. The energy use and CO_2-equivalent emissions are given in Table 4.32. Total CO_2-equivalent emissions are dominated by the loss of soil carbon (assumed to be

Table 4.30 *Energy balance of a loblolly pine plantation at a site in Louisiana with different treatments*

System	New biomass growth after 10 years (tonne/ha)	Energy value of harvested residues (GJ/ha)[a]	Energy value of the difference in biomass after 10 years (GJ/ha)	Embodied energy in herbicides or fertilizer (GJ/ha)	Net gain (GJ/ha)
Stem-only harvest	35.6	0.0	0.0	0.0	0.0
Whole-tree harvest (WTH)	28.0	416.2	−154.3[b]	0.0	261.9
WTH + herbicide	42.5	416.2	140.1[c]	1.3[d]	554.9
WTH + fertilizer	52.3	416.2	339.0[c]	4.3[e]	750.8

Note: [a] Based on harvested residues of 20.5t/ha times an assumed energy content of 20.3GJ/tonne. [b] Based on differences in column 2 relative to the stem-only case, times an assumed energy content of 20.3GJ/tonne. [c] Based on differences in column 2 relative to the whole-tree harvest case, times an assumed energy content of 20.3GJ/tonne. [d] Based on herbicide use of 1kg/ha/yr for five years times a herbicide embodied energy of 263MJ/kg. [e] Based on P and N fertilizer application rates of 56 and 50kg/ha/yr, respectively, and embodied energies of 7.6 and 78MJ/kg, respectively.
Source: Based on Scott and Dean (2006)

Table 4.31 *Energy used in producing various forms of woody biomass for energy in Norway and delivering it to consumers in Norway, EROEI and CO_2-equivalent emissions*

Type of biomass	Distances transported (km)	Energy content (GJ)	Energy used in harvesting, transport and processing				EROEI	CO_2-equivalent emission (kgC/GJ)	
			Electricity (kWh)	Fossil fuel (MJ)	Bioenergy (MJ)	Total (MJ)		Combustion only	Total
Pine wood, 1m³	25, 104	8.5	0	338	0	338	25	1.22	1.96
Birch wood, 1m³	25, 104	9.5	0	364	0	364	26	1.40	2.09
Spruce sawdust, 1m³	50, 64	10	37	212	0	346	29	0.30	0.68
Bark, 1m³	50, 64	2.6	37	173	0	306	8	0.74	2.11
Briquettes, 1t	0, 50	20	340	443	245	1912	10	0.30	0.74
Pellets, 1t	119, 15	20	199	504	335	1555	13	0.30	0.80

Note: Combustion emissions are based on CH_4 and N_2O only. The two transport distances given are for (1) transport of raw biomass to the biomass processing facility, and (2) transport of processed biomass to the consumer. Any fertilizer or other inputs during the growth of the biomass are excluded.
Source: Data given in or computed from Raymer (2006)

Table 4.32 *Energy use and CO_2-equivalent emissions associated with the production of biomass chips from forestry residues in Finland*

Step	Energy use (% of delivered chip energy)	CO_2-equivalent emission (kgC/GJ chips)
Collection, transport and chipping	1.9–2.6	0.33–0.57
Storage (N_2O and CH_4 emission per month)		0.38–0.76
Combustion (N_2O and CH_4 emission only)		0.15
Return of ash to forest soil	0.1	0.02
Nitrogen fertilization	1.4	0.53
Loss of soil carbon		3.03–3.41
Total (for one month storage)	3.4–4.1	4.4–5.4

Source: Wihersaari (2005)

11 per cent over 100 years, based on simulation models). The total CO_2-equivalent emission (about 60kgCO_2/MWh$_{chips}$) is still about five to six times smaller than that of coal (330kgCO_2/MWh$_{coal}$ plus up to another 15 per cent due to leakage of methane from coal mines).[26]

The total energy used for collecting, processing and long-distance transport of biomass fuels to central Europe from Northern Europe, Eastern Europe or Latin America, and biomass energy lost during transport due to decomposition, are indicated in Table 4.33. The EROEI is computed assuming a biomass energy content of 20GJ/tonne and based on the energy inputs excluding biomass energy losses; it ranges from 8 to 17.

De Mira and Kroeze (2006) have assessed the N_2O emissions associated with the production of willow biomass to generate electricity. Nitrous oxide emissions

Table 4.33 *Biomass energy supply chains to central Europe, energy use and EROEI*

Biomass location	Biomass source	Transport mode	Distance (km)	Biomass loss	Energy input	EROEI
				GJ/tonne$_{dry}$ delivered		
Scandinavia	Forestry residues	Chips by ship	1100	1.6	1.6	12.6
Scandinavia	Forestry residues	Chips by trains	1100	1.6	4.6	4.3
Scandinavia	Forestry residues	Pellets by ship	1100	0.5	1.2	16.7
Scandinavia	Forestry residues	Pellets by train	1100	0.5	2.5	8.0
Eastern Europe	Short-rotation willow coppice	Pellets by ship	2000	0.1	1.0	9.4
Latin America	*Eucalyptus* plantation	Pellets by ship	11,500	0.4	1.4	14.1

Source: Hamelinck et al (2005b)

Table 4.34 *N_2O emissions, energy yield and equivalent CO_2 emissions (due to N_2O emissions) per TJ of willow energy yield for soils with different rates of N fertilization in The Netherlands*

Case	Fertilizer input (kg N/ha/yr)	N emission factor (kg N_2O-N/kg N input)			Yield (TJ/ha/yr)	CO_2-equivalent emission (kgC/GJ)	
		Direct	Indirect	Total		Mineral soil	Organic soil
No fertilizer	0	0.0000	0.000	0.0000	0.075	0	8
Low fertilizer	60	0.0025	0.004	0.0065	0.20	0	3
Average fertilizer	100	0.0125	0.035	0.0475	0.10	3	9
High fertilizer, poor soils	120	0.0225	0.14	0.1625	0.04	22	38

Source: Data from de Mira and Kroeze (2006)

arise through direct emissions from fertilized soils (due to nitrification and denitrification of nitrogen fertilizer), from indirect emissions at remote sites due to leaching and runoff of the applied fertilizer, and from indirect emissions through fertilizer-induced atmospheric deposition of nitrogen compounds. Nitrous oxide emissions were computed using the methods recommended by the Intergovernmental Panel on Climate Change (IPCC) and converted to equivalent CO_2 emissions using an N_2O global warming potential of 296. Four cases with different amounts of fertilizer application were considered for willow grown in The Netherlands. The resulting CO_2-equivalent emissions per unit of biomass energy are given in Table 4.34 and exceed that of natural gas (15kgC/GJ) and sometimes coal (23.5kgC/GJ) for the case of high fertilizer use on poor soil.

For the production of *Brassica carinata*, a herbaceous woody crop suitable for the hot and dry conditions of Southern Europe, Gasol et al (2007) compute an EROEI of 8.1. The gross energy yield in Spain is estimated to be about 84GJ/ha/yr, while the required primary energy input for fertilizers, herbicides, equipment and fuel is about 10GJ/ha/yr. Interestingly, delivery of 83GJ of energy to Spain in the form of natural gas requires an upstream input of about 15GJ, so the losses in supplying energy as biomass are less than for natural gas. The biomass, like natural gas, can be used for heating and/or electricity generation.

Production of wood pellets

The energy required to produce and deliver various fuels that could be used for heating in Sweden, as estimated by Gustavsson and Karlsson (2002), and the associated EROEI are given in Figure 4.40. Energy inputs for biofuels are 10–18 per cent of the heating value of the fuel and the EROEI ranges from about six to ten. For these results, it is assumed that wood chips are transported 50km to pellet plants, and that pellets are transported 100km to the end-user.

Magelli et al (2009) estimate the total energy inputs for the production of biomass pellets in western Canada and their delivery to Sweden to be about 40 per cent of the energy content of the delivered fuel when wood residues are used for drying (so the EROEI is about 2.5) and about 35 per cent when natural gas is used for drying. Greater energy is required for drying using wood residues than using natural gas because of the lower efficiency combusting biomass compared to natural gas. However, the fossil fuel requirement is only about 20 per cent of the biomass energy content in this case (giving an EROEI of about five). About a third of the total energy input is for transportation.

Grasses

Among the grasses that can be used for bioenergy are switchgrass, *Miscanthus* and napier grass. Grasses have the advantage that they do not need to be seeded and ploughed every year, thereby reducing energy inputs and the risk of erosion.

Table 4.35 gives the energy inputs and outputs estimated for the production of switchgrass pellets in Canada. The estimated EROEI is 14.6. For the production and harvesting of reed canary grass in Lithuania, Jasinskas et al (2008) estimate an EROEI of about 8 to 14. Table 4.36 presents the energy used and energy yield of napier grass in Brazil. The EROEI is 21.3 if N fertilizer is used and 25.4 if N fertilizer can be

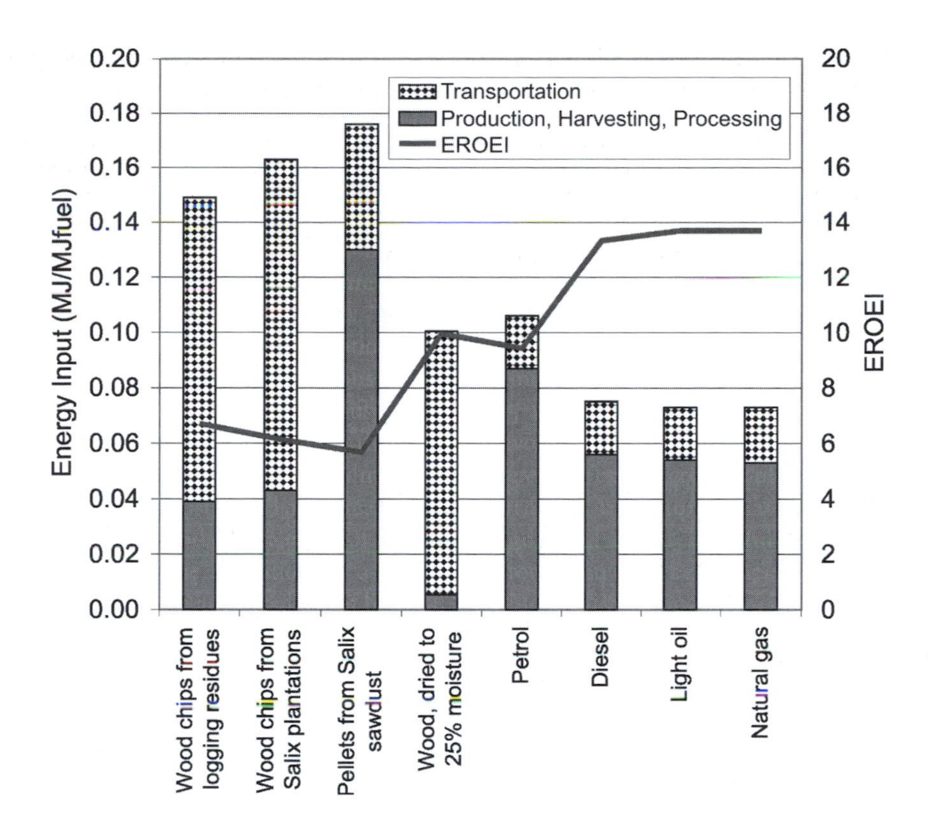

Source: Gustavsson and Karlsson (2002)

Figure 4.40 *The energy required to produce and deliver various fuels for heating detached houses in Sweden, and the EROEI*

Table 4.35 *Energy terms in the production of switchgrass pellets in Canada*

Activity	Energy (GJ/tonne)
Establishment	0.028
Fertilizer + application	0.460
Mowing	0.034
Baling	0.197
Transport	0.072
Biofuel production total	**0.791**
Pellet mill construction	0.043
Pellet mill operation	0.244
Delivery	0.193
Pellet production total	**0.480**
Overall total	**1.27**
Energy supplied	**18.5**
EROEI	**14.6**

Source: Samson et al (2005)

Table 4.36 *Energy terms in the production of 30 oven-dry tonnes/ha/yr of napier grass in Brazil*

Activity	Energy (GJ/ha/yr)
Establishment (every 3 yrs)	0.18
N fertilizer	3.69
Other fertilizer	2.29
Lime	4.37
Pesticides	0.02
Mechanical weeding	0.18
Harvest	7.21
Transport (16km) for processing	1.23
Machinery embodied energy	2.27
Total input	**23.1 (19.4)**
Energy yield @ 16.44GJ/odt	493.2
EROEI	**21.3 (25.4)**
Densification (0.1–0.3GJ/t)	3–9
Drying (2.6–5.2GJ/t)	78–156
Total input	**94–188**
EROEI	**2.6–5.2**

Note: Terms in brackets are for the case without application of N fertilizer.
Source: Samson et al (2005)

avoided. Not included is the energy used for drying and densification, the latter requiring 0.1–0.3GJ/tonne. Grasses commonly have a moisture content of 40–70 per cent if cut at the end of the growing season. In temperate regions, grasses left to overwinter have 12–14 per cent moisture at the time of harvest and require no further drying. Artificial drying from 45–75 per cent moisture to 7–9 per cent requires 2.6–5.2GJ/oven-dry tonne (odt), compared to an energy content of 18–19GJ/odt (Samson et al, 2005). Thus, the net energy yield will be significantly increased if artificial drying can be avoided, or if waste heat from CHP is available. Formation of good-quality briquettes requires a moisture content below 10 per cent.

Cultivation of a mixture of 16 species of perennial grasses on agriculturally degraded, nutrient-poor sandy soils in the US prairies yielded 68GJ/ha/yr of biomass with inputs of 4GJ/ha/yr, giving an EROEI of 17 (Tilman et al, 2006). Yields on these poor soils with only one species of grass ranged from about 5 to 50GJ/ha/yr, indicating a synergistic interaction when multiple species are intermixed.

Reducing the N content of grass biomass improves the net energy balance by requiring less N fertilizer and reducing the potential for spoilage, and reduces potential NO_x emissions during combustion. As discussed in subsection 4.5.1, this can be achieved through harvesting after the late-season translocation of N to the root system.

Comparison of agricultural crops and woody biomass for heating and cogeneration

Table 4.37 compares the energy and CO_2 balance for two agricultural crops (winter wheat and hemp) and for short-rotation poplar grown in The Netherlands and in Poland, and used for heating or cogeneration of heat and electricity, as estimated by Dornburg et al (2005). The embodied energy of farm equipment is not included in the original analysis, so a rough value taken from Note 22 for farms in Canada has been added here for winter wheat and hemp, and one quarter this value has been arbitrarily added for poplar. CO_2- equivalent emissions include an estimate of N_2O emissions related to the use of N fertilizer in The Netherlands and applied to Poland in proportion to fertilizer use, but (unlike EROEI) have not been adjusted to account for the embodied energy in machines. Wheat provides the largest EROEI (about 9–14 in both The Netherlands and Poland) and the greatest net avoided CO_2 emission. Avoided emissions are greater in The Netherlands than in Poland, as the better yield more

Table 4.37 *Energy and carbon balance of energy crops grown in The Netherlands and Poland, and used for heating or for cogeneration of heat and electricity*

	The Netherlands			Poland		
	Winter wheat	Hemp	SR poplar	Winter wheat	Hemp	SR poplar
Yield (tonnes/ha/yr)	11–14	6–9	6–10	6–7	2.3–5.6	3–5
Yield (GJ/ha/yr)	165–252	90–162	90–180	90–126	40–100	45–90
Energy invested (GJ/ha/yr)						
Operation of farm machinery	8.7	8.7	6.0	3.5	6.0	5.2
Seeds	0.6	1.6	0.3	0.6	1.6	0.4
Fertilizer	7.9	3.1	2.5	4.6	4.3	1.1
Pest control	0.3	–	0.0	0.2	–	0.1
Chipping and drying	–	–	–	–	7.2	6.4
Total	17.5	13.4	16	8.9	11.9	13.2
Machine embodied energy	0.4	0.4	0.1	0.4	0.4	0.1
EROEI	**9.3–14.1**	**6.6–11.8**	**5.6–11.2**	**9.8–13.7**	**3.2–8.2**	**3.4–6.8**
CO_2-equivalent emission (kgC/GJ)	4.05	3.65	3.35	4.21	6.08	5.53
Avoided CO_2-equivalent emission (kgC/GJ)	20.1	20.3	21.8	25.1	24.8	26.8
Net avoided CO_2-eq (kgC/GJ)	16.0	16.6	18.4	20.9	18.7	21.3

Note: Yields as GJ/ha/yr are computed by multiplying the lower and upper mass yields (t/ha/yr) by minimum and maximum heating values (LHV) of 15GJ/t and 18GJ/t, respectively. SR = short rotation.
Source: Based on Dornburg et al (2005)

Table 4.38 *Comparison of fertilizer use (kg/ha/yr) for different agricultural systems and bioenergy crops in Ireland*

Fertilizer	Dairy	Beef	Sheep	Sugar Beets	SRCW	*Miscanthus*
N	160	52	35	160	43	67 & 100
P	13	8	5	49	9	13 & 20
K	35	20	11	165	66	67 & 100
GHG	3.29	1.42	1.04	1.01	0.53	0.37

Note: Rates of fertilizer use for dairy and beef cattle and for sheep refer to average of rates for pasture and silage. For *Miscanthus*, the first number pertains to years 1 to 2 and the second to years 3 to 15 of a 15-year production cycle. Also given is the CO_2-equivalent GHG emission (tC/ha/yr) associated with each product. SRCW = short-rotation coppice willow.
Source: Styles and Jones (2007)

than compensates for the greater energy use and N_2O emission associated with greater use of N fertilizer. Short-rotation poplar has a comparatively low EROEI (6–11 in The Netherlands) because of the energy use associated with chipping and drying, which does not apply to the other crops.

Table 4.38 compares the fertilizer requirements for different agricultural products and for production of short-rotation coppice willow (SRCW) and *Miscanthus* in Ireland. Also given are total GHG emissions associated with the production of these commodities. As shown in Table 4.38, the switch from production of feed for cattle or sheep to woody bioenergy crops, or from sugar beets to woody bioenergy crops does not necessarily reduce fertilizer use in Ireland. Replacing beef farming with the bioenergy crops reduces farm-related CO_2-equivalent GHG emissions by about 1tC/ha/yr. Using the biomass so produced to generate electricity that displaces coal would reduce CO_2 emissions by a further 4–6tC/ha/yr (given biomass yields for SRCW and *Miscanthus* of about 9 and 12t/ha/yr, respectively, assuming that biomass is 50 per cent carbon by mass and assuming the same electricity-generation efficiencies using coal and biomass). Finally, sequestration of soil C over a period of 100 years at annual average rates of 0.5 and 1.2tC/ha/yr is thought to be possible for SRCW and *Miscanthus*, respectively, which equals or exceeds the emission associated with the energy inputs required to produce the biomass.

Use of sawmill residues to generate electricity

Cowie and Gardner (2007) examined the impact on GHG emissions of using sawmill residues to generate electricity in Australia, instead of using the residues to manufacture particleboard (as at present). If the residues are used to generate electricity and fresh plantation biomass is used to produce particle board, then 0.56tC of CO_2-equivalent GHG emissions is avoided per tonne of sawmill residue used, but an extra 0.10tC is emitted in the manufacture of particle board, giving an 18 per cent smaller net reduction.

4.6.4 Biogas

Berglund and Börjesson (2006) have investigated the net energy inputs used to produce biogas from eight different raw materials in Sweden. Energy is used for the cultivation and harvesting of energy crops, the recovery of harvest residues or the collection of municipal organic waste, the transportation of raw materials to the biogas plant, the operation of the biogas plant, and the dewatering, transport and spreading of digestate (material remaining after the production of biogas) on arable land. Results of their analysis for a transport distance of 10km are given in Figure 4.41. The energy used in producing biogas in Sweden ranges from 13 to 40 per cent of the energy content of the product biomass. This corresponds to an EROEI of 2.5–8.0.

4.6.5 Synthesis

The studies discussed in the preceding pages can be summarized in two ways: in terms of the EROEI and in terms of the efficiency in converting from biomass to liquid or solid fuels ready for use. EROEI values are summarized in Table 4.39. The lowest EROEI values (1.5–1.8) pertain to ethanol from corn, followed by ethanol from sugarcane and from cellulosic crops, biodiesel from oily crops, biogas from various sources,

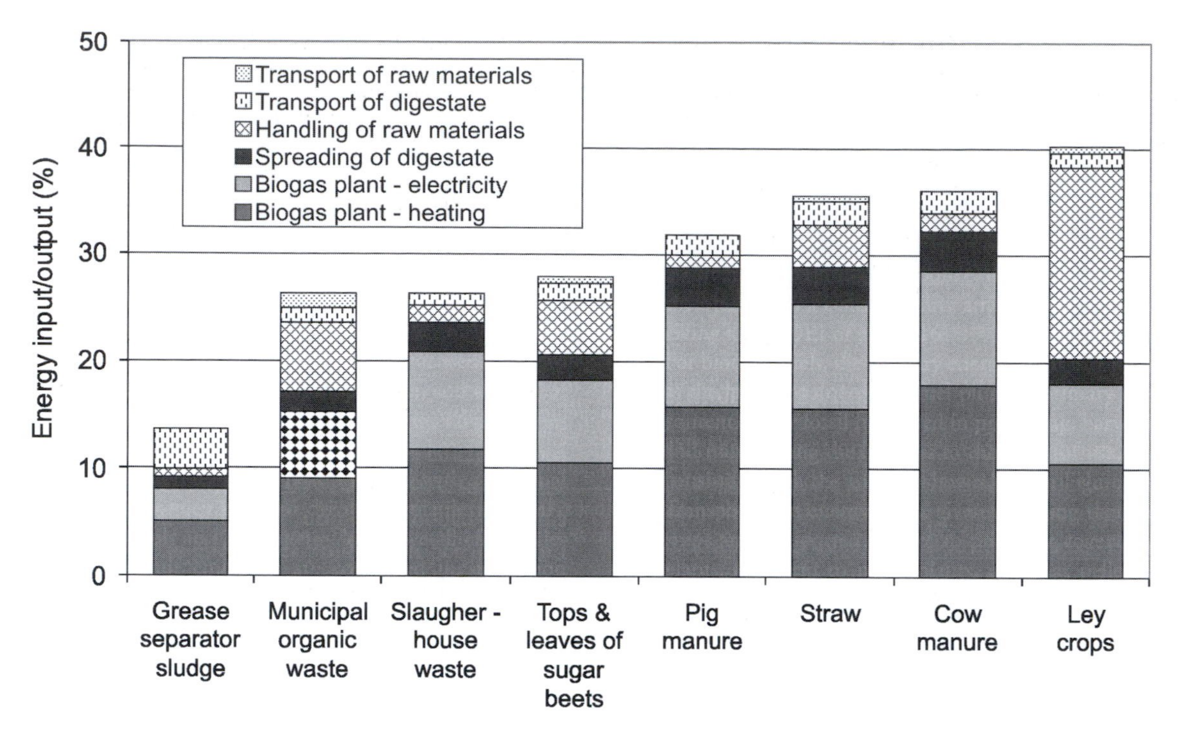

Source: Berglund and Börjesson (2006)

Figure 4.41 *Inputs required for the production of biogas from various biomass sources, for a transport distance of 10km, in relation to the energy provided by the biogas*

and finally, solid fuels for heating and cogeneration of heat and electricity. EROEI values as high as 56 have been calculated.

When primary biomass energy is transformed to another form (such as from wood to biogas, or from sugarcane to ethanol), some of the initial biomass energy is lost. As well, external energy inputs are usually required (unless biomass not converted to the energy product can be used to power the conversion process). The conversion efficiency can thus be defined in terms of secondary energy provided (including the energy saved through the use of co-products) over initial biomass energy, or secondary energy over initial biomass energy plus non-biomass energy inputs. Both of these efficiencies are shown in Figure 4.42. The first efficiencies are generally in the 40–60 per cent range, with the second efficiencies up to 15 per cent smaller but generally only a few per cent smaller. Converting plant oils into biodiesel is simpler than converting starch or sugar into ethanol, usually resulting in greater conversion efficiency. Figure 4.43 shows the breakdown of the total energy inputs for making fuel: the bioenergy feedstock itself, the energy used in producing the bioenergy feedstock and transporting it to the processing plant, and the energy used at the processing plant. Biomass energy is the dominant energy input. Also shown is the total energy of the products produced as a percentage of the total input (this is the second efficiency shown in Figure 4.42). This total energy divided by the non-biomass energy input gives the EROEI.

4.7 Biomass as a chemical feedstock

Biomass can be used as a feedstock in place of petroleum (and other fossil fuels) for the production of all of the products (or close substitutes) that are currently made from fossil fuels. As most of these products (such as plastics, tyres and solvents) are eventually incinerated or decompose, releasing CO_2 to the atmosphere, the replacement of fossil fuel feedstocks with biomass feedstocks will reduce net CO_2 emissions (as well as stretch out conventional fossil fuel supplies). Geiser

Table 4.39 *Summary of estimates of EROEI for bioenergy crops given here*

Energy product	Biomass source	Location	EROEI	Comment	Source
Ethanol	Corn	US	1.5–1.8	Range today	Table 4.19
	Sugarcane	Brazil	8	Average today[a]	Table 4.22
	Sugarcane	Brazil	10	Best today[a]	Table 4.22
	Sugarcane	Brazil	44	Maximum future	Table 4.22
	Corn stover	US	6.8–8.2	Future	Table 4.24
	Woody biomass	US	11	Future	Table 4.19
Biodiesel	Soybean	US, Canada	1.4–2.2		Table 4.26
	Canola	US, Canada	1.0–2.0		Table 4.26
	Jatropha	India	4.6–5.3	EROEI ~ 1 without use of co-products	Subsection 4.6.1
	Jatropha	Thailand	1.9–12.0	With sizeable co-product credits	Subsection 4.6.1
Ethanol & biodiesel	Corn–soybean rotation	US	3.7	Without credits	Table 4.24
			3.9	With credits	
Solid fuel	Coppice willow	New York	36	Chemical fertilizer	Table 4.29
	Coppice willow	New York	48	Biosolids as fertilizer	Table 4.29
	Coppice willow	UK	20–64	No fertilization	Subsection 4.6.3
	Pine, birch	Norway	25	No fertilizer or other inputs	Table 4.31
	Spruce sawdust	Norway	29	during growth are included	Table 4.31
	Bark	Norway	8	here.	Table 4.31
	Briquettes	Norway	10	Transport is included	Table 4.31
	Pellets	Norway	13		Table 4.31
	Forestry residues	Finland	25–30	Includes fertilization	Table 4.32
	Brassica carinata	Spain	8.1	Includes fertilization	Subsection 4.6.3
	Switchgrass pellets	Canada	14.6		Table 4.35
	Pine pellets	western Canada	2.5	Drying with NG	Magelli et al (2009)
		to Sweden	5.0	Drying using residues	
	Napier grass	Brazil	21	Without drying or densification	Table 4.36
	Napier grass	Brazil	2.6–5.2	With drying and densification	Table 4.36
	Reed canary grass	Lithuania	8–14	No transport	Jasinskas et al (2008)
	Miscanthus	Europe today	23–40	Production, storage, and	Smeets et al (2009)
	Miscanthus	Europe in 2030	32–56	transport 100km.	Smeets et al (2009)
	Switchgrass	Europe today	25–47	CO_2-equiv emission of	Smeets et al (2009)
	Switchgrass	Europe in 2030	27–49	1–2kgC/GJ	Smeets et al (2009)
	16 mixed grasses	US prairies	17	Production only on agriculturally-degraded, nutrient-poor soils	Tilman et al (2006)
	Wood chips	Sc to CE	4.3–12.6	Transport is included,	Table 4.33
	Wood pellets	Sc to CE	8–17	based on source and	Table 4.33
	Wood pellets	EE to CE	9	destination given in	Table 4.33
	Wood pellets	LA to CE	14	column 3	Table 4.33
	Winter wheat	Netherlands	9–14	Transport is excluded	Table 4.37
	Hemp	Netherlands	6–12		Table 4.37
	Poplar	Netherlands	6–11		Table 4.37
	Winter wheat	Poland	10–14		Table 4.37
	Hemp	Poland	3.2–8.2		Table 4.37
	Poplar	Poland	3.4–6.8		Table 4.37
	Dried wood, pellets or chips	Sweden	6–10	Local production and use	Figure 4.40
Biogas	Various	Sweden	2.5–8	Local materials	Figure 4.41

Note: Future EROEIs for ethanol are hypothetical. Sc = Scandinavia, CE = Central Europe, EE = Eastern Europe, LA = Latin America. [a] For mechanical harvesting, the expected future norm.

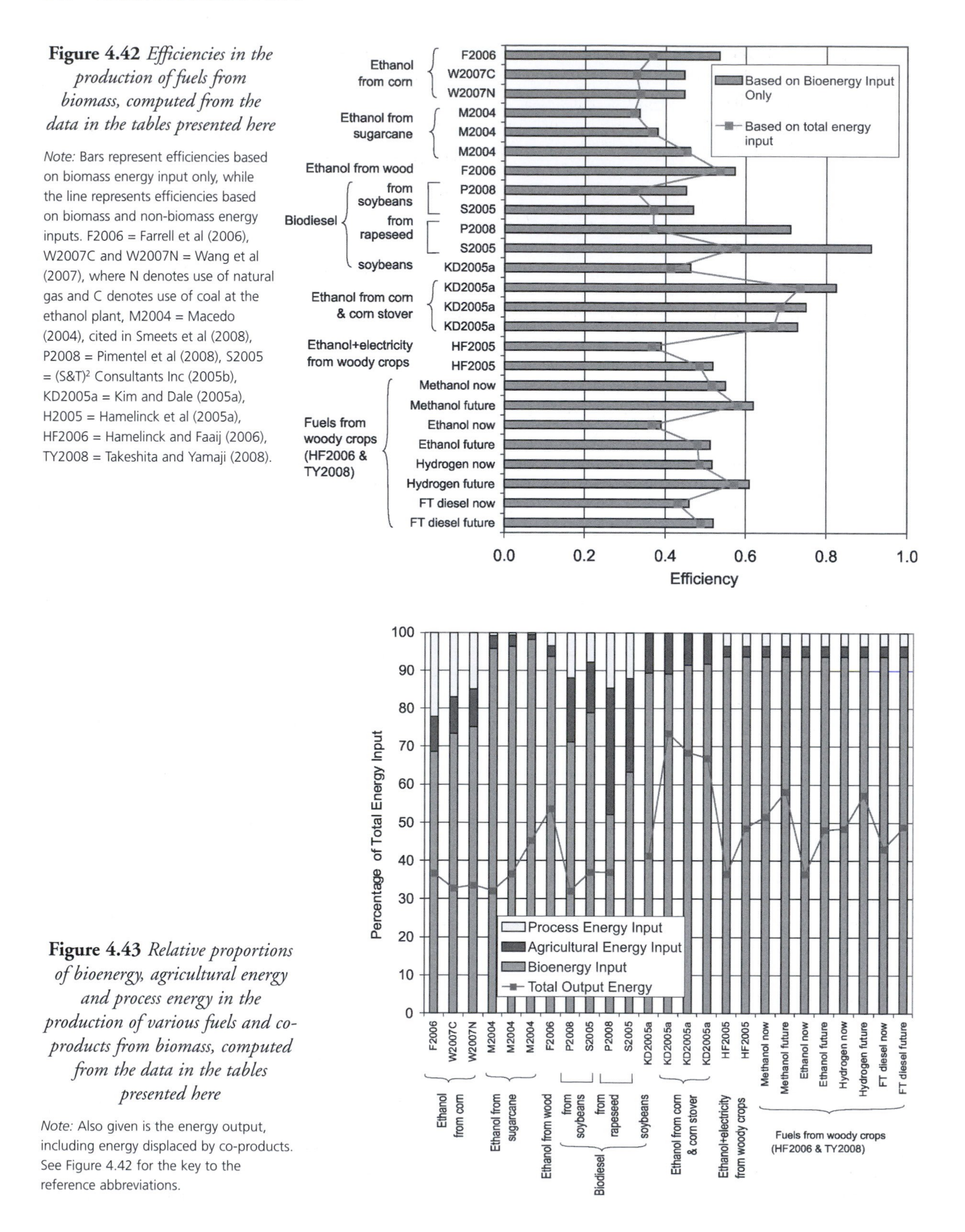

Figure 4.42 *Efficiencies in the production of fuels from biomass, computed from the data in the tables presented here*

Note: Bars represent efficiencies based on biomass energy input only, while the line represents efficiencies based on biomass and non-biomass energy inputs. F2006 = Farrell et al (2006), W2007C and W2007N = Wang et al (2007), where N denotes use of natural gas and C denotes use of coal at the ethanol plant, M2004 = Macedo (2004), cited in Smeets et al (2008), P2008 = Pimentel et al (2008), S2005 = (S&T)^2 Consultants Inc (2005b), KD2005a = Kim and Dale (2005a), H2005 = Hamelinck et al (2005a), HF2006 = Hamelinck and Faaij (2006), TY2008 = Takeshita and Yamaji (2008).

Figure 4.43 *Relative proportions of bioenergy, agricultural energy and process energy in the production of various fuels and co-products from biomass, computed from the data in the tables presented here*

Note: Also given is the energy output, including energy displaced by co-products. See Figure 4.42 for the key to the reference abbreviations.

(2001, Chapter 11) contains an overview of the many petroleum-based products that can be made from biomass feedstocks. Table 4.40 gives the percentage of different classes of industrial materials that are currently produced from biomass in the US, while Figure 4.44 shows the major routes for the synthesis of organic chemicals from biomass feedstocks.

Although all materials currently made from petroleum, such as plastics, can also be made from biomass (with gasification or fermentation as a first step, followed by various catalysed chemical reactions), there are some important differences (Cho, 2007): the hydrocarbons in petroleum are molecules with few functional groups (such as hydroxyls) that provide sites for reactions, so they are relatively stable. Carbohydrates, by contrast, have many functional groups that make them less stable and less able to withstand the high temperatures used to process petrochemicals. Hence, the functional groups must be stripped off first.

In the following subsections, biomass is discussed as a feedstock in the following ways:

- as a feedstock to the existing petrochemical industry in place of fossil fuel feedstocks;
- as a source of biopolymers for new materials with properties similar to existing plastics;

Table 4.40 *Proportion of various industrial products produced from biomass feedstocks in the US in 1996*

Product	% from biomass
Wall paints	9.0
Special paints	4.5
Pigments	9.0
Dyes	15.0
Inks	16.0
Detergents	18.0
Surfactants[a]	50.0
Adhesives	48.0
Plastics	1.8
Plasticizers	15.0
Acetic acid	17.5
Furfural	17.0
Fatty acids	40.0
Carbon black	12.0

Note: The trend from 1992 was upward in every case. [a] Surface-active agents that which remove dirt and grease from various materials by holding them in solution or suspension in water. Natural surfactants are based on fats and oils from plant and animal sources. See Saouter et al (2006) for further information.
Source: Moser (1998)

- as a source of natural fibres that can be used in novel ways in place of synthetic fibres.

4.7.1 Biomass as a feedstock to the existing petrochemical industry

There are a number of routes by which biomass could be used as a feedstock for organic chemicals:

- production of naphtha via Fischer-Tropsch (FT) synthesis;
- production of naphtha via high thermal upgrading (HTU);
- production of ethylene from biomass-based ethanol;
- production of ethylene and propylene from biomass-based methanol.

In the production of naphtha via HTU, biomass is treated with water in a reactor at temperatures of 300–350°C and 120–180atm pressure for five to ten minutes, producing a crude oil product that can be upgraded to naphtha or diesel. The conversion efficiency is 64 per cent (IEA, 2006a). As naphtha is currently produced from petroleum feedstocks and in turn is used as the starting point for a broad array of plastics and other organic chemicals (see Volume 1, Chapter 6, section 6.10), the production of naphtha from biomass would allow much of the existing petrochemical infrastructure to be used. However, naptha would be produced from biomass on a small scale at dispersed sites due to the dispersed nature of biomass resources. For biomass at $3/GJ, crude oil from HTU would cost $9/GJ ($50/barrel) and HTU-naphtha would cost $12–14/GJ (IEA, 2006a).

4.7.2 Production of biopolymers

A variety of biopolymers has been investigated and could be used in some products built from petrochemical polymers. An example is polylactic acid (PLA), which is produced from lactic acid, which in turn is produced from corn and sugar beets. It can be used to make bottles, films, fibre products and extruded containers. It has mechanical properties similar to PET (polyethylene terephthalate) and PP (polypropylene), and the waste stream can be processed through composting or anaerobic digestion. Another example is polytrimethylene terephthalate (PTT),

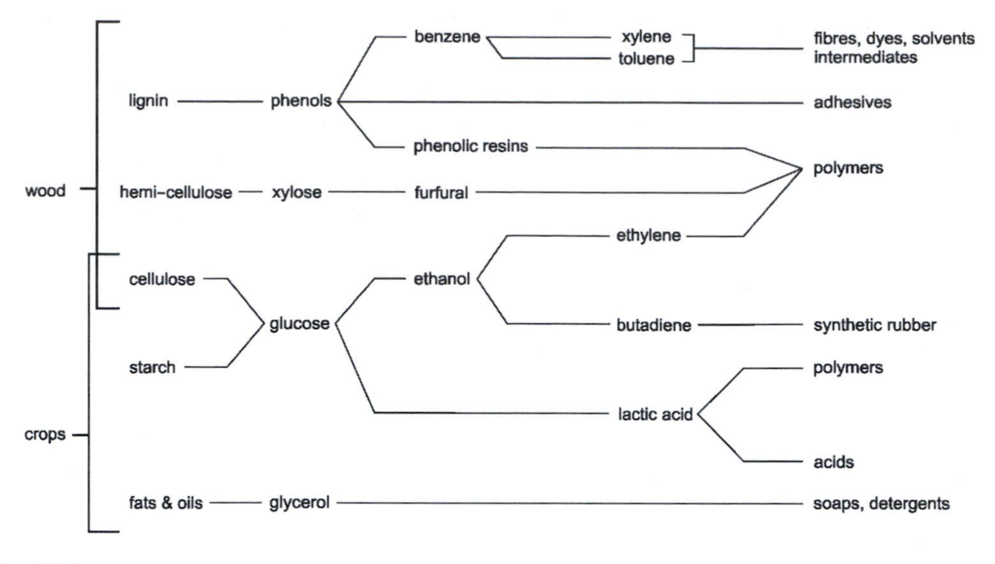

Source: Geiser (2001)

Figure 4.44 *Principle routes for the synthesis of organic chemicals from biomass feedstocks*

which is produced from 1,3-propanediol, which in turn is produced from corn using a genetically modified bacterium. PTT can be used in carpets. Figure 4.45 compares the energy required to manufacture PET and PLA. PLA requires more process energy at present, but requires no fossil fuel feedstock, so that total energy requirement is about 25 per cent less. However, it is expected that the process energy requirement for PLA can be reduced by about 50 per cent, largely by a production sequence using bio-refinery concepts (Detzel et al, 2006). A third biopolymer of interest is polyhydroxyalkanate (PHA). If produced as a co-product of cellulosic ethanol, the fossil fuel energy requirement is reduced to 44MJ/kg from 77–88MJ/kg for its petrochemical counterpart according to an analysis based on laboratory tests (Yu and Chen, 2008).

Hermann et al (2007) have assessed the CO_2 savings resulting from the production of ten different bulk chemicals (many of them used in the manufacture of plastics) through fermentation or other enzymatic conversions of biomass. They performed a cradle-to-grave analysis (from agricultural production to management of post-consumer waste), taking into account the energy and GHG emission savings from

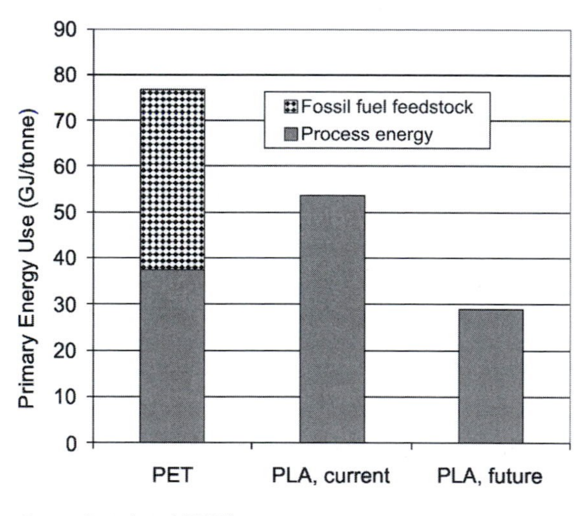

Source: Detzel et al (2006)

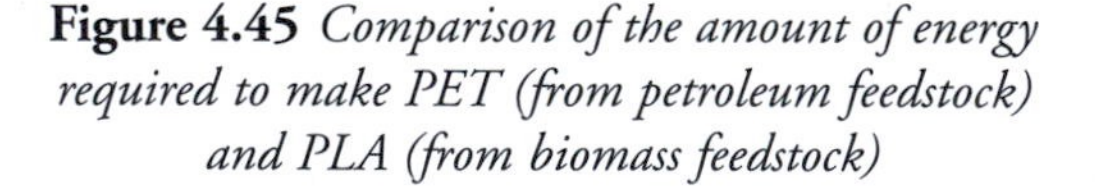
Figure 4.45 *Comparison of the amount of energy required to make PET (from petroleum feedstock) and PLA (from biomass feedstock)*

co-products. CO_2 emission savings of up to 2tC per tonne of chemical product are possible assuming process parameter values thought to be achievable in 20–30 years. In terms of CO_2 emission reduction per tonne of chemical product, sugarcane is preferred to corn stover as a biomass source, which in turn is preferred to corn starch, while in terms of CO_2 emission reduction per hectare of land area used to produce the crop, corn stover is preferred to sugarcane. For some products and biomass sources, the emission savings compared to production from fossil fuel feedstocks are greater than 100 per cent, because of displaced emissions elsewhere due to co-products.

Dornburg et al (2004) have compared the effectiveness of using a unit of land to produce energy crops, natural fibre composites and biopolymers in place of functionally equivalent petroleum-derived polymers. Included in their analysis are direct energy inputs for crop production, indirect energy inputs (for fertilizers, pesticides and machines) and energy inputs for crop processing to make usable energy products or biopolymers. The energy savings per unit of land dedicated to the production of natural fibres is substantially greater than that saved when the unit of land is used to grow crops for CHP production, and is greater than the savings with CHP for most biopolymers that could be produced from the unit of land. The advantage in using land for natural fibres or biopolymers is more pronounced if the alternative is to use the land to produce ethanol for transportation via short-rotation crops, and even more so if the ethanol is produced from corn. This analysis does not take into account the possibility to extract energy from biopolymers through incineration at the end of their life, which further increases the benefit of using land to produce biopolymers.

4.7.3 Novel uses of natural fibres

A particularly large energy saving arises from the replacement of automotive plastics (typically having 30 per cent glass fibre reinforcement) with plastics containing 65 per cent hemp fibre for reinforcement. The saving arises from the replacement of an energy-intensive material (glass) with a natural product, the reduction in automobile weight (which reduces fuel consumption) and the ability to incinerate hemp fibre plastics with energy recovery. Replacement of 50 per cent of glass fibre plastics used in the automotive sector in North America yields a net reduction in CO_2 emissions equal to 4.3 per cent of total US industrial CO_2 emissions (Pervaiz and Sain, 2003). Corbière-Nicollier et al (2001) calculate that the substitution of glass fibre with Chinese reed (*Miscanthus sinensis)* fibre as reinforcement in plastic transportation pallets reduces lifecycle energy use associated with the manufacture, use and disposal of the pallets by almost 50 per cent. They also find that the energy savings per unit of biomass used in transport pallets is four to ten times larger than if the biomass were burned to supply energy.

4.7.4 Summary

In summary, the most effective uses of biomass in reducing energy use, in order of decreasing effectiveness, are: (1) replacement of glass fibres with natural fibres, (2) replacement of polymers made from fossil fuel feedstock with biopolymers, (3) CHP generation, (4) production of ethanol from sugarcane, (5) production of biodiesel from oily crops, and (6) production of ethanol from corn, the last two uses providing very little net benefit.

4.8 Comparison of CO_2 emission savings for alternative uses of land

The analysis to this point shows that, with appropriate safeguards to maintain soil fertility, biomass can be grown on a given parcel of land, harvested and used in place of fossil fuels indefinitely. Conversely, deforested land can by reforested, causing a one-time increase in the amount of carbon stored (sequestered from the atmosphere) that offsets an equal fossil fuel emission. This will be referred to here as *biological carbon sequestration* so as to distinguish it from the capture of CO_2 from combustion of fossil fuels and its compression and injection into geological formations (as discussed in Chapter 9), which shall be referred to as *geological carbon sequestration.* It is possible that soil carbon can be built up while continuously harvesting above-ground biomass for energy purposes, or that CO_2 released from the combustion or gasification of biomass could be sequestered geologically. In both cases, the result would be a *negative* CO_2 emission. In this section the impacts on atmospheric CO_2 of using land for biological sequestration of carbon, for production of bioenergy, and for concurrent biological or geological carbon sequestration and bioenergy are compared.

It should be noted, however, that sequestration of carbon in forests is not equivalent to avoiding an equal emission of carbon from the combustion of fossil fuels. This is because the restoration of forests generally causes a decrease in the *albedo* or reflectivity of the land surface to solar radiation, which in turn causes an increase in the absorption of solar radiation by the land surface. In the case of reforestation in middle and high latitudes, the net effect of cooling due to less CO_2 in the atmosphere and warming due to greater absorption of solar radiation appears to be one of warming, while tropical reforestation has a net cooling effect (Marland et al, 2003; Schaeffer et al, 2006). Indeed, computer simulations indicate that the effect of albedo changes due to the total deforestation since pre-historical times is to cool the climate by 0.13–0.4°C in the global mean and by up to 2°C regionally due to induced changes in atmospheric circulation (Feddema et al, 2005; Brovkin et al, 2006) (local temperature changes would be further altered due to changes in land surface hydrology and in evaporative cooling). This of course is not to imply that reforestation should not be carried out and existing forests not protected, but rather that the purely climatic benefits are smaller than or even opposite to what would be expected based on the change in atmospheric CO_2. Where carbon is sequestered by building up soil carbon, however, there would be no change in the surface albedo and hence no offsetting heating effect from the absorption of more solar radiation.

4.8.1 CO_2 emission offset through biological carbon sequestration

Carbon can be biologically sequestered by converting degraded agricultural land to forested land or to perennial grasses and through improved management of existing land uses. With regard to the former, Lemus and Lal (2005) indicate that there are about 1750Mha of land worldwide that have been degraded due to deforestation, overgrazing or overly intensive farming. These lands have lost 30 per cent or more of their original soil carbon, but much of this can be restored over a period of 40–60 years by planting perennial crops (such as switchgrass, willow or poplar). An accumulation of 10–20tC/ha, or about 0.2–0.3tC/ha/yr over a period of 40–60 years, is a realistic global mean average for degraded lands. Applying this to all 1750Mha of degraded land gives a global sequestration rate of 0.35–0.51GtC/yr. With regard to improved management of existing lands, Table 4.41 summarizes estimates by Lemus and Lal (2005) of the rates at which C can be accumulated. These rates range from as low as 0.3tC/ha/yr in croplands and agro-forestry systems, to as much as 3.0tC/ha/yr in forests and degraded lands. The potential global rate of C sequestration sums to about 0.3–0.9GtC/yr. Thus, the potential sequestration from rehabilitation of degraded land and improvement management of land that is still productive sums to about 0.7–1.4GtC/yr. By comparison, the global fossil fuel CO_2 emission in 2005 was about 8GtC.

Table 4.41 *Estimated rates of accumulation of soil carbon due to improved land management that can be sustained over a period of 50 years*

System	Global land area (Mha)	Rate of C accumulation (tC/ha/yr)	Global total (GtC/yr)
Cropland	150	0.3	0.045
Biofuel croplands	13	0.6–1.2	0.008–0.016
Grasslands	47	0.5–0.7	0.024–0.033
Forests	164	1.0–3.0	0.164–0.492
Degraded lands	109	0.8–3.0	0.087–0.262
Agro-forestry	21	0.3–0.5	0.006–0.011
Total	504		0.33–0.86

Source: Lemus and Lal (2005)

A similar global estimate is found in the IPCC *Special Report on Land Use and Forestry* (Watson et al, 2000): 1.0GtC/yr sequestration from improved land management within existing land uses and from agro-forestry and reforestation after a ten-year ramp up. Nabuurs et al (2007) estimate the potential C sequestration from reforestation and improved forest management to be 0.5–1.5GtC/yr by 2040, about 25–35 per cent of which is in middle and high latitudes and the balance in the tropics. Strengers et al (2008) assessed the C sequestration potential of abandoned cropland as a function of the increasing cost of successive additional units of sequestration. To do this they used the land use scenarios developed by the IPCC in its *Special Report on Emission Scenarios* (the so-called 'SRES' scenarios) (Nakicenovic and Swart, 2000) combined with detailed information on a 0.5° × 0.5° global grid concerning soils, climate and other relevant information as embedded in the Dutch IMAGE-2

(Integrated Model to Assess the Global Environment) framework. They conclude that a sequestration rate of about 0.7GtC/yr could be achieved by 2050 at marginal cost of $200/tC, with the cost of additional sequestration increasing sharply as the total sequestration rate increases further.

Richards and Stokes (2004) review the previous ten years of studies of the potential for carbon sequestration in biomass and soils. Estimated sequestration rates per unit land area are:

- 2–10tC/ha/yr with the establishment of forest plantations on deforested (but not necessarily degraded) land, and cumulative storage of 15–40t/ha in boreal regions, 30–175t/ha in temperate regions and 25–125t/ha in tropical regions;
- up to 7.6tC/ha/yr and up to 125t/ha cumulative storage from improved forest management;
- cumulative storage of 50–200t/ha from agro-forestry.

In summary, the typical rates of biological C sequestration that could occur over a period of several decades with careful management are:

- 0.2–0.3tC/ha/yr through restoration of degraded lands;
- 0.3–3.0tC/ha/yr through better management of still-productive lands;
- 2–10tC/ha/yr through re-establishment of forests on deforested but not necessarily degraded land.

The total global potential is a biological sequestration rate of 0.5–1.5GtC/yr.

4.8.2 Avoided CO_2 emissions from bioenergy crops

To deduce the avoided CO_2 emission per hectare of land devoted to bioenergy, first consider the substitution of coal or natural gas for the generation of electricity. We will make the following assumptions:

- biomass yield is 10t/ha/yr;
- biomass is 50 per cent C by mass;
- the heating value of biomass is 19GJ/t;
- the emission factors for biomass, coal and natural gas are 26.3kgC/GJ, 26.3kgC/GJ and 16.9kgC/GJ, respectively;[27]
- the efficiencies of electricity generation using biomass, coal and natural gas are 50 per cent, 50 per cent and 70 per cent, respectively (these all represent future advanced technologies);
- the energy required to cultivate, harvest and deliver biomass energy is 1GJ/t (out of a range of 0.5–1.5GJ/t given by Cannell (2003)), supplied as oil with an emission factor of 22kgC/GJ.

Then:

- the net reduction in CO_2 emission per tonne of biomass used is 478kgC if coal is displaced and 207kgC if natural gas is displaced;
- the avoided CO_2 emissions per unit land area are 4.8tC/ha/yr and 2.1tC/ha/yr if coal and natural gas, respectively, are displaced;
- 100Mha of land with a biomass yield of 10t/ha/yr saves 0.48GtC/yr and 0.21GtC/yr if coal and natural gas, respectively, are displaced, while providing 19EJ/yr of primary energy.

Thus, rates of avoided C emission when biomass displaces coal and even highly efficient natural gas for the generation of electricity are comparable to the rates at which C can be sequestered through reforestation. However, the displacement of fossil fuel emissions can continue indefinitely and the avoided emission in meeting the energy demand at a given point in time is permanent, whereas the C accumulated in ecosystems can be released due to disturbances or due to adverse climatic change in the future.

To estimate the avoided C emission when bioethanol is used to displace gasoline, we make the following assumptions:

- The biofuel yield is 5000 litres/ha/yr, which is the midrange of what is expected from lignocellulosic biomass in the future (see Table 4.10).
- The fuel automobile consumption is 7 litres/100km using gasoline and 9 litres/100km using biofuel.
- The combustion of 1 litre of gasoline releases 0.63kgC, which is multiplied by 1.2 to account for upstream emissions (associated with exploration for and extraction and refining of crude petroleum).
- The net reduction in CO_2 emissions per km travelled using cellulosic ethanol is 40–80 per cent (Table 4.27).

The result of these assumptions is an avoided CO_2 emission of 1.18–2.35tC/ha/yr. The upper limit is better than the rate at which CO_2 emission can be avoided if biomass is instead used to displace highly efficient generation of electricity with natural gas, but is less than if highly efficient generation of electricity using coal is displaced. If instead we assume an ethanol yield of 13,000 litres/ha/yr and a 92 per cent reduction in CO_2 emissions, as should be achieved by Brazilian sugarcane ethanol in the future, the avoided CO_2 emission is 7.0tC/ha/yr.

To estimate the avoided C emission when biodiesel is used to displace diesel fuel, we will make the following assumptions:

- The biofuel yield is 1200 litres/ha/yr, which pertains to biodiesel from rapeseed in Europe (the best for any crop in Europe) (see Table 4.10).
- The fuel automobile consumption is 6 litres/100km using diesel and 7 litres/100km using biodiesel.
- The combustion of 1 litre of diesel releases 0.74kgC, which is multiplied by 1.2 to account for upstream emissions.
- The net reduction in CO_2 emissions per km travelled using biodiesel is 20–50 per cent (Table 4.27).

The result of these assumptions is an avoided CO_2 emission of only 0.18–0.46tC/ha/yr. Assuming 6000 litres/ha/yr (pertaining to biodiesel from Malaysian palm oil), and assuming the same 20–50 per cent net reduction in CO_2 emissions, the avoided CO_2 emission is 0.79–1.98tC/ha/yr.

In summary, current production of biodiesel from Malaysian palm oil or of ethanol from Brazilian sugarcane provides the best avoided CO_2 emission per unit of land area (0.8–2.0tC/ha/yr for palm oil, 1.2–2.4tC/ha/yr for sugarcane ethanol), *assuming* that cultivation of palm oil or sugarcane does not lead to direct or indirect deforestation or drainage of peatlands to provide land for the bioenergy crops. Future ethanol from Brazilian sugarcane will probably avoid emissions at a rate of 7tC/ha/yr, again ignoring possible induced deforestation. When biomass is used to produce electricity at 50 per cent efficiency (as expected with future technologies), the avoided CO_2 emission is about 2tC/ha/yr if advanced natural gas generation of electricity is displaced and about 5tC/ha/yr if advanced coal generation is displaced.

4.8.3 Simultaneous displacement of fossil fuels and biological carbon sequestration

As discussed in subsection 4.5.1, it should be possible with careful management to simultaneously harvest above-ground biomass for energy purposes while building up soil carbon through the root systems of bioenergy crops. Reported rates of soil carbon sequestration under short-rotation woody crops in various US locations are frequently in the range 0.5–2.0tC/ha/yr, while the most diverse mixtures of perennial grasses on degraded soils reported by Tilman et al (2006) sequestered an average of 0.75tC/ha/yr during the first decade even with yearly removal of the above-ground biomass, a rate that they believe can be maintained for a century. Sartori et al (2006) estimate that 60Mha of land (devoted equally to short-rotation woody crops and herbaceous crops) used to produce biomass for the generation of 5EJ of electricity could displace 0.162GtC/yr of fossil fuel emissions if high-efficiency natural gas is replaced (and much more if coal is displaced), while sequestering 0.024GtC/yr in soils (an average of only 0.4tC/ha/yr). Metting et al (2001) estimate that 0.5–0.8GtC per year could be sequestered worldwide in the soils of biomass cropland while harvesting biomass crops for energy use.

Biochar

Due to the relatively rapid decomposition of most soil C fractions (within years to decades), a continuous flux of organic matter to the soil is required in order to maintain an elevated amount of soil C. If carbon could instead be built up in very long-lasting (hundreds or thousands of years) pools, then continuous addition of new C would not be needed in order to maintain the C pool. *Biochar* is one such form of carbon. It is produced through slow pyrolysis – the slow burning of biomass in the absence of oxygen. Volatile materials are driven off as gases and can be collected and used as a fuel, while leaving behind a solid that is resistant to decomposition (this is how charcoal is made, except that the volatile materials are not collected). Addition of biochar to soil therefore produces a long-lasting carbon sink. By contrast, only 10–20 per cent of unaltered organic matter that is added to soil remains after five to ten years. The pre-European populations of the Amazon Basin appear to have deliberately applied biochar in Amazonian soils, producing the so-called Amazonian Dark Earths or *terra preta* that have a carbon storage of up to 250tC/ha per m of depth, hundreds or thousands of years after having been abandoned. By comparison, unaltered soils from similar parent materials contain about 100tC/ha/m (Lehmann et al, 2006). For specific examples of the production and use of biochar today, see Ogawa et al (2006).

McHenry (2009) reviews the production and properties of biochar. Pyrolysis can be used to produce three products: biochar, bio-oils and hydrogen. The relative amounts of these products depend on the conditions during pyrolysis:

- Maximization of biochar production requires a low temperature and a low rate of heating.
- Maximization of bio-oil production requires a low temperature and a high rate of heating.
- Maximization of H_2 production requires a high temperature and a high rate of heating.

The amount of carbon that remains in biochar produced at 500–550°C ranges from 20–65 per cent of the original carbon in the biomass, and tends to be larger the greater the heating value of the biomass.

The addition of biochar may indirectly increase or decrease GHG emissions. It increases soil fertility by: (1) increasing the cation exchange capacity of the soil; (2) reducing leaching of nutrients (effectively increasing the productivity of fertilizer); (3) serving as a host for nitrogen-fixing organisms; (4) improving the water-holding capacity of the soil; and (5) neutralizing acidity (Fowles, 2007). Improved soil fertility should lead to more vigorous plant growth, with associated C uptake. Biochar also reduces emissions of CH_4 and N_2O from soils. However, it might also accelerate the decomposition of C that is already in the soil, a possibility that needs further investigation (Wardle et al, 2008).

About 30 per cent of the original biomass energy content remains in the biochar, and about two thirds of the remainder can be captured, so only about 80 per cent of the original biomass energy can be used. However, biochar can be made from various agricultural and forestry residues which, today, are often not used as an energy source and indeed often create disposal problems. Production of biochar thus serves to produce some useful energy and eliminate disposal problems while also creating a product that improves soil fertility.

There are some restrictions on the production and use of biochar:

- Biochar produced at temperature less than 400°C has a low surface area and so may not be useful as a soil improver.
- Toxic and possibly carcinogenic organic compounds can be produced from some feedstocks under certain conditions.
- Heavy metal contaminants in some feedstocks can limit the safe addition of biochar to soil to less than 250t/ha (~200tC/ha).
- Biochar should be added to the soil in a way that minimizes disturbance to soil structure or avoids an increase in erosion.

Potential C sequestration with biochar

Gaunt and Lehmann (2008) have assessed the net reduction in CO_2-equivalent emissions when various bioenergy crops (switchgrass, *Miscanthus*, corn, wheat straw and corn stover) are converted to biochar with capture of the offgases and their use to displace electricity generated from coal or natural gas. They considered two strategies: one in which pyrolysis is carried out in such a way as to maximize the production of energy in gases (in which case no biochar is produced) and one in which the production of biochar is maximized (in which case less energy is provided). They also took into account estimated energy inputs for the various crops, and estimates of the reduction in soil N_2O emissions resulting from the application of biochar to the soil. The result is that there is a factor of two to three larger net reduction in GHG emissions when production of biochar rather than of energy is maximized (depending on whether coal or natural gas is displaced by whatever energy-containing gases are produced). When the choice is between pyrolysis of agricultural residues (rather than of dedicated bioenergy crops) with maximization of either biochar or energy production, maximization of biochar production again gives a substantially larger net reduction in GHG emissions.

Day et al (2005) carried out experimental work involving the co-production of hydrogen and biochar from biomass at a small (laboratory) scale. They envisage using heat in part from powerplant flue gas to drive the charring reaction and production of H_2, with the biochar then used as a filter to absorb CO_2, NO_x and SO_x (the last two reacting with some of the H_2 to produce N and S fertilizer within the biochar), then added to soil. From 100kg of biomass, 22kgC would be sequestered as biochar and an additional 3kgC permanently taken up from the powerplant flue gas for geological sequestration, while 6.78kg H_2 beyond that needed to react with NO_x and SO_x would be produced. Assuming a biomass heating value of 19MJ/kg, the efficiency of hydrogen production would be (6.78kg × 120MJ/kg)/(100kg × 19MJ/kg) = 0.43. By comparison, the efficiency of dedicated production of H_2 from biomass is about 55–70 per cent (Larson, 1993). However, co-production of H_2 and biochar, even at lower efficiency than dedicated production of H_2 from biomass, would lead to a larger net reduction in CO_2 emission even if the H_2 produced in both cases displaces coal.[28] This is without considering the added benefits of reduced soil CH_4 and N_2O emission associated with biochar.

The above analysis compares the CO_2 benefit of production of biochar with the benefit that would be obtained if the additional energy required for co-production of biochar were instead used to displace fossil fuels. Once fossil fuels are eliminated from the energy system, the applicable alternative would be dedicated use of land for carbon storage through reforestation. Suppose that 100EJ of biomass primary energy are produced from 500Mha of plantations. This could produce: (1) 60EJ H_2/yr and no biochar; or (2) 40EJ H_2/yr with sequestration of 1.16GtC/yr in biochar; or (3) 38EJ H_2/yr with sequestration of 1.16GtC/yr in biochar and 1.14GtC/yr through carbon capture and storage.[29] To produce 60EJ H_2/yr with co-production of biochar and capture of CO_2 would require an additional 289Mha but now 1.83GtC/yr would be sequestered in biochar and 1.80GtC/yr captured. This is equivalent to C sequestration at the rate of (3.63GtC/yr)/(289Mha) = 12.6tC/ha/yr, which exceeds the range (5–10tC/ha/yr) at which C could be sequestered through reforestation during the mid-growth phase of the newly planted seedlings. As well, sequestration in biochar would be essentially forever and could continue indefinitely, whereas sequestration of C in forest biomass could continue for only a few decades (until the forest is fully grown) and would be reversible (due to human disturbance, fires or dieback due to adverse climatic change). Thus, if the only goal is to maximize the reduction in net CO_2 emission to the atmosphere, use of greater amounts of land for bioenergy plantations with subsequent co-production of hydrogen and biochar is better than using less land, producing only hydrogen from biomass, and reforesting the extra land.

4.8.4 Displacement of fossil fuels with geological sequestration

As shown in Chapter 9 (Table 9.3), 98 per cent of the CO_2 that is released during the gasification of biomass to produce H_2 fuel can be captured. The net CO_2 emission is −18.4kgC/GJ H_2 or approximately −12kgC/GJ biomass (assuming 65 per cent conversion efficiency). 1GJ of H_2 would displace 1.5–2.2GJ of gasoline (due to the greater efficiency with which hydrogen can be used in automobiles; see Volume 1, Chapter 5, subsection 5.4.8), which corresponds to 1.9–2.7GJ petroleum and an avoided CO_2 emission of 23–33kgC/GJ biomass (given a petroleum emission factor of 18.5kgC/GJ). Thus, the total reduction in net emission is 35–45kgC/GJ. For a biomass yield of 200GJ/ha/yr (10.5t/ha/yr at 19GJ/t), the avoided emission is 4.6–6.6tC/ha/yr.

CO_2 could also be captured during the gasification of biomass to generate electricity. According to Table 9.6, 44 per cent of the exhaust CO_2 could be captured with an electrical generation efficiency of 28 per cent (instead of 34 per cent without capture), giving a net emission of −0.14kgC/kWh. If advanced coal (50 per cent efficiency) is the alternative to biomass, the avoided emission is 0.18kgC/kWh, giving a net benefit of 0.32kgC/kWh (25kgC/GJ biomass), or 5tC/ha/yr if the biomass yield is 200GJ/ha/yr.

Box 4.2 Summary of direct avoided CO_2 emissions from alternative uses of land

Dedicated biological C sequestration can achieve rates of CO_2 uptake of:

- 0.2–0.3tC/ha/yr on degraded lands;
- 0.3–3.0tC/ha/yr through improved management within existing land uses;
- 2–10tC/ha/yr through reforestation of deforested lands;
- 0.5–1.5GtC/yr conservative total global potential.

Dedicated bioenergy crops with a yield of 10t/ha/yr and an energy value of 19GJ/t saves CO_2 emissions at a rate of:

- 4.8tC/ha/yr if coal for electricity is displaced;
- 2.1tC/ha/yr if natural gas for electricity is displaced.

Also, 0.4–2.0tC/ha/yr could be concurrently sequestered through buildup of soil C with careful management.

Production of biofuels for transportation saves:

- 1.2–2.4tC/ha/yr if North American cellulosic ethanol displaces gasoline;
- ~7tC/ha/yr if future Brazilian sugarcane ethanol displaces gasoline;
- 0.2–0.5tC/ha/yr if European rapeseed biodiesel displaces diesel;
- 0.8–2tC/ha/yr if Malaysian palm oil biodiesel displaces diesel.

Examining bioenergy and biochar, 500Mha of bioenergy plantations produces:

- 60EJ/yr H_2, or
- 40EJ/yr H_2 + 1.16GtC/yr in biochar, or
- 38EJ/yr H_2 + 1.16GtC/yr in biochar + 1.14GtC/yr CO_2 captured and stored.

789Mha of bioenergy plantations produces 60EJ/yr H_2 + 3.63GtC/yr in biochar and in captured and stored CO_2. Carbon is sequestered on the margin at a rate of (3.63GtC/yr)/(289Mha) ~14tC/ha/yr.

For bioenergy and geological sequestration, production of H_2 by gasification gives 4.6–6.6tC/ha/yr. Generation of electricity with BIGCC saves 5tC/ha/yr compared to advanced coal as the alternative.

By comparison, deforestation of tropical secondary forests entails an emission of up to 150tC/ha while deforestation of primary forest entails an emission of up to 300tC/ha (Righelato and Spracklen, 2007), so avoiding tropical deforestation provides a one-time savings of 150–300 tC/ha. Not included in the emission savings from the production of bioenergy crops are emissions associated with direct or indirect clearing of land due to impacts on prices.

4.8.5 Summary

Box 4.2 compares the rate at which C emissions are avoided or offset, per unit of land area, when land is used for dedicated carbon sequestration in biomass and soils or used for various bioenergy crops with or without biological or geological C sequestration. Also given for comparative purposes is the loss of C that occurs when tropical primary or secondary forest is destroyed. With the possible exception of future ethanol from Brazilian sugarcane, use of land for reforestation is more effective over a 30-year period than using the same land to produce cellulosic ethanol or biodiesel. Conversely, use of land to produce biomass that is then used to generate electricity in place of coal is potentially more effective than reforestation and, unlike reforestation, can continue indefinitely on the same parcel of land. The most important near-term measure by far for reducing net C emissions is to preserve existing forests.

Sequestration of carbon through the buildup of biomass incurs a cost, whereas continuous harvesting of biomass for energy generates a revenue stream and will become increasingly profitable as the cost of conventional fuels increases. However, buildup of biomass can be done as part of ecological restoration, something that is urgently needed in many parts of the world.

4.9 Global biomass energy potential

As noted in the introduction, biomass currently supplies almost 10 per cent of world primary energy needs – 48EJ out of a total supply of 483EJ/yr in 2005. Much of this biomass energy is produced and used far less efficiently than it could be. Potential future biomass energy sources are:

- plantation energy crops (woody or herbaceous) on surplus agricultural land and on degraded lands that are no longer suitable for agriculture;
- biomass residues from forestry, agriculture and food processing;
- organic animal and human waste streams; and (4) biomaterials cascaded to energy use at the end of their life.

Table 4.42 presents on overview of estimates of the global biomass energy potential from these categories. The total global potential ranges from 40EJ/yr (i.e. about the same as current bioenergy supply) to 1100EJ/yr (i.e. more than twice total world primary energy demand today). The single largest component of this bioenergy supply is projected to be from dedicated bioenergy plantations (up to 850EJ/yr), and most of that (up to 700EJ/yr) on surplus agricultural land that is expected to become available over the coming decades due to expected continuing increases in agricultural yields. Large increases in the yields of bioenergy crops themselves are also expected. Because so much of the projected future biomass energy comes from land dedicated to that purpose, and depends on a growing availability of surplus agricultural land, it is pertinent to first examine in some detail the relative land area requirements using different kinds of renewable energy, then to examine the determinants of surplus agricultural land. This will be followed by a discussion of estimates of the potential bioenergy supply in each of the categories listed in Table 4.42.

4.9.1 Sunlight-to-biomass conversion efficiency

The amount of primary energy that can be supplied by biomass is limited by the efficiency of photosynthesis and the availability of land. In most regions, the growing season will be constrained by temperature and/or moisture. Assume that half of the incident solar radiation

Table 4.42 *Overview of the range of estimated potential bioenergy supply from different resource categories in 2050*

Resource type	Bioenergy potential (EJ/yr)	Comment
Energy crops on surplus agricultural land	0–700 (typically 100–300)	Potential available land area of 0–4Gha (typically 1–2 Gha), yields of 8–12 dry tonnes/ha/yr
Energy crops on degraded or deforested land	60–150 (could be 0)	Max potential available land area of 1.7 Gha, yields of 2–5 dry tonnes/ha/yr
Agricultural residues	15–70	More residues are available with intensive (high-input) systems
Forest residues	30–150 (could be 0)	Includes primary and secondary
Animal manure	5–55	Low value is current use
Organic wastes	5–50	
Biomaterials	40–150	An additional initial claim on biomass, later cascaded to energy use. Required land area of 0.2–0.8Gha at 5 dry tonnes/ha/yr average yield
Total	40–1100 (typically 250–500)	

Note: By comparison, world primary energy demand in 2005 was 483EJ.
Source: WWI (2006)

occurs during the growing season and that half of the incident radiation reaches the leaves. Of this, 80 per cent is absorbed. Of the absorbed radiation, about 50 per cent is photosynthetically active and 30 per cent of that will be converted to stored energy. However, much of the stored energy will be used to meet the plant's own metabolic needs; 40 per cent might be available for growth of new biomass. The overall efficiency in converting sunlight to biomass is thus $(0.5 \times 0.5 \times 0.8 \times 0.5 \times 0.3 \times 0.4) \times 100$ per cent ≈ 1.0 per cent. Further losses occur if the biomass is converted to electricity or to liquid or gaseous fuels. On marginal land, with limited nutrients and water, energy yields and the resulting efficiencies are 20–40 times less. For example, a biomass yield of 5t/ha/yr in a region with a mean annual incident solar radiation of 200W/m^2 corresponds to an efficiency in converting sunlight to biomass of 0.07 per cent (assuming the biomass to have an energy value of 18GJ/tonne). Even highly productive sugarcane (at 70t/ha/yr, 35 per cent dry matter and 18GJ/tonne of dry matter) corresponds to an efficiency of only 0.7 per cent. By comparison, it should be possible to mass-produce low-cost PV modules with an efficiency of 10 per cent or better. If we allow for the modules covering only 50 per cent of the land surface area in a central PV facility, we can see that the land area required to produce electricity photovoltaically is several times less than that needed by growing biomass. The sustainability of large-scale biomass production is also an important consideration, and it is not proven that large-scale production of biomass for energy is truly sustainable.

However, there are several important considerations that favour biomass energy, such as the ease of implementation, simplicity, low cost and the ability to store and transport the energy. The cost itself already takes into account the low conversion efficiencies and resulting land area requirement for a given production of biomass.

4.9.2 Determinants of the amount of surplus agricultural land

As noted above, surplus agricultural land figures prominently in most estimates of future bioenergy potential. In many regions, there has already been a growing surplus of agricultural land. For example, according to the 2002 Natural Resources Inventory of the USDA (www.nrcs.usda.gov/technical/land/nrio2/nrio2lu.html), US cropland dropped from 420 million acres in 1982 to 368 million acres in 2002 (−12 per cent). In spite of the decrease in cropland, the percentage of non-cultivated cropland grew from 11 per cent (44 million acres) in 1982 to 15 per cent (57 million acres) in 2002. The area of non-federal grazing lands (rangeland, pastureland, grazed forests) fell from 611 million acres (415.5, 131.0, 64.6) in 1982 to 578 million (405.3, 117.3, 55.1) in 2002 (a reduction of 5 per cent). There has been a growing surplus of agricultural land in Europe too, and more land is expected to become available in the Eastern European countries that have recently joined the European Union. A large potential for yield increases still exists at present in much of the developing world.

The amount of surplus agricultural land available in the future, including both surplus cropland and surplus pasture land, depends on:

- population;
- crop yield;
- diet (proportion of energy from meat and the kinds of meat consumed);
- efficiency in building animal mass from animal feed.

These factors are briefly discussed below, beginning with crop yield.

Yields of agricultural crops

Figure 4.46 shows the trend in the mean grain yield from 1961 to 2005, averaged over various regions. More detailed information can be found in Hafner (2003). A continuous increase in yield, with no sign of slowing down, is evident for all regions except Africa. In Africa, yields have hardly changed since 1961. Yields in Africa are low due to lack of inputs (fertilizers, improved crop varieties, irrigation) and lack of application of modern crop management techniques. If current yield trends outside Africa continue, and if African yields can be brought up to even the current world average, then, even after allowing for a human population peaking at 50 per cent above the current population (as in some scenarios), vast amounts of agricultural could be freed up for other purposes.

Further increases in agricultural yields depend on the extent to which yield can be increased through crop breeding, genetic engineering, fertilization and weed control, and the impact of future changes in climate and in atmospheric CO_2 concentration (which directly

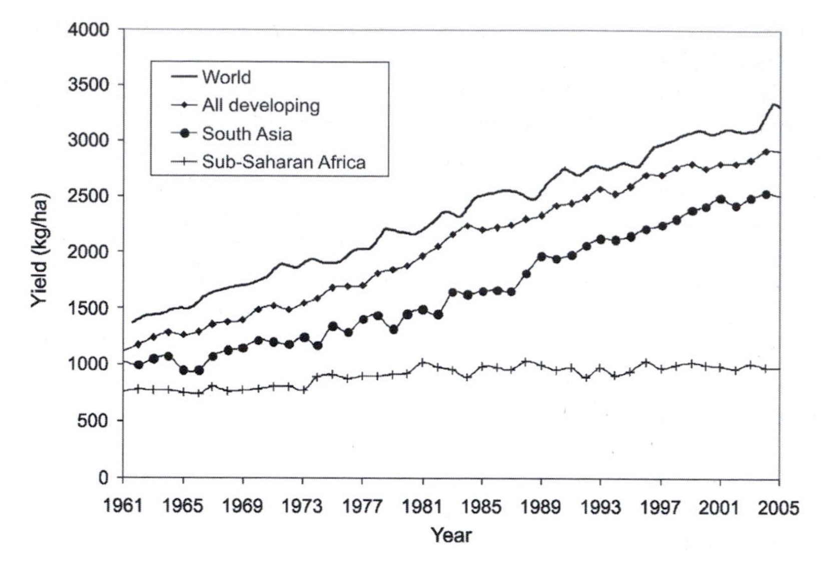

Source: Hazell and Wood (2008)

Figure 4.46 *Trend in grain yield averaged over sub-Saharan Africa, South Asia, all developing countries and the entire world*

stimulates photosynthesis and tends to reduce water loss by plants). Most scenarios generally assume that climatic change has no net effect on agricultural yields in the global mean during the next few decades (although large regional changes are expected), as the beneficial stimulatory effects of higher CO_2 combined with adaptation are believed to be able to offset the generally negative effects expected from changes in climate. However, recent field experiments indicate that the beneficial effects of higher CO_2 will be only about half that previously assumed (Schimel, 2006; Long et al, 2006), so that the net effect of higher CO_2 and climatic change will probably adversely affect agricultural yields in the global aggregate. By the time atmospheric CO_2 concentration doubles, the impact of climatic change could be in the order of a 10 per cent decrease in global food production. This is a small change compared to projected increases in yields, but could nevertheless be critical if future yield increases are not as large as hoped, or for high-population scenarios with high-meat diets (as explained later). Substantially worse effects are expected with larger CO_2 increases.

Table 4.43 provides information on the expected yield increase by 2050 compared to 1998 for various countries and regions. The greatest potential yield increases (by factors of up to six to seven) are expected in regions where yields are currently very low due to low inputs, particularly Africa. Mozambique, for example, has an average grain yield of 1t/ha/yr, compared to a potential yield of 7t/ha/yr with moderate use of fertilizers, pest control, selected seeds and large-scale harvesting practices, and 8–12t/ha/yr elsewhere (WWI, 2006).

Table 4.43 *Projected ratio of agricultural yields in 2050 compared to 1998 according to a range of scenarios, in the absence of increasing CO_2 and climatic change*

Region	Yield ratio
North America	1.6–3.2
Western Europe	0.9–1.9
Eastern Europe	2.1–4.1
Former Soviet Union	3.2–6.7
Japan	2.7–3.0
East Asia	2.3–3.2
South Asia	3.7–5.6
Oceania	2.4–4.6
sub-Saharan Africa	5.6–7.7
Caribbean and Latin America	2.8–4.5
World	**2.9–4.6**

Source: WWI (2006)

If the optimistic projections of increasing agricultural yields presented above are realized then, depending on population and diet, there could continue to be a steady increase in the amount of surplus agricultural land. However, in practice it may be difficult to stimulate demand for bioenergy just sufficient to absorb this amount of land without creating competition with land for food production. If bioenergy competes with land for food production, then food prices will increase and so too will the incidence of hunger and malnutrition. Sometimes it is assumed that biomass for energy will be grown on degraded land, but land reported to be degraded is often used for subsistence agriculture by the rural poor. The transition to cultivation of bioenergy crops on this land (perhaps by these same poor people) would have to be carefully managed. Unless existing forested areas are protected, economic forces arising from a growing demand for bioenergy could provoke (and already are provoking) clearing of existing forests (for example, tropical forests have been cleared and peatlands drained and burned in Malaysia in order to produce biodiesel for Europe).

Role of diet and animal feed intensity

Total grain (or grain equivalent) requirements per person depend on the total food energy intake, the proportion of food energy provided by animal products and the mix of animal products consumed. Food energy is lost when grain is fed to animals that in turn are fed to humans, so a meat-rich diet requires more grain than a vegetarian diet. The overall flow of phytomass in the food system, in energy terms, was presented in Volume 1 (Figure 7.12). As shown there, the world average efficiency in converting phytomass energy into processed vegetable food products is about 62 per cent, whereas the average efficiency in converting phytomass energy into meat products is about 4 per cent. However, the efficiency in converting grain to animal products varies enormously between different animal products, as well as with animal husbandry techniques (animal breeding; disease prevention and treatment; use of feed supplements such as minerals, enzymes and bacterial inoculates; and use of dedicated animal housing). Thus, the impact of alternative diets depends on the assumptions made concerning animal systems.

Table 4.44 *Kilograms of dry phytomass required per kg of wet animal product in various world regions and (in the global average) for various agricultural systems (defined in the main text)*

	Animal product (kg)				
	Bovine meat	**Pork**	**Lamb and goats**	**Poultry**	**Milk**
By region					
North America	26	6.2	58	3.1	1.0
Oceania	36	6.2	106	3.1	1.2
Japan	15	6.2	221	3.1	1.3
Western Europe	24	6.2	71	3.1	1.1
Eastern Europe	19	7.0	86	3.9	1.2
Former Soviet Union	21	7.4	69	3.9	1.5
sub-Saharan Africa	99	6.6	108	4.1	3.7
Caribbean and Latin America	62	6.6	148	4.2	2.6
Middle East and North Africa	28	7.5	62	4.1	1.7
East Asia	62	6.9	66	3.6	2.4
South Asia	72	6.6	64	4.1	1.9
By agricultural system					
Landless and mixed high	15	6.2	46	3.1	1.0
Mixed low	60	7.5	125	4.1	3.0
Pastoral high	37		58		1.4
Pastoral low	125		150		4.5
World	45	6.7	79	3.6	1.6

Source: Smeets et al (2006)

Table 4.44 compares estimates of the animal feed intensity (kg of animal feed per kg of animal product) for different animal products in different world regions and for different production systems (landless, mixed and pastoral) with low and high levels of agricultural technology. The 'landless' system involves animals confined to pens and fed entirely with feed crops and plant residues. The mixed systems for cattle, sheep and goats involve 50–90 per cent grasses and fodder (with the upper end of this range applying to the low-technology variant), while for pigs and poultry they involve feed crops and residues (as in the landless system). The pastoral system involves animals feeding on pastoral land but with 5 per cent feed crops in the high-technology variant and scavenging in the low-technology variant.

For bovine meat, feed intensity varies regionally by more than a factor of six and by more than a factor of eight between different systems. Feed intensity for pig meat is a factor of 2.5–15 less than that for bovine meat (depending on the region) and the feed intensity of poultry and eggs is about half that of pig meat. According to Table 4.44, high yields of animal products (comparable to landless systems) are possible in mixed systems (in which animals are not always confined to pens) with high levels of technology.

Table 4.45 compares the global mean feed requirements and the MJ of plant energy required per MJ and per kg of protein in animal products. Dairy products and poultry require about the same phytomass per MJ of animal product food energy, although poultry requires substantially less phytomass per kg of protein supplied in animal products. However, protein-rich plant foods require about a quarter the phytomass as poultry for supplying the same amount of protein (the typical adult daily protein requirement is 2.6gm per kg of non-fat body mass per day, or 2.2gm/day per kg of total body mass for those who are not overweight – an amount that can be easily supplied from plant foods without exceeding daily required energy intake). Thus, substantial reductions in the grain requirements can be obtained by shifting from bovine meat to pig meat and especially to poultry, eggs and milk, and even more so by shifting to a vegetarian diet.

Table 4.46 provides information on the amount of plant biomass energy required to supply a vegan, vegetarian, moderate and affluent diet for illustrative cases of high and low feed intensity. For the current world average diet and conversion efficiencies, the average phytomass energy requirement is 70MJ/person/day. A universal vegetarian diet would require about 26MJ/person/day, while a universal high-meat diet would require about 140MJ/person/day – almost a factor of six higher. The phytomass requirements for diets with meat will drop in the future if the average feed intensity decreases; the average feed intensity depends on both production efficiencies and the mix of different

Table 4.45 *Properties of different animal products, protein as a percentage of dry weight, kg of phytomass per kg of animal product, and MJ of phytomass per MJ and per kg of protein in animal products*

	Per kg of animal product			Protein as % of dry weight	kg phytomass per kg of wet animal product	MJ phytomass per MJ of animal product	MJ phytomass per kg of animal protein
	MJ	kg protein	kg water				
Bovine meat	10.7	0.186	0.49	36	45	80	4597
Pork	9.8	0.134	0.57	31	6.7	13	950
Lamb	10.6	0.22	0.47	42	79	142	6823
Poultry	5.7	0.238	0.71	82	3.6	12	287
Milk	2.7	0.034	0.87	26	1.6	11	894
Eggs	6.1	0.116	0.74	45	3.6	11	590
Plant food, mid protein	19	0.13	0.15	15	1	1	146
Plant food, high protein	19	0.26	0.15	31	1	1	73

Source: First three data columns from Pimentel (2004); protein as a percentage of dry weight, author's calculations; kg of phytomass per kg of animal product from Smeets et al (2006); and MJ of phytomass per MJ and per kg of protein in animal products, author's calculations assuming a phytomass energy value of 19MJ/kg

Table 4.46 *Impact of diet and animal feed intensity on required plant energy inputs per person per day*

	Current average	Future diet scenario			
		Vegan	Vegetarian	Moderate	Affluent
Energy intake (MJ/day)	13.10	13.10	13.10	13.10	13.10
Plant energy (MJ/day)	10.95	13.10	11.9	9.44	6.84
Land meat energy (MJ/day)	1.04	0.00	0.00	0.60	2.20
Seafood energy	0.12	0.00	0.00	0.06	0.06
Dairy energy (MJ/day)	0.99	0.00	1.20	3.00	4.00
Required land phytomass at current feed intensity (1.5 for plant food, 44.6 for meat, 7.7 for dairy products)					
Plants (MJ//person/day)	16.4	19.7	17.9	14.2	10.3
Meat (MJ/person/day)	46.3	0.0	0.0	26.8	98.1
Dairy (MJ/person/day)	7.6	0.0	9.2	23.1	30.8
Total	**70.3**	**19.7**	**27.1**	**64.0**	**139.2**
Required phytomass at high future feed intensity (1.5 for plant food, 35 for meat, 6 for dairy products)					
Plants (MJ//person/day)	16.4	20.1	17.9	14.2	10.3
Meat (MJ/person/day)	36.3	0.0	0.0	21.0	77.0
Dairy (MJ/person/day)	5.9	0.0	7.2	18.0	24.0
Total	**58.7**	**20.1**	**25.1**	**53.2**	**111.3**
Required phytomass at low future feed intensity (1.5 for plant food, 30 for meat, 5 for dairy products)					
Plants (MJ//person/day)	16.4	20.1	17.9	14.2	10.3
Meat (MJ/person/day)	31.1	0.0	0.0	18.0	66.0
Dairy (MJ/person/day)	5.0	0.0	6.0	15.0	20.0
Total	**52.5**	**20.1**	**23.9**	**47.2**	**96.3**

Note: Food energy intakes are based on food supplied at the wholesale level, and for today are converted to plant energy inputs using inverse conversion efficiencies for plant, meat and dairy foods of 1.5, 44.6 and 7.7.

Source: Current average per capita food intakes and inverse efficiencies were derived from data in Wirsenius (2000)

animal products consumed. Even for the higher future grain-to-meat conversion efficiencies assumed here, the vegetarian diet still requires less than one quarter the phytomass of an affluent, high-meat diet.

Scenarios for surplus agricultural land area

The information in these tables is used in Table 4.47 to illustrate the impact of alternative assumptions concerning population in 2050, diet, agriculture yield and feed intensity on the amount of agriculture land available for energy plantations. Required land areas are computed assuming the current world average cropland food phytomass yield of 43.8GJ/ha/yr and for future yields of 60 and 90GJ/ha/yr. The various combinations of population, diet, crop yield and feed intensity result in required agricultural land areas in 2050 ranging from 0.57 to 22.3Gha.[30] The present world areas of cropland and pastureland are 1452Mha and 3405Mha, respectively (4.85Gha total), so many of the combinations of population, diet, yields and conversion efficiencies used in Table 4.47 are clearly impossible. For the moderate population scenario, moderate diet with low feed intensity (achieved in part through a partial shift from bovine to pig meat and poultry) and high agriculture yield, the total required cropland plus pastureland area is by coincidence equal to the total cropland area alone today (1452Mha). For a moderate diet and medium food yields, the available plantation area is 1600–2960Mha, depending on population and feed intensity. Assuming an average biomass energy yield of 200GJ/ha/yr (which is quite low compared to the summary data given in Table 4.4), the available plantation biomass energy in 2050 is 25–305EJ/yr for the moderate diet and middle crop yield scenarios. Note also, however, that for some combinations of population and diet, the required agricultural land area increases, even assuming the high crop yield and low feed intensity scenario. In this case, land for bioenergy would be competing with land for agriculture, and the two together would represent significant additional pressure on the world's remaining forests.

Although an increase in agricultural yields frees up land that can be used for other purposes, such as the

Table 4.47 *Impact of population (in 2050), diet and average productivity of agricultural land on the amount of land available for energy plantations*

	Present day	Diet																	
		Vegetarian						Moderate						Affluent					
		Population (billions)						Population (billions)						Population (billions)					
		7.8		8.8		9.9		7.8		8.8		9.9		7.8		8.8		9.9	
		Feed efficiency		Feed efficiency		Feed efficiency		Feed efficiency		Feed efficiency		Feed efficiency		Feed efficiency		Feed efficiency		Feed efficiency	
		Low	High	Low	High	Low	High	Low	High	Low	High	Low	High	Low	High	Low	High	Low	High
Crop phytomass (EJ/yr)	33.2	50.8	50.8	57.3	57.3	64.5	64.5	40.3	40.3	45.5	45.5	51.2	51.2	29.2	29.2	33.0	33.0	37.1	37.1
Animal feed phytomass (EJ/yr)	109	20.5	17.1	23.1	19.3	26.0	21.7	111	94.0	125	106	141	119	288	245	324	276	365	311
Required cropland area for	1452	1148	1148	1295	1295	1452	1452	1452	1452	1452	1452	1452	1452	1452	1452	1452	1452	1452	1452
human and animal	1452	847	847	956	956	1075	1075	1452	1452	1452	1452	1452	1452	1452	1452	1452	1452	1452	1452
food (Mha)	1452	565	565	637	637	717	717	1452	1452	1452	1452	1452	1452	1452	1452	1452	1452	1452	1452
Required pastureland	3405	0	0	0	0	0	0	3974	2758	5297	3924	6752	5207	15,887	12,845	18,737	15,304	21,871	18,010
area (Mha)	809	0	0	0	0	0	0	1209	355	2137	1173	3158	2074	9570	7435	11,570	9161	13,770	11,060
	0	0	0	0	0	0	0	0	0	0	0	95	0	4370	2946	5703	4097	7170	5363
Available cropland plus	1	3710	3710	3563	3563	3405	3405	0	648	0	0	0	0	0	0	0	0	0	0
pastureland area (Mha)	2596	4011	4011	3902	3902	3783	3783	2196	3050	1268	2232	247	1331	0	0	0	0	0	0
	3405	4293	4293	4221	4221	4141	4141	3405	3405	3405	3405	3310	3405	0	459	0	0	0	0
Plantation bioenergy	0	401	401	372	372	341	341	0	65	0	0	0	0	0	0	0	0	0	0
supply (EJ/year)	260	462	462	440	440	416	416	220	305	127	223	25	133	0	0	0	0	0	0
	341	518	518	504	504	488	488	341	341	341	341	331	341	0	46	0	0	0	0

Note: The three rows each for required cropland, required pastureland, available cropland plus pastureland and plantation bioenergy supply correspond to the present world average cropland energy yield of 44.3GJ/ha/yr and future yields of 60GJ/ha/yr and 90GJ/ha/yr, in that order. The cropland area for production of food for direct human consumption and to feed to animals is not allowed to exceed the current cropland area of 1452Mha. A value of 1452Mha in the above table invariably indicates a requirement for a greater cropland area.

Calculation procedure:

1 Required cropland phytomass is computed as population × per capita daily plant phytomass consumption from Table 4.46 (16.43MJ/person/day at present) × 365 days/year.

2 Required animal feed phytomass is computed as population × per capita daily plant phytomass consumption from Table 4.46 for meat plus dairy consumption (46.28 + 7.62MJ/person/day at present) × 365 days/year.

3 Cropland area required to grow food for humans is given by the required cropland phytomass divided by cropland food energy yields (after accounting for losses) of 44.27GJ/ha/yr (computed from Wirsenius, 2000), 60GJ/ha/yr and 90GJ/ha/yr for present-day, high future and low future agricultural yield scenarios.

4 Phytomass available for animals as co-products from the production of crops for humans and as food-processing waste and wasted food is given by the crop phytomass (33.2EJ/yr today) times a factor of 0.547, deduced from Figure 3.3 of Wirsenius (2000).

5 The difference between the current cropland area of 1452Mha and that calculated in step 3 is available to provide phytomass to be fed to animals. The same three phytomass yields that were used in step 3 to compute the production of food for humans are used to compute the production of food for animals, times a factor of 1.384 (derived from Wirsenius, 2000) to account for crop residues that can also be fed to animals but not to humans.

6 The total animal feed phytomass required (109EJ/yr at present) minus the phytomass computed from steps 4 and 5 gives the phytomass for animals that must be supplied from pastureland. This divided by pastureland animal feed yields of 14.04GJ/ha/yr (computed from Wirsenius, 2000), 20GJ/ha/yr and 30GJ/ha/yr for present-day, high future and low future scenarios, respectively, gives the required pastureland area.

7 Any excess cropland available after satisfying the requirements of steps 3 and 5 is assumed to provide bioenergy crops at a rate of 200GJ/ha/yr, while any excess pastureland is assumed to provide bioenergy crops at a rate of 100GJ/ha/yr. Present yields are already substantially greater than this in many regions.

production of biomass energy crops, it often does so at the cost of heavy energy, fertilizer and pesticide inputs, which have adverse environmental impacts (see Volume 1, Chapter 7, section 7.2, with regard to the environmental impacts of fertilizers). Low-input farming, which is growing in popularity, will require additional agricultural land while providing environmental and health benefits. Similarly, 'landless' animal food systems raise a number of environmental, food quality and ethical issues related to the treatment of animals. Due to the concentration of animals in large feedlot operations in such systems, the recycling of nutrients in animal wastes back to the soil that produced the animal feed is not possible, and indeed, such wastes represent a disposal problem. Thus, to the extent that large-scale production of bioenergy crops requires greater intensification of agricultural and animal production, it will conflict with other environmental goals. The combination of moderate improvements in agricultural yields and in feed intensity combined with a vegetarian or moderate diet would have less environmental impact than the combination of high agricultural yields, low feed intensity and a high-meat diet.

4.9.3 Potential biomass yield increases

For energy plantations, a critical parameter is the energy crop yield (tonnes of biomass produced per hectare per year (t/ha/yr)). This is a major factor in determining the cost of biomass feedstock, which in turn is a major factor in determining the cost of biomass-derived energy. Yields will be higher on better land, but better land is more costly or might already be used for production of food or livestock. In developed countries, the balance between land cost and yield tends to favour use of better quality land (Perlack et al, 1997, cited in Kartha and Leach, 2001). In developing countries, population pressure for food and fibre will probably force energy crops onto marginal land. These lands require significant preparation and have low yields.

It is widely believed that genetic improvements in energy crops and use of fertilizers and irrigation could greatly increase yields. Woody bioenergy crops are largely undomesticated and have not undergone the centuries of improvement that our major food crops have experienced (Tuskan, 1998). Estimates of the potential improvement by Ravindranath and Hall (1996), which must be regarded as somewhat speculative, are summarized in Table 4.48. In semi-arid regions, it might be possible to increase the baseline yield of 2–5t/ha/yr to 6–12t/ha/yr with genetic improvement and fertilizer, or to 20–30t/ha/yr if irrigation is also used. This is a factor of six to ten improvement. In sub-humid regions, the potential improvement is a factor of four (from 5–10t/ha/yr to 20–35t/ha/yr). For switchgrass in the US, Greene (2004) projects a yield increase of 40–100 per cent by 2025 and a doubling to tripling of yield by 2050 (Table 4.49). A major research effort is needed to determine how to achieve these yield increases in practice on a sustainable basis and to determine how to minimize costs. Yield increases in bioenergy crops are expected to be easier than for food crops because there will be fewer constraints (such as maintaining a given taste).

4.9.4 Estimates of bioenergy potential from bioenergy plantations

Berndes et al (2003) examined 17 projections of potential future global biomass energy supply from plantations, whether on surplus cropland or pastureland, or on degraded lands. For 2050, the projected potential biomass energy supply ranges from 47 to 450EJ/yr, of which

Table 4.48 *Potential productivity (dry tonnes per hectare per year) of tropical tree plantations on degraded forest and non-agricultural lands*

Genetic improvement	Fertilizer	Irrigation	Semi-arid (500–1000mm)	Sub-humid (1000–2000mm)
No	no	no	2–5	5–10
Yes	no	no	4–10	10–22
Yes	yes	no	6–12	12–30
Yes	no	yes	8–18	11–25
Yes	yes	yes	20–30	20–35

Source: Ravindranath and Hall (1996)

Table 4.49 *Current and projected future yields of switchgrass in different regions of the US, assuming no change in climate*

Region	Yield in 2000 (dry tonne/ ha/yr)	Projected yield (dry tonne/ha/yr)	
		2025	2050
Northern Plains	7.8	10.2	13.2
Southern Plains	9.7	19.8	31.9
Great Lakes and northeast	10.8	14.3	18.4
Corn belt	13.4	21.9	31.9
Appalachia	13.1	26.9	43.3

Source: Greene (2004), cited in Larson (2006)

developed countries produce 35–150EJ/yr and developing countries 15–300EJ/yr. Energy crops supply 50–200EJ/yr, with most of the balance (0–250EJ/yr) from various kinds of residues. Table 4.50 gives the assumed biomass plantation area, yield and total global energy supply from plantations for six of these studies and three subsequent studies. With the exception of one study (and the upper limits of two other studies), the assumed plantations areas (185–870Mha) are below the limits deduced here for a moderate diet and high agricultural yields, with total global energy supply from plantations ranging from 45–410 EJ/yr. The plantation areas are also below the global area of degraded land as estimated by Lemus and Lal (2005), namely, 1757Mha (of which 563Mha are deforested land, 662Mha overgrazed land and 532Mha marginal cropland), although financial, policy and social factors could greatly reduce what can be used in practice. Most studies assume increases in the yield of bioenergy crops by 50–100 per cent by 2050, with the average yields being as high as 12–14t/ha/yr (216–252GJ/ha/yr at 18GJ/t). The study by Hoogwijk et al (2005) deduced a feasible energy supply of 130–410EJ/yr from woody energy crops on excess agricultural land in 2050. Energy plantations could be planted with grasses rather than trees, with large yield increases also expected – reaching 17 dry t/ha/yr by 2025 and 28 dry t/ha/yr by 2050 for switchgrass without using genetically modified plants (WWI, 2006). By comparison, present switchgrass yields are in the 8–13t/ha/yr range.

The last three studies listed in Table 4.50 make use of the Dutch IMAGE model, which was also employed in an assessment of the potential to sequester carbon through reforestation (discussed in subsection 4.8.1). This model contains a global data set on a 0.5° × 0.5° latitude–longitude grid of soil and climatic conditions, and is used to determine the potential yields of agricultural and bioenergy crops for various land use scenarios. None of the studies listed in Table 4.50 take into account future changes in climate.

The required rates of establishment of new plantations (given in Table 4.50) are several times the current rate of 4.5Mha/yr, of which only about 3Mha/yr are successful. Reforestation will often meet strong local resistance unless it is tied to local needs. Nilsson and Schopfhauser (1995) concluded that only 41, 27 and 29Mha out of 535, 740 and 162Mha of suitable land in Latin America, Africa and Asia, respectively, could be available to plantations under current conditions. The total (97Mha) is dramatically less than the plantation area assumed to be available in any of the studies listed in Table 4.50.

4.9.5 Potential biomass energy from agricultural and forestry residues and organic wastes

The availability of agricultural residues depends on the ratio of non-crop to crop biomass production for agricultural plants; one way of improving agricultural yields is to breed plants that channel more of their net primary production (NPP) into food parts, thereby reducing the amount of residue available for a given food production. Not all residues can be easily collected, and some portion is usually left on the fields. Most studies assume that only about 25 per cent of the available agricultural and logging residues can be used for energy. Hoogwijk et al (2003) conclude that 5–27EJ/yr could be supplied from primary agricultural residues and another 5EJ/yr from secondary residues (mainly bagasse and rice husks). For forestry residues, they conclude that 10–16EJ/yr could be provided. As for animal manure and municipal organic waste, Hoogwijk et al (2003) conclude that 9–25EJ/yr and 1–3EJ/yr, respectively, could become available. Altogether, 30–76EJ/yr could be available from residues and organic waste.

4.9.6 Energy from biomaterials

Minimization of biomass resource demands requires maximum cascading of biomass. In a biomass energy and material products system, some biomass would be

Table 4.50 *Global plantation bioenergy potential according to various scenarios*

Source	Year	Plantation area (Mha)	Rate of plantation establishment (Mha/yr)	Global average yield (t/ha/yr)	Biomass energy supply (EJ/yr)	Comment
Swisher and Wilson (1993)	2030	870	21.8	3	50	10% of global crop, forest and woodland area assumed
Johansson et al (1993)	2050	429	10.7	14.9	120	Degraded land in developing countries, excess cropland in industrialized countries.
Battjes (1994)	2050 2050	185, low 395, high	4.6 9.9	12.75 12.75	45 96	Yield increases from 4.6t/ha/yr in 2025. Residues not included in energy supply
IIASA-WEC (Nakicenovic et al, 1998)	2050 2050	390 610	9.8 15.3	12.6 8.4	93 97	Based on projected bioenergy demand with post-scenario feasibility check. Includes 75% of mill residues, 50% of forest residues, together assumed to be equal to 65% of roundwood production
Sørensen (1999)	2050	752	18.8	6	86	10% of cropland in regions with surpluses, 50% of rangeland everywhere, 30% of forests except for rainforests and other forests that are assumed to be protected
Fischer and Schrattenholzer (2001)	2050 2050	2165 2388	54.1 59.7	3.8 4.8	156 218	Yields estimated for 1990, then assumed to grow at same rate as assumed by others for agricultural yields (leading to a factor of 1.6–2.1 greater yield by 2050)
Hoogwijk et al (2005)	2050 2100	600–1300 1200–2600			130–410 240–850	Considers only SRWCs on abandoned agricultural lands. Impact of potential loss of soil C on productivity not considered
Smeets et al (2006)	2050 2050	729 358	18.2 89.6	16 20	215 1272	Considers woody bioenergy crops on surplus agricultural land only
Hoogwijk et al (2009)	2050				130–270EJ/yr at ≤\$2GJ, 180–440EJ/yr at ≤\$4GJ, 300–675EJ/yr total potential	Considers only SRWCs on abandoned agricultural lands and rest lands for various land use scenarios. Focus on costs, accounting for increase in labour costs with development

used for products such as wood and various paper products. When the wood products reach the end of the useful lives, or after paper products have been recycled to the extent that can be justified, they would then be converted into synthetic products that are currently made from petroleum and/or natural gas (as discussed in section 4.7). After these have reached the end of their useful lives, they would be combusted in order to produce useful heat and electricity. Ash residues would be returned to the land where the biomass was originally grown so as to recycle nutrients. Hoogwijk et al (2003) estimate that by 2050, production of biomaterials not cascaded from prior uses of biomass would require a land area of 230–325Mha and provide 32EJ/yr at the end of their life. Production of these biomaterials would generate organic wastes that could provide an additional 32EJ/yr, which is in addition to the energy extracted from the used biomaterials through incineration to produce heat and power.

4.10 Cost of biomass and biomass products

The cost of solid or liquid biofuels or of electricity produced from biomass depends on the cost of the biomass feedstock, the capital cost of the processing equipment and financing conditions, the cost of non-biomass energy inputs (which can be important for liquid biofuels) and the conversion efficiency.

4.10.1 Cost of solid biofuels

Table 4.51 summarizes various estimates of the present or projected future costs of various biomass feedstocks. Coppice willow biomass grown in New York State can be delivered to nearby coal-fired powerplants (for co-firing with coal) at a delivered cost of about $3/GJ according to Tharakan et al (2005), given present-day yields of 9.8odt/ha/yr on the first three-year rotation and 14.8odt/ha/yr on subsequent rotations. With increased yields (11.8 and 17.8odt/ha/yr on the first and subsequent rotations, respectively), the price is expected to drop to $2.6/GJ. Yemshanov and McKenney (2008) estimate the cost of supplying fast-growing poplar to small biomass powerplants in rural communities in Canada, using existing agricultural land. Altogether, they identified 241 settlements in Canada with sufficient agricultural land within a short enough distance to supply the minimum size powerplant possible. The computed cost of biomass ranges from $4/GJ to $10/GJ, which is much higher than the estimated cost of coppice willow in New York or other estimates for Canada. However, the assumed yield is generally only 5–6odt/ha/yr (depending on the region).

Smeets et al (2009) estimate the cost of producing *Miscanthus* and switchgrass in a variety of different European countries both in 2004 and as projected for 2030. Their cost estimates take into account regional yields, labour and machinery costs, land rent and fuel costs today and as projected to 2030. The present costs of production and storage of *Miscanthus* and switchgrass amount to €2.3–4.8/GJ and €1.6–4.4/GJ, respectively. Transport a distance of 100km by truck adds €0.6–1.1/GJ, bringing the total cost to €2.2–5.9/GJ today. The net effect of projected changes in yield and in land, labour and fuel costs is a 15–27 per cent reduction in the cost of *Miscanthus* by 2030 but very little change in the cost of switchgrass.[31] These results are summarized in Table 4.51.

The cost per unit of biomass energy is strongly influenced by the biomass yield (t/ha/yr), because fixed costs (crop establishment, land, maintenance and overhead) can be amortized over more biomass if yields are higher. Only harvesting, transport and post-harvest processing costs per unit of biomass are independent of yield. Higher yield also means that the average distance that biomass must be transported to supply a bioenergy facility of a given size will be smaller, or that a larger facility (usually with smaller unit costs) can be supplied with biomass from a given area. Thus, there will be a preference to locate biomass production on better quality land, that is, in competition with food production or with

Table 4.51 *Summary of published estimates of present or projected possible future costs of solid biomass suitable for electricity generation or for heating*

Commodity	Cost	Reference
	Tree crops	
Woody crops in NW Europe	€3–6/GJ	Faaij (2006)
Willow in Central Europe	€1.6–8GJ	van Dam et al (2007)
Coppice willow in New York state	$2.6–3.0/GJ, delivered	Tharakan et al (2005)
Poplar in Canada	$4–10/GJ, delivered	Yemshanov and McKenney (2008)
Willow biomass	$3/GJ	Volk et al (2006)
	Grasses	
Miscanthus in Illinois today	$2.3/GJ average, farmgate	Khanna et al (2008)
Switchgrass in Illinois today	$3.2/GJ average, farmgate	Khanna et al (2008)
Miscanthus in Europe today	€2.9–5.9/GJ, delivered 100km	Smeets et al (2009)
Switchgrass in Europe today	€2.2–5.5/GJ, delivered 100km	Smeets et al (2009)
Miscanthus in Europe in 2030	15–27% lower than today	Smeets et al (2009)
Switchgrass in Europe in 2030	3–7% higher than today	Smeets et al (2009)

Note: By comparison, coal typically costs $1–2/GJ and natural gas $4–6/GJ. Transportation costs are not included for Khanna et al (2008).

commercial forestry. The biomass-intensive scenarios considered earlier assume that biomass is produced on degraded land (perhaps serving multiple purposes, such as erosion and flood control as well as provision of energy). It is therefore important to check that overly optimistic assumptions about yields and costs are not being made.

4.10.2 Cost of liquid biofuels

Figure 4.47 compares the cost of gasoline and bioethanol from various sources, and the cost of diesel fuel and biodiesel from various sources all in 2006. Bioethanol from food crops and sugarcane was just barely competitive with gasoline in 2006, although not without subsidies in most regions. Ethanol from lignocellulose – which represents a larger potential bioenergy resource – was and still is much more expensive than gasoline. Table 4.52 presents information on estimated short-term and long-term conversion efficiencies and costs for the production of various biomass-based transportation fuels (distribution, storage and dispensing would represent additional costs). One barrel of crude oil contains 5.75EJ, so oil at \$70/barrel corresponds to a cost of \$12.2/GJ, while 1 litre of gasoline contains about 0.035GJ, so gasoline at \$1.0/litre (\$3.79/gallon) is equivalent to \$18/GJ. Methanol and ethanol from woody biomass both appear to have good prospects of being competitive with gasoline. Ethanol from sugarcane in Brazil is already generally competitive with gasoline and can be confidently projected to decrease in cost further. Credits from the sale of the protein byproduct as animal feed are important to the final cost of ethanol from corn. A large increase in the supply of animal feed, through vastly increased production of ethanol from corn, would reduce the price of animal feed, thereby worsening the economics of ethanol from corn.

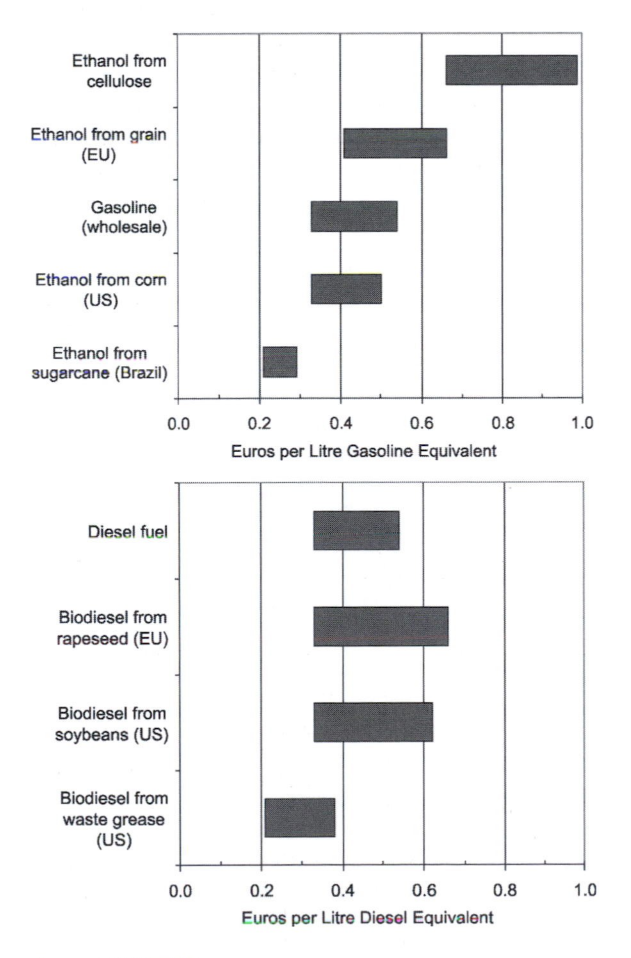

Source: WWI (2006)

Figure 4.47 *Costs of various liquid fuels in 2006. Oil at the time cost about \$60/barrel*

4.10.3 Cost of biomass powerplants, biogas facilities and biomass-derived electricity

Table 4.53 provides recent capital cost, efficiency and operating cost data for direct combustion steam turbines of various sizes as provided by the US Western Governors Association (WGA, 2006). Also given is the resulting cost of electricity assuming a fuel cost of \$0/GJ (applicable when the fuel consists of residues at a pulp and paper mill or sugar mill that need to be disposed of) or \$2/GJ, 7 per cent interest and 20-year financing.[32] Current capital costs range from about \$4200/kW (at a 3.4MW scale) to about \$2000/kW (at a 60MW scale). Electricity costs when no credit is given for heat production range from 4.3 cents/kWh (60MW scale, fuel at \$0/GJ) to 6.9 cents/kWh (3.4MW scale, fuel at \$2/GJ). Also shown in Table 4.53 is the credit for heat production, per kWh of electricity generated, assuming an overall efficiency of 0.75 (from which the heat to electricity production ratio R can be deduced) and that the alternative fuel for heating costs C_h = \$4/GJ and can be used with an efficiency η_h of 0.8

Table 4.52 *Short-term and long-term efficiencies (based on biomass energy input only) in converting biomass to various fuels, and production cost*

Fuel	Efficiency (fuel or electrical energy/ biomass energy, HHV) (%)		Production cost (€/GJ)	
	Short term	Long term	Short term	Long term
Hydrogen	60 (fuel)	55 (fuel) + 6 (electricity)	9–12	4–8
Methanol	55 (fuel)	48 (fuel) + 12 (electricity)	10–15	6–8
FT liquids	45 (fuel)	45 (fuel) + 10 (electricity)	12–17	7–9
Ethanol from wood	46 (fuel) + 4 (electricity)	53 (fuel) + 8 (electricity)	12–17	4–7
Ethanol from wood[a]	35	48	22	11[b]
Ethanol from sugarcane	43 (fuel)	43 (fuel)	16[c]	9.4–11.2[d]
Biodiesel from oilseeds	88 (fuel)	88 (fuel)	25–40	20–30

Note: Gasoline at €1/litre is equivalent to €30.8/GJ. [b] For biomass feedstock at €3/GJ.
Source: Faaij (2006) except [a] Hamelinck et al (2005a), [c] current production cost in Brazil according to van den Wall Bake et al (2009), and [d] cost in Brazil in 2020 as projected by van den Wall Bake et al (2009).

Table 4.53 *Costs and efficiency of direct combustion steam turbines for generating electricity from biomass, and the resulting cost of electricity assuming 7 per cent financing over 20 years*

Size (MW)	Boiler technology	Capacity factor	Efficiency	Capital cost ($/kW)	O&M costs		Cost of electricity (cents/kWh)		Heat credit (cents/kWh)
					Fixed ($/kW/yr)	Variable (c/kWh)	Fuel at $0/GJ	Fuel at $2/GJ	
3.4	Pile	0.9	0.16	4235	274	0	8.5	12.9	6.4
10	Stoker	0.9	0.13	2875	270	0	6.9	12.5	8.7
15	FB	0.9	0.13	3116	254	0	7.0	12.5	8.7
50	Stoker	0.8	0.24	2191	81	0.95	5.1	8.1	3.9
60	Stoker	0.8	0.28	1946	67	0.75	4.3	6.9	3.1

Note: FB = fluidized bed. The heat credit is per kWh of electricity produced, computed as explained in the main text.
Source: Data from WGA (2006)

(so the heat credit is computed as $0.0036RC_h/\eta_h$). The calculated credit can be directly subtracted from the cost of electricity given in Table 4.53, and although the computation is quite crude (and excludes the avoided cost of a boiler for heating), it can nevertheless be seen that the heat credit is substantial and serves to make electricity generation from biomass attractive (if all of the waste heat can be used) in spite of the high capital cost and low electricity generation efficiency.

Figure 4.48 gives the investment cost of wood-fired powerplants in Finland between 1994 and 1997 as a function of capacity. Unit investment cost at a 150MW capacity was about half that at a 2MW capacity. The investment cost at a capacity of 50–60MW (€1200–1500/kW) is roughly comparable with the cost given in Table 4.53 ($2000/kW).

As noted in subsection 4.4.3, BIGCC promises the highest efficiency in generating electricity from biomass (45–50 per cent), but has not yet been commercialized as challenges related to cleaning the biogas before it enters the gas turbine remain (Sims et al, 2006). Capital costs in 2030 are projected to fall between US$2000/kW and US$1100/kW (REN21, 2005), but such distant projections are highly speculative.

Table 4.54 compares estimates of the future performance and costs (including O&M) for various

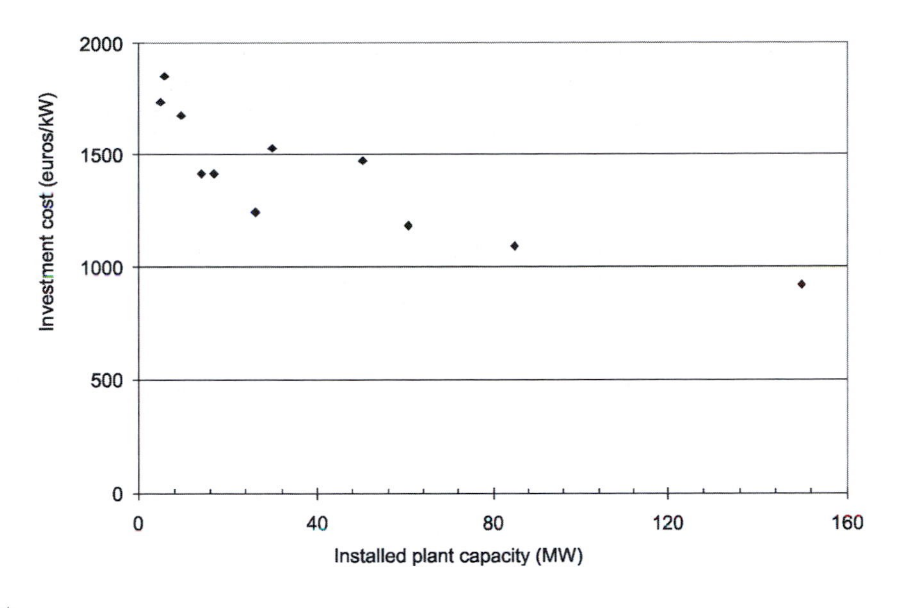

Source: IEA (2003)

Figure 4.48 *Investment cost of wood-fired powerplants in Finland between 1994 and 1997 as a function of capacity*

technologies for producing heat and/or electricity from biomass and natural gas. The capital cost of biomass steam turbine or gasification combined cycle cogeneration systems at the 36–60MW_e scale is about twice that of natural gas combined cycle at the 50MW_e scale, but the feedstock cost is lower (about \$2–4/GJ for biomass, versus \$6–10/GJ for natural gas, and possibly more in the long term in some regions). Biomass heating plants at the 50MW_{th} scale are over three times as expensive as natural gas plants at the same scale.

Amigun and von Blottnitz (2007) gathered data on the cost of 21 biogas plants in Africa ranging in volumetric capacity from 4 to 100m^3. They found that the unit capital cost *increases* with the capacity of the biogas plant (unit cost varies with capacity to the power 1.2). Thus, smaller-scale, decentralized facilities are more economical. The average distance that the fuel must be transported will be shorter with smaller plants, further improving the economics for these plants.

Table 4.55 compares the cost of electricity from various small-scale biomass-based systems and from diesel generators or the power grid in India, and for biomass and kerosene stoves for cooking. All costs are based on actual projects in India. Some forms of small-scale biomass electricity are competitive with grid electricity, and all are competitive with diesel generators. Efficient wood stoves are about five times more expensive than traditional stoves but are 40 per cent less expensive on a lifecycle basis using a 10 per cent discount rate. Community biogas stove systems are 45 per cent more expensive than traditional stoves but are more convenient (fuelwood does not need to be collected) and cleaner burning.

Finally, Dwivedi and Alavalapati (2009) assess the cost of electricity produced by gasification of biomass grown on community-managed, degraded forest land in Madhya Pradesh, India. The energy crops considered are *Eucalyptus* and two species of *Acacia*, which would be grown over a period of three years and then harvested. There have been many pilot projects involving biomass-gasification to generate electricity in India, so there is a good data set based on real operating experience. The estimated costs of electricity leaving the powerplant are \$0.33/kWh, \$0.14/kWh and \$0.10/kWh for 5kW, 40kW and 100kW gasifiers, respectively. By comparison, the cost of electricity supplied to the grid is \$0.08/kWh.

Table 4.54 *Comparison of efficiency, investment cost and O&M for different bioenergy conversion technologies and for technologies using natural gas, and the computed cost of electricity or heat*

Technology	Capacity (MW_{th}/MW_e)	HHV efficiency			Investment cost	O&M		Energy cost (cents/kWh)	Average cost (cents) of $1kWh_e$ and $1kWh_{th}$	
		Heat	Elec	Overall		Fixed	Variable			
Cogeneration					($/kW$_e$)	($/kW$_e$/yr)	($/MWh$_e$)		$6/GJ	$9/GJ
Biomass ST	72/36	0.48	0.24	0.72	1600	43	5.1	5.22	5.46	5.46
BIGCC	60/60	0.34	0.34	0.68	1600	35	7.6	5.41	5.41	5.41
NGCC	50/50	0.40	0.40	0.80	810	16	1.5	4.37	4.37	5.72
Power (condensing) plants					($/kW$_e$)	($/kW$_e$/yr)	($/MWh$_e$)			
Biomass ST	0/200		0.32		1400	16	3.8	6.18	4.26	4.26
BIGCC	0/100		0.38		1300	38	3.8	5.88	4.11	4.11
NGCC	0/300		0.47		700	11	1.3	5.99	4.34	6.12
Heating plants					($/kW$_{th}$)	($/kW$_{th}$/yr)	($/MWh$_{th}$)			
Biomass	50/0	0.72			320	6.3	2.3	2.33	4.11	4.11
Natural gas	50/0	0.85			95	0	0	2.69	4.34	6.12

Note: The energy cost is computed without distinguishing between heat and electricity in the case of cogeneration, and assumes biomass fuel at a cost of $3/GJ and natural gas at $6/GJ. Average costs take into account additional electricity needed in the case of Biomass ST so as to give equal electricity and heat production, the supply of heat from heating plants to match the electricity production from condensing powerplants, and the supply of electricity from condensing powerplants to match the heat production from heating plants. Average costs are shown for biomass at $3/GJ and for natural gas at $6/GJ and $9/GJ. All costs assume 6 per cent/year interest, a 20-year equipment lifespan, and a 65 per cent capacity factor. ST = steam turbine.

Source: Efficiencies, investment cost, and O&M costs are from Gustavsson and Madlener (2003)

Table 4.55 *Comparison of various biomass and non-biomass options for providing electricity and for cooking in India*

Generation of electricity

Fuel	Technology	Capacity	Capacity factor	Cost of fuel (Rs/kg)	Cost of electricity (Rs/kWh)
Biomass + diesel	Biomass gasifier	20kW	0.46	1.0	5.05
Biomass only	Biomass gasifier	20kW	0.46	1.0	4.17
Biomass	Combustion, steam turbine	10MW	0.68	1.0	2.15
Dung, diesel	Biogas	120kW	0.46	Free	3.90
Bagasse	Cogeneration	10MW	0.50	0.66	2.31
Diesel	Diesel engine	20kW	0.46	19[a]	9.24
Coal	Grid electricity	500MW	0.75	1.42	3.25

Cooking

Fuel	Technology	Stove efficiency (%)	Cost of fuel (Rs/kg)	Cost of heat (Rs/GJ)
Firewood	Traditional stove	15	0.6	271
Firewood	Efficient stove	35	0.8	164
Dung	Community biogas plant + stove	55	1.0	394
Kerosene	Kerosene stove	50	7.6[a]	460

Note: All costs are based on real projects and a 10 per cent discount rate. Rs1 = $0.023 in mid-2008. [a] Cost per litre.
Source: Ravindranath et al (2006)

4.10.4 Learning curves for biomass energy

Progress ratios (the fraction by which the cost of a technology is multiplied with each doubling in cumulative production) based on learning through experience have already been discussed with regard to solar energy (Chapter 2, subsection 2.2.12) and wind energy (Chapter 3, subsection 3.13.6). There are fewer data available for developing progress ratios for biomass energy systems, due in part to the wide range of capacities, fuels and plant layouts and due to regional differences. Junginger et al (2006) have computed progress ratios for a few cases: 0.77 for the investment cost of Swedish biomass-fuelled cogeneration plants installed between 1991 and 2002; 0.87 for the cost of woodchips from forest residues in Sweden between 1975 and 2003; 0.92 for the cost of electricity produced from biomass-fuelled cogeneration plants in Sweden between 1990 and 2002; 0.88 for the investment cost of biogas plants in Denmark between 1985 and 1998; and 0.77 for the cost of biogas in Denmark between 1984 and 1997.

4.10.5 Global supply cost curves

Hoogwijk et al (2009) use the IMAGE-2 model to assess the cost of producing increasing amounts of liquid fuels and electricity from biomass. As previously noted, this model contains a global data set of soil and climatic conditions on a 0.5° × 0.5° latitude–longitude grid, and so can be used to estimate the yield of bioenergy crops in specific locations. Hoogwijk et al (2009) estimate the cost of producing bioenergy crops based on labour and capital costs in 17 different world regions and based on geographically varying land costs. In poor countries, labour costs are assumed to increase due to economic development, leading to some substitution of capital for labour in the production process. A progress ratio of 0.9 is applied to some technological factors, leading to a decrease in cost with increasing experience. Several distinct steps of the production chain are considered: land preparation, planting, crop management during growth, harvesting and transport. Only short-rotation commercial woody crops on large-scale plantations on abandoned agricultural land and non-productive land that is suitable for bioenergy crops (excluding bio-reserves and forests) are considered.

The costs of two biomass end products were computed: diesel fuel produced through FT synthesis and electricity from BIGCC. The conversion efficiency from biomass to electricity is assumed to increase from 40 per cent today (in pre-commercial projects) to 55 per cent by 2050, while the conversion efficiency from biomass to diesel is assumed to increase from 40 per cent today (in pre-commercial projects) to 56 per cent by 2050. Costs with today's technology but applied at a large scale are assumed to be $1370/kW for BIGCC and $160/kW for production of diesel, and are assumed to decrease over time.

Figure 4.49 shows the resulting global supply cost curves for biodiesel and for electricity. Shown is the cost of successive increments of secondary energy supply as the total supply increases. Results are shown for four coupled scenarios of land use and socio-economic development, as developed by the IPCC in its *Special Report on Emission Scenarios* (Nakicenovic and Swart, 2000). Between about 100 and 240EJ/yr of biodiesel is estimated to be potentially available in 2050 at a cost of up to $15/GJ (by comparison, gasoline at $1/litre or $4/gallon corresponds to a cost of about $32/GJ). In addition, between about 18,000TWh/yr and 50,000TWh/yr of electricity could be available from biomass at a cost of 5 cents/kWh or less (excluding transmission, distribution and administration costs). By comparison, total world oil primary energy demand

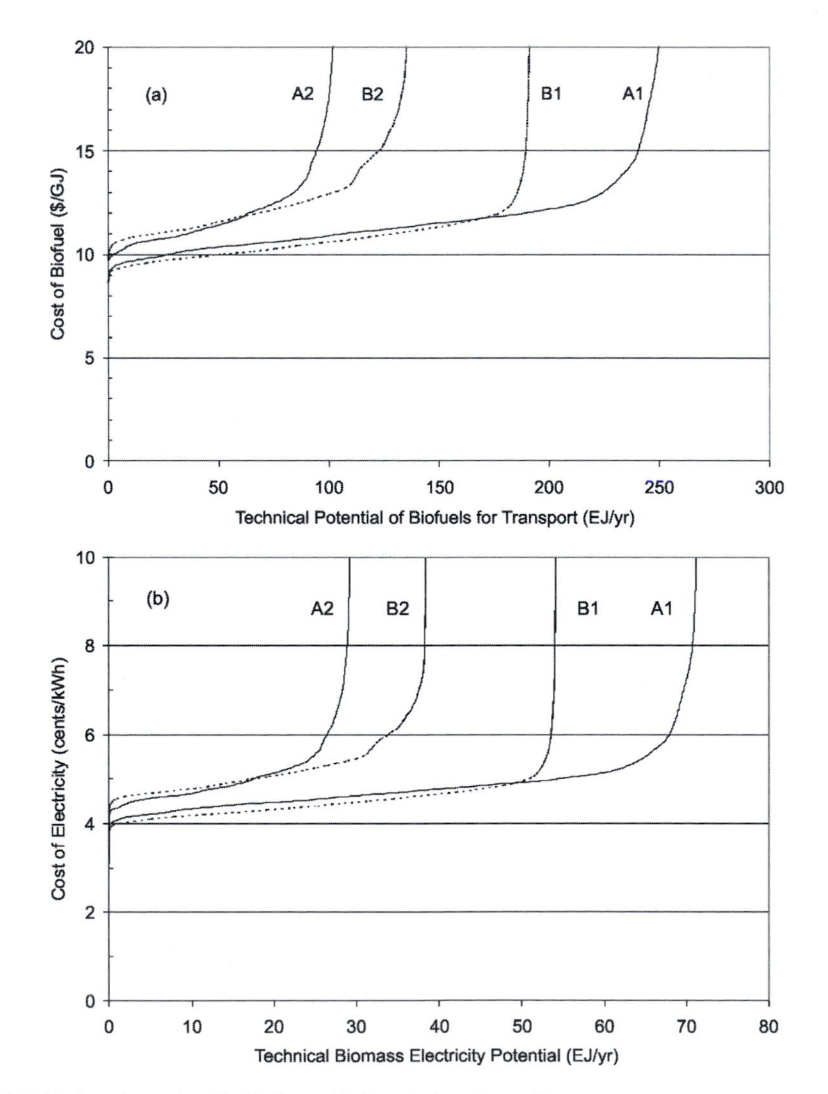

Source: Hoogwijk et al (2009), based on a data file kindly provided by Monique Hoogwijk

Figure 4.49 *Cost of supplying increasing amounts worldwide of (a) biofuels using FT synthesis and (b) electricity from biomass using BIGCC*

in 2005 was only 160EJ and global electricity demand was only about 18,500TWh. The resource potentials identified by Hoogwijk et al (2009) are so large, and the costs so low, that even if much of the projected surplus land cannot be used for bioenergy crops for various reasons, and even if costs turn out to be substantially more expensive than projected, biomass could still provide a significant fraction of future energy demands at competitive costs.

This conclusion rests on two key assumptions: (1) that the growth in total energy demand is constrained through a strong emphasis on efficient end-uses, and (2) that climatic change is limited enough that any adverse impacts do not substantially reduce the bioenergy potential or substantially increase the amount of land needed for food production. Of course, the sooner that we begin the transition to a C-free energy system, in part through advanced use of biomass coupled with strong improvements in end-use efficiency, the greater the likelihood that climatic change can be limited to levels such that impacts on bioenergy potentials are small.

4.11 Vulnerability of biomass energy production to climatic change

A disadvantage of biomass energy is that its availability will fluctuate from year to year with fluctuations in weather. These fluctuations will be particularly marked for annual energy crops (such as sugarcane or soybeans), but short-term fluctuations will have less impact on the average yield of perennial crops. However, long-term climatic change runs the risk of adversely affecting bioenergy (and food) yields. As yields decrease, crop prices will increase, thereby increasing the final cost of energy produced from biomass (or forcing some bioenergy producers to shut down).

Bachelet et al (2001) have assessed the impact of climatic change on biomass production in the US over the next century, as projected by seven different coupled atmosphere–ocean climate models. They find that moderate temperature increases (2.8–3.7K national and annual mean temperature change) are associated with an increase or decrease in overall plant growth over most of the US, but that larger temperature increases are associated with significant reductions in plant growth (by up to 35 per cent for the model giving 6.5K average warming by the end of the century). Particularly vulnerable are the forests of the southeast US, with catastrophic fires potentially causing a rapid conversion from forest to savannah. Tuck et al (2006) find that Mediterranean oil and solid biofuel crops, currently restricted to southern Europe, could extend northward. However, bioenergy crops in Spain are particularly vulnerable, with many crops expected to decline dramatically by the 2080s based on the climate simulated by four global climate models for the 2080s.

Adverse climatic change could in some cases act to favour bioenergy crops, in that hardy, deep-rooted perennial energy crops could become an alternative for farmers in regions that become unsuitable for traditional crops. Hopefully, the loss of agricultural production will be compensated by increased yields (due to climatic and non-climatic factors) elsewhere or by reductions in meat consumption.

4.12 Summary and synthesis

Biomass energy is perhaps the most complicated of the (net) C-free energy options. Complications arise from the many different possible biomass energy feedstocks (primary and secondary agricultural and forestry residues, dedicated bioenergy plantations on surplus agricultural land or on degraded land, municipal solid and sewage waste, biomaterials at the end of their useful life), the different forms in which biomass can be used (as solid, liquid or gaseous fuels), the variety of energy conversion processes available (combustion, gasification, pyrolysis, anaerobic digestion and conversion to ethanol, methanol, biodiesel or hydrogen), the many potential end-use applications (heating, cooking, transportation, generation of electricity and as a feedstock in the production of various chemicals and plastics), the potential adverse and beneficial environmental impacts (including impacts on soil fertility, biodiversity, water supplies, atmospheric and water pollution and erosion), social considerations through competition with land for food and impacts on the price of food crops, the difficulty in assessing the net energy gain using biomass, and uncertainties related to long-term sustainability and the impacts of climatic change during the coming century and beyond.

One of the most difficult issues with regard to biomass energy is the determination of the net energy gain or yield using biomass. Calculation of the net energy yield of biomass requires accounting for all of the direct and indirect energy inputs at all stages in the production of bioenergy crops and their conversion to usable solid, liquid or gaseous fuels. These inputs include the energy required to make the fertilizers, pesticides, herbicides and farm machinery used in the cultivation of bioenergy crops. In the production of liquid or gaseous fuels from biomass, significant additional energy inputs are needed to power the conversion process. Energy credits need to be assigned to account for co-products that can displace commodities that otherwise require energy to produce. Credit should also be assigned for the recycling of farm machinery and production facilities at the end of their useful life. Proper accounting requires determining the sources of electricity used in the production of the various inputs (such as fertilizers and herbicides) and the efficiencies with which the electricity is generated. Even when there is a net energy gain using biomass, the net greenhouse balance can be unfavourable due to emissions of N_2O (nitrous oxide, a powerful GHG) associated with the use of nitrogen fertilizer.

In spite of these uncertainties, there are clear differences in the relative benefits of using biomass in different applications. These benefits can be expressed in terms of EROEI – the ratio of energy provided by the biomass to the non-solar energy inputs required to supply the biomass in the form in which and at the location where it is used. The EROEI of different uses of biomass is summarized in Table 4.56. Use of biomass as a solid fuel, rather than its conversion to liquid fuels, is much superior from an EROEI point of view. However, liquid fuels from biomass represent one of the few options for displacing petroleum-based fuels in the transportation sector.

Given limitations on the available land area that can be devoted to the production of biomass for energy purposes, and potential competition between land for food and land for bioenergy or nature reserves, it is essential that biomass be used as effectively as possible. The use of corn to produce ethanol is the least effective of the possible uses of biomass, and is also the least effective of the possible ways of reducing transportation energy use (with at best a 30 per cent savings potential, while improved vehicle efficiency and shifts to less energy-intensive modes of travel have far greater potential, as discussed in Volume 1, Chapter 5). Ethanol from sugarcane is much more attractive, and ethanol from lignocellulosic materials (grasses such as switchgrass or *Miscanthus,* corn stover and wood) is potentially much more attractive, but this technology is not yet developed, so it remains to be seen if it can perform as projected. However, preparation and delivery of solid biomass generally yields an EROEI of 30–50. This biomass can be used for heating and/or generation of electricity. If biomass used for electricity generation displaces coal, this would provide the largest CO_2 emission reduction per unit of biomass grown. Biomass can be gasified, and if gasified biomass is used in a CAES system to provide baseload or dispatchable electricity from wind (as discussed in Chapter 3,

Table 4.56 *Summary of EROEI for various uses of biomass*

Biomass product	EROEI
Ethanol from corn today	1.5–1.8
Ethanol from Brazilian sugarcane today	8–10
Ethanol from sugarcane in the future	much higher
Ethanol from ligno-cellulosic plants or materials, future expectations:	
– corn stover	7–8
– wood	11
Biodiesel from oily plants today:	
– soybeans	1.4–2.2
– canola (rapeseed)	1.0–2.0
– *Jatropha*	2–12
Local production of wood fuels	8–55
Production and delivery of wood pellets	8–17 for long-distance transport
Production and delivery of wood chips	4–13 for long-distance transport

subsection 3.11.2), this would permit the complete replacement of coal with renewable energy for electricity generation. In effect, 1GJ of biomass energy would leverage the displacement of several GJs of coal energy by permitting wind to assume a much larger role than it otherwise could, with the balance supplied by biomass.[33]

The difference in GHG emissions between using biomass to make fuels and using conventional petroleum fuels depends not only on the net energy yield of biomass but also on how the biomass is produced. This is because nitrous oxide (N_2O) emissions are associated with the use of nitrogen fertilizers, and the greenhouse effect of such emissions can be a significant factor in the overall GHG balance, in some cases more than offsetting the reduction in CO_2 emissions due to reductions in the use of fossil fuels. Conversely, substantial amounts of soil carbon can sometimes build up over time with the repeated growth and harvesting of above-ground biomass, which will augment the reduction in net CO_2 emissions due to reduced use of fossil fuels (where there is indeed a net reduction to begin with).

The use of plantation-scale biomass energy is fraught with numerous potential environmental and social problems. Its development should therefore proceed in a gradual, well-planned manner and with minimal government subsidies so as to avoid distortions that create unexpected and undesirable side-effects. Enforceable restrictions in the development of bioenergy resources will be needed so as to prevent destruction of forests or other natural areas that are better left intact. However, there can be substantial indirect emissions through market interactions if bioenergy development does not proceed slowly enough that the land requirements can be offset through reductions in the need for land for food or forest products.

Biomass energy could supply a significant fraction of future energy needs (up to several hundred EJ per year, compared to world primary energy demand in 2005 of 483EJ) and at attractive costs in the long run (generally \$3–6/GJ for solid fuels, less than \$15/GJ for liquid fuels and about 5 cents/kWh for electricity) Most of this biomass will have to come from bioenergy plantations, but if existing forests are to be protected, the plantations will have to be limited to surplus agricultural and grazing lands. Whether or not there are surplus lands in the future will depend on:

- the future human population,
- future diet (in particular, the proportion of food energy provided by meat and the kinds of meat consumed),
- future agricultural productivity, and
- the future efficiency in converting animal feed into animal food products.

Because up to 100kg or more of phytomass are required to produce 1kg of meat, a diet rich in meat today requires about five times the land area of a vegetarian diet and about seven times the land area of a vegan diet. High agricultural productivity means that less land will be required to produce a given amount of food, so that more land is available for bioenergy crops, but very high productivity generally requires high rates of use of chemical fertilizers, pesticides and herbicides, with adverse environmental consequences. However, in many parts of the world there is still the possibility of large (up to a factor of seven) increases in agricultural yield with minimal inputs. High animal feed efficiency means that less animal feed and hence less land is required in order to produce a given amount of food, but the highest animal feed efficiencies are obtained when animals are confined to pens their entire lives and injected with chemical growth stimulants, with adverse effects on animal well-being prior to slaughter, on food quality and probably on human health. Such systems also preclude recycling of nutrients back to the soil that supplies the food to the animals. For a future widespread meat-rich diet, it is unlikely that surplus agricultural land will be available for bioenergy crops, even with high agricultural yields and high animal feed efficiencies. Conversely, a hypothetical universal vegetarian diet would free up enough land to potentially meet the world's entire future energy needs with biomass energy alone, while a low-meat diet would allow a large fraction of future energy needs to be met with biomass while permitting less extreme agricultural yields and animal feed efficiencies, as well as providing a buffer against adverse effects of a warming climate. Thus, diet emerges as a significant factor in determining both the future potential of biomass energy and the environmental impacts associated with the food system.

Notes

1 Mha = mega-hectare (millions of hectares), where 1 hectare (ha) = 100m × 100m = 10000m^2 = 0.01km^2

2 Grows over a period of several years.
3 One of two major pathways by which plants carry out photosynthesis.
4 Nitrogen as it exists in the air (N_2) cannot be used by plants. Instead, it must be converted to NH_4^+ – a process called nitrogen fixation, which can be done by certain bacteria only in association with certain plant species.
5 Straw is the dry stalk of cereal plants after the grain has been removed. It makes up about half of the yield of the cereals barley, oats, rice, rye and wheat.
6 No-till cultivation refers to cultivation without ploughing or 'tilling' the soil. It is practised so as to conserve soil moisture, minimize erosion and the loss of soil carbon and to reduce costs, but generally requires greater use of herbicides for weed control.
7 Charcoal is produced by heating wood with a minimal supply of oxygen so as to drive off volatile materials, and so entails an initial energy loss. Overall efficiencies and pollutant emissions are compared in subsection 4.5.11.
8 Four to ten tonnes of wood (with a typical heating value of 18MJ/kg) are used to produce one tonne of charcoal (with a heating value of about 29MJ/kg, inferred from Bailis et al, 2003), implying a conversion efficiency of about 16–40 per cent. Further energy losses occur during the use of charcoal.
9 Glycerine (also spelt 'glycerin') is used in the cosmetics, ink, lubrication and preservative industries, and is an ingredient in soap.
10 Makhijani (2007), quoting unpublished sources, gives the algae biomass yield in desert areas when fertilized with powerplant CO_2 as 250 tonnes dry mass/ha/yr, compared to typical bioenergy crop yields of 10–20t/ha/yr. This corresponds to an efficiency of solar energy capture of about 5 per cent, or about 10 times that of corn (based on the energy content of the entire plant) or 20 times that of corn (based on the energy content of the corn kernels – the only part used to make ethanol). A yield of 0.5kg biodiesel/kg dry matter would give the 142,000 litres/ha/yr biofuel yield given in the NREL study. Water hyacinth can also achieve 5 per cent solar energy conversion efficiency but will not grow in brackish water. It would be used to produce biogas and can also absorb toxic metals, which it may be possible to harvest from the residue associated with the production of biogas.
11 These yields conserve C mass, as biomass is typically 50 per cent C by mass (see Table 4.2), while ethanol (C_2H_5OH) is only 39 per cent C by mass.
12 Nm^3 = normal cubic metres, pertaining to a temperature of 0°C and a pressure of 101.325kPa (1atm).
13 Senescence is last phase in the growth cycle, during which nutrients are withdrawn from leaves and stored in roots or stems (in the case of woody plants).
14 Many nutrients are part of the photosynthetic machinery, so shifting nutrients toward the upper part of the plant, where there is more sunlight, maximizes overall photosynthesis.
15 The available species consisted of four C_4 grasses, four C_3 grasses, three herbaceous legumes, one woody/shrubby legume, four non-legume herbaceous forbs and two savannah oak species.
16 Fossil energy required to produce, harvest and transport the biomass to a central facility amounts to 4 GJ/ha/yr, giving an EROEI of about 16.
17 For biomass fuels, the primary energy input is computed as the heating value of the fuel plus the primary energy content of any fossil fuels used in producing, harvesting, and transporting the biomass to the point of use, and is consistent with the EROEI shown later in Figure 4.40.
18 Transportation energy use is given assuming that wood chips are transported 100km to power or cogeneration plants, or 50km to pellet plants, and that pellets and wood are transported 100km or 20km, respectively, to the end-user.
19 The current area of sugarcane plantations in Brazil used for ethanol production is about 3.5 million ha (Fischer et al, 2009) while the area used for cattle is 2.4 million ha (Goldemberg et al, 2008). Thus, doubling the cattle stocking density would free up enough land for about a 35 per cent increase in the sugarcane for ethanol plantation area.
20 The warming impact of CO and NMHCs pollutants arises largely from their reaction, in the presence of sunlight, to produce ozone, a GHG. Emissions of CH_4 also contribute to ozone formation, but CH_4 itself is a GHG and its direct effect dominates.
21 Global warming potential is a measure of the heating effect of a mass of a given gas compared to the heating effect of the same mass of CO_2, integrated over some arbitrary time horizon (typically 100 years) (see Volume 1, Chapter 2, subsection 2.6.5).
22 One of the main sources of disagreement between various assessments of the EROEI for corn ethanol has been the embodied energy assigned to farm machinery, with most workers ignoring it altogether and Pimentel and Patzek (2005) assigning a value (4.3GJ/ha/yr) equal to almost 25 per cent of the total energy input to corn production found in other studies. F2006 estimate a value of 0.32GJ/ha/yr based on a rough input–output analysis of dollar expenditures (see Appendix B for an explanation of input–output analysis). Bottom-up data for tractors on Canadian farms presented in Dyer and Desjardins (2006) support the relatively small embodied energy estimated by F2006. In particular, Dyer and Desjardins (2006) indicate a tractor power ranging from 15kW to 120kW on farms ranging in size from about

20 to 300ha. Lee et al (2000) indicate that 85 per cent of a tractor mass is iron and steel, with the balance being mainly aluminium, copper and glass. From Tables 6.1 and 6.4 of Volume 1, the world average primary energy requirements are 35.6MJ/kg, 193.3MJ/kg, 88MJ/kg and 21.1MJ/kg for primary (that is, made from virgin materials) steel, aluminium, copper and flat glass. Using an 85:5:5:5 weighting, the average primary energy requirement is 45MJ/kg (this is much less than in the old studies cited by Dyer and Desjardins (2006)). Assuming an eight-year average lifespan and multiplying by 3.6 to take into account other farm machinery (as suggested by Dyer and Desjardins, 2006) results in a farm machinery embodied energy of 1.10GJ/ha/yr for a 20ha farm and 0.59GJ/ha/yr for a 300ha farm. These account only for the materials in the machinery, and need to be multiplied by a factor of no more than 2.0 (and probably closer to 1.5) to account for the energy used in assembling the farm machinery, building the facilities where it is built, and other upstream inputs, as well as in delivering the farm machinery. However, if all the materials in the machinery are assumed to be recycled at the end of their life, then the energy saved through recycling can be credited to the machinery. The net embodied energy in that case is the same as the energy required to recycle old materials into new materials. From Volume 1, these are roughly 7MJ/kg for steel, 17MJ/kg for aluminium, 56MJ/kg for copper and 18MJ/kg for glass. These give a mean tractor embodied energy of 10.5MJ/kg and farm machinery embodied energies of 0.14–0.26MJ/ha/yr, which again need to be multiplied by a factor of up to two in order to account for other energy inputs.

23 The mass of ethanol that is produced per kg of dry corn is equal to the ratio of glucose mass to corn mass (about 0.66) times the ratio of ethanol to glucose mass (0.511, a fixed property of the conversion reaction $C_6H_{12}O_6 \rightarrow 2C_2H_5OH + 2CO_2$) times the mass conversion efficiency.

24 N_2O is a powerful GHG that is produced in association with the use of N fertilizer and must be taken into account. It is released from the soil through two microbial activities: *nitrification* (the oxidation of NH_4^+ to NO_3^- with NO_x and N_2O produced as intermediates or byproducts); and *denitrification* (the anaerobic reduction of NO_3^- to N_2, again with production of NO_x and N_2O). The emission of N_2O depends on the concentrations of NH_4^+ and NO_3^-, the soil water content, temperature and texture, and the amount of labile (reactive) carbon. For this reason, N_2O emissions will vary significantly from place to place, but can be significant.

25 According to Smeets et al (2008), extensive areas in Brazil currently used for cattle ranching are ideal for sugarcane production.

26 As indicated in Volume 1 (Table 2.12), a typical methane leakage from underground coal mines is about 0.5kgCH_4 per GJ of coal mined. This has to be multiplied by the global warming potential of methane (~25 on a 100-year timescale) to give the supposedly equivalent CO_2 emission. Methane emissions from surface mines are about a factor of four smaller.

27 The emission factor for biomass is calculated from the assumed carbon fraction and biomass heating value, while the emission factor for coal is 25kgC/GJ × 1.05 and that of natural gas is 13.5kgC/GJ × 1.25, with the factors of 1.05 and 1.25 included so as to account for typical emissions associated with extracting and supplying the fuel.

28 Production of 1GJ of H_2 requires 2.33GJ biomass with co-production of biochar and sequesters 30.6kgC, while production of 1GJ of H_2 without biochar production and C sequestration requires 1.54GJ. The extra 0.8GJ biomass could displace 0.8GJ coal (if both are used to generate electricity at the same efficiency), which would avoid the emission of only 20kgC.

29 Assuming capture of 90 per cent of CO_2 in offgases with a 5 per cent energy penalty, as given in Table 9.8 for capture of CO_2 during the production of H_2 from coal.

30 Gha = giga-hectares, or billions of hectares.

31 Labour costs are projected to increase most strongly in Eastern Europe, but would still remain below today's labour costs in Italy and the UK.

32 The reader can extrapolate the electricity cost results given in Table 4.53 to higher fuel costs.

33 In the long run, even this small amount of biomass would not be necessary for handling fluctuations in wind energy with CAES if advanced adiabatic CAES systems develop as hoped, as discussed in subsection 3.11.2.

5

Geothermal Energy

5.1 Introduction

Heat flows outward from the interior of the earth due to the fact that the temperature at the centre of the earth is around 7000°C. This heat is partly left over from the formation of the earth about 4.6 billion years ago, and partly a result of heat released from the radioactive decay of long-lived radioactive isotopes, principally thorium-232, uranium-238 and potassium-40. The average geothermal heat flux is about 0.06W/m^2 under the continents and 0.10W/m^2 under the oceans (Barbier, 2002). This heat flux is equal to the vertical temperature gradient times the thermal conductivity. Given a thermal conductivity of crustal rocks of 2.5–3.5W/m/K, this implies a typical vertical temperature gradient under continents of about 15–20K/km. Mudrocks (clays and shales) have a smaller thermal conductivity of (1–2W/m/K), resulting in a more rapid increase of temperature with depth for a given geothermal heat flux. Under most of the densely populated parts of the continents, the geothermal heat flux is less than 0.05W/m^2. A few regions attain heat fluxes of 0.1W/m^2. In local volcanic regions where molten or partially molten rocks occur within a few kilometres of the surface (and at temperatures of up to 1000°C), much larger heat fluxes can be found. At present, geothermal energy supplies only a small fraction of the world's energy needs – about 0.12 per cent of global electricity generation and 0.6 per cent of global residential and commercial space heating and hot water requirements.

5.2 Geothermal resources

Geothermal energy occurs in several different useful forms, which are described here.

5.2.1 Hydrothermal

Hydrothermal resources consist of steam or liquid water at temperatures of up to 500°C at a depth of up to 4km in permeable rock.[1] Three conditions are required for the formation of hydrothermal resources: an *aquifer* (a porous and permeable layer of rock filled with water), an overlying impermeable layer and a heat source. The relationships between these ingredients and surface features are shown in Figure 5.1.

Porosity refers to the fraction of the rock volume that is filled with cavities, while *permeability* refers to the ease with which a fluid will flow through the rock. Porosity is important for the utilization of a hydrothermal resource because it determines how much hot water is present in a given volume of rock, while permeability is important because it determines how rapidly the hot water that is extracted at a given point can be replenished through flow from the surrounding rocks. The volume of water (Q) flowing across a cross-section of area A depends on the cross-sectional area, the *hydraulic conductivity* (K_w, a measure of permeability) and the *hydraulic gradient*, as given by Darcy's Law:

$$Q = K_w A\left(\frac{\Delta H}{\Delta L}\right) \tag{5.1}$$

H is the so-called *pressure head* of the water. Its value at a given point in the aquifer is equal to the height to which water in a tube would rise if the base of the tube were placed at that point. ΔL is the distance over which the variation ΔH is measured.

Table 5.1 gives typical values of porosity and hydraulic conductivity for different kinds of rock. Rocks such as silt and clay have very high porosity but very low hydraulic conductivity. Unconsolidated

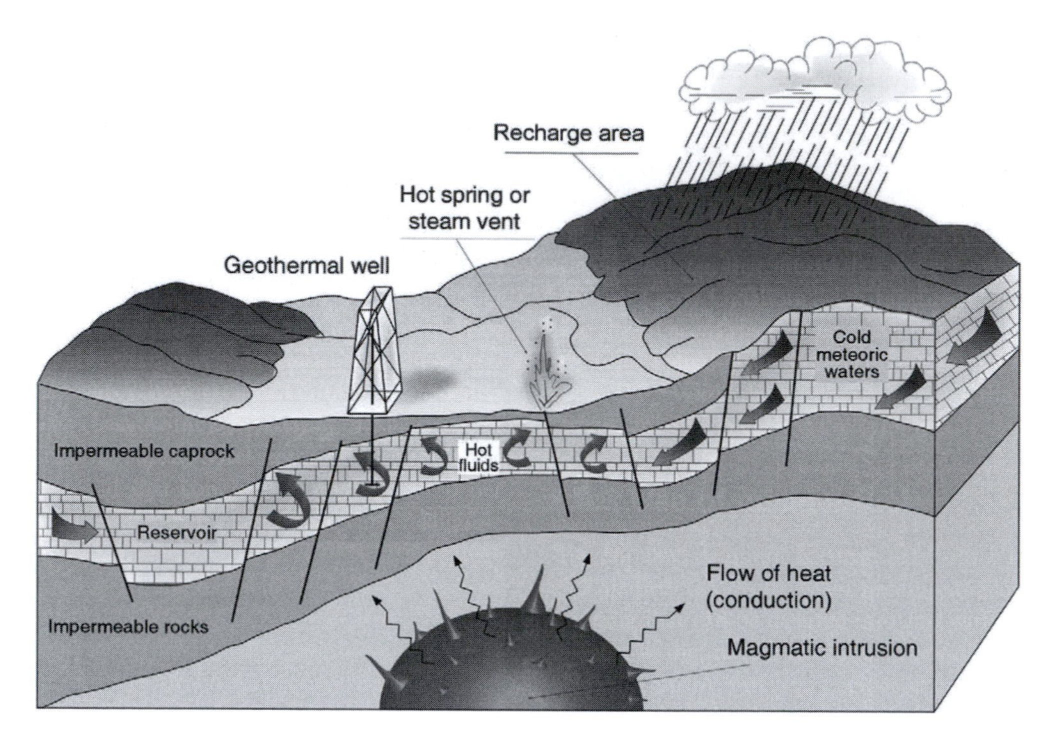

Source: Barbier (2002)

Figure 5.1 *A hydrothermal geothermal resource, showing the aquifer (labelled 'reservoir'), impermeable caprock, heat source and surface water recharge zone*

Table 5.1 *Typical porosities and hydraulic conductivities of common sediments and rocks*

Material	Porosity (%)	Hydraulic conductivity (m/day)
	Unconsolidated sediments	
Clay	46–60	<10^{-2}
Silt	40–50	10^{-2}–1
Sand, volcanic ash	30–40	1–500
Gravel	25–35	500–10000
	Consolidated sedimentary rocks	
Mudrock	5–15	10^{-8}–10^{-6}
Sandstone	5–30	10^{-4}–10
Limestone	0.1–30	10^{-5}–10
	Crystalline rocks	
Solidified lava	0.001–1	0.0003–3
Granite	0.0001–1	0.003–0.03
Slate	0.001–1	10^{-8}–10^{-5}

Source: Brown (1996)

(loose) sediments such as sand and gravel have high porosity and hydraulic conductivity. In unconsolidated sediments, permeability arises through the interconnection of the pore spaces between grains. Sedimentary rocks (consolidated sediments) form as crystals precipitate from the water flowing through the sediments, filling in the pore spaces and 'cementing' the grains together. In these rocks, permeability arises through incomplete cementation, fracturing of the rock or subsequent partial dissolution in the case of calcareous sedimentary rocks (limestone).

The most useful hydrothermal resources largely contain steam rather than hot water, but are rare. Occurrences include the Larderello field in Italy, the

Geysers field in California and the Matsukawa field in Japan (Mock et al, 1997). The next best hydrothermal resources contain hot water at temperatures of up to 350°C, and occur in many regions, including the western US, Mexico, Nicaragua, New Zealand, Iceland, China, Japan, Indonesia, the Philippines, Italy, Turkey and several countries in eastern Africa. Vapour-dominated resources occur where the temperature at depth is greater than the local boiling point of water (the boiling point increases with pressure and hence with increasing depth), while liquid-dominated resources occur where the temperature is less than the local boiling point. The water may flash into steam as it rises through a well.

Low-temperature geothermal resources are common in sedimentary basins having higher than average geothermal heat flux. The association of higher than average heat fluxes with sedimentary basins is not coincidental: extension and thinning of the earth's crustal plates results in greater than average heat flux (as warmer mantle rock is brought closer to the surface), while the thinning creates topographic depressions into which sediments can accumulate. Beneath the south Hungarian Plain, for example, the geothermal gradient exceeds 100K/km, such that water at 120°C occurs at a depth of 1km.

In some crystalline rocks the geothermal heat flux is significantly enhanced by the in situ generation of heat by the decay of radioactive isotopes. For example, a granite body in south-western England contains enough potassium-40, thorium-232 and uranium-238 to add a heat flux of 0.056W/m^2 to the background flux (in that region) of 0.04W/m^2 (Brown, 1996, Box 9.5).

5.2.2 Geopressurized

Geopressurized resources consist of hot high-pressure brines (saline water) containing dissolved methane (which forms 92–98 per cent of natural gas). They occur in petroleum basins worldwide. Along the Gulf of Mexico coast of Texas and Louisiana, temperatures of 150–180°C and pressures of 1400 atmospheres occur. Geopressurized resources thus contain three energy resources that can be exploited: thermal energy, hydraulic energy because of the high pressure and chemical energy (as methane). Typical proportions are 60 per cent thermal, 10 per cent hydraulic and 30 per cent chemical (Mock et al, 1997).

5.2.3 Hot dry rock

The term 'hot dry rock' refers to rocks at high temperatures that do not contain fluids. Hot dry rock systems can be found anywhere if one drills deep enough. Inasmuch as temperatures in excess of 150°C are needed for producing electricity, this requires drilling 4–8km in low thermal gradient regions (20–40K/km) and 2–5km in high-gradient (>60K/km) regions.

5.2.4 Magma

Magma resources consist of partially or completely molten rock at accessible depths (<7km) in regions of recent volcanic activity. Examples include the Kilauea Iki volcano in Hawaii and the Krafla field in northern Iceland. Magma intrusions can take several million years to cool back to ambient conditions, so even extinct volcanic fields can provide good geothermal resources (an example being in Yellowstone National Park in the US).

5.3 Technologies for utilizing geothermal energy

Geothermal heat can be used directly to provide domestic hot water and space heating, or can be used indirectly to produce electricity by first generating steam that is used in a steam turbine. Taylor (2007) reviews technologies for geothermal exploration, drilling, reservoir management (maintaining energy supply rates) and for enhanced geothermal systems, while Kagel (2008) reviews surface technologies used in the conversion of geothermal heat into useful energy.

5.3.1 Direct use of geothermal heat, with or without heat pumps

Low-temperature geothermal resources can be directly used for heating. This requires a hot water distribution network, consisting of insulated pipes of varies sizes, pumps, valves, meters, expansion joints and controls. The geothermal water is typically too saline and corrosive to be directly used in heating systems, so corrosion-resistant heat exchangers are used to transfer heat to a second, closed circulation loop. Figure 5.2 shows the temperatures that are useful for different purposes, while Figure 5.3 shows the temperatures that are encountered at depths of 6.5km in the US and 5km in Europe. Many agricultural and food-processing uses require heat at temperatures of 50–90°C, which are generally available at a depth of 3.5km or less in the western US and at a depth of 3.5km or somewhat greater in the eastern US.

The breakdown of thermal direct uses of geothermal heat worldwide in 2005 is given in Table 5.2. This compilation counts heat from ground-source heat pumps (GSHPs) as geothermal heat, but GSHPs extract largely seasonal heat with the aid of electricity and so can be excluded. This leaves bathing and heating of swimming pools as the single largest direct use of geothermal heat (at 45 per cent of the total), followed by space heating (at 30 per cent) and heating of greenhouses (at 11 per cent). Also given in Table 5.2 are the average capacity factors for the different direct utilizations of geothermal heat. Excluding snow melting, these range from a low of 40 per cent for space heating to a high of about 70 per cent for industrial uses.

The French are at the forefront in the utilization of low-temperature geothermal resources in district heating systems, having installed 55 such systems in the Paris region and several more in south-western France. Paris is situated over a 200km-wide sedimentary basin, with hot water at 55–70°C and depths of 1–2km. One such district heating system is illustrated in Figure 5.4. Note that a heat pump is used to extract additional useful heat from the geothermal water after it has already heated 2000 apartment units, thereby permitting heating of an additional 2000 units.

Heat pumps can also be used to utilize geothermal heat sources that are not quite warm enough to use directly for heating. Substantial savings in electricity use (by a factor of four or more compared to electric-resistance heating) will be possible due to the small required temperature lift (see Volume 1, Chapter 4, subsection 4.4.2).

5.3.2 Electricity production from high-pressure steam fields

Four main technologies have been used to generate electricity from geothermal steam, the choice of technology being dependent on the nature of the geothermal resource. These technologies are illustrated in Figure 5.5 and described below. The efficiency of geothermal powerplants is limited by the relatively low temperature (generally less than 350°C) of geothermal steam compared to the temperatures (around 600°C today, eventually reaching around 700°C) used in a fossil fuel-powered steam turbines (see Volume 1, Chapter 3, section 3.2). Geothermal steam generally contains non-condensable gases (CO_2, H_2S, NH_3, CH_4, N_2 and H_2) that have to be extracted from the

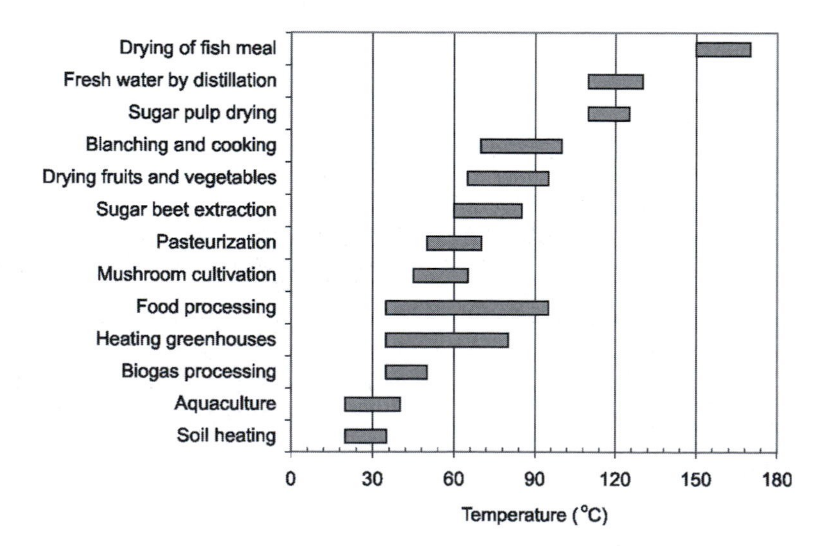

Source: Data from Popovski and Popovska-Vasilevska (2003)

Figure 5.2 *Temperature ranges that are useful for different agricultural purposes*

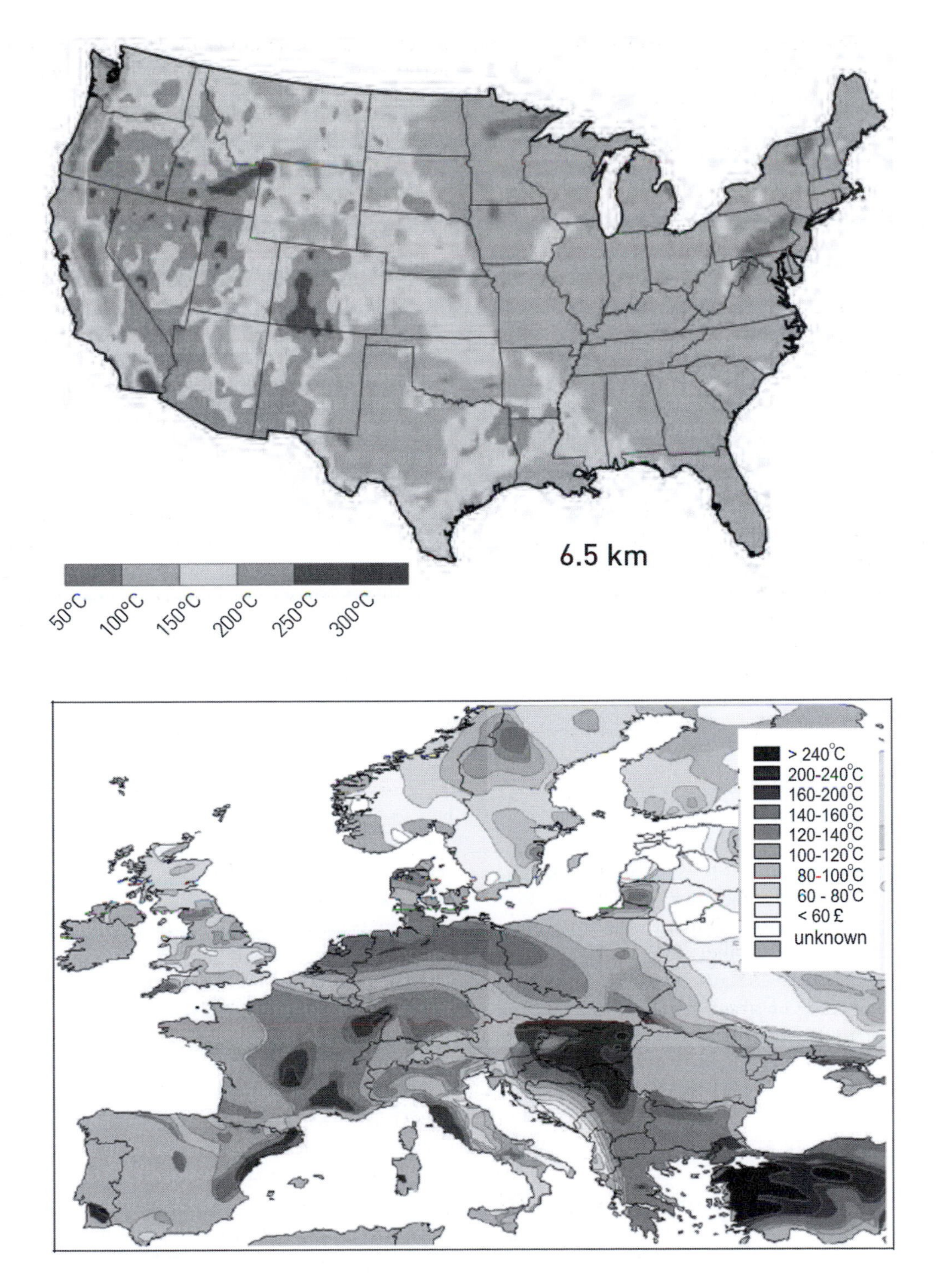

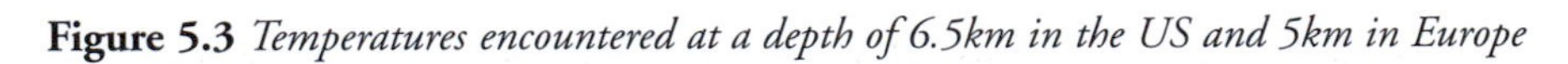
Source: MIT (2006) for the US and GAC (2006) for Europe

Figure 5.3 *Temperatures encountered at a depth of 6.5km in the US and 5km in Europe*

Table 5.2 *Direct uses of geothermal heat and comparison with heat supplied by GSHPs in 2005*

Application	Capacity (MW_{th})	Energy supplied (TJ/yr)	Capacity factor	Per cent of non-GSHP supply
GSHPs	15,384	87,503	0.18	
Bathing and swimming	5401	83,018	0.49	44.7
Space heating	4366	55,256	0.40	29.7
Greenhouse heating	1404	20,661	0.47	11.1
Aquaculture pond heating	616	10,976	0.57	5.9
Industrial uses	484	10,868	0.71	5.8
Cooling/snow melting	371	2032	0.17	1.1
Agricultural drying	157	2013	0.41	1.1
Others	86	1045	0.39	0.6
Total	28,269	273,372	0.31	

Source: Lund et al (2005)

condensers of powerplants, which reduces the efficiency compared to that which would be obtained with pure water. Overall efficiencies in generating electricity are generally in the 10–17 per cent range (Barbier, 2002). Geothermal powerplants can be built in relatively small units (5–50MW), so the powerplant capacity can be gradually built up as demand grows. Further information on the four technologies is given below.

Dry steam powerplants

This is used where dry steam (unaccompanied by liquid water) at a temperature of up to 350°C and a pressure of eight to nine atmospheres is found. A steam turbine is used, with steam expanding and cooling as it passes through the turbine. The low-pressure steam can then be vented to the atmosphere (in a 'back-pressure' turbine) or condensed (in a 'condensing' turbine) (see Figure 5.5a). Modern plants require about 6.5kg of steam per kWh of electricity generated, or 100kg/steam for a 55MW plant (Brown, 1996). The best thermal efficiency achieved is around 30 per cent.

Single-flash steam powerplant

Where the geothermal resource consists of a mixture of steam and hot liquid water, the steam and liquid water must be separated. If the resource consists of hot water, one wishes to avoid flashing of the water into steam as it rises (so as to avoid deposition of minerals that will eventually clog the well), and this is done by pressurizing the well. Steam is produced at the surface in a separator and fed into a conventional turbine. The steam temperature is typically 155–165°C, so more steam is required than in the dry steam powerplant (around 8kg/kWh). Most of the fluid remains as unflashed brine, which is either re-injected into the aquifer or used in a local district heating system (see Figure 5.5b).

Binary cycle powerplant

At temperatures below about 200°C, steam cannot be effectively used. Instead, heat is transferred through a heat exchanger to a second fluid (hence, the term 'binary') that has a lower boiling point than water, and is used to generate electricity in a thermodynamic cycle known as the Organic Rankine Cycle. Typical secondary fluids are pentane and butane. The geothermal fluid is kept under pressure and re-injected into the aquifer, while the secondary fluid is condensed with a water-cooled or air-cooled condenser after expanding through the turbine, then cycled back to the heat exchanger to be reheated (see Figure 5.5c). Chemically impure and low-temperature geothermal resources (warmer than 85°C) can be used, but with very low efficiency (3–6 per cent). Nevertheless, binary cycle powerplants have emerged as the most cost-effective and reliable way to convert large amounts of low-temperature geothermal resources into electricity, and large low-temperature reservoirs exist at accessible depths almost everywhere (Barbier, 2002). The Mammoth (Casa Diablo) geothermal plant in California uses 700kg/s of steam to produce 30MW of electricity (Brown, 1996).

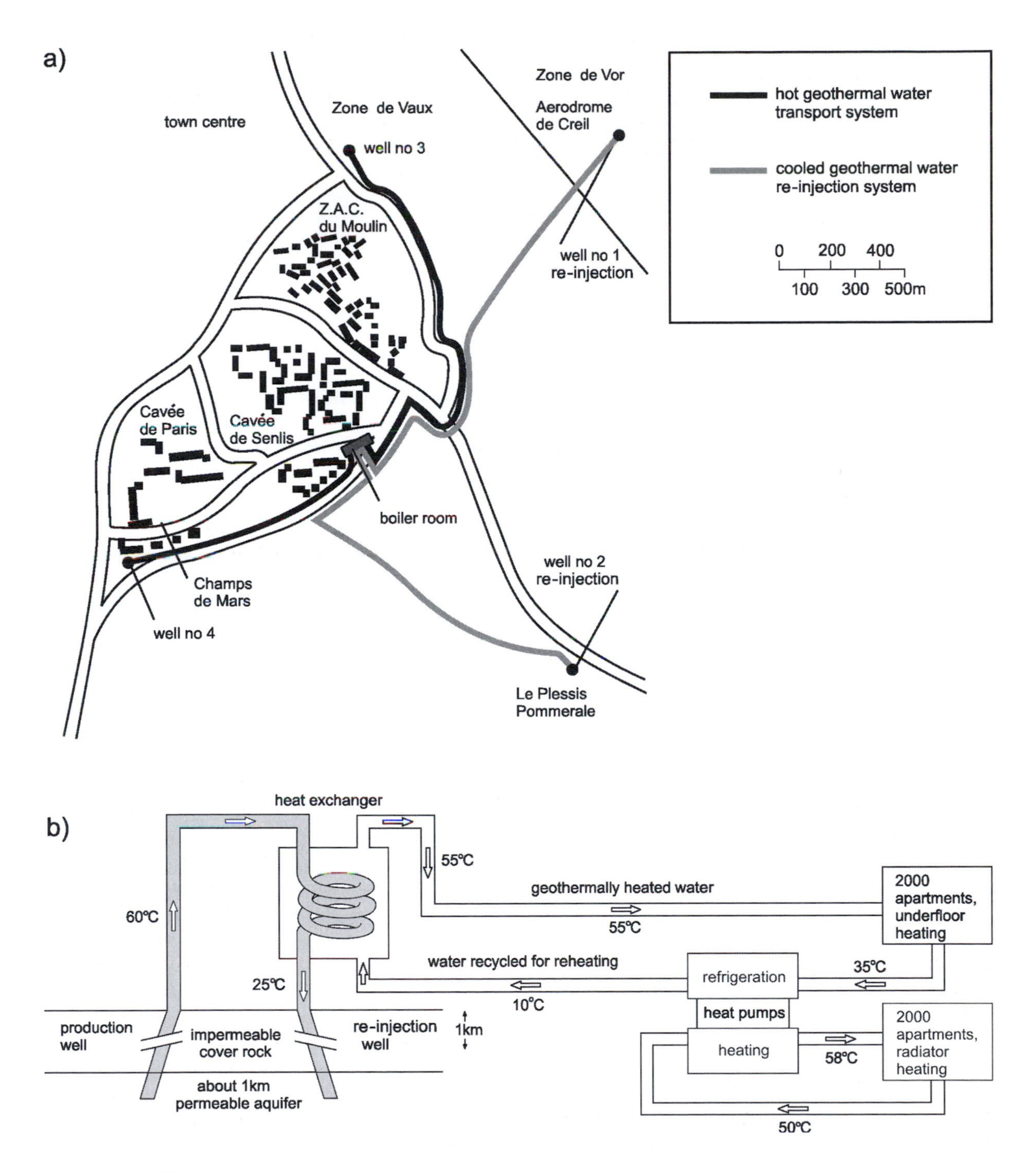

Source: Brown (1996)

Figure 5.4 *The Creil district heating system, installed north of Paris in 1976*

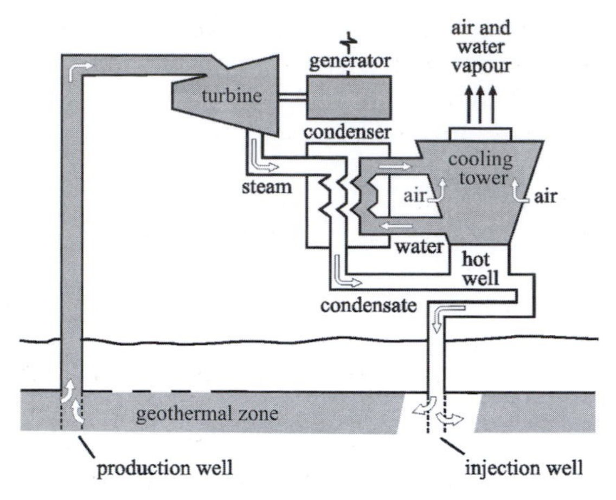

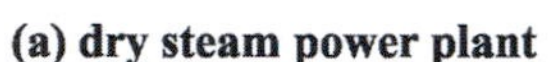
(a) dry steam power plant

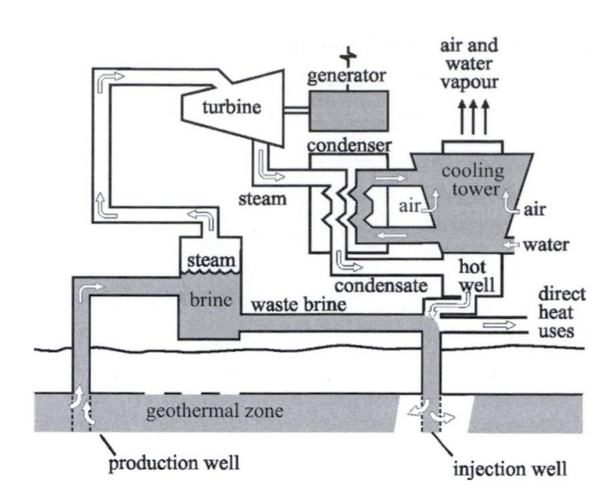

(b) single-flash steam power plant

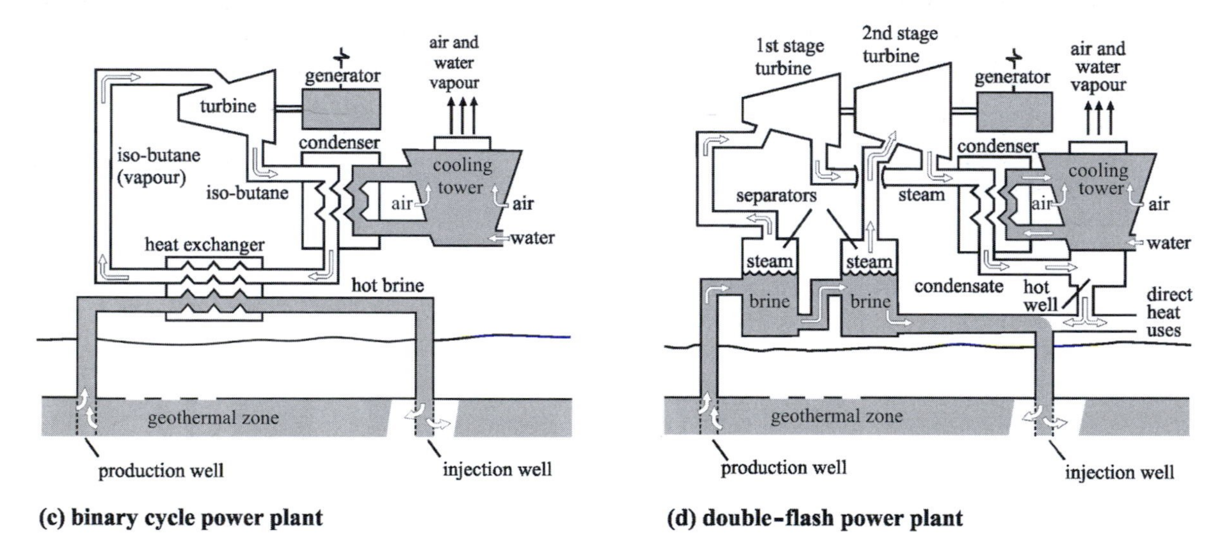

(c) binary cycle power plant

(d) double-flash power plant

Source: Brown (1996)

Figure 5.5 *Simplified flow diagrams for the four major systems for generating electricity from geothermal energy*

Double-flash steam powerplant

In a double-flash powerplant, unflashed liquid from the first flashing flows to a low-pressure tank where another pressure drop provides additional steam. This steam is mixed with exhaust from the high-pressure turbine to drive a second turbine at lower pressure (see Figure 5.5d). Conversely, the low-pressure steam can be fed into the same turbine as the high-pressure steam, but at a different point, where the pressure is lower. Power output can be increased by 20–25 per cent for an extra 5 per cent capital cost (Brown, 1996).

5.3.3 Hot dry rock technologies or enhanced geothermal systems

The regions with the hottest rocks at accessible depths (within 10km of the surface) tend to have very low porosity and permeability. Thus, techniques are required to enhance the porosity and permeability so that water can be injected into the hot rocks and heated. Such systems are referred to as enhanced geothermal systems (EGSs). An EGS involves:

1 drilling an injection borehole, down which water will be injected;
2 fracturing the rock in the region at the bottom of the injection borehole, so that water can flow through the rock and become hot in the process; and
3 drilling an extraction or 'production' borehole in the vicinity of the injection borehole, from which the heated water will be drawn.

This arrangement is illustrated in Figure 5.6. The major technical challenge is in fracturing hard, crystalline rock at depths of 3–6km enough to create an adequate hydraulic conductivity. At the same time, one wants to create a system where the loss of water during the entire injection/extraction circuit will be no more than 10 per cent, particularly in regions where water is scarce. Fracturing is achieved by pumping water down the production borehole at increasing pressure until the rock fractures (this technique, known as hydro-fracturing, is used in the oil industry).

An 18-member panel led by the Massachusetts Institute of Technology (MIT) recently assessed the potential for power generation from EGSs in the US (MIT, 2006). Their conclusion is that, with modest research and development support ($800 million to $1 billion over 15 years) from a combination of private and government sources, EGSs and other unconventional geothermal resources could provide 100GW of cost-competitive baseload electric generating capacity in the US by 2050. The total EGS resource base in the US is estimated to be 13 million EJ, of which 200,000EJ is thought to be extractable with current technology. This is 1000 times the total US primary energy use in 2005. The geothermal option has been largely ignored in the US up to the present, but EGS research has been actively pursued in France and Australia, and significant progress in EGS techniques has been demonstrated. For example, at Soultz, France, a connected reservoir well system with an active volume of 2km^3 at a 4–5km depth has been created, and fluid flow rates are within a factor

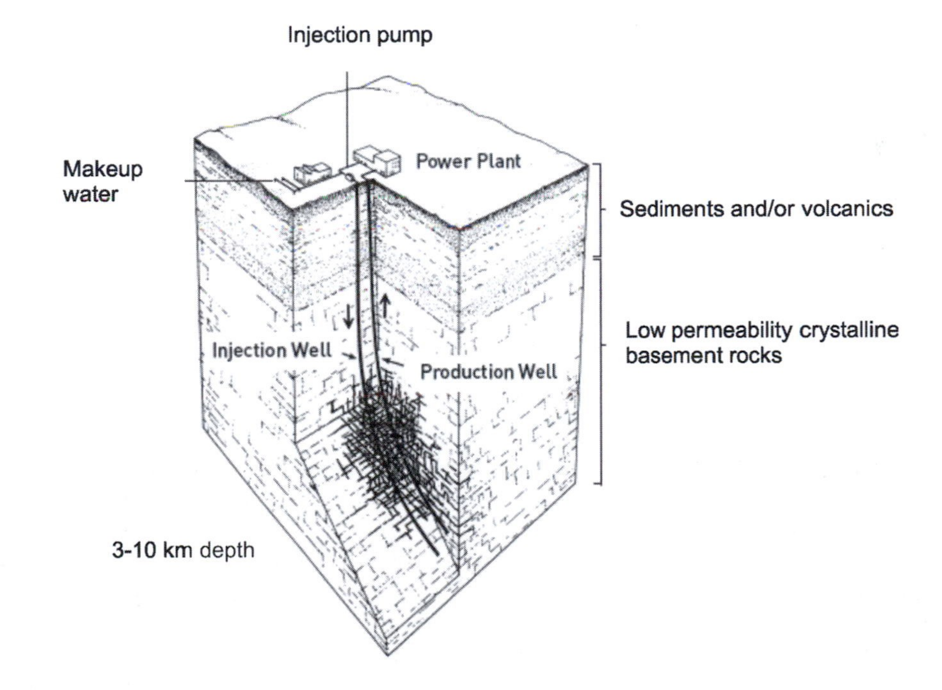

Note: The injection and production boreholes are separated by a horizontal distance of about 300m.
Source: Mock et al (1997)

Figure 5.6 *Production of electricity from a hot dry rock geothermal resource*

Table 5.3 *Estimated land area for powerplant and auxiliaries, and subsurface volume*

Plant size (MW_e)	Area (km^2)	MW_e/km^2	Volume (km^3)	MW_e/km^3
25	1	25	1.5	16.7
50	1.4	36	2.7	18.5
75	1.8	42	3.9	19.2
100	2.1	48	5.0	20.0

Note: Beyond a plant size of 100MW, area and volume vary in proportion to plant size.
Source: MIT (2006)

of two to three of what is required for commercial viability. Improvements in heat transfer performance for low-temperature fluids and further evolution in plant designs are expected to lead to a factor of ten increase in power generation from a given thermal reservoir.

EGSs are inherently modular and scalable from 1 to $50MW_e$, and could provide widely dispersed power, district heating and combined heat and power systems. Table 5.3 gives the land area needed for the powerplant and auxiliaries and the subsurface reservoir volume needed for EGS plants of various sizes.

5.3.4 Cascading of geothermal energy

To make maximal use of geothermal energy, it should be cascaded from high-temperature to successively lower-temperature applications. Thus, waste heat from a relatively high-temperature steam flash powerplant could be used in a lower-pressure turbine as part of a binary cycle, and waste heat from that in turn could be used to supply space heating needs in a district heating system or for aquaculture or agriculture (with further heat extracted from the return flow via heat pumps). Metals such as zinc, arsenic, manganese and lithium can be extracted from geothermal fluids returning from a binary cycle plant.

5.4 Environmental considerations

As noted above, steam from geothermal fields contains a number of non-condensable gases (CO_2, H_2S, NH_3, CH_4, N_2 and H_2), as well as a number of toxic chemicals in suspension and in solution (arsenic, mercury, lead, boron and sulphur). Potentially serious water pollution problems can therefore occur if geothermal water is discharged into surface lakes or rivers, or if steam is vented to the atmosphere. Non-negligible CO_2 emissions can also occur. Figure 5.7 compares the CO_2 and sulphur emissions (per kWh of electricity generated) for various geothermal powerplants in the

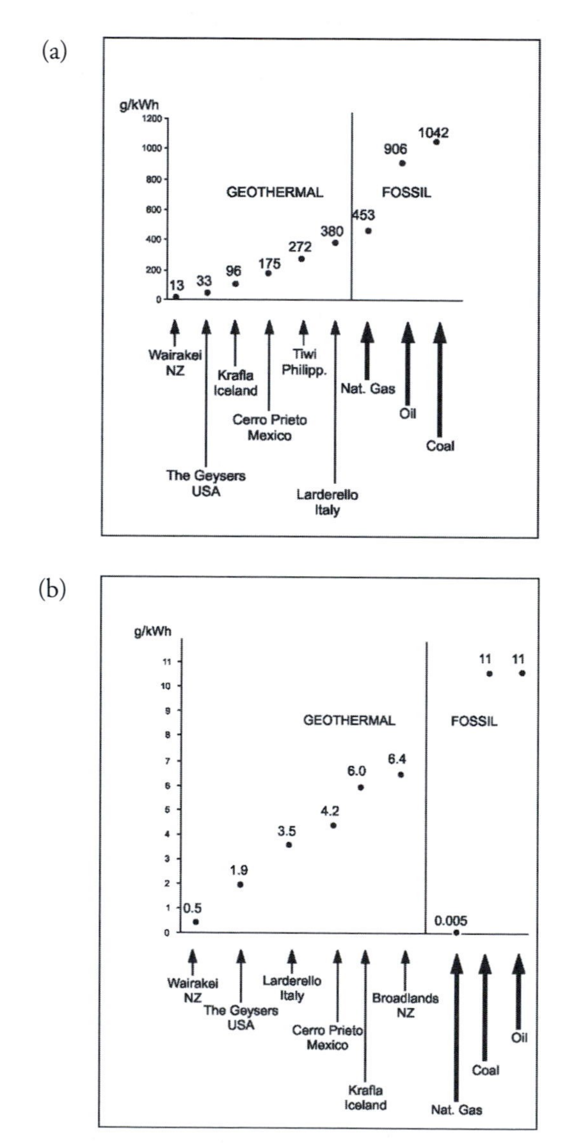

Source: Barbier (2002)

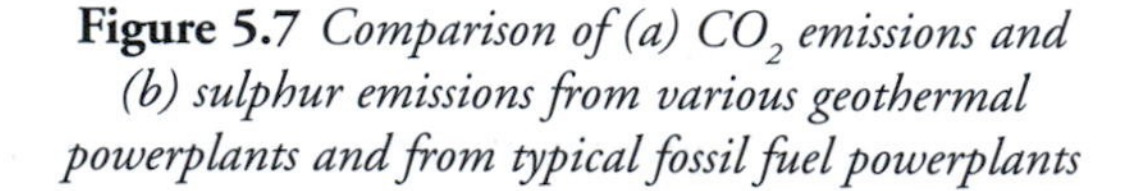

Figure 5.7 *Comparison of (a) CO_2 emissions and (b) sulphur emissions from various geothermal powerplants and from typical fossil fuel powerplants*

world, and for conventional fossil fuel powerplants. In some cases, CO_2 emissions are not substantially better than from natural gas powerplants, while S emissions can be over half of those from coal powerplants. Mercury emissions range from 45 to 900µgm/kWh, which is comparable to those from coal powerplants (Barbier, 2002). Boron vapours near geothermal powerplants can have a serious impact on vegetation. Clearly, a case-by-case environmental impact assessment is needed for geothermal energy, with appropriate control measures taken. In most cases, some or all of the CO_2 emitted during power production would have been gradually emitted anyway (Kristmannsdóttir and Ármannsson, 2003). A proper assessment therefore requires measuring the rate of emission prior to building the geothermal energy facility – not an easy task.

Binary cycle powerplants, in which heat is transferred to a secondary fluid without exposure to the atmosphere, and then re-injected into the ground, will not discharge gases or fluids to the environment during normal operation. Hot dry rock systems, in which surface water is circulated through freshly fractured rock, will probably not pose a pollution problem either.

Large-scale extraction of geothermal water can lead to subsidence of the ground or could trigger induced micro-earthquakes. Re-injection of geothermal fluids back into the aquifer using re-injection wells is therefore essential. This also serves to lengthen the lifespan of the geothermal energy resource project by maintaining pressure within the aquifer, and avoids pollution of surface waters.

5.5 Current and potential utilization and costs

5.5.1 Current utilization

Figure 5.8 gives the variation in worldwide geothermal electricity generation capacity and heating capacity from 1975 to 2005. Worldwide electrical capacity reached 8.9GW by 2005 while the heating capacity reached 28GW. To put this in context, Table 5.4 compares the electrical and (where applicable) heating capacities of solar, wind, commercial biomass, hydro and geothermal energy. Geothermal electricity generation is the smallest of the five, and provides less heat than solar thermal collectors or commercial biomass.

Table 5.5 lists the 15 countries with the greatest utilization of geothermal energy for electricity

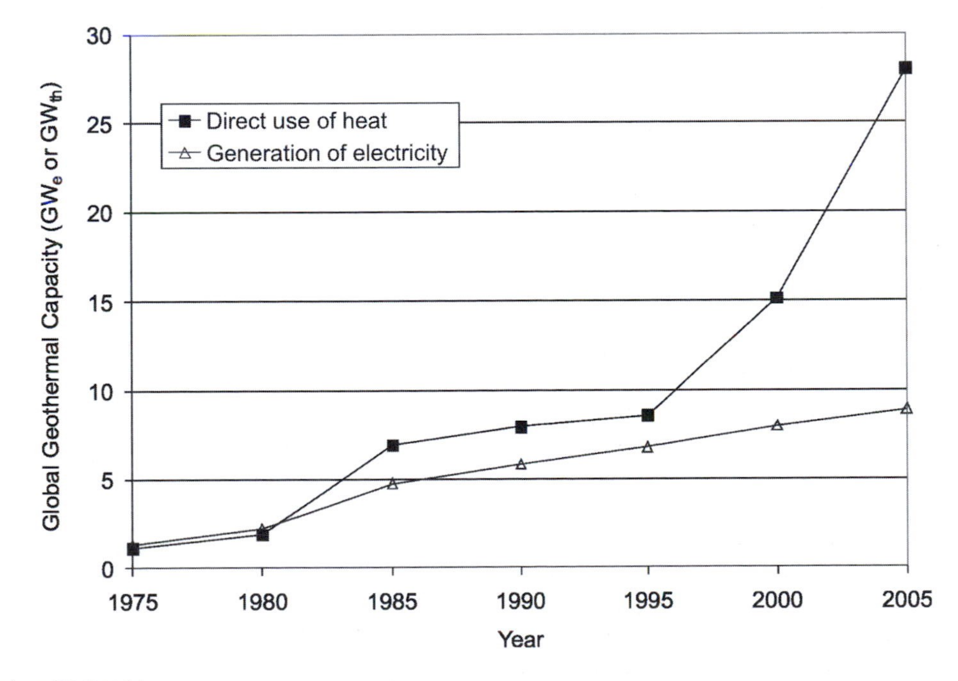

Source: Data from IEA (2006c)

Figure 5.8 *Variation in geothermal electricity power capacity from 1975 to 2005*

Table 5.4 *Electrical and heating capacity (GW), electricity generation (TWh/yr) and heating (EJ/yr) supplied by geothermal and other renewable energy sources*

	Electricity (capacity or generation)				Heating (capacity or supply)		
	GW	CF	TWh	Year	GW	EJ	Year
Solar	15	0.20	26	2008	148	0.93	2007
Wind	121	0.23	244	2008			
Commercial biomass	32	0.60	167	2007		3.2	2007
Geothermal	9.1	0.28	22.2	2005	28	0.25	2005
Hydro	867	0.39	2996	2005			
Global total	4267.0	0.49	18,454	2005		40	2005

Note: Global totals for electricity pertain to all sources of electricity, while the global total for heat is the sum of oil and natural gas use by residential and commercial buildings × 0.7 (the assumed average supply efficiency) plus heat supplied through district heating systems (all as given in Table 2.3 of Volume 1). CF = global mean capacity factor, taken from Table 2.5 of Volume 1 and applied (in the case of wind and biomass) to thermal energy production as well as to electricity production. Solar, wind, biomass and hydro data are from other chapters in this volume or from Table 2.4 of Volume 1.

Table 5.5 *Installed geothermal electricity generating capacity and energy production in 2004 for the 15 countries with the largest geothermal generating capacity*

Country	Running capacity (MW)	GWh generated	Capacity factor	% of national electrical energy
US	2228	17,917	1.057	0.5
Philippines	1909	9253	0.575	19.1
Mexico	953	6282	0.752	3.1
Indonesia	838	6085	0.829	6.7
Italy	699	5340	0.872	1.9
Japan	530	3467	0.747	0.3
New Zealand	403	2774	0.786	7.1
Iceland	202	1483	0.838	17.2
Costa Rica	163	1145	0.802	15
Kenya	129	1088	0.963	19.2
El Salvador	119	967	0.928	22
Nicaragua	38	271	0.814	9.8
Guatemala	29	212	0.835	3
Turkey	18	105	0.666	0
Guadeloupe	15	102	0.776	9

Source: Bertani (2005)

generation, while Figure 5.9 gives the countries with the highest share of geothermal in their electricity production. In absolute terms, the leading countries for geothermal electricity are the US, Philippines, Mexico and Indonesia, while in terms of percentage of total electricity production, the leading countries are El Salvador, Kenya, Philippines and Iceland. Table 5.6 lists the 15 countries with the greatest direct use of geothermal heat plus ground-source heat pumps. The leading countries are China, Sweden, the US and Turkey.

5.5.2 Future potential

Geothermal energy is not a renewable energy resource, since the rate of heat extraction in geothermal fields is many orders of magnitude greater than the typical geothermal heat flux of 0.05–0.1W/m^2. However, it could supply steady baseload electricity and heat for many centuries. Table 5.7 gives an estimate of the size of the geothermal resource that could eventually become available technically. The resource available worldwide down to depths of 10km in regions where the

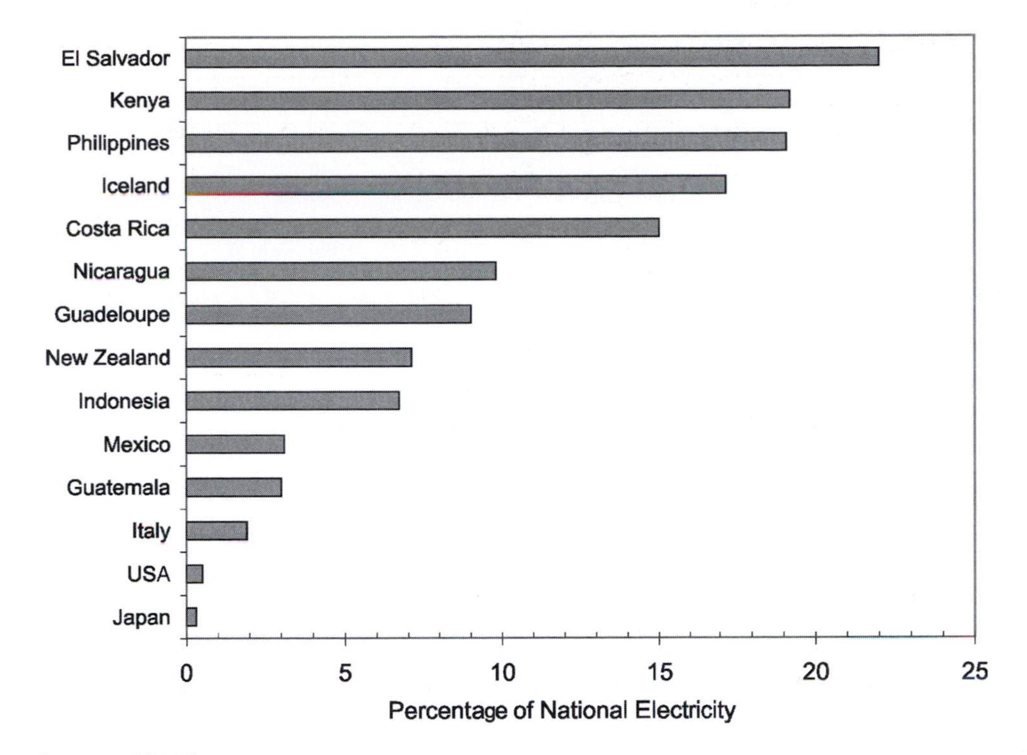

Source: Data from Bertani (2005)

Figure 5.9 *Countries with the largest share of total 2004 electricity production as geothermal electricity*

Table 5.6 *The world's top 15 direct users of geothermal heat in 2004*

Country	Capacity (MW_{th})	Energy supply (GWh/yr)	Capacity factor
China	3687	12,605	0.39
Sweden	3840	10,000	0.30
US	7817	8678	0.13
Turkey	1495	6900	0.53
Iceland	1844	6806	0.42
Japan	822	2862	0.40
Hungary	694	2206	0.36
Italy	607	2098	0.39
New Zealand	308	1968	0.73
Brazil	360	1840	0.58
Georgia	250	1752	0.80
Russia	308	1707	0.63
France	308	1443	0.53
Denmark	330	1222	0.42
Switzerland	582	1175	0.23
World	28,268	75,942	0.31

Source: Lund et al (2005)

geothermal temperature gradient exceeds 40K/km is around 32,000EJ. By comparison, total primary energy demand in 2005 was around 480EJ. The resource available in regions with a geothermal temperature gradient less than 40K/km, which would require deeper drilling and thus greater expense to exploit, is another 78,500EJ, bringing the total resource to about 110,000EJ (this resource estimate is, however, less than

Table 5.7 *Estimated global geothermal resource*

Resource type	Resource size (EJ)
Hydrothermal	130
Geopressurized[a]	540
Magma[b]	5000
Hot dry rock[c]	
Moderate to high grade (Δ T>40K/km)	26,500
Low grade (ΔT<40K/km)	78,500
Total	105,000
Total	**110,500**

Note: [a] Includes hydraulic and methane energy content. [b] To depths of 10km and rock temperatures greater than 650°C. [c] To depths of 10km and rock temperatures greater than 85°C.
Source: Mock et al (1997)

the 200,000EJ estimated by MIT to be recoverable in the US alone). If utilized at a rate of 100EJ/yr, it would last over 1000 years. The vast majority of this resource, however, occurs as hot dry rock, so development of the technology for using this resource is a key to long-term use of geothermal energy. Very little (1–2 per cent) of this resource is available at depths of less than 3km. More conventional geothermal resources could, however, provide 100EJ/yr of primary energy for about 60 years according to Table 5.7.

Figure 5.10 gives an estimate of the amount of heat available at various temperatures and in various depth intervals below the US land surface. One of the smallest bins shown in Figure 5.10 – temperature at 150°C at a depth of 3–4 km – contains almost 92,000EJ of available heat.

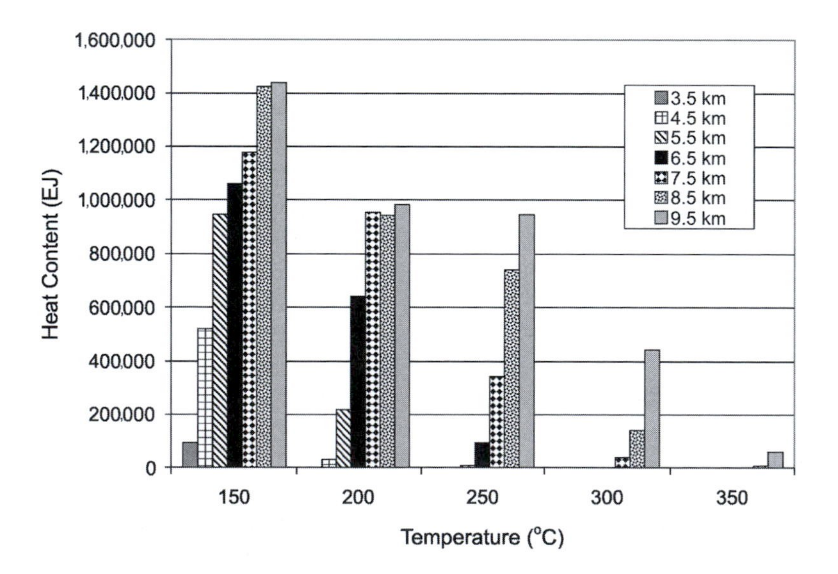

Note: The heat resource is large but non-renewable in practical terms.
Source: MIT (2006)

Figure 5.10 *Amount of heat available at different temperatures and at different depth intervals below the US land surface*

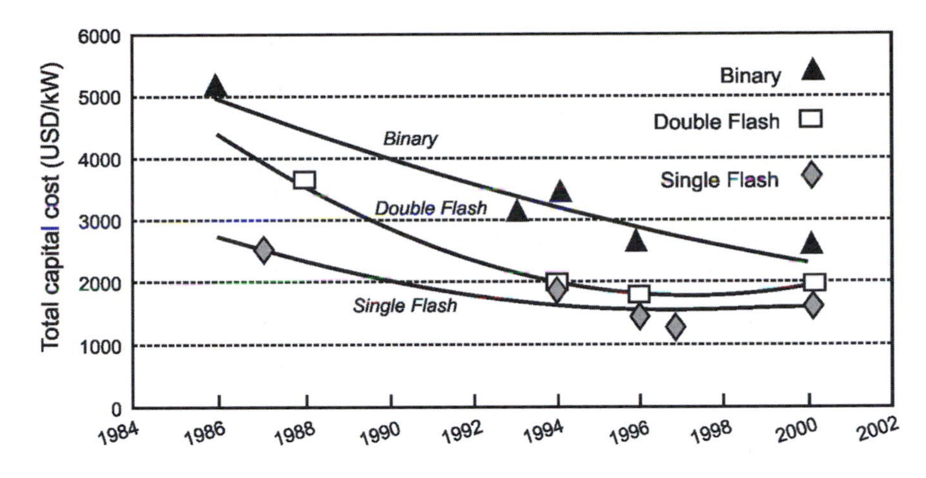

Source: IEA (2006c)

Figure 5.11 *Variation in capital cost for single-flash, double-flash and binary geothermal power systems from 1986 to 2000*

Table 5.8 *Cost of single-flash geothermal electricity (year 2000 $) as function of resource quality and plant size*

Plant size (MW)	Direct capital costs ($/kW)			Electricity cost (cents/kWh)		
	High-quality resource (T>250°C)	Medium-quality resource (T=150–250°C)	Low-quality resource (T<150°C)	High-quality resource (T>250°C)	Medium-quality resource (T=150–250°C)	Low-quality resource (T<150°C)
Small (< 5)	1600–2300	1800–3000	2000–3700	5–7	5.5–8.5	6–10.5
Medium (5–30)	1300–2100	1600–2500	NNS	4–6	4.5–7	NNS
Large (> 30)	1150–1750	1350–2200	NNS	2.5–5	4–6	NNS

Note: NNS = normally not suitable
Source: IEA (2006c)

5.5.3 Present and projected future costs

Figure 5.11 shows the variation from 1986 to 2000 in the capital cost of single-flash, double-flash and binary-cycle geothermal power systems. Costs for binary systems (the most expensive type) declined from about $5000/kW in the mid-1980s to about $2500/kW by 2000. Single-flash systems declined from about $2500/kW to $1500/kW. All unit costs depend on the quality of the resource and the size of the project. Table 5.8 gives capital costs and electricity costs for single-flash systems as a function of resource quality and plant size. Costs in 2000 ranged from $1150/kW for large (>30MW) plants with high-quality (>250°C) resources, to up to $3700/kW for small (<5MW) plants with low-quality (<150°C) resources.

Table 5.9 gives the costs of recent hydrothermal electricity generation and anticipated costs of hot dry rock electricity generation, broken down into powerplant capital costs, O&M costs and drilling costs. Also given are the costs of supplying heat through geothermal-powered district heating systems. Hydrothermal electricity costs range from about 5 to 17 cents/kWh, while projected high-grade hot dry rock (EGS) costs range from about 6 to 8 cents/kWh. Thermal energy costs about $4–7/GJ (by comparison, natural gas prices have fluctuated largely between $4 and $10/GJ in recent years).

Table 5.9 *Costs of existing hydrogeothermal energy and anticipated future costs of hot dry rock geothermal energy (1996 $)*

Resource	Powerplant cost ($/kW)	Powerplant cost (cents/kWh)	O&M cost (cents/kWh)	Drilling cost (cents/kWh)	Total cost (cents/kWh)
Electricity					
Hydrothermal					
– high grade	1000–1500	2.4–3.6	0.3	2–3	4.7–6.9
– low grade	2000–2500	5.1–6.3	0.4	4–10	9.5–16.7
Hot dry rock					
– high grade[a]	1000–1500	2.4–3.6	0.3	3–4	5.7–7.9
– low grade[b]	2000–2500	5.1–6.3	0.4	20	25.5–26.7

Resource	Fuel cost ($/GJ)	Heat distribution cost ($/GJ)	Total cost ($GJ)
Heat supply			
Hydrothermal			
– high grade	1.7	2.1	3.8
– low grade	2.1	2.6	4.7
Hot dry rock			
– high grade[a]	2.6	2.6	5.2
– low grade[b]	4.7	2.1	6.8

Note: [a] Thermal gradient >60K/km. [b] Thermal gradient of 30K/km.
Source: Mock et al (1997)

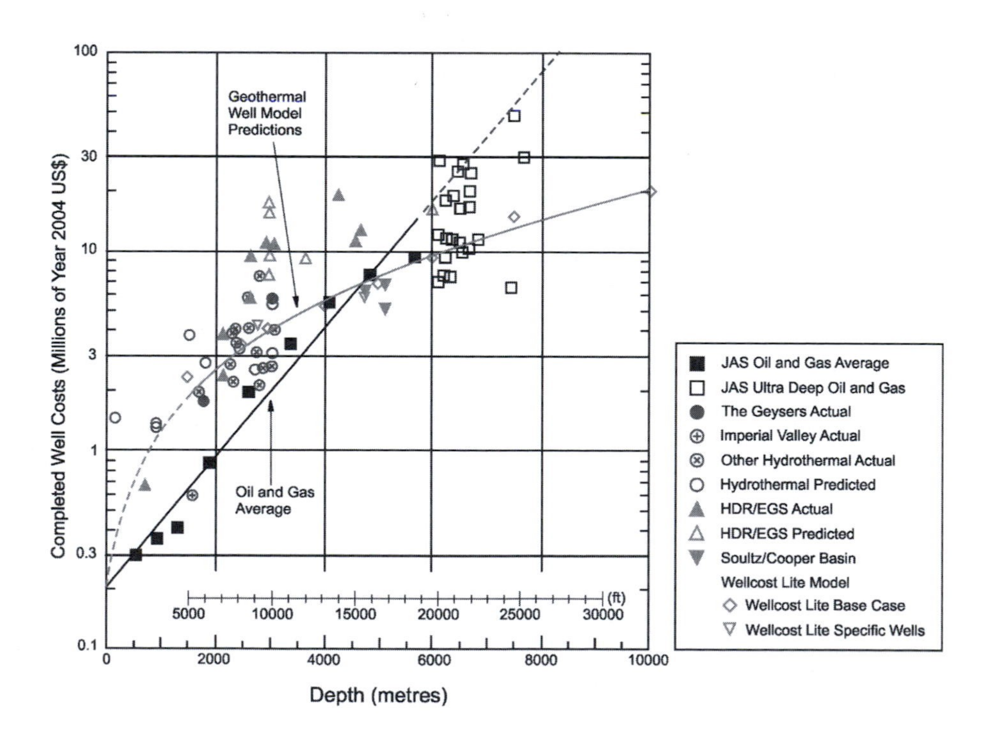

Note: The logarithmic scale on the y axis.
Source: MIT (2006)

Figure 5.12 *Costs of oil, gas and hot dry rock geothermal wells*

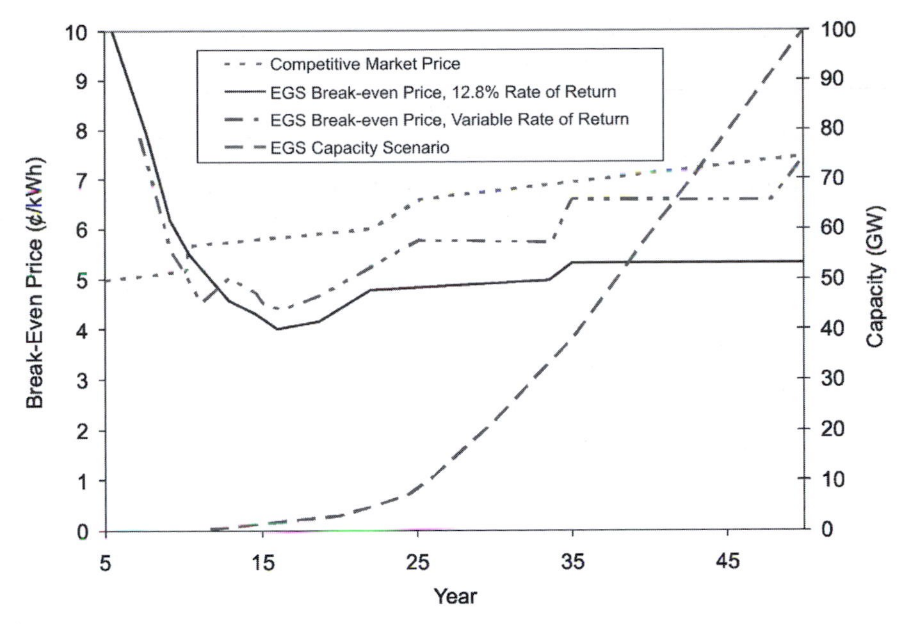

Source: MIT (2006)

Figure 5.13 *Scenario for the variation in the cost of electricity from EGS as the EGS power capacity in the US increases to 100GW by 2050*

Costs of EGS are highly uncertain, especially for depths approaching 10km, but the aforementioned MIT panel identified a number of emerging drilling technologies as well as potential long-term revolutionary technologies (hydrothermal flame spallation, fusion drilling, chemically enhanced drilling and metal shot abrasive-assisted drilling) that could dramatically reduce drilling costs. Figure 5.12 shows data on actual and expected drilling costs as a function of depth; costs and uncertainties increases dramatically with increasing depth. Figure 5.13 shows a scenario of low and high electricity costs from EGS as the EGS capacity increases to 100GW by 2050. Costs initially decrease due to technological advances and learning-by-doing, then gradually increase as less attractive geothermal resources are tapped. Long-term electricity costs of 4–7 cents/kWh are foreseen, which would make EGS a very attractive energy option (and, unlike wind and solar energy, would provide continuous baseload power without the need for storage). These cost projections must be regarded as highly uncertain at present.

5.6 Concluding comments

The key for geothermal energy to make a significant contribution to world energy needs is the development of hot dry rock technology, also known as EGS. If this technology can be developed according to the projections cited here, geothermal energy could provide one to several hundred EJ/yr for many centuries. Geothermal energy can be used directly for heating and for the generation of electricity. In heating applications, the resource could be significantly extended through the use of heat pumps to extract additional heat from the already-used water that is returned to the ground.

Note

1 Steam is vaporized liquid water in equilibrium with liquid water, although in common parlance, 'steam' refers to condensed water itself.

6

Hydroelectric Power

This chapter reviews the physical principles involved in the generation of hydroelectric power, the current hydroelectric power capacity, potential future increases in this capacity and potential GHG emissions associated with hydroelectric power.

Hydroelectric power can be divided into run-of-the-river power, in which a river flow is directed through a turbine without holding back water in a reservoir, and reservoir-based power. Hydroelectric power is often regarded as C free, entailing no GHG emissions. This may be true for run-of-the river power, but it is not true for reservoir-based power. The latter involves emission of CO_2 during the production of cement, with some of the CO_2 produced as a byproduct of the chemical reactions that produce cement and therefore independent of the type of energy used to make the cement. A potentially much greater source of GHG emissions is the release of CO_2 and especially methane associated with the decomposition of organic matter in the region flooded by the creation of the reservoir.

6.1 Physical principles

In a hydroelectric powerplant, the gravitational potential energy of water in the reservoir is converted to electrical energy. The efficiency of electricity generation is the ratio of electric energy generated to the loss of gravitational potential energy of the water flowing from the reservoir through the turbine. The loss of gravitational potential energy depends on the difference in elevation between of the water surface of the reservoir and the turbine.[1] This difference in elevation is referred to as the *head*. The gravitational potential energy of the reservoir water is ultimately provided by solar energy, as it is solar energy that drives the evaporation of water from the oceans and its ascent into the atmosphere, from which it eventually falls as rain.

The mechanical power P_w of a volumetric rate of water flow Q is given by:

$$P_w = \rho g Q H \tag{6.1}$$

where ρ is the density of water, g is the acceleration due to gravity and H is the effective head (m) across the turbine (the effective head is 75–95 per cent of the real head, the difference being due to frictional energy losses as the water flows through channels and pipes to the turbine). The mechanical power P_t produced at a turbine shaft is equal to the turbine efficiency η_t times the mechanical power of the water, while the electrical power P_e produced by the generator is equal to the shaft power times the electrical generator efficiency η_e. Thus:

$$P_t = \eta_t \rho g Q H \tag{6.2}$$

and:

$$P_e = \eta_e \eta_t \rho g Q H \tag{6.3}$$

Hydroelectric powerplants involve mature technologies, with typical turbine efficiencies of 75–85 per cent, but as high as 90 per cent in new large turbines and as low as 60 per cent in small old turbines (IEA, 2002). Generator efficiencies are typically about 98 per cent.

The maximum turbine efficiency occurs under optimal conditions (rotation speed and direction of incoming water relative to the turbine blades). When demand falls, the output can be reduced by reducing the water flow (in a Francis turbine, described below, this is done by turning the guide vanes to constrict the flow), but the generator must retain the same rotation speed so as to produce electricity at the same frequency. The change in flow velocity with a constant turbine speed means that the water hits the blades at a different and therefore suboptimal angle, so the efficiency falls. In some propeller type turbines the angle of the blades can be altered.

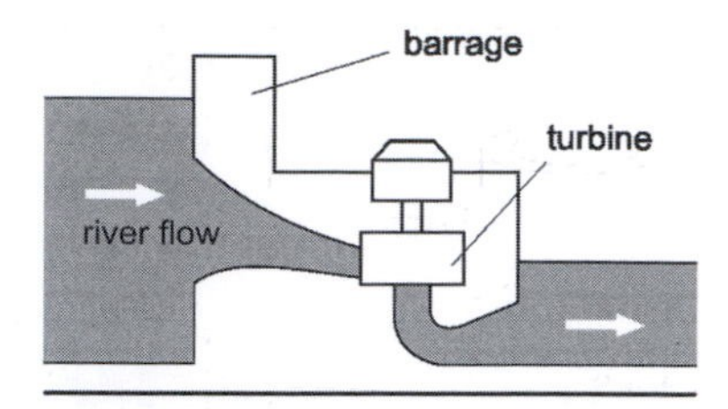

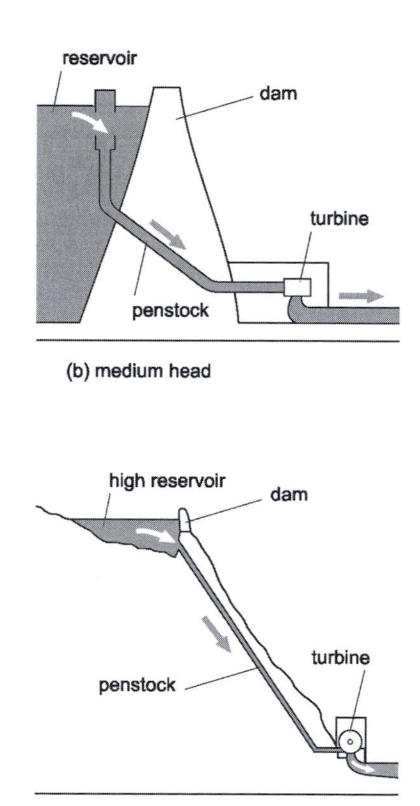

Source: Ramage (1996)

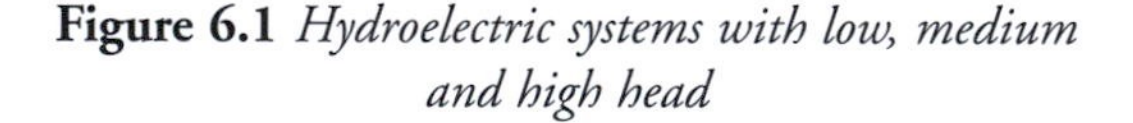

Figure 6.1 *Hydroelectric systems with low, medium and high head*

6.2 Hydroelectric configurations and types of runners

Figure 6.1 illustrates different configurations of hydroelectric reservoirs and turbines, with low, medium and high hydraulic head. In high-head systems, the bottom of the reservoir is above the turbine. The channel that directs the water to the turbine in medium- and high-head configurations is referred to as the *penstock*. The same power output could be achieved through a combination of low head and large flow rate, or high head and low flow rate. The former case is typical of hydroelectric reservoirs and requires larger engineering works and machinery to accommodate the large water flow, which increases cost (some of the cost of hydroelectric reservoirs might be assigned to flood control and irrigation benefits). Conversely, high-head systems can use more compact machinery, but it must be able to withstand greater pressures.

There are several different turbine designs, four of which – the propeller-type, Francis type, Pelton wheel and cross-flow turbine – are illustrated in Figure 6.2. The combination of blades and support structure is referred to as the *runner*. Different types of turbine are most suitable for different combinations of flow rate and head, as illustrated in Figure 6.3.

With the propeller turbine, water enters along the axis of the turbine and through the entire area swept by the blades. This makes it suitable for very large flows with only a few metres head. The speed of the tip of the blade is about twice the water speed, allowing for rapid rotation even with low water speeds. Propeller turbines in which the angle of the blade can be altered are referred to as Kaplan turbines.

With the Francis turbine, water arrives at the tip of the blades and flows inward to a channel along the axis of the turbine. It is suitable for heads of 2–200m. This is a *reaction* turbine, as water is deflected by the blades, pushing the blades in the opposite direction. The water arrives under pressure, with a pressure drop through the turbine accounting for a significant part of the delivered energy. The blade speed is comparable to the water speed.

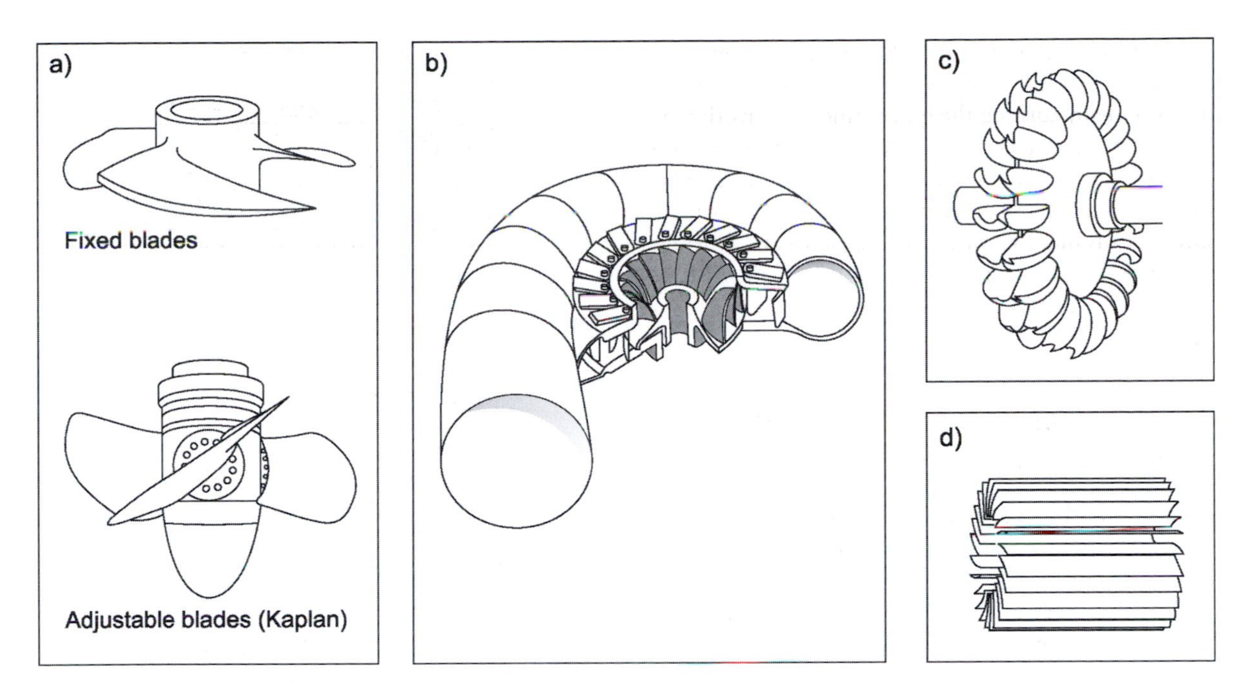

Source: Ramage (1996)

Figure 6.2 *Four runner designs used in hydroelectric turbines: (a) propeller type, with either fixed blades or adjustable blades (the Kaplan turbine), (b) Francis type, (c) Pelton wheel type, and (d) cross-flow type*

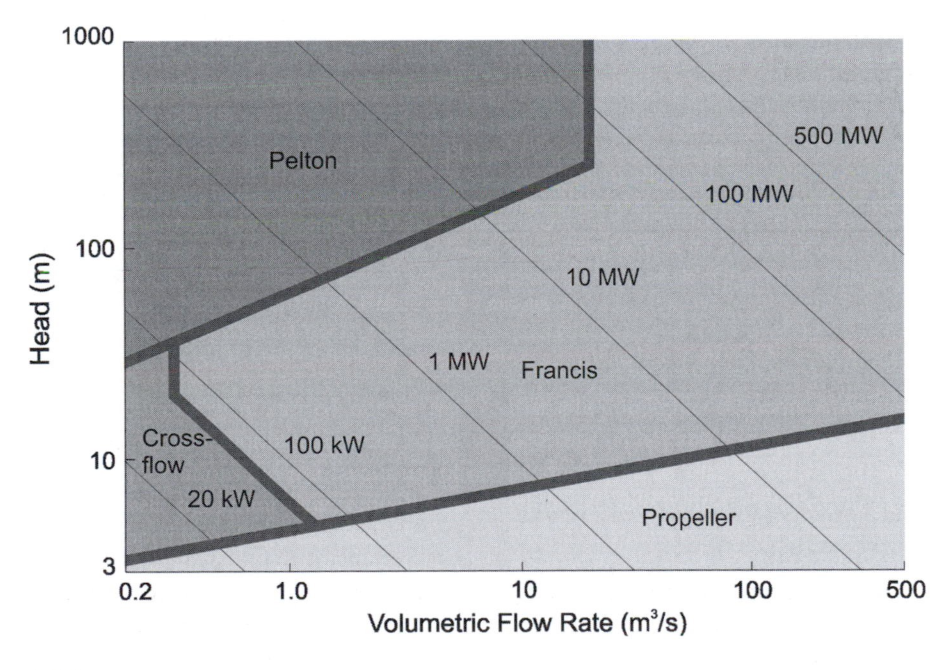

Source: Ramage (1996)

Figure 6.3 *Combinations of water flow and head where different types of turbines are used*

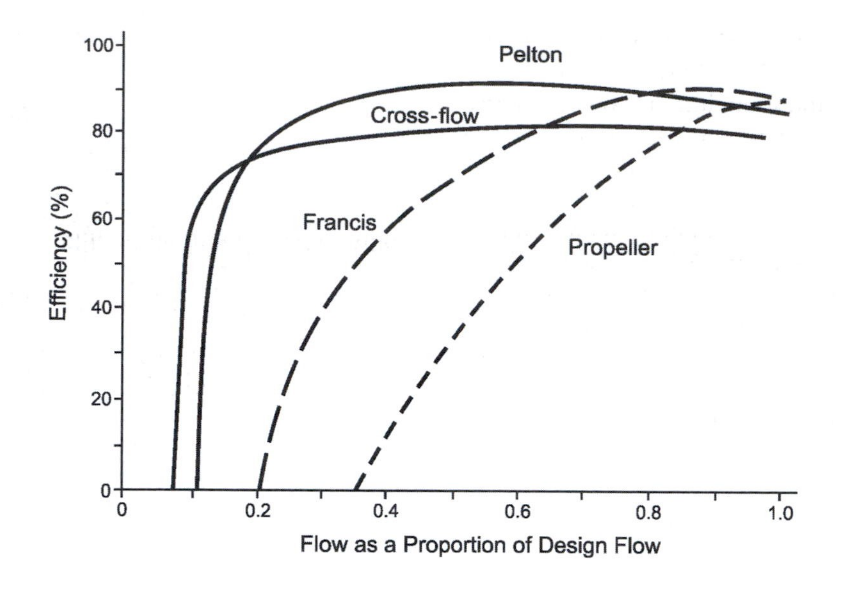

Source: Paish (2002)

Figure 6.4 *Variation of hydroturbine efficiency as a function of the flow rate*

The Pelton wheel consists of a wheel with a set of cups mounted around the rim. It is driven by a jet or jets of water hitting the cups. The energy is delivered in short pulses, so this is referred to as an *impulse* turbine. Unlike the propeller and Francis turbines, which are entirely submerged in water, the Pelton wheel largely operates in air. Maximum efficiency occurs when the speed of the cups is half the speed of the water jet or jets. It is restricted to relatively large heads (greater than 250m for large systems, less for smaller systems).

Turbine efficiencies are roughly constant down to about 30 per cent of the peak flow for the Pelton and cross-flow propellers, but then drop abruptly at lower flows, whereas efficiency drops continuously with decreasing flow for the Francis propeller. This is illustrated in Figure 6.4. This compounds the decrease in electricity production when water flow decreases.

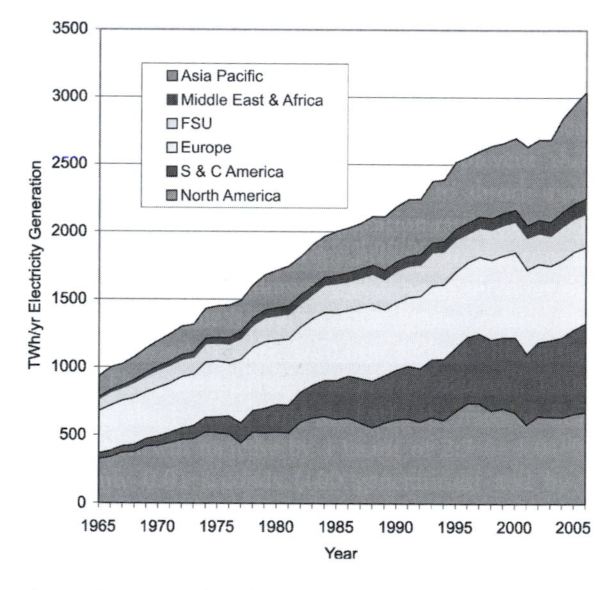

Source: Data from BP (2007)

Figure 6.5 *Growth in the generation of hydroelectricity in major world regions over the period 1965–2006*

6.3 Existing hydroelectric capacity and electricity generation

Figure 6.5 shows the variation in the production of electricity from hydroelectric powerplants in major world regions over the period 1965–2006. Over this 41-year period, the average growth rate was 2.9 per cent/year, although during the last ten years the growth has averaged only 1.7 per cent/year. In 2005, hydropower accounted for about 20 per cent of world electricity-generating capacity (867GW out of 4267GW) and 16 per cent of world electricity production (2996TWh out of 18,454TWh) (see Volume 1, Table 2.4). This translates into an average capacity factor (average power output divided by capacity) of 39 per cent, compared to an average capacity factor of about 50 per cent for the world's fossil fuel powerplants and 83 per cent for the world's nuclear powerplants (see Volume 1, Table 2.5). Figures 6.6, 6.7 and 6.8 give the top ten countries in terms of hydroelectric power capacity, hydroelectric energy production and fraction of total electricity produced from hydropower in 2005, respectively.

The hydropower potential can be divided into 'large' hydro and 'small' hydro. Different countries use different boundaries between the two, but a common boundary is at 10MW. Large hydro generally involves reservoirs and flooding of land, whereas small hydro either involves small reservoirs or 'run-of-the-river' powerplants. Table 6.1 gives the small hydro capacity and energy production for the top ten countries as given in WEC (2007), along with the fraction of total hydroelectric power capacity in the form of small hydro. China is the leader in the development of small hydropower stations, with 9.6GW of capacity, equal to about 5 per cent of its total hydropower capacity.

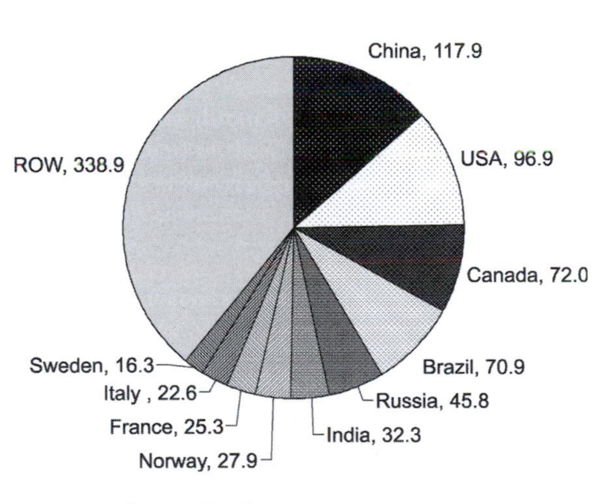

Source: Data from UN (2008)

Figure 6.6 *The top ten countries and rest of the world in terms of hydroelectric power capacity (GW) in 2005*

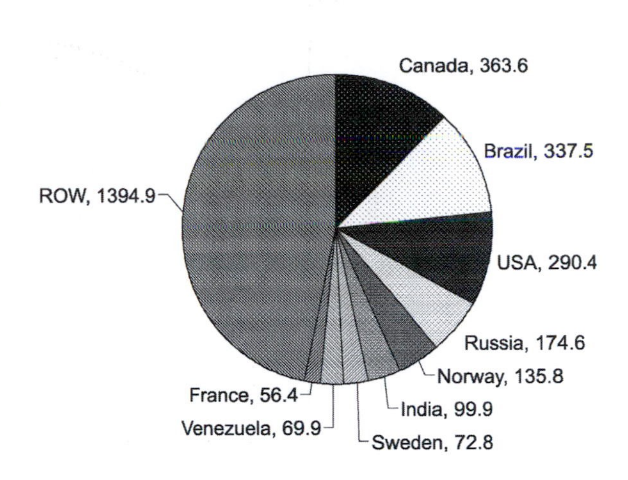

Source: Data from UN (2008)

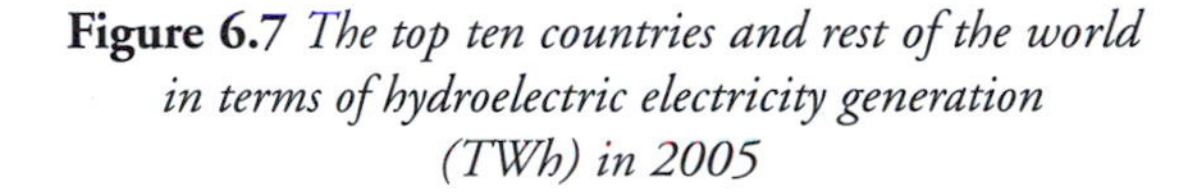

Figure 6.7 *The top ten countries and rest of the world in terms of hydroelectric electricity generation (TWh) in 2005*

6.4 Technical and economic potential for further hydropower

The gravitational potential energy of a parcel of mass M is given by:

$$PE = gMH \tag{6.4}$$

where g is the acceleration due to gravity (9.8m/s^2) and H is the height of the parcel. The worldwide rate of precipitation over land is about 1.06×10^{17}kg/yr, of which 0.69×10^{17}kg/yr evaporates and 0.37×10^{17}kg/yr reaches the ocean as runoff (Harvey, 2000). The average height of the land surface is about 800m, so the total potential energy of rainfall runoff is about 2.9×10^{20}J or 106,000TWh/yr. By comparison, the worldwide generation of electricity in 2005 amounted to about 18,500TWh – about six times less. Of course, it will never be possible to capture every drop of rain nor to extract the potential energy in its entire descent from where it falls on land to sea level. In the absence of hydropower turbines, the potential energy of river flow will be entirely converted to heat, through frictional dissipation of kinetic energy created as the water descends. That portion of the water kinetic energy that is converted to electricity is also ultimately dissipated as

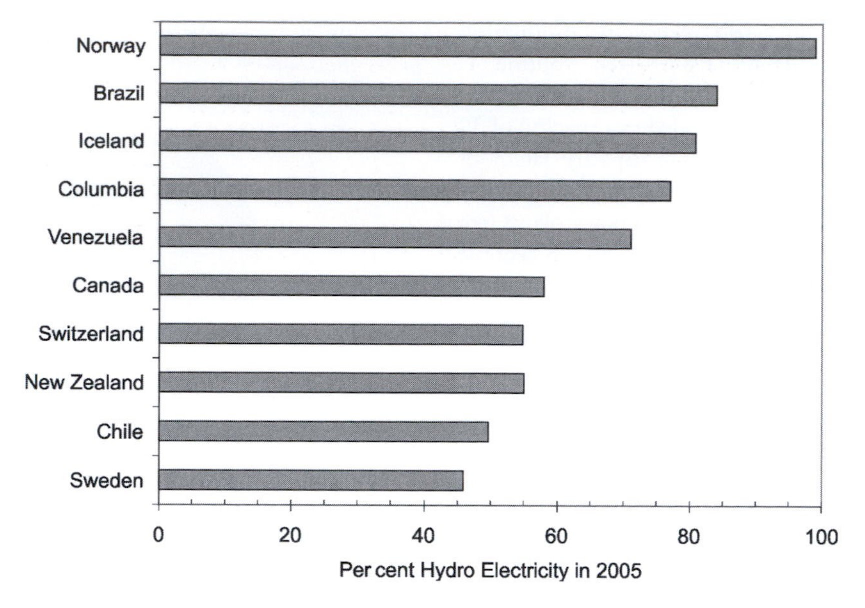

Source: Computed from data in Volume 1, Table 2.4

Figure 6.8 *The ten countries where hydroelectric power provides the largest fraction of total electricity generation*

Table 6.1 *Capacity (GW) and energy production (TWh/yr) from small hydro (facilities <10MW in size) in 2005 for the ten countries with the largest absolute contribution from small hydro, supplemented with year 2000 data for China, Europe and North America*

Country	Actual		Small hydro as % of total hydro	
	GW	TWh/yr	Capacity	Energy
China	9.5		5	
Japan	3.48		12.5	0.0
USA	2.84	10.7	3.7	4.0
Italy	2.41	7.6	13.9	21.1
France	2.02	5.8	7.9	10.3
Spain	1.79	4.7	9.6	20.4
Brazil	1.43	6.7	2.0	2.0
Austria	0.99	4.0	8.4	10.3
Sweden	0.99	3.8	6.1	5.3
Czech Republic	0.28	1.1	27.5	44.6
Peru	0.23	1.0	7.2	5.6
Europe	12.5		9.7	
North America	5.1			

Source: Ten countries with largest absolute contribution from small hydro listed in WEC (2007); supplementary data on China, Europe and North America from IEA (2003)

heat, partly during transmission but mostly at the point of use.

Table 6.2 gives the theoretical potential to generate electricity, along with the technical and economic potentials by continent and for the countries with the greatest potentials in each continent. The theoretical potential is defined as the electricity that could be generated if all natural flows down to sea level or to the border of the country were captured with 100 per cent efficiency. The technical potential is that portion of the theoretical potential that could be captured with current technology. The economic potential is that portion of

Table 6.2 *Potential electricity generation (TWh/yr), existing (end of 2005) or future generation (TWh/yr); total electricity generation (TWh) in 2005, and percentage of total electricity generation met by hydropower in various continents and selected countries (listed for each continent in order of decreasing technical potential)*

Region	Potential hydro generation			Actual or hydro future generation				Total generation	% hydro
	Theoretical	Technical	Economic	Existing	UC	Planned	Total		
Africa	>3884	>1852	1007	90.7	22	300	413	558.2	16.2
Congo	1397	774	419	7.4		271	278	7.42	99.7
Ethiopia	650	>260	160	2.9	10.5	5.0	18	2.87	99.3
Cameroon	294	115	103	3.9		2.4	6.3	4.15	94.2
Asia	>16,703	>5793	3818	824.5	286.4	730	1841	6640.2	12.4
China	6083	2474	1753	397	169	270	836	2500	15.9
India	2638	660	600	99.9	53	33	186	705.2	14.2
Indonesia	2147	402	40	10.8	0.4	2.4	14	127.4	8.5
Tajikistan	527	>264	264	16.9	2.2	90	109	17.1	98.8
Australasia	495	>189	69	49.7	0.1	1.0	51	305.1	16.3
Australia	265	100	30	15.7			16	252.5	6.2
Papua New Guinea	175	49	15	0.5			0.5		
New Zealand	46	37	24	23.5	0.1	1.0	25	43	54.7
Europe	4945	2714	1830	735.6	31	51	818	4843.5	15.2
Russia	2295	1670	852	174.6	20	37	232	953	18.3
Norway	560	200	187	135.8	1.5	4.2	142	137.3	98.9
Italy	340	105	65	33.6	0.1	0.6	34	294.4	11.4
Sweden	130	100	85	72.8			73	158.4	46.0
France	270	100	70	56.4			56	576.8	9.8
North America	8054	3012	1114	702.3	12	83	797	5284.4	13.3
US	4485	1752	501	290.4			290	4304.8	6.7
Canada	2216	981	536	363.6	6.9	55	426	629.5	57.8
Mexico	135	49	32	27.7	1.9	3.8	33	235.4	11.8
South America	>7121	>3036	1725	601.3	42	303	946	822.6	73.1
Brazil	3040	1488	811	337.5	23	169	530	403.1	83.7
Peru	1577	395	260	20	1.1	6.1	27	25.5	78.4
Venezuela	320	246	130	69.9	12	13	95	98.5	71.0
Columbia	1000	200	140	39.8	2.7	41	84	51.6	77.1
Total	41,202	16,596	9563	2996	393.5	1468	4858	18,454	16.2

Note: UC = under construction.

Source: WEC (2007) for potential hydro generation; UN (2008) for existing hydro and total generation (with total generation adjusted to include wind, as described in Volume 1, Table 2.4)

the technical potential that could be exploited with current technology under current economic conditions (and may or may not exclude economic potential that would be unacceptable for social or environmental reasons). Also given in Table 6.2 is the current (2005) hydroelectrical generation, the projected annual generation due to projects under construction and planned, the total existing plus projected generation, current total electricity generation and the percentage of current generation met by hydro in 2005. Worldwide, existing generation amounts to about 3000TWh/yr, projects under construction or planned add another 1900TWh/yr (bringing the total to 4900TWh/yr), the economic potential is about 9600TWh/yr and global electricity demand in 2005 was about 18,500TWh/yr. The global technical potential is close to total current global electricity demand and the economic potential is about half of current demand, but the realizable potential (taking into account social and environmental constraints) is much less. Continental totals are summarized in Figure 6.9.

Table 6.3 gives existing hydroelectric capacities (GW) and capacity under construction or planned as of 2005 in each continent and in various countries. The world hydroelectric capacity in 2005 was 867GW and another 531GW were under construction or planned. This makes hydroelectric power the fastest-growing renewable energy source in absolute terms. By comparison, world electric generating capacity was about 4300GW in 2005.

In the US, no new hydroelectric reservoirs have been built since the mid-1970s and no new ones are likely to be built. However, there is still the potential to almost double the existing US hydropower capacity of 97GW. According to Kosnik (2008), 3.9GW could be added by upgrading existing hydropower facilities (with, for example, more efficient turbines), 17.1GW could be added to dams that were built for purposes other than hydropower generation, and 58.9GW of what he considers to be environmentally friendly small (<30MW) hydro could be added out of a technical potential of 431GW and an available potential of 275GW. This brings the total addition to 80GW. For China, Zhou et al (2009) estimate a potential of 128GW from small hydro (defined as <50MW), compared to 35GW in place at present. About 64 per cent of the small hydro potential is in the western part of the country, away from major demand centres.

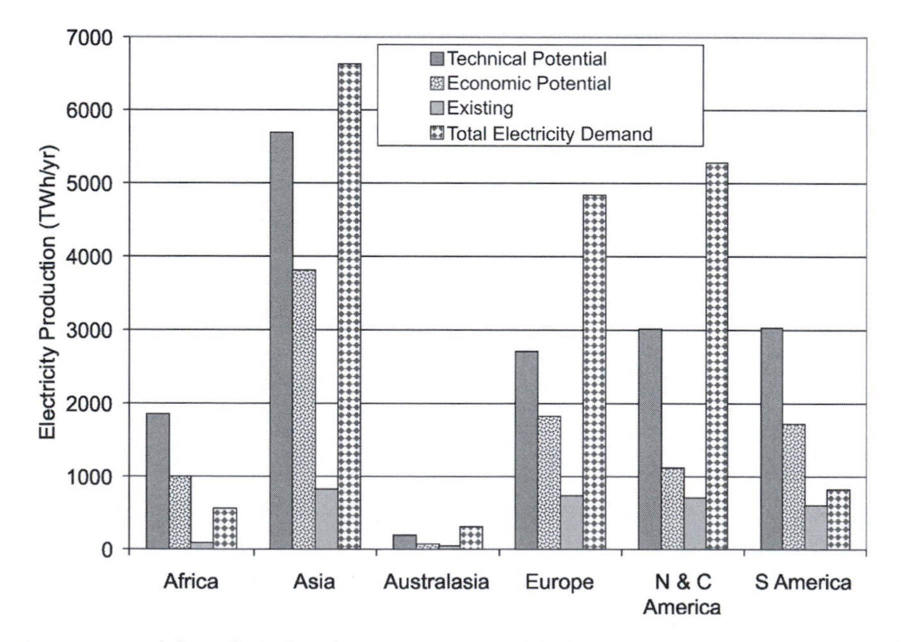

Figure 6.9 *Comparison of the technical and economic potential hydroelectric energy generation, and hydroelectric energy generation and total electricity demand in 2005, given in Table 6.1 for each continent*

Table 6.3 *Existing hydroelectric power capacity (GW), hydro capacity under construction or planned, and total electric power capacity*

Region	Actual (2005) or future capacity			
	Existing	UC	Planned	Total
Africa	23.1	5.7	85	113.8
Congo	2.4		42.8	45.2
Ethiopia	0.7	3.1	1.3	5.1
Cameroon	0.7		0.5	1.2
Asia	276.4	96.3	227	599.7
China	117.9	50	80	247.9
India	32.3	13.2	8.9	54.4
Indonesia	3.2	0.1	0.8	4.1
Tajikistan	4.5	0.7	27	32.2
Australasia	15.1	0	0.2	15.3
Australia	9.3			9.3
Papua New Guinea	0.2			0.2
New Zealand	5.3	0	0.2	5.5
Europe	244.8	10.1	12.3	267.2
Russia	45.8	5.6	8	59.4
Norway	27.7	0.3	0.9	28.9
Italy	22.6	0	0.2	22.8
Sweden	16.1			16.1
France	25.3			25.3
North America	184.1	2.9	19.8	206.8
US	96.9	0	0	96.9
Canada	72.0	1.5	11.6	85.1
Mexico	10.6	0.8	2.1	13.5
South America	123.3	9.1	62.9	195.3
Brazil	70.9	5	36.6	112.5
Peru	3.2	0.2	1.1	4.5
Venezuela	14.4	2.3	3	19.7
Columbia	9	0.7	10	19.7
Total	866.8	124.1	407.2	1398.1

Source: WEC (2007) for hydro capacities planned or under construction, UN (2008) for existing capacities (with total capacity adjusted to include wind)

6.5 Upgrading existing facilities

Relicensing in the US results in an average 8 per cent loss in generation due to the imposition of new environmental constraints on operation. However, there are two ways in which hydroelectric power production could be increased if there is a sustained research and development effort to improve the design of turbine systems and minimize adverse environmental impacts (Brown et al, 1998).(1) Modernizing and upgrading existing turbines could result in a 5–10 per cent increase in electricity production in the US (counteracting the impact of operational restrictions in new licences). New technologies provide opportunities for concurrent improvement of fish passage, water quality and downstream aquatic habitats. (2) Adding new generation capacity at existing dams could amount to a 20–25 per cent increase (20–25GW) in the US.

As noted above, Kosnik (2008) estimates a potential increase of about 21GW from upgrades of existing hydroelectric facilities and additions of power capacity to existing non-hydropower dams.

6.6 Social and environmental considerations

Not all of the economically viable technical potential for hydropower will be exploited, due to other constraints on the construction of dams. These include displacement of populations living on lands flooded by reservoirs and a variety of environmental impacts, including loss of biodiversity due to flooding, disruption or prevention of fish migration along rivers, changes in the downstream hydrology and supply of sediment to coastal regions (where it sometimes serves to maintain beaches and shore areas in the face of erosion by waves), changes in water temperature and quality, and mobilization of mercury in rivers (which is a problem with many of the hydroelectric developments in Quebec, Canada). Many hydroelectric projects have occurred with little or no consultation with the affected people, inadequate compensation of displaced people and little or no consideration of environmental impacts. Sometimes the lifespan and power production from a reservoir have been far less than expected due to the accumulation of sediment behind the dam. Often the benefits have flowed to people and national-level governments far removed from the hydroelectric project, with little benefit to local communities. GHG emissions can also be significant, as discussed in section 6.7. In response to these problems, the *World Commission on Dams* was established in May 1998 as an international, multi-stakeholder process that comprehensively reviewed 125 dams and commissioned over 100 papers, resulting in the release in November 2000 of its final report, *Dams and Development: A New Framework for Decision-Making* (WCD, 2000). Among the Commission's key conclusions are:

- Dams have made an important and significant contribution to human development...
- In too many cases an unacceptable and often unnecessary price has been paid to secure those benefits
- Lack of equity in the distribution of benefits has called into question the value of dams in meeting water and energy development needs when compared with the alternatives
- Negotiating outcomes will greatly improve the development effectiveness of water and energy projects by eliminating unfavourable projects at an early stage...

Most of the negative impacts of hydroelectric power can be eliminated through small, run-of-the-river projects, which do not entail building dams and flooding land.

6.7 GHG emissions

Hydroelectric power would appear to qualify as a C-free power source, as no fossil fuel energy is needed. However, plant material in the reservoirs flooded by hydroelectric dams decomposes anaerobically, producing methane, some of which is oxidized as it rises to the water surface, but some of which escapes to the atmosphere. As methane is 26 times more effective at trapping infrared radiation than CO_2 on a molecule-per-molecule basis, the impact on climate of methane emissions could be substantial (even after accounting for the relatively short lifespan of methane molecules in the atmosphere, as embodied in the global warming potential index that is discussed in Volume 1, Chapter 2, subsection 2.6.5). An assessment of the net effect of hydropower on GHG emissions requires accounting for the following:

- pre-dam emissions of CH_4 from the river and from periodically flooded wetlands, or pre-dam sinks of CH_4 due to oxidation of atmospheric methane in soil;
- emissions of CH_4 from the reservoir itself;
- emissions of CH_4 from the water as at flows through the turbines and spillways (relative losses to the atmosphere probably being different for the two);
- emissions of CO_2 from organic matter that decomposes as a result of the flooding that created the reservoir behind the dam.

For present reservoirs, there generally are no measurements of pre-dam CH_4 emissions. Emissions from the reservoir itself are highly variable in space and time, so a comprehensive long-term monitoring programme would be needed. To measure the loss of CH_4 from the water as it flows through the turbines and spillways, one needs site-specific measurements of the methane concentration in water entering the spillways and turbines, and some distance downstream. The methane concentration in entering water depends on the depth from which the water is drawn and the vertical profile of methane concentration in the reservoir, something that will vary from reservoir to reservoir and also over time (there being both a seasonal cycle and long-term trends). Not all the emission of CO_2 from the reservoir should be counted as an impact of the reservoir; some of the emission will be from decomposition of organic matter swept into the reservoir from elsewhere within the watershed (which would decompose anyway), and some will be from the respiration of organic matter produced by photosynthetic organisms within the reservoir (this CO_2 does not represent a net flux). Needless to say, these processes pose difficult accounting and measurement problems, as exemplified by the heated exchange between Fearnside (2004) and Rosa et al (2004).

In spite of these uncertainties, various workers have estimated the CO_2 and CH_4 emissions from hydroelectric reservoirs. Figure 6.10a shows the annual

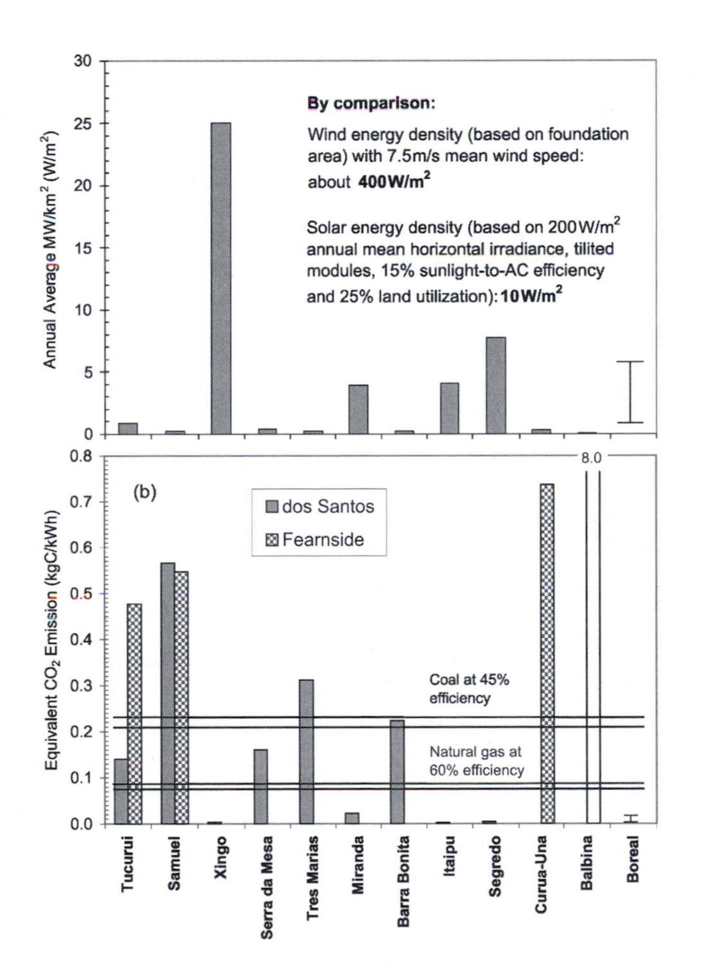

Source: Data from dos Santos et al (2006) and Fearnside (2002, 2005a, 2005b) for Brazilian reservoirs; Duchemin et al (2002) for boreal reservoirs

Figure 6.10 *(a) Annual average power output per unit of flooded land area for reservoirs in Brazil and boreal reservoirs in Quebec, assuming a capacity factor of 0.5, and (b) CO_2-equivalent GHG emissions per kWh of generated electricity for the reservoirs shown in (a), in comparison to emissions from coal and natural gas powerplants*

average power density (the ratio of annual average power to flooded area) for 11 hydroelectric reservoirs in Brazil, and the range of power densities for 5 reservoirs in Quebec (annual average power was computed as powerplant capacity times an assumed capacity factor of 0.5). It ranges from 0.38W/m^2 to 50W/m^2, with all except two of the Brazilian projects having a power density of less than 5W/m^2. By comparison, the power output from wind turbines per unit of land area that cannot be used for other purposes is about 400/m^2 in regions with a mean wind speed of 7.5m/s (a very good onshore wind regime) (see Chapter 3, section 3.9), while that from tilted PV arrays with 200W/m^2 mean annual irradiance, 15 per cent conversion efficiency and an array/land area ratio of 0.25 is about 10W/m^2. Thus, the power production per unit of flooded land is generally less than the power production per unit land area used for tilted PV arrays, and substantially less than for wind farms (however, the timing of the electricity production from hydro reservoirs is controllable at short timescales and to some extent at seasonal timescales).

Figure 6.10b shows the equivalent CO_2 emission (based on estimated CO_2 and CH_4 emissions) in comparison to that from a state-of-the-art coal powerplant at 45 per cent efficiency or natural gas combined cycle powerplant at 60 per cent efficiency (for the latter, allowance has been made for CH_4 emissions as given in Volume 1, Table 2.10). The emissions as computed by dos Santos et al (2006) are gross emissions from the reservoir (i.e. pre-reservoir emissions have not been estimated and subtracted), but CH_4 emissions from degassing downstream from the reservoir have not been included. The estimates by Fearnside are meant to be net emissions, including degassing emissions. Fearnside also takes into account the elimination of N_2O emissions from flooded soils and of methane from forest termites. According to the estimates shown in Figure 6.10b, some of the Brazilian reservoirs have greater emissions than from coal powerplants, while other Brazilian reservoirs and all of the Canadian reservoirs have much smaller emissions than natural gas powerplants. Emissions related to the decay of biomass originally in the reservoir will decrease over time, but some portion of the methane emissions is related to organic matter that washes into the reservoir and then decomposes anaerobically; this emission will persist over time.

Figure 6.11 plots CO_2-equivalent emissions per kWh against power density for the Brazilian and Canadian hydroelectric reservoirs. In both cases, lower emissions per kWh are associated with greater power densities. Thus, proposed new reservoirs should go forward only if, as a minimum, they have relatively high power densities.

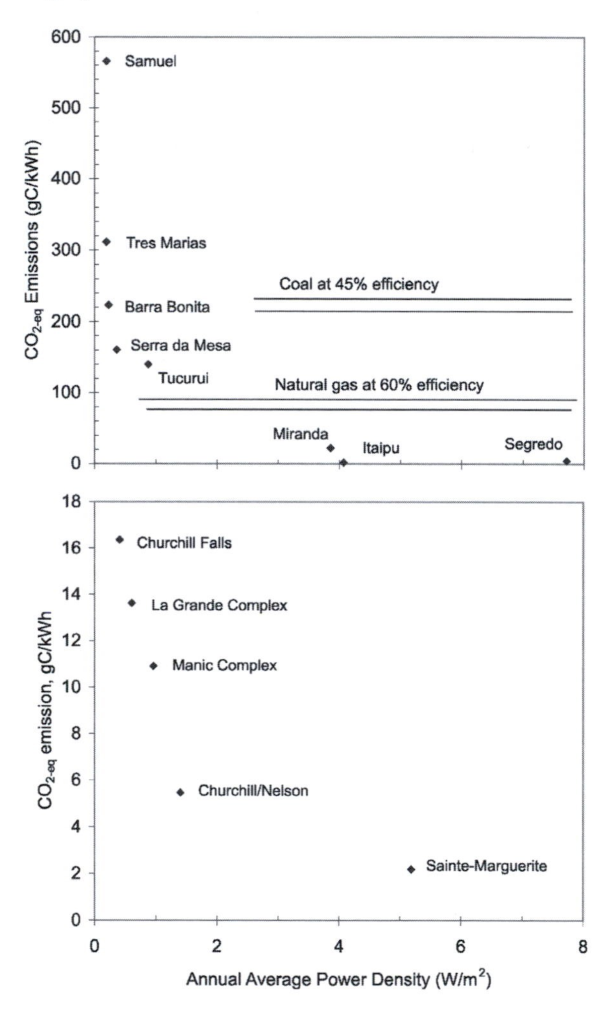

Note: There is a difference in the vertical scale between the two panels
Source: Hydro data from Duchemin et al (2002)

Figure 6.11 *CO_2-equivalent GHG emissions (in terms of kgC) per kWh of electricity generated versus annual average power density for hydroelectric reservoirs in Brazil (upper panel) and Quebec (lower panel), along with emissions from coal and natural gas powerplants in the upper panel*

6.8 Cost

REN21 (2005) indicates the following investment costs for hydroelectric power:

- small hydro in China, \$900/kW;
- small hydro elsewhere, \$1300/kW;
- large hydro in China, \$1400/kW;
- large hydro elsewhere, \$2000/kW.

There will, however, be many exceptions to these estimates. For example, the Rupert River project in Quebec, currently under construction, is expected to cost Cdn\$5 billion but will add only 900MW of capacity, equivalent to about Cdn\$5500/kW (~US\$4800), while the proposed Romaine River project in Quebec is projected to cost \$8 billion for 1550MW, or Cdn\$5200/kW (these costs include the cost of construction of access roads and transmission lines).[2] At the opposite extreme, Carvalho and Sauer (2009) cite a Brazilian assessment indicating that 75GW of additional hydropower in Brazil (compared to 71GW existing) could be developed for costs of up to \$1500/kW, with transmission costs presumably extra.[3]

Figure 6.12 gives a range of capital costs for various capacities up to 2MW. Costs outside developing countries range from \$4000–8000/kW at the 100kW scale to \$2000–2500/kW at the 2MW scale according to this figure, while costs in developing countries at scales of less than 500kW are given as \$1500–3000/kW.

Table 6.4 gives the costs of hydroelectricity for investment costs (including possible new transmission costs) of \$1000/kW to \$8000/kW, a 50-year

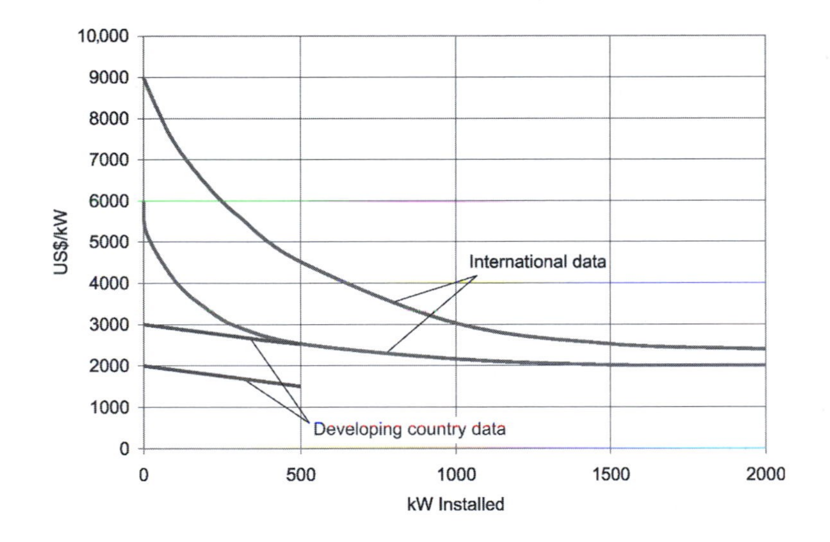

Source: Paish (2002)

Figure 6.12 *Variation of capital cost with capacity for small hydro (<2MW)*

Table 6.4 *Cost of hydroelectric energy (cents/kWh) for various capital costs, interest rates and capacity factors, assuming amortization of the initial investment over a 50-year period*

Interest rate	0.03	0.03	0.06	0.06	0.09	0.09
Capacity factor	0.3	0.6	0.3	0.6	0.3	0.6
Capital cost						
\$1000/kW	1.48	0.74	2.41	1.21	3.47	1.74
\$2000/kW	2.96	1.48	4.83	2.41	6.94	3.47
\$4000/kW	5.92	2.96	9.66	4.83	13.89	6.94
\$8000/kW	11.83	5.92	19.31	9.66	27.77	13.89

Note: O&M, insurance, water rent, transmission and administrative costs are not included.

amortization period *N*, real interest rates *i* (excluding inflation) of 3 per cent, 6 per cent and 9 per cent and capacity factors *CF* of 0.3 and 0.6. The cost is computed as:

$$C = \frac{CRF \cdot CC}{8760 \cdot CF} \tag{6.5}$$

where *CC* is the capital cost, 8760 is the number of hours in a year and the *CRF* is the cost recovery factor, computed using Equation (C.1) of Appendix C. For an interest rate of 3 per cent, electricity costs range from 0.7 to 5.9 cents/kWh for a capacity factor of 0.6 and capital costs ranging from \$1000 to \$8000/kW (O&M, administrative and water charges are not included). This makes hydroelectricity one of the least expensive forms of electricity, even with substantial transmission costs. Because hydropower is a mature technology, the potential for cost reduction is small.

6.9 Concluding comments

Existing hydroelectric powerplants already constitute a large, renewable source of electricity – about 16 per cent of current global electricity demand. Projects under construction or planned will increase electricity production by about 50 per cent, while the economic potential is more than three times current global hydroelectric electricity supply under current conditions. However, not all of the economic potential can be developed, because of adverse environmental or social impacts. Further hydro development should be critically scrutinized to make sure that there is a large benefit–cost ratio, including minimal GHG emissions from decomposition of organic matter in flooded terrain. Hydropower projects should only go forward based on negotiated and legally binding agreements with the affected people based on their free and informed consent. Displaced peoples should be fully and fairly compensated, and the cost of such compensation should be included in the projected cost of hydroelectric energy and taken into account, along with the cost of other renewable or C-free supply alternatives and energy efficiency measures, when deciding whether or not to proceed. Hydroelectric developments are particularly advantageous when they serve to compensate for variable wind or solar electricity production, as they then serve to leverage greater renewable energy contributions without requiring greater fossil fuel spinning reserve (as discussed Chapter 3, section 3.11).

Notes

1 Although the inlet to the turbine will be placed some distance below the water surface, water entering the inlet is replaced by water in the layer above the inlet, extending to the water surface.

2 The levelized cost of electricity from the Romaine River project is expected to be 9.2 cents/kWh – about twice what it will be sold for to aluminium smelters and other larger industrial consumers, so it would represent a large subsidy to these industries.

3 For two of the projects (Santo Antônio and Jirau), totalling 6.45GW, the power density would be about 6.5W/m^2 if the average capacity factor is 0.5 – better than most Brazilian hydroelectric developments to date but still not impressive.

7
Ocean Energy

7.1 Introduction

The potential forms of ocean energy that can be converted to electricity are ocean wave energy, tidal energy and energy from vertical thermal and salinity gradients. All of these forms of energy are at a very early stage of development, and intensive research would be needed to bring them to commercial viability. The greatest challenge is to build systems that work reliably for 10–20 years in a continuously harsh environment. The smallest technical glitch can undermine otherwise promising designs. However, the potential energy resources from the ocean, particularly wave and thermal gradient energy, are enormous.

A review of ocean energy technologies carried out as part of the IEA's *Implementing Agreement on Ocean Energy Systems* identified 135 different ocean energy technologies or projects that are under development, most beginning in the 1990s (Khan and Bhuyan, 2009). These include 76 ocean wave technologies, 40 tidal current devices, 8 tidal barrage projects, 8 ocean thermal energy conversion (OTEC) projects and 3 salinity gradient concepts. This tally excludes technologies that have been discontinued due to poor cost and performance. Of the 135 technologies, 23 were still at the concept stage, 42 involved part-scale testing in tanks, 38 involved part-scale testing in the ocean, 18 were at the full-scale ocean-testing stage and 13 were at a pre-commercial demonstration stage (the last of these consisted of five ocean wave projects, four tidal current projects and four tidal barrage projects). The largest share of technologies under development is in the UK (40 projects), followed by the US (23 projects), Canada (11 projects) and Norway (8 projects). Current designs are still far from optimal (so there is room for improvement) and there has been inadequate testing to determine which concepts will perform as hoped for in harsh marine environments.

7.2 Wave energy

The power in waves varies with the square of the wave height and linearly with the wave period. Figure 7.1 gives the variation in average wave power (kW per m of coastline) along the coastlines of the world. The greatest wave power densities occur off north-western Europe, western Canada and Alaska, and southernmost South America and Australia. Wave energy is the most concentrated of the forms of renewable energy, with energy fluxes of 4–11kW/m width of incoming wave along the European Mediterranean coast, 25kW/m off the southern Atlantic coast of Europe and up to 75kW/m off the coast of Ireland and Scotland (Clément et al, 2002). Average wave power is less in the region within 25° of the equator, but is steady due to the steady trade winds, and so can still be attractive. Studies in the early 1990s indicated that the total wave energy resource for Europe is 120–190TWh/yr offshore and 34–46TWh/yr onshore (Boud, 2002), compared to total European electricity demand of about 4700TWh/yr (see Table 2.4). The US wave energy potential that could be harnessed is estimated to be 260TWh/yr, which is roughly equal to current hydroelectricity generation (6.5 per cent of the total) in the US (Musial, 2006).

Table 7.1 outlines the various wave energy technologies under development and the number of projects of each type identified by IEA (2006d). The large number of different technologies still under development indicates that the level of understanding is not sufficiently advanced to show which concepts will be the most cost effective.

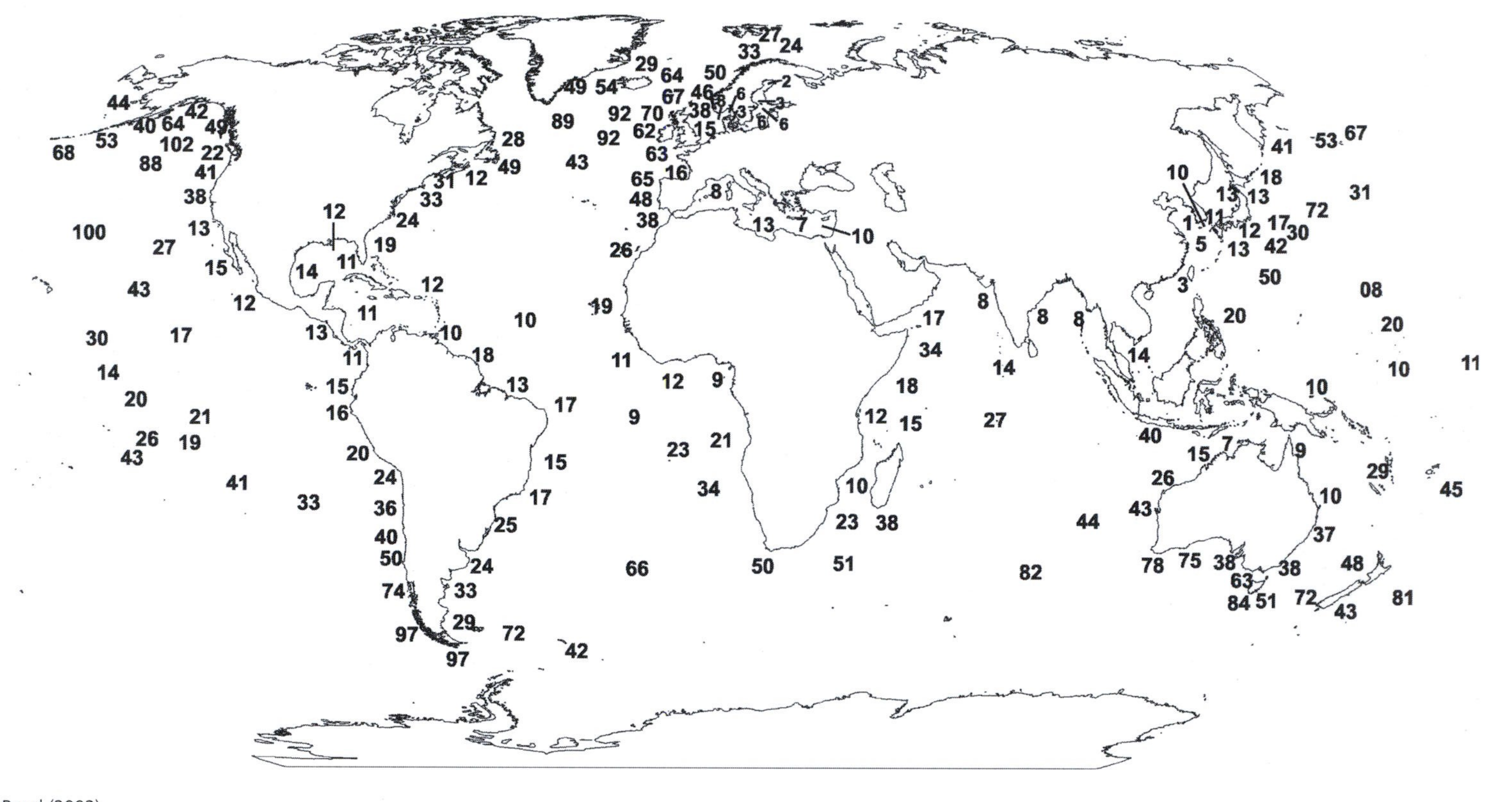

Source: Boud (2002)

Figure 7.1 *Wave power density (kW per metre of coastline) along the world's coastlines*

Table 7.1 *Types of ocean wave energy technologies*

Technology	Description	Number of projects
Overtopping	A wave surge/focusing device, in which waves topple over a wall into a reservoir. The reservoir creates a head of water that is released through hydroturbines as the water flows back to the sea	3
Collector	A floating device that captures waves to concentrate energy into various power systems	2
Oscillating water column (OWC)	A submerged, resonantly tuned column, sealed at the top and open to the sea below the water surface and with air trapped in a column above the water. The water in the column moves up and down with wave movements, compressing and decompressing the air. The air is channelled through an air turbine that makes use of the airflow in both directions	12
Oscillating wave surge converter (OWSC)	A wave surge/focusing device that consists of a near-surface collector mounted on an arm that oscillates with the movement of water in the waves	4
Point absorber	A floating structure that absorbs energy from all directions through its movement at or near the water. It can be designed to resonate	21
Terminator	A near-surface floating structure similar to a point absorber, but absorbing energy only from one direction. It extends perpendicularly to the wave direction, restraining waves as they arrive. It can be designed to resonate	1
Submerged pressure differential	Located near the shore and attached to the seabed. Waves cause sea level to rise and fall above the device, inducing a pressure differential that pumps a fluid through the system to generate useful energy	2
Attenuator	A long floating structure aligned parallel with the wave direction. Movements along its length can be selectively constrained to produce energy. It has a smaller area perpendicular to the waves than a terminator, so it experiences smaller forces	4
Other		5

Source: IEA (2006d)

Figure 7.2 illustrates two possible wave energy conversion devices: an onshore oscillating water column device and a floating ocean wave converter buoy. There are many concepts for floating offshore devices, all of which would be able to take advantage of the greater energy in waves in deeper water. These include the point absorber and various overtopping devices (in which waves spill into a slightly elevated floating basin, then drain through a low-head turbine).

Wave energy, like wind energy, is variable in time. However, the available wave energy can be accurately predicted 24–48 hours in advance through the use of satellite remote sensing of surface winds and wave heights over entire ocean basins. This in turn reduces the fossil fuel spinning reserve that would otherwise be required in most jurisdictions to rapidly compensate for fluctuations in power production (as discussed in Volume 1, Chapter 3, subsections 3.1.6 and 3.2.6).

7.3 Tidal energy

Tides are produced by the gravitational attraction of the moon and sun on the waters of the ocean. In fact, with regard to the moon, there are two factors at work: gravitational attraction as well as a centrifugal effect.

7.3.1 Physics of tides

The centrifugal effect arises from the tendency of an object to move in a straight line, so if the object is on a spinning globe, it tends to be continuously thrown outward from the axis of rotation. The earth does not travel in a smooth trajectory in its orbit around the sun as the moon rotates around the earth; rather, the earth and the moon rotate around a point located just below the earth's surface as the moon rotates around the earth, as illustrated in Figure 7.3. The net result is that there is an outward-directed centrifugal force related to the rotation of the earth around this centre of rotation. A centrifugal force is strongest for points furthest from the axis of rotation, so the centrifugal force related to the rotation of the earth and moon around a common centre is strongest on the side of the earth *furthest* from the moon and weakest on the side facing the moon (this centrifugal force is in addition to that associated with the rotation of the earth around its own axis, but

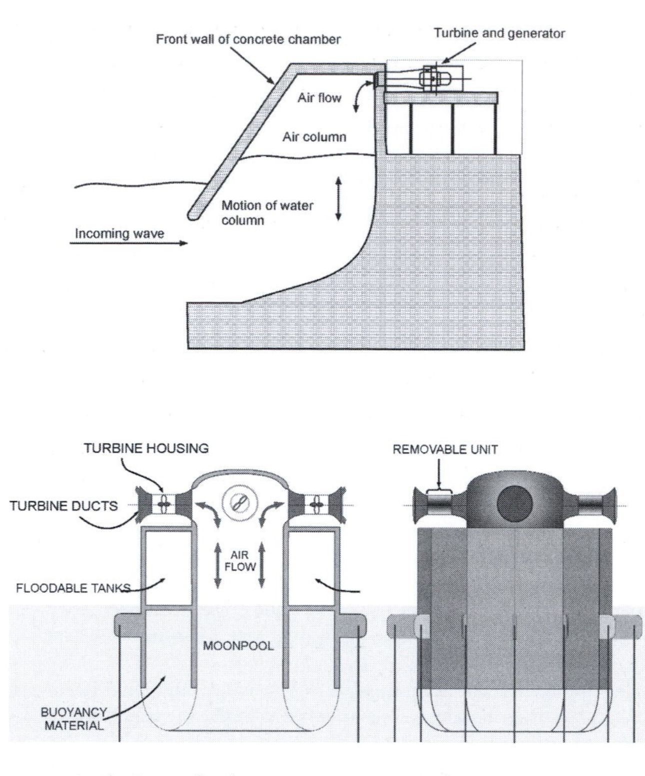

Note: Top, a shoreline ocean wave converter; bottom, a floating ocean wave convertor buoy
Source: Khan and Bhuyan (2009)

Figure 7.2 *Wave energy conversion devices*

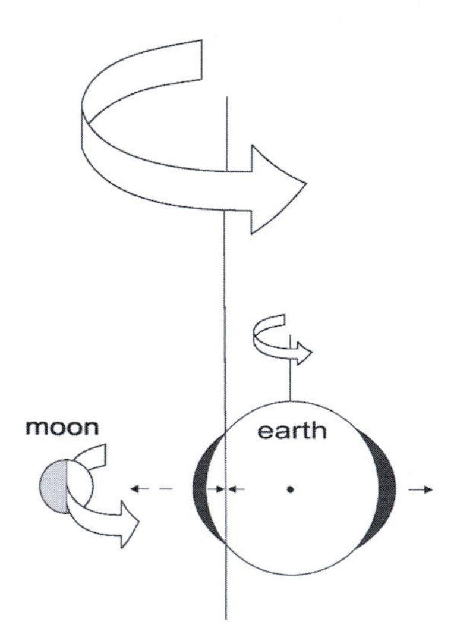

Source: Modified from Elliott (1996)

Figure 7.3 *Rotation of the earth and the moon around a common centre and the resulting bulge in the ocean surface due to the associated centrifugal force*

is ignored here because it is unrelated to the presence of the moon and is the same at all longitudes). This in turn creates an outward bulge in the ocean surface that is largest on the side furthest from the moon, as illustrated in the lower part of Figure 7.3.

The second lunar effect, the gravitational attraction of the moon, draws the ocean water on the side of the earth facing the moon toward the moon, creating a bulge toward the moon. The ocean water on the side of the earth opposite the moon experiences less than average gravitational attraction by the moon and so bulges slightly outward. The net result of a centrifugal effect that is strong on the side of the earth furthest from the moon and weak on the side facing the moon, and vice versa for the gravitational effect, is to create tidal bulges that, in theory, are the same on the sides facing and opposite the moon. In reality, there are slight differences between the two bulges related to topographic effects and the tilt of the earth's axis in relation to the orbit of the moon.

Because the earth rotates on its axis once per day, the bulges remain fixed in position 'under' and opposite the moon. Thus, a given location will experience two high tides and two low tides per day. As the moon rotates around the earth in the same direction as the earth's rotation once every 29.5 days, the time interval from one occurrence of a given longitude directly below the moon to the next is $24.0 \times (1 + 1/29.5) = 24.8$ hours rather than 24 hours. Thus, lunar high tides occur once every 12.4 hours.

This pattern is modified by the gravitational attraction of the sun. The gravitational attraction of the sun on the seas of the earth is 177 times greater than that of the moon, but the effect on tides is much less. The reason for this is that the effect of both the sun and the moon on the tides depends on the ratio of the earth's diameter to the distance between the earth and the sun or moon. When the sun and the moon are aligned and pull on the oceans together, the result is the very high *spring tide*, whereas when the sun and the moon are at 90° to each other, the result is the lower *neap tide*. Spring tides can be more than twice as high as neap tides. The basic pattern of two spring tides and two neap tides per month is modified by the fact that the moon's orbit is elliptical rather than circular, and by the fact that the plane of the moon's orbit is tilted relative to the plane of the earth's orbit around the sun, which causes a semi-annual cycle with a 10 per cent variation in the height of the tides.

In the middle of the ocean, the variation in the height of the water surface due to the tides is about 0.5m. However, *resonances* from one side of an ocean basin to another can amplify the tidal range considerably near the shore. In music, resonance occurs when the dimensions of a sound cavity match the wavelength of an incoming vibration or signal, so that waves trapped in the cavity reflect off the walls in such a way as to reinforce or amplify the original signal. The cyclic rising and falling of the tides is analogous to an incoming sound signal, and these can be reflected off the boundaries of an ocean basin so as to reinforce the original signal. This happens between the two sides of the North Atlantic Ocean, where the distance of 4000km between Europe and North America is just right to create a resonance with the 12-hour tidal cycle. The result is that the tidal range is about 3m rather than 0.5m on either side of the North Atlantic Ocean. Resonance effects around the Pacific Ocean are weaker and less regular. In some parts of the world there is only one significant tide per day because resonance effects occur only over a 24-hour period.

Resonances can also occur between the edges of shallow coastal seas and in bays, increasing the tidal range from its typical coastal value of 3m to over 10m. In the case of estuaries, a combination of a funnelling effect (as the width of the incoming tidal flow decreases up-estuary) and resonance effects can be at work. A good example is the increase in tidal range as one moves into the Severn Estuary in the UK, illustrated in Figure 7.4. The funnelling effect is offset to some extent by frictional losses as the tide flows into the estuary. Friction increases as the water becomes more concentrated, such that the tidal range will reach a maximum at some point as one moves up an estuary and not increase further thereafter. In the Severn Estuary, the maximum tidal range occurs at the Severn Bridge (shown in Figure 7.4).

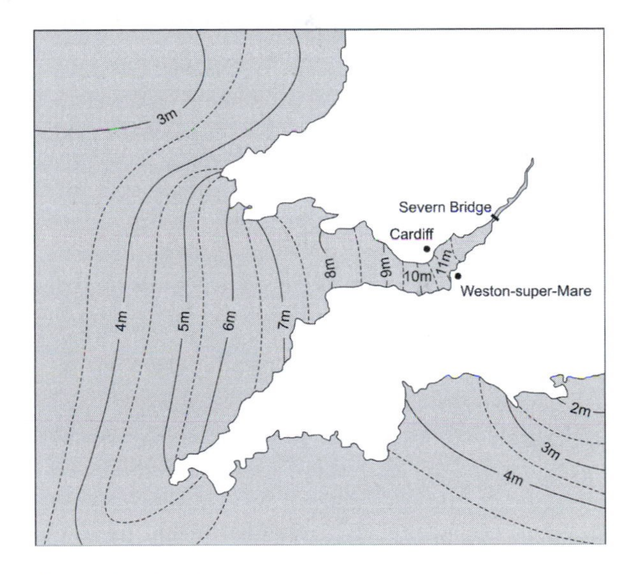

Source: Elliott (1996)

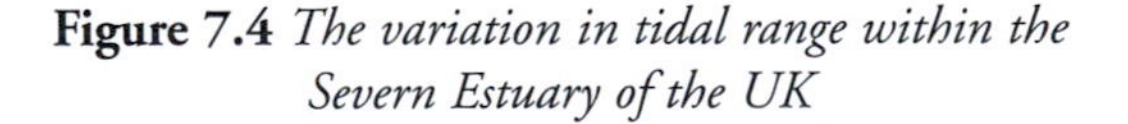

Figure 7.4 *The variation in tidal range within the Severn Estuary of the UK*

7.3.2 Power generation from tides

To generate power from tidal energy requires constructing a dam or tidal *barrage* across a suitable estuary and either holding back the incoming water or holding back the outgoing water (or both) until a sufficient difference in water on the two sides of the barrage has been created that it can be used to generate electricity with a turbine (analogous to hydroelectric power generation). Water is held back and then released using *sluice gates* that can be lowered into a channel between the two sides of the barrage and later raised. Power can be generated during the *flood* (incoming) tide or during the *ebb* (outgoing tide). The sea level outside the barrage must be allowed to rise first, then water is allowed to flow through the turbines into the basin, but the basin water level will not have had time to rise to the peak water level outside the barrage before sea level starts to decrease. This is illustrated in Figure 7.5a. The net result is that the tidal range and maximum water level inside the barrage are reduced compared to before. Conversely, water flow can be initially blocked during the ebb, resulting in a reduced tidal range with a higher minimum water level inside the barrage compare to before, as illustrated in Figure 7.5b. Ebb generation is the most popular, perhaps because maximum water levels are unaffected.

It is possible to generate electricity during both the flood and ebb phases, with variation in basin level and power output as illustrated in Figure 7.5c. In this case, the design of the turbine blade would not be simultaneously optimized for flow in both directions. Total energy production is not increased substantially compared to ebb or flood generation alone, but the electricity production would be spread over a longer time period and peak power would be reduced, thereby reducing the cost of generators (WEC, 2007).

Figure 7.6 illustrates some different possible configurations of the water channel, turbine and generator. In the *bulb* design, the generator is sealed inside an enclosure that is mounted within the flow. This requires cutting off the flow of water during maintenance. In the *stratflo* design (used at Annapolis Royal in the Bay of Fundy, Canada), the generator is mounted around the rim and only the turbine is in the water flow. In the *tubular* design, the channel curves such that the turbine shaft extends outside the channel, so the generator set can be placed next to the channel. An advantage of tidal energy is that it is entirely predictable, as the ebb and flow of tides are governed entirely by the gravitational attraction of the sun and moon on the earth. Although predictable, some form of energy storage (or backup) is needed, as the power

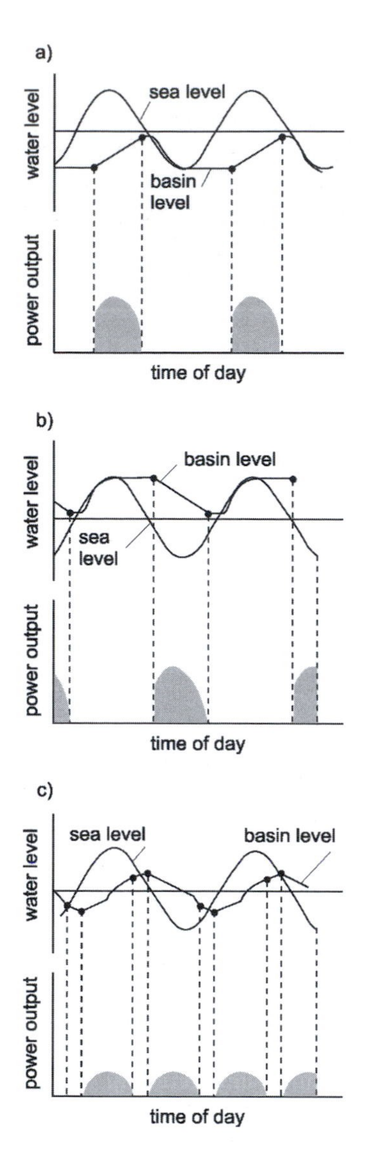

Source: Elliott (1996)

Figure 7.5 *Variation of sea level and of water level inside a tidal barrage during the course of one day, and power output for generation during (a) flood flow only, (b) ebb flow only and (c) flood and ebb flow*

output from tidal energy in general will not coincide with times of peak demand.

7.3.3 Tidal energy potential

The rise and fall of the tides dissipates about 3000GW of energy in shallow seas worldwide (Charlier, 2003).

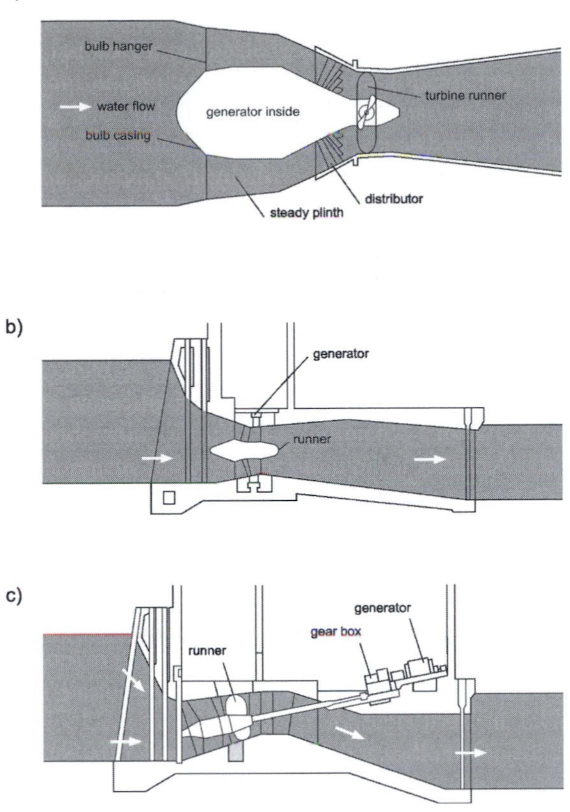

Source: Elliott (1996)

Figure 7.6 *Possible configurations of the water channel, turbine blades and generator in tidal barrages for (a) bulb turbine, (b) stratflo turbine and (c) tubular turbine*

Early schemes to capture tidal energy required a rather large vertical variation, but recent technological developments of low-flow turbines have changed the picture. The largest tidal power station in the world is the La Rance River plant on the coast of Brittany (France), built between 1961 and 1967 and consisting of a 750m long dam with a fish pathway across it. Peak power output is 240MW with an annual average capacity factor of 0.26. The other significant tidal powerplants today are the 18MW single generator at Annapolis Royal in the Bay of Fundy, Nova Scotia, installed in 1984; a 400kW unit in the Bay of Kislaya (100km from Murmansk), completed in 1968; a 500kW unit at Jangxia Creek in the East China Sea; and a 300kW horizontal-axis turbine mounted on a 30m high foundation, with 20m diameter propeller blades installed in Kvalsundet, Norway in 2003. The annual capacity factor at the last of these is 0.26. There have been many proposals for large tidal power projects, such as a 12GW plant across the Severn Estuary and a 15GW scheme to enclose a vast area of the sea near the existing La Rance project in France, but they have not been pursued.

Figure 7.7 shows the locations of possible tidal powerplants around the world, along with the tidal range and the potential peak power generation. The total potential worldwide sums to about 239GW (the total global electrical generation capacity was about 4300GW in 2005).

7.3.4 Environmental considerations

Environmental impacts of tidal energy arise from the inhibition to the migration of fish and other marine animals into and out of the basin behind the barrage, and the reduction in the variation in water level in the basin behind the barrage. This would be important for mud flats, home to many migratory birds. Salinity would be reduced behind the barrage due to the smaller volume of ocean water available to mix with river inflow. The ability of currents to transport and suspend sediments would be reduced, resulting in less turbid water. This could have a beneficial effect, as more sunlight could penetrate into the water in the basin behind the barrage, leading to increased biological productivity. Conversely, sediment that today nourishes beaches outside the barrage could instead be trapped behind the barrage.

Artificial lagoons have been proposed as an alternative to barrages across estuaries. This would entail enclosing an offshore basin with levees, thereby avoiding impacts on coastlines and the inter-tidal zone. However, there has been no in-depth peer review of this concept according to WEC (2007).

7.3.5 Economics

The capital cost of tidal power projects will, like that of hydropower, depend strongly on the site characteristics. Elliott (1996) indicates a typical estimated cost of around \$2000/kW, while a typical capacity factor is around 0.25. The resulting cost of levelized electricity as a function of the interest rate and financing period is given in Table 7.2. Inasmuch as the lifespan of the tidal barrage is indefinite, a very long financing period (such

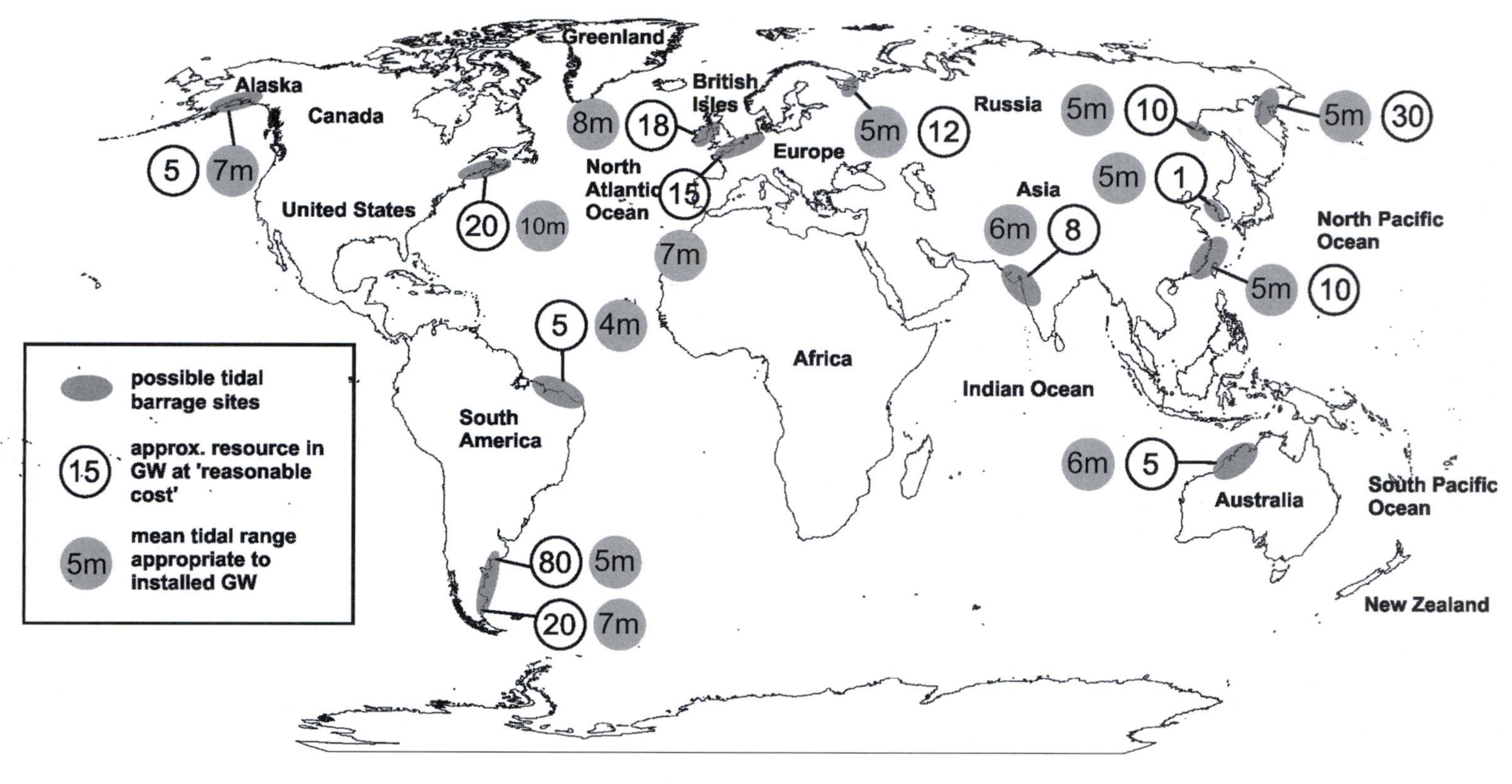

Source: Elliott (1996)

Figure 7.7 *Locations of possible sites for tidal barrages, showing the tidal range (m) and potential installed power capacity (GW)*

Table 7.2 *Illustrative levelized costs of tidal power for various interest rates and financing periods, assuming a capital cost of $2000/kW and an annual capacity factor of 0.25*

Interest rate	0.03	0.03	0.03	0.09	0.09	0.09	0.15	0.15	0.15
Financing lifespan (yrs)	20	40	100	20	40	100	20	40	100
CRF (yr^{-1})	0.067	0.043	0.032	0.110	0.093	0.090	0.160	0.151	0.150
Cost of electricity (cents/kWh)	6.14	3.95	2.89	10.00	8.49	8.22	14.59	13.75	13.70

as 100 years) is justified but would probably not be attractive to the private sector. With long-term (100 year) financing by government at a social discount rate of 3 per cent, the cost of tidal power is 2.9 cents/kWh. If financed over a 40-year period, the cost of power is 4.0 cents/kWh but essentially free after that (except for minor maintenance and operational costs).

7.4 Tidal currents and other marine currents

Energy can be extracted from steady ocean currents or from oscillating currents created by the flood and ebb tides. For steady currents, a velocity of 1m/s is likely to be economically worth exploiting, while for tidal currents, velocities must likely exceed 1.5m/s. Capacity factors for electricity generated from tidal currents are 40–50 per cent, while capacity factors for marine currents can be as large as 80 per cent (Boud, 2002). There are promising marine current resources around the UK, Greece, France and Italy, with a total potential of 48TWh/yr (compared to European electricity demand of 4840TWh in 2005). In China, 7000MW of tidal current power could be installed, which would translate into 25–30TWh/yr.

Table 7.3 lists the various tidal current technologies under development and the number of projects identified by IEA (2006d) for each type of technology. Most of the tidal current projects surveyed by IEA (2006d) utilize a horizontal-axis turbine or vertical-axis rotor. Other concepts besides rotary devices are ducted Venturi choke systems, an oscillating hydrofoil system and air-injected pressure amplifiers. Figure 7.8 gives schematic illustrations of a horizontal-axis turbine and a proposed reciprocating wing concept, while Figure 7.9 shows prototype tidal current turbines in a fjord in Norway. Due to the greater density of water than of air, a relatively large power can be produced with a relatively small current velocity and rotor diameter (see Equation (3.4)). Thus, a rotor with a diameter of 10–15m can generate 200–700kW of power, whereas a 600kW wind turbine requires a rotor diameter of about 45m.

A variety of techniques have been or could be used to hold the tidal current device in place: mounting on a very heavy base, held in place by gravity (three projects); attached to a pile (steel column) driven into the seabed (allowing horizontal-axis devices to yaw about the pile) (ten projects); attached to moorings (cables) that are attached to the seabed and to floats,

Table 7.3 *Types of ocean tidal current energy technologies*

Technology	Description	Number of projects
Horizontal-axis turbine	A rotary device on a horizontal axis that extracts energy from moving water much like a wind turbine extracts energy from moving air. Can be housed inside ducts that concentrate the flow	14
Vertical-axis turbine	A rotary device on a vertical axis	6
Venturi	A funnel-like collecting device submerged in the tidal current. The water flow can directly drive a turbine, or the induced pressure differential in the system can drive an air turbine	2
Oscillating hydrofoil	A hydrofoil (wing) is attached to an oscillating arm. Oscillatory motion is caused by the tidal current flowing along either side of the wing, causing lift. The motion of the oscillating arm drives fluid in a hydraulic system that is converted into useful energy	1

Source: IEA (2006d)

Source: Boud (2002)

Figure 7.8 *Schematic illustration of a horizontal-axis turbine (left) and a proposed reciprocating wing concept (right) for extracting energy from marine currents*

Source: www.e-tidevannsenergi.com

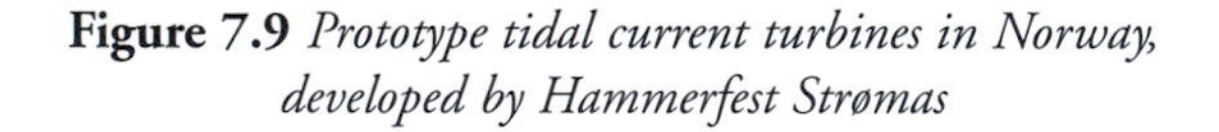

Figure 7.9 *Prototype tidal current turbines in Norway, developed by Hammerfest Strømas*

allowing the device to swing as the tidal current changes direction (four projects) or attached to moorings with minimal leeway (one project); using a hydrofoil to create a downward force from the tidal current to hold the device into place; attached to a floating platform (one project); or fixed to the seabed in other ways (two projects).

Boud (2002) indicates an estimated cost of electricity from early (and relatively simple) tidal current devices of 4.5–13.5 cents/kWh. The main potential environmental impact would be through the reduction in current velocities when large arrays of devices are installed, which would affect the transport of sediments and affect downstream ecosystems.

7.5 Ocean thermal energy conversion

OTEC would make use of the temperature difference between surface water in tropical regions (at 26–28°C) and water at a depth of 1km or so (at about 4–6°C) to generate electricity. Figure 7.10 plots the vertical variation of temperature in the upper ocean at selected tropical locations, while Figure 7.11 shows the temperature difference between surface water and water at depth in the low-latitude oceans. The temperature difference is small, so the efficiency in generating electricity (which cannot exceed the Carnot efficiency, given by Equation (3.16) of Volume 1) would be small. As well, large volumes of water would need to be pumped from the 1km ocean depth. There was a lot of interest in OTEC in the 1970s and through to the mid-1990s, but only one ongoing project, developed by the Institute of Ocean Energy in Japan and operating off the coast of India, was identified by IEA (2006d). Some analysts consider OTEC to be logistically near impossible, in part because of the large volumes of water that must be pumped over great distances. A 1.0–1.2MW demonstration project is under construction in Hawaii at the Pacific International

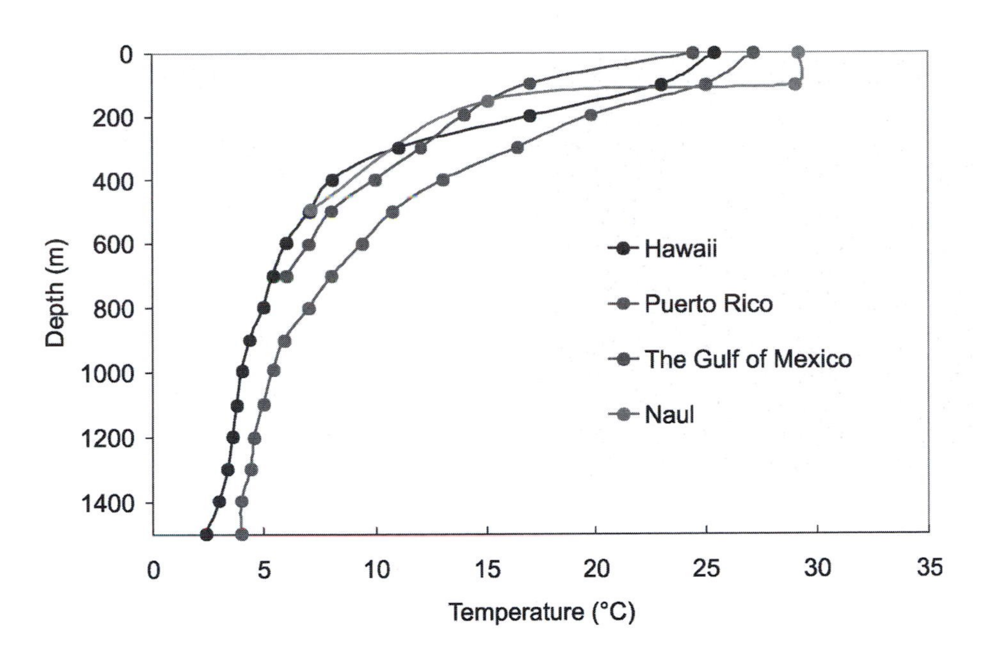

Source: Xenesys Inc, www.xenesys.com

Figure 7.10 *Typical vertical variation of temperature in the upper 1.5 km of the ocean in tropical and subtropical regions*

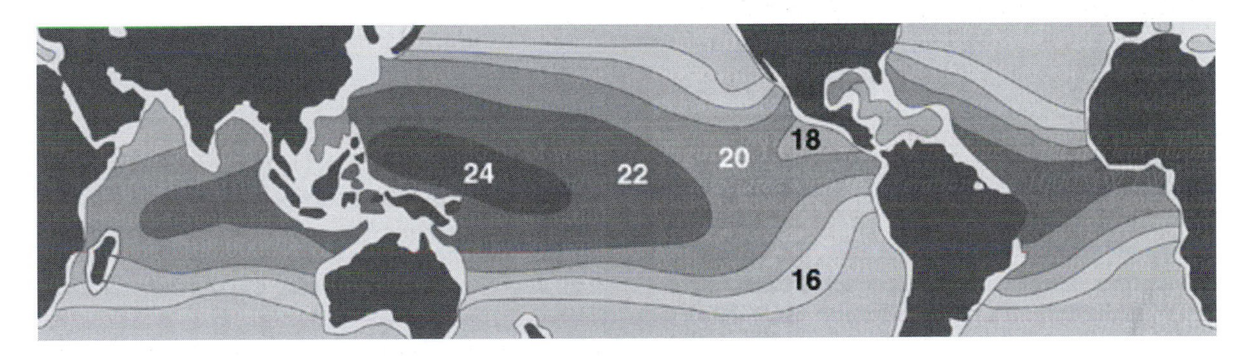

Source: Xenesys Inc, www.xenesys.com

Figure 7.11 *Variation in the temperature difference between ocean surface water and deep water (typically at a depth of 1000m)*

Center of High Technology Research (PICHTR). It will provide much-needed information using modern materials and methods. Research is also continuing in Japan (see www.xenesys.com).

OTEC powerplants would operate either as a closed-cycle system, an open-cycle system or a combination of the two. In a closed-cycle system, heat is transferred from warm seawater to a separate working fluid (such as ammonia, which boils at −33°C and atmospheric pressure) to turn it into a vapour. A closed-cycle system is illustrated in Figure 7.12. The expanding vapour drives a turbine. Cold seawater passing through a condenser containing the working fluid turns the vapour back into a liquid which is then recycled through the system. In an open-cycle system the warm surface water itself is the working fluid. It vaporizes in a near-vacuum at surface water temperatures, then drives a low-pressure turbine. The vapour loses its salt during evaporation, so when it condenses by exposure to cold deep water, it is almost pure fresh water. If condensation occurs without contact with the cold deep water (via a heat exchanger), the condensate can be used as a source of fresh water. If condensation occurs through direct contact, more electricity is produced but useful fresh water is not produced.

Given a surface water temperature of about 27°C in tropical regions, and a deep-water temperature of about 7°C, the Carnot efficiency is only 7.1 per cent, and efficiencies achieved in practice are around 3 per cent. Straatman and van Sark (2008) propose use of a floating solar pond (constructed from heat-resistant plastic foils) that would create a temperature difference of 50–80K rather than 20K, thereby greatly increasing the efficiency (and reducing the cost of electricity). Yamada et al (2009) propose using flat-plate solar collectors to boost the temperature of the surface water, thereby increasing the power output by about 50 per cent.

There are a number of practical considerations in the deployment of OTEC:

- distance from shore to the thermal resource;
- depth of the thermal resource;
- size of the resource within the exclusive economic zone;
- the ability of ocean currents to replenish both warm and cold water;
- waves and hurricanes;
- seabed conditions for power cables.

Rather than producing electricity for connection to the land-based power grid, OTEC plants could produce hydrogen (which could be transported by ship as part of the hydrogen economy, discussed in Chapter 10) or nitrogen fertilizer (using atmospheric nitrogen and hydrogen that would be generated onsite using OTEC electricity).

A simpler use of ocean thermal gradients is for desalination of seawater (Sistla, 2006; Bhattacharjee, 2007). Warm seawater is evaporated at low pressure, then the water vapour is condensed to pure water using the cold, deep seawater in a heat exchanger. A 1000m^3/day demonstration plant operates on a barge 40km off the coast of India. A 1m diameter, 700m long high-density polyethylene pipe draws 10°C water from a depth of 600m. Warm water in this case is at 28°C.

Whether electricity or fresh water is produced, the cold deep water that would be discharged at the surface will be nutrient rich. It could thus be used in aquaculture to produce food for human consumption, or to grow algae that could be harvested to produce a combination of biodiesel and ethanol fuels.

WEC (2007) indicates capital costs for OTEC of $7000–15,000/kW, costs which had risen substantially during the preceding three years due to the increased cost of materials. More effective and hence smaller and less costly heat exchangers are critical to reducing the capital cost. The above capital costs are several times those of wind energy ($1300–1900/kW onshore,

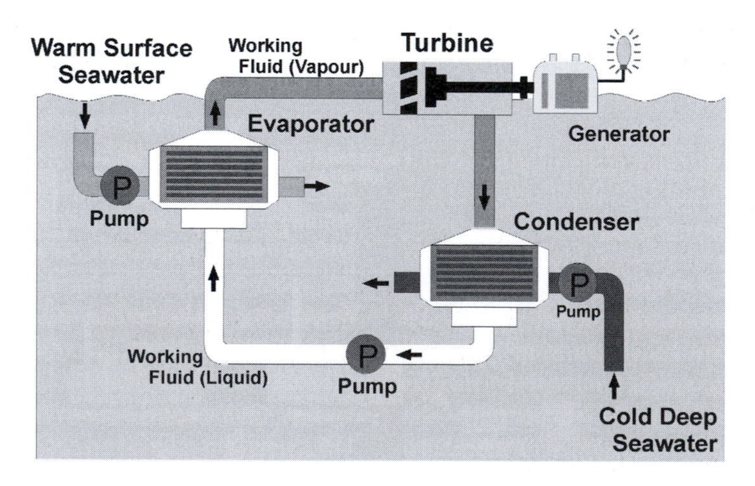

Source: Khan and Bhuyan (2009)

Figure 7.12 *A closed-cycle OTEC process based on the Rankine Cycle*

\$1800–2600/kW offshore), but this is mitigated to a large extent by the high expected capacity factor (0.9 versus 0.3–0.4 for good wind sites). At 5 per cent interest, 25-year financing and annual O&M equal to 2 per cent of the capital cost (compared to 1.0–1.5 per cent for most other power systems), a capital cost of \$7000–15,000/kW translates into an electricity cost of 8.1–17.3 cents/kWh. This would be competitive with the cost of electricity in remote island locations, where electricity is produced from diesel fuel that is brought in by ship.

Apart from possible niche applications (such as remote islands), the sale of co-products will probably be crucial to the economic viability of OTEC. WEC (2007) indicates that the sale of potable water at 40 cents/m^3 would provide a credit of 4 cents/kWh, implying that 0.1m^3 of fresh water could be produced per kWh of electricity that is generated. On remote Pacific islands, fresh water can cost as much as \$1.6/m^3. In the future, with global warming, fresh water is likely to become increasingly scarce and hence valuable in many subtropical parts of the world.

7.6 Energy from ocean salinity gradients

There are two ways in which salinity gradients can be used to produce energy. The first is the solar pond technique, in which large amounts of salt are dissolved in a layer of water at the bottom of a shallow body of water. This makes it too dense to rise as it is heated by absorption of solar radiation, so heat can be stored in the salty layer. The thermal stratification can be used to generate electricity using a thermodynamic cycle similar to that used in OTEC. The second technique makes use of the osmotic pressure difference across a semi-permeable membrane between salty and fresh water that in turn is used to drive a turbine and generate electricity. The European Union is investigating this concept under a project entitled Salinity Power. The best-performing membranes produce power at a rate of 1.5W/m^2, while a commercially viable powerplant is estimated to require 6W/m^2 and have a power output of 10MW, corresponding to about 1.7 million m^2 of membranes (IEA, 2006c). The concept is illustrated in Figure 7.13.

Salinity gradient systems make use of the salinity difference between fresh water and salt water. They could be applied where rivers discharge into the ocean, or could tap municipal fresh water before it is discharged into the oceans.

7.7 Resource potential

Table 7.4 provides estimates of the potential annual electricity production from various ocean energy resources. The greatest potential is thought to lie with ocean wave energy (potentially many times current world electricity demand), followed by OTEC (up to two thirds of current world electricity demand). How much of this potential can be achieved in practice or at a cost that is competitive with other renewable energy options is unknown (OTEC has been largely abandoned, for example). Proponents suggest that electricity could be produced for about 10–14 cents/kWh from ocean waves, 6–17 cents/kWh from

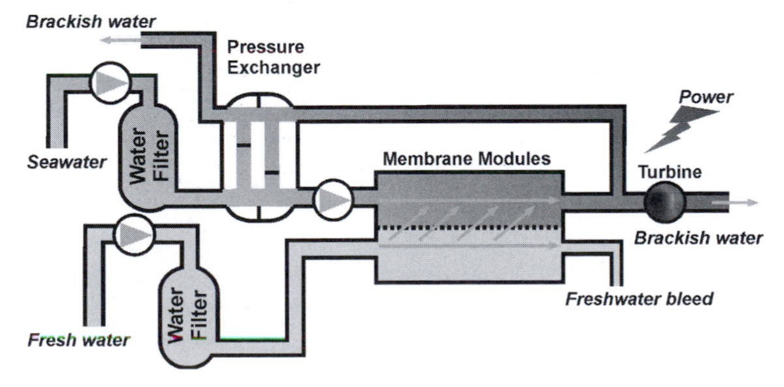

Source: Khan and Bhuyan (2009)

Figure 7.13 *A pressure-retarded osmosis (PRO) process*

Table 7.4 *Overview of ocean energy technologies and global electricity generation potential*

Ocean energy type	Technologies	Global resource	Potential cost
Ocean wave	Attenuator, collector, overtopping, OWC, point absorbed, submerged pressure differential, terminator, rotor	8000–80,000TWh/yr	10–14 cents/kWh
Tidal and tidal current	Horizontal/vertical-axis turbine, oscillating hydrofoil, venturi	≥800TWh/yr	6–17 cents/kWh
Salinity gradient	Semi-permeable osmotic membrane	2000TWh/yr	
OTEC	Rankine Cycle turbine	10,000TWh/yr	8–17 cents/kWh[a]

Note: Global electricity production in 2005 was about 18,000TWh. [a] Based on the analysis presented in section 7.5, without taking into account a credit from the sale of the freshwater co-product, which could easily amount to 4–8 cents/kWh.
Source: Based on IEA (2006c, 2006d)

tidal currents and 8–17 cents/kWh from OTEC (or less if co-products such as fresh water can be produced and sold), but it is too early to suggest costs for salinity gradient systems.

Ocean energy could be important on a regional basis and for remote islands. China has 7000 islands and 18,000km of coastline, and Zhang and Sun (2007) estimate the technical resource potential to be as follows: tidal, 110GW; tidal current, 14GW; wave, 13GW; salinity gradient, 125GW. The total (262GW) is large compared to China's current generation capacity from all sources (509GW in 2005) but the economic and practical potential could be much less. Ocean energy could also make an important contribution in northwest Europe.

7.8 Concluding comments

The various forms of renewable energy from the oceans are not likely to form a large fraction of the renewable energy mix, except locally where there are particularly good and economical energy resources. The technical potential is largest (up to several times current global electricity demand) for wave energy, but the realizable potential is likely to be a very small fraction of the technical potential. OTEC could be broadly applicable in low-latitude regions and could make an important contribution, but requires significant further research, development and testing before a clearer picture of its potential emerges.

8

Nuclear Energy

8.1 Introduction

Nuclear energy does not entail any direct emissions of GHGs and has been promoted by advocates as an at-least partial solution to the problem of global warming. However, it is plagued by a number of ongoing questions and uncertainties. Nuclear energy may be a largely C-free power source, but it is not renewable energy. Global supplies of accessible uranium, the element currently used to fuel nuclear reactors, are limited, and efforts to extend the supply through reprocessing of spent fuel would be fraught with enormous risks for the proliferation of nuclear weapons, either by terrorists or by states, although some reprocessing systems that are thought to be proliferation resistant have been proposed. The issue of long-term isolation of nuclear waste from the biosphere has not been settled, more than 50 years after the first use of nuclear power for civilian purposes. Finally, debris left over from uranium mining operations needs to be carefully managed essentially forever (1 million years) in order to prevent contamination of the regional groundwater supply.

8.2 Basics of nuclear energy physics

Heat is produced in the fuel of a nuclear reactor through the emission of energetic particles from the decay of radioactive nuclei within the fuel; as the radiation passes through the fuel it imparts some of its kinetic energy to the surrounding material. The heated fuel is then used to generate steam, which is used in a steam turbine similar to that used in a coal-fired powerplant. In the following pages, the fuels that can be used as a source of radiation, the forms of radiation and how the radioactivity of the spent fuel and other wastes changes over time, are outlined. Much of the discussion in this section is based on the excellent and more detailed treatment found in Wolfson (1993). Readers not interested in the background physics can jump to section 8.3.

8.2.1 Nuclei, isotopes and radioactive decay

The nucleus of an atom consists of *protons* (each having a positive electric charge) and *neutrons* (having no charge), which together are referred to as *nucleons*. The nucleus is surrounded by *electrons*, which have a negative electric charge. The number of protons in a nucleus is called the *atomic number*, while the number of protons plus neutrons is referred to as the *atomic mass number*. The chemical properties of a particular element (such as how it can combine with other atoms and the energy contained in the bonds with other atoms) depend on the number of electrons surrounding the nucleus, which in turn is equal to the number of protons in the nucleus. Thus, specifying the number of protons in an atom is equivalent to specifying what element it is. However, the number of neutrons in the nucleus of a given element can vary. Atoms with the same number of protons but different numbers of neutrons all belong to the same element but are referred to as *isotopes* of that element. For example, the element carbon contains six protons but can have six, seven or eight neutrons in its nucleus. The nuclei of carbon thus have a total of 12, 13 or 14 nucleons. The different isotopes of carbon would be written as ${}^{12}_{6}C$, ${}^{13}_{6}C$ and ${}^{14}_{6}C$ respectively, where the superscript refers to the number of protons plus neutrons and the subscript refers to the number of protons. Alternatively, since specifying the element and the number of protons is redundant, it is common to write the isotopes of carbon as ^{12}C, ^{13}C and ^{14}C. Hydrogen, which contains one proton in its nucleus, occurs as three different isotopes: hydrogen (with no neutrons), deuterium (with one neutron) and tritium (with two neutrons). Hydrogen is the only element where the different isotopes have different names.

Some isotopes are stable, meaning that the nucleus remains as it is forever, while other isotopes are unstable, meaning that the nucleus can change into the nucleus of another isotope or element while emitting on energetic (and therefore damaging) particle. The reasons for this are sketched out in Box 8.1.

The breakup of an unstable nucleus is called *radioactive decay*, and radioactive isotopes are called *radio-nuclides*. The radioactive decay of an unstable nucleus involves the emission of one of three kinds of energetic particles:

- *alpha particles* (consisting of two protons and two neutrons, as in the nucleus of helium), the weakest form of radiation, capable of being stopped by a sheet of paper or a few millimetres of air but harmful if ingested or introduced into the human body through even a small break in the skin;
- *beta particles* (electrons), able to travel through a few centimetres of human tissue but stopped by a few millimetres of metal;
- *neutrons*, highly penetrating particles released when a nucleus splits in two and stopped by thick shields of concrete or water.

In addition to high-energy particles, there is another kind of radiation, *gamma radiation*, consisting of very short wavelength electromagnetic radiation that is produced when the emission of alpha or beta particles is accompanied by the transition of the nucleus to an 'excited' or high-energy state rather to the 'low' or ground state; gamma radiation is emitted as the nucleus drops to its ground state. It is also produced when a nucleus is struck by another particle and enters an excited high energy state for a while before dropping back to its normal level. Gamma radiation is able to penetrate flesh, bone and metal, but is stopped by 1m of concrete or 3m of water.

With emission of an alpha particle (containing two protons), the emitting element drops down two elements on the periodic table and initially has two excess electrons, which are quickly stripped away. The emission of a beta particle is preceded by the transformation of a

Box 8.1 Protons, neutrons and radioactive decay

The protons and neutrons in the nucleus of an atom are referred to as *nucleons*. As is well known, an *electric force* operates between charged particles, such that opposite charges attract each other and like charges repel each other. The strength of the electric force varies inversely with the square of the distance between the charges. However, there is a second force that acts inside a nucleus, an attractive force that acts between any two nucleons (whether charged or not), called the *nuclear force*. Its strength varies much more strongly with distance than the electric force. As a result, the attractive nuclear force – which attracts protons or neutrons to each other or attracts protons to neutrons – is stronger than the repulsive electric force at the very short distances found inside the nucleus of an atom. Thus, by packing neutrons among the protons of a nucleus, the nucleus is held together in spite of the repulsive electric force between the protons.

The stability of a nucleus represents a fine balance between the repulsive electric forces between the protons and the attractive nuclear forces between the nucleons. The nuclei of light elements contain nearly equal numbers of protons and neutrons, but for progressively heavier elements, more neutrons than protons are needed in order to form a stable nucleus. This is because two widely spaced protons in a large nucleus experience a relatively strong repulsive force (which decreases with distance squared) but a relatively weak attractive nuclear force (due to its much more rapid decrease with distance) and so would fly apart if there were not an excess of neutrons over protons. Thus, the ratio of neutrons to protons gradually increases as we consider elements with successively higher atomic number. However, if too many extra neutrons are added or are subtracted, the nucleus is unstable – it will eventually come apart (for many elements, there is only one stable isotope). Furthermore, no isotope of any element with atomic number greater than 82 (lead) is stable, because in this case the nucleus is so large that the repulsive electric force always exceeds the attractive nuclear force. There is no naturally occurring isotope heavier than $^{238}{}_{92}U$.

neutron to a proton and an electron (thereby conserving charge), so the emitting element goes up by one on the periodic table. The atom now has a positive charge, which is quickly balanced by taking an electron from the surrounding medium.

An unstable isotope can also decay by splitting into two nuclei while also emitting neutrons. The splitting of a nucleus is call *fission*. The neutrons in a nucleus serve as a 'glue', holding the nucleus together through the nuclear force, but with the emission of neutrons and the splitting of the original nucleus in two, the repulsive force between the two positively charged nuclei produced by fission causes the two nuclei to fly apart at high speed. Fission can occur spontaneously, or it can occur as the result of the absorption of a neutron. The latter is called *neutron-induced fission*, and nuclei that can split this way are said to be *fissionable*. Most fissionable isotopes, including the predominant isotope of uranium, U-238, require neutrons with a substantial amount of energy. However, for U-233, U-235 and Pu-239, fission can occur with neutrons having arbitrarily low energy. These isotopes are said to be *fissile*.

Although a particular isotope of a particular element might be unstable, the unstable atoms do not all immediately decay. Rather, it is observed that half of the atoms existing at any given time will decay after a fixed length of time, referred to as the *half-life* of the isotope. Thus, if we begin with 1000 atoms of an isotope with a half-life of one day, 500 atoms will remain after one day, 250 after two days, 125 after three days and so on with the amount remaining decreasing by half for each additional day. Given equal quantities of different radioactive materials, those with the shorter half-lives will be more radioactive (that is, a greater number of particles will be emitted per second, which is what leads to the shorter half-life).

Figure 8.1 is a plot of the number of neutrons versus the number of protons for all of the stable elements (solid squares) (lead being the heaviest element with a stable isotope) and for all the unstable nuclei of elements up to and including lead (open squares). Straight lines for different ratios are shown, and the increase in neutron to proton ratio with increasing atomic number can be seen. If a heavy element splits into two lighter elements, the lighter elements will on average have the same neutron to proton ratio as the original element, and so will have an excess of neutrons (as indicated in Figure 8.1). They will therefore be unstable and so will decay through emission of a beta particle, which is accompanied by the conversion of a neutron to a proton, thereby reducing the neutron to proton ratio.

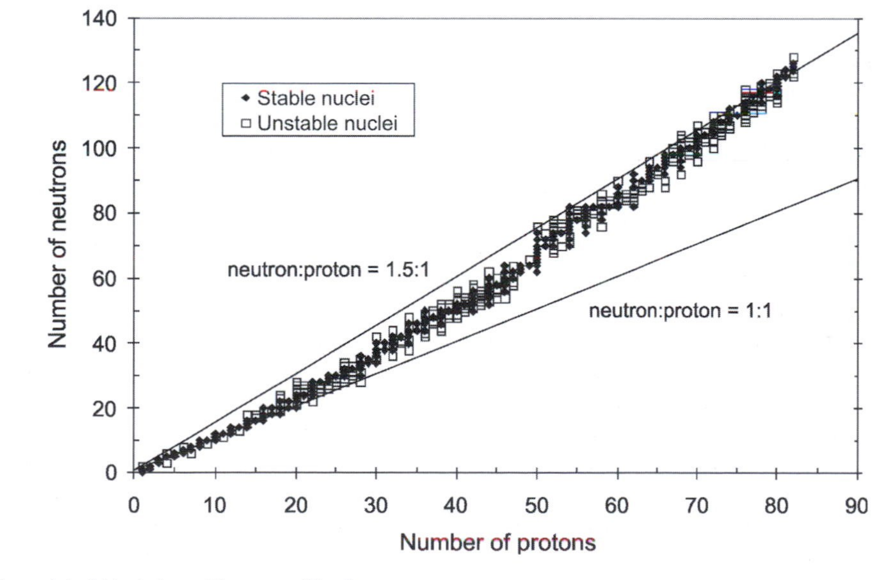

Source: Entries for each individual element in www.wikipedia.org

Figure 8.1 *Ratio of neutrons to protons in the stable isotopes (solid squares) and unstable isotopes (open triangles)*

8.2.2 Fuels used and decay reactions inside a nuclear power reactor

Nuclear power reactors today are overwhelmingly fuelled with uranium, but some are fuelled in part with plutonium derived from spent fuel, and nuclear reactors could in principle be fuelled with a mixture of thorium and uranium. Natural uranium consists of 0.0055 per cent U-234, 0.7205 per cent U-235 and 99.275 per cent U-238 (where percentages are in terms of numbers of atoms). As noted above, U-235 is fissile, meaning that it is capable of undergoing fission in response to absorbing a neutron with arbitrarily low energy. All three isotopes of natural uranium (U-234, U-235 and U-238) are radioactive, but U-238 emits less penetrating alpha particles than other alpha emitters and only weak gamma rays, so it is not a health hazard unless ingested or inhaled. If ingested, it is hazardous both due to its radiation and chemical toxicity.

When U-235 absorbs a neutron, it first turns into U-236, which in turn splits into lighter elements while releasing two or three neutrons. The general reaction is:

$$^{235}\mathrm{U}+n \rightarrow {}^{236}\mathrm{U} \rightarrow (A_1,M_1)+(A_2,M_2)+Nn \quad (8.1)$$

where the two lighter elements have atomic numbers A_1 and A_2 (with $A_1 + A_2 = 92$, the atomic number of uranium) and atomic mass numbers M_1 and M_2 (with $M_1 + M_2 + N = 236$, the mass number of the parent uranium atom), and N is the number of neutrons emitted (usually two or three). There is a probability distribution of different elements produced by fission, rather than a fixed pair of fission products for the fission of U-236 (or of any other element). The probability distribution for the fission of U-236 is given in Figure 8.2. If $M_1 = M_2$, the fission is called *symmetric fission*. Otherwise, it is *asymmetric fission*. As seen from Figure 8.2, the fission process is overwhelming asymmetric, usually with one fission product having an atomic mass number around 95 and another around 140.

One of the two or three neutrons emitted in reaction (8.1) might be absorbed by the cooling water or structural materials, one might be absorbed by U-235 and initiate reaction (8.1) again and one might be absorbed by U-238. When U-238 absorbs a neutron it turns into U-239 that then decays to neptunium-239 with a half-life of 23 minutes, which in turn decays to plutonium-239 with a half-life of 56 hours, as follows:

$$^{238}\mathrm{U}+n \rightarrow {}^{239}\mathrm{U} \rightarrow {}^{239}\mathrm{Np}+\beta \quad (8.2)$$

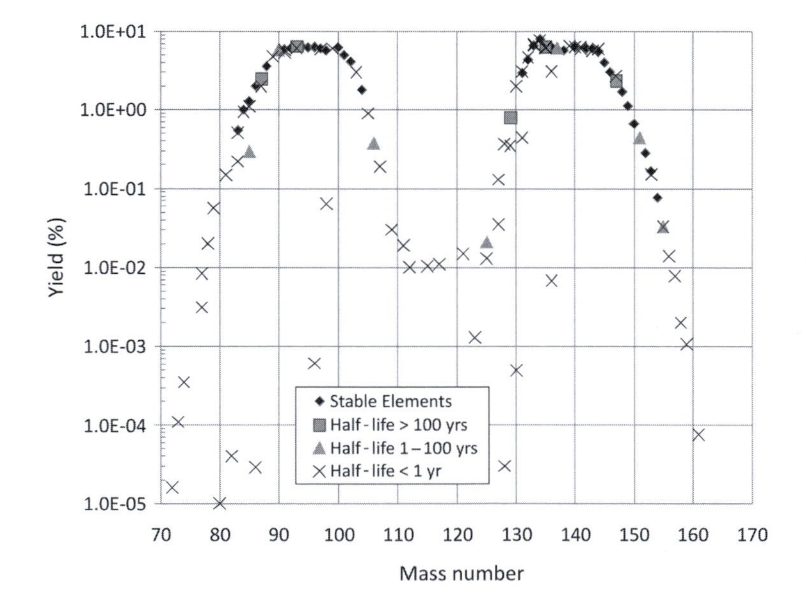

Note: As the fission of one U-236 nucleus produces two fission products, the sum of the probabilities is 200 per cent.
Source: Based on data in Appendix 3 of Wilson (1996b)

Figure 8.2 *The probability that fission of U-236 nucleus produces a given fission product as a function of the atomic mass number of the fission product*

$$^{239}Np \rightarrow {}^{239}Pu + \beta \qquad (8.3)$$

U-238 is not fissile (it does not split into two lighter elements) but it is *fertile*, meaning that it can absorb a neutron to form new fissile material. Pu-239 is both fissile and fertile – it can eventually release more neutrons as it splits into lighter elements (with a half-life of about 24,000 years), but it can also absorb further neutrons if available, forming successively heavier isotopes (Pu-240, Pu-241 and Pu-242). Pu-241 can release a beta particle to form americium-241, which in turn can form successively heavier isotopes of americium through absorption of neutrons. Am-242 can release a beta particle to form curium-242, which can also absorb neutrons and form heavier isotopes of curium. A typical sequence of transmutation by absorption of neutrons and emission of beta particles is shown in Figure 8.3. Other sequences involve production of the elements berkelium and californium, in addition to curium. The elements produced through transmutation, beginning with neptunium, are referred to as *transuranic* ('beyond' uranium) elements.

Over time, the amount of U-235 in the nuclear fuel decreases while the amount of transuranic elements increases. The transuranic elements can be split by neutrons into two energetic lighter elements and further neutrons. The energy released during the fission of transuranic elements compensates to some extent for the reduction in the amount of U-235 remaining in the fuel (and the resulting reduction in the rate of energy release from fission of U-236) toward the end of the fuel's residence inside the reactor.

The stable isotopes of lighter elements require a smaller neutron to proton ratio than heavy elements (see Figure 8.1), so the fission of a heavy element such as U-236 or the transuranic elements will produce lighter elements with an excess of neutrons. The fission products therefore undergo further decay, through the emission of beta particles, with each beta emission converting a neutron to a proton. Many of these transitions occur rapidly, but some transitions have half-lives of a few years to decades, while several fission products decay with half-lives of several million years. Because of the ongoing decay of the fission products, heat continues to be generated long after the fission reactions have stopped and long after the spent fuel has been removed from the reactor. This in turn makes the storage of radioactive waste difficult, as discussed later.

The half-lives of some of the more important fission products and of other radioactive elements are given in Table 8.1. The fission products of particular

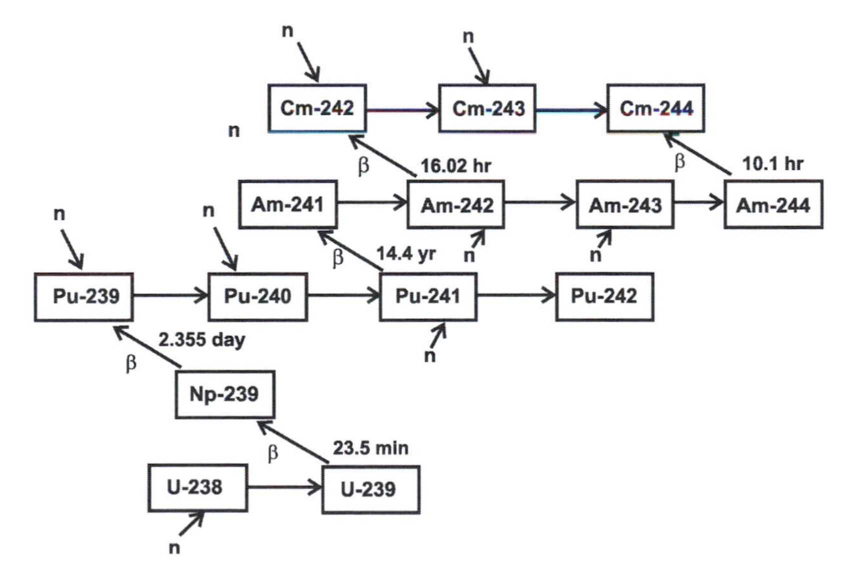

Source: Modified from Wilson (1996a)

Figure 8.3 *A transmutation sequence beginning with the absorption of a neutron by U-238*

concern are I-131 (with a half-life of eight days), which becomes concentrated in milk and is absorbed by the thyroid gland; Cs-137 (half-life of 30 years), which behaves like potassium and so is distributed throughout the body, and decays to Ba-137 with the release of gamma radiation; and Sr-90 (half-life of 29 years), which mimics calcium and so becomes concentrated in bones. I-131, with its high activity and short pathway to milk, is important in the period just following any release to the atmosphere. Cs-137 and Sr-90, with their half-lives of about 30 years, determine the length of time that the land would need to be evacuated (several half-lives). The transuranic radio-isotopes, mostly with half-lives of 60–150 thousand years, are the major reason why reactor waste must be isolated from the environment for 200,000 to 1 million years.

Thorium is a fertile material (it can absorb neutrons but is not fissile) that can also be used to produce U-233 (which is fissile) in combination with a neutron source such as U-235. The reactions are:

$$^{232}_{90}Th + n \rightarrow {}^{233}_{90}Th \rightarrow {}^{233}_{91}Pa + \beta + \nu_e \qquad (8.4)$$

and:

$$^{233}_{91}Pa \rightarrow {}^{233}_{92}U + \beta + \nu_e \qquad (8.5)$$

where ν_e is an anti-neutrino. After some time the irradiated materials can be removed from the reactor and the U-233 easily separated from the thorium (through chemical rather than isotopic separation) and fed into another reactor as part of a closed cycle. When U-233 absorbs a neutron, it will either fission into lighter elements or release another neutron. Like Pu-239 (produced from U-238), U-233 can be used in weapons.

Use of thorium in combination with uranium can increase the energy output per tonne of mined uranium by 85 per cent compared to a once-through uranium cycle, and by 22 per cent compared to recycling plutonium once, while the production of plutonium is reduced by a factor of five (Kazimi and Todreas, 1999). Reprocessing spent fuel when thorium is used in the fuel would be more hazardous (due to greater radiation levels) than reprocessing spent fuel from the uranium-only cycle, which makes diversion into weapons more difficult, but radioactivity after 200 years would be reduced by

Table 8.1 *Major sources of radiation in spent nuclear fuel*

Radio-isotope	Yield during fission (%)	Half-life	Particle energy (keV)	Type of radiation
		Fission products (half-life <1 year)		
Sr-89	4.79	50.5 days		
Sr-91	5.81	9.5 hrs		
Sr-92	5.3	2.7 hrs		
Y-93	6.1	10.2 hrs		
Zr-95	6.2	64 days	1.125	β, γ
Zr-97	5.9	16.8 hrs		
Mo-99	6.06	2.8 days		
Ru-103	3	39.3 days	0.764	β, γ
I-131	3.1	8.04 days	0.971	β, γ
Te-132	4.7	3.3 days		
Xe-133	6.62	5.243 days	0.427	β, γ
I-133	6.9	20.8 hrs	1.77	β, γ
I-134	7.8	52.6 min		
I-135	6.1	6.6 hrs		
Xe-135	6.3	9.1 hrs		
Ba-139	6.55	1.4 hrs		
Ba-140	6.35	12.8 days		
La-141	6.4	3.9 hrs		
Ce-141	6	32.5 days		
Ce-143	5.7	1.4 days		
Ce-144	6	284.6 days	0.319	β, γ
Nd-147	2.7	11 days		
Subtotal	124.4%			

Table 8.1 *Major sources of radiation in spent nuclear fuel* (Cont'd)

Radio-isotope	Yield during fission (%)	Half-life	Particle energy (keV)	Type of radiation
		Fission products (half-life of 1–100 years)		
Kr-85	0.29	10.7 yrs	0.687	β, γ
Sr-90	5.77	29.1 yrs	0.546	
Cs-137	6.15	30.2 yrs		
		Primary fission products (half-life >100 years)		
Zr-93	6.45	1.5 Ma		β, γ
I-129	0.8	15.7 Ma	0.191	β, γ
Cs-135	6.41	2.3 Ma		β
Sm-147	2.36	106 Ga		
		Secondary fission products (half-life >100 years)		
Se-79[a]	0.04	0.295 Ma		β
Tc-99m[b]	6.06	0.211 Ma		β
Pd-107[c]	0.19	6.5 Ma		β
Sum	158.9%			
		Thorium and uranium (particle energies in MeV)		
Th-228		1.91 yrs	5.423	α, γ
Th-232		14 Ga	4.013	α
U-233		159.2 ka	4.824	α
U-234		246 ka	4.776	α
U-235		704 Ma	4.4	α, γ
U-238		4.47 Ga	4.197	α
		Transuranic elements (particle energies in MeV)		
Np-237		2.14 Ma	4.788	
Pu-238		87.7 yrs	5.4992	
Pu-239		24.1 ka	5.156	
Pu-240		6.56 ka	5.1683	
Pu-242		375 ka	4.901	
Am-241		432.7 yrs	5.4857	
Am-242m		150 ka		
Am-243		7.37 ka	5.276	
Cm-242		162.8 days	6.1127	
Cm-244		18.1 yrs	5.8048	
Cm-245		8.5 ka		

Note: Yields of fission products are for fission of U-235 by thermal neutrons. The total fission yield is 200 per cent, as two nuclei are produced for every fission that occurs. ka = thousands of years, Ma = millions of years, Ga = billions of years, k_eV = thousands of electron volts, MeV = millions of electron volts. [a] From decay of As-79, half-life of 9 minutes. [b] From decay of Mo-99, half-life of 2.8 days. [c] From decay of Rh-107, half-life of 21.7 minutes.
Source: Appendices 3 and 4 of Wilson (1996b)

75 per cent. Demonstrations were carried out in the 1960s and 1970s with various thorium-uranium oxide fuels, but economic considerations did not favour use of the thorium cycle. As well, significant new technologies would need to be developed in order to use this cycle.

In summary, there are two reactions that initiate a nuclear chain reaction – the initial absorption of a neutron by U-235 (reaction 8.1) and the initial absorption of a neutron by U-238 (reaction 8.2). The absorption of a neutron by U-238 leads eventually to the production of plutonium and other transuranic elements that split into lighter elements along with the emission of neutrons. The absorption of a neutron by U-235 produces U-236 that in turn splits into lighter nuclei while releasing neutrons that can trigger more reactions with U-235 or U-238 in a potentially self-sustaining process. There are two main fissile radio-nuclides in the nuclear fuel: U-235 and Pu-239. When the fuel is first loaded into the reactor, the only fissile material is U-235. Over time, the concentration of U-235 decreases while that of Pu-239 increases, but the increase in Pu-239 does not entirely offset the decrease in U-235, so at some point the fuel needs to be replaced with fresh fuel. Some initial neutrons are required to begin the fission process, and this comes from the spontaneous decay of the enriched U-235 (which has a 704 million year half-life). The heat in a nuclear reactor arises from the kinetic energy of the energetic lighter elements that are produced by fission of U-235 and the transuranic elements (mostly Pu-239), from the kinetic energy of emitted neutrons, from the kinetic energy of beta particles that are emitted during the transmutation chain that produces transuranic elements, and during the decay of the lighter elements produced by fission.

8.2.3 Sustaining and stabilizing the nuclear decay reactions

As noted above, the absorption of a neutron by U-235 can lead to fission of U-235 and the release of up to three more neutrons, one of which would need to be absorbed by another U-235 nucleus to induce further fission and neutron release, thereby sustaining the process. However, the neutrons released by fission are moving so fast that they have a very low probability of inducing fission of further U-235 nuclei. In order to greatly increase the probability of absorption, the neutrons must be slowed down. This can be done using a *moderator*. Collisions between the nuclei in the moderator and the neutrons slow them down. Ideally, the moderator should absorb neutron energy but not neutrons themselves. When two objects collide, greater energy transfer from the fast-moving to the slow-moving object will occur if the two objects have similar masses. As the proton in the nucleus of the hydrogen atom has essentially the same mass as a neutron, this makes water (H_2O) a good moderator. However, water also absorbs some of the neutrons, and as a result, there are not enough U-235 atoms in natural uranium to sustain a reaction. Thus, nuclear reactors using water as a moderator (the vast majority in the world) require that the U-235 be enriched from its natural concentration of 0.72 per cent to about 3 per cent.

Water having ${}^{2}_{1}H$ (deuterium) rather than ${}^{1}_{1}H$ (ordinary hydrogen), referred to as *heavy water*, can also be used as moderator. Such reactors are referred to as *heavy-water reactors* (HWRs), while those using normal water are referred to as *light-water reactors* (LWRs). Heavy water is less effective in slowing down neutrons (because the mass of the deuterium nucleus is twice that of the neutron), but absorbs almost no neutrons. Thus, reactors using heavy water as the moderator can operate with unenriched uranium. Most of the world's nuclear power reactors are LWRs, while the Canadian CANDU (Canadian deuterium uranium) reactor is the only major HWR used for electricity generation.

In order to maintain a steady rate of nuclear reactions and hence a steady rate of heat production, exactly one neutron from each fission must cause a subsequent fission. Since not all neutrons released from fission will be absorbed by U-235 and cause a subsequent fission, the ratio of neutrons produced by a nuclear reaction to neutrons absorbed (η) must exceed one in order to sustain a chain reaction. As noted above, this is the case for the absorption of slow neutrons by U-235 (as well as for absorption of slow neutrons by U-233 and Pu-239). The ratio of neutrons produced to neutrons absorbed with subsequent fission is called the *multiplication ratio*, and to ensure that this ratio is exactly 1.0, neutron-absorbing materials in the form of *control rods* are inserted between uranium fuel rods and boron is added to the cooling/moderating water. If the multiplication ratio exceeds 1.0, the rate of nuclear reactions would increase exponentially, eventually producing so much heat as to cause a violent explosion inside the reactor building (but *not* a nuclear explosion as in nuclear weapons). To prevent this, control rods are inserted further so as to absorb more neutrons and reduce the multiplication ratio, whereas if the multiplication ratio drops below 1.0, the control rods are partially withdrawn. The time required for the neutrons from one fission to cause another fission (the *generation time*) is only 1/10,000 second (Bodansky, 2004). These are referred to as *prompt neutrons*. Thus, if the multiplication ratio is 1.01, the rate of heat production will increase by a factor of 2.7 (= 1.01^{100}) after only 0.01 seconds (100 generations) and by a factor of 145 after only 0.1 seconds. No mechanical control system can respond this quickly.

However, some of the neutrons involved in the chain reaction are produced during decay of fission products rather than during fission. The generation time for these *delayed neutrons* is such that the average generation time is 0.1 seconds. This gives enough time for mechanical control systems to operate. The key is to maintain the multiplication ratio so close to 1.0 that the chain reaction cannot be sustained without the delayed neutrons. In the case of the explosion at the Chernobyl reactor, the reactor was inadvertently put in a state where the multiplication ratio rose to the point where the prompt neutrons alone were sufficient to sustain the reaction, at which point control was lost and the reactor power rose to 500 times its intended value in five seconds (Wolfson, 1993).

8.2.4 Thermal reactors, fast reactors and fast breeder reactors

As noted above, a sustained nuclear reaction requires that the ratio η of neutrons produced to neutrons absorbed exceed 1.0 (do not confuse this ratio with the multiplication ratio). If $\eta > 2$, we can 'breed' new fuel. For a chain reaction to occur, the speed of the neutrons must be reduced or there must be a minimum density of fissile nuclei. Neutrons that have been slowed down are called *thermal* neutrons, and reactors (such as the LWRs and HWRs mentioned above) using such neutrons are called *thermal reactors*. Table 8.2 gives η values for thermal and fast neutrons produced by the decay of U-233, U-235 and Pu-239 (U-233 is produced from thorium, while Pu-239 is produced from U-238, as explained above). The value of η for fast neutrons is significantly greater than for slow neutrons. Reactors with a high enough density of fissile material that the neutrons do not need to be slowed down are therefore called *fast reactors*. Fast reactors could be fuelled with U-233, U-235 or Pu-239. Pu-239 would be used instead of U-235 as a fuel as this provides a means of using the Pu-239 that is produced from a thermal reactor, rather than treating it as a waste.

If the Pu content reaches 10–20 per cent, more Pu will be produced than is consumed. Such reactors are referred to as *fast breeder reactors* (FBRs). Through repeated cycling through FBRs, a significant fraction (30–60 per cent) of the U-238 in natural uranium can be converted to Pu (via reactions 8.2 and 8.3) and used as a fuel. Otherwise, only the small portion of U-235 (0.72 per cent of the total times the utilization fraction, typically 0.7) can be used as a fuel, and the vast majority of uranium remains unused. Taking into account that only 60–70 per cent of the Pu produced from U-238 is fissile (Pu-239 and Pu-241) and that the fissile Pu is 10 per cent less effective than U-235 as a fuel, the fast breeder reactors would multiply the uranium energy resource by a factor of about 30–75.

In both thermal reactors and fast reactors, the neutron source can be thought of as the 'fuel'. This fuel (U-235 in the case of thermal reactors, Pu-239 in the case of fast reactors) is consumed over time, so these reactors are called *burner* reactors. The basic conceptual difference between burner reactors and FBRs is that, in a burner reactor, the rate of conversion of U-238 to fissile Pu is less than the rate of fission of the fuel, so the fissile content decreases over time, whereas in a breeder reactor, more fissile material (Pu-239 and Pu-241) would be produced than is consumed.

As discussed later (subsection 8.3.1), getting a breeder reactor (and the associated reprocessing facility) to operate reliably has proven to be far more difficult and expensive than expected.

Table 8.2 *Number of neutrons produced per neutron absorbed for common fertile nuclear fuel isotopes*

Fuel isotope	Absorption of thermal neutrons (KE ≈ 0.25eV)	Absorption of fast neutrons (KE > 0.5MeV)
U-233	2.28	2.60
U-235	2.10	2.65
Pu-239	2.09	3.04

Note: KE = kinetic energy.
Source: Hoffert and Potter (1997)

8.2.5 Measures of radioactivity

The metric unit for radioactivity is the *becquerel* (Bq), where 1Bq = a rate of one decay per second. One *curie* (Ci), defined as the rate of decay from 1gm of radium, is equal to 3.7×10^{10}Bq. The rate of radiation per gm of radio-nuclide is referred to as the *specific activity*.

As for human health effects, radioactivity is measured in terms of the amount of energy deposited in a unit mass of tissue. The units are the *gray* (equal to 1J/kg) and the *rad* (equal to 100ergs/gm) (1 gray = 100 rad). However, biological harm depends not only on the total energy deposition, but also on the *linear energy transfer* (LET) rate, which is defined as the energy lost by a particle (ΔE) divided by the distance (Δx) traversed. A larger LET causes greater but more localized damage for the same total energy deposition. Alpha particles and neutrons have a large LET while gamma rays and beta particles have a small LET.

In order to account for the differing biological effectiveness of different kinds of radiation, the *dose equivalent* (H) has been introduced. It is given by:

$$H = QD \tag{8.6}$$

where D is the absorbed dose (in grays or rads) and Q is the *quality factor* (equal to 1 for gamma rays and beta particles, 5–20 (depending on energy) for neutrons and 20 for alpha particles). If D is in grays, the H is in *sievert* (Sv), while if D is in rads, H is in *rem* (1 sievert = 100 rem). Further information about health-related measures of radiation doses can be found in Bodansky (2004).

8.2.6 Variation of nuclear radioactivity over time

Nuclear radiation consists of the following:

- beta particles emitted during the transmutation of one of the actinides[2] to an element with a higher atomic number (as shown in Figure 8.3);
- neutrons emitted during the fission of an actinide element;
- the fission products themselves;
- beta particles produced by the decay of the fission products;
- alpha and beta particles produced from four *radioactive decay chains* that proceed spontaneously without the absorption of a neutron. The four decay chains are:
  - the *thorium series*, beginning with Th-232 and ending with Pb-208;
  - the *uranium series* (also referred to as the *radium series*), beginning with U-238 and ending with Pb-206;
  - the *actinium series*, beginning with Pu-239, passing to U-235 and ending with Pb-207;
  - the *neptunium series*, beginning with Pu-241 and ending with Tl-205;
- gamma radiation produced when a nucleus drops from an excited to a low-energy state.

As an example of a radioactive decay chain, the uranium series is shown in Figure 8.4 (the other decay chains are given in Appendix 2 of Wilson, 1996b).

The radioactivity associated with unstable isotopes that decay directly into a stable isotope decreases continuously over time. This is the case for fission products, which are the dominant radio-nuclides in spent fuel. Thus, the radioactivity of reactor fuel one year after removal from the reactor core is about 3 million curies/tonne and heat is produced at a rate of 13kW/tonne. After ten years these quantities have decreased to 0.6 million curies/tonne and 2kW/tonne (MIT, 2003).

The radioactivity of the four radioactive decay chains, in contrast, initially increases over time, then decreases. This is shown in Figure 8.5 for the uranium series. U-238 has a half-life of 4.46 billion years, which means that the rate of decay (and the associated radiation) is quite low. The next two radio-nuclides in the chain decay essentially instantly (see Figure 8.4), so there will be no appreciable buildup. The fourth and fifth elements in the chain have half-lives of 245,000 and 77,000 years, respectively, and so will build up to a quasi-equilibrium concentration on a timescale equal to a few of their half-lives. That is, as they build up in concentration, the associated radioactivity increases, until a balance is achieved between the rate of production from the decay of upstream radio-nuclides, and decay to subsequent members of the series. Once this equilibrium is reached, the amounts and radioactivities associated with all the members of the series will decrease in proportion to the slow decrease in U-238 (governed by its half-life). However, even the peak radioactivity associated with these series is about five orders of magnitude smaller than the peak radioactivity associated with fission products.

Figure 8.6 shows the contribution of different radioactive elements to the radioactivity of spent fuel as a function of time since the fuel was removed from the reactor, beginning one year after removal. Total radioactivity is extremely high at the time of removal, and is dominated by the decay of the many fission products with half-lives of hours to a few days (many of which are listed in Table 8.1). After one year the radioactivity has fallen to about 1.3 per cent of the activity at discharge (Bodansky, 2004), and by year ten it has fallen by another factor of six (as seen in Figure 8.6). During the next 170 years the total radioactivity is largely produced by the fission products strontium-90 and caesium-137, each of which have half-lives of about 30 years. Both Sr-90 and Cs-137 decay to short-lived unstable products that emit further radiation, thereby roughly doubling the radioactivity associated with Sr-90 and Cs-137 to the levels shown in Figure 8.6. The radioactivity associated with Am-241 (half-life 432.7 years), which is a decay product of Pu-241 (half-life 14.3ka), gradually increases while that associated with Sr-90 and Cs-137 decreases, and so dominates the total radioactivity between about 180 years and 1800 years after discharge. Thereafter, the dominant contributors to total radioactivity are Pu-240 (until year 5000, due to its 6.56ka half-life), then Pu-239 (until year 100,000,

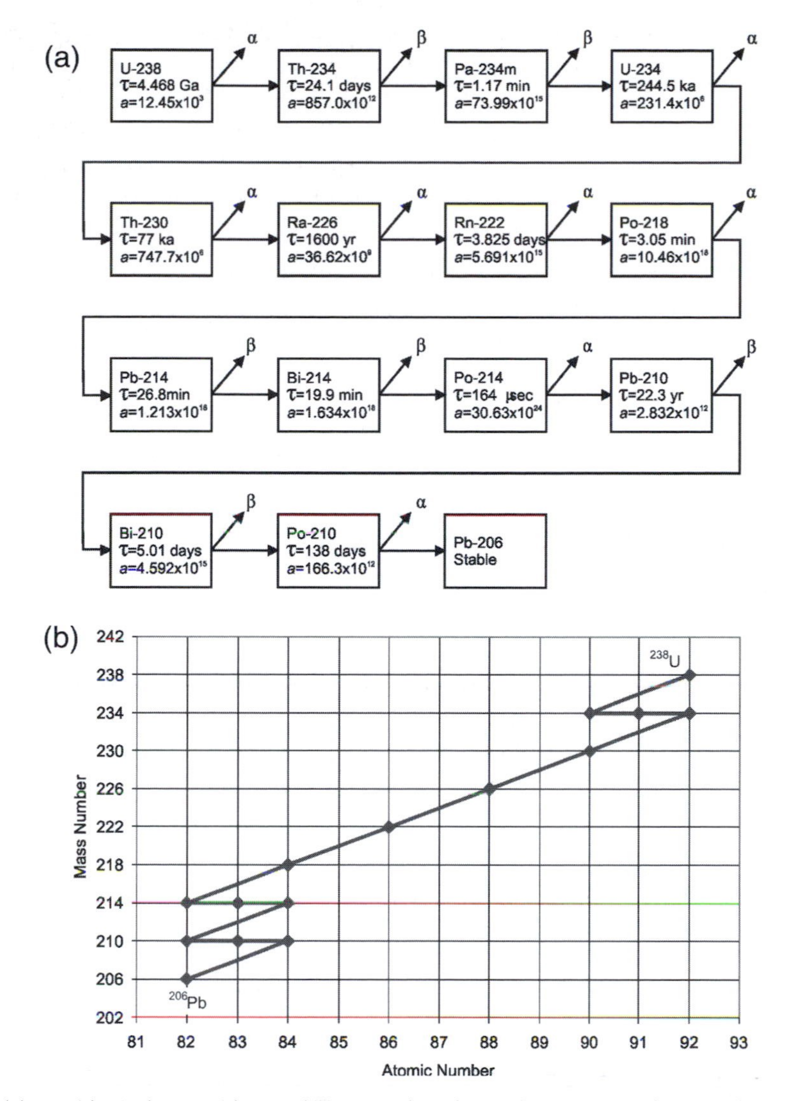

Note: τ = half-life, α = alpha particle, β = beta particle, Ga = billion years, ka = thousand years, *a* = specific activity (Bq/gm).
Source: Data from Shultis and Faw (2008) for half-lives, www.wise-uranium.org for specific activities

Figure 8.4 *(a) The uranium series radioactive decay chain, beginning with U-238, and (b) a plot of the uranium series on an atomic mass–atomic number diagram*

due to its 24.1ka half-life) and finally the uranium and neptunium series, whose radioactivity experiences a much delayed increase for the reasons explained above. Wastes from uranium mining and depleted uranium from the enrichment process consist largely of U-238, so the radioactivity from these wastes also increases over time for the first 200ka, although the peak radioactivity is several orders of magnitude less than that of spent fuel one year after its removal from the reactor.

8.2.7 Nuclear fusion

Energy is released (in the form of energetic particles) not only when the nuclei of heavy elements are split (fission), but also when the nuclei of lighter elements (such as deuterium and lithium) are fused together. The radioactive products of fusion decay with half-lives of decades, and an explosive runaway reaction or meltdown of a fusion energy system is not possible

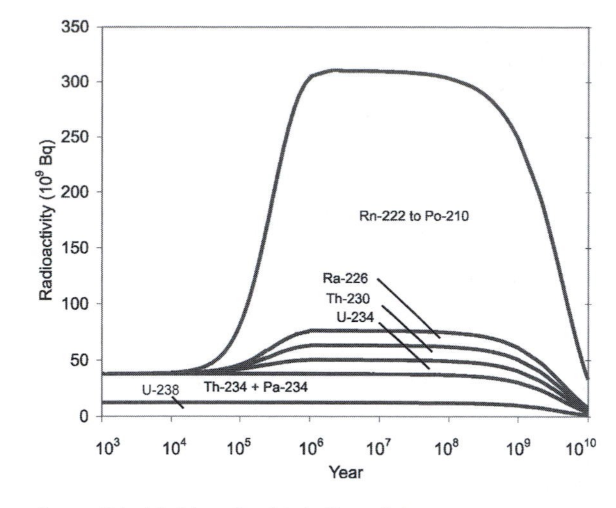

Source: Calculated from the data in Figure 8.4

Figure 8.5 *Variation over time in radioactivity associated with an initial tonne of U-238*

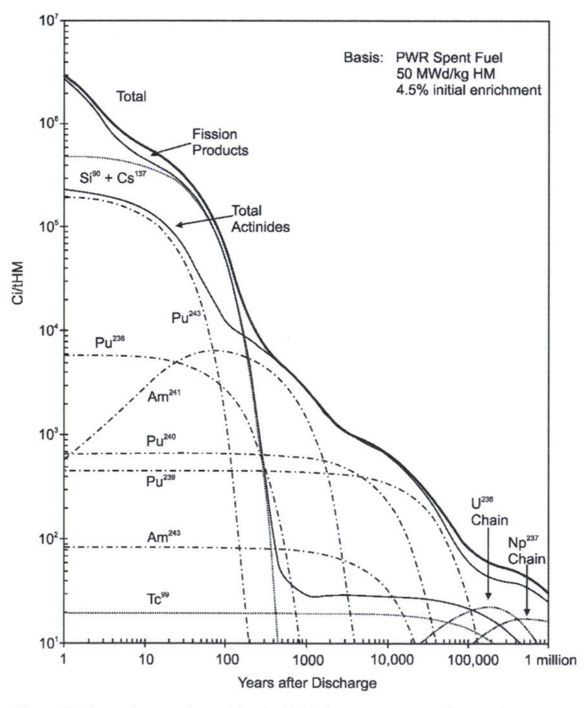

Note: Both scales are logarithmic. MWd = megawatt-day and HM = heavy metal.
Source: MIT (2003)

Figure 8.6 *Contributions of different radioactive elements to the total radioactivity of spent fuel, as a function of time since the fuel is produced*

(von Hippel et al, 2010). The fuel needed to operate a 1GW fusion powerplant for one year would be the deuterium in 5000 tonnes of natural water and the ^{6}Li in 4 tonnes of natural lithium. The lithium reserve is estimated to be 12 million tonnes, which would allow the operation of 2000GW of fusion powerplants for 1500 years. An additional 200 billion tonnes of lithium is thought to be economically extractable from seawater, while the supply of deuterium in seawater is effectively unlimited (von Hippel et al, 2010).

8.3 Nuclear powerplant reactor technologies

In this section the major nuclear reactor technologies are described, followed by a breakdown of the past and projected future progression of nuclear energy through four generations.

8.3.1 Reactor technologies

The major nuclear reactor technologies are described below, based on the more extensive discussion found in Wolfson (1993) and Hesketh (1996). An overview of the major reactor types and their relationship to the concepts discussed in previous sections is presented in Figure 8.7. These reactor concepts are:

- plants enriched in U-235 (to 3–5 per cent) and using thermal neutrons and normal water as a moderator and coolant (the LWRs);
- plants with natural uranium and using thermal neutrons and heavy water as a moderator and coolant (the HWRs);
- plants enriched to 95 per cent U-235 and using thermal neutrons, graphite as a moderator and helium gas or water as a coolant;
- fast reactors using Pu-239 as a fuel, having no moderator and using liquid metal as a coolant.

A brief elaboration on each reactor concept is given below.

LWRs

LWRs can be classified as either boiling-water reactors or pressurized-water reactors. The simplest method of transferring heat from the reactor to the steam turbine

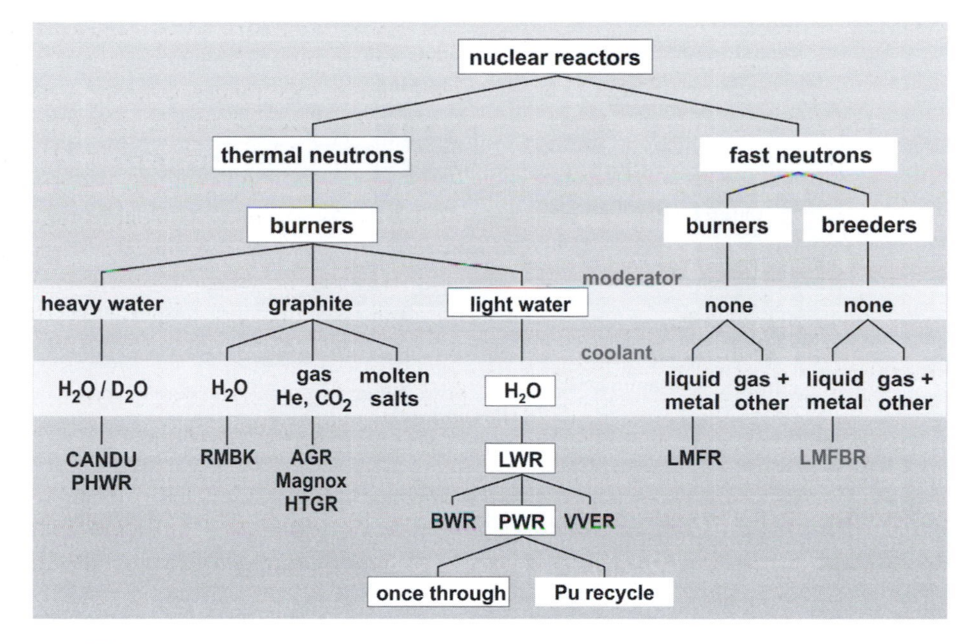

Source: van Leeuwen (2007)

Figure 8.7 *Nuclear reactor technologies in relation to type of neutrons (thermal or fast), absence or presence of a moderator and type of moderator if present*

is to circulate the water to be turned into steam through the reactor core and then directly through the steam turbine, as illustrated in Figure 8.8a. This gives the *boiling-water reactor.* The reactor temperature is about 300°C but the pressure is low enough that the water can boil into steam. This water passes through the steam turbine but also serves as the moderator and reactor coolant. If there is a loss of coolant flow, there is also a loss of moderator, so the fission reactions shut down (heat will still be produced from the decay of fission products, but this is a less intense heat source). However, the water picks up some fission products and absorbs neutrons (which in turn produce radioactive tritium), so the steam circulating through the turbine is radioactive (leading to 'turbine shine'). An alternative is the *pressurized-water reactor,* in which the water circulating through the reactor is kept under sufficiently high pressure to remain in the liquid state rather than turning into steam. This water exchanges heat through a heat exchanger with a separate water loop that passes through the steam turbine, as illustrated in Figure 8.8b. Water in the primary loop (passing through the reactor core) serves as both a coolant and a moderator (as in the boiling-water reactors), while water passing through the turbine is non-radioactive. The reactor core of both boiling-water and pressurized-water reactors is placed inside a pressurized heavy steel vessel, and so needs to be shut down in order to replace spent fuel rods with new fuel rods, a process that occurs every 12–24 months and takes six to eight weeks.

HWRs

In the CANDU HWR, bundles of unenriched uranium are placed inside channels through which pressurized heavy water circulates. This heavy water serves as a coolant, taking heat to a heat exchanger, where ordinary water is turned into steam. The entire assembly of coolant channels with fuel bundles is placed inside an unpressurized tank of heavy water, which serves as the moderator. Because the entire assembly is inside an unpressurized vessel, individual pressurized fuel and coolant channels can be removed and the fuel replaced without shutting down the reactor. This is done continuously, with about 15 fuel

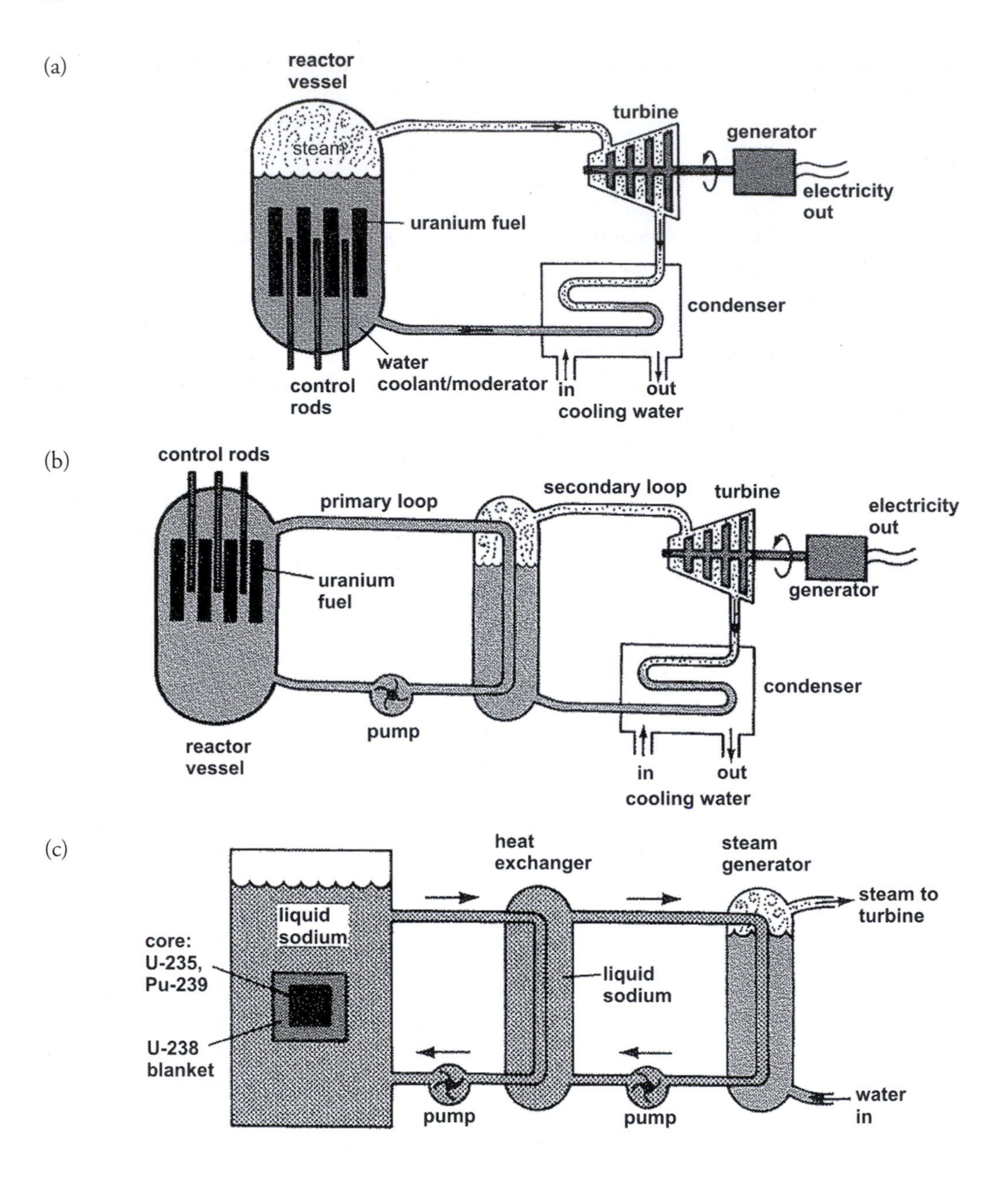

Source: Wolfson (1993)

Figure 8.8 *Schematic diagram of nuclear powerplants using (a) a boiling-water reactor, (b) a pressurized-water reactor and (c) a liquid-metal FBR*

bundles replaced per day. As a result of continuous refuelling, average conditions inside the reactor are more uniform over time. However, it is also easier to use the fuel just long enough to optimize the production of plutonium for use in nuclear weapons and it is less practical for international inspection teams to monitor the spent fuel for possible diversion to nuclear weapons (see section 8.6 for connections between nuclear power and nuclear weapons proliferation and potential nuclear terrorism).[3]

Gas-cooled reactors

Graphite-moderated reactors have been developed in Russia, the UK and the US. The US *high-temperature gas-cooled reactor* (HTGR) uses uranium fuel enriched to 93 per cent U-235 and uses helium gas as a coolant. Its high temperature leads to more efficient electricity generation by the steam turbine, and helium does not boil or react chemically with other substances (so steam explosions are impossible). However, such high enrichment makes the uranium already weapons grade (see subsection 8.6.1), and graphite is flammable. The single commercial US HTGR was retired in 1989 after years of mixed performance. The UK's graphite reactors, the Magnox and the advanced graphite reactor (AGR), are still in operation and use CO_2 as a coolant, while the Russian version (the RBMK) uses water as a coolant (there were four such reactors at Chernobyl, and 13 remain in operation elsewhere) (Bodansky, 2004).

FBRs

FBRs use fast neutrons (as produced by fission), without being slowed down with a moderator, to produce ('breed') Pu-239 from U-238. Fission of Pu-239 by absorption of fast neutrons produces 3.04 neutrons on average, whereas fission of U-235 by absorption of fast neutrons produces 2.65 neutrons on average (see Table 8.2). One of the neutrons produced by fission is needed for the next fission (so as to sustain the process), so the remaining two or three can produce new Pu-239 once absorbed by U-238. The core of the FBR is very compact because it has no moderator, which in turn means that it is so hot that water or gas cannot be used as a coolant. Liquid sodium is used as a coolant because it is very effective in removing heat without pressurization and because it absorbs very few neutrons. However, sodium burns spontaneously on contact with air and reacts violently with water. For this reason, the sodium-cooled FBR has two separate liquid sodium loops and the usual steam loop so that radioactive liquid sodium cannot come into contact with the steam loop. These loops are illustrated in Figure 8.8c.

Getting FBRs to operate reliably has been difficult and expensive, with sodium fires and leaks being a particular problem. France built a 233MW experimental FBR called Phenix in 1973, and built a larger (1200MW) FBR called SuperPhenix in 1986, but the latter was shut down in 1998 (Bodansky, 2004). A 246MW FBR called Monju was operated for only a few months in Japan in 1995 before being shut down (and not subsequently restarted) after a sodium fire. The US was first to build a FBR, in 1951, but cancelled its FBR programme in 1983. A subsequent fast reactor programme was cancelled in the mid-1990s. Small breeders in the UK and Kazakhstan were also shut down in the 1990s. Germany also briefly worked on FBRs. Thus, there are only two FBR reactors still operating today – the Phenix in France and a 560MW reactor in Russia. India recently began pursuing a FBR programme (Rosenkranz, 2006).

8.3.2 Reactor generations

The progression of nuclear reactors for the generation of electricity has been divided into four generations: Generation I, early prototype reactors; Generation II, the larger central station reactors of today; Generation III, advanced LWRs with improved safety features; and Generation IV, the next generation of reactors, to be designed and built during the next two to four decades. These are described next.

Generation I reactors

There are still a few Generation I reactors in operation – the Soviet VVER (voda-vodyanoi energetichesky reactor) 440-230, which is a pressurized-water reactor using water as a coolant, and the UK Magnox.

Generation II reactors

Today's reactors are largely Generation II reactors. The technology of today's nuclear industry is based largely on military research of the 1940s and 1950s (Beck, 1999). There have been improvements but few radical changes. As of September 2009, there were the 264 pressurized-water reactors in operation worldwide, 92 boiling-water reactors, 44 CANDU HWRs in seven countries and 18 gas-cooled reactors (only in the UK). This information is summarized in Table 8.3.

Table 8.3 *Number of different types of nuclear power reactors in the world as of September 2009*

Type of reactor	Number of units	Total capacity (MW_e)
Fast breeder reactor	2	690
Light-water gas reactor	16	11,404
Gas-cooled reactor	18	8909
Pressurized HWR	44	22,441
Boiling-water reactor (an LWR)	92	83,656
Pressurized-water reactor (an LWR)	264	243,160
Total	436	370,260

Source: www.iaea.org/programmes/a2/index.html

Generation III reactors

About 20 different designs of Generation III reactors are under development. Most represent incremental rather than major changes to Generation II reactors. Among these are the *European pressurized-water reactor* and the *pebble bed modular reactor* (which is sometimes classified as Generation IV and is so discussed later). The latter was pursued by several countries up until the late 1980s, but only prototypes were ever built (in the US, UK and Germany) and these were decommissioned after at most 12 years of operation. Generation III reactors are supposed to have:

- a standardized design for each type;
- a simpler and more rugged design;
- a higher availability and longer operator life (typically 60 years);
- reduced possibility of core-melt accidents;
- a higher burn-up (defined in subsection 8.4.3) so as to reduce fuel use and the amount of waste.

The development, demonstration and commercialization of even incremental changes to existing reactors, as in the Generation III concepts, would still require two or more decades.

Generation IV reactors

In 2001, the Generation IV International Forum (GIF) was formed in order to pool resources in the initial development of six advanced reactor concepts. The GIF consists of Argentina, Brazil, Canada, France, Japan, South Africa, South Korea, Switzerland, UK and US. The six reactor concepts, three of which are fast reactors but not FBRs, are listed in Table 8.4. No more than three of the six concepts are expected to survive the feasibility testing phase and become research prototypes (at a cost of $1 billion each). All would run at a temperature of at least 510°C, whereas today's water-cooled reactors run at about 300°C. The result is up to 50 per cent higher thermal efficiency (47 per cent instead of 33 per cent), and the capability to co-produce hydrogen and electricity, increasing yet further the overall efficiency.[4] All except the supercritical water concept use coolants other than water, and all depend on untested engineering and the development of new ultra-hard materials. The goal is to reach technical

Table 8.4 *The six Generation IV reactor concepts under development*

Concept	Coolant	Temp (°C)	Pressure	Waste recycling	Output	Earliest delivery
Gas-cooled fast reactor (or modular helium reactor)	Helium	850	High (75atm)	Yes	Electricity Hydrogen	2025
Lead-cooled fast reactor	Lead-bismuth	550–850	Low	Yes	Electricity Hydrogen	2025
Molten salt reactor	Fluoride salts	700–800	Low	Yes	Electricity Hydrogen	2025
Sodium-cooled fast reactor	Sodium	550	Low	Yes	Electricity	2015
Supercritical water-cooled reactor	Water	510–550	Very high	Optional	Electricity	2025
Very-high-temperature reactor	Helium	1000	High	No	Electricity Hydrogen	2020

Source: Butler (2004)

maturity by 2030, but many regard this as highly optimistic. A substantial funding effort would be required, but there may not (perhaps with good reason) be the will to provide the funding. All Generation IV reactors exist only on paper at present.

Most Generation IV concepts involve a closed fuel cycle. This means that reprocessing of spent fuel in order to extract and use the plutonium would occur, such that the wastes would flow back into nuclear reactors (thereby 'closing' the cycle). The result would be an enormous increase in the risk of nuclear proliferation if these reactors were used, as dozens of sites with separated plutonium would replace a few centralized and (presumably) highly monitored sites (see section 8.4). In experienced hands, only 1–2kg of Pu-239 would be required to make a bomb. According to a study by MIT (Deutch et al, 2003), the fuel cost with a closed cycle, including waste storage and isolation costs, would be 4.5 times that of a cycle in which fuel is used only once, without reprocessing, although this is a trivial consideration compared to the overall cost of nuclear energy and compared to the risks of nuclear weapons proliferation.

The additional time needed to develop a new reactor, obtain an operating licence and then have several years of operation by one or more demonstration plants in order to gain real-world experience, adds up to at least 20 years before any Generation IV reactors could begin to be deployed commercially. Another two decades or more would then be required for a new reactor to reach a significant fraction of the reactor fleet. Thus, none of the Generation IV concepts is likely to occupy a significant fraction (25 per cent or more) of the nuclear fleet before 2050, if ever.[5]

8.3.3 Fusion reactors

The technology for nuclear fusion is still under development. Current fusion energy research is focused on the confinement of hot ionized gases, called plasma, in toroidal (doughnut-shaped) magnetic fields. About 10MW of heat have been produced for about one second. A research facility is under construction in France by a consortium of countries (China, the European Union, India, Japan, Russia, South Korea and the US) and is expected to produce 500MW of heat for a few hundreds of seconds. It is still a long way from a demonstration fusion powerplant, with the earliest projected date for the start-up of such a plant being 2035–2040 (von Hippel et al, 2010). Fusion energy thus has no prospect of making a non-negligible contribution to the world's energy supply before 2050, and it may never turn out to be technically and economically viable.

8.4 The nuclear fuel chain

An assessment of the desirability of nuclear energy relative to other energy sources requires a consideration of the environmental impacts, hazards, costs and energy inputs associated with the entire fuel chain, beginning with the mining and processing of uranium, its use in nuclear reactors and the ultimate recycling or isolation of wastes from the biosphere. This is referred to as the nuclear fuel 'cycle', but it is really a nuclear fuel 'chain', as even recycled uranium is reused only once before being isolated (the isolation of nuclear wastes from the biosphere is commonly and incorrectly referred to as 'disposal'). The present-day uranium fuel chain is discussed here, followed by a depiction of the mass flows of uranium and other elements associated with the present and various possible future nuclear fuel chains. Wilson (1996b) provides further technical information on the nuclear fuel chain, from mining to eventual isolation of wastes and decommissioning of nuclear powerplants.

8.4.1 Mining and milling of primary uranium

Uranium ores contain many U-bearing minerals, the most common being uraninite, which is largely UO_2 (uranium dioxide) and UO_3, mixed with oxides of lead, thorium and the rare earth metals. The ratio of the mass of uranium oxide (in the standardized form U_3O_8) to the mass of ore is called the ore *grade*. The highest grade of uranium ore currently being mined is 17.96 per cent (at the McArthur River underground mine in Saskatchewan, Canada) and the lowest being mined on a large scale is 0.029 per cent (at the Rössing open-pit mine in Namibia) (Kreusch et al, 2006). Data presented later in Figure 8.30 suggest that the world average grade is about 0.2–0.3 per cent. All uraninite minerals contain a small amount of radium as a radioactive decay product of uranium. Radium in turn decays to produce radon, which has adverse health effects.

As with most other minerals, there are two kinds of mine: open-pit and underground. Open-pit mines, as the name suggests, involve removing the layers of rock that lie on top of uranium-bearing layers, creating a large open pit but allowing easy access to the uranium ores. This is practical if the deepest layers are not more than about 1km deep (the proposed open-pit expansion of the Olympic Dam underground mine in Australia, described later in subsection 8.8.1, would entail excavating 450m of overburden, then excavating the ore body down to a depth of 1200m). Uranium can also be extracted through leaching of uranium minerals from uranium-bearing ores with chemicals in situ (that is, without excavating the ore), as described later. A fourth alternative is to extract ores from the ground, pile them in a heap near the mine, and then use chemicals to leach the uranium from the ores. This is referred to as heap leaching and is done in association with open-pit mining (NEA/IAEA, 2008).

Figure 8.9 shows the breakdown of uranium extraction techniques worldwide in 2007. Open-pit mining (including heap leaching) accounted for about 26 per cent of the total, underground mining 38 per cent and in situ leaching 28 per cent. The other category, co-product mining, refers to uranium mining in association with the mining of another mineral (typically copper or zinc) and accounted for about 8 per cent the total uranium extraction in 2007.

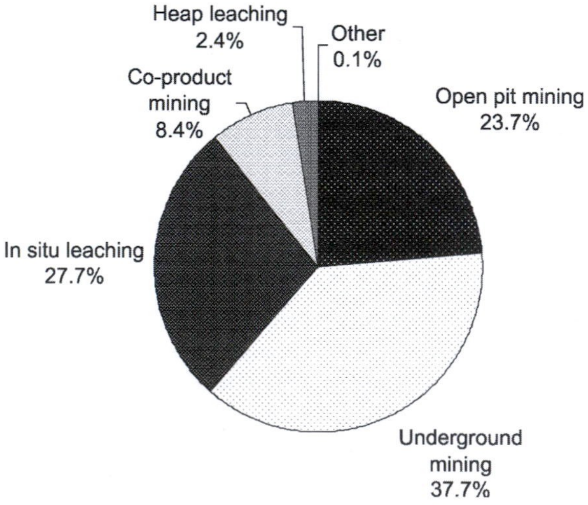

Heap leaching 2.4%
Other 0.1%
Co-product mining 8.4%
Open pit mining 23.7%
In situ leaching 27.7%
Underground mining 37.7%

Source: NEA/IAEA (2008)

Figure 8.9 *Breakdown of world uranium extraction techniques in 2007*

During mining, large volumes of contaminated water are pumped out of the mines and released to rivers and lakes. Piles of waste rock that usually have elevated concentrations of radio-nuclides compared to normal rock are also created. For open-pit mining, 40 tonnes or more of waste rock can be generated for every tonne of ore extracted, while for underground mines the ratio may be 1:1 or less (Winfield et al, 2006). Radioactive and toxic minerals seep into the groundwater, while radon is released to the air from the debris.

Mining is followed by milling, during which the recovered ore is crushed and leached in a uranium mill, which is usually located close to the mines. Sulphuric acid or alkaline fluids can be used to leach the uranium from the crushed ore. Other minerals (such as molybdenum, vanadium, selenium, iron, lead and arsenic) are removed as well and must be separated from the uranium. The final product is U_3O_8 plus impurities and is referred to as *yellowcake*, illustrated in Figure 8.10. The residue from the milling process (the mill *tailings*) is a slurry, which is usually pumped to settling ponds for final disposal. Up to 85 per cent of the radio-nuclides in the original uranium ore end up in the tailings, as well as heavy metals and chemicals from the milling process. The tailings are a source of radiation (typically 20–100 times the normal background levels), radon, fine sand lifted by wind when the tailings dry out, and contaminants that leach into the ground and surface water. Tailings cannot

Source: www.wise-uranium.org

Figure 8.10 *An open drum of yellowcake (U_3O_8)*

simply be dumped into abandoned mines because the contaminants are much more mobile than in their original state, and so would enter the groundwater.

Assuming that no uranium is lost during processing, the ratio of the mass of ore that needs to be mined and processed to the mass of uranium recovered would be equal to the reciprocal of the ore grade. Thus, for an ore grade equal to the current word average of about 0.002 (0.2 per cent), 500 tonnes of ore would need to be mined per tonne of uranium extracted from the ore. In reality, some uranium ends up in the tailings rather than captured during the milling and chemical processing. For ore with a grade of 0.2 per cent, the captured fraction might be 0.95, but could drop to 0.6 for ore with a grade of 0.02 per cent (van Leeuwen, 2007). Thus, the mass of ore that would need to be mined per tonne of captured uranium would increase from (1t)/(0.002 × 0.95) to (1t)/(0.0002 × 0.6), or from 526t to 8333t, as the ore grade drops from 0.2 per cent to 0.02 per cent. This in turn has significant implications for the net energy yield of nuclear reactors as increasingly lower grade ores are exploited, as discussed in section 8.8. The volume of the tailings is essentially that of the original mined ore, so the volume per unit mass of uranium extracted also increases rapidly as the ore grade declines.

In newer mining operations, the tailings are placed in specially constructed pits that are kept in a continuously dewatered state so that groundwater will flow into the tailings pits rather than out of or through the pit, or a zone of high hydraulic conductivity may be constructed around the tailings site so as to channel groundwater around rather than through the tailings. Measures may be taken to contain radon gas. Management of tailings will need to continue in perpetuity (radon is produced by decay of radium-226, which has a half-life of 1600 years and is in turn produced by decay of thorium-230, which has a life-life of 80,000 years, while the radioactivity associated with decay of Th-230 would require many Th-230 half-lives to decay to that associated with the residual uranium).

Uranium has been extracted from ores through in situ leaching in North America, Australia and the former Soviet Union. This method involves circulating dilute sulphuric acid or an alkali solution through porous underground ores and dissolving the uranium into a slurry that is pumped to the surface. Uranium daughters such as radium remain in the ground, thereby avoiding the release of radon to the atmosphere (Vattenfall, 2007). However, this technique involves irreversible contamination of aquifers with both the added chemicals and mobilized radioactive and toxic elements.

8.4.2 Enrichment of primary uranium

With the exception of HWRs, uranium needs to be enriched in the U-235 isotope in order to be used as a fuel. The different isotopes of uranium differ slightly in mass, and if in gaseous form, will diffuse under pressure across a membrane at different rates. They can also be partly separated in a centrifuge. By applying membrane separation or a centrifuge thousands of times, a stream of gases containing U-235 and U-238 can be separated into two streams, one enriched in U-235 and the other depleted in U-235. Enrichment technology is explained further by Upson (1996).

The enrichment process for a light-water power reactor typically produces an enriched stream with 3.6 per cent U-235 and a depleted stream with 0.2–0.3 per cent U-235 (compared to a starting concentration of 0.72 per cent). If C_i is the initial concentration of U-235 and C_e and C_d are the concentrations in the enriched and depleted streams, respectively, then by writing equations expressing conservation of the total mass of uranium and conservation of U-235, it follows that the mass M_n of natural uranium that needs to be processed to produce a mass M_e of enriched uranium is given by:

$$M_n = \left(\frac{C_e - C_d}{C_i - C_d}\right) M_e \qquad (8.7)$$

Thus, if uranium is depleted to 0.2 per cent in one stream, 6.7kg of natural uranium are needed to produce 1kg of uranium enriched to 3.6 per cent, whereas if uranium is depleted to only 0.3 per cent, 8.0kg of natural uranium are needed. However, fewer enrichment units are needed if the uranium is depleted to a lesser extent in the waste stream. Enrichment by gaseous diffusion consumes about 3–4 per cent of the electricity produced by the reactor that uses the enriched uranium, while enrichment with centrifuges requires less than 0.1 per cent of the electricity generated by the nuclear plant. The depleted uranium (DU) is referred to as *tails* (not to be confused with tailings).

In order to separate U-235 and U-238, the uranium has to be converted to a gaseous form, namely, uranium hexafluoride (UF_6). After enrichment, the depleted uranium (DUF_6) must be stored somewhere forever (at temperatures less than 56.4°C it is a solid). In the US there are 560,000 tonnes of DUF_6 in 14-tonne steel cylinders at three different sites, one of which is shown in Figure 8.11. The storage of DUF_6 poses a health and safety hazard, as UF_6 exposed to air reacts with water in the air to produce uranyl fluoride (UO_2F_2) and hydrogen fluoride (HF), both of which are toxic. As well, DUF_6 is a radioactive waste, although it has not been declared as such – perhaps because the radioactivity of DU starts off relatively small and peaks after approximately 500,000 years (as illustrated in Figure 8.5 for the uranium series). The main concerns are Ra-226 and Pb-210 (from the decay of U-238) and Pa-231 (from the decay of U-235) (van Leeuwen, 2007).

Source: www.wise-uranium.org/stk.html?src=stkd02e

Figure 8.11 *Containers of depleted uranium at the Portsmouth, Ohio storage site (top) and corroded canisters after a few years of storage (bottom)*

In 1984, a French ship with 30 UF_6 containers on board sank off the Belgian coast after colliding with a car ferry. Fortunately, the UF_6 containers were retrieved without leaking. In 1986 a UF_6 cylinder at a conversion plant in Gore, Oklahoma, US ruptured, spilling waste into the adjacent river. The accident happened when an overfilled cylinder was heated in an attempt to remove excess UF_6. When the solid UF_6 liquefied, the associated volume increase breached the cylinder.

8.4.3 Use of nuclear fuel

The enriched uranium is converted from UF_6 to UO_2 and shaped into rods with zirconium alloy or steel cladding that can be inserted into and withdrawn from the core of the nuclear reactor. A given rod is used for a period of up to four years before it must be replaced. The effectiveness of fuel use is expressed as *burn-up*, which is the amount of thermal energy produced per unit mass of uranium in the fuel. Thermal energy produced times the thermal efficiency of the steam cycle gives the gross amount of electrical energy generated.

Box 8.2 lays out the calculation of fuel burn-up in terms of the initial and final concentration of U-235 in the fuel and in terms of the average energy production per fission event. Greater burn-ups are achieved by enriching the uranium fuel to a greater degree in U-235, which means that more U-235 is extracted from the original uranium, less remains in the discarded uranium, and so slightly less uranium is needed. Burn-ups increased from a typical value of 20GWd/t in the 1970s to about 40GWd/t by the late 1990s (Bodansky, 2004, Chapter 9). Table 8.5 shows the calculated burn-up for various enrichment factors, as well as the fraction of the U-235 in the fuel that is used, the mass of uranium needed per tonne of enriched uranium fuel, the tonnes of rock that need to be mined per tonne of fuel produced (assuming an ore grade of 0.2 per cent), the energy provided per tonne of rock mined and the energy provided per gm of U-235 in the original fuel. Doubling the burn-up from 40GWd/t to 80GWd/t requires increasing the enrichment from 4.5 per cent to 8.3 per cent, but reduces the uranium requirement for the same energy production by 6.9 per cent. It also

Box 8.2 Energy yield from uranium fuel

The average energy produced from the fission of U-235 (after its absorption of a neutron) depends on the energy produced by the different fission reactions, weighted by the probability of each reaction, plus the energy released from the decay of the fission products of U-235, from the fission of transuranics produced by the initial absorption of a neutron by U-238, from the decay of the fission products of the transuranics, and from gamma rays that are produced from neutrons that are absorbed by U-235, U-238 or other materials without causing fission. Some of the energy released from the decay of fission products consists of neutrinos that are lost from the reactor. The average energy retained in the reactor per fission of U-235, based on Bodansky (2004, Chapters 6 and 9), is:

- 180.5MeV from the fission of U-235, of which 168.2MeV is the kinetic energy of the fission fragments, 4.8MeV is the kinetic energy of prompt neutrons and 7.5MeV the energy of prompt gamma rays;
- 14.6MeV from the decay of the fission products themselves;
- 3.6MeV from gamma rays produced by neutron capture by U-235 and U-238;
- ~20MeV from the fission of transuranic elements;
- ~5MeV from gamma rays emitted following neutron capture by transuranic elements.

Thus, the average total energy yield per U-235 fission is about 225MeV. Given that 1eV = 1.6×10^{-19}J, this corresponds to an energy yield of 92.16GJ per *gram* of U-235 that fissions. By comparison, the heat released from burning coal is about 28.5GJ *per tonne*. The burn-up per tonne of fuel depends on the initial U-235 concentration and the fraction of the initial U-235 that it consumed before the fuel is removed. Thus, the burn-up per tonne of fuel (counting the mass of U only) is given by:

$$\textit{Burn-up}(\text{GWd/t\_fuel}) = C_e\left(\frac{C_e - C_f}{C_e}\right)\left(\frac{92.16\times10^6\text{GJ/t}}{86400\text{GJ/GWd}}\right) = (C_e - C_f)1067(\text{GWd/t}) \quad (8.8)$$

where C_e is the concentration of U-235 in enriched uranium (as fed to the reactor) and C_f is the final concentration (when the fuel leaves the reactor). For $C_f = 0.008$, the burn-up increases from 40GWd/t at an enrichment of 4.6 per cent to 80GWd/t at an enrichment of 8.3 per cent.

Table 8.5 *Fuel burn-up and other quantities as a function of the initial enrichment of U-235 in the fuel, assuming an ore grade of 0.2 per cent, 100 per cent recovery of the uranium in the ore and a U-235 concentration in spent fuel of 0.8 per cent*

U-235 concentration in enriched fuel (%)	Fraction of U-235 used	Burn-up (GWd/t U)	Tonnes of U mined per tonne of U in fuel	Tonnes of rock processed per tonne of U in fuel	GJ of heat per tonne of rock processed	GJ of heat per gm of U-235
3.6	0.778	29.9	6.7	3939	655	71.7
4.6	0.824	40.0	8.5	5039	686	76.0
6.0	0.867	55.5	11.4	6719	713	79.9
7.5	0.893	71.5	14.3	8456	730	82.3
8.3	0.904	80.0	15.9	9383	737	83.3

Note: Tonnes of rock processed per tonne of U in fuel is given by the ratio in the previous column divided by (ore grade × 0.8463), where 0.8463 is the mass fraction of U in U_3O_8.

reduces the mass of spent fuel by 50 per cent. The amount of plutonium and other actinides per unit of electricity generated is reduced only modestly, and so is more concentrated in the discarded fuel assemblies. As a result, they would generate more decay heat and so would have to be spaced further apart in the repository. Increasing the burn-up beyond the current rate requires that possible problems due to excessive corrosion, degradation of the cladding around the fuel, distortion of the fuel rod assembly and other problems are addressed (Kazimi and Todreas, 1999).

8.4.4 Reprocessing of spent fuel

Spent reactor fuel consists of some of the original U-235 and U-238 plus fission products, plutonium and minor actinides.[6] The important minor actinides in spent nuclear fuel are neptunium-237, americium-241 and -243, curium-242–248 and californium-249–252. All of these have half-lives of 300 to 20,000 years, while the important fission products either have very short half-lives (<50 years) or very long half-lives (≥64,000 years) (see Table 8.1). It was originally expected that spent nuclear fuel would be routinely reprocessed in order to make new fuel for use by LWRs, or to supply Pu for fast reactors or FBRs. Reprocessing involves separating some of the residual U-235 and the Pu-239 from the fuel rods and concentrating it in new fuel. This is done with a process called PUREX (plutonium uranium recovery by extraction), which involves dissolving the spent fuel in an acid. Reprocessed fuel is a mixture of oxides of uranium and plutonium, and is referred to simply as mixed oxide (MOX). About 5–6 kg of spent fuel must be processed in order to produce 1kg of MOX. Fresh uranium fuel is referred as uranium oxide (UOX) fuel.

With reprocessing, much of the radioactivity that had been locked in spent fuel is instead made freely available in large tanks that must be actively cooled so as not to evaporate and spread radioactivity over the surroundings (in the case of the UK reprocessing facility at Sellafield, this involves a triply redundant cooling system). As well, the recycling facility will itself have been rendered radioactive and will need to undergo a decommissioning process and cleanup after its useful life similar to that of nuclear powerplants, with long-term isolation of the materials from the dismantled facilities.

It is expected that reprocessing to produce MOX would occur only once, that is, the spent MOX fuel would not be reprocessed again. If there are not enough fast reactors to use reprocessed Pu fuel, then the spent Pu fuel from the fast reactors will have to be purified and then used in MOX fuel for use by LWRs. Thus, the possible chains are:

- once-through uranium fuel;
- spent uranium fuel (1 per cent Pu) → reprocessing → Pu-239 and U-235 used in a 6:1 ratio (for example) in MOX fuel (~4.5 per cent total, the balance being U-238) for a LWR → waste;
- spent uranium fuel (1 per cent Pu) → reprocessing → Pu-239 used in MOX fuel for a fast reactor (10 per cent Pu) → clean the spent Pu → Pu used in MOX fuel for LWR → waste;
- spent uranium fuel (1 per cent Pu) → reprocessing → Pu-239 used in MOX fuel for a FBR (10–20 per cent Pu) → spent Pu used in FBR → continue indefinitely.

If uranium is used only once, then at most only 0.7 per cent of the potential energy in the original uranium ore is used (corresponding to the U-235 fraction). As noted in subsection 8.2.4, 30–75 times more electricity can be produced with reprocessing.

Recycling for use in LWRs

Figure 8.12 compares the composition of fresh nuclear fuel for LWRs and spent fuel after a burn-up of 33GWd/t. The fresh fuel has been enriched to 3.1 per cent U-235 (31kg out of 1000kg), which is necessary in order for the chain reaction to occur. Recycling involves separating the U-235 and Pu from the spent fuel. Because of the presence of U-234 and U-236 (which absorb neutrons) in the recycled uranium, the value of the recycled U-235 is reduced by about 20 per cent compared to freshly enriched uranium. The recycled uranium and plutonium can be mixed with natural or depleted uranium oxide to form a MOX that is then used as fresh fuel in the reactor. Table 8.6 gives the composition of Pu isotopes in spent fuel; only about 70 per cent of the Pu is fissile (Pu-239 and Pu-241), and in MOX the fissile Pu is only about 90 per cent as effective as U-235 (Albright and Feiveson, 1988). Thus, the 9kg of extracted Pu can substitute for about 5.7kg of

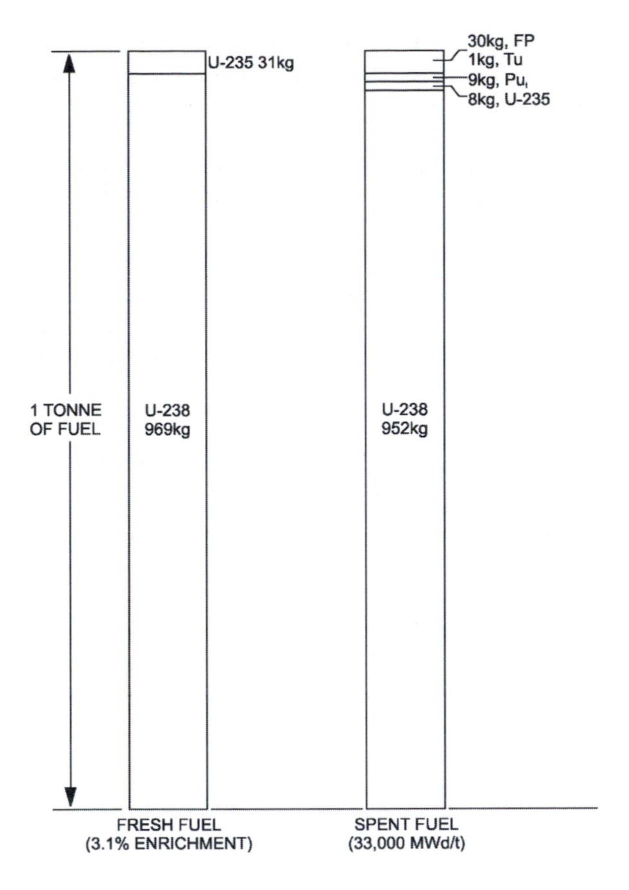

Note: FP = fission products, Tu = transuranic products, Pu_t = total plutonium (including non-fissile Pu).
Source: Albright and Feiveson (1988)

Figure 8.12 *Composition of fresh and spent uranium fuel*

U-235. The Pu would probably be recycled only once, due to the buildup of unwanted isotopes each time that the fuel is used. Altogether, a total of 8kg × 0.8 + 5.7kg = 12.1kg of U-235 are saved, corresponding to about 40 per cent of the U-235 required using fresh fuel. Additional U-235 would be provided from the natural or depleted uranium that would be mixed with the recycled uranium and plutonium. To get the equivalent of 31kg U-235 (as in the original fuel), the spent fuel from several LWRs could be processed, or supplemental plutonium from dismantled nuclear weapons could be used. About 94 per cent of the plutonium from nuclear weapons would be fissile (see Table 8.6).

Table 8.7 gives the amounts of natural or depleted uranium that would be required per 1000kg of MOX fuel, assuming that either fresh natural uranium or depleted uranium (from enrichment activities) is combined with reprocessed uranium and Pu, with or without the addition of weapons Pu (also given and discussed later are cases where only weapons Pu is used along with natural or depleted uranium). When supplemental weapons plutonium is used, all of the spent fuel from one LWR reactor could supply the fuel for another LWR running on 100 per cent reprocessed fuel. However, current LWRs using reprocessed fuel would take only a third of their fuel as MOX, the other two thirds being normal U-235-enriched UOX fuel. Thus, one LWR using fresh fuel would supply three LWRs using a third MOX fuel. This would be in place of four LWRs using fresh fuel. If the UOX is enriched to 3.1 per cent U-235, then (from Equation (8.7)), 5.8t of natural uranium are needed per tonne of UOX.

Table 8.6 *Plutonium composition and rate of heat production in spent nuclear fuel and in materials that can be used to make nuclear weapons*

Isotope	Weapons materials		Uranium spent fuel		Thorium-uranium spent fuel	
	Super grade	Weapons grade	4.5 years use, 45MWd/kg	6 years use, 72MWd/kg	Th-U 72MWd/kg	Th-U 100MWd/kg
			Composition (per cent of total Pu)			
Pu-238	0.0	0.012	2.3	4.4	10.8	13.9
Pu-239	98.0	93.80	55.8	54.6	38.9	36.4
Pu-240	2.0	5.80	21.2	19.6	19.8	19.4
Pu-241	0.0	0.35	14.9	15.4	14.9	14.0
Pu-242	0.0	0.022	5.8	5.9	15.6	16.3
			Decay heat			
(W/kg)	2.0	2.3	16.1	27.8	63.4	80.7
Relative	1.0	1.1	8.1	13.9	31.7	40.4

Source: Kazimi and Todreas (1999)

Table 8.7 *Combinations of mass of enriched U (Case 1) or mass of U-235 and P-239 from spent fuel, mass of P-239 from nuclear weapons and mass of natural or depleted uranium (Cases 2–7) that give an effective mass of 31kg U-235 in 1000kg of MOX*

	Enriched U fuel	Reprocessed U-235 and P-239 with weapons P-239		Reprocessed U-235 & P-239, no weapons P-239		Weapons P-239, no processing of spent fuel	
	Case 1	Case 2	Case 3	Case 4	Case 5	Case 6	Case 7
kg enriched U	1000.0	0.0	0.0	0.0	0.0	0.0	0.0
			Composition of 1000kg of MOX fuel				
kg reprocessed U-235	0	8.0	8.0	16.1	19.3	0.0	0.0
kg reprocessed Pu	0	9.0	9.0	18.1	21.7	0.0	0.0
kg fresh natural U	0	969.8	0.0	965.9	0.0	973.7	0.0
kg depleted U	0	0.0	964.5	0.0	959.0	0.0	968.5
kg weapons Pu	0.0	13.2	18.5	0.0	0.0	26.3	31.5
			Numbers of reactors				
L_{UOX}, number using UOX	1	1	1	1	1	0	0
L_{MOX}, number using one third MOX	0	3	3	3	3	3	3
			Uranium mining requirements and savings compared to Case 1				
t U per t fuel	5.8	4.6	4.4	4.6	4.4	4.2	3.9
Per cent savings		20.8	25.0	20.8	25.0	27.8	33.3

Note: Also given are the tonnes of U that need to be mined per average tonne of fuel supplied using the proportions of reactors fuelled with UOX and with a third MOX given below. Additional assumptions: reprocessed U-235 is 80 per cent as effective as pure U-235 (Albright and Feiveson, 1988); reprocessed and weapons Pu consist of 70 per cent and 94 per cent fissile isotopes, respectively (Table 8.6); the fissile Pu in reprocessed Pu is 90 per cent as effective as U-235 in a uranium fuel element due to the presence of 30 per cent non-fissile Pu (Albright and Feiveson, 1988); the fissile Pu in nuclear weapons Pu is 98 per cent as effective as U-235 (due to the presence of 5 times less non-fissile Pu as in reprocessed Pu); natural and depleted uranium contain 0.7 per cent and 0.2 per cent U-235, respectively.

MOX would use either natural uranium (to which the 5.8 multiplier would not apply) or depleted uranium (in which case no new uranium mining would be required for the MOX portion of the fuel supply).

From the above, it follows that the mass M of uranium that needs to be mined per average tonne of fuel supplied is given by:

$$M = (E_R(1+R_{MOX}(1-F_{MOX})) + R_{MOX}F_{MOX}M_{MOX})/(1+R_{MOX}) \quad (8.9)$$

where E_R is the ratio of natural to enriched uranium (5.8 here), R_{MOX} is the number of reactors using MOX for every reactor using pure UOX, F_{MOX} is the fraction of fuel as MOX in the reactors using MOX, and M_{MOX} is tonnes of natural U per tonne of MOX fuel ($R_{MOX} = 1/F_{MOX}$). The lower rows of Table 8.7 show the uranium mining requirements and percentage savings. The overall savings in uranium mining through reprocessing of spent fuel in LWRs is 20–25 per cent. For some countries, recycling would replace dependence on foreign supplies of uranium with dependence on foreign reprocessing facilities.

Recycling in HWRs

The CANDU reactor uses unenriched uranium and so could also use the spent fuel from LWRs, which has a U-235 concentration close to that of natural uranium (0.8 per cent verus 0.7 per cent). For the same total energy production, using the spent fuel in a CANDU reactor rather than supplying fresh fuel to an LWR reactor reduces the amount of waste for the entire system by 30 per cent (Kazimi and Todreas, 1999).

Recycling in fast reactors and FBRs

The fast reactor can be designed to exactly consume the amount of Pu that is supplied from a thermal reactor, along with other transuranic elements, rather than being a net source of Pu. All that would be left are the fission products themselves, the most radioactive of which have half-lives of 30 years or less (see Table 8.1 and Figure 8.6). Alternatively, fast reactors can be designed to produce more than they consume. As previously noted, such reactors are referred to as FBRs. As discussed in subsection 8.2.4, the amount of energy that could be extracted from a gram of uranium with FBRs would be increased by a factor of about 30–75. Considerable effort was directed toward the development and commercialization of sodium-cooled FBRs during the 1960s and 1970s, but as discussed in subsection 8.3.1, the technology turned out to be far more expensive and difficult to maintain than expected, and most of the world's FBR programmes have been shut down. One difficulty is the use of sodium as a coolant, as liquid sodium burns on contact with air and explodes on contact with water. Three of the six Generation IV reactor concepts listed in Table 8.4 are fast reactors, one of which is sodium-cooled.

Current status of reprocessing

In the early days of nuclear power, it was expected that reprocessing would consume the steady supply of plutonium created by light- and heavy-water nuclear reactors. However, the US suspended development of reprocessing due, in part, to proliferation concerns after India tested a bomb made from extracted plutonium in 1974. As well, development of the FBR – which was to use the plutonium from reprocessing – has not proceeded as originally expected. As a result, reprocessing is now carried out in only a handful of countries, listed in Table 8.8 along with the capacity of the reprocessing facilities. Most of the reprocessing had been done in the UK and France, but the UK facilities were shut down in April 2005 after the discovery that 25 tonnes of spent fuel dissolved in 83m^3 of acid had leaked over a period of months into a stainless-steel-lined concrete enclosure (Garwin, 2007). Much of the spent fuel processed in France is of German and Japanese origin, and the separated Pu, fission products and other radioactive wastes are, under contractual agreement, to be returned to the country of origin. France uses its own Pu to make MOX, reducing its own uranium demand by 20 per cent. Extraordinary security measures have been taken when wastes have been transported internationally. The time for a round trip, from shipment of waste from the country of origin, reprocessing and return of the separated materials, is expected to be in the order of ten years.

Table 8.8 *Countries with the capacity to reprocess spent nuclear fuel (and their capacity) and countries considering constructing reprocessing facilities*

Country	Capacity (tonnes/year)
UK	2400, inoperative since April 2005 after the discovery of a leaking pipe
France	1700
Russia	400
India	275
Japan	40 and 800 under construction
Sweden	Under consideration
Australia	Under consideration
US	Under consideration

Source: Marris (2006)

Currently, there are only 27GW of nuclear powerplants (out of a total of 370GW in 2009) that use MOX fuel, but MOX provides only a third of the fuel to these plants, the balance being UOX (fresh uranium oxide fuel). Thus, the equivalent MOX-fuelled nuclear fleet is 9GW out of 370GW. As a result, there is a growing stockpile of plutonium (about 270 tonnes in separated form from civilian reactors, 700 tonnes in spent LWR fuel and 100 tonnes discarded from each of the US and Russian nuclear weapon programmes, for a total of about 1170 tonnes by 1999 according to Kazimi and Todreas, 1999). The US is now reconsidering reprocessing and proposes a method called UREX+ that reduces the possibility of reprocessed plutonium being used in weapons by leaving it mixed with other highly radioactive metals (Marris, 2006).

8.4.5 Decommissioning of nuclear powerplants

Decommissioning is the process of dismantling a facility and cleaning the site. The decommissioning of nuclear powerplants is rendered difficult by the fact that many of the plant materials are rendered radioactive during the operation of the plant, through the absorption of neutrons that produce radio-nuclides called *activation products*. Table 8.9 lists some of the activation products and their occurrence, half-lives and mode of decay. Co-60 is derived from neutron bombardment of steel; tritium and C-14, from lithium and nitrogen, respectively; Ca-41 from Ca-40; and Cl-36 from Cl-35 (Buck, 1996). As well as activation products, some fission products and small amounts of transuranic elements can be expected in the scrap metal and dismantling debris.

Many of the activation and fission products have half-lives of a few years or less, so decommissioning is therefore easier after a 'cooling off' time of a few years to decades, during which time the plant must be kept under surveillance while the most intensive radioactivity decays to lower levels. Japan requires decommissioning to be completed within ten years of shutting down a plant, the US allows up to 60 years between shutdown and decommissioning and the UK allows up to 135 years before decommissioning of the Magnox and AGR stations (Buck, 1996).

Decommissioning involves: (1) removing fuel; (2) cleaning and decontaminating the plant as thoroughly as possible; (3) altering ventilation systems; (4) removing all ancillary equipment and buildings; and (5) dismantling or reduction in size and removal of the remaining plant parts. In situ decontamination simplifies the subsequent dismantling by removing residual radioactive species and chemicals (from equipment such as heat exchangers). Very strong chemical reagents may be used for this process. Decontamination is beneficial only if the radioactivity is less of a problem in its new home. The ventilation systems that are appropriate during the operation of a nuclear powerplant are different from those appropriate after the plant had been shut down, so modifications (in terms of airflow routes, airflow volumes and filtration) are required for the current generation of nuclear powerplants (Buck, 1996). Dismantling involves taking apart equipment in such a way that it could be reassembled, whereas size reduction entails cutting the equipment into smaller pieces. Robotic systems with special tools are required in the areas with highest radioactivity. Even after all this, the site is not likely to be returned to the condition that existed prior to construction of the nuclear powerplant.

8.4.6 Isolation ('disposal') of nuclear wastes

Radioactive wastes are generated during the mining of uranium ores, the milling of the ore and concentration of uranium into uranium oxide, during the enrichment

Table 8.9 *Activation products found in nuclear powerplants after power production*

Radio-nuclide	Material found in	Half-life	Decay mode
Tritium	Concrete, steel	12.32 yr	β^-
C-14	Concrete, steel	5715 yr	β^-
Cl-36	Concrete, steel	3.0 ka	β^-, β^+, EC, A
Ca-41	Concrete, steel	1.3 ka	EC
Mn-54	Steel	312.5 days	EC, γ
Fe-55	Steel	2.73 yr	EC
Co-57	Steel	271.8 days	EC, γ
Co-60	Steel	5.27 yr	β^-, γ
Ni-59	Steel	76 ka	EC
Ni-63	Steel	100 yr	β^-
Zn-65	Steel	243.8 days	β^+, EC, A
Nb-94	Steel	24 ka	β^-, γ
Ag-108m	Steel	130 yr	EC, γ
Eu-152	Concrete	13.48 yr	β^-, EC, γ
Eu-154	Concrete	8.59 yr	β^-, EC, γ

Note: β^-, β^+ = negative and positive beta emission, γ = gamma radiation, EC = orbital electron capture, A = annihilation radiation, ka = thousands of years.
Source: van Leeuwen (2007)

process (as required for LWRs), during the use of the fuel, during fuel reprocessing (if carried out) and eventually when the nuclear powerplant is decommissioned. These wastes fall into the following categories (Ahearne, 1997):

- high-level radioactive waste, consisting of spent fuel, liquid wastes resulting from the production or reprocessing of reactor fuel, possibly some structural materials of the plant that are loaded with activation products (such as the pressure vessel), and solids into which the liquid wastes have been converted;
- transuranic or intermediate-level waste, produced during reprocessing of nuclear fuel;
- low-level waste, produced during most steps in the nuclear fuel chain.

Spent nuclear fuel contains by far the most radioactivity of stored nuclear wastes in the US, followed by other high-level waste, transuranic waste and low-level waste. High-level wastes, which include all spent fuel, are generated from the production of reactor fuel, reprocessing of spent fuel and from the eventual decommissioning of powerplants. These wastes are stored as liquids, sludge and solids. Low-level waste consists of everything that is not spent fuel, other high-level waste or transuranic waste. There are large volumes of it but with low radioactivity. It consists of contaminated equipment such as gloves, ventilation ducts, protective clothing and packaging. Decommissioning of fuel processing facilities will generate large volumes of low- and medium-level wastes.

When the spent fuel is first removed from a reactor, the bulk of the radioactivity is produced from the decay of short-lived fission products and actinides that were produced in the fuel while it was irradiated with neutrons inside the reactor (see Figure 8.6). The intense radiation from the short-lived fission products and actinides requires that the fuel be initially stored in such a way that it can cool, either in ventilated racks or in cooling ponds. Spent fuel also consists of radioactive materials with very long half-lives (such as the long-lived fission products listed in Table 8.1) that contribute to long-term lower (but still hazardous) levels of radiation.

About 7000 tonnes of spent fuel are currently accumulating per year in temporary storage depots on power station sites throughout the world. When reprocessing was disallowed in the US under the Carter administration due to concerns about the possible diversion of plutonium waste to weapons, US utilities made efforts to expand the storage capacity of the storage ponds by storing the spent fuel more compactly, something that could be done without exceeding safe temperature limits. However, several powerplants have already reached their storage capacity.

The term 'disposal' is usually used in reference to the process of isolating nuclear waste from the biosphere. The term 'disposal' implies that the waste has somehow been eliminated, which is not the case. There are four options with regard to the isolation of spent fuel:

1. processing recycling, which requires separating and recovering reusable U and Pu from the spent fuel, then consuming the U-235 and P-239 as a fuel;
2. isolation of used fuel without reprocessing in deep geological formations after a period of surface storage;
3. surface storage of waste until a decision concerning long-term storage is reached;
4. transmutation.

These options are briefly discussed below.

Reprocessing

Reprocessing was discussed in subsection 8.4.4 in the context of extending uranium supplies. Reprocessing would only slightly reduce the mass of waste (U-235 and Pu-239 would be taken out), and it would not significantly increase the capacity of long-term underground storage repositories because the capacity of such repositories is not limited by the bulk of the fuel but rather by the heat still produced by the fission products and transuranics in the spent fuel. Inert waste in the form of spent fuel would be converted to liquids and sludges that would seriously complicate high-level waste isolation. More importantly, by separating Pu from the spent fuel, reprocessing opens the possibility that some Pu could be diverted into nuclear weapons (see subsection 8.6.2). It is also expensive and accident-prone (see subsection 8.4.4), and would require frequent transport of radioactive materials, thereby increasing the number of possible targets for terrorist groups.

Deep geological storage

Another option is to place high-level nuclear waste in mined structures called *geologic repositories* hundreds of metres below the earth's surface. Among the countries considering deep geological storage, only Sweden has made a final site selection (at Forsmark, 200km north of Stockholm and the site of a nuclear powerplant), while Finland has made a preliminary site selection (next to the Olkiluoto nuclear powerplant, on the west coast).[7] Both the Swedish and Finnish plans call for placing 2t batches of spent fuel in 25t cast-iron canisters coated with 5cm of copper. The canisters will be placed in tunnels 500m below the surface and surrounded with bentonite, which swells when mixed with water to form a supposedly watertight barrier that would also provide protection against earthquakes.

The Swedish and Finnish plans are both counting on engineered barriers to prevent the release of radioactive material and its eventual migration to where the water might be used as drinking water or for agriculture. The assumption is that the 5cm copper coating will not corrode at depth, which in turn is based on the assumption that copper will not corrode in water lacking oxygen. However, recent work challenges this assumption (Szakálos et al, 2007; Hultquist et al, 2008, 2009). It is recognized that corrosion of copper at depth can occur, and evidence from 333-year-old copper coins recovered from a shipwreck in an anoxic environment indicates that up to 1m of copper could be corroded over a 100,000-year time period. The corrosion produces hydrogen, but if the hydrogen builds up next to the copper shell, corrosion will stop. The hope is that the bentonite clay layer will prevent diffusion of hydrogen away from the canister, thereby inhibiting further corrosion (Johan Swahn, personal communication, September 2009).

The US, at Yucca Mountain, had been counting on the geology rather than engineered barriers, the assumption being that groundwater would not enter the repository. Conditions for groundwater entry must not only be unfavourable today, but must remain unfavourable following large human-induced or natural climatic changes over the next million years. Plans for the Yucca site were recently shelved. Uncertainty associated with the true rate of groundwater flow in the Yucca Mountain region had been an ongoing issue (Metlay, 2000).

Transmutation

In principle, all of the unwanted radioactive products of fission can be transformed (transmuted) into non-radiative elements by absorption of neutrons in fast reactors. Transmutation would involve constructing special reactors to split the most radioactive and long-lived isotopes in nuclear waste into isotopes that are radioactive for only a few hundred years. Innovative processing plants would be needed that could separate the various chemical elements in the waste; this would be far more complex than anything in existing plants. A series of reactors would then be needed to bombard the various separated isotopes with appropriate rapid neutrons – one neutron source would have to be built for every five to eight operating reactors, and this would probably cost as much as a nuclear reactor. It is not clear that the whole process is technically feasible, nor that its development could be funded. The whole process is likely to be several times more expensive than deep geological burial.

Current national plans

The options currently preferred by all the countries with nuclear powerplants are listed in Table 8.10. Högselius (2009) presents an analysis of the factors that have led to the adoption of specific options in many of these countries. Almost half of the countries with nuclear powerplants (14 out of 32) intend on deep burial of their nuclear wastes, but only Finland and Sweden have selected a site. Political paralysis characterizes many other democratic countries. Plans for the proposed Yucca Mountain site in the US were shelved by the Obama administration after 20 years of study, but no other site had been under investigation, so the site selection process must begin anew (Ewing and von Hippel, 2009).

8.4.7 Summary of waste generated during normal operation

Figure 8.13 summarizes the quantities of solid and liquid wastes generated during each stage of the nuclear fuel chain for the generation of 1GW-year of electricity. The quantities given are for a waste rock/ore ratio of 5, ore at 0.2 per cent uranium, enrichment to 3.6 per cent U-235, a burn-up of 42GWd/tU and a thermal

Table 8.10 *Options for the isolation of nuclear wastes preferred at present for each country with nuclear powerplants*

Reprocessing	*Deep burial (continued)*
China	Lithuania
France	Romania
India	Slovakia
Japan	Slovenia (or possible export)
Russia (for VVER reactors)	South Korea
UK	Spain
Export to other countries with return of wastes	Sweden
	Taiwan
Belgium (to France)	UK (possibly)
Italy (to France)	US
Netherlands (to France)	*No long-term decision yet*
Ukraine (to Russia)	Argentina
for VVER reactors	Armenia
Export to other countries without return of wastes	Brazil
	Czech Republic
Bulgaria (to Russia)	Mexico
Deep burial	Pakistan
Canada (with retrievability)	South Africa
Finland	Switzerland
Germany	Ukraine (for fuel not exported to Russia)
Hungary	

Source: Högselius (2009)

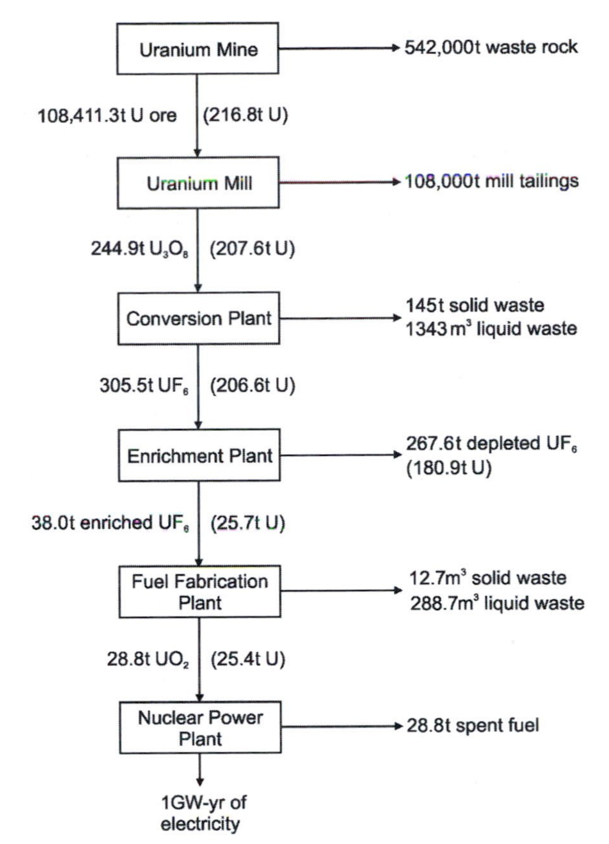

Source: Based on Beck (1999)

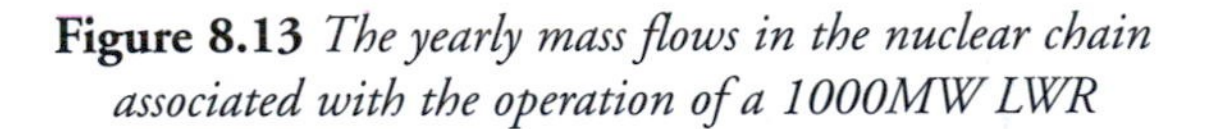

Figure 8.13 *The yearly mass flows in the nuclear chain associated with the operation of a 1000MW LWR*

efficiency of 34.2 per cent. The greatest volume of waste by far is generating during uranium mining (542,000t/yr of overburden removed for a 5:1 overburden to ore ratio) and milling (108,000t/yr of radioactive tailings), with subsequent steps each generating a few tens or hundreds of tonnes of waste per year.

8.4.8 Alternative long-term nuclear fuel chains

Figure 8.14 shows four hypothetical fuel chains associated with a 1500GW fleet of nuclear reactors in 2050, developed by MIT's Nuclear Energy Study Advisory Committee (MIT, 2003). The first two chains (1a and 1b) are once-through chains but with differing amounts of initial U-235 enrichment and burn-ups. Doubling the burn-up (from 50GWd/t to 100GWd/t) reduces the uranium requirement by about 7 per cent and cuts the mass of spent fuel in half but does not change the mass of fission products, plutonium or minor actinides produced.

Option 2 involves separating and recycling Pu (but not U-235) in thermal reactors, which reduces the uranium requirements by 16 per cent. 334t/yr of separated plutonium are combined with depleted uranium to produce MOX fuel. The UOX to MOX ratio (5.27:1.0) is chosen so as to exactly use up the plutonium that is separated from spent UOX fuel. Altogether, 59 per cent of the plutonium that is generated is used as a fuel, the balance ending up in spent MOX fuel.

Option 3 uses reprocessed spent fuel in fast reactors, but it is a fully closed cycle, in that the number of fast reactors deployed is sufficient to consume essentially all of the actinides (including plutonium) produced in once-through thermal reactors. Only the fast reactor fuel is reprocessed, ideally in a developed country in a secure

Option 1a, Low Burn-up: 50GWd/t, n=0.33

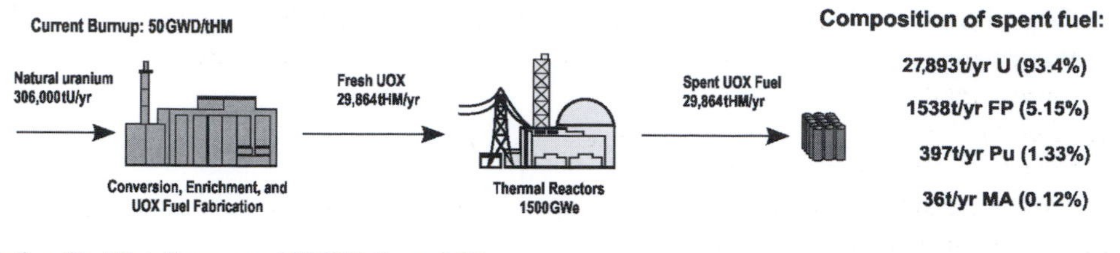

Option 1b, High Burn-up: 100GWd/t, n=0.33

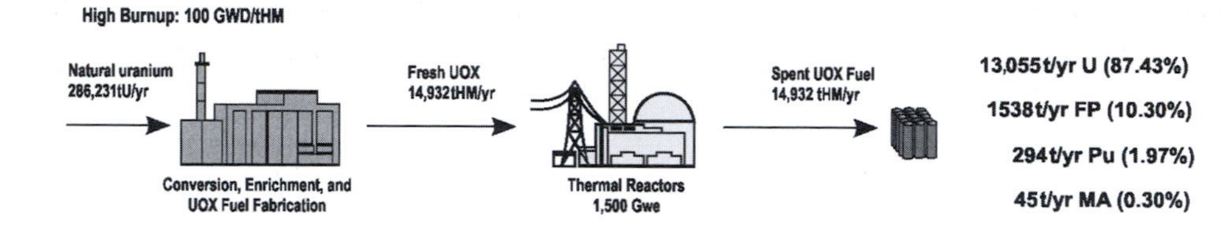

Option 2, Burn-up =50GWd/t, n=0.33

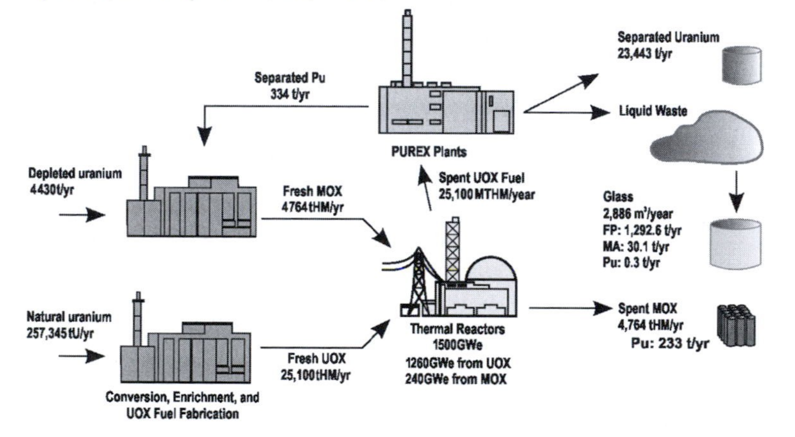

Option 3, LWR burn-up = 50GWd/t, n=0.33, CF=0.9; FR burn-up = 120GWd/t, n=0.44, CF=0.9

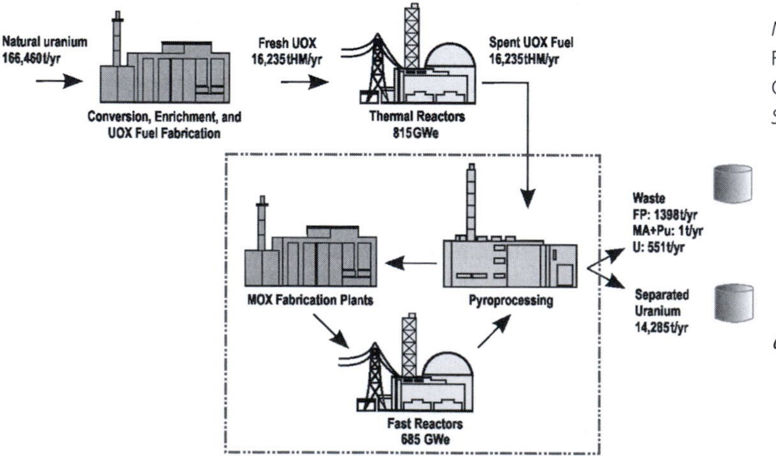

Note: LWR = light-water reactor, FR = fast reactor, FP = fission products, MA = minor actinides, CF = capacity factor, n = thermal efficiency.
Source: MIT (2003)

Figure 8.14 *Possible uranium fuel chains associated with 1500GW of nuclear powerplants operating on average at 90 per cent of capacity: Options 1a and 1b, once-through chains; Option 2, a chain with recycling of plutonium once using thermal reactors; Option 3, a chain with recycling of plutonium and actinides*

energy park. The thermal reactors, operating on a once-through cycle, could be located anywhere. This option is different from the FBR scheme, in which the fast reactors produce excess plutonium that serves as the fuel for the thermal reactors, thereby eliminating the need for a continuous supply of fresh uranium. Option 3 requires a continuous supply of fresh uranium, but it stretches out the uranium supply (uranium requirements are reduced by 45 per cent compared to present once-through systems) while making it more difficult to divert plutonium to the production of nuclear bombs. Both once-through and partly closed cycles can operate on uranium or Th fuel using different reactor types.

Table 8.11 compares the material flows for a global nuclear fleet of 1500GW at 90 per cent capacity factor using option 3 and for the 2007 fleet of 370GW modelled entirely at option 1a. During each cycle of reprocessed fuel through the fast reactors in option 3, the transuranic content of the fuel decreases by 20 per cent by irradiation, so this amount must be supplied by transuranics in the spent UOX from the LWRs. The relative capacities of LWRs and fast reactors are chosen so as to give a perfectly balanced flow (supply = consumption), except that 0.1 per cent is assumed to be lost during reprocessing and appears in the waste stream. However, the net production of transuranics (Pu + minor actinides) in a hypothetical future nuclear fleet of 1500GW is over 100 times smaller than for the present fleet of 370GW. The rate of consumption of uranium resources is 2.2 times greater and the rate of production of fission products is 3.7 greater while producing 4.1 times more electricity.

8.5 Safety

The safety issues related to nuclear energy are:

- accidental release of radiation to the environment due to a malfunction of some part of the nuclear powerplant due to, for example, design error, construction error, operator error, maintenance failure or 'Act of God';
- intentional release of radiation due to a terrorist attack on a nuclear powerplant (for example, by striking the plant with a fully fuelled large commercial aircraft, or through internal sabotage);
- intentional release of radiation through a military strike with conventional weapons;
- accidents, sabotage or military strikes at reprocessing plants.

These risks are elaborated upon below.

8.5.1 Risks associated with routine operation of nuclear reactors

The likelihood of an accident that proceeds all the way to core damage can be estimated by analysing data on

Table 8.11 *Mass flows (tonnes/yr) associated with the current world fleet, modelled using option 1a of Figure 8.14, and for a fleet of 1500GW modelled using option 3 of Figure 8.14*

	Present fleet	Future fleet	Future/present
Capacity (GW)	370	1500	4.05
Natural U	75,527	166,460	2.20
Fresh UOX	7366	16,235	2.20
Spent UOX	7366		
FP	379		
Pu	98		
MA	8.9		
Separated U		14,285	
Reprocessing wastes		1950	
FP		1398	
Pu+MA		1	
U		551	
Total FP	379	1398	3.69
Total Pu+MA	106.8	1	0.0094

Note: FP = fission products, MA = minor actinides.
Source: Based on estimates in MIT (2003)

the occurrence of individual system malfunctions. Such an analysis for the US indicates a drop by a factor of 100 between 1974–1978 (just before the Three Mile Island accident) and 1994–1998 (Sailor et al, 2000). Recent analyses for the US indicate a likely average frequency of core damage of 1 in 10,000 reactor-years. Thus, for a fleet of 200 reactors (twice the current US fleet), core damage to a reactor would be expected once every 50 years. MIT (2003) suggests that a reasonable goal would be to reduce this frequency to 1 in 100,000 reactor years, and furthermore, designers of advanced LWRs claim that their design already meets this goal. Core damage does not necessarily imply release of radiation to the surrounding environment, so the risk to people living near nuclear powerplants would be smaller still. A key caveat is that the frequency estimates depend on proper O&M of nuclear facilities, which in turn requires a persistent commitment to a safety 'culture'.

The average age of the global reactor fleet is 25 years. Initial operating licences are typically for 20–25 years, but reactor operators have been applying for licence extensions, which will increase the risk of an accident due to the effects of aging, unless there are increasingly stringent and frequent safety inspections. Aging involves changes in material properties over time due to the simultaneous effects of irradiation, corrosion and thermal and mechanical loads, and is inherently difficult to detect (prior to a failure) because it involves complicated systems of pipes that cannot be easily accessed and/or are subject to high radiation levels (Frogatt, 2005). Serious accidents have been avoided because the damage had been detected in time by monitoring systems or by routine checks during downtime and repairs, but sometimes the discoveries were made by chance (see Rosenkranz, 2006, for examples). Deregulation and a greater emphasis on minimizing costs – at the same time that the reactor fleet is aging – may compromise safety, especially in jurisdictions with weak enforcement of standards or lax standards to begin with.

8.5.2 Risks associated with terrorist attacks

The Electric Power Research Institute in the US evaluated the effect of a direct strike on a nuclear powerplant by a commercial aircraft for the US Nuclear Energy Institute. Their conclusion is that the containment structure for US plants, used-fuel storage pools and used-fuel dry storage facilities, as well as used-fuel transportation containers, would not be breached by a direct hit with a Boeing 767-400 aircraft (EPRI, 2002). However, this may not mean the same as surviving the full aftermath of a direct hit. Engineering analysis had indicated that the World Trade Center towers would survive the kinetic energy of an airliner crashing into them, which they did, but they nevertheless succumbed to the resulting fires.

8.5.3 Risks associated with reprocessing plants

There have been a number of accidents at reprocessing plants (MIT, 2003): the explosion of a waste tank at Chelyabinsk in the former Soviet Union, leakage from a plant at Hanford, US and discharges to the environment at the Sellafield plant in the UK. At any given time a commercial reprocessing plant would have on the site the waste from tens or even hundreds of reactor years.

8.5.4 Risks associated with a military strike

Nuclear reactors and spent fuel stored onsite (even in the more secure systems used in new plants) could be vulnerable to a military attack. A commercial nuclear powerplant contains an order of magnitude more radioactivity than is released by exploding an atomic bomb, so the threat of attacking a nuclear powerplant or reprocessing plant militarily could be as dangerous as an adversary acquiring nuclear weapons themselves (Rosenkranz, 2006).

8.6 Nuclear weapons, arms proliferation and terrorism risks

There are two distinct risks related to the proliferation of nuclear weapons from the use of nuclear energy: first, the facilities used to enrich natural uranium to reactor-grade uranium can also be used to produce bomb-grade uranium. Second, the reprocessing of spent fuel would separate plutonium from the other, more radioactive materials in the spent fuel, thereby making it relatively easy and safe to use the separated plutonium to make nuclear weapons. Unless it is inhaled, ingested or comes into contact with a skin abrasion or cut, plutonium itself is not particularly harmful, as the radiation from it cannot even penetrate human skin. However, even after 50 years, the radiation from other elements in spent fuel would be lethal in less than an hour at a distance of 1m. Thus, the separation of Pu from other radioactive elements by reprocessing makes it readily usable in nuclear weapons.

8.6.1 Risks associated with the enrichment of uranium for power generation

Uranium requires a U-235 concentration of at least 20 per cent in order to make a bomb, but such a bomb would be far too heavy, so the U-235 is typically enriched to 90 per cent. Uranium enriched to 20 per cent or more U-235 is defined as *high-enriched uranium* (HEU), although in practice the term means uranium enriched to 90 per cent U-235. Anything over 20 per cent U-235 is classified as 'weapons useable'. Conversely, uranium enriched to less then 20 per cent U-235 is defined as *low-enriched uranium* (LEU), although in practice the term means enriched to the 3–5 per cent level required for nuclear powerplants.

The same processes are used to enrich uranium for power generation and for bombs, the only difference being one of degree. As discussed in subsection 8.4.2, enrichment of one stream of uranium in U-235 leaves another stream that is depleted in U-235. The inherent effort required to enrich uranium by a given amount is represented by the number of *separative work units* (SWUs) N_{swu}, which is calculated as (Bodansky, 2004, Chapter 9):

$$N_{swu} = M_n V_n + M_d V_d - M_e V_e \tag{8.10}$$

where M_n, M_d and M_e are the masses of the natural, depleted and enriched streams, and the V_x are value functions calculated as:

$$V_x = (1 - 2C_x)\ln\left(\frac{1 - C_x}{C_x}\right) \tag{8.11}$$

where C_x is the concentration of U-235 in stream x. M_n is given by Equation (8.7), while M_d is given by:

$$M_d = \left(\frac{C_e - C_n}{C_n - C_d}\right) M_e \tag{8.12}$$

N_{swu} as given in Equation (8.10) has units of mass and is per unit of enriched uranium produced. To get the effort per unit of U-235 produced in the enriched fuel, the N_{swu} given by Equation (8.10) should be divided by the concentration of U-235 in the enriched fuel. Figure 8.15 shows (N_{swu}/C_e) as a function of the U-235 concentration in the enriched stream, assuming C_d to be 0.002 or 0.003. For C_d= 0.002, about 65 per cent of the effort needed to produce bomb uranium at 90 per cent U-235 is expended in producing reactor uranium at 4 per cent U-235, and only one third is used to go from reactor to bomb uranium. For C_d = 0.003, the relative effort needed to reach 4 per cent enrichment is about 62 per cent of that required to reach 90 per cent enrichment. Thus, all uranium enrichment facilities are potential sources of uranium for bombs.

As of 2004, uranium enrichment facilities existed in Argentina, Brazil, China, France, Germany, India, Iran, Japan, Netherlands, Pakistan, Russia, the UK, the US and possibly a few other countries. Two issues related to the use of centrifuges for nuclear weapons programmes are: (1) the speed with which any peaceful-use plant could be converted to non-peaceful purposes, and (2) the potential for clandestine plants. Both issues are discussed at some length by Wood et al (2008). The centrifuge cascade used for peaceful purposes could be converted to a cascade capable of producing HEU within days, and a centrifuge plant built to fuel just one commercial-sized reactor could produce enough HEU for dozens of nuclear weapons per year.

8.6.2 Risks associated with the recycling of spent fuel

It has recently been estimated that 1.9 million kg of HEU and 1.83 million kg of plutonium exist worldwide (Hecker, 2006). Approximately 1.4 million kg of the plutonium are currently in highly radioactive spent fuel and so could not be easily used by terrorists. Although the plutonium normally used in nuclear weapons and the plutonium in spent reactor fuel have quite different isotopic compositions (as shown in

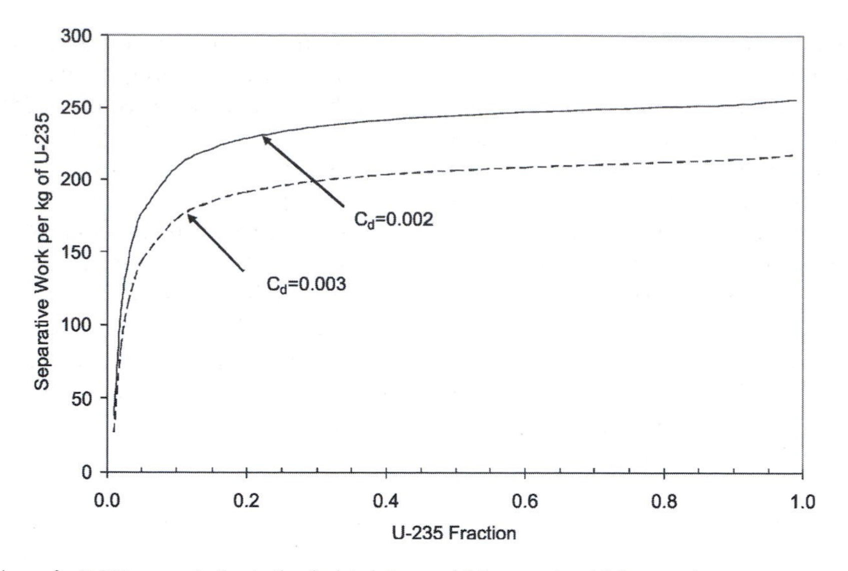

Note: Amounts are shown for U-235 concentration in the depleted stream of 0.2 per cent and 0.3 per cent.
Source: Calculated from Equations (8.7) and (8.10) to (8.12)

Figure 8.15 *The number of SWUs required to enrich U-235, per kg of enriched U-235, as a function of the degree of enrichment*

Table 8.6), plutonium of virtually any isotopic composition can still be used to make nuclear weapons according to the US Committee on International Security and Arms Control (CISAC, 1994). Reactor-grade plutonium could be used in relatively simple designs such as that used in the Nagasaki bomb – which are within the capabilities of many nations and possibly of sub-national groups – with an assured minimum nuclear explosive yield of at least 1 or 2 kilotons (compared to 20kt for the Nagasaki bomb). Using more sophisticated designs, reactor-grade plutonium could be used for weapons with much higher explosive yields. Thus, according to CISAC (1994), 'the difference in proliferation risk posed by separated weapons-grade plutonium and separated reactor-grade plutonium is small in comparison to the difference between separated plutonium of any grade and unseparated material in spent fuel'.

A major expansion of nuclear power, and in particular, an increase in the number of countries using nuclear energy, would require a significant strengthening of international supervision (Hecker, 2006). However, accurate accounting of fissile materials is difficult because fissile materials occur in many locations (including 120 research reactors using HEU in 40 countries) and plutonium occurs in seven different crystal structures with differing densities (making mass inventories difficult). As little as 1kg of plutonium or 3kg of HEU could be used to build a bomb with a 2kt yield, with the exact Pu or HEU requirements depending on the technical skills of those making the bomb (Cochran and Paine, 1995).

8.6.3 Reducing the proliferation risk associated with reprocessing

Current reprocessing (PUREX/MOX) does not provide significant resistance against proliferation of nuclear weapons. The US embarked on three initiatives in order to reduce the risks of proliferation of nuclear weapons associated with reprocessing:

1 Research and development into reprocessing of spent fuel by 'pyroprocessing' or electro-refining in a molten salt electrolyte as an alternative to the PUREX nitric-acid dissolution, organic solvent-extraction cycle used in most commercial reprocessing plants (von Hippel, 2001). This does not fully separate plutonium from other

transuranic elements, so there would be some proliferation resistance.

2 Research and development into the UREX+ (uranium extraction) process, in which an acid dissolves the fuel (as in PUREX) while generating uranium as low-level waste or for use in a fuel. Cs and possibly Sr are also removed. The remainder is separated into the fission products (which become high-level waste) and the Pu and other transuranics, which are incorporated into new reactor fuel. The minor actinides are kept with the Pu, making it dangerous to handle. The US is probably at least a few decades away from the large-scale implementation of either UREX+ or pyrochemical processing (Bodansky, 2006).
3 Design and construction of a prototype *integrated fast reactor* (IFR), in which a reprocessing and fuel-recycle plant would be integrated into the reactor plant. It would use high-energy fast neutrons but would consume plutonium rather than breeding it. Because fast neutrons can cause many more types of elements to undergo fission, it would not be limited to uranium and plutonium as fuels (Hoffert and Potter, 1997). It could thus consume many other highly radioactive elements. The IFR programme was terminated in the mid-1990s, but parts of it were later revived.

The key to any system with reprocessing that is proliferation resistant would be to have a secure fuel cycle, in which spent fuel is sent to a small number of facilities in existing nuclear power countries, the fuel is reprocessed, and the reprocessed fuel is sent back to the original country for reuse. This would be economically attractive from the point of view of the country supplying spent fuel for reprocessing as long as reprocessed fuel is supplied at the same price as fresh fuel (in spite of its much greater cost), and would be attractive from the point of view of the countries doing the reprocessing because of the greatly reduced risk of diversion of nuclear materials to nuclear weapons.

8.6.4 Use of excess nuclear weapons Pu or HEU and depleted U from enrichment activities

Rather than reprocessing spent nuclear fuel to produce new fuel, with attendant large risks of the diversion of separated Pu for military purposes or to terrorist groups, the world's nuclear fleet could be used to consume excess weapons-grade Pu. The US and Russia alone have built up an inventory of 250,000kg of Pu in over 50,000 nuclear warheads. In a hypothetical future where the nuclear powers largely disarm, there will be a need to dispose of the excess plutonium.

In September 2000 the US and Russia agreed to each dispose of 34 tonnes of excess plutonium from downsizing of their nuclear arsenals, with the option of using the discarded plutonium to produce MOX fuel (see www.state.gov/documents/organization/18557.pdf). For this purpose, the US is planning to build an MOX fuel plant at the Savannah River site in South Carolina while Russia will build a plant at Seversk. Some fuel with US plutonium was manufactured in France and delivered for testing at the Catawba nuclear station in South Carolina in April 2005. Each tonne of separated plutonium can replace about 1.2 tonnes of natural uranium. The mass of Pu in the spent fuel would be comparable to the mass of Pu in the MOX, but would now be in a form that could not easily be used to make nuclear weapons.

The last two columns of Table 8.7 show the composition of MOX using weapons Pu and either natural uranium or depleted uranium to give the equivalent of 3.1 per cent U-235. If natural uranium is blended with weapons Pu (Case 6 in Table 8.7), the ratio of mined U to uranium fuel would drop from between six and eight (for each unit of enriched fuel produced) to one. If depleted U is used (Case 7), more weapons U would be needed per tonne of fuel (thereby using up the excess weapons Pu faster) and a hazardous waste (depleted U) would be consumed. If reactors are fuelled with a third MOX and two thirds UOX (as in current designs) using weapons Pu and enriched U, the saving in U mining is about 28 per cent, whereas if depleted U is used, the saving is 33.3 per cent, with a similar reduction in landscape impacts from mining. If reactors are designed to use 100 per cent MOX, and this MOX is made using depleted U, then no new U mining would be needed.

As seen from Table 8.7, about 30kg of Pu would be required per tonne of MOX fuel produced with an equivalent U-235 concentration of 3.1 per cent. From Box 8.2, the burn-up would be about 31GWd/t. Given a weapons arsenal of 250,000t Pu and assuming a powerplant thermal efficiency of 35 per cent, about 750GW-yr of electricity could be produced while using up the nuclear weapons Pu arsenal. Given a current world nuclear capacity of 370GW, only a small fraction of the current nuclear fleet would need to be fuelled with weapons Pu over a period of two decades in order to repackage all this Pu into spent nuclear fuel (in the event that the world were to largely disarm).

The world's nuclear weapons arsenals also contain HEU. This uranium is enriched to a much greater concentration of U-235 than is required for power reactors, and so can be blended with uranium that is depleted in U-235 to produce U enriched to the degree needed by nuclear reactors. In 1993, the US and Russia concluded an agreement whereby Russia supplies 500 tonnes of HEU per year for 20 years, the equivalent of 153,000 tonnes of natural uranium. Depleted uranium, which contains about 0.2 per cent U-235 (compared to 0.7 per cent in natural U), can be re-enriched (while co-producing uranium that is even more depleted in U-235). At present, some European depleted U is sent to Russia for re-enrichment. The re-enriched uranium is then either sent back to Europe or used as a feedstock for blending with HEU from nuclear weapons. In this way, Russia can supply downblended HEU to the US without using its own limited uranium resources. The US has also been downblending its own HEU, with a total of 153 tonnes expected to be downblended by 2016.

Because of the supply of secondary uranium, the price of uranium fell to around \$40/kg in 1995 after peaking at around \$100/kg in 1985, and fluctuated between \$30 and 40/kg until 2005. This drop in prices led to a sharp drop in exploration effort. Very little uranium is mined in the US at present; instead, 60 per cent of current extraction occurs in three countries: Canada, Australia and Kazakhstan. Since 2005, prices have risen sharply due to a resurgent demand. By 2003, 46 per cent of the global uranium supply for civilian reactors came from secondary sources (Nassauer, 2005), but eventually this will have to be replaced by new primary supplies.

8.7 Cost

It is very difficult to get information about the true cost of building recent nuclear reactors, so the information that is available is quite limited. This makes future costs quite uncertain, which in turn has resulted in little credible commercial interest in new nuclear powerplants.

The levelized cost of electricity is the constant price of electricity that needs to be charged in order to pay back the initial investment cost with interest, as well as to cover yearly operating costs (fuel, insurance, O&M and, in the case of nuclear reactors, contributions to funds for waste disposal and decommissioning). The levelized cost, C_{elec} (\$/kWh), is given by:

$$C_{elec} = \frac{(CRF + I)C_{cap} + OM_{fixed}}{8760CF} + OM_{variable} + C_{fuel} \qquad (8.13)$$

where CRF is the cost recovery factor, I is the annual insurance payment as a fraction of the capital cost, C_{cap} is the capital cost of the powerplant, OM_{fixed} is the fixed annual O&M cost per kW of power capacity (\$/kW/yr), $OM_{variable}$ is the variable O&M cost (\$/kWh), CF is the capacity factor and C_{fuel} is the contribution of fuel to the cost of electricity. The cost recovery factor depends on the real interest rate expressed as a fraction, i, and the powerplant lifespan in years, N, as given by Equation (C.1) of Appendix C.

Capital costs of powerplants are given as the cost per kW of peak output (per kW of capacity). If the powerplant were to run at full output all year, it would produce 8760kWh of electricity per year for each kW of capacity (8760 being the number of hours in a year). CF is the average output as a fraction of the peak output, so $8760CF$ is the number of kWh produced per year for kW of generating capacity. The capital cost includes the direct costs of equipment and construction as well as interest on the expenditures made during construction. The latter can be quite substantial, as construction can require ten years or more. As well, the discounted future cost of decommissioning the powerplant and building the waste repository should be included in the capital cost. Unlike other kinds of powerplants, nuclear powerplants also incur substantial costs after decades of operation: isolation of radioactive waste, guarding closed reactors and eventually decommissioning the reactors after a cool-down period that could exceed 30 years.

8.7.1 Capital cost

There are several different ways in which the capital cost of nuclear powerplants can be reported, and comparison of costs that are reported in different ways is one reason for differences in the quoted cost (Du and Parsons, 2009). The costs in building a nuclear powerplant are incurred over a period of five to ten years (depending on how long it takes to build the powerplant), during which time interest accumulates on the costs already incurred while costs not yet incurred increase due to general inflation. One method of quoting costs is in terms of the costs from the vendor using prices at the start of the construction, as if everything could be supplied at once (that is, without interest charges). This is referred to as the 'overnight' cost. These are costs for engineering, procurement and construction (EPC). In addition to purchase costs, the utility will entail costs of its own that are typically 20 per cent of EPC costs. There may be additional costs for transmission upgrades so that the grid is able to absorb a large new source of power. With inflation and accumulated interest during construction, the final cost can be up to twice the overnight EPC cost and is frequently 50–75 per cent greater.

Figure 8.16 gives the variation in the cost of completed nuclear powerplants from 1971 to 1997 (the first era of nuclear powerplant construction) and of projections of the cost for the hoped-for next generation of nuclear powerplants made at various times and by various types of organizations, as compiled by Cooper (2009). Between 2001 and 2003, various university and government groups were projecting overnight EPC costs of \$1400–2350/kW. In 2007–2008, a number of power utilities projected costs of \$2500–5500/kW, and in 2008–2009, a number of financial firms (including Moody's and Standard and Poors) projected overnight costs of about \$5000–10,000/kW.

A group at MIT (2003) had initially projected a cost of about \$2000/kW, based on overnight costs (in 2002\$) of \$1611–2536/kW for seven reactors completed in Japan and South Korea between 1994 and 2002 (local costs were converted to dollars using a purchasing power parity (PPP) conversion factor, not the official exchange rate). Increasing this by the general inflation of 15 per cent between 2002 and 2007 brings the costs to \$1853–2916/kW in 2007 dollars. During the period 2004–2006, five more reactors were completed in Japan and South Korea. The overnight costs in 2007 PPP dollars were \$2357–3357/kW.

A new reactor is under construction at the Olkiluoto site in Finland, the first of the new European evolutionary pressurized-water (EPR) reactors. The original contract, signed in 2004, was for a 1600MW unit at a cost of €2000/kW (about \$2800/kW). By March 2009, construction was three years behind schedule and the projected cost had risen to €3190/kW (about \$4500/kW) (Schneider et al 2009). A second

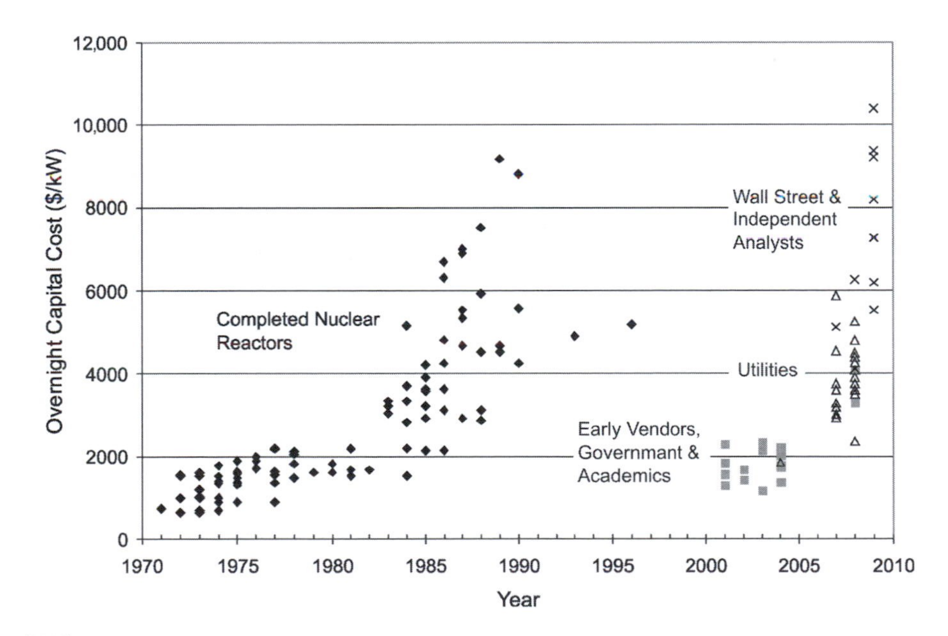

Source: Cooper (2009)

Figure 8.16 *Costs of nuclear powerplants completed between 1971 and 1997, and projections of the cost of future nuclear powerplants made from 2001 to 2009*

EPR was ordered in January 2007, this time by the French utility Électricité de France, with a revised overnight cost estimate in December 2008 of about €2500/kW (up from €2000/kW in early 2007).

A number of US utilities filed cost estimates in 2007 and 2008 (shown in Figure 8.16) that averaged about $4000/kW in 2007 (these estimates are discussed by Du and Parsons, 2009). In light of this, the MIT revised its earlier cost estimate of $2000/kW to $4000/kW (MIT, 2009). Compared to the cost (in PPP terms) of about $2000/kW in 2002 for the Japanese and South Korean reactors, this implies a nuclear cost inflation of about 15 per cent/year (about five times the general rate of inflation). Part of this is due to sharp increases in commodity prices and in engineering services due to demand outstripping supply, and may very well decrease following the economic slowdown that began in late 2008. However, all of the recent utility estimates are just that – estimates, not real after-the-fact costs. As noted above, independent financial analysts have been very sceptical of utility estimates, projecting instead costs in the $5000–10,000/kW range.

In the past, major new technologies such as nuclear energy have experienced start-up problems that greatly increased costs. It is unlikely that entirely new nuclear energy concepts would not also entail much greater costs than forecast by their proponents, so they pose substantial financial risk to would-be investors (whether private or public). The risk of design changes for new reactor concepts cannot be ruled out; design changes would occur if unanticipated problems arise during construction, or if the initial operating experience of operating units for a new type of reactor leads to the need for a change in design after construction of subsequent units has started.

Unlike most other technologies, costs of nuclear energy have not been falling over time. This could be due in part to a paucity of orders in recent decades, and in part due to the lack of mass production of a standardized design. Current reactors are built largely on site, which is considerably more expensive than factory production.

8.7.2 Capacity factor

The average capacity factor of nuclear powerplants (average output divided by capacity) has gradually increased over time, rising from about 70 per cent (with wide variation) in the early 1970s to an average of about 83 per cent during 2003–2005 (Cooper, 2009). The maximum possible capacity factor for LWRs is 85–94 per cent, given that they need to be shut down for six to eight weeks every 12–24 months for refuelling. Additional downtime is due to malfunctions or unscheduled maintenance. However, if a new generation of nuclear reactors is built based on new designs, one can expect the capacity factors to be initially lower than at present, as the inevitable problems with new (and especially with complicated) technologies are worked out. Loss of output during the early years of a powerplant has a disproportionately large effect on the cost of electricity over the lifetime of the plant.

8.7.3 Operating costs

The operating costs of a nuclear powerplant include: fuel cost, fixed O&M, variable O&M, contributions to waste and decommissioning funds and insurance. The cost of fuel is given by the cost of the ore and cost of the conversion, enrichment and fabrication steps, plus carrying costs (interest on the initial expenditures) associated with these steps due to the fact that costs for the early steps are incurred up to several years before the fuel is used. The cost of fuel ($/kg) is computed as:

$$C_f = \sum_i M_i C_i + \sum_i M_i C_i \phi \Delta T \tag{8.14}$$

where M_i is the mass of material used in stage i, C_i is the cost per unit mass at stage i, φ is the carrying charge per year as a fraction of the initial investment and ΔT is the number of years from when a cost is incurred to the halfway point in the use of the fuel. The first summation involves costs incurred year by year and the second summation involves one-time costs incurred prior to the generation of electricity. Table 8.12 gives the inputs to Equation (8.14) for the fabrication of UOX, as given by MIT (2003). The final cost is $2040/kg of fuel for uranium ore at a cost of $30/kg. For uranium ore at $150/kg (five times more expensive), the cost of fuel is $3788/kg – less than twice as expensive.

The fuel cost per kWh of electricity generated can be computed as:

$$\frac{\$}{\text{kWh}} = \frac{C_f(\$/\text{kg})}{B(\text{MWd/kg})\eta} \frac{1\text{MW}}{1000\text{kW}} \frac{1\text{day}}{24\,\text{hours}} \tag{8.15}$$

Table 8.12 *Inputs to Equation (8.14) for the computation of the cost of 1kg of UOX fuel, and final cost assuming a carrying charge factor φ of 0.1*

	Input			Direct cost	Carrying charge
Item	M_i	C_i	ΔT_i (yr)	$M_i C_i$ (\$)	$M_i C_i \varphi \Delta T_i$ (\$)
Ore purchase	10.2kg/kg	\$30/kg	4.25	307	130
Conversion	10.2kg/kg	\$8/kg	4.25	82	35
Enrichment	6.23kgSWU/kg	\$100/kgSWU	3.25	623	202
Fabrication	1kg/kg	\$275/kg	2.75	275	76
Storage and isolation	1kg/kg	\$400/kg	−2.75	400	−90
Total				1686	353
Grand total				\$2040/kg	

Source: MIT (2003)

where B is the burn-up and η is the thermal efficiency. For B = 40MWd/kg and η = 0.33, we obtain a fuel component to the cost of electricity of 0.64 cents/kWh and 1.20 cents/kWh for C_f= \$2040/kg and \$3788/kg, respectively.

Table 8.13 gives the inputs needed to compute the cost of MOX fuel using the PUREX method. Reprocessed fuel is over four times as costly (\$8890/kg) as fuel made from fresh uranium at \$30/kg (MIT, 2003), and even at a hypothetical future uranium cost of \$150/kg, MOX is still over twice as expensive. Thus, where PUREX/MOX reprocessing is carried out today, it is heavily subsidized. However, even with the much greater cost of recycled fuel, the fuel cost would still represent a very small fraction of the total cost of nuclear-generated electricity.

Du and Parsons (2009) indicate a current fixed O&M cost for new nuclear reactors of \$56/kW/yr and a variable cost (excluding fuel) of 0.042 cents/kWh.

8.7.4 Decommissioning and long-term isolation of wastes

Given the complexity and difficulty of the decommissioning process (subsection 8.4.5), the decommissioning process could easily be as or more

Table 8.13 *Inputs to Equation (8.15) for the computation of the cost of 1kg of MOX fuel, and final cost assuming a carrying charge factor φ of 0.1*

	Input			Direct cost	Carrying charge
Item	M_i (kg/kg fuel)	C_i (\$/kg)	ΔT_i (yr)	$M_i C_i$ (\$)	$M_i C_i \varphi \Delta T_i$ (\$)
Credit for avoided isolation of UOX spent fuel	5.26	–400	4.25	–2105	–895
Reprocessing	5.26	1000	4.25	5263	2237
High-level waste storage and isolation	5.26	300	3.25	1579	513
MOX fabrication	1	1500	3.25	1500	488
MOX storage and isolation	1	400	–2.25	400	–90
Total				6637	2253
Grand total				\$8890/kg	

Source: MIT (2003)

expensive than the construction of the powerplant in the first place. The British Nuclear Decommissioning Authority (NDA, 2006, Appendix 4) presents estimates of the cost of decommissioning the graphite-cooled reactors (and one HWR and one FBR) that have been permanently shut down in the UK. Using data on the powerplant net capacities from the *Power Reactor Information System* (PRIS) database, the cost per kW has been computed here and plotted against powerplant size in Figure 8.17. The estimated cost of decommissioning decreases with increasing reactor size, ranging from over $10,000/kW for a 200MW unit to about $1600/kW for a 1000MW unit. In the same way that the costs of nuclear powerplants have invariably been greater than initial estimates, the actual costs of decommissioning are likely to be greater than the estimated costs.

There are a few examples of nuclear reactors that have been partly decommissioned, summarized in Wilson and Burgh (2008). These are a 72MW PWR in Shippingport, Pennsylvania, which was partly dismantled at a cost of $1267/kW; a 100MW LWR in Niederaichbach, Germany, which was partly decommissioned after only 18 days operation at a cost of $1910/kW; and a 45MW boiling-water reactor BWR) in Japan, partly decommissioned at a cost of $3180/kW.

The proposed, and now shelved, Yucca Mountain isolation site in Nevada would have stored 70,000 tonnes of spent fuel waste, at an estimated cost of $40 billion by the time it would have been sealed (Beck, 1999), but any such estimates are little more than educated guesses. At the current worldwide rate of production of spent fuel waste of 7000 tonnes per year, a new Yucca Mountain site would be required every ten years. Given that 2768TWh of electricity were generated in 2005 by the world nuclear fleet (see Volume 1, Table 2.4), a $40 billion repository every ten years corresponds to a cost of 0.15 cents/kWh.

8.7.5 Financing costs

Powerplants can be financed either through debt (borrowing money and paying interest) or through equity (selling shares in the powerplant). Equity holders

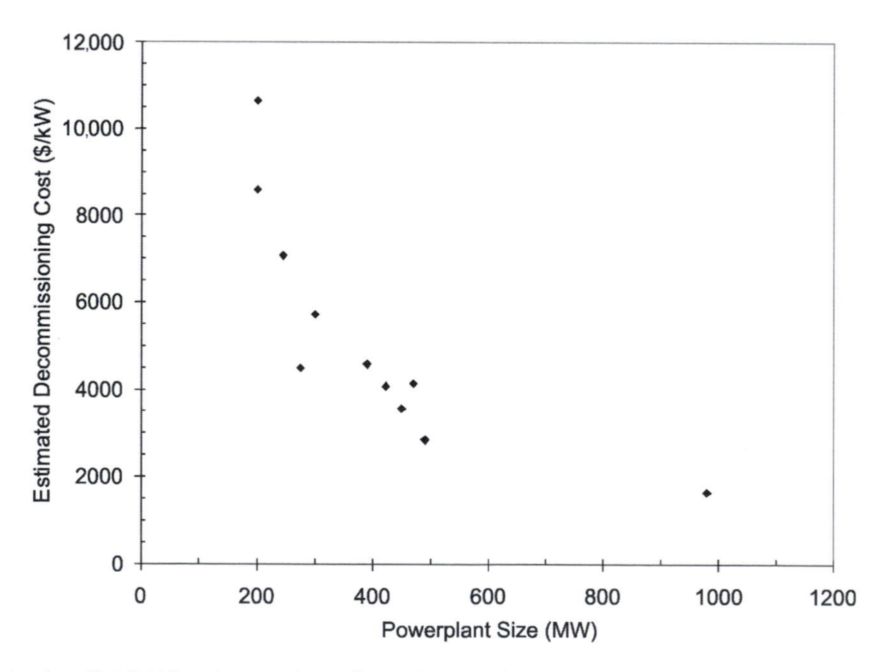

Source: Total cost data from NDA (2006) and converted to US$ assuming an exchange rate of £1 = $1.6, powerplant capacity data from the PRIS database (www.iaea.org/programmes/a2/)

Figure 8.17 *Estimated cost of decommissioning graphite-moderated nuclear powerplants that have been permanently shut down in the UK*

own part of the powerplant and expect a higher rate of return than the interest rate on debt to credit-worthy borrowers. The real interest rate on debt (i.e. after subtracting the general inflation rate) might be about 4 per cent for publicly owned utilities, whereas the expected rate of return for equity investments would typically be about 16 per cent/year (Cooper, 2009). Because nuclear powerplants are subject to higher risks than some other kinds of powerplants (including financial, regulatory and performance risks), private lenders or equity holders will demand a higher rate of return than for other investments. This has a disproportionately large impact on the final cost of nuclear electricity because capital costs are large and fuel (and other operating) costs are lower for nuclear power than for other sources of power. As well, the very long construction periods mean that accumulated interest on expenditures during the construction period can be large, particularly since interest rates are large.

8.7.6 Cost of nuclear electricity

To illustrate the impact of the various factors discussed above on the cost of nuclear electricity, representative results are shown in Table 8.14, as computed using Equations (8.13) and (8.15), subject to the assumptions given in Table 8.15. The plant is assumed to be constructed over a five-year period (something achieved for recent plants in India, China, Japan and South Korea (Schneider et al, 2009)). Results are shown with and without decommissioning costs, where decommissioning the powerplant is assumed to cost as much as building it but these costs are discounted at 3 per cent per year (from the end of the powerplant lifespan at year 40 to the start of electricity production) and added to the construction costs. Fuel contributes very little to the cost of electricity (only 0.6 cents/kWh for fuel bundles at $2000/kg). Depending on the capital cost and rate of return on the investment, nuclear electricity costs anywhere from 5 cents/kWh to 34 cents/kWh.

Figure 8.18 shows the calculated cost of nuclear electricity from the same set of studies that provided the capital cost estimates shown in Figure 8.16. The estimated costs range from 4 cents/kWh to 24 cents/kWh, with the more recent estimates tending to be higher.

These costs are merely the estimated internalized cost of nuclear energy. Not included are the external costs, including environmental impacts throughout the entire nuclear fuel chain (as discussed in section 8.4) and monetization of risks – all privately owned nuclear

Table 8.14 *Illustrative costs of electricity from nuclear powerplants for various capital costs and costs of financing, subject to the additional assumptions given in Table 8.15*

Overnight EPC cost ($/kW)	EPC + financing + owner's costs ($/kW)	Discounted decommissioning cost ($/kW)	Cost of electricity (cents/kWh)	
			Without decommissioning	With decommissioning
		5%/yr cost of capital		
2000	2696	591	4.0	4.5
4000	5392	1183	6.1	7.0
6000	8088	1774	8.2	9.6
8000	10,785	2366	10.3	12.2
		10%/yr cost of capital		
2000	3022	591	6.1	6.9
4000	6044	1183	10.2	11.8
6000	9066	1774	14.4	16.8
8000	12,088	2366	18.5	21.8
		15%/yr cost of capital		
2000	3380	591	8.7	9.9
4000	6760	1183	15.6	18.0
6000	10,140	1774	22.4	26.0
8000	13,520	2366	29.2	34.0

Table 8.15 *Assumptions adopted here in computing the cost of electricity from future nuclear powerplants*

Parameter	Assumed value
Construction period	5 years, distributed as 10%, 20%, 30%, 25% and 15% of the total cost over years 1 to 5, respectively[a]
Owner's cost	20% of EPC cost
Decommissioning cost	Equal to construction cost, discounted at 3%/yr
Plant lifespan	40 years
Capacity factor	0.85
Fixed O&M	$60/kW/year
Variable O&M	0.05 cents/kWh
Insurance & liability	Zero
Fuel cost	$2000/kg
Burn-up	40MWd/kg
Thermal efficiency	0.328

Note: [a]Taken from Table 3.7 of Volume 1.

facilities in the world have very limited liability in the event of a major accident, so that government rather than the owner of the nuclear powerplant bears the overwhelming majority of the financial risk.

8.8 Lifecycle energy balance and GHG emissions from nuclear powerplants

The lifecycle energy use of nuclear powerplants and related GHG emissions depend on the energy used throughout the entire nuclear fuel chain, from mining through to disposal of spent fuel, and including construction and decommissioning of the powerplant. In order to determine the overall net gain or loss in primary energy, the energy inputs should all be converted to primary energy and compared with the savings in fossil fuel primary energy if nuclear displaces fossil fuel powerplants.

8.8.1 Terms in the lifecycle energy budget of nuclear powerplants

The full nuclear lifecycle energy use and associated GHG emissions involve:

- uranium exploration, mining and land reclamation after mining;
- uranium milling to produce yellowcake (U_3O_8);

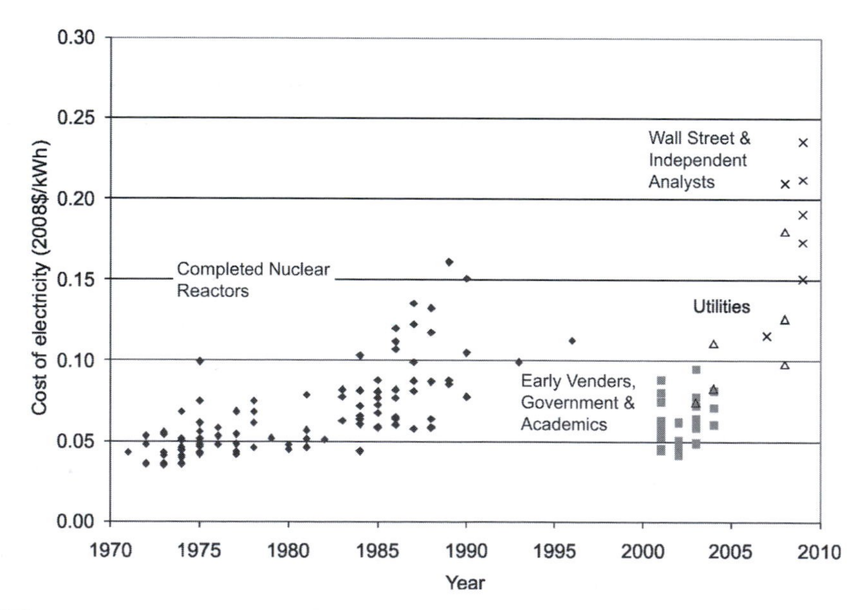

Source: Cooper (2009)

Figure 8.18 *Costs of nuclear electricity from powerplants completed between 1971 and 1997, and projections of the cost of future nuclear electricity from future powerplants made from 2001 to 2009*

- conversion of (U_3O_8) to UF_6 and enrichment (as required for LWRs), or production of heavy water (as required for HWRs);
- reconversion of UF_6 back to U_3O_8 prior to final isolation;
- fabrication of fuel;
- transportation at various stages of fuel production;
- construction of the powerplant and production of the materials used in the plant;
- operation, maintenance and refurbishment of the powerplant;
- processing and long-term isolation of spent fuel and other nuclear wastes;
- decommissioning of the powerplant.

Sovacool (2008) and Lenzen (2008) reviewed recent studies of the energy use and GHG emissions associated with the entire nuclear energy lifecycle. The estimated CO_2-equivalent emissions range from <1gC/kWh to about 80gC/kWh (that using natural gas at 60 per cent efficiency is 54gC/kWh). However, the lowest values do not include all elements of the lifecycle, while the highest values appear to have greatly overestimated some of the energy inputs. The average GHG emission is about 18gC/kWh (one third that of state-of-the-art natural gas power generation), of which about 7gC/kWh is from uranium mining and milling.

However, the energy use associated with mining and milling (and land reclamation) increases strongly with decreasing grade. As we move from reasonably assured resources to inferred resources and speculative resources, the grade of the ore in question tends to decrease. Higher-cost uranium tends to be associated with lower ore grades. The current average ore grade is about 0.2 per cent, but by the time the majority of the uranium resources available at $130/t U are exploited, the average grade will be closer to 0.01 per cent (see subsection 8.11.4).

To explore this issue further, we make use (with modifications) of the detailed, thorough and well-documented analysis of van Leeuwen (2007, henceforth vL2007) to work out the variation in lifecycle energy use and the net energy balance as a function of the ore grade. In the following analysis, all energy inputs are converted to primary energy by assuming electricity to be generated and transmitted to the point of use with an overall efficiency of 40 per cent and assuming the primary energy equivalent of fuels to be 1.2 times the onsite fuel use. The parameters of the powerplant under consideration, the required fuel inputs and the unit and lifecycle energy use are given in Table 8.16. Some brief comments on the terms in Table 8.16 are given below.

Mining and milling

The energy requirement per unit of fuel fed to the reactor depends on the grade of uranium ore, the hardness of the rock containing uranium minerals and the type of mining operation. The ratio of mass of ore that must be mined and processed to the mass of uranium that is extracted from the ore is given by:

$$R = \frac{1}{0.848GY} \qquad (8.16)$$

where 0.848 is the ratio of the molecular weight of U_3 to U_3O_8, G is the ore grade (the fraction of the ore as U_3O_8) and Y is the extraction yield – the fraction of the uranium in the ore that is extracted (the balance ends up in the mine tailings). The extraction yield decreases with decreasing ore grade, as shown in Figure 8.19, falling from about 0.95 at an ore grade of 1 per cent to about 0.6 (optimistically) at an ore grade of 0.01 per cent. The amount of ore that must be mined thus increases by at least factor of 167 as the ore grade falls from 1 per cent to 0.01 per cent. The mining energy used per tonne of natural uranium produced is given by the energy use per tonne of ore times R.

Additional material (the overburden) would need to be removed and transported some distance in an open-pit mining operation. Figure 8.20 is a cross-section of a proposed conversion of the Olympic Dam uranium mine in Australia from an underground mine to an open-pit mine. Once the starter pit is completed, the ratio of mass of overburden removed to mass of ore extracted would be about 8.0, falling to 3.3 by the end of the operation. About 450m of overburden would be removed before reaching the top of the ore body, which extends to a total depth of 1200m. For underground mines, very little non-ore material would be removed, and no non-ore material is removed with in situ leaching.

Mining is followed by milling, which produces a natural uranium material called yellowcake (see Figure 8.10). The energy used for milling involves the energy required for crushing and grinding down to a final size

Table 8.16 *Characteristics of the LWR used for the lifecycle energy analysis that is presented here, and unit energy requirements for various processes as assumed by van Leeuwen (2007) and as adopted here (where different)*

	Units	Van Leeuwen	Alternative value
	Reactor characteristics		
Capacity	GW	1	
Capacity factor		0.82	
Lifespan	years	40	
Natural uranium used	tonnes	6804	
Depleted uranium produced	tonnes	5918	
Enriched uranium used	tonnes	873	
Spent fuel produced	tonnes	873	
Million SWUs used		6.06[a]	4.19
	Energy requirements related to mining, milling and mine reclamation		
Mining	GJ/t ore	1.06	0.96
Milling (soft ores)	GJ/t ore	1.27	1.01
Tailings reclamation	GJ/t ore	4.2	0.60
	Energy requirements not related to ore grade		
Conversion to yellowcake	TJ/t U	1.48	1.32
Enrichment	TJ/MSWU	3100	1348
Fuel element fabrication	TJ/t U	3.792	2.97
Reactor OMR	PJ/GW-yr	3.44	0.55
Reconversion of DU	TJ/t DU	1.48	1.32
	Packaging of waste other than spent fuel or from decommissioning		
DU waste	TJ/t DU	0.102	0.04
Enrichment waste	TJ/MSWU	129	48.4
Operational waste	TJ/yr	223.9	84.0
	Sequestration of waste other than spent fuel or from decommissioning		
DU waste	TJ/t DU	0.103	0.06
Enrichment waste	TJ/MSWU	0.13	81.0
Operational waste	TJ/yr	225.5	129.6
	Energy requirements related to spent fuel		
Interim storage	TJ/t SF	3.3	3.30
Packaging	TJ/t SF	2.0	0.75
Sequestration	TJ/t SF	26.2	16.3
	Construction and decommissioning of reactor		
Construction of reactor	PJ	80	15
Decommissioning of reactor	PJ	100	11
Packaging decommissioning waste	PJ	10.8	4.0
Sequestration of decommissioning waste	PJ	9.5	8.3

Note: SWU = separative work unit, OMR = operation, maintenance and refurbishment, DU = depleted uranium. [a] This implies 6.94SWU/kg fuel, whereas the value computed from Equations (8.9–8.11) (and given by WNA, 2006), is 4.81SWU/kg if the enrichment and tails concentrations are 3.5 per cent and 0.25 per cent, respectively. I have therefore adjusted the lifecycle SWUs accordingly.

of 30–300μm, the energy used for separation of uranium and non-uranium grains, the energy used to produce the chemicals used and the energy used for roasting if needed.

The energy used per tonne of ore for mining plus milling is one of the most important terms in the overall energy balance of nuclear energy once the ore grade drops substantially below 0.1 per cent, but is subject to substantial disagreement. vL2007 uses a secondary energy intensity of 1.06 GJ/t of ore for mining based on Rotty et al (1975) (henceforth RPR75) and a milling energy intensity similar to that

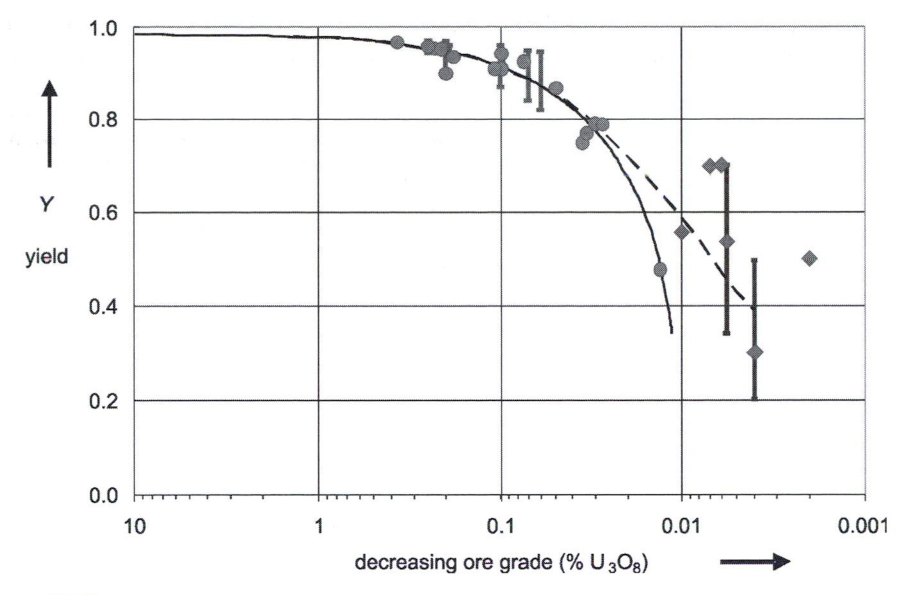

Source: van Leeuwen (2007)

Figure 8.19 *Extraction yield (the fraction of uranium in an ore that is extracted) as a function of the ore grade*

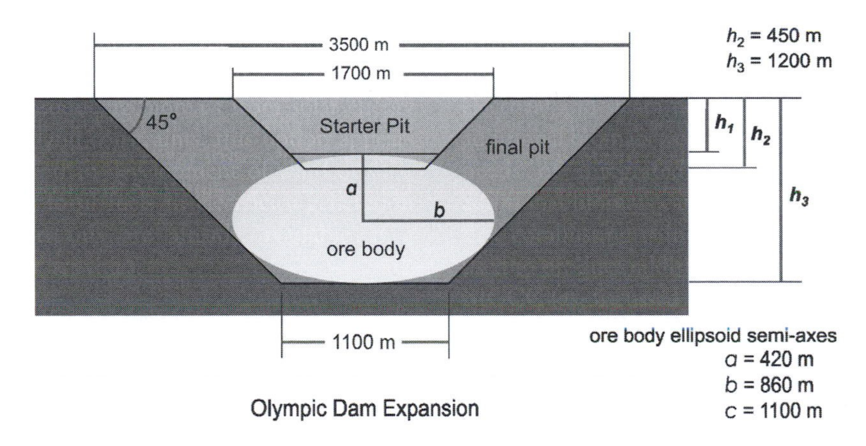

Source: van Leeuwen (2007)

Figure 8.20 *Cross-section of the Olympic Dam uranium deposit in Australia, showing the starter pit and final pit that would be excavated if the mine (currently an underground mine) is converted to an open-pit mine*

given by RPR75. The energy requirements given in RPR75 for both mining and milling are among the highest in the literature, but their report is also the most detailed among those published. For this reason, a detailed breakdown of the energy use for mining and milling, as given by RPR75, is presented in Table 8.17. Shown is direct energy use at the mine site per tonne of ore mined, as well as the embodied energy in materials (most of which is the embodied energy of explosives) and in mining machinery. For milling, most of the material embodied energy is the energy required to make sulphuric acid. Direct energy use was calculated by RPR75 based on surveys of uranium production and energy use by mines in the US, while the embodied energy in materials and machinery was estimated from the dollar value of these inputs combined with an

Table 8.17 *Energy use in uranium mining and milling in the US, per tonne of ore mined*

Item	Electricity (kWh/t)	Fuel energy use (GJ/t)				Total secondary (GJ/t)
		Petroleum products	Natural gas	Coal	Total	
			Mining			
Direct, sandstone	20.4	0.45	0.01	0.00	0.46	0.53
Direct, shale	7.8	0.02	0.00	0.00	0.02	0.04
Process materials						
- Surface sandstone	10.9	0.32	0.00	0.00	0.32	0.36
- Subsurface sandstone	7.5	0.23	0.00	0.00	0.23	0.25
- Subsurface shale	1.0	0.04	0.00	0.00	0.04	0.04
Machinery						
- Surface sandstone	1.7	0.13	0.00	0.00	0.13	0.13
- Subsurface sandstone	4.1	0.31	0.00	0.00	0.31	0.32
- Subsurface shale	2.7	0.10	0.00	0.00	0.10	0.11
Total mining						
- Surface sandstone	33.0	0.89	0.01	0.00	0.90	1.02
- Subsurface sandstone	32.0	0.98	0.01	0.00	0.99	1.10
- Subsurface shale	11.4	0.16	0.00	0.00	0.16	0.20
			Milling			
Direct, sandstone	29.0	0.15	0.44	0.00	0.59	0.70
Direct, shale	18.3	0.00	0.00	0.56	0.56	0.63
Process, sandstone	6.9	0.09	0.19	0.00	0.28	0.31
Process, shale	1.0	0.19	0.00	0.00	0.19	0.19
Machinery, sandstone	1.2	0.06	0.00	0.00	0.06	0.06
Machinery, shale	0.8	0.02	0.00	0.00	0.02	0.03
Total milling, sandstone	37.1	0.30	0.63	0.00	0.93	1.06
Total milling, shale	20.1	0.21	0.00	0.56	0.77	0.85

Source: Rotty et al (1975)

economic input–output table for the US economy. There is little difference in the estimated overall electricity or fuels requirements for surface or subsurface mining of sandstone ores, both requiring about 1.0GJ/t in total according to Table 8.17, while the energy requirement for mining shale ores is about five times smaller. However, uranium-containing shales typically have a grade of only 0.002–0.006 per cent, whereas uranium-containing sandstones typically have a grade of 0.03–0.3 per cent (according to sources cited in vL2005), so the shale case is not of interest. The energy requirements for mining are largely as petroleum products (to operate transportation and excavating machinery), while about two thirds of the total energy input for milling is in the form of heat, which is supplied either from natural gas or from coal in the US.

The estimated energy intensities for mining and milling of uranium, as given by RPR75, are compared with estimates of the energy requirement for mining and processing of other materials in Table 8.18. Some of the estimates of mining energy use, or of mining plus milling energy use, are much smaller than given by RPR75 for uranium. However, all of these exclude the embodied energy of chemical inputs and of machinery (which were estimated by RPR75 using input–output tables). As an indication of the importance of these terms, Table 8.19 presents data on chemical inputs per tonne of U_3O_8 produced at the Ranger uranium mine, Australia. These inputs are converted to inputs per tonne of ore and multiplied by the embodied energies per unit mass of input (from Volume 1, Chapter 6) to give the energy input per tonne of ore mined. This ranges from 0.2–0.3GJ/t (depending on the ore grade to which the input data pertain), but these are process-based estimates (estimates from input–output analysis would be larger, probably by a factor of two).[8] In light of this, the overall direct plus indirect milling energy uses of about 1.0GJ/t of shale ore and 2.1GJ/t of sandstone ore

Table 8.18 *Comparison of secondary energy use (fuels plus electricity) for mining of different materials or for mining plus milling of uranium*

Material	Energy use	Comment	Reference
Bauxite	0.08–0.25MJ/kg	0.7–4.4kg diesel/t, 1.4kg other oil/t, 0.4kWh/t electricity.	Green (2007)
Coal	0.23MJ/kg	Mining energy use is 0.81% the energy value of the coal	Cleveland and Ruth (1995)
Coal	0.39MJ/kg	Deduced from mining energy per GW-year of electricity, assuming an efficiency of 34%	White and Kulcinski (2000)
Copper	0.62MJ/kg-ore	Olympic Dam subsurface mine	BHP Billiton (2004)
Copper Nickel	0.6MJ/kg-ore 0.7MJ/kg-ore	Derived from variation of energy use per kg of metal with grade of ore	Norgate et al (2007), their Figure 11
Copper Nickel	0.14MJ/kg-ore	2kg diesel and 13kWh electricity per tonne of ore (surface mining)	Norgate and Rankin (2000)
Copper	0.08–0.16MJ/kg-ore	Global mean copper mining energy use of 7.8MJ/kg Cu, converted to a kg-ore basis assuming an average copper ore grade of 1–2%	Kuckshinrichs et al (2007)
Copper	0.065MJ/kg-ore	5.7MJ/kg Cu for open-pit mining in Chile and an average grade of 1.14%	Alvarado et al (2002)
Limestone	0.046MJ/kg	This is the fuel energy use for mining + grinding, assumed to be entirely for mining. Electricity use is 0.123MJ/kg but is assumed to be entirely for grinding	West and Marland (2002)
Uranium	0.11MJ/kg-ore	Subsurface mine in Namibia	Rössing (2004)
Uranium, direct	0.53MJ/kg-ore	For mining of sandstone ores in the US	Rotty et al (1975)
	0.04MJ/kg-ore	For mining of shale ores in the US	
Uranium, total	1.06MJ/kg-ore	For mining of sandstone ores in the US	Rotty et al (1975)
	0.20MJ/kg-ore	For mining of shale ores in the US	
		Uranium mining + milling energy use	
Uranium	1.0MJ/kg-ore	Energy use in the US of 5 × 10^{15} Btu per 10^3t of U_3O_8. This is about 500MJ/kg U_3O_3, or 1.0MJ/kg-ore if the average ore grade is 0.2%.	DeLucchi (1991)
Rössing Olympic Dam Ranger McClean Lake Cluff Lake	0.14MJ/kg-ore 0.26MJ/kg-ore 0.66MJ/kg-ore 3.78MJ/kg-ore 5.26MJ/kg-ore	Based on 1980s data Calculated here from midpoint of energy use data (GJ/t U_3O_8) times the midpoint of the ore grade (as a U_3O_8 fraction) given for various mines	Various sources in Mudd and Diesendorf (2008)

Table 8.18 *Comparison of secondary energy use (fuels plus electricity) for mining of different materials or for mining plus milling of uranium* (Cont'd)

Material	Energy use	Comment	Reference
Soft ore	2.33MJ/kg-ore	From separate mining + milling terms	vL2007
Hard ore	5.55MJ/kg-ore	From separate mining + milling terms	vL2007
Direct, sandstone ore	1.23MJ/kg-ore	From separate mining + milling terms	Rotty et al (1975)
Direct, shale ore	0.67MJ/kg-ore	From separate mining + milling terms	Rotty et al (1975)
Total, sandstone ore	2.12MJ/kg-ore	From separate mining + milling terms	Rotty et al (1975)
Total, shale ore	1.05MJ/kg-ore	From separate mining + milling terms	Rotty et al (1975)

Note: Energy use given by Rotty et al (1975) includes direct energy use and the energy embodied in explosives or chemicals and machinery used during mining and/or milling operations. The energy use in other cases is or is probably the energy use at the mine site only.

Table 8.19 *Chemicals used in the production uranium from uranium ores at the Ranger mine*

Material	t/t U_3O_8	GJ/t input or GJ/l input	GJ/t ore	
			0.28% grade	0.42% grade
Sulphuric acid	15	2.29	0.10	0.14
Ammonia	0.46	42	0.05	0.08
MnO_2	1.75	?		
Lime	5.9	1.20	0.02	0.03
Kerosene (L)	320	36.6	0.03	0.05
Amine (L)	12.7	?		
Total			0.21	0.32

Note: Original data are input per tonne of U_3O_8 produced, but have been converted here to input per tonne of ore mined assuming that the ore grade is either 0.28 per cent or 0.42 per cent (this being the given range) and then multiplied by the embodied energies per tonne of input to give the embodied energy per tonne of ore.

Source: Inputs from Mudd and Diesendorf (2008); embodied energies from Volume 1, Chapter 6

given by RPR75 seem to be quite reasonable. However, because these estimates depend in part on the overall energy intensity of the US economy (through the input–output tables), I shall assume a reduction by a third compared to RPR75 to account for energy efficiency improvements since the early 1970s. This results in a primary energy intensity of 0.96GJ/t-ore for mining and 1.01GJ/t-ore for milling.

Land reclamation

The practice at uranium mines today is to retain the tailings in specially designed retaining ponds. As the tailings will remain radioactive for many thousands of years, a more responsible approach would be to return the tailings to the ground from which they were taken but with measures to neutralize the acidity and isolate the radioactive materials from the surrounding environment. This option has been considered by vL2007. The mass of tailings is equal to the mass of ore mined plus chemicals added during milling minus the mass of uranium extracted, and so will be about 8 per cent greater than the original ore mass. During reclamation, the tailings would first be mixed with powdered limestone to neutralize the acids and with a phosphate (such as sodium phosphate) to render the radioactive nuclides and other heavy metals insoluble in water. The resulting material would then be poured between layers of bentonite (a clay) in order to isolate the tailings from groundwater. The limestone, bentonite and phosphate (whose total mass will be comparable to that of the tailings mass) must all be mined and transported to the mine site and then placed in the mine. vL2007 assumes the total energy for this process to be four times that for the original excavation of the ore. This could arise as one unit for mining the materials added to the tailings, one unit for transporting the added materials to the mine site and mixing the materials, and two units for putting twice the original mined mass back into and around the mine. However, the total energy used for mining limestone given in Table 8.18 (0.046GJ/t) is much smaller than that given by RPR75 for surface mining of uranium sandstone (1.02GJ/t-ore), and the energy required for mining bentonite is presumably also much less than that for mining sandstone.

Here, two components of land reclamation will be considered, the first involving the mine tailings and the second (for open-pit mines) involving the placement of the overburden over the neutralized tailing. The energy required for reclamation of tailings is assumed to be three times the direct plus indirect energy required (0.2GJ/t) for mining shale (a relatively soft rock), derived as follows: 0.2GJ/t for mining the materials that need to be mixing with the tailings, 0.2GJ/t for transporting these materials a distance of 200km to the mine by truck (using an energy intensity of 0.9MJ/t-km as given in Table 5.31 of Volume 1), and an energy intensity for putting material back into and around the mine of 0.2GJ/t (derived by assuming that the energy requirement for putting materials back is half what was required to mine it, as no explosives are needed to put it back, and assuming that the volume put back is about twice the volume of uranium ore that is mined). Thus, I adopt an energy requirement for tailings reclamation of 0.6GJ/t-ore. For the removal and later return of the overburden, I adopt an energy requirement equal to that for mining the ore (0.96GJ/t-overburden). These are clearly very crude estimates but are useful for illustrative purposes and hopefully are within a factor of two of the correct typical values.

Conversion, enrichment and reconversion

Energy is used to convert U_3O_8 to UF_6, which is necessary prior to enrichment. RPR75 give energy requirements of 16.1kWh of electricity and 1571GJ fuels per t U. As with their mining and milling energy, I multiple these intensities by 0.7 to crudely account for overall improvements in efficiency since then. The result is a secondary energy requirement of 1.10GJ/t U (compared to 1.48GJ/kg U used by vL2007 and 1.27GJ/kg U given by the World Nuclear Association (WNA, 2006)) and a primary energy requirement of 1.32GJ/t U.

The energy required for enrichment is given by the number of SWUs (see subsection 8.6.1) times the energy requirement per SWU. Enrichment by gaseous diffusion (which is used in the US) requires 2400–3000kWh of electricity per SWU, but enrichment by centrifuge (widely used in Europe) requires only 100kWh/SWU according to Lenzen (2008) and about 50kWh/SWU according to Bodansky (2004). Any new enrichment facilities will use centrifuges, and to reflect technical progress, an electricity requirement of 50kWh/SWU will be adopted here. However, one must also take into account thermal energy used during operation of the centrifuges, as well as the energy used to build the centrifuges. According to a 1984 source cited by Lenzen (2008), this amounts to a total of $0.936GJ_{th}$/SWU, but again the energy requirement should have fallen over time, so I multiply this value by 0.8. The total primary energy requirement would then be 1348TJ/million SWU.

The depleted fuel stream is in the form of UF_6, which is toxic and flammable, so sound environmental principles require that it be converted back to U_3O_8 and stored in perpetuity (except for some small portion that could be used in a nuclear reactor by blending it with highly enriched uranium from dismantled nuclear weapons). vL2007 indicate the same energy requirement per unit mass for reconversion as far the initial conversion to U_3O_8, whereas WNA (2006) neglects this step altogether.

Fuel fabrication

Nuclear fuel is shaped into pellets about 0.8cm in diameter and 1.35cm long, which are then clad in a 0.06cm thick layer of zircaloy, a zirconium alloy (98 per cent Zr, 1.5 per cent Sn) (Bodansky, 2004). vL2007 cites a source giving an overall energy requirement for fuel fabrication of 3.79GJ/kg U, whereas WNA (2006) gives a value of 0.79GJ/kg U. Lenzen (2008) indicates an energy requirement for fuel fabrication as reported in various studies ranging from 0.64–8.0GJ/kg U, with an average of 2.97GJ/kg U. The average value is adopted here.

Construction of the nuclear powerplant

The two approaches used for estimating the amount of energy required to construct nuclear powerplants, process chain analysis (PCA) and an input–output analysis (IOA), are described in Appendix B. The PCA approach accounts for the energy used in the first two or three steps in the process chain (onsite energy use, energy directly used to manufacture the materials that become part of the powerplant, and possibly the energy used to used to make machinery and other inputs used in making the materials), but neglects higher-order energy inputs. A large amount of 'soft' services (such as engineering design) goes into the construction of a nuclear powerplant, entailing energy use in the buildings where the design teams work. This will not be captured in PCA, thereby probably causing a particularly large underestimation of the true energy input. IOA implicitly sums over all orders of inputs and includes tangible as well as intangible inputs, but relies on economic energy intensities (MJ/\$) combined with monetary inputs (such as \$) rather than physical energy intensities (MJ/kg) combined with physical inputs (such as kg of steel). The economic energy intensities used pertain to broad sectors (such as steel making) rather than to the specific products used in nuclear powerplants. A cruder approach is to multiply the total monetary cost of a powerplant times the average energy intensity (AEI) of the national economy in the country where the plant is built.

Lenzen (2006) presents a detailed listing of published estimates of the embodied energy of various kinds of 1GW nuclear powerplant, as estimated by each of these three methods. Results are plotted in Figure 8.21. For pressurized-water reactors (which make up the bulk of the world's nuclear reactor fleet), IOA-based estimates of the embodied energy of a 1GW powerplant range from 5 to 15PJ. Average estimated energy requirements are 4.2PJ, 12.6PJ and 61.9PJ using the PCA, IOA and AEI approaches, respectively. The IOA approach is probably the most reliable, but it ultimately depends on the estimated monetary cost of the powerplant. As estimates of the cost have doubled to quadrupled during the past few years, the corresponding effort of all kinds and associated energy inputs would have increased, although not necessarily

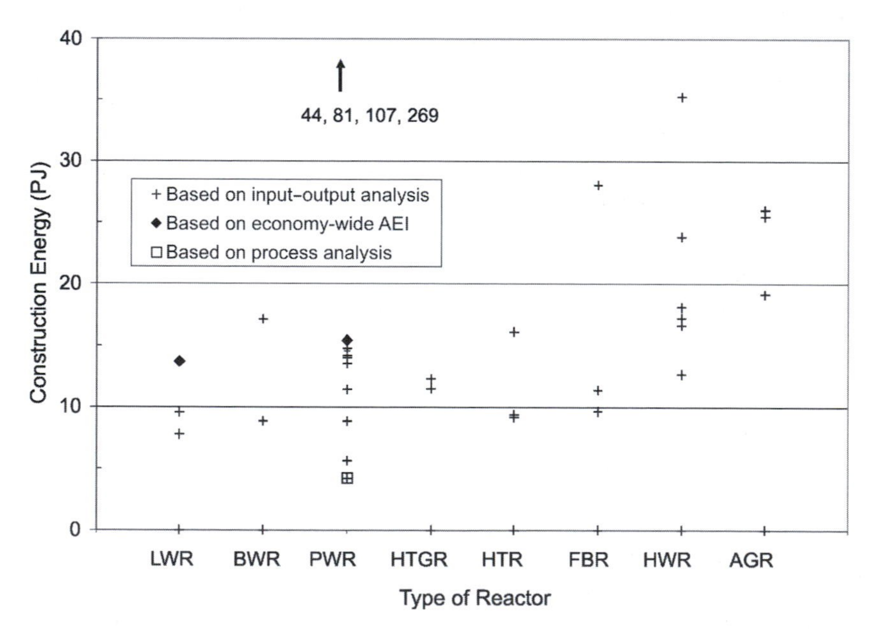

Note: LWR = light-water reactor, BWR = boiling-water reactor, PWR = pressurized-water reactor, HTGR = high-temperature gas-cooled reactor, HTR = high-temperature reactor, FBR = fast breeder reactor, HWR = heavy-water reactor, AGR = advanced gas-cooled graphite reactor.
Source: Lenzen (2006)

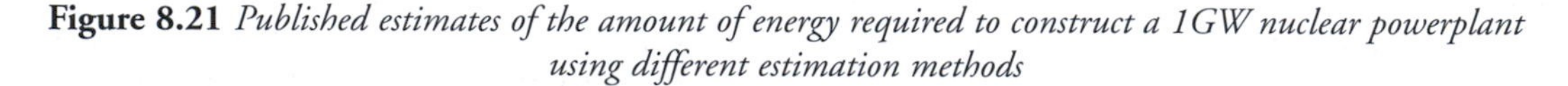

Figure 8.21 *Published estimates of the amount of energy required to construct a 1GW nuclear powerplant using different estimation methods*

Table 8.20 *Comparison of material intensities (tonnes/MW) for nuclear powerplants in and around Belgium, along with the plant embodied energy per kWh of electricity generated*

PCA		IOA	
Material (t/MW)		*Economic input (millions 1996 Belgium francs/MW)*	
Concrete	560	Machinery	23
Steel	60	Services	23
Plastic	1.27	Building	14.2
Copper	0.75	Other	8.8
Glass	0.69		
Oil	0.69		
Aluminium	0.20		
Primary embodied energy in a 1GW plant			
6.9PJ			19.5PJ

Note: PCA = process chain analysis, IOA = input–output analysis.
Source: Voorspools et al (2000)

by the same factor.[9] I shall adopt a value of 15PJ here, although other values could be easily justified.

Table 8.20 compares the PCA and IOA estimates of the energy required to build nuclear powerplants in Belgium and the surrounding region, as computed by Voorspools et al (2000). The primary energy required to build a 1GW nuclear powerplant is 6.9PJ using the PCA method (but only including the materials listed in Table 8.20 and not accounting for their fabrication into final products) and 19.5PJ based on IOA (about one third of which is associated with the 'services' input).

Operation, maintenance and refurbishment of the nuclear powerplant

Energy inputs during the operation of a nuclear powerplant include various chemicals, small amounts of diesel fuel and maintenance. RPR75 indicate energy requirements of about 8.55GWh/yr of electricity and 287TJ/yr of fuels for a 1GW powerplant operating at 75 per cent capacity factor. This translates into a primary energy requirement of 0.563PJ/GW-yr. Adjusting to an 82 per cent capacity factor (as in the calculations here) and multiplication by 0.7 to allow for efficiency improvements reduces this to 0.323PJ/yr. vL2007 cites recent sources indicating an energy use for routine O&M of about 0.7–1.2PJ/GW-yr (or 0.6–1.0PJ/yr for a plant with an 82 per cent capacity factor), but one or more major refurbishments (costing 20–80 per cent of the original powerplant cost) are also required during the plant lifespan. Assuming the refurbishment energy input to be proportional to cost, and assuming a 40-year powerplant lifespan, the annual average refurbishment energy input is 0.5–2.0 per cent of the construction energy input, or 0.075–0.3PJ/yr (assuming a construction energy requirement of 15PJ). Adding the midpoint of this range to the energy use for routine O&M gives a total operation, maintenance and refurbishment (OMR) energy requirement of 0.55PJ/yr. Over 40 years, the OMR energy requirement here is 22PJ. WNA (2006) gives a total construction plus OMR energy use of 24.7PJ over a 40-year lifetime, which is comparable to the IOA-based estimate of construction energy use alone in Voorspools et al (2000), and so is likely to be too small.

Decommissioning

vL2007 gives a decommissioning energy use (excluding packaging and isolation of wastes) of 100PJ for a 1GW powerplant, compared to 80PJ for the construction of the powerplant. Although the cost of decommissioning may approach or even greatly exceed the cost of constructing the powerplant, it seems unreasonable that the energy requirements would be greater (as new, energy-intensive materials do not need to be manufactured). WNA (2006) gives a decommissioning energy use of 6PJ/GW, based on a 1970s estimate for the CANDU nuclear reactor. I adopt a value of 11PJ (about 75 per cent of my adopted construction energy input).

Packaging and isolation of nuclear wastes

Except for mill tailings, all radioactive wastes need to be placed in containers. vL2007 considered five different types of containers, depending on the type of waste. The different wastes considered are depleted uranium, spent fuel and wastes from uranium conversion, uranium enrichment, fuel fabrication, reactor operation and reactor decommissioning and dismantling. The container assumed for spent fuel is the 25t steel and copper canister (each holding 2t of spent fuel) that is intended for the Finnish and Swedish geological repositories. The calculated embodied energy in waste containers is computed as the mass of each type of container times a uniform material energy intensity of 80GJ/t, which is meant to represent not only the energy required to make the materials, but also to process them into containers and to transport and handle the containers. Based on data in Table F.23 of vL2005 on the total number of containers of different types, the mass-weighted average composition of the containers would be about 60 per cent fibre-reinforced concrete and 40 per cent steel plus copper plus lead (mostly steel). Given current energy intensities for cement, steel and copper of 4.8GJ/t, 26.3GJ/t and 88GJ/t, respectively (see Volume 1, Table 6.1), and given transportation energy requirements of about 0.6–1.0GJ/t/1000km using heavy trucks (see Volume 1, Table 5.31), a total embodied energy of 80GJ/t seems to be excessive. Keeping in mind that concrete is only about 10 per cent cement (the rest being aggregate with negligible energy input requirement), a more appropriate container energy intensity is likely to be 10–15GJ/t at present and less in the future. A value of 10GJ/t will be adopted here.

With regard to the construction of the repository for spent fuel and other wastes, vL2007 indicates that 7.2m^3 of rock would need to be excavated for every m^3 of waste, based on the Swedish repository concept. Later, the repository would be backfilled with bentonite, which has to be mined, prepared and transported. He derives an energy requirement per m^3 of excavated repository of 20.2GJ, derived as an energy for excavating granite of 5.2GJ/t plus twice the energy used in mining uranium sandstone (1.06GJ/t), times the density of granite of 2.76t/m^3. Here, I use 5.2GJ/t × 0.8 (to reflect efficiency improvements) plus twice the shale mining energy use (0.2GJ/t) to represent mining, transporting and backfilling of bentonite. vL2007 gives lifecycle energy requirements for packaging and burial of 27.5PJ for spent fuel, 20.4PJ for decommissioning wastes and 20.6PJ for other wastes, or a total of 68PJ. WNA (2006) gives only one term with no justification, 1.5PJ, for storage of wastes. My assumptions result in a requirement of 39.9PJ for all forms of waste.

8.8.2 Net energy gain and EROEI of nuclear powerplants

Table 8.21 compares the lifecycle energy requirements for each of the terms discussed above, as given by vL2007 for an ore grade of 0.15 per cent, as assumed here for the same grade and as given by WNA (2006) for an unspecified ore grade. The lifecycle energy input given by WNA (2006) is about one ninth that based on vL2007, whereas my estimate is three times that of WNA (2006) and one third that of vL2007. Figure 8.22 shows the lifecycle energy use calculated for a 1GW nuclear powerplant as a function of the ore grade, using the alternative parameters given in Table 8.16. Results are shown for an underground mine with zero non-ore material excavated, and for an open-pit mine with the final overburden mass removed equal to 2.9 times the mass of ore excavated (the 2.9:1 ratio corresponds to that of the proposed Olympic Dam mine, illustrated in Figure 8.20).[10] The lifecycle energy input begins to rise rapidly as the ore grade drops below 0.03 per cent.

Inasmuch as the nuclear powerplant is generating electricity that would otherwise be produced using fossil fuels, the appropriate EROEI is the ratio of the total primary energy input divided by the primary energy that would otherwise be required to produce the electricity generated by the nuclear powerplant. For consistency with the calculation of the primary energy input, it is assumed that 1 unit electricity requires 2.5 units of primary energy. Figure 8.23 shows the EROEI computed this way for subsurface mines (neglecting any non-ore materials that need to be excavated), open-pit mines with a 2.9:1 overburden to ore mass ratio,

Table 8.21 *Comparison of total lifecycle energy inputs (PJ) for a 1GW nuclear powerplant that operates for 40 years with an average capacity factor of 0.82, based on the energy intensities given in vL2007, as adopted here and as given in WNA (2006)*

	Source of input data		
	vL2007	**Here**	**WNA (2006)**
Mining and milling	13.4	11.4	1.6
Land reclamation	24.3	18.3	0.0
Conversion	10.1	9.0	9.2
Enrichment	18.8	5.7	3.1
Reconversion	8.8	7.8	0.0
Fuel fabrication	25.8	20.3	5.8
Construction	80.0	15.0	24.7
Reactor OMR	137.6	22.0	
Decommissioning	99.7	11.0	6.0
Spent fuel waste	27.5	17.8	
Decommissioning waste	20.3	12.4	1.5
Other waste	20.1	9.7	
Total	486.3	160.2	51.9

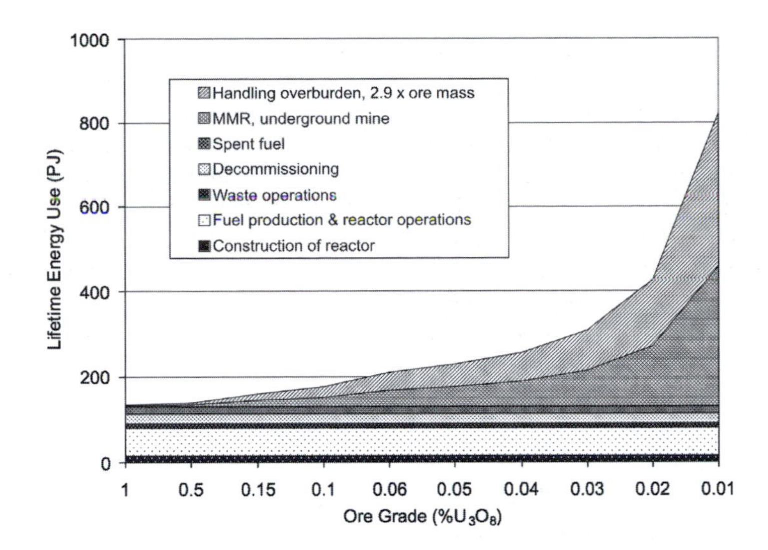

Note: MMR = mining, milling and reclamation

Figure 8.22 *Energy use over the 40-year life of a 1GW LWR nuclear powerplant with a capacity factor of 87 per cent, as computed using the alternative parameter values from Table 8.16*

and for in situ leaching (ISL). With regard to ISL, vL2007 cites a single 1977 source that gives the ISL energy requirement as 0.15–10.9GJ/t-ore. For illustrative purposes, the high end of this energy intensity range is used here but reduced by 30 per cent to reflect plausible improvements in energy efficiency during the past 30 years. However, based on the discussion in vL2007, a reasonable lower limit to the

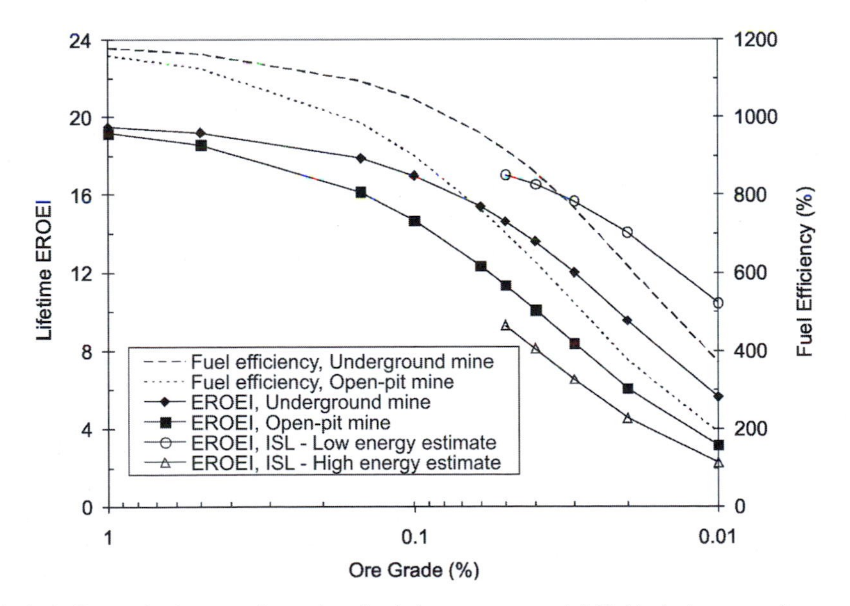

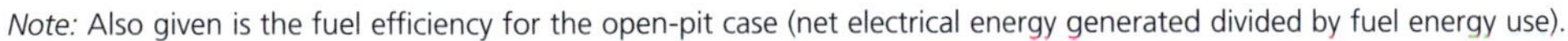

Note: Also given is the fuel efficiency for the open-pit case (net electrical energy generated divided by fuel energy use).

Figure 8.23 *Energy return over energy invested for nuclear powerplants with the lifecycle energy use given in Figure 8.22 (open-pit case) and when uranium is extracted through in situ leaching with an energy requirement of 0.3 or 7.0GJ/t-ore*

energy required for ISL seems to be very roughly about 0.3GJ/t-ore, so that value is use here for the case labelled 'ISL – Low energy estimate'. Results for ISL are shown only for ore grades of 0.05 per cent or less, as ISL is today generally applied only to ore grades of 0.03–0.05 per cent.

The resulting EROEIs drop from about 19.5 at an ore grade of 1 per cent to 17–19 at the current world average grade of 0.2–0.3 per cent. For an ore grade of 0.01 per cent, the EROEI computed here drops to 5.6 for underground mining and to 3.2 for open-pit mining, but could be as low as two or as high as ten for ISL. However, ISL involves significant and irreversible chemical and radioactive contamination of underground aquifers. Of the 5.4Mt of identified uranium resources listed later in Table 8.25, 1.1Mt are amenable to open-pit mining, 2.0Mt are amenable to underground mining, 0.6Mt are amenable to ISL, and 1.1 would be mined as co-products with other minerals (in which case only a portion of the mining energy would be attributed to uranium).

The EROEIs based on primary energy for the major renewable energy sources are often better than the EROEIs for nuclear energy calculated here for the present world average ore grade, and will be decidedly better at lower grades (with the possible exception of some uranium from co-product mines). In particular, the EROEI for PV modules is 10–20 in good locations today and will be >20 in the near future, given payback times of <1 year and lifespans ≥20 years (see subsection 2.2.9), while current EROEIs are 10–20 for hybrid PV/thermal modules (see section 2.6), 8–13 for concentrating solar thermal power generation (see subsection 2.3.5) and up to 50 for wind turbines (see section 3.15).

8.8.3 Efficiency in using fossil fuels to generate electricity

Nuclear powerplants can be thought of as a very complicated way of using fossil fuels (primarily oil) to generate electricity, with the assistance of uranium. The efficiency in using fossil fuels to generate electricity in this way is the ratio of net electrical energy generated to fossil fuel energy inputs. This has been computed here as a function of ore grade, based on the separately computed lifecycle electricity and fuel requirements, and is shown using the right axis in Figure 8.23. The fossil fuel efficiency ranges from about 1200 per cent at a grade of 1 per cent to about 100–300 per cent at a grade of 0.01 per cent. This of course is much better than the efficiency with which oil and natural gas can be directly used to generate electricity. However, this is at the cost of enormous landscape impacts associated with mining (especially as ores with very low grade are exploited) and with a radioactive waste legacy not associated with the direct use of oil and gas. One may also object to using dwindling oil supplies to generate electricity rather than for transportation, although if the electricity from the nuclear powerplants is used to charge plug-in hybrid or all-electric vehicles (discussed in Volume 1, Chapter 5, subsection 5.4.6), then nuclear power becomes a complicated way of using oil for transportation purposes but with higher efficiency (at least initially).

8.9 Operational Constraints

Due to the thermal mass of the reactor core and restraints arising from radioactive decay chains, nuclear power reactors have a very limited ability to produce variable heat output. Thus, they normally run at full and constant electricity output. The electricity output from the steam turbines can be reduced by simply not making use of all of the steam produced by the reactor, but this increases the cost of electricity.

One of the fission products produced in large quantity (~6 per cent of all fission products) during operation of the reactor is I-135. I-135 decays to Xe-135 with a half-life of seven hours, which in turn decays with a half-life of nine hours, but while it lasts it is a powerful neutron absorber (Hesketh, 1996). During normal operation of the reactor, Xe-135 reaches a steady concentration that is governed by the balance between production from I-135 and losses from its own decay to Cs-135 or transmutation (through absorption of neutrons) to Xe-136 (neither Cs-135 nor Xe-136 absorb neutrons). However, when the reactor is shut down, loss by transmutation ceases while production from decay of I-135 initially continues, so the Xe-135 concentration builds up, peaking about ten hours after shutdown. As neutron absorption inhibits the chain reaction needed to sustain the fission process, the reactor cannot be restarted until after the Xe-135 concentration has dropped sufficiently – about 24 hours after the initial shutdown.

Because of these constraints, nuclear powerplants can be operated only as baseload powerplants (that is, with steady output that matches the minimum electricity demand encountered). Other power sources are required to meet the fluctuating component of electricity demand, unless excess nuclear thermal energy is discarded at certain times so as to permit reduced nuclear generation of electricity. In a hypothetical future hydrogen economy, excess nuclear-generated electricity could be used to produce hydrogen through electrolysis of water, thereby providing for a steady electricity demand. In this case, excess nuclear electricity-generated potential would be used (thereby improving the economics of the nuclear powerplant), but the electrolyser would not be used at its full capacity (as it would be used only when there is excess nuclear electricity), which would increase the price of the hydrogen so produced. The economics of intermittent operation of electrolysers are discussed in Chapter 10 (subsection 10.7.1).

8.10 Current nuclear capacity

As of September 2009 there were 436 nuclear reactors in the world used to generate electricity, with a total capacity of 370.3GW, distributed as shown in Table 8.22. These reactors supplied about 16 per cent of total electricity production in 2006, but contributed

Table 8.22 *Number of reactors in the world and total capacity as of September 2009*

Country	Number of reactors	Total capacity (MW_e)	Average capacity factor (%), 2003–2005[a]	Electricity production (TWh/yr)	Technologies
Argentina	2	935	84	6.9	HWR
Armenia	1	376	70	2.3	VVER
Belgium	7	5863	90	46.2	PWR
Brazil	2	1766	76	11.8	PWR
Bulgaria	2	1906	78	13.0	VVER
Canada[b]	18	12,577	82	90.3	HWR
China	11	8438	86	63.6	PWR, HWR, VVER
Czech Rep	6	3634	78	24.8	VVER
Finland	4	2696	95	22.4	VVER, BWR, PWR
France	59	63,260	83	460.0	PWR, FBR
Germany	17	20,470	88	157.8	PWR, BWR
Hungary	4	1859	76	12.4	VVER
India	17	3782	81	26.8	HWR, FBR, VVER
Japan[c]	53	45,957	65	261.7	BWR, PWR
South Korea	20	17,647	90	139.1	PWR, HWR
Lithuania	1	1185	71	7.4	RBMK
Mexico	2	1300	88	10.0	BWR
Netherlands	1	482	94	4.0	PWR
Pakistan	2	425	58	2.2	HWR, PWR
Romania	2	1300	89	10.1	HWR
Russia	31	21,743	77	146.7	VVER, RBMK, FBR
Slovak Rep	4	1711	84	12.6	VVER
Slovenia	1	666	94	5.5	PWR
South Africa	2	1800	85	13.4	PWR
Spain	8	7450	89	58.1	PWR, BWR
Sweden	10	8958	88	69.1	PWR, BWR
Switzerland	5	3238	87	24.7	PWR, BWR
Ukraine	15	13,107	82	94.2	VVER
UK	19	10,097	74	65.5	HTGR, PWR
US[d]	104	100,683	91	802.6	PWR, BWR
Total	436	370,260	83	2665.0	

Note: Electricity production given here is computed from the given capacity and average capacity factor for 2003–2005. PWR = pressurized-water reactor, BWR = boiling-water reactor, HWR = heavy-water reactor, VVER = voda-vodyanoi energetichesky reactor (a Russian PWR), RBMK = reaktor bolshoy moshchnosti kanalniy (a Russian design using graphite and water), FBR = fast breeder reactor, HTGR = high-temperature gas-cooled reactor. [a] Includes only operational reactors. [b] Includes 4 units with 2568MW_e total capacity in long-term shutdown. [c] Includes 1 unit of 246MW_e capacity in long-term shutdown. [d] Includes 1 unit of 1046MW_e capacity in long-term shutdown.

Source: www.iaea.org/programmes/a2/index.html and www.world-nuclear.org/info/reactors.htm

over 30 per cent of national electricity use in many countries, as shown in Figure 8.24. Figure 8.25 shows the growth in annual electricity generation from nuclear energy in major world regions between 1965 and 2006. The world average capacity factor (average output over peak capacity) steadily increased from 1970 until about 2000, contributing in part to the increase in annual electricity production over that time period (see Figure 8.26). In 2009, 31 nuclear reactors with a total capacity of 24.1GW were under construction worldwide (see Table 8.23). Almost half have been in progress for 18–30 years (Rosenkranz, 2006), and construction has been halted on a total of 11 nuclear powerplants with a planned total capacity of 8642MW (Thomas, 2005). A nuclear powerplant may contain many reactors. Reactor size has grown from 60MW_e in the 1950s to more than 1300MW_e today for the largest units, but there is a move back to smaller units as these permit a more gradual increase in capacity and easier integration into smaller grids. Figure 8.27 gives the age distribution of nuclear reactors in the world in mid-2009. The average age then was 25 years, but inasmuch as very few new reactors are being built, the average age of the fleet is increasing by about one year per year.

8.11 Potential contribution of nuclear energy to future world energy needs

The potential contribution of nuclear energy to the world's energy supply is severely restrained in terms of: (1) the rate at which the nuclear fleet can be expanded, (2) the rate at which depositories for long-term storage of nuclear waste can be developed, (3) the rate at which

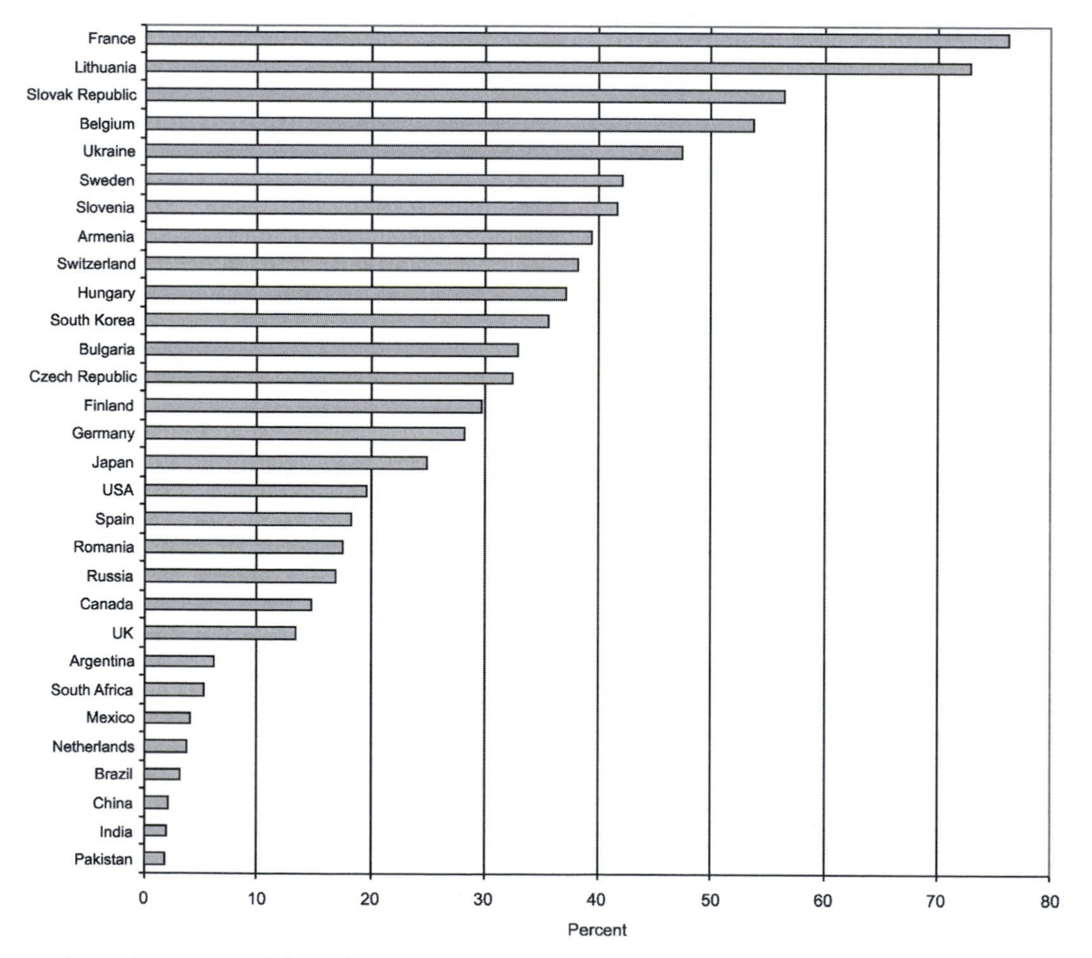

Source: www.iaea.org/programmes/a2/index.html

Figure 8.24 *Percentage of electricity generated by nuclear power in 2008*

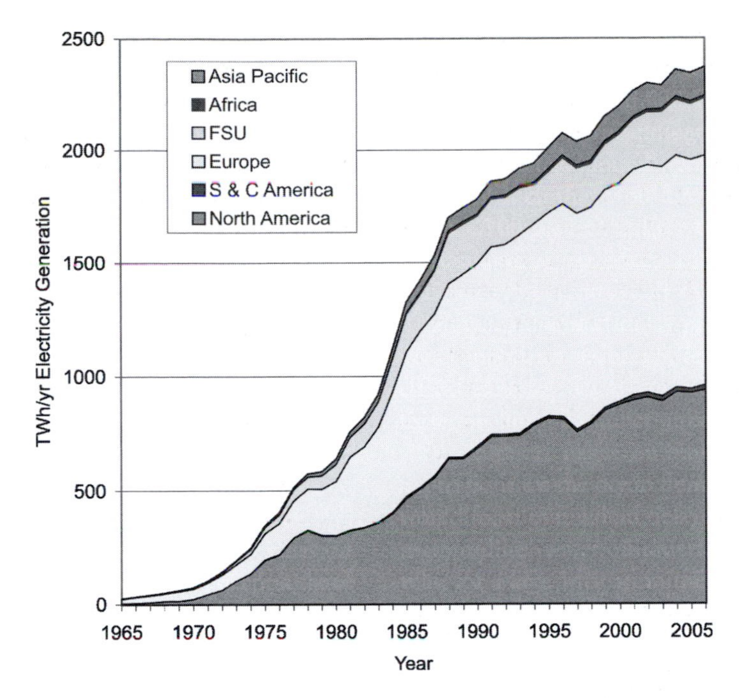

Source: Data from BP (2007)

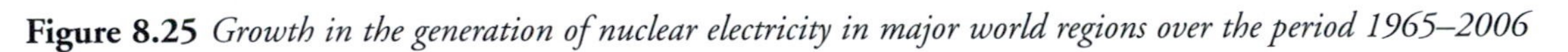
Figure 8.25 *Growth in the generation of nuclear electricity in major world regions over the period 1965–2006*

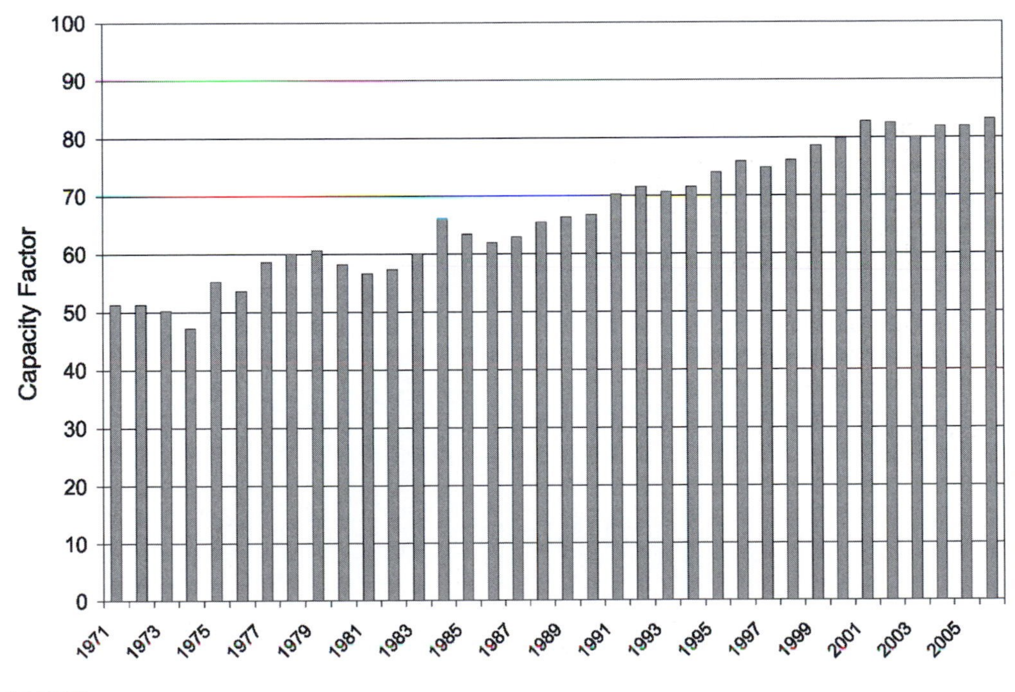

Source: WEC (2007)

Figure 8.26 *Variation in the world average capacity factor for nuclear reactors from 1971 to 2006*

Table 8.23 *Nuclear reactors under construction as of September 2009*

Country	Number of reactors	Total capacity (MW_e)
Argentina	1	692
Bulgaria	2	1906
China	16	15,220
Finland	1	1600
France	1	1600
India	6	2910
Iran	1	915
Japan	2	2191
South Korea	5	5180
Pakistan	1	300
Russian Federation	9	6894
Slovak Republic	2	810
Ukraine	2	1900
US	1	1165
Total	52	45,883

Source: www.iaea.org/programmes/a2/index.html

a closed cycle using plutonium and fast breeder reactors can be built up (if this path were to be chosen), and (4) limitations in the supply of uranium. There are also non-technical constraints, such as public opposition and the need to develop a meaningful anti-proliferation regime. Here, only the technical constraints are briefly discussed.

8.11.1 Constraints on the rate at which the nuclear fleet can be expanded

As of September 2009, 127 reactors worldwide had been on the grid for more than 30 years, but the oldest are only 40 years old (Figure 8.27). Assuming that reactors are retired after 40 years, 127 new reactors will have to be planned and built over the next ten years – an average of one every four weeks – just to maintain the current worldwide production of electricity from nuclear power.[11] During the following decade, 212 new reactors would have to be built – an average of one every 17 days – just to maintain the current nuclear capacity.

If the current world electricity nuclear capacity were to quadruple by 2050 (from 370GW to 1500GW) and electricity demand were to double, then, neglecting possible changes in the average powerplant capacity factor, the share of total electricity generated by nuclear reactors would increase from 16 per cent in 2005 to 30 per cent in 2050.[12] Assuming

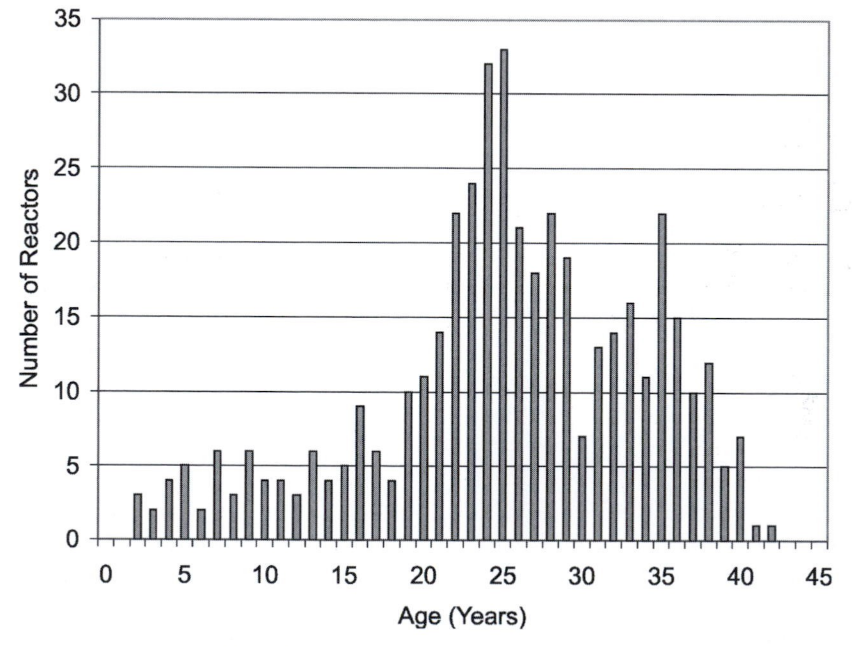

Source: Data from www.iaea.org/programmes/a2/index.html

Figure 8.27 *The age distribution of nuclear reactors as of September 2009*

that this expansion begins in 2020 (it could not start any sooner due to the ten-year permitting and construction process), 44.3 reactors would need to be built per year on average. During the decade 2020–2030, 61 new reactors would need to be built per year when replacement of existing reactors that would be retired is taken into account – an average of one every six days. If the nuclear fleet doubles rather than quadruples by 2050, a new reactor would be needed every 11.4 days during the decade 2020–2030. Such a rapid rate of expansion appears to be highly unlikely.

8.11.2 Constraints on the development of repositories for long-term storage of wastes

A fleet of 1500GW of LWRs, with an average capacity factor of 0.9 and a burn-up of 50–100GWd/t, would generate 13,000–27,000t/yr of spent fuel (based on options 1a and 1b Figure 8.14). This in turn would require a new Yucca Mountain type repository (70,000t storage, $40 billion cost) every two to four years. This is probably an impossible target, because of the difficulty of finding suitable sites. As noted earlier, plans for the Yucca Mountain site in the US were shelved in 2009 and the process of site selection started anew, and most other countries are in a state of political paralysis with regard to site selection.

8.11.3 Constraints on the rate of establishment of reprocessing facilities and FBRs

An alternative to long-term storage of spent fuel is to base expansion of the nuclear fleet on reprocessing and FBRs. Reprocessing of the 13,000–27,000t of annual worldwide production of spent fuel given above for a 1500GW fleet would require 8–16 times the current reprocessing capacity in France (1700t/yr). The reprocessing plants are complex, would have large stocks of plutonium in temporary storage and would involve thousands of shipments of plutonium between reactors, reprocessing plants and fresh-fuel production centres, with attendant risks of weapons proliferation. The present process generates 14m^3 of waste per tonne of reprocessed fuel (Beck, 1999).

8.11.4 Constraints based on the size of the recoverable uranium resource

Estimates of the amount of the size of the uranium resource in various countries and worldwide are published every two years by the Nuclear Energy Agency and International Atomic Energy Agency in a publication that is know within the industry as the 'Red Book'. Uranium resources are classified in the Red Book as:

- reasonably assured resources (RAR) – resources thought to exist with sufficient confidence that decisions to invest in mining operations can proceed;
- inferred resources – resources that require further direct measurements before making an investment decision;
- prognosticated resources – resources that can be expected to occur in known uranium provinces based on geological knowledge of previously discovered deposits and regional geological characteristics as determined by geological mapping;
- speculative resources – resources expected to occur in geological provinces that may host uranium deposits.

Tables 8.24 and 8.25 summarize the resource estimates from the 2008 edition of the Red Book (NEA/IAEA, 2008). Table 8.24 gives the breakdown of RAR and inferred resources by extraction technique and cost category. The total over all extraction techniques and cost categories is 5.36Mt, of which 1.1Mt are amenable to surface mining (with greater energy requirements for a given ore grade due to the need to remove the overburden), 2.0Mt are amenable to underground mining, and 1.1Mt will be mined as co-products of other minerals (so some of the mining energy can be assigned to the other minerals). The grand total including all resource categories is 15.9Mt. Some of the inferred and probably more of the prognosticated and speculative resources may turn out not to exist, but further exploration will probably add some uranium to all of the resource categories. It is reasonable to expect that 6–12Mt of uranium are available at costs of up to $130/kg. Schneider and Sailor (2008) review other estimates of the size of the uranium resource, which are

Table 8.24 *Estimated size of the global reasonably assured uranium resource (RAR) and inferred uranium resource (kt U) in different cost (<$40/kg, $40–80/kg, $80–130/kg) and extraction-technique categories*

Extraction technique	Reasonably assured resources			Inferred			Total (RAR + inferred)			Overall total
	<40	40–80	80–130	<40	40–80	80–130	<40	40–80	80–130	
Open pit	338	172	342	215	7	54	552	179	396	1127
Underground	541	403	281	266	427	75	807	830	356	1993
ISL	37	59	57	346	57	12	383	116	69	567
Co-product	547	59	0	367	79	47	914	138	47	1100
Other	28	139	227	10	85	85	39	224	312	574
Total	1491	832	907	1204	655	272	2695	1486	1180	5361

Source: NEA/IAEA (2008)

Table 8.25 *Estimated size of the global uranium resource (kt U) in different cost and resource categories*

Resource confidence class	Cost category ($/kg)				Overall total
	<40	40–80	80–130	Unsure	
Reasonably assured	1491	832	907		3230
Inferred	1204	655	272		2131
Prognosticated		1946[a]	823		2769
Speculative			4798[b]	2973	7771
Total	2695	3433	6800	2973	15,901

Note: [a]Listed as <$80/kg. [b]Listed as <$130/kg.
Source: NEA/IAEA (2008)

based on various speculative mathematical models that are rather poorly constrained by data.

The reasonably assured and inferred resources together are referred to as 'identified' resources. Figure 8.28 shows the distribution of identified resources at $130/kg or less among the world's countries; Australia, Kazakhstan and Russia account for almost half of the this resource.

Figure 8.29 shows how an earlier estimate of 5.7Mt of identified resources is distributed according to ore grade. Most of the remaining uranium is in ore with a grade between 0.2 and 0.02 per cent. Figure 8.30 shows how uranium production costs vary with the ore grade. Standard economic thinking would indicate that increasing quantities of uranium will become available as the market price of uranium increases. This may be true but higher-cost uranium tends to be lower-grade uranium. As the cost reaches about $130/kg U, the grade of the remaining ore drops to about 0.02 per cent, at which point the EROEI

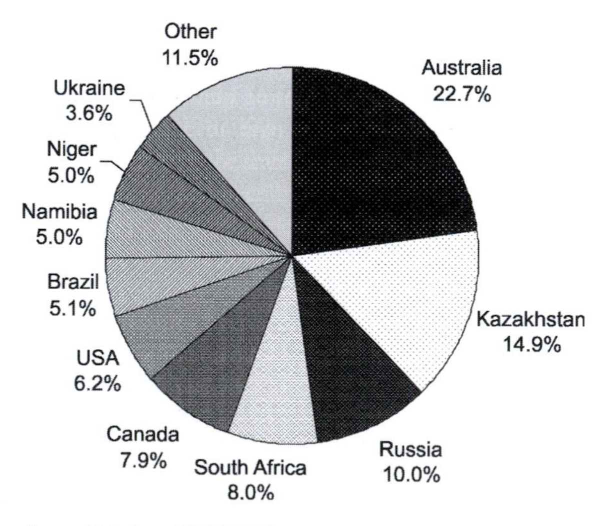

Source: Data from WEC (2007)

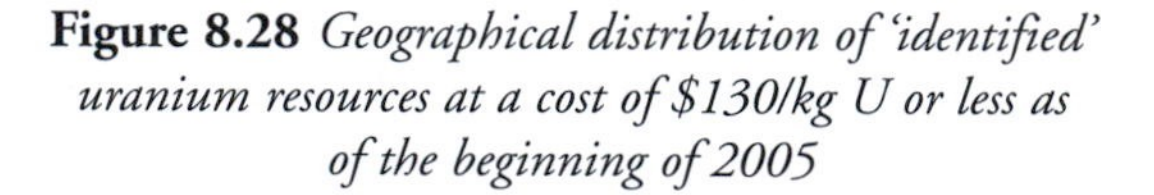

Figure 8.28 *Geographical distribution of 'identified' uranium resources at a cost of $130/kg U or less as of the beginning of 2005*

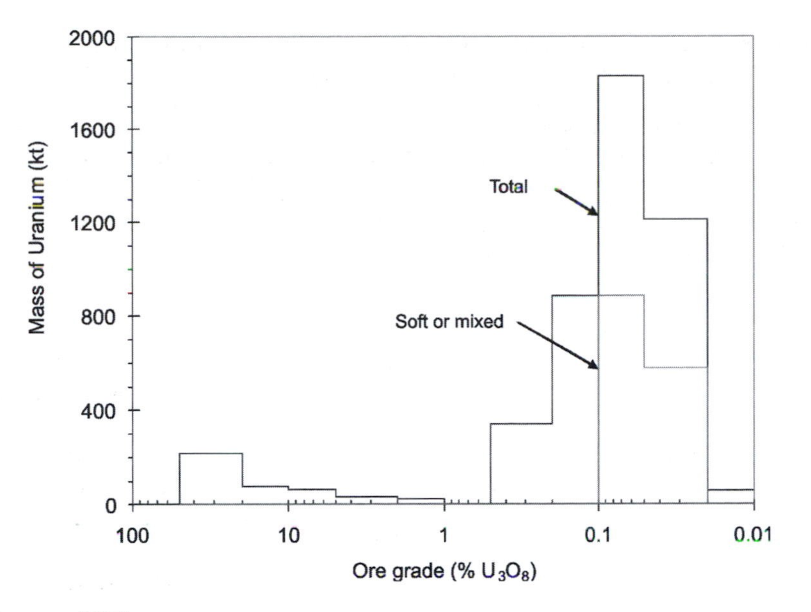

Source: Data from van Leeuwen (2007)

Figure 8.29 *Distribution of 'identified' uranium resources at a cost of $130/kg U or less with regard to the grade of ore*

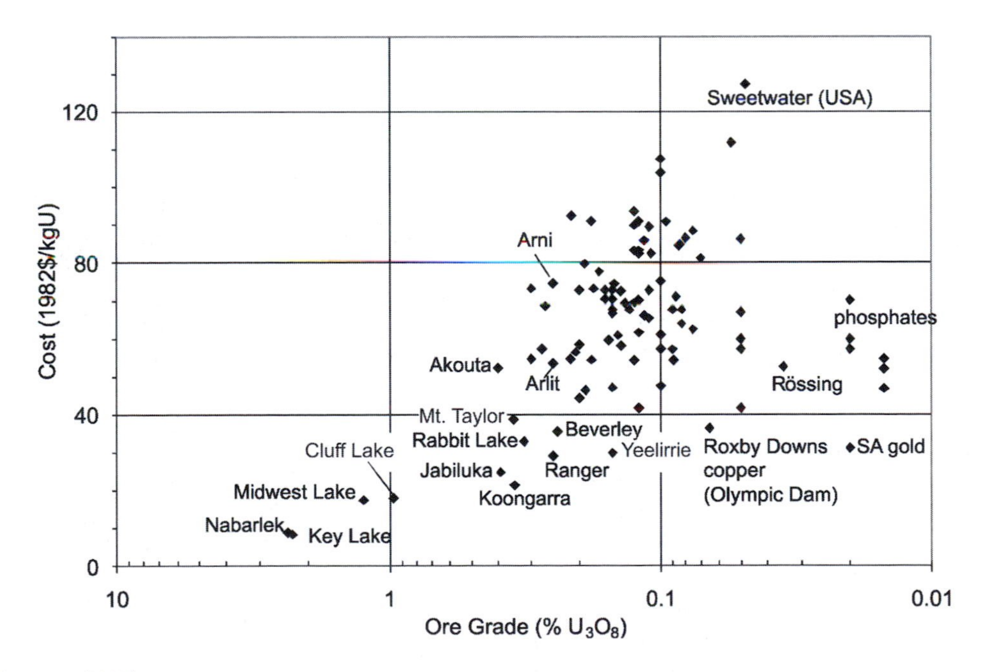

Source: van Leeuwen (2007)

Figure 8.30 *Cost of uranium versus the grade of ore supplying the uranium*

would have dropped to about seven to ten based on the calculations shown in Figure 8.23. By the time the ore grade drops to 0.0036 per cent, the EROEI would have dropped to about one and the fuel efficiency would have dropped to about 50 per cent if the mining energy use per tonne of rock processed is unchanged. To be viable, nuclear energy needs to do much better than break even. The amount of uranium that it is worth using in energy terms is likely to fall somewhere between the 5.4Mt of 'identified' uranium resource available at up to $130/kg U and the 15.9Mt of estimated total resources including as-yet undiscovered uranium. Most of whatever uranium remains to be discovered is almost certain to be of low grade, small ore body size and low accessibility.

A fleet of 1500GW of nuclear reactors would use about 306,000t/yr under option 1a of Figure 8.14, and 166,000t/yr under option 3. Including the entire estimated known and undiscovered resource (15.9Mt), there would be only a 52-year supply under option 1a (the more likely technology scenario) and a 96-year supply under option 3 (a much less likely scenario, given the difficulties experienced so far with both fast reactors and reprocessing). Unless FBRs are used (an even less likely scenario), nuclear power is likely to serve as only a short-term (<100 years) transitional fuel to a fully renewably based energy system if it were to supply a significant fraction of future world electricity demand. In the meantime, large quantities of radioactive wastes that will endure up to 1 million years will have been generated.

Seawater contains uranium at a concentration of 3mg/tonne (3×10^{-6}) and so represents an enormous resource (about 6×10^{12}t), but a volume of at least 285km^3 of seawater would need to be processed per year for a 1GW plant running at full capacity, or 9040m^3 per second per GW, assuming (as in van Leeuwen, 2007) that 17 per cent of the uranium in the processed volume could be extracted. To power the current world nuclear fleet of 370GW would require processing about 2.8 million m^3 per second. Estimated costs from seven different studies reviewed by van Leeuwen (2007) range from $280/kg U to $23,700/kg U, while van Leeuwen (2007) himself estimates the minimum energy requirement to be 195–250TJ/t U. The electricity yield of uranium (in the absence of reprocessing of spent fuel) is only about 140TJ/t U (assuming 75 per cent utilization of the U-235, 65GJ heat per gm of U-235 and a powerplant thermal efficiency of 40 per cent) and the net energy yield (after subtracting powerplant construction and decommissioning, fuel fabrication and waste disposal energy costs) is smaller still. Thus, if van Leeuwen's estimate of the energy requirement holds up, seawater is unviable as a source of uranium, both in terms of net energy yield and almost certainly in economic terms too.

8.12 Concluding comments

Nuclear energy is beset with a number of adverse environmental impacts and risks, such as:

- contamination of soils, groundwater and surface water at uranium mines and uranium milling sites due to the disposal of large volumes of radioactive and toxic wastes, and the need to monitor and manage these wastes in perpetuity;
- the accumulation of large volumes of low-level nuclear wastes from uranium enrichment, which will need to be managed in perpetuity;
- risks of accidental release of radioactivity during routine operation of nuclear powerplants or of facilities for reprocessing of spent fuel;
- risks of nuclear weapons proliferation by facilitating enrichment of uranium to weapons-grade uranium using facilities claimed to be only for enriching natural uranium to reactor-grade uranium;
- risks of nuclear weapons proliferation through the diversion of plutonium that had been separated from spent fuel – something that would need to occur on a large scale if nuclear energy were to supply a significant fraction of future electricity demand on a long-term basis;
- the continuing inability to find an acceptable long-term method for the isolation of high-level nuclear waste, and the growing burden of storing wastes on site until and if a long-term solution is found.

The probability of a serious accident during the operation of nuclear powerplants is thought to be extremely low as long as proper procedures are followed and there is a culture of safety. However, these conditions may break down somewhere at some point in time.

It will be difficult for nuclear energy to play a significant role in reducing GHG emissions at a global scale. Indeed, the industry will be hard pressed to maintain the current electricity production over the coming two decades, while total electricity demand continues to grow. Currently, nuclear energy provides 16 per cent of world electricity demand, with a capacity in mid-2009 of 370GW out of a total global capacity (in 2005) of about 4300GW. However, the current fleet is old (average age is 25 years), and a new reactor would have to be built once every five weeks over the next ten years and once every 22 days over the following decade just to maintain the current capacity.

If, in spite of the above, the nuclear power capacity were to increase to 1500GW (almost four times the current capacity) by 2050, known and speculative uranium supplies available at a cost of $130/kg U or less would last about 40 years with once-through use of uranium fuel. The supply could be expanded by up to a factor of 75 with reprocessing of spent fuel to separate and use plutonium in FBRs, but this would create enormous risks of nuclear materials sufficient to build a crude bomb getting into the hands of terrorists. Alternative reprocessing schemes that would provide some resistance to proliferation of nuclear weapons and that would stretch the existing uranium supply (to perhaps the end of this century for a nuclear capacity of 1500GW) have been proposed. However, these would require significant technological development.

The savings in primary energy when nuclear-generated electricity displaces fossil fuel-generated electricity at 40 per cent efficiency, divided by the primary energy inputs throughout the nuclear lifecycle from mining to decommissioning and isolation of wastes, averages about 15 at present. This ratio, referred to as the EROEI, is quite uncertain. However, the EROEI decreases as the grade of ore used decreases because the mass of ore that must by mined and processed, and the mass of tailing that must be treated for land reclamation, increases faster than the inverse of the ore grade. By the time uranium prices have reached $130/kg U, the grade of much of the remaining ore will probably have dropped to about 0.02 per cent and the EROEI will have dropped to about three to five (except for uranium that is co-mined with other minerals). Further decreases in ore grade would see an accelerating drop in the EROEI.

The cost of new nuclear powerplants is also uncertain. Recent assessments indicate a cost of about $4000/kW to over $10,000/kW. Decommissioning costs are uncertain but are likely to be at least as expensive as the lower estimates of the cost of constructing a nuclear powerplant.

Development of a new generation of nuclear reactors will require another 15–20 years or more to reach the pilot demonstration stage, and then perhaps ten years of operation before such reactors could begin to be produced on a large scale. Another 20 years or more would be required before the new generation could provide a significant fraction of the total nuclear fleet. Thus, a major change in the technology of the operating fleet cannot be expected before 2050 at the earliest – a timescale much longer than that needed for significant reductions in CO_2 emissions if the risk of catastrophic climatic change is going to be reduced to a meaningful extent.

Notes

1 Some might regard bismuth (atomic number 83) as having a stable isotope, as ^{209}Bi decays only with an extremely long half-life (about 19×10^{18}yr).

2 The elements in the periodic table from atomic number 89 (actinium) and upward are referred to as *actinides.* These include thorium, uranium and plutonium (atomic numbers 90, 92 and 94). As previously noted, those elements higher than uranium are called transuranic elements, so the transuranic elements are a subset of the actinides.

3 The first plutonium isotope to be produced in a nuclear reactor is Pu-239, which is most effective in making nuclear weapons. As the fuel is used for a longer period of time, other isotopes of plutonium build up that are prone to earlier fission, thereby reducing explosive power, or that absorb neutrons. A sign that fuel in a reactor is being used to produce plutonium for military purposes would be if the fuel throughput is an order of magnitude greater than is normal for commercial power reactors.

4 Hydrogen would not be produced by electrolysis using the generated electricity, but rather, through the thermo-chemical decomposition of water into H_2 and O_2 using the high-temperature waste heat from the reactor.

5 The first nuclear reactor in the world went critical in 1942 and the first commercial reactor did not go online until about 1970, and that was with enormous financial support (Dean Abrahamson, personal communication, June 2009).

6 Defined as actinide elements in the periodic table from neptunium and upward. Uranium and plutonium are called major actinides.

7 A definitive decision in Finland is supposed to wait until testing after excavation of an initial tunnel to the 500m storage depth.

8 Energy use associated with two items has been left out due to lack of data, but is probably small.

9 To the extent that the cost increase is due to increases in the costs of raw materials without an increase in the underlying inputs, the estimated energy would not go up because the MJ/$ ratios for the commodities in the input–output matrix would go down in proportion to the increase in commodity costs. However, to the extent that the increase in the estimated costs of nuclear powerplants represents real increases in the effort and inputs needed to make them, IOA estimates that are based on powerplant costs that are too small will also be too small.

10 This ratio is derived as follows: the ore body is assumed to be an ellipsoid with horizontal semi-axis lengths of 860m and 1100m and a vertical semi-axis of 420m. The bottom of the pit is assumed to be an ellipsoid that is concentric within the outer edge of the ore body, with semi-axis lengths of 550m and 790m. The top of the ore body is also an ellipse with semi axes that are each 1200m greater (consistent with Figure 8.20), namely, 1750m and 1990m. Letting A_1 and A_2 be the areas of the lower and upper ellipses and assuming the upper ellipse to have the same proportions as the lower ellipse, the volume of the pit would be that of a frustum (truncated cone), namely $V=1/3h(A_1+\sqrt{(A_1A_2)}+A_2)$ while the volume of the ore body is $4/3\pi abc$ (where a, b and c are the three semi-axes). This approximation is adopted here.

11 Perhaps a little less often if the capacity factor of new reactors is greater than that of the reactors that they replace.

12 In the most aggressive energy efficiency scenarios presented in Volume 1 (Figure 10.14), global electricity demand increases by about 50 per cent between 2005 and 2050 for the low population and GDP growth scenario, and approximately doubles for the high population and GDP growth scenario.

9 Carbon Capture and Storage

9.1 Introduction

The climatic hazard from the combustion of fossil fuels arises from the *emission* of CO_2 into the atmosphere, not from the use of fossil fuels per se. If the CO_2 produced from combustion of fossil fuels could be prevented from entering the atmosphere, these energy sources would qualify as 'carbon-free' as far as the climate is concerned. To prevent CO_2 from entering the atmosphere, it is first necessary to capture the CO_2 at the point of combustion, then compress and transport the CO_2 to some disposal reservoir and inject the compressed CO_2 into the disposal reservoir. The process of capturing and disposing of CO_2 so that it cannot enter the atmosphere is referred to as *carbon sequestration* or as *carbon capture and storage (CCS)*. The process of building up biomass (through reforestation) or of building up soil carbon is also referred to as carbon sequestration. In the case of CCS, it is specifically CO_2 that is captured and stored (so a more appropriate term would be *carbon dioxide capture and storage* or *CDCS*, as carbon in the form of soot or methane is not included), while in the case of biomass or soils it is indeed carbon and not carbon dioxide that is being stored, although if the carbon returns to the atmosphere it does so in the form of CO_2 after oxidation with atmospheric O_2. Potential CCS disposal reservoirs include coal beds, deep saline aquifers, depleted oil and gas fields, sub-seabed sediments and the deep ocean itself. In the case of disposal in the ocean or in sub-seabed sediments, the captured CO_2 would probably be liquefied first. Thus, except for the possibility of disposal in the deep ocean (which seems unlikely now), CCS involves disposal in geological formations and so can be referred to as *geological carbon sequestration*, in contrast to sequestration in biomass and soils, which can be referred to as *biological carbon sequestration*.

Capturing CO_2 from numerous, dispersed, small-scale sources such as automobiles and individual household furnaces will never be practical, but capturing it from large, centralized facilities such as fossil fuel-powered electricity powerplants, district heating systems and industrial facilities such as integrated steel mills and cement kilns could be practical. This would allow about 50 per cent of current emissions to be sequestered. However, if the transportation fleet is eventually powered by hydrogen, and this hydrogen is produced from fossil fuels, the CO_2 that is produced in this process could also be sequestered. In this way, another 30 per cent of fossil fuel-related CO_2 emissions that would otherwise occur could be sequestered.

In this chapter we discuss potential sources of CO_2 for geological or deep-ocean sequestration and the technical options, costs, hazards and storage limits associated with the capture of CO_2 and its geological or deep-ocean sequestration. The possibilities with regard to biological carbon sequestration are discussed in Chapter 4 (subsections 4.8.1 and 4.8.3).

9.2 Sources of CO_2 for capture and storage

Practical capture of CO_2 requires individual sources of CO_2 with a large enough CO_2 flux to make capture worthwhile. Capture is easier the higher the concentration of CO_2 in the candidate gas stream. For a given concentration, capture will be easier the greater the total pressure of the gas stream (which will give a greater CO_2 partial pressure). Table 9.1 gives the CO_2 concentration and pressure for gas streams that are good candidates for capture of CO_2: powerplant flue gases and gas streams from oil refineries, petrochemical

Table 9.1 *Properties of gas streams large enough to make CO_2 capture worthwhile*

Source	CO_2 concentration (% dry volume)	Pressure of gas stream (atmospheres)	CO_2 partial pressure (atmospheres)
	Streams where CO_2 is not normally separated		
Powerplant flue gases			
Natural gas-fired boilers	7–10	1.0	0.07–0.10
Natural gas turbines	3–4	1.0	0.03–0.04
Oil-fired boilers	11–13	1.0	0.11–0.13
Coal-fired boilers	12–14	1.0	0.12–0.14
IGCC before combustion	8–20	20–70	1.6–14
IGCC after combustion	12–14		0.12–0.14
Streams from industrial processes			
Oil refineries	8	1.0	0.08
Blast furnace gas before combustion	20	2–3	0.4–0.6
Blast furnace gas after combustion	27	1.0	0.27
Cement kiln offgas	14–33	1.0	0.14–0.33
	Streams from which CO_2 is separated as part of the production process		
Ammonia production	18	28	5
Ethylene oxide production	8	25	2
Hydrogen production	15–20	22–27	3–5
Methanol production	10	27	2.7
Processing of natural gas	2–65	9–80	0.5–44

Source: Gale et al (2005)

plants, blast furnaces and cement kilns. Gas streams from fuel combustion are at atmospheric pressure, but the gas streams from blast furnace gas before combustion and the synthesis gas produced at integrated gasification combined cycle (IGCC) powerplants are at substantially greater pressure. Blast furnace gas typically contains 20 per cent CO_2 by volume and 21 per cent CO, the balance being N_2, and is at a pressure of 2–3atm. CO_2 could be captured before or after combustion, but would be somewhat higher in concentration (27 per cent) after combustion due to the conversion of CO to CO_2. In either case, the concentration and pressure would be significantly greater than from powerplants (3–4 per cent for natural gas plants, 12–14 per cent for coal plants). The flue gas from cement kilns contains 14–33 per cent CO_2, depending on the production process and type of cement produced. Again, this is substantially higher than in the flue gas from powerplants. In ammonia plants, H_2 is produced from natural gas and used as a feedstock. The CO_2 is already separated from the H_2 as part of the production process, and so could be captured at relatively little cost.

Table 9.2 gives the numbers of different kinds of sources worldwide with a flux of more than 0.1MtCO_2 per year, the total emissions from these sources and the average emission per source. The total of all listed sources (3.64GtC) amounts to just over half of the worldwide fossil fuel CO_2 emission at the time of the estimate (6.74GtC in 2000). The top 25 per cent of all the stationary CO_2 sources listed in Table 9.2 accounts for 85 per cent of the emissions from all sources listed in Table 9.2. Most of the existing US fossil fuel powerplants are close to deep saline formations suitable for storage of CO_2, as seen from maps presented by Dahowski and Dooley (2004). High-concentration/ high-partial-pressure sources such as ammonia and hydrogen production account for a small share (<2 per cent) of the total stationary emissions but would be good candidates for the initial implementation of CO_2 capture and storage. Meng et al (2007) identified a number of Chinese coal-fed ammonia plants that are close to suitable deep saline aquifers. Many of these produce relatively pure streams of CO_2 in excess of 500,000 tonnes/yr that are currently vented to the atmosphere.

The *Carbon Sequestration Atlas of the United States and Canada* contains a map of CO_2 sources according to the type of facility (such as cement, fertilizer, industrial and electrical plants) and according to the magnitude of the annual source (NETL, 2007).

Table 9.2 *Profile of stationary CO_2 sources with a minimum emission per source of 0.1MtCO$_2$/year*

Process	CO_2 concentration in gas stream (% by volume)	Number of sources	Total emission ($MtCO_2$/year)	% of total emission from sources considered here	Average emission per source ($MtCO_2$/year)
CO_2 from electricity generation					
Coal	12–15	2025	7984	59.69	3.94
Natural gas	3	985	759	5.68	0.77
Natural gas	7–10	743	752	5.62	1.01
Fuel oil	8	515	654	4.89	1.27
Fuel oil	3	593	326	2.43	0.55
Other fuels		79	61	0.45	0.77
Cement production	20	1175	932	6.97	0.79
Refineries	3–13	638	798	5.97	1.25
Iron and steel industry					
Integrated mills	15	180	630	4.71	3.50
Other processes		89	16	0.12	0.17
Petrochemical industry					
Ethylene	12	240	258	1.93	1.08
Ammonia: process	100	194	113	0.84	0.58
Ammonia: combustion	8	19	5	0.04	0.26
Ethylene oxide	100	17	3	0.02	0.15
Hydrogen		2	3	0.02	1.27
Other sources		90	83	0.52	0.37
Total		7584	13,375	100	1.76
CO_2 from biomass					
Bioenergy	3–8	213	73		0.34
Fermentation	100	90	17.6		0.20

Source: Gale et al (2005)

9.3 Capture of CO_2

Carbon dioxide can be captured during the use of fossil fuels or biomass in four different ways:

1 from the exhaust gases of electricity powerplants after combustion in air, where CO_2 constitutes about 3–14 per cent of the exhaust gas;
2 from the exhaust gases of powerplants where the fuel is burned in close to pure oxygen so that CO_2 and H_2O are the only combustion products and the CO_2 is separated by condensing the water vapour;
3 prior to combustion, during the gasification of solid fuels or transformation of natural gas to make hydrogen;
4 during the operation of fossil fuel-powered fuel cells.

Each of these options is discussed below.

9.3.1 Post-combustion CO_2 separation processes

Capturing CO_2 from a powerplant is already a commercial process, with more than a dozen capture plants worldwide (Herzog, 2001). Captured CO_2 is used for industrial purposes, including by the soft-drink industry and for injection into oil fields to increase the amount of oil that can be extracted from the field. There are four major kinds of processes that can be used for separating CO_2 from flue gases: *absorption, adsorption,* through *membranes* and by liquefying the CO_2 (*cryogenic* separation). A fifth method is absorption with chilled ammonia. Below is a brief overview, based on IEA (2001a). An absorption CO_2 capture facility is illustrated in Figure 9.1.

Source: Thambimuthu et al (2005)

Figure 9.1 *A chemical solvent-based plant that captures a mere 200tCO_2/day from a flue gas stream*

Absorption systems

Absorption systems can be based on chemical solvents, physical solvents or a combination of the two. The exhaust gas passes through a tower filled with solvent that absorbs CO_2. The CO_2 is periodically driven from the solvent in concentrated form, thereby regenerating the solvent for reuse. Regeneration requires that heat be added to the solvent because CO_2 forms weak chemical bonds with the solvent when it reacts with the solvent. The heat is extracted from the steam turbine in the powerplant, which reduces the power output. Chemical solvents are generally good when the CO_2 concentration in the exhaust gas is relatively low, such as for natural gas combined cycle (NGCC) and pulverized coal powerplants. A common chemical solvent is MEA (monoethylamine), which was developed over 60 years ago in order to separate CO_2 from natural gas (which contains up to 20 per cent CO_2 by volume). With a physical solvent, CO_2 dissolves into the solvent (like a gas dissolving into water). Physical solvents are good when the CO_2 concentration (and hence pressure) in the exhaust gas is relatively high, such as for IGCC powerplants or hydrogen production plants. The CO_2 can be separated and the solvent regenerated either through a pressure drop or with heat. A common physical solvent is Selexol (dimethylether of polyethylene glycol). Most commercial CO_2 capture plants today use chemical absorption with MEA.

The major energy penalty in using chemical solvents is the thermal energy required to regenerate the solvent. The conventional regeneration process requires 4.2MJ/kg CO_2. A new process using altered amines (KS-1) has been commercialized and used in ammonia plants; the energy requirement is 3.2MJ/kg CO_2 and might be reduced to 2.3MJ/kg CO_2 – a saving of 45 per cent compared to the MEA process (Damen et al, 2006).

Both chemical and physical solvents require that concentrations of SO_x be very low, as SO_x reacts with the solvent. Chemical solvents also require very low concentrations of NO_x. In purely economic terms, how low the SO_x and NO_x should be involves a tradeoff between the cost of reducing SO_x and NO_x concentrations and the cost of the solvent consumed by reaction with SO_x and NO_x and treating the waste stream created by the reaction. The required concentrations are lower than most or all current emission limits. Thus, in achieving CO_2 removal by absorption, air pollution benefits will occur that would not otherwise occur, unless pollution emission limits would have been tightened anyway.

Adsorption systems

In *adsorption*-based systems, CO_2 binds to the outer surfaces of small particles (either alumina, zeolite or activated carbon). The adsorbing material is regenerated either through pressure release, which requires seconds, or through high temperatures (requiring hours). Adsorption processes are used today to separate CO_2 from H_2 that is produced by steam reforming of natural gas.

Membrane systems

Two kinds of membranes are possible: gas-separation membranes and gas-absorption membranes. With gas-separation membranes, one gas in a mixture passes more readily through the membrane than another, allowing it to be concentrated. The two key parameters for membranes are the *permeability* for the various gases and the *selectivity* of CO_2 – the ratio of the permeability of CO_2 to that of the other gases in the mixture. Ideally, one would like a membrane with both a high permeability and a high selectivity for CO_2. Usually the differential permeability of the membrane is not enough to give adequate concentration of CO_2, so multiple membrane stages are needed, with recompression of the transmitted gas between each stage.

With gas-absorption membranes, gas flow occurs on one side of the membrane and an absorbing liquid is on the other side. The membrane itself does not need to selectively transmit CO_2; the selectivity of the processes arises from the absorbing liquid. Membrane technology is still at an early stage of development, and often requires more energy (up to 14 per cent of power output for compression) than is used to regenerate MEA (Corti et al, 2004).

Liquefaction

CO_2 can be liquefied by lowering its temperature and raising its pressure. The higher the pressure, the less it needs to be cooled. Figure 9.2 is a phase diagram for CO_2, which shows the temperature and pressure combinations at which it occurs as a gas, liquid, or solid.

Absorption with chilled ammonia

A 5MW pulverized coal pilot plant began operation in Wisconsin in 2007 using chilled ammonia to absorb

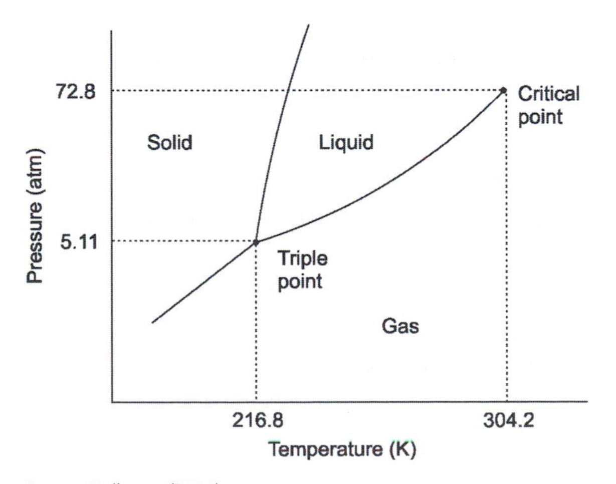

Source: Holloway (2001)

Figure 9.2 *The phase diagram for CO_2*

CO_2 after combustion (Blankinship, 2007). The process is said to have the potential to capture more the 90 per cent of the flue gas CO_2 less expensively than other techniques, and does not require extremely low SO_2 levels to work. Instead, the process can remove residual high levels of SO_2, condensable particulate matter ($PM_{2.5}$) and other pollutants. It is thus a promising technique for retrofitting onto existing coal powerplants.

9.3.2 Capturing CO_2 from flue gases after combustion in oxygen

Combustion in pure oxygen (oxy-fuel combustion) eliminates nitrogen from the flue gas and thus the need for any of the techniques described in the preceding section. The only products of combustion would be CO_2 and water vapour, which can be easily separated by condensing the water vapour – without the cycling of chemicals or the generation of chemical waste products (see subsection 9.7.1). Combustion in pure oxygen would result in combustion at a temperature of 3500°C, which is far too hot for most materials, so the temperature is limited by recirculating some of the CO_2 from the exhaust back into the combustion chamber. Very close to 100 per cent of the CO_2 from combustion can be captured.

All of the components of oxy-fuel combustion are already in use in the aluminium, iron and steel, and glass-melting industries, but they have not yet been deployed on a commercial scale for capture of CO_2. The main energy-consuming step is the separation of oxygen from air for subsequent use in oxy-fuel combustion. This is done using cryogenic techniques (cooling the air to 89K). In the long run, ion transport membranes may prove to be more attractive for separating oxygen. Future coal powerplants should achieve efficiencies of 45 per cent with carbon capture (after subtracting energy output used to separate oxygen), while plant efficiencies of >50 per cent with natural gas combined cycle and CO_2 capture are expected to be achieved by 2020 (Thambimuthu et al, 2005).

There are no technical barriers to retrofitting the boilers of existing coal-fired powerplants with oxygen combustion and CO_2 capture (Thambimuthu et al, 2005). In one design study, it was found to be more economical to design the air separation units for only 95 per cent O_2 purity rather than 99.5 per cent, given air leakage into the boilers, and to remove the associated argon and nitrogen after combustion. Combustion in this case produces a stream consisting of 60 per cent CO_2, 30 per cent H_2O and 10 per cent other gases. After cooling to condense out water vapour, the result is 96 per cent CO_2, 2 per cent nitrogen, 1 per cent argon and 1 per cent O_2 and SO_2, which can then be further purified through distillation. However, in order to justify the expense of retrofitting an existing plant, a long remaining life would be needed. Old plants have low efficiency compared to new plants, such that new plants with CO_2 capture can be as or more efficient than old plants without capture.

In natural gas powerplants, combustion does not occur in a boiler but in the combustion chamber of the turbine. The combustion products then expand through the turbine, but a CO_2/H_2O mixture (produced by combustion in oxygen) has quite different expansion characteristics than a $CO_2/H_2O/N_2$ mixture (produced by combustion in air) (Damen et al, 2006). The optimal pressure ratio is higher than for conventional turbines, so a whole new gas turbine (compressor and combustor) would need to be developed.

Two advanced concepts under development merit brief mention: the *mixed conducting membrane (MCM) reactor* and *chemical looping combustion (CLC)*. In the former, compressed air enters a reactor, from which oxygen passes through a membrane to the combustion chamber. The heat of combustion is transferred to the oxygen-depleted air, which expands through a conventional turbine. Heat from the turbine exhaust and from the CO_2/H_2O stream generated in the combustion chamber is used to make steam for a steam turbine. Conventional turbines can be used. In CLC, the oxidation and reduction parts of combustion are separated. The fuel is oxidized by a metal oxide, which is highly exothermic, providing high-temperature exhaust for power generation. The reduced metal oxide is returned (in particulate form) to an oxidation reactor where it reacts with oxygen to complete the cycle. Further information on MCM and CLC can be found in Thambimuthu et al (2005) and Damen et al (2006).

9.3.3 Capturing CO_2 from coal IGCC powerplants

The IGCC powerplant using coal is described in Volume 1 (Chapter 3, subsection 3.2.2). Coal is gasified to a mixture of CO_2, CO and H_2 called *syngas* by heating it in the presence of 95 per cent purity oxygen. A typical gas composition will be about 30 per cent CO_2, 40 per cent CO and 30 per cent H_2. A number of options are available from this point onward:

1 The CO can be largely converted to CO_2 and H_2 by reaction with water and removed prior to combustion. Conversion of CO to CO_2 produces a gas with a typical composition of 33 per cent CO_2 and 67 per cent H_2 by volume. The high CO_2 concentration lends itself to removal with a physical solvent such as Selexol prior to combustion of the H_2 and any residual CO in a gas turbine using air. The only CO_2 in the exhaust would be from combustion of the residual CO and from CO_2 not removed by the physical solvent prior to combustion.
2 The CO and H_2 can be fed to the gas turbine without the water-shift reaction, and the CO_2 removed after combustion in air. In this case, the CO_2 concentration in the exhaust gas will be relatively low, and so will need to be removed using a chemical solvent.
3 The CO and H_2 can be fed to the gas turbine without the water-shift reaction, and the CO_2 removed by condensation of water vapour after combustion in pure oxygen. As with the combustion of natural gas in pure oxygen, the working fluid passing through the turbine is a CO_2/H_2O mixture for which current turbines are not designed, so this option requires the development of new gas turbines. As well, additional separation of oxygen (beyond that used for gasification) would be required, but essentially 100 per cent of the CO_2 could be captured.

Selexol (used for pre-combustion physical absorption) can be used to remove both sulphur (as H_2S) and CO_2 from the syngas, followed by a two-step regeneration to produce separate sulphur and CO_2 streams (Davison, 2007). Co-capture of H_2S and CO_2 significantly reduces the energy penalty associated with CO_2 capture, compared to separate capture of CO_2 alone. Theoretical calculations by Ordorica-Garcia et al (2006) indicate that separate capture of 80 per cent of the CO_2 from a 577MW IGCC powerplant along with H_2S capture reduces the net power output to 488MW – a penalty of 89MW (15.4 per cent). However, co-capture of H_2S and CO_2 reduces the output by only 25MW (a penalty of 4.3 per cent).

It should be noted that the few IGCC demonstration plants in operation (none of which capture CO_2) have had great difficulty achieving semi-reliable operation, requiring three to five years to reach an average availability of only 70–80 per cent (MIT, 2007). This is because IGCC powerplants are considerably more complex than conventional pulverized coal powerplants.

Another potential concept – still theoretical at this stage – is to use an advanced ceramics proton membrane to separate the syngas into an H_2 gas (99.99 per cent purity) and a C-rich gas without the water-shift reaction, then to direct each gas stream to a gas turbine, where it is combusted in pure oxygen (Duan et al, 2004). Exhaust heat from the two turbines is used to produce steam, some of which is used to generate additional electricity, and some of which is fed into the H_2 turbine so as to limit the turbine temperature. Exhaust from the steam turbine is used to cool the other gas turbine. Compared to a conventional IGCC scheme, about twice as much oxygen needs to be separated from air, but the electrical efficiency is calculated to decrease from 46.0 per cent without carbon capture to 45.2 per cent with carbon capture – a remarkably small efficiency penalty.

9.3.4 Capturing CO_2 from the operation of pressurized fossil fuel-powered fuel cells

Fuel cells require hydrogen or a hydrogen-rich gas as the input fuel. They could be powered with hydrogen that is produced electrolytically from electricity that is produced from renewable energy sources (such as wind and solar energy), or with hydrogen that is produced from fossil fuels – either natural gas or coal. In the latter case, the hydrogen could be produced separately from its use in a fuel cell, or as part of a process that is integrated with the operation of high-temperature fuel cells. Either way, a CO_2-rich gas stream would be produced. Pressurized high-temperature fuel cells may thus provide a low-cost method for capturing CO_2.

The reactions to produce H_2 from methane are:

$$CH_4 + H_2O \leftrightarrow CO + 3H_2 \quad \text{(steam reforming)} \quad (9.1)$$

$$CO + H_2O \leftrightarrow CO_2 + H_2 \quad \text{(water shift reaction)} \quad (9.2)$$

giving the net reaction:

$$CH_4 + 2H_2O \leftrightarrow CO_2 + 4H_2 \quad (9.3)$$

The reactions to produce H_2 from coal or biomass are:

$$3C + 2H_2O \leftrightarrow CH_4 + 2CO \quad \text{(partial oxidation)} \quad (9.4)$$

$$CH_4 + H_2O \leftrightarrow CO + 3H_2 \quad \text{(steam reforming)} \quad (9.5)$$

$$3CO + 3H_2O \leftrightarrow 3CO_2 + 3H_2 \quad \text{(water shift reaction)} \quad (9.6)$$

giving the net reaction:

$$C + 2H_2O \leftrightarrow CO_2 + 2H_2 \quad (9.7)$$

Thus, production of hydrogen from methane involves reaction with steam (steam reforming) followed by the water-CO shift reaction to make additional hydrogen. Production of hydrogen from coal (or biomass) begins with an extra step (partial oxidation) followed by the same steps as in the production of hydrogen from natural gas (steam reforming and water shift). The water-CO shift reaction entails an energy loss. However, both CO and H_2 can serve as fuel for solid oxide fuel cells (SOFCs) and molten carbonate fuel cells (MCFCs), so the CO + H_2 gas could be directly supplied to the fuel cell anode (without energy loss). Air can be directly supplied to the fuel cell cathode, but only the oxygen passes through the electrolyte (either as O_2^{2-} or CO_3^{2-}) to the anode, where it reacts with the CO and H_2 (see the MCFC and SOFC rows in Table 10.7). Unreacted CO and H_2 in the anode exhaust could be burned in pure O_2 to produce additional electricity via a gas (or steam) turbine (as illustrated in Volume 1, Figure 3.10). However, the quantity of pure O_2 needed (and the associated energy used to separate it from air) would be much less than in powerplants based on combustion of the fuel (discussed in subsection 9.3.2) – about 1 per cent of the thermal output. In effect, the fuel cell serves as a membrane for separating oxygen from air as a byproduct of its operation! The final exhaust gas would consist of CO_2 and H_2O only, and the H_2O would be separated from the CO_2 through simple cooling and condensation.

Inui et al (2005) foresee an overall efficiency of 64.9 per cent (higher heating value, HHV, basis) for a SOFC using natural gas that captures and liquefies essentially all of the CO_2. With further development, they foresee coupling of a SOFC to an MHD (magneto-hydrodynamic) generator (which is suited to operating at a temperature of 2200K using helium as a working fluid) and a net efficiency (after accounting for liquefaction of CO_2) of 67.5 per cent (HHV basis).

9.3.5 Comparison of the cost and performance of alternative methods of CO_2 capture from fossil fuel powerplants

All of the CO_2 removal processes require energy inputs, which reduce the net energy output and efficiency of the powerplant. This in turn increases the fuel consumption. Summary information on the energy penalty, effective CO_2 fraction captured and cost from pulverized coal, IGCC, NGCC and fuel cell powerplants is given in Tables 9.3–9.5, with more detailed information in Supplemental Tables S9.1–S9.3. One set of estimates is given for existing technology and two sets for advanced technology – one based on the 2005 IPCC assessment of carbon sequestration (Thambimuthu et al, 2005), the other based on a 2006 review (Damen et al, 2006) that tends to be a bit more optimistic than the IPCC assessment.

All of the technologies entail a substantial efficiency loss at present. As seen from Table 9.3, the retrofit of existing coal powerplants for carbon capture is expected to entail a larger efficiency penalty (increasing fuel use by 43–77 per cent) than construction of new powerplants with post-combustion carbon capture (24–40 per cent increase in fuel use). In the case of an MEA retrofit, steam is taken from the boiler to regenerate the solvent. The steam turbine thus operates at part load and hence at lower efficiency, which amplifies the loss of electricity production due to diversion of some steam away from the turbine. A better approach (which is no longer a retrofit) is to rebuild the core of a subcritical (less efficient) PC (pulverized coal) unit, installing more efficient supercritical or ultra-supercritical technology along with MEA CO_2 capture. Existing subcritical PC plants have efficiencies of around 34 per cent, which would decrease to around 25 per cent after a retrofit. Ultra-supercritical units

Table 9.3 *Energy penalties associated with CO_2 capture only*

Process	Using existing technology (%)	Using advanced technology	
		IPCC (2005) (%)	Damen et al (2006) (%)
Retrofitting existing PC powerplants	43–77		
New PC powerplant, post-combustion	24–40	15–43	9
– oxyfuel		25–33	9–12
New IGCC, pre-combustion	14–25	21–24	5–9
– oxyfuel			8
New NGCC, post-combustion	11–22	16–25	6
– pre-combustion with membrane			5–6
– oxyfuel MCM			2–8
Fuel cell/turbine hybrid		13–44	
Biomass IGCC			36[a]
Pulp and paper mill using black liquor waste			19[b]
H_2 production from coal		2.2	

Note: PC = pulverized coal, IGCC = integrated gasification combined cycle, NGCC = natural gas combined cycle, MCM = mixed conducting membrane. [a] Rhodes and Keith (2005); [b] Möllersten et al (2004).
Source: Tables S9.1 to S9.3 and sources cited therein

would be at 43.4 per cent without capture and 34 per cent with capture (MIT, 2007). Thus, the efficiency would be unchanged with this strategy. However, the energy penalty – and increase in coal consumption – is still significant compared to upgrading without installing CO_2 capture. MIT (2007) projects a cost of electricity from a rebuilt subcritical PC plant with carbon capture to be about 7 cents/kWh, although we are nowhere near that at present.

The contributions to the efficiency loss for specific alternative PC, NGCC and IGCC powerplants are compared in Figure 9.3. The IGCC plants considered are two different designs with a water shift reaction and pre-combustion capture using Selexol (Option 1 of subsection 9.3.3). Two cases of post-combustion separation and one case of combustion in oxygen are shown for PC and NGCC plants. Combustion in pure oxygen, although facilitating easier and more complete removal of CO_2, does not reduce the energy penalty, due to the substantial energy requirement for separation of oxygen and for compression and purification of the CO_2. The oxygen plant for PC produces O_2 at atmospheric pressure, while the oxygen plant for the NGCC must produce it at 30atm pressure, thereby increasing the energy penalty. Although the IGCC plant uses separated oxygen for the gasification process, this does not contribute to an efficiency penalty when capturing CO_2 because the oxygen for gasification is needed whether or not CO_2 is captured.

A distinction needs to be made between the percentage of CO_2 captured and the effective percentage of CO_2 captured. The former refers to the percentage of CO_2 in the exhaust gas stream that is removed. However, because of greater fuel use due to the efficiency penalty caused by CO_2 capture, the exhaust flow is greater. The effective CO_2 capture is based on the absolute emission in the case with capture compared to the absolute emission for the case without capture. Suppose, for example, that 80 per cent of the exhaust CO_2 is captured but that fuel use increases by 20 per cent (from 100 to 120 units). Then the emission is 120 × (1.0–0.8) = 24 units, so the effective CO_2 percentage captured is only 76 per cent. Table 9.4 shows the expected effective CO_2 capture fraction for existing and advanced technologies after taking into account the extra fuel use.

All of the carbon capture technologies are subject to considerable cost uncertainty. Table 9.5 presents expected future costs for various powerplants using

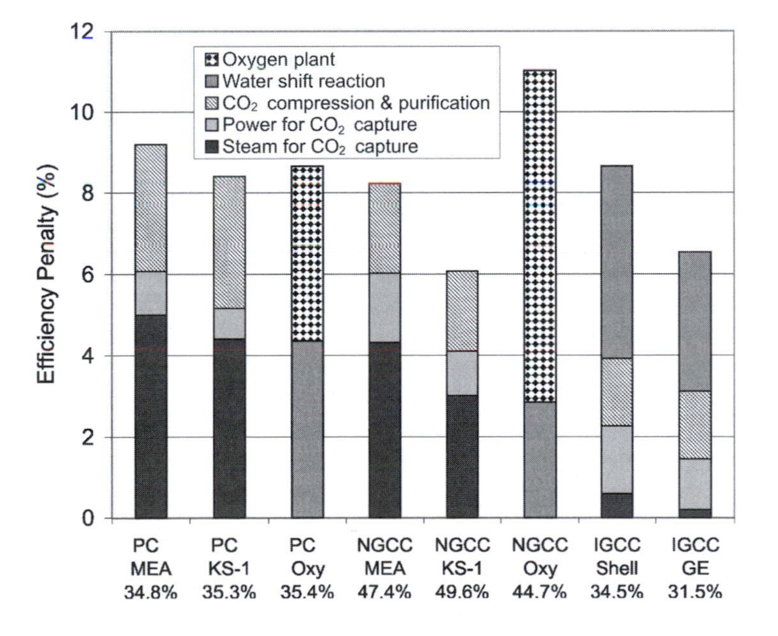

Source: Davison (2007)

Figure 9.3 *Breakdown of the sources of the efficiency penalty associated with the capture of CO_2 from different PC, NGCC or IGCC powerplants*

Table 9.4 *Effective fraction of CO_2 captured*

Process	Using existing technology (%)	Using advanced technology	
		IPCC (2005) (%)	Damen et al (2006) (%)
Retrofitting existing PC powerplants	63–94		
New PC powerplant, post-combustion	81–88		85
– oxyfuel		88–99.5	90–100
New IGCC, pre-combustion	81–91	90–91	85–90
– oxyfuel			100
New NGCC, post-combustion	83–88		85–90
– pre-combustion with membrane		82–100	100
– oxyfuel MCM			85–100
Fuel cell/turbine hybrid		86–92	80–100
Biomass IGCC			39[a]
Pulp and paper mill using black liquor waste			88[b]
H_2 produced from coal		98	

Note: PC = pulverized coal, IGCC = integrated gasification combined cycle, NGCC = natural gas combined cycle, MCM = mixed conducting membrane. [a] Rhodes and Keith (2005); [b] Möllersten et al (2004).
Source: Tables S9.1 to S9.3 and sources cited therein

Table 9.5 *Capital cost (\$/kW) of powerplants equipped with technologies for capture of CO_2*

Process	Using existing technology	Using advanced technology	
		IPCC (2005)	Damen et al (2006)
Reference powerplants (no C capture, actual costs today for NGCC and PC, projected for IGCC)			
NGCC	515–724		
New PC	1200–1500		
IGCC	1200–1600		
Powerplants with C capture (all costs are projected)			
Retrofitting existing PC powerplants	650–1950		
New PC powerplant, post-combustion	1900–2600	1700–1800	1520
– oxyfuel		1850–2850	1800–2200
New IGCC, pre-combustion	1500–2300	1450–2200	1450–2200
– oxyfuel			1420–1550
New NGCC, post-combustion	900–1300	950–1225	700–1010
– pre-combustion with membrane			940
– oxyfuel MCM			820–1250
Fuel cell/turbine hybrid		1800	990–2060
Biomass IGCC			1980[a]

Note: PC = pulverized coal, IGCC = integrated gasification combined cycle, NGCC = natural gas combined cycle, MCM = mixed conducting membrane. Costs are projected costs after some period of learning. [a] Rhodes and Keith (2005).
Source: Tables S9.1 to S9.3 and sources cited therein

either present or future CO_2 capture technologies. The cost estimates with both present and advanced technologies pertain to the plants built after a period of learning. With advanced technologies, NGCC with carbon capture is expected to cost \$700–1250/kW (compared to \$515–724/kW at present without capture), IGCC about \$1400–2200/kW and PC plants about \$1500–2850/kW (compared to \$1200–1500/kW at present). Figure 9.4 shows projections published in 2005 of the expected cost of electricity in 2020 from various powerplants with and without carbon capture. Shown are the separate contributions of capital, fuel and O&M to the cost of electricity, as well as the assumed capital costs (about €2000/kW for PC with carbon capture and about €1900/kW for IGCC with carbon capture). The cases with carbon capture include the additional cost of transporting CO_2 100km and injecting it into an aquifer. Fuel costs are assumed to be €1.55/GJ for coal and have been adjusted here to €6/GJ for natural gas from €3.35/GJ in the original study. Projected costs of electricity with CCS are about 6 eurocents/kWh from PC, IGCC or natural gas powerplants, but over half of the final cost for natural gas is the fuel cost, the future of which is highly uncertain.

However, current capital costs are nowhere near those assumed in Figure 9.4 or assumed in Table 9.5. The US Department of Energy cancelled its *FutureGen* project in 2008 after the initial cost estimate doubled from about \$3250/kW to \$6500/kW, although the project was revived by the Obama administration. This project aimed to design, construct and operate a 275MW plant that would produce electricity and hydrogen. A proposed 450MW IGCC plant with carbon capture in Saskatchewan was shelved in September 2007 after the estimated cost soared from Cdn\$3778/kW to Cdn\$8444/kW (the high cost due in part to labour shortages associated with uncontrolled tar sands projects in Alberta).

As discussed in Chapter 2 (subsection 2.2.12) with regard to solar energy and in Chapter 3 (subsection 3.13.6) with regard to wind energy, most technologies exhibit an approximately constant fractional decrease in cost with each doubling in cumulative production. Cost typically varies over time as:

$$C(t) = C_o PR^{\left(\frac{\ln R(t)}{\ln 2}\right)} \tag{9.8}$$

where C_o is the initial cost, PR is the progress ratio and $R(t)$ is the ratio of cumulative production at time t to

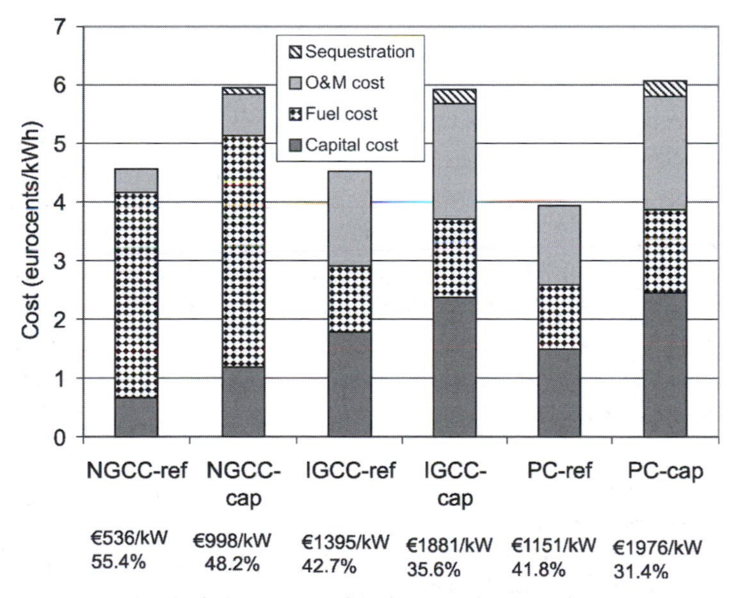

Source: Tzimas and Peteves (2005), except that the fuel component of the final cost of electricity from NGCC has been adjusted to reflect a natural gas cost of €6/GJ rather than €3.35/GJ

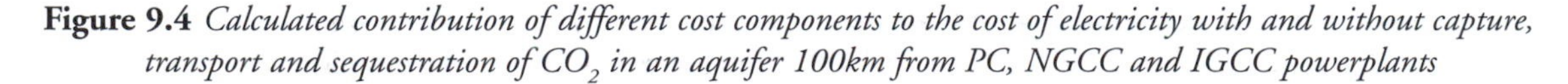

Figure 9.4 *Calculated contribution of different cost components to the cost of electricity with and without capture, transport and sequestration of CO_2 in an aquifer 100km from PC, NGCC and IGCC powerplants*

the cumulative production at the time when the cost was C_o. As shown by Rubin et al (2007), many new technologies initially increase in price with increasing use, due perhaps to the discovery of defects in the original design, the correction of which entails greater cost. Later, however, costs begin to follow a learning curve given by Equation (9.8). This was the case for four technologies studied by Rubin et al (2007) that are related to CO_2 capture. The four technologies and the eventual progress ratios are: flue gas desulphurization, 0.89; selective catalytic reduction, 0.88; gas turbine combined cycle, 0.90; and production of liquefied natural gas, 0.86. The cost of another three related technologies fell into a learning curve without an initial increase; they and their progress ratios are: pulverized coal boilers, 0.95; production of oxygen, 0.90; and steam methane reforming to produce hydrogen, 0.73.

9.3.6 Capturing CO_2 from standalone biomass powerplants

The most efficient method to generate electricity from biomass would be to gasify the biomass and use the gas so produced in combined gas and steam turbines. This technology is referred to as *biomass integrated gasification combined cycle* (BIGCC). Rhodes and Keith (2005) model the capture of CO_2 produced by BIGCC powerplants. Gasification in pure oxygen produces a mixture of CH_4, CO_2, CO and H_2 called *syngas* (just as does gasification of coal) and leaves a char residue (30 per cent of the original carbon) that is combusted to provide the heat for the gasification process. The CO_2 produced at this stage can be captured, but combustion of the CH_4 and CO produces further CO_2 that is not readily captured. Instead, the CO can be reacted with water vapour in the aforementioned water shift reaction (Equation (9.2)) to convert most of the CO to CO_2 (which can be captured) while producing more H_2, but this requires combusting some of the syngas to produce heat to drive the water shift reaction. A flow diagram for the scheme with the water shift reaction is shown in Figure 9.5, and the expected performance of BIGCC without carbon capture, with carbon capture but without the water shift reaction, and with carbon capture and the water shift reaction is summarized in Table 9.6. Cost and performance characteristics reflect possibilities that could be achieved after ten years of aggressive research and development. The GTCC powerplant is assumed to have a gross efficiency of 60 per cent (lower heating value, LHV), and when multiplied by the gasification efficiency (74 per cent), gives an overall net efficiency (after subtracting electrical output used by the operation itself) of 34 per cent (HHV) in the absence of carbon capture. With carbon capture (which is performed using the Selexol process) and compression, the net efficiency drops to 25 per cent (HHV). The rather substantial energy penalty associated with capture and compression (31 per cent

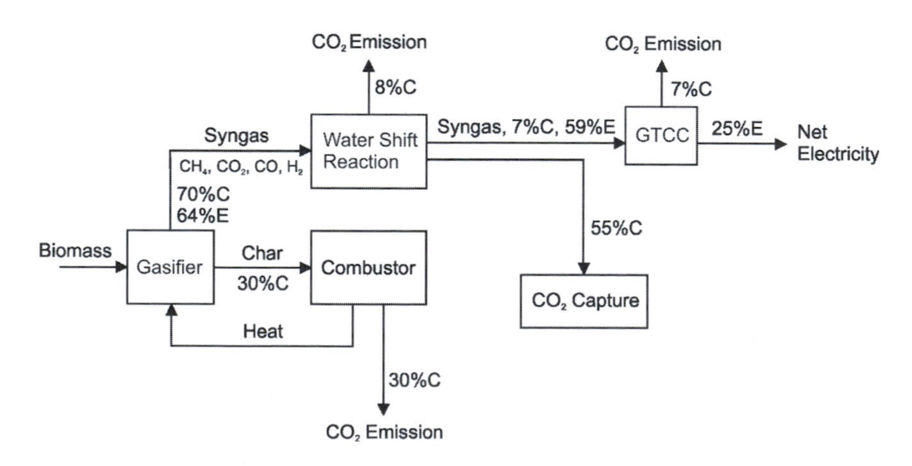

Source: Derived from information in Rhodes and Keith (2005)

Figure 9.5 *Flow diagram for carbon (C) and energy (E) associated with the BIGCC scheme with a water shift reaction*

Table 9.6 *Characteristics of capture of CO_2 from BIGCC powerplants (with or without the water shift reaction) that could be available after ten years of intensive research and development*

Parameter	BIGCC without capture	BIGCC with capture	
		without water shift	with water shift
Capital cost ($/kW)	1250	1730	1980
Net efficiency (HHV basis)(%)	34	28	25
Emission (kgC/kWh)	0	−0.14	−0.20
Carbon capture fraction		0.44	0.55
Effective CO_2 capture fraction		0.32	0.39
Non-fuel O&M ($/kW-yr)	100	131	146
Cost of electricity (cents/kWh)	5.9	8.2	9.3
Cost of CO_2 removed[a] ($/tC)	102	123	135

Note [a] Removed CO_2 for BIGCC.
Source: Estimated by Rhodes and Keith (2005)

extra energy) combined with a small capture fraction (53 per cent) results in a rather small effective CO_2 capture (39 per cent). Because of the large energy penalty with CO_2 capture, about one third more biomass would be needed for the same electricity production as without capture. Given the underlying low efficiency in producing biomass from solar energy (~1 per cent) and land constraints, this option appears to be less attractive than development of wind or solar energy to displace fossil fuels in electricity generation unless biomass streams are already available as a byproduct of other activities.

9.3.7 Capturing CO_2 from pulp mills and from integrated pulp and paper mills with cogeneration of heat and power

Möllersten et al (2004) investigate the performance of biomass-based combined heat and power with carbon capture from standalone pulp mills and from integrated pulp and paper mills using the chemical pulping process. Chemical pulp mills produce a biomass waste called black liquor, which is combusted to produce steam and electricity (referred to as CHP). In existing mills with modern CHP systems, the electrical efficiency is deliberately low (≤15 per cent) so as to provide the required steam. The development of systems with higher electrical output (which implies less steam output) is dependent on the use of new technologies that require less steam (see Volume 1, Chapter 6, subsection 6.9.4). A black liquor gasifier with combined cycle (BLGCC) could achieve an electrical efficiency of 28–31 per cent, but this technology is not yet commercially available. The output from BLGCC would be sufficient to meet the electrical and steam needs of advanced pulp mills, but additional biomass would be needed to meet the energy needs of integrated pulp and paper mills. The additional biomass can produce heat and electricity by first combusting it in a boiler, or can be gasified and used to produce additional electricity and heat through BIGCC (as discussed in subsection 9.3.6).

Möllersten et al (2004) analysed the cases of (1) pulp mills powered by black liquor without CO_2 capture, with CO_2 capture but no water shift reaction, and with CO_2 capture with a water shift reaction (see subsection 9.3.6); and (2) integrated (combined) pulp and paper mills using energy from black liquor plus additional biomass from which energy is extracted using a boiler with no CO_2 capture, or using BIGCC with or without CO_2 capture.

In all cases, energy is extracted from black liquor using BLGCC. CO_2 is captured using selexol and compressed to 80atm. Results are presented in Table 9.7. Note that efficiencies are lower with the shift reaction, as any change in energy from one form to another entails losses. However, the fraction of CO_2 that can be captured increases by a factor of three (from about 30 per cent to 90 per cent). Less electricity is

Table 9.7 *Performance of pulp mills and of integrated pulp and paper mills with and without CO_2 capture*

	Pulp mills			Integrated pulp and paper mills			
	No CO_2 Capture	With CO_2 capture		No CO_2 capture		With CO_2 capture	
		No water shift	With water shift	Boiler for non-BL biomass	BIGCC for non-BL biomass	BIGCC for non-BL biomass	
						No water shift	With water shift
Black liquor input (MW)	338	338	338	338	338	338	338
Woody biomass input (MW)	0	0	0	97	114	114	184
CO_2 capture (per cent)	0	31	90	0	0	33	90
CO_2 capture rate (kg/s)	0	2.6	6.2	0	0	3.7	12.2
CO_2 absorption power consumption (MW)	0	1.9	4.0	0	0	2.8	6.1
CO_2 compressor (MW) (to 80atm)	0	4.2	12.5	0	0	6.1	20.2
Air separation power consumption (MW)	4.5	4.5	4.5	4.5	6.2	6.2	6.8
Other internal power consumption (MW)	10.1	10.1	10.1	13.1	13.6	13.6	15.7
Gross electrical output (MW)	120.2	124.5	102.4	131.5	135.4	135.4	162.7
Net electrical output (MW)	105.6	93.8	71.3	114.0	115.5	106.7	113.9
Electrical surplus (MW)	66.5	54.6	32.1	40.0	41.8	32.9	40.1
Steam output (MW)	238.7	238.7	238.7	375.2	375.2	375.2	375.2
Electrical efficiency (LHV basis) (%)	31	28	21	26	26	24	22
Overall efficiency (LHV basis) (%)	75	72	65	80	77	76	67
Avoided CO_2 emission (kgC/kWh sacrificed)	0.00	0.21	0.41	0.00	0.00	0.09	0.18

Note: BLGCC (black liquor gasification combined cycle) is used in all cases to produce steam and electricity from black liquor (BL), while energy is extracted from the additional biomass (mostly bark) available from pulp and paper mills using either a boiler or BIGCC.
Source: Möllersten et al (2004)

produced with greater CO_2 capture, so there is a smaller surplus that could otherwise potentially displace coal-fired electricity elsewhere. For the integrated pulp and paper mill, the electricity surplus with CO_2 capture falls by less, but extra biomass is used. The relevant parameter is the avoided CO_2 emission per kWh of sacrificed electricity production (including the electricity that could hypothetically be produced using the extra required biomass in a dedicated biomass powerplant). If this parameter is larger than the emission associated with the power source that would be used to replace the sacrificed electricity production, then capture of CO_2 from the pulp mill gives a larger net reduction in CO_2 emission than not capturing CO_2 and using any extra biomass to produce electricity (assuming that this is feasible in practice). The avoided C emission per kWh of sacrificed electricity is shown in the last row of Table 9.7, assuming that the extra biomass could otherwise be used to generate electricity at 40 per cent efficiency (as indicated in Chapter 4, subsection 4.4.3 for BIGCC). The avoided emission is generally less than the emission incurred if advanced coal (45 per cent efficiency) has to replace the sacrificed electricity (0.2kgC/kWh) but greater than if advanced natural gas (60 per cent efficiency) has to replace the sacrificed electricity (0.09kgC/kWh). Thus, CO_2 capture is not a good option if the sacrificed electricity would be generated from coal, but it is a good option if natural gas (or C-free energy sources) would otherwise be used.

Hektor and Berntsson (2007) computed the performance of a standalone chemical pulp mill in which energy is extracted from black liquor using a boiler and steam turbine, rather than using BLGCC. For the same black liquor input (338MW) as in the analysis discussed above, the heat and power requirements of the mill are met while providing an additional 14MW of electricity for export to the grid, while for the case with 90 per cent CO_2 capture by chemical absorption, there is a small net electricity requirement of 1MW from outside the mill. The surplus without capture (14MW) is substantially less than the surplus using BLGCC with 90 per cent capture shown in Table 9.7 for the pulp mill (32MW), but uses conventional technology. The CO_2 capture rate is about five times that shown in Table 9.7 for the pulp mill with BLGCC and water shift (about 27kg/s versus 6.2kg/s). Combined with the net electricity requirement of 15MW, the avoided C emission is an impressive 1.8kgC/kWh (many times the emission entailed even if inefficient coal replaces the lost electricity production). The proposed scheme, like that of Möllersten et al (2004), still requires substantial development work, in this case related to the integration of various process heat flows.

9.3.8 Capturing CO_2 during the production of hydrogen and other fuels from coal and natural gas

In the post-fossil fuel era, hydrogen could be produced by splitting water using renewable energy sources, and then used in all applications where fossil fuels are used today (as discussed in Chapter 10). Today, hydrogen is not used as a fuel (except for the space programme), but it is used as a feedstock in the chemical industry. It is generally produced by steam reforming of methane (Equations (9.1)–(9.3)), but it can also be produced by gasification of coal or biomass (Equations (9.4)–(9.7)). A typical efficiency (hydrogen energy output over fuel energy input) is 75 per cent using natural gas and 50 per cent using coal. Although not sustainable in the long run, production of hydrogen from fossil fuels for energy uses could serve as a step in the transition to a hydrogen economy (in which only renewable energy sources are used to make hydrogen). If the CO_2 produced as a byproduct of hydrogen production were captured and sequestered, emissions to the atmosphere could be eliminated even during the transitional stage to a renewably based hydrogen system. Similarly, Fischer-Tropsch (FT) liquids (gasoline, diesel fuel), methanol, ethanol and dimethyl ether (DME) could be produced from carbonaceous fuels with sequestration of the CO_2 produced during the production of these fuels (although the CO_2 emission associated with their end use could not be sequestered). FT liquids produced in this way would be relatively clean, as they would not contain sulphur or nitrogen and would have a very low content of aromatic compounds (Yamashita and Barreto, 2005).

The characteristics of these CO_2 capture systems are summarized in Table 9.8. Energy penalties are generally small using coal (2–8 per cent) but larger using natural gas (10–22 per cent). The fraction of CO_2 captured is around 90 per cent for all fuels produced except methanol, where the fraction captured is only 58–62 per cent. The extra capital cost for carbon capture is largest in relative terms for hydrogen production from natural gas, while the capture of CO_2 associated with the production of methanol or DME plus electricity from coal is projected in some studies to result in smaller total capital cost than without capture.

Figure 9.6 shows the mass flows for production of H_2 from natural gas. A two-step process is assumed: steam methane reforming ($CH_4+H_2O \rightarrow CO+3H_2$) followed by the water shift reaction ($CO+H_2O \rightarrow CO_2+H_2$). About 30 per cent of the input fuel is burned to produce steam, and the CO_2 so produced is mixed with exhaust gases and vented to the atmosphere. The remaining input fuel produces a stream of CO_2 mixed with H_2, and is easily separated from the H_2 using a pressure swing absorption unit. Thus, 70 per cent of the CO_2 released during the production of H_2 from natural gas can be captured at low cost and with no energy penalty (it needs to be separated from the H_2 in any case). Removal of a greater fraction of CO_2 would entail some energy penalty. Energy would be required for the subsequent compression of CO_2 and transport to disposal sites.

Figure 9.7 compares the estimated cost of producing hydrogen from natural gas and from coal, with and without capture and sequestration of CO_2. Capture of CO_2 increases the cost of hydrogen by about €1/GJ according to the analysis shown in Figure 9.7: from €10/GJ to €11/GJ using natural gas at €6/GJ, and from €8/GJ to €9/GJ using coal at €1.55/GJ. By comparison, oil at \$100/barrel is equivalent to about \$17/GJ or ~€14/GJ (one barrel of oil contains 5.75GJ).

9.3.9 Capturing CO_2 produced during the production of nitrogen fertilizer

Currently, nitrogen fertilizers are produced by reacting H_2 (produced through steam reforming of natural gas or gasification of coal) with N_2, to produce ammonia (NH_3), which can either be directly used as a N fertilizer or reacted to produce ammonium nitrate

Table 9.8 *Summary of the estimated eventual cost and performance of CO_2 capture associated with the production of hydrogen, hydrogen plus electricity, or other fuels plus electricity from new facilities*

	Energy products produced					
	Hydrogen	**Hydrogen**	**Hydrogen + electricity**	**FT liquids + electricity**	**Methanol + electricity**	**DME + electricity**
			Reference plant (without capture)			
Feedstock	Natural gas	Coal	Coal	Coal	Coal	Coal
Efficiency (LHV basis)	74.5	62.9	56–62	60	46	48–50
Carbon released (kg CO_2/GJ products)	78–81	168	145–174	163	203	185–198
			Capture plant			
Capture technology	Amine or MEA scrubber (autothermal reforming)	Not reported (gasifier)	Selexol	Amine scrubber	Selexol	Selexol, rectisol
Efficiency (LHV basis)	61–68 (78.3)	60.2 (70)	52–60	56	44	43–49
Carbon released (kg CO_2/GJ products)	14–23 (1.46)	28 (19.5)	7–25	72	101–109	43–145
			Energy costs			
Reference plant electricity (cents/kWh)	4.5	4.5	3.1–4.6	3.6	4.3	4.4
Capture plant electricity (cents/kWh)	4.5	4.5	3.1–6.2	5.4	4.3	5.8
Reference plant fuel ($/GJ)	8.6–10.0	8.0	6.5–7.3	5.6	9.1	7.4–8.7
Capture plant fuel ($/GJ)	10.1–13.3	8.6	7.5–8.3	5.4	8.4–10.4	6.7–8.1
			Impact of CO_2 capture			
Per cent extra fuel use	10–22	4.5	4–8	6.5	3.6–4.0	2–15
Per cent extra capital cost	38–54	2.2	−2–17	4	−10–3	−10 to 2
Per cent CO_2 captured	90 (95)	90 (90)	87–95	91	58–63	32–97
Per cent effective CO_2 captured	72–83	83	86–96	56	46–50	27–97
Cost of CO_2 captured (US$/t$CO_2$)	21–39	4.1	2.2–8.7		−6–12	−18 to 12
Cost of CO_2 avoided (US$/t$CO_2$)	24–56	4.4	2.3–9.2		−7–13	−18 to 13

Note: Costs of CO_2 captured do not include costs of transport and storage. LHV/HHV = 0.96 for coal, 0.90 for natural gas. Cost of electricity is computed assuming fuel costs of $1.0–1.5/GJ for coal, $2.8–4.4/GJ for natural gas.
Source: Condensed from Tables 3.11 and 3.14 of Thambimuthu et al (2005)

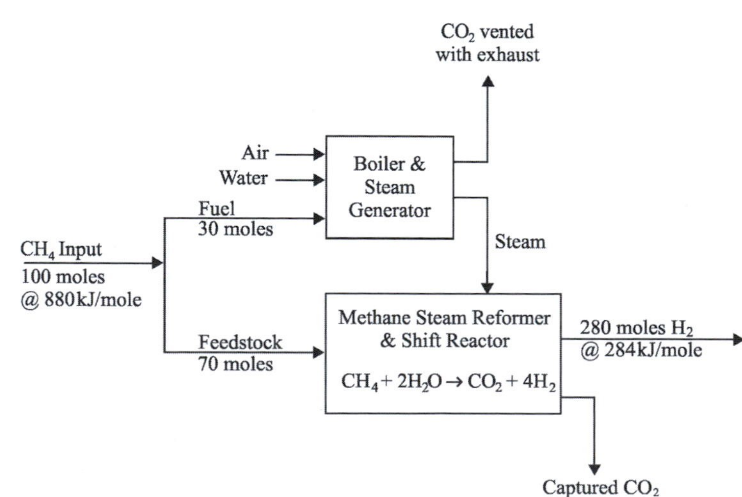

Source: Derived from text

Figure 9.6 *Mass and energy flow (HHV basis) associated with production of hydrogen from natural gas*

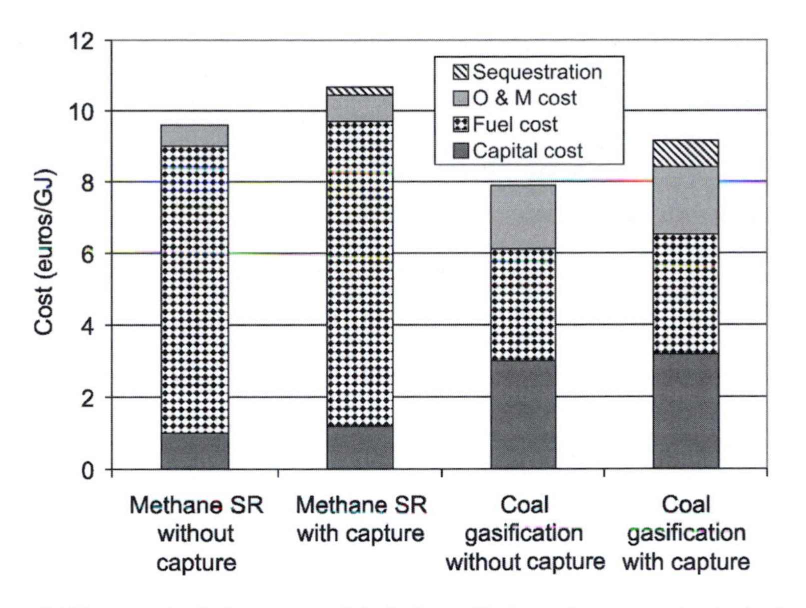

Source: Tzimas and Peteves (2005), except that fuel component of the final cost of hydrogen from steam reforming has been adjusted to reflect a natural gas cost of €6/GJ rather than €3.35/GJ

Figure 9.7 *Calculated contribution of different cost components to the cost of hydrogen produced by steam reforming (SR) of natural gas and gasification of coal with and without capture, transport and sequestration of CO_2 in an aquifer 100km from hydrogen production plants*

(NH_4NO_3) or ammonium bicarbonate (NH_4HCO_3). The net reactions are:

$$3CH_4 + 4N_2 + 2H_2O + 8O_2 \rightarrow 4NH_4NO_3 + 3CO_2 \uparrow \qquad (9.9)$$

and:

$$3CH_4 + 4N_2 + 14H_2O + 5CO_2 \rightarrow 8NH_4HCO_3 \downarrow \qquad (9.10)$$

respectively. The production of ammonium nitrate releases CO_2 through the chemical reaction, as well as through the combustion of fossil fuel to provide the heat needed to drive the reaction. The production of ammonium bicarbonate is a net sink for CO_2 but still consumes natural gas.

Alternatively, flue gases from combustion of fossil fuels or biomass can be used as a source of carbon for the production of ammonium bicarbonate. The net reaction is:

$$2CO_2 + N_2 + 3H_2 + 2H_2O \rightarrow 2NH_4HCO_3 \downarrow \qquad (9.11)$$

This reaction releases heat and so should be able to occur spontaneously at ambient temperature and pressure. With a suitable catalyst and heat from flue gases, it should be able to occur very rapidly. According to Lee and Li (2003), 90 per cent of flue gas CO_2 could be removed in this way. The required hydrogen could be produced electrolytically using renewable energy sources (see Chapter 10, subsection 10.3.3). Through the use of nickel and platinum catalysts, NO_x and SO_x in the flue gases can be converted to HNO_3 and H_2SO_4 and then would react with ammonia to produce the additional fertilizers NH_4NO_3 and $(NH_4)_2SO_4$.

When ammonium nitrate is added as a soil fertilizer, the nitrate frequently is lost as runoff, contaminating groundwater and surface water. Furthermore, NO_3^- reacts to form N_2O, a GHG that is released to the atmosphere. Use of ammonium bicarbonate fertilizer, by contrast, releases HCO_3^- that percolates deep into the ground and reacts with Ca^{2+}, Mg^{2+} and Na^+ (if present) to form stable carbonates, essentially permanently locking up the CO_2 removed from flue gases through reaction (9.11).

An alternative scheme would use volatile gases given off from the pyrolysis of solid biomass to produce H_2 which would then be used to make ammonia. The ammonia would be combined with the residual solid biomass (char) and flue-gas CO_2 to produce a nitrogen-enriched char (containing ammonium bicarbonate) that could be used as a soil fertilizer. The heat to drive the pyrolysis reaction would be provided by the reaction to produce ammonium bicarbonate, while the heat required to make steam for the water shift reaction would come from the powerplant exhaust.

9.3.10 Capturing CO_2 from ambient air

Zeman (2007) has outlined a process for the direct removal of CO_2 from ambient air. CO_2 would be absorbed by an alkaline NaOH solution to produce dissolved sodium carbonate, a strongly exothermic (heat-releasing) reaction. The carbonate ion would be removed from the solution by reaction with calcium hydroxide ($Ca(OH)_2$), causing the precipitation of calcite ($CaCO_3$), a reaction that is mildly exothermic. The calcite would then be decomposed to CaO (lime) and a concentrated stream of gaseous CO_2 through the addition of heat. Hydration of lime back to $Ca(OH)_2$ would complete the cycle. The reaction cycle is thus:

$$2NaOH_{(aq)} + CO_{2(g)} \rightarrow Na_2CO_{3(aq)} + H_2O_{(l)} \qquad \Delta H° = -109.4 \text{ kJ/mol} \quad (9.12)$$

$$Na_2CO_{3(aq)} + Ca(OH)_{2(s)} \rightarrow 2NaOH_{(aq)} + CaCO_{3(s)} \qquad \Delta H° = -5.3 \text{ kJ/mol} \quad (9.13)$$

$$\mathrm{CaCO_{3(s)} \rightarrow CaO_{(s)} + CO_{2(g)}} \qquad \Delta H° = 179.2 \text{ kJ/mol} \quad (9.14)$$

and:

$$\mathrm{CaO_{(s)} + H_2O_{(l)} \rightarrow Ca(OH)_{2(s)}} \qquad \Delta H° = -64.5 \text{ kJ/mol} \quad (9.15)$$

Reaction (9.14) is the calcination reaction, which is a key step in the production of cement (see Volume 1, Chapter 6, section 6.7.1). Here, it would be performed in a lime kiln fired with pure oxygen rather than air, so as to produce a relatively pure stream of CO_2. The reactions' enthalpies add up to zero, as the enthalpy changes are all given for the same (standard) temperatures and pressures. In reality, CO_2 would enter and exit the cycle at different temperatures and pressures. For CO_2 input at 380ppmv and output at 1atm pressure, both at 300K, the minimum energy requirement is 19.6kJ/mol. The actual energy requirement is much larger because heat released during the exothermic steps ($\Delta H° < 0$) cannot be fully captured and used by the endothermic step. As well, about 3 million cubic metres of air would need to circulate through the system per tonne of CO_2 removed, and this would require electrical energy for fans. Zeman (2007) calculates energy requirements of 214kJ thermal and 123kJ electrical per mole of CO_2 captured, broken down as shown in Table 9.9. If the overall energy requirement of 337kJ/mol were supplied through cogeneration of heat and electricity using coal with an overall efficiency of 90 per cent, then 32kgC would be captured per GJ of coal used, which is only modestly larger than the typical emission of 25kgC per GJ of coal combusted. If the required air movement could be provided passively, as suggested by Zeman (2007), the overall energy requirement would be 257kJ/mol, which translates into a capture of 42kgC/GJ. The net removal would then be 17kgC/GJ of coal used.

Table 9.9 *Energy requirements for air capture using the NaOH cycle*

Step	Energy requirement (kJ/mol)	Energy requirement (% of total)	Type of energy
Airflow	88	19.9	Electrical
Cryogenic separation of oxygen from air	16	3.6	Electrical
Drying of $CaCO_3$ precipitate	63	14.3	Thermal
Calcination	256	57.9	Thermal
Compression of CO_2	19	4.3	Electrical
Total	442	100.0	
Available heat of hydration	105		Thermal
Net energy requirement	337		

Source: Zeman (2007)

A more interesting option would be to use concentrated solar thermal energy (discussed in Chapter 2, section 2.3) to produce both the electricity and heat required for the process. Solar thermal energy, once collected, can be used to produce electricity with an efficiency of about 30 per cent, so there would be about two units of waste heat available per unit of electricity generated. This matches the ratio of required heat (214kJ/mol) to required electricity (123kJ/mol) given in Table 9.9. Thus, 1GJ of solar thermal energy would provide 0.3GJ of electricity along with all of the required heat, and this would allow the capture of about 29kgC. Conversely, using the solar thermal energy to generate electricity at 30 per cent efficiency that in turn displaces coal-based electricity that would be generated at 40 per cent efficiency would avoid an emission of only 19kgC. Thus, if the air capture scheme can be made to work, powering it with electricity and heat from concentrating solar thermal energy would appear to be worthwhile from a CO_2 emission point of view even while coal is still part of the electricity supply mix.

9.4 Compression or liquefaction and transport

Captured CO_2 would be compressed for transport by pipeline, something that is already done for industrial uses of CO_2. The longest pipeline today for carrying CO_2 is over 800km in length (it provides CO_2 for enhanced oil recovery). A 1m-diameter pipeline would be sufficient for transporting the CO_2 from a 3GW powerplant operating at full output (IEA, 2001c). A major break in a CO_2 pipeline, particularly in low-lying areas, would have serious consequences, so careful routing of the pipelines would be necessary.

Williams (1998) indicates that compression from 1.3 to 110 atmospheres pressure requires about 300–400kWh/tC (1.08–1.44GJ/tC), depending on the size of the operation, and adds $15–20/tC to the cost of sequestration for electricity at 5 cents/kWh. Compressors add capital costs of around $8/tC, while compressors plus a 250km pipeline add $20–55/tC to the cost of sequestration. The compressed CO_2 needs to be dried (reducing the water content from as high as 1000ppm down to 10ppm) and cooled (to 10°C) prior to transport by pipeline. Drying is done with a desiccant and requires 0.029GJ/tC as thermal energy, while cooling would require 0.029GJ/tC as electrical energy using a chiller with a COP of five (Farla et al, 1995). Haugen and Eide (1996) indicate an electrical energy requirement of about 0.08GJ/tC/1000km for large-diameter (0.864m) pipelines. Thus, the energy requirements for drying, cooling and transporting CO_2 are small compared to the energy required for compression. The cost of transporting natural gas is less than that of transporting H_2, which in turn is less than the cost of transporting CO_2 (Blok et al, 1997). This will create a preference for locating H_2 production facilities with CO_2 capture close to the CO_2 sequestration sites.

If, instead of compression to 110 atmospheres, CO_2 is liquefied for injection into the ocean, then the energy requirement is about 400–440kWh/tC.[1] This is comparable to the energy required for compression, but reduces the volume of CO_2 by a factor of 600 (compared to a factor of 110 if pure CO_2 gas is compressed to 110atm pressure). Additional energy would be required in order to transport the liquefied CO_2 to the oceanic injection sites. For natural gas powerplants fuelled by LNG (liquefied natural gas, at 110K), Zhang and Lior (2006) propose using the coldness of the LNG to condense CO_2 to a liquid state under pressure while vaporizing the LNG for use by the gas turbine.

Doctor et al (2005) provide information on safety and legal issues associated with transport of CO_2 by ships and by pipeline, as well as costs. Figure 9.8 shows the variation in the cost of transporting CO_2 by offshore and onshore pipelines, as a function of the flow rate. Transport costs by pipeline vary in direct proportion to the distance transported, but decline exponentially with the mass flow rate. For a flow rate of 6MtCO_2/yr, costs are $2–3/t$CO_2$/250km for onshore pipelines and $3–4/t$CO_2$/250km for offshore pipelines. In the case of transport by ship, costs increase at a decreasing rate as the transport distance increases, as shown in Figure 9.9. This is because loading and unloading costs are fixed and only the relatively small operating component of the cost increases with increasing distance. Similarly, CO_2 emissions associated with energy use by ships are initially relatively large (2 per cent of the transported CO_2 for a distance of 200km) but are relatively small for long-distance transport (18 per cent for a distance of 12,000km).

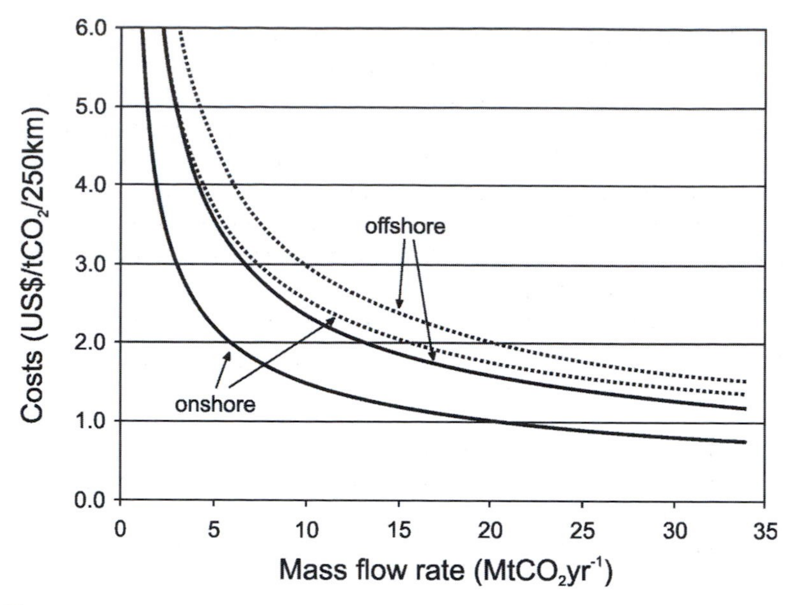

Source: Doctor et al (2005)

Figure 9.8 *Cost of transport of CO_2 by offshore and onshore pipeline as a function of the mass flow rate*

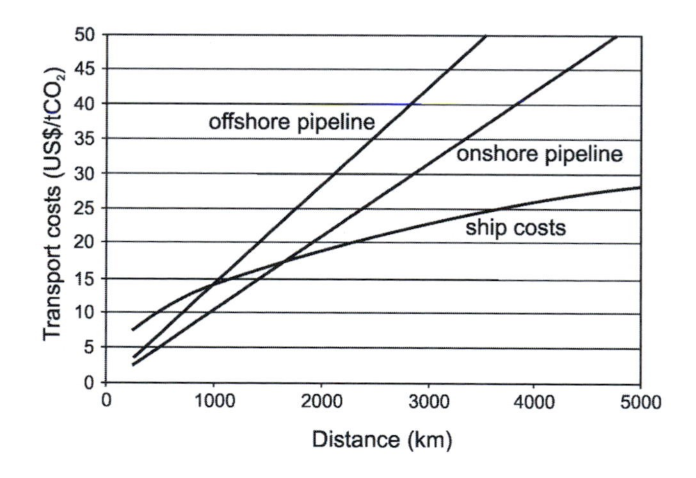

Source: Doctor et al (2005)

Figure 9.9 *Cost of transport of CO_2 by offshore and onshore pipeline and by ship for a mass flow rate of 6.5MtCO_2/yr as a function of the distance transported*

9.5 Compounded energy penalties for production of electricity and hydrogen

Table 9.10 gives the energy penalties associated with CO_2 capture, compression and transport. For CO_2 capture from electric powerplants, energy penalties estimated based on representative efficiencies of electricity generation with and without carbon sequestration from Tables S9.1–S9.3 are used. Electrical requirements for compression are assumed to be 300kWh/tC, and are converted to (GJ electricity)/(GJ of input fuel) using the carbon capture rate (kgC/GJ of input fuel) given in Table 9.10. Division by the efficiency for generating electricity with capture of CO_2 gives (GJ primary energy)/(GJ primary energy), that is, the fraction of the input fuel (primary energy) that is needed to meet the compression energy requirement. For hydrogen production, the energy penalty associated with CO_2 capture is zero because CO_2 needs to be separated in any case. Division by the efficiency in generating electricity from hydrogen elsewhere, and by the efficiency in generating hydrogen from input fuel, gives the fraction of the hydrogen production used for compression. Of the extra x units of fuel needed to supply the electricity for compression of CO_2, a fraction x is needed to compress the extra CO_2 produced by the extra x units, so another x^2 units have to be added, and so on. The resulting fuel requirement as a fraction of the fuel requirement without compression is the sum of the infinite series $1+x^2+x^3$ ..., namely, $x/(1-x)$. If y is the fractional increase in fuel use due to the loss in electricity generation efficiency, the overall fuel requirement is multiplied by $(1+x/(1-x))(1+y)$.

As seen from Table 9.10, the overall increase in fuel use is 15–30 per cent for electricity generation but only 3–7 per cent for hydrogen production. The reasons for the much smaller energy penalty in converting coal to H_2 than to electricity are (1) there is no energy cost in capturing CO_2 associated with hydrogen production, and (2) less primary energy is used to produce a unit of energy as hydrogen than as electricity.

9.6 Methods for and costs of long-term storage of CO_2

Carbon dioxide could be stored for varying lengths of time in subsurface rock reservoirs, in depleted oil and gas fields, in coal beds, in sub-seabed sediments and in the deep ocean. The *Carbon Sequestration Atlas of the United States and Canada* contains maps showing potential regions in North America were CO_2 could be stored in oil and gas reservoirs, unmineable coal seams and deep saline aquifers (NETL, 2007). Geological carbon sequestration below the land surface is critically reviewed in Holloway (2005), Haszeldine (2006) and Tsang et al (2008), while some of the key issues are outlined here. Carbon dioxide can also be temporarily

Table 9.10 *Compounded energy penalty for capture and subsequent sequestration of carbon produced during the generation of electricity using different fuels and technologies, or from the production of hydrogen using different feedstock*

	Electricity production			Hydrogen production		
	NGCC	PC	IGCC	NG	Coal	Biomass
	CO_2 emission and capture					
Fuel emission factor (kgC/GJ)	14	25	25	14	25	18
Hydrogen production efficiency				0.874	0.718	0.692
Electricity generation efficiency without capture	0.601	0.424	0.478	0.600	0.500	0.500
Electricity generation efficiency with capture	0.541	0.361	0.435			
Energy penalty for capture[a]	0.111	0.175	0.099			
Capture rate (kgC/GJ input fuel)	12.6	22.5	22.5	12.6	22.5	16.2
	Energy use for compression and transport					
Electricity (GJ/tC)	1.454	1.454	1.454	1.455	1.125	1.479
Electricity (GJ/GJ input fuel)	0.018	0.032	0.032	0.018	0.032	0.023
Primary energy (GJ/GJ fuel)	0.034	0.090	0.074	0.030	0.065	0.047
Increase in fuel use (fraction)[b]	0.035	0.099	0.080	0.031	0.069	0.049
	Compounded energy use for capture, compression and transport					
As a fraction of energy use without sequestration	0.149	0.290	0.187	0.031	0.069	0.049

Note: PC = pulverized coal powerplant, NGCC = natural gas combined cycle powerplant, IGCC = integrated gasification combined cycle powerplant (using coal). [a] Additional energy used for capture as a fraction of energy used without capture. [b] Equal to $x/(1-x)$, where x is the primary energy requirement given in the line above.

stored in a variety of chemical products. This is briefly discussed here, as the use of captured CO_2 in such products would displace CO_2 produced from the dedicated combustion of natural gas or other fossil fuels.

9.6.1 Storage in deep saline aquifers on land and below the ocean bed

Carbon dioxide can be stored in a subsurface rock reservoir if the reservoir is overlain by an impermeable layer called a caprock. Sedimentary basins that have undergone only minor tectonic deformation and are at least 1000m thick are ideal. The pore spaces of rocks very close to the ground surface are normally filled with a mixture of air and fresh water. Below this the pore spaces will be filled with fresh water, which percolates down from the ground surface. At very great depth or beneath the sea, however, rock pore spaces are filled with saline water. If a well is drilled into a reservoir rock and CO_2 is injected, the CO_2 will be stored in one of three ways:

1 as a gas, displacing some of the pore water and retained in the pores by capillary forces;
2 dissolved into the pore water;
3 as a carbonate, produced through chemical reaction between the CO_2 and the host rock.

The amount of CO_2 that can dissolve into pore water corresponds to about 7 per cent of the pore volume, but the rate at which injected CO_2 dissolves into pore water is expected to be quite slow – requiring thousands of years (Holloway, 2005). Most of the CO_2 will migrate upward, under pressure, until it reaches an impermeable layer, but in some cases all of the CO_2 can be trapped in the storage layer without reaching the caprock. Thick slate layers are essentially impermeable, whereas shale and mudstone permit extremely slow migration. Pre-existing fractures in the caprock or imperfectly sealed drill holes would permit CO_2 to leak. It is important not to induce fracturing of the caprock by injecting too much CO_2 at too high a pressure. Some CO_2 may migrate laterally out of the reservoir if the caprock is not horizontal or if there are horizontal pressure gradients in the reservoir. If a small amount of CO_2 is injected into a large reservoir, the combination of many small traps at the base of the caprock and dissolution of CO_2 into the pore fluids may mean that none of the CO_2 will ever escape, even if there is not a continuous large-scale trap structure.

Both pressure and temperature increase with increasing depth. Whether CO_2 exists as a gas, solid or liquid depends on the pressure and temperature, as shown in Figure 9.2. Above a temperature of 31.1°C and a pressure of 72.8 atmospheres (which typically occurs near a depth of 700m), CO_2 occurs as a supercritical fluid: it is still compressible, like a gas but has liquid-like densities (around 600–700kg/m^3). The density increases sharply as the CO_2 approaches a supercritical state, and this greatly reduces the storage volume required for a given mass of CO_2. Even as a supercritical fluid, however, the density of CO_2 is still less than that of water (1000kg/m^3), so it is more buoyant than water (but comparable in buoyancy to that of crude oil) and will tend to rise.

For aquifers in carbonate rocks, the presence of CO_2 in the pore fluids will promote dissolution of the rock through the reaction:

$$CaCO_3 + CO_2 + H_2O \rightarrow 2HCO_3^- + Ca^{2+} \qquad (9.16)$$

This, however, has a negligible effect on the storage of CO_2 by the aquifer. If the host rock is a silicate rock, the CO_2 uptake will be enhanced by up to a factor of two through dissolution of the silicate rock and precipitation of solid carbonates (this mimics what happens through the weathering of silicate rocks on land, delivery of the solute to the oceans by rivers and the eventual precipitation of carbonate sediments and burial in the ocean floor). Maximum CO_2 uptake would occur in a calcium-magnesium-silicate rock, with the reaction:

$$\begin{aligned} &3MgCa(SiO_3)_2 + 3CO_2 + 2H_2O \\ &\rightarrow H_4Mg_3Si_2O_9 + 3CaCO_3 + 4SiO_2 \end{aligned} \qquad (9.17)$$

The precipitation of $CaCO_3$ would reduce the permeability of the aquifer, thereby possibly creating a permanent trap for the injected CO_2 in cases where none existed before (IEA, 2001b). Aquifers rich in calcium and magnesium silicates would therefore be preferred for CO_2 sequestration. Basalts are particularly good for absorbing injected CO_2 through the chemical reaction to form carbonates, and there is interest in both the Columbia River flood basalts in the north-western US (with a potential storage of more than 50GtCO_2) and the Deccan Traps in India (with a potential storage of 150GtCO_2) (Jayaraman, 2007). The CO_2 would be injected into the ground below the basalts, so that it would react with the basalt, forming a cap as it rose into the basalt layers.

André et al (2007) performed detailed computer simulations of the fate of CO_2 injected into the carbonate rocks of the Paris Basin. Assuming the CO_2 to be injected in a supercritical state, the CO_2 acts as a piston, pushing water laterally away from the injection well. As the CO_2 dissolves into the aquifer water at the interface between the supercritical CO_2 and water, the pH of the water initially drops to 3.6, causing dissolution of calcite (which constitutes 70 per cent of the aquifer rock) sufficient to increase the porosity by up to 7 per cent of the original porosity by the end of a ten-year injection period (some dissolution continues after this time). No chemical reactions occur between the rocks and supercritical CO_2. Residual water remaining in the pores after the supercritical CO_2 has passed through slowly evaporates. Complete drying occurs near the well region, leading to precipitation of salts such as dolomite and anhydrite, leading to a decrease in porosity by about 10 per cent of the original porosity in the first 2m around the injection well. This would cause only a minor inhibition of further injection of CO_2 into the aquifer.

The first commercial-scale project dedicated to the geological storage of CO_2 in a saline rock formation is the Sleipner Project, about 250km off the coast of Norway in the North Sea. The Sleipner West gas field contains CO_2-rich natural gas (4–9.5 per cent CO_2). Sales-quality natural gas should have no more than 2.5 per cent CO_2, so the excess CO_2 must be removed. This is done next to the production rig. Rather than venting the excess CO_2 (272,000tC/yr, or 3 per cent of Norway's total emissions) to the atmosphere, it is injected into a 150–200m thick sandstone layer at a depth of 800–1000m below the seabed. The sandstone is overlain by more than 100m of shale. CO_2 injection began in 1996, at a cost of about US$55/tC (Holloway, 2001). Norway had a carbon tax of about $180/tC beginning in 1991, which is avoided if CO_2 is not emitted (this tax was lowered to about $140/tC on 1 January 2000) (Herzog, 2001).

Other projects involving storage in deep aquifers are underway in Japan and US, and more are planned for Australia, Germany and Norway (summarized in Benson et al, 2005). A major project that has been discussed would involve capturing more than 27Mt of C per year from the Natuna natural gas field in the South China Sea beginning in 2010 (Williams et al, 2000). The reservoir gas is 71 per cent CO_2, and the potentially sequestered CO_2 is equivalent to almost 0.5 per cent of present total global fossil fuel emissions.

9.6.2 Storage in depleted oil and gas fields for enhanced recovery of oil and natural gas

Carbon dioxide is often injected into oil fields in order to force more oil out of the field than would otherwise be possible. This is referred to as enhanced oil recovery (EOR). In the US, 12 million tonnes of carbon were injected at 67 commercial EOR projects in 1998 (Herzog, 2001). Since oil (or natural gas, which often occurs along with oil) has been stored in the reservoir for millions of years, the reservoir is a proven long-term storage medium. The maximum amount of CO_2 that can be stored in this way is 1tC for every 2.8tC extracted as oil (Holloway, 2001). This is equivalent to reducing the net emission for every tonne of oil extracted to 64 per cent of the original emission, not counting the energy used in collecting, transporting and injecting the CO_2 into the reservoir. However, to the extent that it permits extraction of oil that could not otherwise be extracted, use of CO_2 in EOR should certainly not count as an emission reduction. An EOR project in Saskatchewan (Canada) is using CO_2 from a coal gasification plant in North Dakota (US), with the expectation of sequestering 5 million tC over the lifespan of the project (Hattenbach et al, 1999). The global storage potential through EOR is estimated to be about 34GtC – about four years worth of global fossil fuel emissions at the present global emission rate of 8GtC/yr. Carbon sequestration through enhanced oil recovery is economically attractive because the extra oil production can largely pay for the cost of sequestering CO_2, but the storage capacity is limited.

In many cases of enhanced oil recovery using CO_2, half to two thirds of the injected CO_2 returns with the extracted oil after a few years, and is usually separated and re-injected (this would be obligatory for sequestration of CO_2). At this point, a smaller stream of fresh CO_2 is needed, so alternative sites for CO_2 injection will need to become available as recycling ramps up.

Carbon dioxide can also be injected into natural gas fields to increase the amount of gas that can be extracted (Blok et al, 1997). Natural gas fields have initial pressures of up to 400 atmospheres, but this pressure decreases as gas is extracted. By the time the pressure has dropped to 20–50 atmospheres there is insufficient pressure to extract more gas. Injection of CO_2 into the depleted gas field raises the pressure, allowing more gas to be extracted. There would be three phases to the exploitation of a natural gas field: (1) initial extraction without CO_2 injection; (2) simultaneous gas extraction and CO_2 injection, until contamination of the remaining natural gas with CO_2 becomes too large; and (3) continued injection of CO_2. The total amount of natural gas that can be extracted from a field increases from about 90 per cent (without CO_2 injection) to 93–96 per cent (with CO_2 injection), depending on the extent of permeable layers within the field and the associated CO_2 contamination. Natural gas from the Krechba Field at In Salah in Algeria contains 10 per cent CO_2 that, like the CO_2 in North Sea gas, must be removed in order for the gas to be marketed. At In Salah, CO_2 removed from the natural gas has been re-injected into a sandstone reservoir at a depth of 800m since 2004 (Benson et al, 2005). The project is expected to store 4.6MtC over its life.

Another option is to store CO_2 in depleted oil and gas fields. The storage potential is larger (200–500GtC) but there are no secondary economic benefits. Costs might range from \$50–100/tC. Since the major oil fields (such as in the Middle East) are not close to where most of the fossil fuels are burned, realization of the full potential of depleted oil fields will require transporting some of the captured CO_2 very long distances – something that would increase costs.

9.6.3 Storage in coal beds

The process of forming coal from peat (burial, compression and heating) drives off volatile matter that forms methane, some of which is retained within the coal bed itself. This is referred to as *coal bed methane* (CBM). Coal contains fractures (cleats) that impart some permeability to the coal. Between cleats, coal has many micropores. Methane molecules are adsorbed onto the surfaces of micropores within the coal bed, and the gas density can exceed that in a conventional sandstone natural gas reservoir of equal thickness. Up to 20m^3 of methane can be extracted per tonne of bituminous coal, with a total heating value (at 38MJ/m^3 of methane) of 0.76GJ/tonne of coal, compared to about 32GJ/tonne for the coal alone. Methane has been extracted from in situ coal beds; this is done by drilling a well into the bed and pumping out water to lower the pressure within the bed. CO_2 has a greater affinity for coal than methane, so simultaneous injection of CO_2 into the coal bed has the double advantage of storing the CO_2 and facilitating the extraction of methane. However, coal swells as it adsorbs CO_2, which can reduce the permeability by orders of magnitude (Shi and Durucan, 2004), although this can be at least partly counteracted by increasing the injection pressure. Conversely, the swelling of coal might induce additional faulting, thereby facilitating migration but also leakage of CO_2 from the coal seam. CO_2 might also react with the coal, further complicating the movement and storage of CO_2. Laboratory experiments suggest that two CO_2 molecules are adsorbed for every CH_4 molecule displaced (Byrer and Guthrie, 1999). Gunter et al (1997) discuss some technical and practical considerations (such as the timing and steadiness of methane release). Pilot projects are currently underway in Canada, China, Poland, Japan and the US (Benson et al, 2005). Wells for the injection of CO_2 into the coal bed and the extraction of methane would begin at the same time, but the extraction well would be sealed after 15–20 years while injection continues for another 20 years (Damen et al, 2005).

The CO_2 storage potential worldwide in CBM is estimated to be 60–200GtCO_2 (6–54GtC), with 0.93GtC stored for every trillion m^3 of CH_4 released (Benson et al, 2005). This implies a CBM energy resource of 17–57 trillion m^3 or about 550–2200EJ. This is significant compared to the cumulative worldwide consumption of natural gas up to the present (about 2100EJ). However, as discussed in Volume 1 (Chapter 2, subsection 2.6.5), a given mass of methane is 72 times as effective as the same mass of CO_2 in warming the climate over a 20-year time period, and 25 times as effective over a 100-year time period, Thus, if even a small fraction of the methane that is displaced by injected CO_2 is not captured but is instead released to the atmosphere, the global warming effect would exceed the benefit from capturing and storing CO_2.

Once CO_2 is stored in a coal bed, it cannot be mined without the stored CO_2 escaping into the atmosphere. However, if the use of coal is to be phased out for climatic (if not other environmental) reasons, this is no constraint. Furthermore, much of the world's coal (90 per cent in the case of the US) is unmineable with current technology, either because the seams are too deep, too thin or unsafe to mine (Byrer and Guthrie, 1999).

Injection of CO_2 into coal beds for methane extraction has been carried out on a pilot scale in the US (Stevens et al, 1999) and in Canada (Wong et al, 1999). A significant density of injection wells would be required, however, which has a negative visual and practical effect on the landscape and would be sure to incite opposition from local residents. Current CBM operations in Alberta (Canada) and various US states (in which methane is extracted without CO_2 injection) have contaminated groundwater (which is the major source of drinking water in many of the regions with significant CBM potential) (Marsden, 2007, Chapter 13).

9.6.4 Storage in the deep ocean

The density of both liquid carbon dioxide and seawater increases with increasing depth in the ocean due to the increase in pressure, but at a much greater rate for liquid CO_2. At depths shallower than 2500m, liquid CO_2 is always less dense than water and so will rise, while at depths greater than 3000m it is always denser than seawater and so will sink. In between, liquid CO_2 can be either denser or less dense than seawater, depending on the temperature (mainly) and salinity of the seawater. Below a depth of about 400m, seawater will tend to be cold enough for CO_2-hydrate ($CO_2{\cdot}6H_2O$) to be stable. Thus, liquid CO_2 will react with seawater to form a solid ice-like crust around the liquid.

Carbon dioxide could be injected into the ocean in three ways. The first would be to inject it as 5mm liquid drops at a depth of 1000–2500m (Ozaki et al, 1999). The drops would be buoyant and so would slowly rise, dissolving well before reaching the surface. For example, a 0.9cm-diameter droplet would rise about 400m over a period of one hour before dissolving completely, with 90 per cent of its mass lost in the first 200m (Brewer et al, 2002). The vertical motion would cause some initial dilution of the CO_2 as it is released through dissolution.

The second option would be to inject 1m-diameter blobs of liquid CO_2 at a temperature of −30°C to −40°C at a depth of 500m. These blobs would be denser than the surrounding water and would remain colder than the surrounding water to a depth of at least 2700m, which they would reach in 25 minutes. Below about 2700m, the blobs would be denser than the surrounding water even if they warmed up to the ambient temperature, and so would continue sinking. This avoids the need to lay pipes to a depth of 3000m, the deepest that has been attempted being only 1600m (IEA, 2001b). CO_2 could be transported to the injection sites by supertanker, at −55°C and 6atm (IEA, 2001c). Lakes of liquid CO_2 would form in depressions on the ocean floor, and a thin hydrate layer would form that would delay the release of CO_2 to the surrounding water. Numerical simulations indicate that, for a lake with an initial depth of 50m, complete dissolution would require 30–400 years, depending on the local conditions (Haugan and Alendal, 2005).

A third option would be to drop projectile-shaped solid CO_2 into the ocean, where it would sink rapidly and penetrate into the sediment layer. However, the energy required to make solid CO_2 is twice that needed to liquefy it (IEA, 2001b).

9.6.5 Storage in chemical products

About 110MtCO_2 (30MtC) are used globally per year for the production of urea (a N fertilizer), methanol, acetic acid, polycarbonates, cyclic carbonates and various other chemicals (Klemeš et al, 2006). The CO_2 used in these chemicals is usually provided by combustion of natural gas, and is eventually released to the atmosphere when the chemicals break down (or when they are used as a fuel, in the case of methanol). If CO_2 that is captured from powerplants or other industrial sources is used instead, there would be a net reduction in CO_2 emissions equal to the amount of captured CO_2 that displaces CO_2 provided to the chemical industry from the dedicated combustion of fossil fuels. The potential reduction is, however, quite small.

9.6.6 Costs of storage

Table 9.11 gives the estimated cost of storage of CO_2 in onshore and offshore geological reservoirs, as summarized Benson et al (2005). These are the costs of

Table 9.11 *Estimates of the cost of storing CO_2*

Storage option	Cost ($/tC)	
	Range	Best estimate
Onshore saline formation	0.7–23	2, 10
Offshore saline formation	1.8–110	12, 28
Onshore depleted oil field	1.8–15	5
Onshore depleted gas field	1.8–45	9
Onshore disused oil or gas field	4.4–14	6
Offshore disused oil or gas field	14–30	22

Source: Benson et al (2005)

developing the storage reservoirs and of injecting CO_2 into the reservoirs, once it has been transported to the reservoir site. There is typically a factor of two to eight difference between the lowest and highest estimated cost, with the lower estimates in the order of $1–4/tC. Eccles et al (2009) indicate that the true cost uncertainty is much greater, with costs per tC varying by over four orders of magnitude for the full range of reservoir characteristics. Estimated costs for deep ocean storage are not available, but the costs of dispersal technologies are expected to be low compared to the cost of capture, compression or liquefaction and transport.

If CO_2 from the combustion of biomass is to be captured and geologically sequestered, then only relatively small amounts of CO_2 will be captured at any one site due to the limited size of biomass plants which in turn is related to the relatively high cost of transporting biomass. Haszeldine (2006) indicates that a 1km deep onshore borehole can be drilled and evaluated for $1–2 million. For CO_2 valued at $75/tC, a minimum of about 25,000tC/yr would need to be injected in order to cover the borehole and related costs over a 20-year period. A flux of this magnitude could be provided by a powerplant in the order of 10MW_e, which would require collecting biomass from land within only an 8km radius if just 10 per cent of the land in the region can be devoted to energy crops. Thus, biomass powerplants would easily be of the size needed to support a single borehole, so there would be no need to collect and pool CO_2 from several powerplants in a local area, although there may be economic advantages in doing so.

9.7 Environmental and safety issues

Environmental and safety issues pertaining to carbon sequestration involve changes in the lifecycle emission of atmospheric pollutants, the long-term impact on atmospheric CO_2 of sequestered carbon that leaks into the atmosphere, hazards associated with the rapid release of CO_2 in underground rock reservoirs (high CO_2 concentrations at the surface can be dangerous), potential effects on groundwater (which could be contaminated through upward displacement of deep saline water or mobilization of toxic elements) and impacts on the chemistry of seawater from oceanic sequestration.

9.7.1 Use of non-energy resources and emission of atmospheric pollutants

Emissions of atmospheric pollutants are associated with the production and transport of materials used for capture of CO_2 (such as chemical or physical solvents, limestone and ammonia), transport and processing of waste produced during the regeneration of solvents (as in the case of MEA-based capture) and in the operation of the powerplant. As well, additional inputs of limestone, ammonia and water are required for some removal processes (water being a scarce resource in many parts of the world). Table 9.12 compares model-based estimates of the amounts of limestone, ammonia and water required per kWh of electricity generated for three different kinds of powerplants (supercritical PC, NGCC and IGCC) with and without capture of CO_2. Also given are emissions of NO_x, SO_2, particulates and ammonia (NH_3).[2] SO_2 emissions are greatly reduced and NO_x emissions substantially increased (by 30–45 per cent) with carbon capture from PC powerplants. Table 9.12 and the work by Tzimas et al (2007) indicate that NO_x emissions from NGCC increase when carbon capture is added. For IGCC, Table 9.12 indicates that SO_2 emissions increase modestly (by 10–15 per cent) and that NO_x emissions decrease modestly. Ammonia emissions go up and particulate emissions go down for pulverized coal. Thus, there is no consistent change in emissions across different pollutants or types of powerplant when carbon capture is added.

Capture systems based on solvents generate some degraded solvent wastes, with greater quantities generated from post-combustion than from pre-combustion removal processes. The expected procedure for waste from MEA is to remove metals and then to incinerate it (Thambimuthu et al, 2005). The associated emissions should be added to the emissions from the powerplant to get a full accounting.

Figure 9.10 compares an estimate of the amount of water used by three different coal powerplant

Table 9.12 *Efficiency (LHV basis), resource use and pollutant emissions of powerplants with and without capture of CO_2*

	Efficiency (%)		Resource use (gm/kWh)					Pollutant emissions (gm/kWh)			
	Plant	Life-cycle	Lime-stone	NH_3	MEA	Selexol	Water (L/kWh)	NO_x	SO_2	Parti-culates	NH_3
PC	39.6	36.3	16.9	0.61			3.1	0.410	1.250	0.058	0.005
PC, C	30.0	27.7	27.2	0.80	3.6		4.1	0.590	0.009	0.030	0.470
NGCC	50.1	41.0		0.20			n/a	0.140	-	-	n/a
NGCC, C	42.8	36.5		0.23	1.33		n/a	0.160	-	-	n/a
IGCC	37.2	35.0				0.02	0.6	0.120	0.300	0.004	–
IGCC, C	32.0	30.2				0.03	0.9	0.100	0.330	0.004	–

Note: C = with capture of CO_2.
Source: Odeh and Cockerill (2008)

technologies with and without CO_2 capture. Water is used in powerplants primarily for cooling purposes in cooling towers. Two water-related parameters are the amount of water withdrawn from rivers, lakes or the ground (all or some of which can be returned to the natural environment) and the amount of water consumed. Figure 9.10 shows the water requirements for recirculating systems, and as seen from this figure, water requirements for PC powerplants more than double when CO_2 capture is added.[3] The water requirements for IGCC powerplants increase by almost 40 per cent when CO_2 capture is added, although the total water use for IGCC with capture is substantially less than that of PC plants without capture.

9.7.2 Potential slow leakage from geological reservoirs

Any leakage from geological reservoirs is likely to be very slow and so would not pose a health or safety

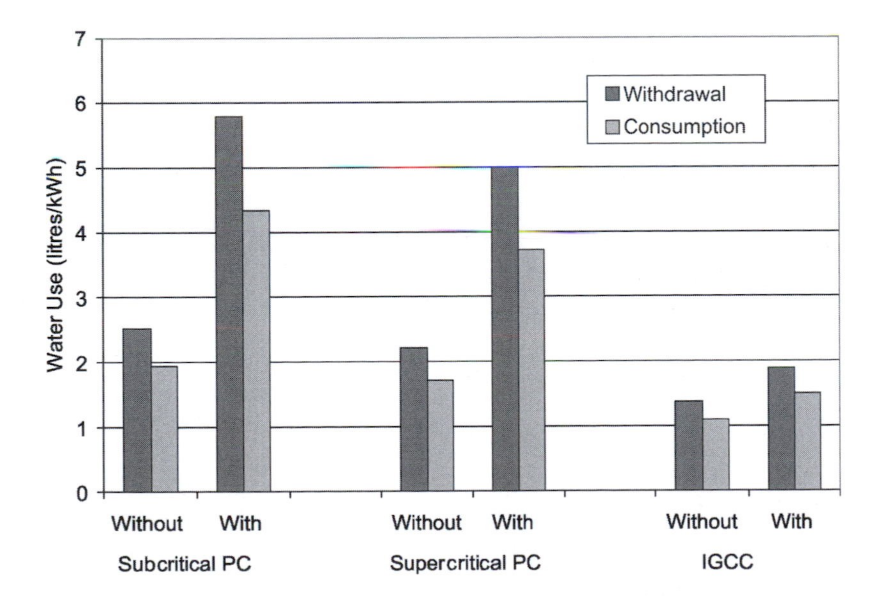

Note: Water consumption is for powerplants with recirculating cooling water systems but includes non-cooling water requirements as well.
Source: Shuster (2008)

Figure 9.10 *Withdrawal and consumptive water use in powerplants without and with capture of CO_2*

hazard to those living nearby. However, very slow but continuous leakage would undermine the benefit (in terms of atmospheric CO_2 concentration) of sequestration. Leakage of CO_2 from depleted oil fields and of CO_2 and methane from coal beds is of particular concern. If cumulative storage reaches 500GtC and the leakage rate were only 1 per cent/year, that would produce an emission of 5GtC/yr, and would compromise any possibility of stabilizing atmospheric CO_2 at 450ppmv or even 550ppmv. Thus, leakage rates have to be very low – in the order of 0.1 per cent/year or less. Storage does not have to be strictly permanent, as long as whatever leakage does occur occurs after fossil fuel emissions have peaked and is very small. Alternatively, if we want to draw down atmospheric CO_2 as quickly as possible after peaking, leakage could be a problem as it will slow down the drawdown of atmospheric CO_2.

There are a number of ways in which CO_2 could leak from depleted oil fields (Haszeldine, 2006). First, the natural top seal of the reservoir could have been fractured by the decrease in pressure during extraction of oil. Second, fracturing could occur during injection of CO_2, which must occur at high pressure. Third, exploration boreholes, oil-extraction boreholes and the injection boreholes themselves could serve as leakage points. Thomson (2009) reports that about 400,000 wells have been drilled in Alberta (a province considered to be ideal for geological storage of CO_2) during the past 70 years, while Tsang et al (2008) report that Texas has about 1 million wells. A CO_2 plume with a radius of 5km in areas of high well density would encounter several hundred wells according to sources cited by Tsang et al (2008), while even in areas of low density, tens of wells would be encountered. The likelihood of finding all or even the majority of these boreholes is very small. Those boreholes that are found could be plugged with cement, but standard Portland cement reacts with CO_2, eventually disintegrating. New cement mixes would need to be developed that remain intact for thousands or tens of thousands of years. However, sulphuric acid produced from SO_2 that contaminates the injected CO_2 would increase the rate of dissolution of cement plugs.

Micro-seismic events (small earthquakes) are well known in the oil and gas industry (Thomson, 2009). They could be triggered by the buildup of pressure associated with large-scale injection of CO_2, and while not posing a direct hazard themselves, they involve movement along fractures that in turn can cause leakage of stored CO_2.

Assuming that the capacity to securely store CO_2 has not been completely compromised by the presence of thousands of drill holes, the integrity of CO_2 storage would be increased by injecting CO_2 into the lower portion of folded structures. Wells in oil and gas fields tend to be near the apex of overlying folded caprocks, so if CO_2 is injected into the lower portion of folded structures, the risk of leakage from old wells will be reduced (but not eliminated) because the injection pressures will have decreased by the time the CO_2 migrates to the wells, significant amounts will be trapped in zones without wells, and some of the CO_2 will have dissolved into pore waters (Bradshaw et al, 2005). The density of CO_2 increases sharply below a depth of about 700m, so storage at greater depths will be safer than shallow storage because upward buoyancy forces will be reduced. Sedimentary basins with a low geothermal heat flux are preferred, so that temperatures at depth will be cooler than otherwise, which will also enhance the security of storage.

There has been no measurable leakage from the Weyburn site in Saskatchewan, Canada after four years of injection and with hundreds of wells, and this is often cited as evidence that underground CO_2 storage is secure. However, four years is far too short a time period on which to base inferences about long-term security of storage, and it would be very dangerous to generalize on the basis of one or even a few sites. As discussed by Thomson (2009), the scale of the operation also matters – the impact of the injection of 1MtC or less per year for a few years in a small region tells us very little about the impact of the injection of hundreds or thousands of MtC per year over a broad region for a century or longer, as high pressures over a broad region in the latter case could induce fracturing. As well, the Weyburn site is an enhanced oil recovery project, in which pressure is released through the withdrawal of oil. Acid gases, often mixed with significant amounts (14–98 per cent by volume) of CO_2, have been injected into deep geological formations in western Canada for several years (into 51 different formations at 44 different locations), and so far there has been no detectable leakage of CO_2 or acid from these sites (Benson et al, 2005). However, many of the same concerns about drawing inferences from the Weyburn site apply here as well.

There are 470 geological natural gas storage facilities in the US with a total storage capacity of 160Mt and a capacity-weighted average age of 25 years. There have been nine documented cases of significant leakage, five of which involved problems of well-bore integrity, three of which involved leaks in caprocks and one of which was related to poor site selection (Benson et al, 2005). Natural gas storage involves rapid pressure cycling, which increases the risk of leakage through caprocks. Storage of CO_2 would not be subject to such cycling, and furthermore, some of the CO_2 will dissolve in pore water and some will react chemically with the rocks. Monitoring techniques are able to detect movement of CO_2 in storage reservoirs, and a number of remedial actions are available to stop or at least reduce any leakage from underground reservoirs, as detailed in Benson et al (2005).

In summary, the risk of slow leakage of CO_2 that is stored underground is a serious issue and could largely negate the climate benefits of CCS. The very large number of drill holes (in the hundreds of thousands to millions) in some regions is a particular problem. Available measures could at best reduce but not eliminate the problem.

9.7.3 Well failures during injection

Perhaps the greatest public safety hazard is the risk of a well blowout during injection of pressurized CO_2 into the ground. This occurred during the drilling of a well to extract CO_2 from a natural field, with chunks of dry ice and rock several centimetres in diameter thrown hundreds of metres into the air (Holloway et al, 2007). Leaks and collapse structures have formed on occasion in association with venting of natural CO_2 when drilling for mineral water. These risks can be reduced or eliminated through careful selection of injection sites and ongoing monitoring of CO_2 concentrations.

9.7.4 Impacts on groundwater supply

Injection of CO_2 into deep saline aquifers is usually not considered to adversely affect potable water sources, as saline water cannot be used for drinking purposes. However, there are a number of issues of concern:

1. the forced upward displacement by the pressurized CO_2 of saline water into previously saline-free aquifers;
2. the mobilization of toxic materials due to the weak acidity produced as injected CO_2 reacts with groundwater;
3. foreclosure of the possibility of using saline groundwater in the future through desalination (which could be powered with renewable energy, as discussed in Chapter 3, subsection 3.14.3); and
4. foreclosure of the possibility of making use of the minerals dissolved in saline groundwater (Li, Zn and Mn in particular).

Birkholzer et al (2009) have simulated the impact on the subsurface pressure field of large-scale injection of CO_2 into saline aquifers. The area affected by high pressure is much larger than the CO_2 plume produced from injection. They find that brine can be displaced into overlying freshwater if the intervening caprock has a relatively high permeability, but as with leakage of CO_2, fracturing induced by pressure buildup over a larger region or unfilled drill holes could provide a conduit for saline water into fresh water even if no CO_2 escapes.

With regard to potential mobilization of toxic elements, Apps et al (2009) find that the reduction in pH associated with CO_2 injection could mobilize arsenic, barium, cadmium, lead, antimony and zinc, with greatest concern over arsenic. This and other studies have led the US Environmental Protection Agency to raise this issue as a concern (EPA, 2009). With regard to potential mineral extraction from saline groundwater, a US company (Simbal Mining) hopes to soon begin extraction of Li from saline water 3000m deep (Wenzel, 2008).

9.7.5 Interaction with compressed air energy storage and geothermal energy

Injection of CO_2 in porous confined underground aquifers would not preclude their use for underground compressed air energy storage, as compressed air would ideally be stored at depths of less than 700m, whereas CO_2 would ideally be injected at depths of greater than 700m (discussed in Chapter 3, subsection 3.14.2). However, it could interfere with exploitation of geothermal energy resources.

9.7.6 Stability of storage in deep-sea sediments

Both temperature and pressure increase with increasing depth in terrestrial and marine sediments, but because the deep sea is bathed everywhere with cold water that originates from sinking at high latitudes, deep sediments at the surface of the ocean floor will be cold compared to terrestrial sediments having the same pore pressure. The pressure and temperature of the top few hundred metres of sediments below ocean beds of 3000m or greater depth are such that CO_2 injected into these regions would be liquid and denser than the overlying pore fluid, and therefore stable (House et al, 2006). This is in contrast to CO_2 injected into terrestrial sediments, where it would be buoyant and would leak to the atmosphere if the overlying caprock were fractured. The top portion of the sediment layer would be in the zone where CO_2 hydrate (ice-like crystals) would form, and this would impede the injection of CO_2. The depth of both the hydrate and buoyancy stability zones depends on the depth and temperature of the sea floor and the geothermal temperature gradient. If CO_2 were injected a few hundred metres below the water–sediment interference it would be outside the hydrate stability zone and may or may not be in the buoyancy stability zone. If the injected CO_2 is buoyant it will tend to rise into the buoyancy stability zone, but the formation of hydrates would impede its upward movement, providing further protection. The potential storage capacity in buoyantly stable sediments within the US economic zone is estimated to be several thousand GtC. However, a modelling study of the injection of CO_2 into seabed sediments indicates that there are large uncertainties at present concerning the amount of leakage during and after injection (Li et al, 2009). Recall (from subsection 9.6.1) that CO_2 is already injected into marine sediments in the North Sea, without difficulty. However, costs are minimized in this case because the injected CO_2 is taken from the natural gas that is withdrawn at the same site.

9.7.7 Impact on atmospheric CO_2 of sequestration in deep ocean water

Most of the CO_2 that is emitted by humans into the atmosphere will eventually end up in the oceans, but over a period of thousands of years (as discussed in detail in Harvey, 2000). This is because, when CO_2 is added to the atmosphere, a balance between the partial pressure of CO_2 in the atmosphere and that in the oceans is achieved with 85–95 per cent of the added CO_2 in the ocean and only 5–15 per cent remaining in the atmosphere (depending on whether partial dissolution of $CaCO_3$ sediments on the ocean floor above about 1–4km depth has had time to occur). Direct injection of CO_2 into the oceans would bypass the slow, natural processes of oceanic absorption of CO_2. However, since the ocean would not absorb all of the CO_2 emitted by humans, not all of the CO_2 injected into the ocean would remain in the ocean either.

The change in the amount of CO_2 in the atmosphere or ocean following a sudden injection of a pulse of carbon into the atmosphere or ocean is called the *impulse response*. The impulse responses as computed by Harvey (2004) for atmospheric and oceanic injection beginning with present conditions are compared in Figure 9.11. 200 years after injection into the atmosphere, 70 per cent of the injected carbon is taken up by the ocean, while after 2000 years (and in steady state), about 87 per cent is taken up by the ocean and 13 per cent remains in the atmosphere (a larger fraction of successively greater pulses would remain in the atmosphere due to the gradual saturation in the ability of the ocean to hold CO_2 as the cumulative emission increases). When carbon is injected directly into the ocean at 3000m, the flux into the atmosphere steadily increases until 13 per cent of the injected CO_2 is in the atmosphere. When CO_2 is injected at a depth of 500m, some of the injected CO_2 diffuses out of the ocean before it has a chance to become diluted by thoroughly spreading through the ocean volume, causing the atmospheric fraction to overshoot its long-term value. Some of this carbon re-enters the ocean later as ongoing mixing within the ocean reduces the surface concentration. The behaviour shown in Figure 9.11 is very close to that obtained by Caldeira et al (2001), who examine in more detail the impact of injection at different depths and in different ocean basins.

The results shown in Figure 9.11 do not account for the partial dissolution of $CaCO_3$ sediments on the ocean floor that would occur as the oceans absorb or receive CO_2. The effect of dissolution on atmospheric CO_2 is negligible during the first 400 years, but would reduce the atmospheric CO_2 fraction by about 4 per cent after 2000 years (Kheshgi and Archer, 1999). That is, the

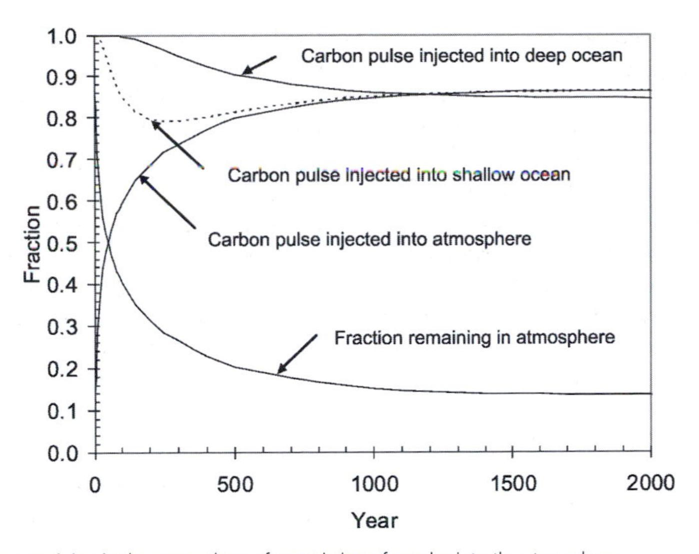

Note: Also given is the fraction remaining in the atmosphere after emission of a pulse into the atmosphere.
Source: Harvey (2004)

Figure 9.11 *Fraction of carbon remaining in the ocean after injection of a pulse into the ocean, or accumulated in the ocean after emission of a pulse into the atmosphere*

13 per cent of CO_2 seen in the atmosphere in Figure 9.11 after 2000 years would be reduced to 9 per cent.

Continuous injection of CO_2 into the oceans over a period of decades or centuries can be thought of as a series of annual injection pulses, each of which leaks according to the response shown in Figure 9.11. However, the cumulative leakage would be somewhat worse than implied by the impulse response shown in Figure 9.11 due to the decreasing ability of the oceans to hold CO_2 as the cumulative emission (or injection) increases. The integrated effect of various scenarios of CO_2 emission and injection can be investigated using a coupled climate–carbon cycle model, which accounts for climate–carbon cycle feedbacks and the response of the underlying oceanic and terrestrial biosphere sinks to altered atmospheric CO_2 concentration. A detailed explanation of how the carbon cycle responds to sequestration of CO_2 is given in Harvey (2004).

Harvey (2003) considers the impact on atmospheric CO_2 for four emission scenarios and three sequestration scenarios, shown in Figure 9.12. Emission Scenario 1 is a business-as-usual scenario in which emissions reach 28GtC/yr by 2100, Scenario 2 entails modest restraints on fossil fuel emissions and in scenarios 3 and 4, emissions decrease to zero by 2100 and 2075, respectively. In Sequestration Scenarios 1–3, an amount of CO_2 equal to the difference in emissions between Emission Scenarios 1–3 and Emission Scenario 4 is injected into the oceans at a depth of 3000m (i.e. no allowance is made for additional CO_2 production to meet the energy needs for sequestration). Figure 9.13 shows the resulting variation in atmospheric CO_2 concentration for the period 1900–3100. In Emission Scenario 4, atmospheric CO_2 peaks at 430ppmv, whereas it peaks at about 1600ppmv in Scenario 1. When the difference in emissions between Scenarios 1 and 4 is sequestered in the oceans, atmospheric CO_2 still rises to over 850ppmv by 3100. By 3100, about 25 per cent of the carbon injected into the oceans has flowed into the atmosphere. Only in Scenario 3, in which fossil fuel use is eliminated by 2100, is sequestration in the ocean able to keep the atmospheric CO_2 concentration from rising above the peak level obtained in Scenario 4 (in which fossil fuel use is eliminated by 2075).

Thus, if the goal is to limit atmospheric CO_2 concentration to very low levels (such as 450ppmv), then sequestration of CO_2 in the oceans as a substitute for early reductions in fossil fuel use is not acceptable because the eventual leakage of the injected CO_2 back to the atmosphere still causes CO_2 to rise to several hundred ppmv above 450ppmv. Oceanic sequestration

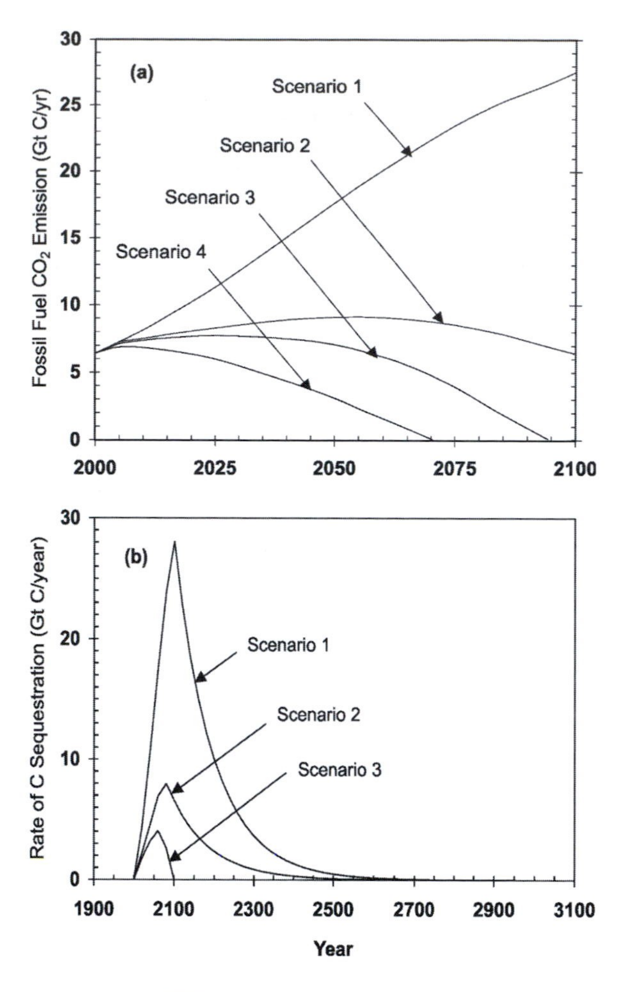

Source: Harvey (2003)

Figure 9.12 *Scenarios of (a) fossil fuel emission, and (b) rates of carbon sequestration, where the carbon sequestration in Scenarios 1, 2 and 3 is equal to the difference in the emission between the corresponding emission scenario and Emission Scenario 4*

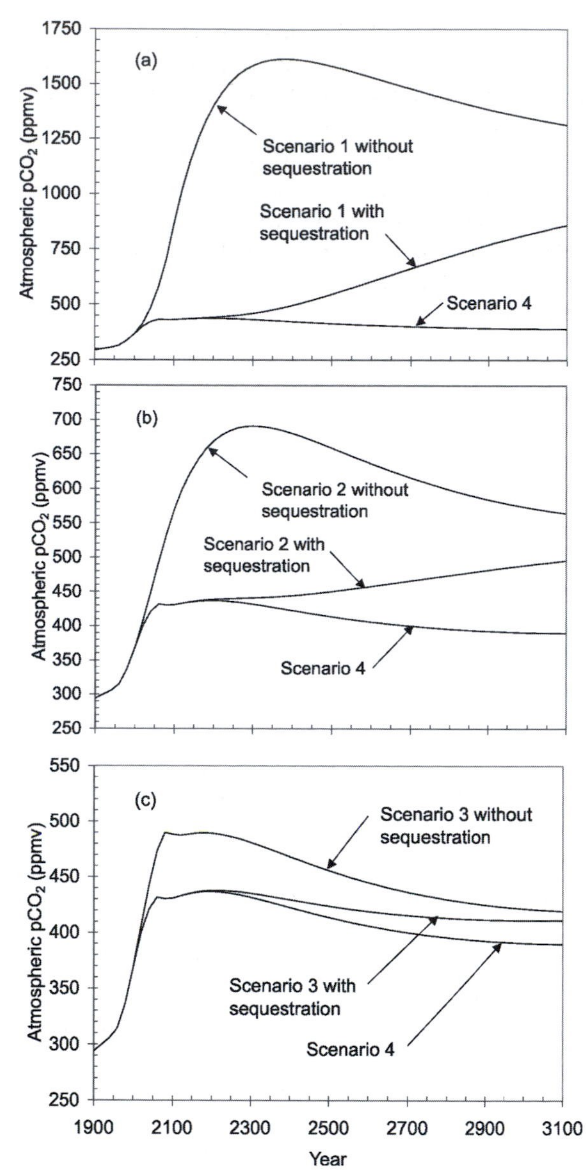

Note: Also given for each comparison is the variation of atmospheric CO_2 concentration for Emission Scenario 4.
Source: Harvey (2003)

Figure 9.13 *Comparison of atmospheric CO_2 concentration for Emission Scenarios 1, 2 and 3 with and without sequestration of the difference between the given emission scenario and Emission Scenario 4*

of CO_2 is acceptable, if at all, only as a complement to rapid reductions in fossil fuel use.

9.7.8 Impact on oceanic chemistry of sequestration in deep ocean water

In addition to eventual leakage into the atmosphere, the absorption of CO_2 by the oceans has two other effects of concern: a reduction in the pH of seawater and a decrease in the degree of supersaturation of seawater with respect to calcite – the structural building material of calcareous micro-organisms (which are scattered throughout the marine food chain) and of coral reefs.

The emission of CO_2 into the atmosphere from human activities leads to an increase in the partial pressure of atmospheric CO_2, and in response to this, there is a net flow of gaseous CO_2 into the surface layer of the ocean. Once CO_2 is dissolved in surface water (either by absorption from the atmosphere or due to direct injection), it combines with water to form a weak acid (carbonic acid, H_2CO_3), which then dissociates to bicarbonate (HCO_3^-) and carbonate (CO_3^{2-}) ions. The reactions are:

$$CO_{2(gas)} + H_2O_{(liquid)} \rightarrow H_2CO_{3\ (aq)} \text{ (carbonic acid)} \quad (9.18)$$

$$H_2CO_3 \rightarrow H^+ + HCO_3^- \quad (9.19)$$

$$CO_3^{2-} + H^+ \rightarrow HCO_3^- \quad (9.20)$$

giving the net reaction:

$$H_2O + CO_2 + CO_3^{2-} \rightarrow 2HCO_3^- \quad (9.21)$$

Reaction (9.19) would tend to increase the acidity of seawater as CO_2 is added, except that CO_3^{2-} consumes the H^+ that is released by reaction (9.19), so that there is no change of pH as long as the occurrence of reaction (9.19) is balanced by reaction (9.20), but the concentration of CO_3^{2-} decreases. The carbonate ion thus acts as a buffer, inhibiting changes in pH to the extent that it is available. However, reaction (9.20) does not proceed fully, so there is some reduction in pH but a smaller reduction in CO_3^{2-} concentration.

Figure 9.14 shows the impact on the pH and calcite supersaturation in non-polar surface waters for the emission and sequestration scenarios shown in Figure 9.12. For Emission Scenarios 1 to 4 (without sequestration), the supersaturation falls from a pre-industrial value of 486 per cent (i.e. 4.86 times saturation) to 155 per cent, 286 per cent, 360 per cent and 381 per cent, respectively. Even the smallest of these decreases is likely to adversely affect coral reef and marine calcareous micro-organisms by reducing calcification rates (Wolf-Gladrow et al, 1999; Langdon et al, 2000; Riebesell et al, 2000). The impact on marine ecology of the more extreme reductions would probably be severe. The corresponding reductions in pH are by 0.66, 0.30, 0.19 and 0.14 units, respectively. The larger of these pH reductions would probably adversely affect

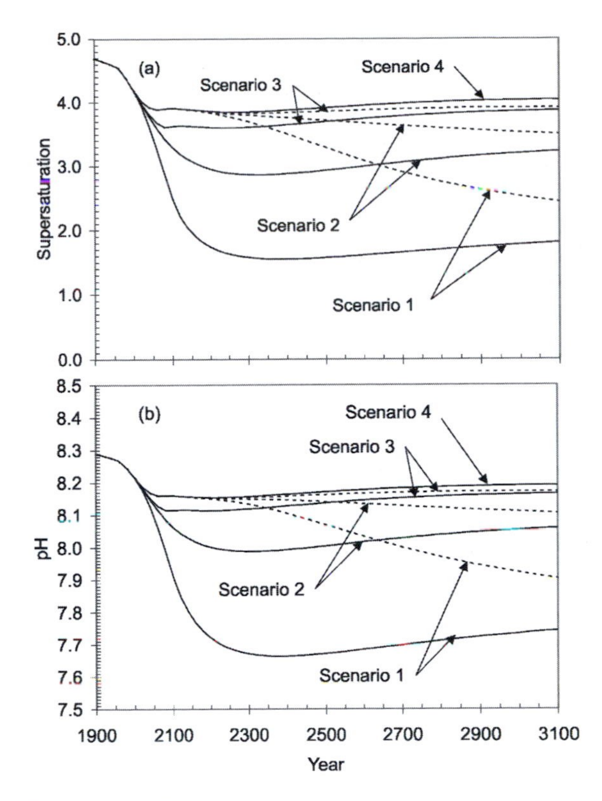

Source: Harvey (2003)

Figure 9.14 *Variation in (a) the non-polar surface water degree of supersaturation with respect to calcite, and (b) the non-polar surface water pH for the emission and carbon sequestration scenarios shown in Figure 9.12*

both calcareous and non-calcareous marine organisms, independently of the decrease in calcite supersaturation. As with the impact on atmospheric CO_2, the response of supersaturation and pH with sequestration converges toward the response without sequestration near the end of the simulation. These responses are a significant fraction of the peak responses seen for the cases without carbon sequestration.

As well as eventual large effects on surface water chemistry, immediate local effects next to the carbon injection sites are also of concern (Adams et al, 1997; Auerbach et al, 1997; Tamburri et al, 2000; Drange et al, 2001). Minimization of these effects requires injecting CO_2 from a large number of dispersed points rather than from a small number of large injection sites, a task that is rendered more difficult for the extreme carbon sequestration scenarios.

9.7.9 Neutralization of CO_2 acidity prior to injection into the ocean

Rau and Caldeira (1999), Caldeira and Rau (2000) and Rau et al (2007) discuss the idea of reacting captured fossil fuel CO_2 with limestone ($CaCO_3$) in reaction chambers on land before being released into the ocean, rather than directly injecting captured CO_2 into the deep ocean, as in the conventional ocean carbon sequestration concept. This replicates what would happen in nature over a period of many thousands of years. Surface waters of the ocean are supersaturated with respect to $CaCO_3$, but deep ocean waters are unsaturated with respect to $CaCO_3$. As a result, carbonate sediments produced by the downward rain of dead calcareous plankton have accumulated on the shallower ocean floor (above depths ranging from about 0.3 to 4km). As the ocean absorbs atmospheric CO_2 (or as captured CO_2 is injected into the ocean), the decrease in CO_3^{2-} concentration (reaction (9.21)) will induce the dissolution of the deepest carbonate sediments over a period of several thousand years, as water that had been supersaturated becomes unsaturated. This will restore the CO_3^{2-} that had been depleted from ocean water, neutralize the increase in acidity of ocean water and allow the ocean to absorb further CO_2 from the atmosphere.

Rau and Caldeira (1999) propose directly reacting captured CO_2 with limestone that has been ground to a size of a few mm, thereby provoking the reaction:

$$CaCO_3 + CO_2 + H_2O \rightarrow 2HCO_3^- + Ca^{2+} \qquad (9.22)$$

This scheme avoids the impacts on ocean acidity and the eventual outgassing of some of the injected CO_2, which would occur with conventional oceanic carbon sequestration, but would require a large infrastructure to mine, crush and transport a mass of limestone several times the mass of captured CO_2, as well as an infrastructure to pump seawater and captured CO_2 to the dissolution sites and to pipe the reaction mixture to the ocean over a sufficiently dispersed area.

9.7.10 Domestic and international legal issues

A rigorous regulatory process that has broad public and political support will be required if CO_2 is to be sequestered underground on a large scale. As well, some sort of international monitoring system would need to be set up to verify that CO_2 was in fact being captured and permanently stored if countries or energy companies were to engage in international trading of CO_2 emission permits or of CO_2 emission credits. It could take many years just to set up such a system, given international concerns over sovereignty.

Conversely, disposal of CO_2 in the oceans will require resolution of questions pertaining to international agreements. The 1972 London Convention prohibits the dumping of industrial waste at sea or in sub-seabed formations, while the 1992 OSPAR Convention reinforces and extends the provisions of the London Convention within the northeast Atlantic region. It therefore needs to be determined whether ocean carbon storage would be an activity that can be regulated under these conventions, and whether fossil fuel CO_2 constitutes an industrial waste. Reiner and Herzog (2004) discuss regulatory issues related to the long-term management of sequestered carbon.

9.8 Regional and global storage potential

As explained above, CO_2 can be stored in deep saline aquifers, in coal beds, in depleted oil and gas reservoirs, in sub-seabed sediments and in the deep ocean. Estimates of the potential storage capacity in various non-oceanic reservoirs are summarized in Table 9.13. To estimate the potential storage capacity of underground reservoirs requires estimates of (1) the available pore volume, and (2) the fraction of the reservoir that can be used. Information on rock pore volumes is generally quite poor outside of oil and gas provinces. The fraction of a reservoir that can be used depends on how extensive and permeable the caprock is, the tilt of the reservoir and the rate of fluid flow within the reservoir. Major uncertainties remain at all stages of the estimation procedure, as reflected by the wide range of estimates given in Table 9.13. The estimates for the storage potential in depleted oil and gas fields do not take into account the fact that not all the pore volume previously occupied by oil or gas will be available to hold CO_2 because of residual water trapped in the pores. In western Canada, the resulting loss of storage capacity is estimated to be about 30 per cent for gas reservoirs and 50 per cent for oil reservoirs (Benson et al, 2005).

Table 9.13 *Estimates of the global carbon sequestration potential, excluding disposal in the deep ocean or in ocean sediments*

Reservoir	Storage potential (GtC)
Oil and gas reservoirs	
NW Europe	13
Western Canada	1.4
US	27
Australia	4.2
Total of above	45.6
Extrapolation to entire world	230
Deep saline aquifers	
Range of 14 assessments	55–15000
Likely minimum	270
Coal beds	
Theoretical potential	16–55
Practical potential for bituminous coal	2
Minimum total	**about 500**

Source: Based on Benson et al (2005)

The global storage capacity excluding oceanic disposal is likely to be at least 500GtC. A critical factor in the global storage capacity is whether or not storage in aquifers should be limited to closed aquifers with structural traps. If, as suggested above, large open aquifers can be used, then the global storage capacity could be many times larger. Other unresolved issues are the extent to which the integrity of aquifers will be compromised by the hundreds of thousands of boreholes present in many regions, the potential displacement of saline water upward into underground freshwater resources and the mobilization of toxic elements due to the reduction of pH caused by injection of CO_2 into groundwater. Given that 1GtC of CO_2 in the atmosphere corresponds to a concentration of 0.469ppmv, and that about half of current fossil fuel emissions are quickly removed from the atmosphere in any case, sequestration of 300GtC corresponds to an avoided CO_2 concentration peak of about 75ppmv. Given strong efforts to reduce the world's dependence on fossil fuels during the next few decades, carbon sequestration could make a significant contribution to limiting the peak atmospheric CO_2 concentration to no more than 450ppmv.

9.9 Current status and scenarios

At least ten years of intensified research are needed to improve and verify many of the technologies needed for carbon capture, and to bring down costs, before attempting large-scale demonstration projects. Especially important is work related to the optimal integration of the many components of CCS systems. Many different technologies would need to be developed and tested if CCS is to be applied to coal, given the wide diversity of coal properties (in terms of heating value and sulphur, water and ash contents). Gibbins and Chalmers (2008) suggest that two generations of carbon capture demonstration projects with learning, each spanning five to six years, would be needed.

MIT (2007) believes that about ten full-scale (1 million tonnes CO_2/yr) demonstration projects of storage techniques would be needed worldwide in order to cover the likely range of accessible geologies that would be encountered, and that each project would need to be thoroughly monitored and analysed over a period of ten years. Such large-scale projects could be part of the second generation of carbon capture projects, but most would probably not be deployed with the first generation. It would be another 20 years after that before CCS could be applied to the majority of fossil fuel powerplants. Thus, CCS could not make a significant contribution to reduction in CO_2 emissions before mid-century. Current and planned carbon sequestration projects are listed in Table 9.14; most of these projects are quite small, with only 2 out of 11 existing projects and one out of seven planned projects exceeding 1MtCO_2 per year.

Damen et al (2005) used a geographical information system (GIS) to look for suitable combinations of point sources of CO_2 and oil and coal beds for sequestration of CO_2. Source/reservoir combinations within 100km of each other were identified, then ranked and screened according to various criteria, then the potential total CO_2 storage and extra oil extraction (in the case of oil fields) or methane extraction (in the case of coal fields) were assessed in a probabilistic manner. Altogether, 64 promising CO_2 sources for oil fields and 59 promising CO_2 sources for coal beds were identified. Meng et al (2007) identified a number of Chinese coal-fed ammonia plants that are close to suitable deep saline aquifers, with estimated total costs of \$15–21/tCO_2 stored (\$55–77/tC).

Building up the infrastructure required to capture and sequester a significant fraction of the CO_2 produced from fossil fuels would be an enormous undertaking. The CO_2 emission associated with world

Table 9.14 *Summary of current and planned geological carbon dioxide sequestration projects*

Project	Country	Type	Lead organizations	Project start date	Approximate daily average injection rate	Total expected storage	Storage type
Sleipner	Norway	Commercial	Statoil, IEA	1996	3000t/day	20Mt	Sub-sea aquifer
Wyeburn	Canada	Commercial	EnCana, IEA	2000	3000–5000t/day	20Mt	EOR
Minami-Nagoaka	Japan	Demo	RIITE[a]	2002	≤40t/day	10,000t	Gas field
Yubari	Japan	Demo	METI[b]	2004	10t/day	200t	ECBM
In Salah	Algeria	Commercial	Sonatrach, BP, Statoil	2004	3000–4000t/day	17Mt	Depleted gas field
Frio	US	Pilot	U of Texas	2004	177t/day for nine days	1600t	Saline aquifer
K12B	Netherlands	Demo	Gaz de France	2004	100–1000t/day	8Mt	EGR
Fenn Big Valley	Canada	Pilot	Alberta Research Council	1998	50t/day	200t	ECBM
Recopol	Poland	Pilot	TNO-NITG	2003	1t/day	10t	ECBM
Qinshui Basin	China	Pilot	Alberta Research Council	2003	30t/day	150t	ECBM
Salt Creek	US	Commercial	Anadarko	2004	5000–6000t/day	27Mt	EOR
Planned projects							
Snøhvit	Norway	Commercial	Statoil	2006	2000t/day		Saline aquifer
Gorgon	Australia	Commercial	Chevron	2009	10,000t/day		Saline aquifer
Ketzin	Germany	Demo	GFZ Potsdam	2006	100t/day	60kt	Saline aquifer
Otway	Australia	Pilot	CO_2CRC	2005	160t/day for two days	100kt	Saline aquifer and depleted oil field
Teapot Dome	USA	Demo	RMOTC	2006	170t/day for three months	10kt	Saline aquifer and EOR
CSEMP	Canada	Pilot	Suncor Energy	2005	50t/day	10kt	ECBM
Pembina	Canada	Pilot	Penn West	2005	50t/day	50kt	ECBM

Note: EOR = enhanced oil recovery, EGR = enhanced gas recovery, ECBM = enhanced coal bed methane. [a] Research Institute of Innovative Technology for the Earth. [b] Japanese Ministry of Economy, Trade and Industry.
Source: Benson et al (2005)

use of oil amounted to 3.1GtC in 2005. To sequester 1GtC/year in geological reservoirs thus amounts to about one third of the current flow of oil out of the ground. Conversely, if the infrastructure for sequestration of CO_2 comparable to that currently in place for the world oil industry were in place by 2050, it would handle only 20–25 per cent of projected business-as-usual emissions in 2050 (12–16GtC/yr).

9.10 Summary

Capture of carbon dioxide from new fossil fuel powerplants using existing technologies would increase fuel requirements by 11–40 per cent according to various estimates, while retrofitting existing coal-fired plants is estimated to increase fuel requirements by 43–77 per cent. With future technologies the energy penalty in new plants might be reduced to 2–12 per cent. The effective CO_2 capture fraction, taking into account the effect of additional fuel use on emissions, would be about 63–94 per cent for retrofits of existing coal powerplants, 80–90 per cent with existing technologies on new plants, but up to about 100 per cent with future technologies on new plants. A new PC powerplant using existing technology for post-combustion capture is projected to cost $1900–2600/kW, compared to real costs at present of $1200–1500/kW without capture. Costs given with existing technology are not costs that have been achieved at present, but rather, are costs that would be expected after learning through widespread implementation. The future costs given using advanced technology are particularly uncertain. As noted earlier, a proposed 450MW IGCC powerplant with carbon capture in Saskatchewan was abandoned after estimated costs ballooned from Cdn$3778/kW to Cdn$8444/kW. Estimated costs of carbon capture, compression, transport and disposal in terms of $ per tonne of carbon sequestered are summarized in Table 9.15. Total costs for all four steps are likely to be at least $75–100/tC ($20–30/tCO_2), and could easily reach $200/tC or more.

Sequestration of at least 500GtC might be feasible in practice. This number includes sequestration in coal beds with extraction of methane from the coal, which is appealing as an alternative to the direct use of coal because it could produce negative net emissions associated with the methane derived from the coal (and avoids other significant negative environmental and social externalities associated with the direct use of coal, discussed in Volume 1, Chapter 2, section 2.7, while raising some environmental problems of its own). This is because more CO_2 carbon might be stored in the coal bed than would be extracted as methane and released to the atmosphere when the methane is combusted. However, the amount of methane that could be extracted from coal beds would be limited by the amount of CO_2 that can be captured from other carbon-containing fuels.

Table 9.15 *Summary of estimated costs of carbon capture, compression, transport and disposal*

Process	Cost ($/tC)
Capture	
From pressurized fuel cells	60–65
From IGCC, pre-combustion	66–85
From BIGCC, post-combustion	114–150
From H_2 production from fossil fuels or biomass	≈0
Compression (to 110 atmospheres) and transport	
250km pipeline	20–60
Disposal	
In land-based aquifers	<10–30
Through enhanced natural gas recovery	<10
Through enhanced oil recovery	<50
In depleted oil and gas fields	50–100
By displacing coal bed methane	<50
In the deep ocean	50–100

There are enormous uncertainties concerning the cost and viability of large-scale storage of CO_2. However, Hansson and Bryngelsson (2009) find that carbon capture and storage experts display a level of confidence that is not justified given the underlying uncertainties. Among the environmental issues associated with sequestration on land are potential leakage through the thousands of drill wells that are found in most populated regions, potential displacement of saline groundwater into freshwater groundwater supplies and mobilization of toxic elements in saline aquifers due to the decrease in groundwater pH associated with CO_2 injection. Other issues are the preclusion of the future use of saline groundwater through desalination and preclusion of future mining of saline groundwater for trace elements.

Disposal of CO_2 in the oceans as anything more than a supplement to a major shift from fossil fuels to renewable energy sources is not acceptable. If the bulk of future business-as-usual emissions were to be avoided through ocean sequestration rather than a shift to non-fossil fuel energy sources, then eventually several 1000GtC would be disposed in the ocean and fossil fuel resources would be depleted faster than would otherwise be the case due to the energy cost of CO_2 capture. For every 1000GtC disposed in the ocean, about 250GtC – corresponding at an increase in atmospheric CO_2 concentration of 125ppmv – would eventually return to the atmosphere. Since the difference between the present concentration of 385ppmv and the upper limit of 450ppmv adopted in this book is only 65ppmv, wholesale reliance on oceanic disposal is not viable. Furthermore, the injection of large amounts of CO_2 into the ocean would have important effects on oceanic pH and the health of marine ecosystems. Alternatively, burial in sub-seabed aquifers (as in the Norwegian Statoil project) would be acceptable.

The most promising methods of CO_2 capture involve production of hydrogen from fossil fuels or direct use of fossil fuels in high-temperature fuel cells. Costs and energy penalties associated with these methods of CO_2 capture are likely to be significantly less (and CO_2 recovery rates better) than capture from powerplant flue gases after combustion of the fuel. Furthermore, CO_2 capture in these ways could be part of the transition from fossil fuels to a hydrogen-based economy – a concept that is thoroughly discussed in Chapter 10.

At least 20 years of demonstration projects involving carbon capture from powerplants using a variety of different types of coal and carbon storage in a variety of different geological settings would be required before large-scale deployment of CCS could begin. Another 20 years would be required before a significant fraction of the world's powerplants would (through normal retirement and replacement) be equipped with CCS. Thus, CCS could not make a significant difference before mid-century. Carbon sequestration could nevertheless be used as an emergency measure to accelerate the later stages in the phase-out of fossil fuel CO_2 emissions. In conjunction with the capture of CO_2 released from the use of biomass, it could create negative CO_2 for many decades, if this is needed in order to reduce atmospheric CO_2 concentration. The potential to draw down atmospheric CO_2 in this way will be discussed further in Chapter 12. An important conclusion is that it is not possible, even with rapid CO_2 emission reductions, maximal sequestration of biomass carbon and sharp reductions in heating from other greenhouse gases, to avoid a century or more of CO_2-doubled climate. There are still benefits in reducing the CO_2 concentration as rapidly as possible during this transient warmth, however, as sea level rise due to thermal expansion would be reduced and it might be possible to avoid the irreversible melting of the Greenland ice sheet and destabilization of the West Antarctic ice sheet, processes that could take hundreds of years to be triggered. Abrupt reorganization of the oceanic circulation might also be averted if the peak warming is relatively short lived.

Notes

1 This is deduced from the statement in Haugen and Eide (1996) that 55–60MW of power would be needed to liquefy the CO_2 produced by a 500MW coal-fired powerplant, and assumes a powerplant efficiency of 33 per cent and a coal emission factor of 25kgC/GJ. Although this is a 10–12 per cent energy penalty for the liquefaction of CO_2 from a coal powerplant, it is less than a 2 per cent penalty for the liquefaction of 71 per cent of the CO_2 released by steam reforming of natural gas. More specifically, for a steam reforming efficiency of 89 per cent, a natural gas emission factor of 13.5kgC/GJ, and 71 per cent capture, the amount of CO_2 that needs to be liquefied is 0.0107tC/GJ H_2, so the energy requirement is 1.6–1.8 per cent of the energy content of the H_2 product.

2 Comparative emissions data are also found in Volume 1, Table 2.19 for coal plants.

3 Recirculating cooling systems are cooling systems in which water is repeatedly circulated through a cooling tower and cooled through partial evaporation. An alternative is a once-through cooling system, in which water is used once for cooling and is then returned to the environment. Recirculating-loop systems require much smaller water withdrawals (and so are used in arid regions) at the expense of much greater water consumption due to evaporative losses.

10
The Hydrogen Economy

10.1 Introduction

The two most promising carbon-free and renewable energy resources are solar and wind energy, but these forms of energy are intermittent and therefore can meet only a portion of primary energy demand. With plug-in hybrid powertrains and advances in battery technology, it may be practical to displace a substantial portion of the liquid fuels that would otherwise be used in automobiles, but batteries are unlikely to be viable as a replacement of carbon-based fuels for long-distance passenger travel or for freight transportation, and certainly not as a replacement for carbon-based fuels for air travel. Furthermore, wind and solar energy often do not coincide with the locations of large energy demand. Geothermal energy can provide continuous heating and electricity production, but the resource is limited at a global scale and is not renewable. It could nevertheless provide a non-negligible fraction of present-day energy demand for many centuries. Biomass can provide continuous electricity generation and heat, can be easily stored and can be converted to liquid fuels (ethanol and biodiesel fuel in particular) that could be produced for transportation uses. However, the biomass energy potential is limited, it does not represent the most efficient use of limited land resources for the production of energy and some emissions of air pollutants will remain (particularly in transportation uses).

One way to resolve the space–time mismatch between the supply of solar and wind renewable energy on the one hand, and energy demand on the other, would be to use hydrogen (H_2) as an *energy currency* that would complement electricity. Electricity is an energy currency, but it cannot be stored on a large scale and cannot be used in all end-use applications. Hydrogen can be readily produced from electricity (by electrolysis), transmitted long distances and stored, and either converted back to electricity (using fuel cells) when and where needed, or used in place of fossil fuels in all other end-use applications where fossil fuels are currently used. Electricity and hydrogen would be the two interchangeable energy currencies in an energy system based entirely on renewable energy.

The term *hydrogen economy* has been coined to describe a system in which hydrogen is produced from renewable energy and used in place of hydrocarbon fuels. In this chapter, the elements and possible configuration of a hydrogen economy are discussed. The production of hydrogen from fossil fuels is also included in the discussion, since the byproduct CO_2 could be sequestered relatively easily, as discussed in Chapter 9 (subsection 9.3.8). Production of hydrogen in this way is seen by some as a potential bridge in the transition to a full-fledged hydrogen economy. A key technology in the hydrogen economy will be the fuel cell, which was discussed in Volume 1 with regard to generation of electricity (Chapter 3, subsection 3.2.5) and with regard to fuel cell vehicles (Chapter 5, subsection 5.4.8). In this chapter, the primary emphasis is on infrastructure and system issues related to the hydrogen economy.

10.2 Properties of hydrogen as an energy carrier

The physical properties of hydrogen are compared with those of gasoline, diesel fuel and jet-A fuel (used in commercial aircraft worldwide) in Table 10.1. The important properties of hydrogen as an energy carrier are (1) for a given amount of energy, hydrogen is about *three times lighter* than gasoline or aircraft fuel, and (2) for a given amount of energy, hydrogen in liquid form is about *four times bulkier* than carbon-based fuels. The low weight of hydrogen fuel will be of particular

Table 10.1 *Physical properties of hydrogen, gasoline, diesel fuel, jet fuel and methane*

	Hydrogen	Gasoline	Diesel fuel	Jet fuel	Methane
Molecular formula[a]	H_2	$C_{7.14}H_{14.28}$	$C_{13.57}H_{27.14}$	$C_{15}H_{30}$	CH_4
Molecular weight[a]	2	100	190	210	16
MJ/kg (LHV)	120.2	43.7	41.8	43.0	50.0
MJ/kg (HHV)	141.9	45.9			55.5
MJ/litre (LHV) of gas at STP[b]	0.0108 (3.73)				0.167 (57.6)
MJ/litre (LHV) of liquid	8.5	32.2	35.8	34.4	
Boiling point	−252.7°C	100°C	230°C	245°C	−161°C
Gas density at STP (kg/m³)	0.0899				0.719
Liquid density (kg/m³)	70.8	737	856	811	423
Specific heat (J/gm/K)	9.69			1.98	3.50
Heat of vaporization (J/gm)	446			360	510

Note: [a] Pertains to surrogates in the case of gasoline, diesel and jet fuel. [b] Values at 345atm pressure are given in brackets
Source: Brewer (1991), Brown (2001) and others

advantage for aircraft, since the aircraft weight during take-off (the most energy-intensive part of a flight) will be greatly reduced, and the aircraft structural weight can also be reduced. The bulkiness of hydrogen, however, could pose a difficulty for its use in automobiles. The most important parameters are the bulk and mass of the stored hydrogen plus container compared to that of a gasoline tank. These are compared in Table 10.2 for three different systems for storing hydrogen and for storing an amount of gasoline sufficient to give the same driving range. Also given are the system energy intensity (stored fuel energy divided by the total volume of the storage device) and the ratios of system masses and volumes. Compressed gas at 700atm has less than one tenth the energy density of a full gasoline tank, but 4.2 times less hydrogen energy is projected to be needed for an advanced hydrogen vehicle compared to a current typical (7.7MJ/100km) mid-sized vehicle (see Table 10.4), so the storage system is only (a still substantial) 2.6 times as bulky. The additional required storage volume is 0.077m³, which should be manageable.

Table 10.2 *Mass and volume of alternative hydrogen storage systems for a mid-sized hydrogen fuel cell automobile, with 3.0kg of stored hydrogen (sufficient for a 610km driving range using the energy intensity given in Table 10.4)*

	Fuel + storage system mass (kg)	Mass % fuel	Volume (litres)		System energy density (MJ/litre)	Ratio of system masses	Ratio of system volumes
			System	Fuel			
Compressed gas @ 345atm	64.5	4.8	175	107	2.07	1.41	3.55
Compressed gas @ 700atm	85.4	3.8	126	55	2.87	1.86	2.56
Liquid H_2	25.5	13.1	99	47.1	3.64	0.56	2.02
Metal hydride ($TiFeH_2$)	230	1.92	40.9		11.36	5.02	0.83
Gasoline	45.9	75	49.1	46.7	30.69	1.00	1.00

Source: Calculated by the author. For compressed H_2, based on the following data from Lasher et al (2009): 5.6kg usable H_2, 5.8kg stored H_2, 17.8kg usable H_2 per m³ of system volume, 0.06 H_2 mass fraction and 258 litres volume for storage at 350atm. The corresponding numbers for storage at 700atm are 5.6kg, 6.0kg, 25.6kg/m³, 0.048 and 14 litres, respectively. For storage of 3.0kg H_2, volume decreases in proportion to the mass of H_2 and the container mass decreases in proportion to volume to the 2/3 power. For storage of 3.9kg of liquid H_2, Ogden (1999) gives a system volume and mass of 116 litres and 28.5kg, respectively, from which the numbers given here can be derived assuming the same scaling relationships as for compressed hydrogen and assuming that 0.9 of the container volume contains usable H_2. Metal hydride results pertain to $TiFeH_2$, which has H_2 and system densities of 105kg/m³ and 5470kg/m³, respectively (Volume 1, Table 5.19). The usable hydrogen is assumed to be 0.9 times the stored H_2 and is multiplied by 1.29 (based on Volume 1, Table 5.29) in order to give the same driving range as the other vehicles in spite of significantly greater mass. Gasoline results are based on the statement in Keoleian et al (1997) that a 117-litre steel tank (assumed to hold 111.2 litres of fuel) has a mass of 21.92kg. Assuming a vehicle energy intensity of 2.47MJ/km (7.67 litres/100km) (from Table 10.4), the required tank storage capacity is 46.7 litres. Tank volume and mass are scaled in the same way as for the other storage technologies.

Another notable feature of hydrogen is that it has an extremely low boiling point: 20K or −253°C. This makes it difficult to store hydrogen in automobiles in liquid form. However, the low boiling point of hydrogen can be used to advantage in hydrogen-fuelled aircraft, as explained in subsection 10.6.4.

10.3 Production of hydrogen

More than 500 billion cubic metres of hydrogen are produced annually, with an energy value of 6.5EJ/yr (about 1.3 per cent of the total world primary energy demand of 483EJ in 2005). About half of all hydrogen produced is used in the manufacture of ammonia, which in turn is mostly used to make fertilizers. A further 37 per cent is used in oil refineries to reduce the C to H ratio in refined petroleum products by adding hydrogen. Another 8 per cent is used in the production of methanol, 4 per cent for various other uses and 1 per cent in the world's space programmes (IEA, 1999). Hydrogen can be produced from natural gas or by gasification of coal. It can also be produced by splitting water into H_2 and O_2 using electrolytic, thermal or chemical methods, or various combinations of these. The vast majority of the hydrogen that is produced today is produced from natural gas (which is 90–95 per cent methane). In special applications where pure hydrogen is required (as in spacecraft), it is generally produced by electrolysis of water. Present-day and potential future methods of producing hydrogen are described below.

10.3.1 Steam methane reforming

Hydrogen can be produced from methane (CH_4) through a process called *steam methane reforming* (SMR). This is the reaction:

$$CH_4 + H_2O \rightarrow CO + 3H_2 \tag{10.1}$$

which requires an external heat input. It is generally carried out at a temperature of 700–850°C and a pressure of 3–25atm (Ogden, 1999). This is followed by the water shift reaction:

$$CO + H_2O \rightarrow CO_2 + H_2 \tag{10.2}$$

The net reaction, if completely carried through, would be:

$$CH_4 + 2H_2O \rightarrow CO_2 + 4H_2 \tag{10.3}$$

The H_2 is separated from the CO_2, and remaining trace amounts of CO_2, CO and CH_4 are removed as the hydrogen is purified to up to 99.999 per cent purity.

The efficiency of large-scale steam reforming of methane can be computed as:

$$\eta = \frac{m_{H_2} LHV_{H_2}}{LHV_{CH_4} + Q_{in}} \tag{10.4}$$

where m_{H_2} is the number of moles of H_2 produced per mole of fuel (4 in Equation (10.3)), LHV_{H_2} and LHV_{CH_4} are the lower heating values of H_2 and CH_4, respectively, and Q_{in} is the amount of heat that must be added per mole of fuel. Equation (10.4) is applicable if Q_{in} is applied externally (by, for example, burning additional fuel) and leads to a theoretical (maximum possible) efficiency of 92 per cent. If Q_{in} is supplied by combusting some of the H_2 that is produced then, as there are some losses in producing the hydrogen, the maximum possible efficiency is slightly smaller – 89 per cent. However, if Q_{in} can be provided as waste heat from elsewhere, then it can be dropped from the denominator of Equation (10.4), and the maximum possible efficiency is then 121 per cent (see Lutz et al, 2003, for further analysis).

These efficiencies assume that reactions (10.1) and (10.2) go to completion, which is overly optimistic because the reactions are reversible. The actual efficiency depends on the composition of gases leaving the reformer, which in turn depends on the operating temperature and pressure. Warmer temperatures drive the reactions to the right, so that there is less CH_4 and CO in the output stream, but thermal losses also increase. Excess steam also drives the reactions further to the right, improving efficiency, but requires additional energy to make the steam. The net effect on efficiency of varying operating temperature and steam use is shown in Figure 10.1. The results here assume that sufficient heat is extracted from the exhaust (and contributes to Q_{in}) to cool the exhaust to 100°C. If the exhaust were cooled to only 600°C, for example, the efficiency would plateau at around 65 per cent instead of 85 per cent. In practice, the efficiency of steam reforming of methane is around 75–80 per cent but could be pushed to 85 per cent with good waste heat recovery and use.

Capital costs for SMR plants depend strongly on capacity, with costs for small plants (5–25kg H_2/hr, suitable for small refilling stations) differing widely between different sources but in the range of

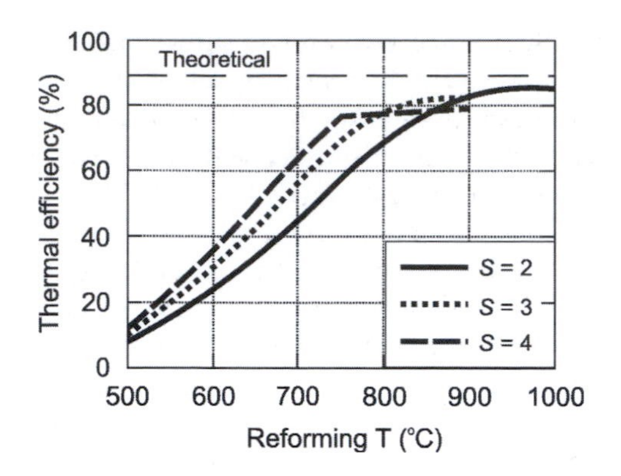

Source: Lutz et al (2003)

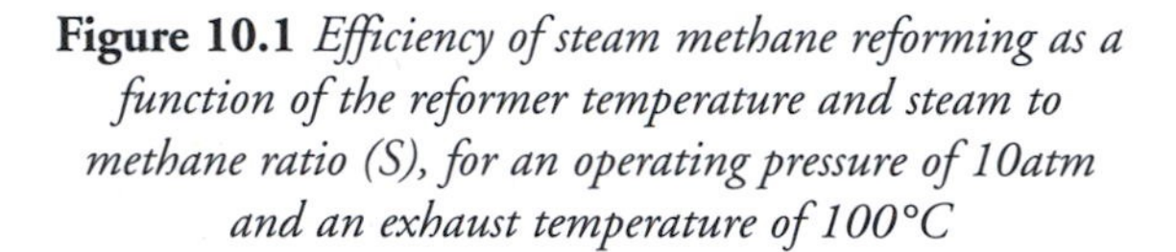

Figure 10.1 *Efficiency of steam methane reforming as a function of the reformer temperature and steam to methane ratio (S), for an operating pressure of 10atm and an exhaust temperature of 100°C*

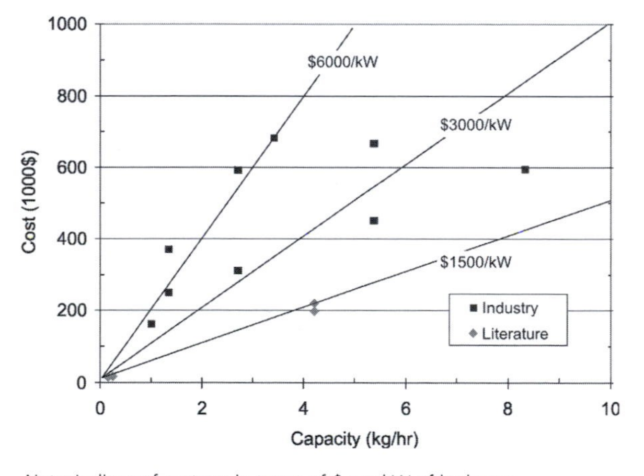

Note: Isolines of cost are in terms of $ per kW of hydrogen production capacity.
Source: Weinert and Lipman (2006)

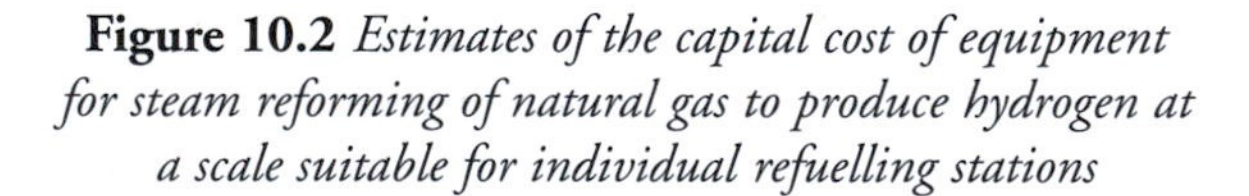

Figure 10.2 *Estimates of the capital cost of equipment for steam reforming of natural gas to produce hydrogen at a scale suitable for individual refuelling stations*

\$1500–6000/kW H_2, as shown in Figure 10.2. Costs given in the academic literature tend to be much lower than costs given by industry sources. In a mature and competitive market, unit costs should decrease with increasing capacity. Future costs for systems producing 40kg H_2/hr could drop to the \$400–600/kW range. For large refinery-scale SMR plants (operating at 850°C and 15–25atm), costs could be much less, ranging from \$80/kW of H_2 for a production of 70,000GJ/day (24,000kg/hr), to \$200/kW H_2 for a production of 7000GJ/day (2400kg/hr) according to Ogden (1999).

10.3.2 Gasification of biomass or coal

As discussed in Chapter 9 (subsection 9.3.8), coal can be gasified to produce H_2 (or other fuels). The first step is partial oxidation (which produces CH_4 and CO), followed by SMR and water shift, as in the steam reforming of methane. A speculative capital cost for large-scale (20,000–150,000m^3_{STP}/hr) systems is about \$850/kW of H_2, with an LHV efficiency of about 55 per cent. Advanced concepts involving the integration of CO_2 separation with the water-gas shift may give an efficiency of 64–68 per cent (Mueller-Langer et al, 2007). The process is easier using biomass than coal, since sulphur is absent from biomass and biomass gasifies at lower temperatures (see Chapter 4, subsection 4.3.2). Capital costs for large-scale biomass gasification systems are estimated to be \$950/kW, with a conversion efficiency of 50–60 per cent. It might be possible to narrow the gap with the capital costs of H_2 production by large-scale steam reforming of methane (\$80–200/kW) through novel membrane separation techniques (Ogden, 1999). Sewage sludge can also be gasified to produce H_2 while leaving residual ash (Midilli et al, 2002).

Both SMR and gasification of biomass or coal followed by the production and separation of hydrogen would produce two streams of CO_2, one of which would be highly concentrated and such that the CO_2 could be collected at minimal cost and with no energy penalty. About 70 per cent of the CO_2 could be captured with no energy penalty using natural gas (see Chapter 9, subsection 9.3.8). Energy would be required for the compression and transport of the captured CO_2 and its injection into underground aquifers, oil and natural gas fields, or the deep ocean. Major issues concern supplies of biomass (Chapter 4, section 4.9) and coal (Volume 1, Chapter 2, subsection 2.5.3), and the size of the available and secure C sequestration capacity (see Chapter 9, section 9.8).

10.3.3 Electrolysis of water

Electrolysis is the splitting of water with electricity through an electrochemical process, carried out using an *electrolyser*. Like a fuel cell, an electrolyser consists of many cells stacked on top of each other, where each cell consists of an anode, a cathode and an electrolyte.

There are three kinds of electrolysers: the alkaline, proton exchange membrane (PEM) and solid oxide electrolyser. The chemical reactions occurring at the cathode and anode, along with the operating temperature and other characteristics for these three kinds of electrolysers, are given in Table 10.3. Almost all hydrogen produced by electrolysis today is produced using alkaline electrolysers, while the PEM electrolyser is in commercial use in laboratories and the solid oxide electrolyser is still at an early stage of development.

PEM fuel cells have the highest specific energy (W/litre) of any fuel cell, thereby making them more compact, an advantage that would carry over to PEM electrolysers. PEM electrolysers permit high-pressure operation, potentially higher than 200atm without mechanical compression, although 14atm is a common operating pressure. The additional electrical energy required to produce hydrogen at high pressure is less than the energy required to separately produce and compress hydrogen, so pressurized operation reduces the overall energy use if compressed hydrogen is needed.

Cost

Capital costs of electrolysers as given by different sources vary widely, as shown in Figure 10.3 for

Table 10.3 *Chemical reactions occurring at the cathode and anode of electrolysers, and other characteristics*

	Type of electrolyser		
	Alkaline	**PEM**	**Solid oxide**
Cathode reaction	$4H_2O + 4e^- \rightarrow 2H_2 + 4OH^-$	$4H^+ + 4e^- \rightarrow 2H_2$	$H_2O(g) + 2e^- \rightarrow H_2(g) + O_2^-$
Anode reaction	$4OH^- \rightarrow O^{2-} + 2H_2O + 4e^-$	$2H_2O \rightarrow 4H^+ + 4e^- + O_2$	$O^{2-} \rightarrow 1/2O_2 + 2e^-$
Electrolyte	20–30% KOH	Nafion (a polymer)	Zirconia with yttria
Charge Carrier	OH^-	H^+	O^{2-}
Operating temp	60–140°°C	43–83°C	600–1000°C
Cell efficiency	80–90%	95%	90%
System efficiency	63–73%	56%	
Cost	Low	High	Medium

Note: Efficiencies are based on the HHV of hydrogen.
Source: Berry et al (2003a) and Ivy (2004)

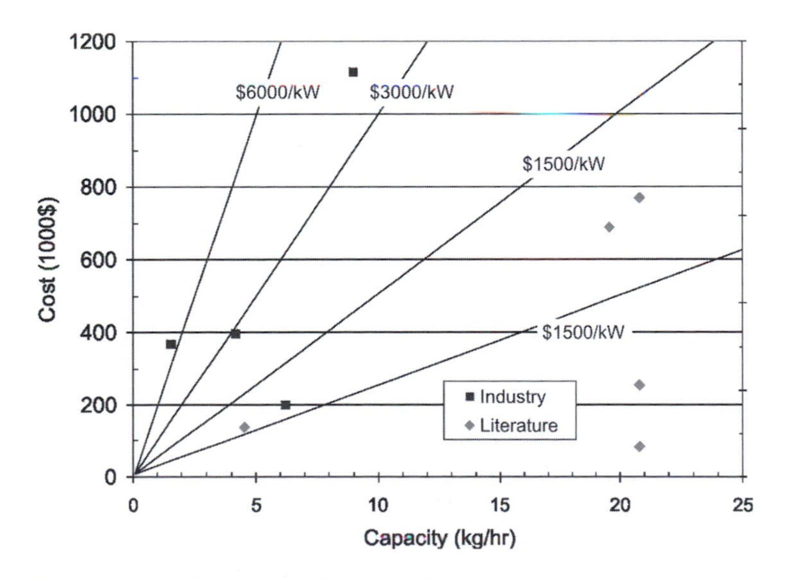

Note: Isolines of cost are in terms of $ per kW of hydrogen production capacity.
Source: Weinert and Lipman, (2006)

Figure 10.3 *Estimates of the capital cost of electrolysers (including water purification)*

alkaline electrolysers. Costs range from \$1500/kW H_2 to over \$6000/kW H_2. As with SMR plants, costs given in the academic literature tend to be much lower than costs given by industry sources. Speculative long-term costs are in the \$700–1000/kW H_2 range.

Efficiency

The theoretical minimum amount of work required to split water is given by the Gibbs free energy for the water-splitting reaction plus a term to account for any difference in pressure between the input (water) and output (H_2 and O_2):

$$Minimum\_work = \Delta G^0_{(H_2O)}(T) + RT\ln\left(\frac{[H_2][O_2]^{1/2}}{[H_2O]}\right) \quad (10.5)$$

where $\Delta G^0_{(H_2O)}$ is the standard Gibbs free energy of water decomposition, R is the universal gas constant (8.314J/mole/K), $[H_2]$, $[O_2]$ and $[H_2O]$ are the H_2, O_2 and H_2O pressures in atm, and T is absolute temperature. If water can be supplied under pressure (as steam), the required voltage decreases. As well, $\Delta G^0_{(H_2O)}$ itself decreases with increasing temperature. For water vapour and product gases at 1atm and 300K, the minimum work for decomposition is 237.141kJ/mole H_2O, which corresponds to an electrolyser voltage of 1.229 volts (Berry et al, 2003a). For production of H_2 from liquid water, the minimum required energy is 286kJ/mole, which corresponds to 1.48 volts.

The efficiency of an electrolyser cell is given by the theoretical minimum voltage divided by the actual voltage. The voltage required for electrolysis is greater than the theoretical minimum given above for three reasons: due to resistance to the flow of ions through the electrolyte and due to the activation overpotential (needed to accelerate the electrochemical reactions) at both the anode and cathode. The contributions of the resistance and of the two activation overpotentials to the total voltage required for electrolysis are shown in Figure 10.4. These contributions increase with increasing current density through the electrolyser. Thus, the cell efficiency – given by the theoretical minimum voltage divided by the actual voltage – decreases with increasing current density. However,

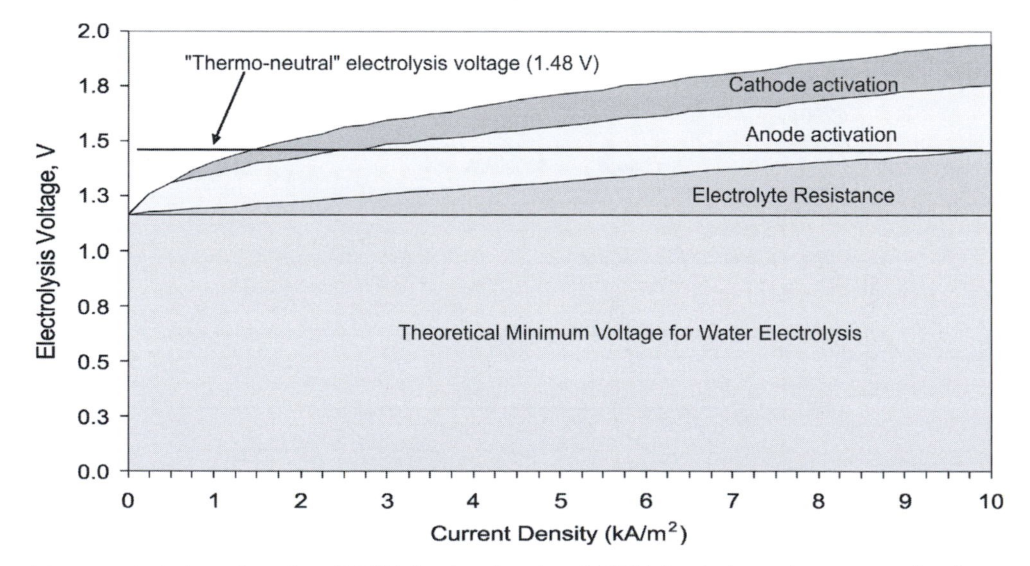

Note: The minimum required voltage shown here (1.18V) is less than the value of 1.229V given in the text because excess heat is assumed to be available.
Source: Berry et al (2003a)

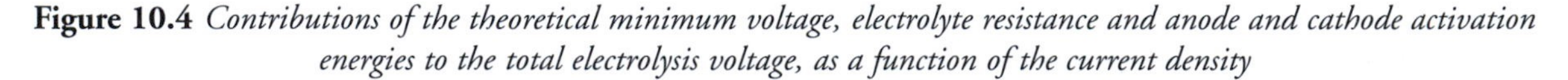
Figure 10.4 *Contributions of the theoretical minimum voltage, electrolyte resistance and anode and cathode activation energies to the total electrolysis voltage, as a function of the current density*

capital costs also decrease with increasing current density, as more power – and a greater rate of hydrogen production – can be forced through a given cell area if the current density is larger.

A distinction must be made between electrolyser cell efficiencies and overall electrolyser efficiencies. The cell efficiency refers to the hydrogen energy produced divided by the energy input to the cell, whereas the overall efficiency accounts for the additional electricity used by auxiliaries such as pumps and water purification units. Unfortunately, efficiencies are usually reported without stating which efficiency is being given. As explained above, the cell efficiency is equal to the theoretical minimum voltage needed to split water divided by the voltage used (voltage times current gives power). Cell efficiencies are 80–90 per cent for commercially available alkaline electrolysers and 90–95 per cent for PEM electrolysers, but the overall efficiency of PEM electrolysers is lower at present – only 56 per cent for the one model discussed by Ivy (2004), compared to 63–73 per cent for the alkaline electrolysers. One goal of current research programmes is to achieve a full-load system efficiency of 78 per cent, including compression to 420atm.

The efficiency of an electrolyser increases with decreasing load, because with decreasing load the current density and related energy losses decrease. This is illustrated in Figure 10.5, which gives the system efficiency for an electrolyser operating at 30atm. The efficiency increases from 78 per cent at full-load to 94 per cent at 10 per cent of full-load.

Use of byproduct oxygen and heat

The oxygen produced as a byproduct of electrolysis could be used in various industries, providing a financial and energy credit against the cost and energy used to produce hydrogen. At 71 per cent efficiency, 5000kWh of electricity would produce 1000Nm^3 of hydrogen and 500Nm^3 of oxygen that would otherwise require 250kWh through cryogenic separation from air (Kato et al, 2005). The energy credit is thus 5 per cent. This credit would of course be applicable only if the byproduct oxygen can be used, but some advanced industrial processes are more efficient if performed in pure oxygen, so there could be a growing demand for oxygen.[1]

If the electrolytic production of hydrogen using renewable sources of electricity is integrated with the pulp and paper industry, so that both byproduct oxygen and waste heat are produced, essentially all of the energy that would be lost in converting electricity to hydrogen can be captured and used (Saxe and Alvfors, 2007). A significant new use of oxygen could be combustion of coal, natural gas and biomass in pure oxygen, or gasification of coal or biomass in pure

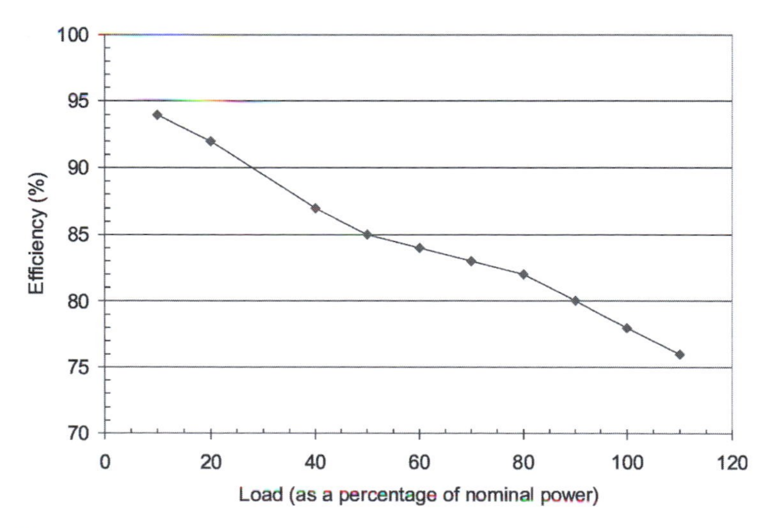

Source: Ntziachristos et al (2005)

Figure 10.5 *Variation of electrolysis efficiency with load for a typical high-pressure (30atm) electrolyser*

oxygen (in all cases with subsequent capture and sequestration of CO_2, as discussed in Chapter 9, subsections 9.3.2, 9.3.3 and 9.3.6).

Electrolysis of seawater

If seawater is electrolysed, chlorine rather than oxygen is generated at the anode, creating disposal and safety problems. If manganese oxide with tungsten is used for the anode, the result is the production of a mixture that is 99.6 per cent oxygen and 0.4 per cent chlorine. Use of a mixture of manganese, molybdenum and iron oxides results in 100 per cent oxygen over the operating temperature range of 30–90°C (Ghany et al, 2002). Thus, there are good prospects for large-scale production of hydrogen in coastal desert regions using solar energy.

Direct coupling to PV modules

Electrolysers require a DC electricity input, while PV power produces DC power. Thus, an electrolyser can be directly coupled to a PV array without intermediate DC/AC and AC/DC power conversion, thereby saving on power conditioning costs and energy losses. As noted in Chapter 2 (subsection 2.2.4), both the PV current I and voltage V increase as the incident solar radiation increases, and power output is maximized if I and V co-vary in such a way as to maximize their product. Increasing the power input to an electrolyser also entails increasing I and V; the characteristic I-V variation of an electrolyser closely matches the I-V solar module variation required for maximum power output. With an optimal series parallel arrangement of PV modules and electrolysers, PV modules and electrolysers can be directly coupled with only 5 per cent energy loss compared to the maximum power output obtained from a PV module, while avoiding the cost (about $700/kW) of active electronic coupling (Paul and Andrews, 2008). Given a PV module efficiency of 15 per cent, 5 per cent coupling losses and an eventual electrolyser efficiency of 80 per cent (averaged over all loads), the overall solar-to-hydrogen conversion efficiency would be about 10 per cent.

Coping with intermittent electricity sources

The power input to an electrolyser would be highly variable if it were coupled to PV arrays or wind turbines, whereas currently available electrolysers are designed to operate with a steady power input. Variable input would (1) reduce efficiency, (2) result in the occurrence of some H_2 in the O_2 stream, and vice versa, and (3) pose problems of corrosion of the electrode of alkaline electrolysers at times of very low or no hydrogen production.

Variable input from a fixed-speed wind turbine reduced the amount of hydrogen produced by 2.6 per cent compared to steady input in tests by Dutton et al (2000), and increased the amount of H_2 in the oxygen stream from 0.53 per cent to 0.72 per cent, for an overall reduction in the production of usable hydrogen by 2.8 per cent (efficiency decreases from 64.5 per cent to 62.7 per cent). Use of variable-speed rather than fixed-speed wind turbines will reduce fluctuations in the order of a few tens of seconds and less in duration. PEM electrolysers might be the ideal choice for fluctuating power sources, given that, like PEM fuel cells, they can start up and shut down quickly and handle fluctuating power supply well.

A relatively small battery could be used to store and release short-term fluctuations in PV or wind electricity output, thereby allowing for smoother variation in the input to the electrolyser. This would entail a small additional loss for that portion of the wind or solar output that passes through the battery. Longer periods (hours) with no wind or solar output could be bridged using the grid. Although some of the power from the grid would come from fossil fuel powerplants, the proportion would be small and, by facilitating operation of electrolysers to absorb excess wind or solar power, would permit a larger solar or wind powerplant without an economic penalty than would otherwise be the case. For isolated systems that are not connected to the grid, and where hydrogen produced by the electrolyser is used in a fuel cell to generate electricity at times when wind is inadequate to meet demand, some of the fuel cell electricity could be used to operate the electrolyser at a very low level (i.e. 10 per cent of peak load) to avoid the need to shut down the electrolyser and the associated energy losses during start-up. This option was included in a system for the Aegean islands simulated by Ntziachristos et al (2005). The overall energy loss in this case in meeting some of the electricity demand through stored hydrogen and fuel cells, and in low-level operation of the electrolyser during times of weak wind, is about 40 per cent of the wind energy output.

Ghosh et al (2003) present the results from ten years of operation of a 30kW PV array whose output is either fed to the grid, fed to a battery, or (when the battery is fully charged and there is excess PV power) fed to an electrolyser to produce H_2 (which is compressed to 120atm) and O_2 (which is compressed to 70atm), then used in a fuel cell when needed. The annual average efficiency (from DC output to AC power) ranges from 64 to 76 per cent. The round-trip efficiency through the battery is 91–95 per cent and that through the electrolyser fuel cell is 34–36 per cent (a result of an 80 per cent electrolyser efficiency, 50 per cent fuel cell efficiency and compressional energy use equal to 10 per cent of the PV output). About 50 per cent of the total electricity demand was met through the battery and 20–25 per cent through the fuel cell, giving an average efficiency in supplying AC electricity from DC electricity of 51–64 per cent.

An option that decidedly is not of interest is the dedicated production of hydrogen by electrolysis using electricity generated at fossil fuel powerplants. The overall efficiency will be far higher, and the associated CO_2 and pollutant emissions far lower, if hydrogen is produced directly from fossil fuels, either through SMR or gasification of coal. As explained in Chapter 9 (subsection 9.3.8), capture of CO_2 during the production of hydrogen from fossil fuels is inherently easier than its capture at fossil fuel powerplants.

10.3.4 High-temperature water vapour electrolysis

As noted above, the energy required for splitting water is smaller in the vapour phase than in the liquid phase. Furthermore, as temperature increases, the minimum voltage (and associated electrical energy) required for electrolysis decreases and the remaining energy can be provided as thermal energy. The decrease in electrical energy required as operating temperature increases from 200°C to 1000°C is shown in Figure 10.6. At an operating temperature of 1000°C, the required electricity input is only 71 per cent of the total required energy input. Solid oxide electrolysers operate at temperatures up to 1000°C, with an overall efficiency (based on electrical and thermal energy inputs) of 70–75 per cent (Ni et al, 2007). Thus, if heat can be supplied externally from some other source (such as industrial waste heat or concentrated solar thermal energy using the methods discussed in Chapter 2, subsection 2.7.2), then the efficiency based on the electricity input only would be about 100 per cent.

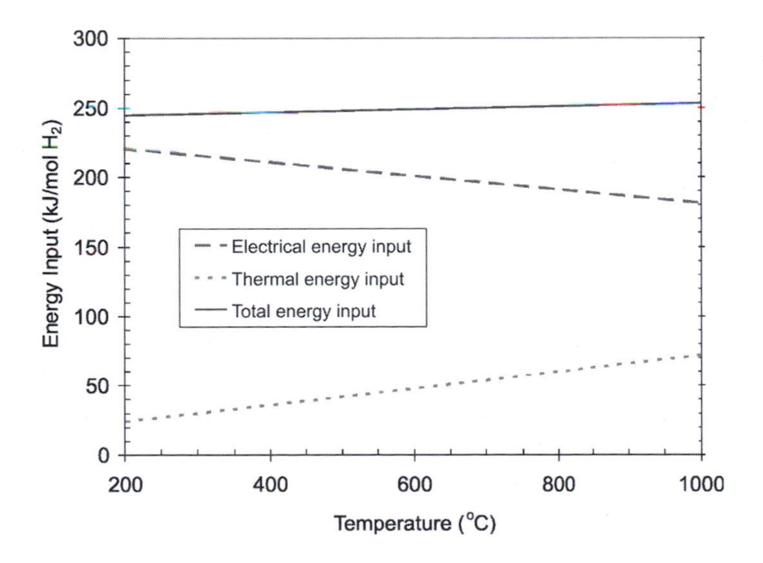

Source: Ni et al (2007)

Figure 10.6 *Variation with operating temperature of the thermal, electrical and total energy required for electrolysis of water*

Alternatively, electricity and heat for high-temperature electrolysis can be produced together from solar energy in an integrated process, as outlined in Figure 10.7. Conceptual design studies indicate that the net conversion efficiency from thermal energy to hydrogen using high-temperature water vapour electrolysis could be 47 per cent (Erdle et al, 1986). Given an effective solar collector efficiency of 50 per cent, this gives a conversion efficiency from solar energy to hydrogen of about 25 per cent. This is much larger than what might be achievable through large-scale PV arrays (15–20 per cent electricity generation × 80 per cent electrolysis efficiency, or 12–16 per cent overall sunlight-to-hydrogen efficiency) and relies largely on easy-to-make mirrors rather than on hard-to-make (and energy-intensive) semi-conductors. Another option is to use joint PV/thermal collectors. According to a computer simulation by Padin et al (2000), a sunlight-to-hydrogen conversion efficiency of 21 per cent would be achieved for various locations in the US (assuming GaAs cells at 25 per cent efficiency, a solar collector thermal efficiency of 40 per cent and an optimistic future electrolyser efficiency of 90 per cent). If excess heat can be used, the overall efficiency in using solar energy would be about 50 per cent.

10.3.5 Thermo-chemical splitting of water

At temperatures greater than 3000K, $\Delta G = 0$, meaning the H_2O will decompose spontaneously into H_2 and O_2. The only energy requirement is sufficient heat. The problem is that H_2 and O_2 also decompose (forming OH and O and H atoms), and most other materials are unstable at these temperatures. A multi-step decomposition process involving reactions with other chemical substances can be used to reduce the required temperature, resulting in a *thermo-chemical* decomposition process. A generic two-step cycle would be (Berry et al, 2003a):

$$heat + 2AO \rightarrow 2A + O_2 \qquad (10.6)$$

and:

$$2H_2O + 2A \rightarrow 2AO + 2H_2 + heat \qquad (10.7)$$

Hydrogen and oxygen are produced in different steps, so there is no need for high-temperature gas separation. In practice, few substances have been proposed that absorb sufficient heat in step one and release sufficient heat in step two. Instead, cycles with three or more steps have been studied, but with increasing complexity and cost. More then 200 thermo-chemical cycles have been reported, but many have problems. The two most promising thermo-chemical cycles are the sulphur-iodine (SI) and calcium-bromine-iron (UT-3) cycles (Berry et al, 2003a; Yildiz and Kazimi, 2006). Steinfeld (2002) proposes using concentrated solar energy to drive the thermal dissociation of ZnO(s) into Zn(g) and O_2 at 2300K. The second step would be the hydrolysis of Zn(l) at 700K to form H_2 and ZnO(s),

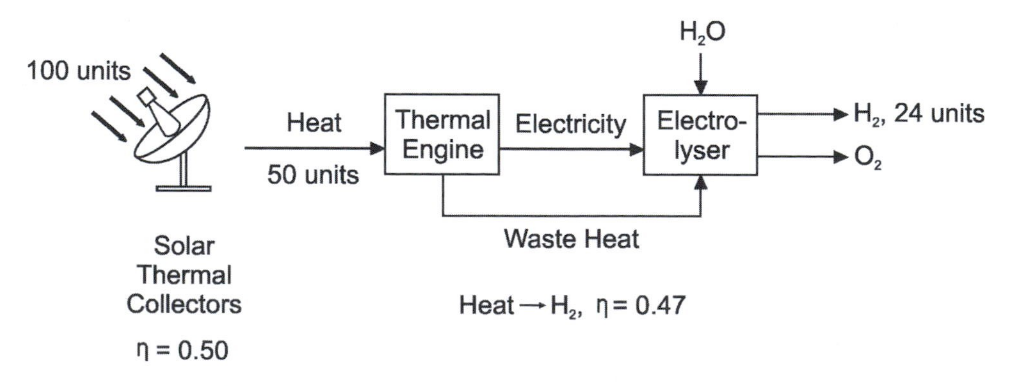

Figure 10.7 *Production of hydrogen through high-temperature electrolysis using solar thermal energy to generate electricity with a heat engine, and using waste heat from the heat engine to supply some of the energy needed to split water, thereby reducing the electricity required and increasing the overall efficiency in using solar energy*

which is recycled to the first step. The maximum feasible efficiency is 29 per cent for a solar-cavity receiver operating at 2300K and a solar flux concentration ratio of 5000.

Kolb and Diver (2008) present a preliminary screening of concepts for using solar energy to drive the thermo-chemical production of hydrogen. They identified solar tower concepts (see Chapter 2, subsection 2.3.3) capable of producing temperatures ≥850°C, ≥1450°C and ≥1800°C, and solar dish technologies capable of producing temperatures ≥900°C and ≥1800°C. The best solar tower concepts are expected to achieve solar-to-hydrogen efficiencies in excess of 20 per cent with a 75 per cent capacity factor.

For any thermo-chemical process to be viable, there would have to be essentially 100 per cent recycling of the chemicals used in each step, given the enormous throughput associated with large-scale production of hydrogen.

10.3.6 Photobiological production

Photobiological production involves the controlled production of hydrogen by certain algae or bacteria in light-driven reactions using potato starch, sugarcane juice or whey as substrates. The organisms use iron and nickel in enzymes that make hydrogen from water; the hydrogen is then used as a fuel by the organisms. The main difficulty with this method is that the byproduct oxygen inhibits the enzymes that produce hydrogen. It might be possible to develop mutants that are more tolerant of oxygen, or to continuously remove the product oxygen. Conversion efficiencies are very low – 1–2 per cent without immediate removal of oxygen, 3–10 per cent with total and immediate removal. A closed system is required so that the hydrogen gas can be collected, but there will be some losses during collection – perhaps 15 per cent. The organisms would be grown in either flat-plate or tubular bioreactors that might track the sun. There are questions concerning the feasibility of scaling up from the experimental scale to the large commercial scale. For a detailed review, see Akkerman et al (2002).

An alternative scheme would be to grow algae high in carbohydrates in open ponds, supposedly at 10 per cent sunlight-to-carbohydrate conversion efficiency (an extraordinarily high efficiency), then to transfer the algae to a chamber where a dark fermentation reaction converts the carbohydrate to hydrogen (at 28 per cent efficiency), followed by transfer of the algae back to the solar ponds to produce new starch. This is an indirect use of photosynthesis to produce hydrogen. Further details concerning this scheme are found in Das and Veziroğlu (2001) and Hallenbeck and Benemann (2002).

Burgess and Fernández-Velasco (2007) have estimated the ratio of energy production to energy inputs for photobiological production of hydrogen. For low-density polyethylene (LDPE) bioreactors (the material that results in the lowest bioreactor embodied energy among those considered), the energy return over energy invested (EROEI) ranges from 2.0 at 1 per cent photosynthetic efficiency (better than achieved at present) to about 7.0 at 10 per cent photosynthetic efficiency (the minimum considered to be necessary for the method to be economically viable). This output to input ratio is quite poor compared to what can be achieved with biomass used to generate electricity (see Chapter 4, subsection 4.6.2). LDPE has a lifespan of less than three years, so substantial amounts of waste would be generated. Glass has only a slightly smaller output to input ratio but a lifespan of at least 20 years.

10.3.7 Photoelectrochemical splitting of water

The production of hydrogen by electrolysis using electricity generated by PV panels is a two step process: first, the production of electron/electron hole pairs by the absorption of photons (and the subsequent production of an electric current); and second, the electrolytic splitting of water molecules. A photoelectrochemical device combines these steps into a single step: photon absorption creates an electron/electron hole pair that directly splits an adjacent water molecule. A photoelectrochemical cell (PEC) involves two electrodes immersed in an aqueous electrolyte, at least one of which is a semi-conductor exposed to light and able to absorb light. Three arrangements are possible (Bak et al, 2002):

- a photo-anode made of an n-type semi-conductor and a cathode made of metal;
- a photo-anode made of an n-type semi-conductor and a photo-cathode made of a p-type semi-conductor;

- an anode made of metal and a photo-cathode made of a p-type semi-conductor.

As discussed in Chapter 2 (subsection 2.2.1), an n-type semi-conductor consists of an oxide semi-conductor (SiO_2 in the case of PV cells) doped with atoms having a valence of five (one more than that of silicon). The valence-5 atoms readily release an electron upon absorption of electromagnetic radiation of sufficient energy, which is then carried away, leaving behind an electron hole h^*. In PECs, TiO_2 is commonly used as a semi-conductor instead of SiO_2, owing to its high corrosion resistance, while the cathode is typically platinum (an expensive and rare metal). The reactions occurring in PECs having an n-type semi-conductor as anode are:

At the anode,

$$h\nu \rightarrow 2e^- + 2h^* \tag{10.8}$$

At the anode/electrolyte interface,

$$2h^* + H_2O_{(liquid)} \rightarrow \frac{1}{2}O_{2(gas)} + 2H^+ \tag{10.9}$$

At the cathode/electrolyte interface,

$$2H^+ + 2e^- \rightarrow H_{2(gas)} \tag{10.10}$$

The energy of the incident photons is equal to $h\nu$, where ν = the frequency of radiation and h = Plank's constant (6.6262×10^{-34} joule sec). Electrons released at the anode flow through an external circuit to the cathode, where they join the H^+ that had migrated through the electrolyte from the anode to the cathode. The net reaction is:

$$2h\nu + H_2O_{(liquid)} \rightarrow \frac{1}{2}O_{2(gas)} + H_{2(gas)} \tag{10.11}$$

The circuit is illustrated in Figure 10.8. H_2 and O_2 are generated at the anode and cathode, respectively, and so are already separated. The minimum required photon energy ($h\nu$) is 1.23eV, which corresponds to a wavelength of 1.01μm.[3] In principle, any photon with energy greater than 1.23eV can dissociate water in the PEC. However, various resistances within the circuit and voltage losses at the contacts increase the required photon energy to about 2.0eV (wavelengths of 0.62μm or less). Undoped TiO_2 has a band gap (the energy difference between the valence and conduction bands) of 3.2eV. Thus, only photons with energy greater than 3.2eV (wavelengths of 0.39μm or less) can boost electrons in undoped TiO_2 to the conduction band, where they can be used in reactions ((10.8)–(10.10)). This excludes photons in the visible part of the spectrum (0.4–0.7μm), where the bulk of the incident solar energy lies. A research objective is to modify TiO_2 so as to bring the band gap closer to the practical energy requirement of 2.0eV, or to find corrosion-resistant oxides with a smaller band gap, so that more of the available photons can be used to split water, thereby increasing the solar-to-hydrogen conversion efficiency. Hybrid PECs have been created, consisting of a TiO_2 thin-film in contact with the liquid electrolyte, next to a semi-conductor such as SiO_2, which has a band gap of

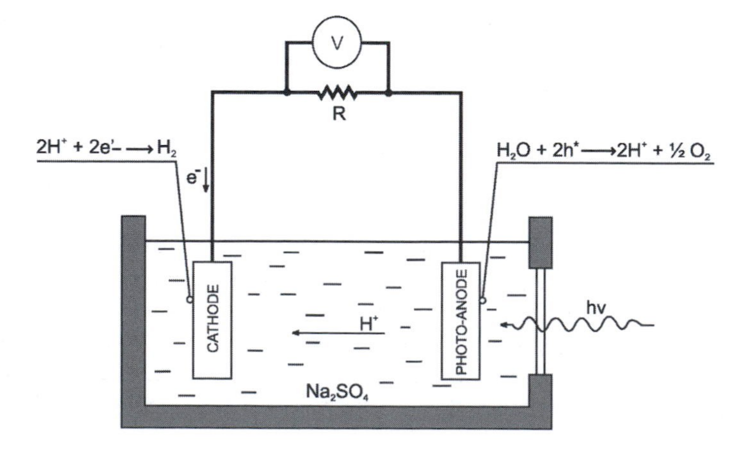

Source: Bak et al (2002)

Figure 10.8 *Structure of a photoelectrochemical cell for the direct electrolysis of water using solar energy*

only 1.2eV. Light passes through the TiO_2 film, which absorbs photons with energy greater then 3.2eV, then passes through the SiO_2, which absorbs photons with energies between 1.2 and 3.2eV.

Optimal performance of a PEC requires the imposition of an external voltage across the cell, which consumes energy. The net result is that the maximum achievable efficiency is 10–12 per cent with current losses, assuming that a corrosion-resistant material with a band gap of 2.3eV can be devised. Measured efficiencies achieved in real PECs using sunlight are in the range of 0.2–2 per cent, excluding PECs with corrosive anodes. However, very little is known about the durability of any kind of PEC, or about the long-term cost prospects. For further information, see Nowotny et al (2007).

10.3.8 Photochemical splitting of water

This involves using sunlight to drive reactions that directly split water into hydrogen and oxygen. A sensitizer is excited by visible light and yields electrons that are used for the reduction of water. The first step in the natural process of photosynthesis also involves splitting water using solar energy, so photochemical splitting of water mimics, to some extent, processes occurring in photosynthetic organisms. A catalyst is usually needed, usually a noble metal in the form of small particles. There has been difficulty getting reproducible results (Momirlan and Veziroǧlu, 2002).

10.4 Storage of hydrogen

In a hydrogen economy, hydrogen would need to be stored in large quantities on a seasonal basis if used for heating. Intermediate-scale storage and possibly seasonal storage would be needed as a buffer against fluctuations in the supply of variable renewable energy sources such as wind and solar, unless fossil fuel or biomass energy backups are available. Intermediate-scale storage would also be needed at hydrogen refuelling stations for cars, trucks and trains, and at airports. Finally, onboard storage would be needed in cars, trucks, trains, aircraft and ships. Here, we discuss the technical characteristics (including energy losses) associated with different ways of storing hydrogen.

10.4.1 Energy required for compression and liquefaction of hydrogen

Hydrogen has a very low density – 0.0899kg/m^3 at 1atm pressure and 273K. Prior to storage, hydrogen would be either compressed or liquefied so as to reduce the required storage volume. Liquid hydrogen has a density of about 70.8kg/m^3, the same as gaseous H_2 at 788atm pressure. In current demonstration vehicles using compressed hydrogen, storage at 700atm is state-of-the-art. So as to permit smooth refuelling, the H_2 at the refuelling station is stored at 880atm (Wietschel et al, 2006).

Compression of a gas tends to increase the temperature of the gas, causing heat to be lost to the surroundings. The slower the compression, the smaller the temperature rise during compression. The limiting case would be isothermal compression, which is impossible in practice. The opposite extreme is adiabatic compression, in which no heat is lost to the surroundings. The mechanical energy required in the two cases is given by the equations:

$$W = p_o V_o \ln(p_1 / p_o) \tag{10.12}$$

for isothermal compression, and:

$$W = \left[\frac{\gamma}{\gamma - 1}\right] p_o V_o \left[\left(\frac{p_1}{p_o}\right)^{(\gamma-1)/\gamma} - 1\right] \tag{10.13}$$

for adiabatic compression (Bossel, 2006). In these equations, γ is the adiabatic coefficient (equal to 1.41 for hydrogen, 1.31 for methane), p_o is the initial pressure (in Pa), V_o is the initial specific volume (equal to 11.11m^3/kg for H_2, 1.39m^3/kg for CH_4) and p_1 is the final pressure. The mechanical energy required for isothermal and adiabatic compression of H_2 from 1atm (0.1MPa) to 1000atm (100MPa) is plotted in Figure 10.9. Also shown in Figure 10.9 is the ratio of the energy required to compress hydrogen adiabatically to that required to compress methane adiabatically; this ratio increases from 8.5 when compression first starts, to 10 with compression to 100atm (the ratio for isothermal compression is constant at 8.0). Actual energy use falls between the isothermal and adiabatic limits, and takes into account the efficiency of the compressor in converting electrical energy into mechanical energy. A

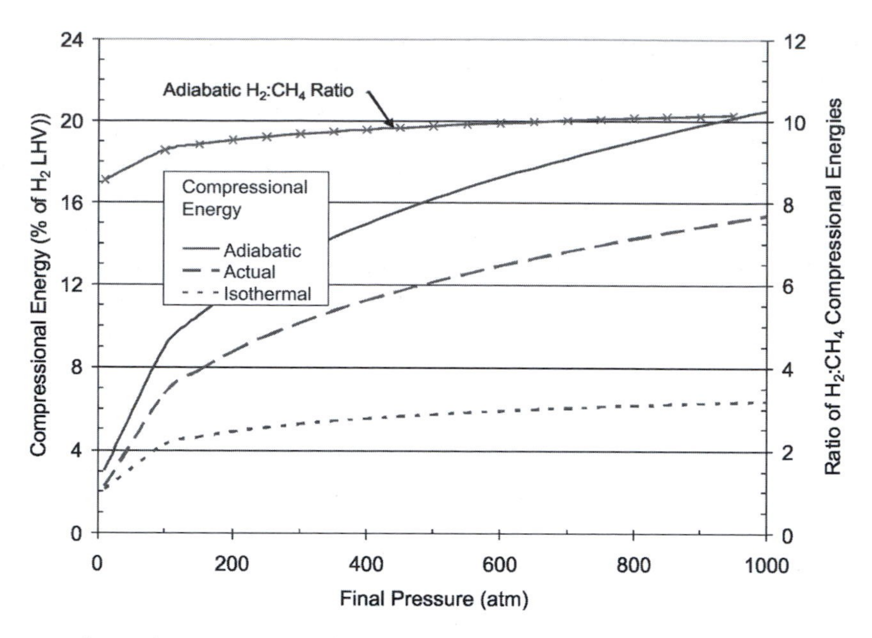

Source: Based on energy use for a real compressor as given by Bossel (2006)

Figure 10.9 *Energy required to compress hydrogen from a pressure of 1atm to the indicated pressure, assuming isothermal and adiabatic compression (computed from Equations (10.11) and (10.12))*

curve based on data from a real compressor is also shown in Figure 10.9. Each additional increment in desired final pressure requires less additional energy than the previous increment. Compression of hydrogen from 1atm to 350atm requires about 10 per cent of the energy content (on an LHV basis) of the hydrogen, whereas compression to 700atm requires about 14 per cent. If the electricity used for compression is generated with an efficiency of 40 per cent, the primary energy required for compression to 700atm is 35 per cent of the energy content of the hydrogen.

If could be argued that, if there is a renewable source of electricity (such as wind, solar or hydro) for electrolysis of water and for compression of the product hydrogen, diversion of some electricity from electrolysis to compression simply reduces the amount of hydrogen produced by 14 per cent (in the above case). However, for the same hydrogen production, the electricity used for compression could instead displace fossil fuel generation elsewhere if there were no need for compression. Thus, as long as there is fossil fuel generation somewhere on the grid that can be displaced, it is appropriate to divide the electricity requirement for compression by the efficiency of the fossil fuel powerplant whose output would be displaced in order to determine the true energy cost of hydrogen compression. However, at times of excess renewable energy there is no displaceable fossil fuel electricity, so the energy cost at those times would be just the direct compressional energy use.

Hydrogen can be liquefied if cooled to 20.3K. Hydrogen molecules, whether liquid or gaseous, exist with two different nuclear spin arrangements. The higher energy arrangement (spins aligned) is called *orthohydrogen*, while the lower energy arrangement (spins opposite) is called *parahydrogen*. The equilibrium fraction of parahydrogen increases from 25 per cent at 300K to 99 per cent at 20K. Spontaneous conversion of orthohydrogen to parahydrogen after liquefaction releases enough heat to vaporize 10 per cent of the LH_2 (liquid hydrogen) within ten hours and 50 per cent within one week. For this reason, the conversion to parahydrogen is intentionally accelerated in LH_2 plants, but this requires removing more heat during the liquefaction process: 11.7MJ/kg for orthohydrogen, 14.1MJ/kg for parahydrogen. These are the theoretical minimum energy requirements for liquefaction. Current industrial liquefaction plants require 10–14kWh of electricity per kg of pure para-LH_2 (36–52MJ/kg), equal to 30–42 per cent of the LHV of the hydrogen (120MJ/kg), with small plants being less efficient than large plants. Magnetocaloric refrigeration might be able to reduce the energy requirement to 7kWh/kg (25KJ/kg) or about 20 per cent of the hydrogen energy content (Berry et al, 2003b). According to Wietshel et al (2006), hydrogen liquefaction today must be done on a relatively large scale in order to be relatively efficient, and so will have to occur in association with centralized production plants.

It is possible to recover about 25 per cent of the energy used to liquefy hydrogen (that is, 9MJ/kg) if it is used to absorb heat from the low-temperature leg of a natural gas turbine using helium as a working fluid and LH_2 as a fuel (Trevisani et al, 2007). Similar ideas are being explored for LNG. If magnetocalorific refrigeration succeeds in reducing the liquefaction requirement to 18MJ (from 36MJ today with conventional liquefaction) and this is combined with recovery of 25 per cent of the remaining energy requirement, the net energy requirement when be only 13.5MJ/kg, or 11 per cent of the energy value of the hydrogen.

Hydrogen can also be stored in compressed form but at very low temperature – 80K. An insulated vessel with an external volume of 250 litres could store 2.5kg of hydrogen at 200atm and 300K (sufficient for a 500km range in an advanced vehicle), but could store 8.5kg at 200atm and 80K. The theoretical minimum energy to compress orthohydrogen to 250atm while cooling it to 80K is 2.5kWh/kg, compared to a theoretical minimum of 3.25kWh/kg for liquefaction of orthohydrogen and 3.92kWh/kg for liquefaction of parahydrogen (Berry et al, 2003b, Figure 4). Further energy would be saved due to the fact that the equilibrium fraction of parahydrogen is only 50 per cent at 80K, so only about half of the 0.7kWh/kg difference between liquid para- and orthohydrogen would be added to the energy requirement. Safety would be enhanced, as the theoretical maximum amount of energy that can be released from hydrogen at 200atm due to a rupture of the storage tank is 7.5 times smaller at 80K than at 300K (Berry et al, 2003b, Figure 7).

As with compression, the liquefaction energy use should be divided by the efficiency of the fossil fuel powerplant whose electricity output would otherwise be displaced in the absence of liquefaction in order to get a true account of the primary energy cost of hydrogen liquefaction. Thus, assuming future advanced powerplants with 50 per cent efficiency, the true energy cost of liquefaction at today's energy requirement of 10–14kWh/kg would be equal to 60–84 per cent of the energy value of the hydrogen. With possible magnetocalorific refrigeration and recovery of 25 per cent of the energy used to liquefy hydrogen, the true energy cost would be 15 per cent of the energy value of the hydrogen.

At fuelling stations, a non-negligible amount of energy is required for storage and for fuelling of vehicles, as discussed later.

10.4.2 Large-scale underground storage

Very large amounts of hydrogen can be stored underground at pressures of up to 70atm (7MPa, or about 1000 pounds per square inch) in depleted oil or gas fields, aquifers and salt or rock caverns and withdrawn as needed. The city of Kiel (Germany) has been storing town gas with 60–65 per cent H_2 at 80–160atm pressure in a 32,000m^3 cavern at a depth of 1330m since about 1971, while Gas de France has stored hydrogen-rich refinery gases in an aquifer and Imperial Chemical Industries has stored hydrogen in salt mine caverns in the UK (Pottier and Blondin, 1995). Seasonal underground storage of natural gas is also practised today, in order to build up local inventories for the winter heating peak. In gas or oil fields and aquifers, only one third to two thirds of the stored hydrogen is accessible, as the remainder must remain as a cushion in order to maintain the reservoir pressure.

10.4.3 Intermediate and small-scale above-ground storage

For small to intermediate amounts of storage, hydrogen can be stored as a liquid or as a compressed gas. As noted above, hydrogen must be cooled to a temperature of −253°C (20K) in order to become a liquid, a process that consumes an amount of energy equal to 30–40 per cent of the energy content of the hydrogen being liquefied using current liquefaction technology. Liquid hydrogen is stored in special containers called *dewars* that have a double skin with a vacuum in between. Dewars range in capacity from a few kilograms to hundreds of tonnes. The main advantage of storage as liquid hydrogen is that higher hydrogen densities (and hence more compact storage) are possible than with compressed hydrogen, unless the hydrogen is compressed to 10,000psi (690atm).

10.4.4 Storage in ground-transportation vehicles

The options for onboard storage of hydrogen in ground-transportation vehicles include metal hydrides (which are heavy and thus incur a weight penalty) or compressed or liquefied hydrogen, all of which are discussed in Volume 1 (Chapter 5, subsection 5.4.8). As discussed earlier, all of the storage systems would be relatively bulky, even in highly efficient vehicles, compared to the size of gasoline tanks required for the same driving range with today's inefficient vehicles.

10.5 Transmission of hydrogen

Most hydrogen that is produced today is used at the production site, but some has been transported by truck as liquefied or compressed hydrogen, and by gas pipelines at pressures of up to 100atm (Ogden, 1999). These pipelines were built to transport hydrogen, but there is already an extensive network of pipes to transport natural gas. However, this network will not be able to be used to transport hydrogen if potential problems of embrittlement of the pipes (discussed in subsection 10.5.3) are not resolved. Instead, a parallel network to transport hydrogen would have to be built before the natural gas network is phased out. This would be feasible for the high-pressure transmission network, but could be very difficult for the local distribution network due to lack of space. During the early stages of the hydrogen economy, it should be possible to inject up to 17 per cent hydrogen by volume into the natural gas network (which of course would make sense only if the hydrogen is produced from energy sources other than natural gas).

Here, we discuss the energy losses associated with, and cost of, transmission of hydrogen by pipeline in comparison to transmission of natural gas.

10.5.1 Large-scale distribution

The energy flux Q (J/s or watts) associated with the flow of a given gas through a pipeline is given by:

$$Q = Av\rho HV \tag{10.14}$$

where A is the cross-sectional area of the pipe, v is the flow velocity, ρ is the gas density and HV is the LHV or the HHV of the gas (depending on one's accounting preference). Given hydrogen and natural gas densities at STP of 0.0899kg/m^3 and 0.719kg/m^3, respectively, and HHVs from Table 10.1, the energy flux for methane would be 3.13 times greater than that for hydrogen for the same pipeline diameter, flow velocity and gas pressure. Thus, for the same energy flux, the hydrogen flow velocity would have to be 3.13 times greater or the pipeline diameter 1.77 times larger. The pumping power needed to transmit a gas depends on the flow velocity and the pressure drop ΔP along the pipe and is given by Bossel (2006) as:

$$P = Av\Delta P = \left(\pi\frac{D^2}{4}\right)v\left(\frac{L}{D}\frac{1}{2}\rho v^2\zeta\right) \tag{10.15}$$

where ζ is a resistance coefficient (assumed to be the same for hydrogen and methane), D is the pipe diameter and L is the length of pipe. Thus, the pumping power varies with v^3D, but since v varies with $1/D^2$ for a given flow volume, the required pumping power for a given flow varies with $1/D^5$.

For the same pressure, pipeline diameter and energy flux (and thus 3.13 times the flow velocity), the ratio of the pumping powers required for hydrogen to that required for methane is:

$$P_{H_2}/P_{CH_4} = (\rho_{H_2}/\rho_{CH_4})(v_{H_2}/v_{CH_4})^3 = 3.83 \tag{10.16}$$

However, for the same pressure, flow velocity and energy flux (but 1.77 times the pipe diameter), the ratio of the pumping powers required for hydrogen to that required for methane is:

$$P_{H_2}/P_{CH_4} = (\rho_{H_2}/\rho_{CH_4})(D_{H_2}/D_{CH_4}) = 0.221 \tag{10.17}$$

Thus, if new pipelines for hydrogen are built with 1.77 times larger diameter than existing natural gas pipelines, the energy required to transmit hydrogen is about one quarter that required to transmit natural gas, and a factor of 17.3 (= 1.77^5) less than if existing natural gas pipelines were to be used. However, as existing natural gas pipelines are up to 1.2m in diameter, this would necessitate a hydrogen pipeline that is 2.1m in diameter, which may be impractical.

The fractional energy use per 100km of pipeline distance is given by Equation (10.15) divided by Equation (10.14) (times 100,000 to convert 100km to metres). That is:

$$\left(\frac{P}{Q}\right)_{\text{per 100km}} = 100{,}000\frac{v^2\varsigma}{2D\cdot HV} \qquad (10.18)$$

where ς is a coefficient that now also takes into account the departure of actual energy use from the theoretical energy use. If D = 1.3m, v = 18.3m/s and ς = 0.02775, then we get an energy loss for methane transport of 2.5 per cent per 1000km (the approximate value given in Volume 1, Chapter 3, subsection 3.2.10). The energy used to push the gas through a pipeline is taken from the gas flow itself and used to power compressors that are spaced every 60–150km along the pipeline. Figure 10.10 shows the extra gas energy that must go into the pipeline at the beginning, as a fraction of the energy delivered at the end of the pipe for methane with the above parameter values, as well as for hydrogen with D = 1.0m and v = 31.3m/s or with D = 1.77m and v = 10m/s. In the first case, the energy consumption for transport a distance of 5000km is 63.5 per cent of the delivered energy, while in the second case it is only 2.9 per cent of the delivered energy.[4] Thus, if the existing natural gas pipeline network were to be used to distribute hydrogen, there would be a very significant energy loss. However, if new pipelines with D only 31 per cent larger were built, then v = 18.3m/s and the energy consumption would be the same as for methane – 13.3 per cent of the delivered energy over a distance of 5000km. This is comparable to the losses associated with DC transmission at 800kV – about 2.5 per cent/1000km (see Table 3.18).

Instead of building larger pipelines, hydrogen could be compressed to a greater pressure than natural gas in order to achieve the same energy flux. This will increase the energy required for compression. For both natural gas and hydrogen, a larger diameter pipe will increase the capital cost of the pipe but will reduce energy losses for a given flow rate, so both compressor capital and operating costs will decrease with wider pipes. There

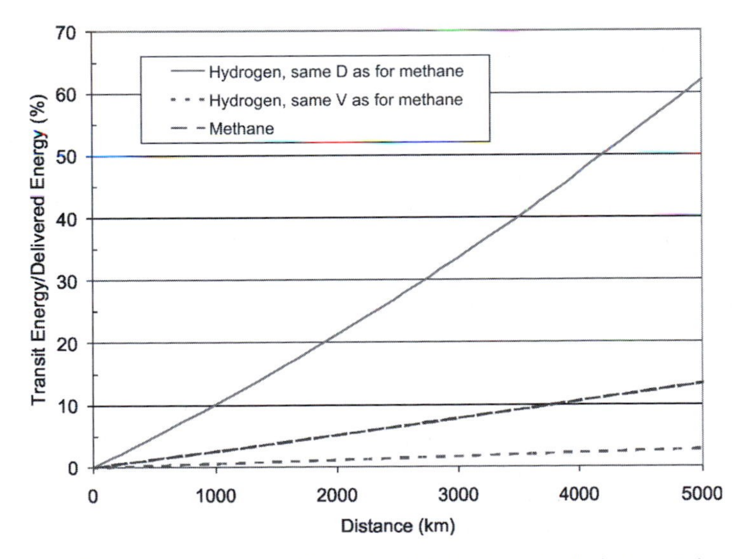

Note: The magnitude of the energy flow is recomputed every 150km, which slightly reduces the absolute energy loss on successive segments.
Source: Computed from Equation (10.18) assuming that the required energy is taken from the gas flow

Figure 10.10 *Energy required to transmit methane and hydrogen for distances of up to 5000km, the latter assuming either the same pipe diameter as for methane (and a 3.13 times greater velocity) or the same velocity as for methane (and a 1.77 times greater pipe diameter)*

will therefore be some optimal pipe diameter that minimizes the total cost of transmitting hydrogen, given the capital costs of pipes and compressors and of the energy used to power the compressors. The optimal diameter depends on the delivery pressure, energy delivery rate and the transport distance. Figure 10.11 shows the variation in the total cost of transmitting natural gas and hydrogen as a function of pipe diameter for one particular set of conditions. There is no fixed ratio between the costs of transmitting hydrogen and natural gas. For the conditions assumed in Figure 10.11, switching from natural gas to hydrogen at the diameter optimized for natural gas (i.e. using existing pipelines, if that were possible), would multiply the cost by a factor of 2.5. The optimal diameter for hydrogen is 60 per cent greater than the optimal diameter for natural gas, and the cost at the optimal diameter for hydrogen is twice that at the optimal diameter for natural gas.

10.5.2 Local distribution

Development of a new pipeline infrastructure to supply individual houses and businesses for heating or combined heat and power generation would be a major undertaking. It might be possible to use the existing natural gas pipe network, if seals and compressors are replaced with units compatible with hydrogen. However, the increasing use of plastic pipes in local natural gas distribution networks is a potential problem, as plastic is four to six times more permeable to hydrogen than to natural gas. Provision of hydrogen to local district heating systems (discussed in Chapter 11, subsection 11.2) would be a much easier undertaking, as hydrogen would need to be supplied only to the central heating facility rather than to every building on the system.

Use of hydrogen in the transportation sector would require a much smaller transmission infrastructure than its direct use in buildings, as only the refuelling stations would need to be supplied with hydrogen. This could be provided by truck (carrying compressed or liquefied hydrogen from a centralized electrolysis facility) or by pipeline. The local distribution network could be fed by the long-distance hydrogen pipeline, or from a centralized hydrogen-production facility. Conversely, hydrogen could be produced onsite at refuelling stations by electrolysis, eliminating the need for both long-distance and local hydrogen transmission. Transport of liquid hydrogen a distance of 300km by

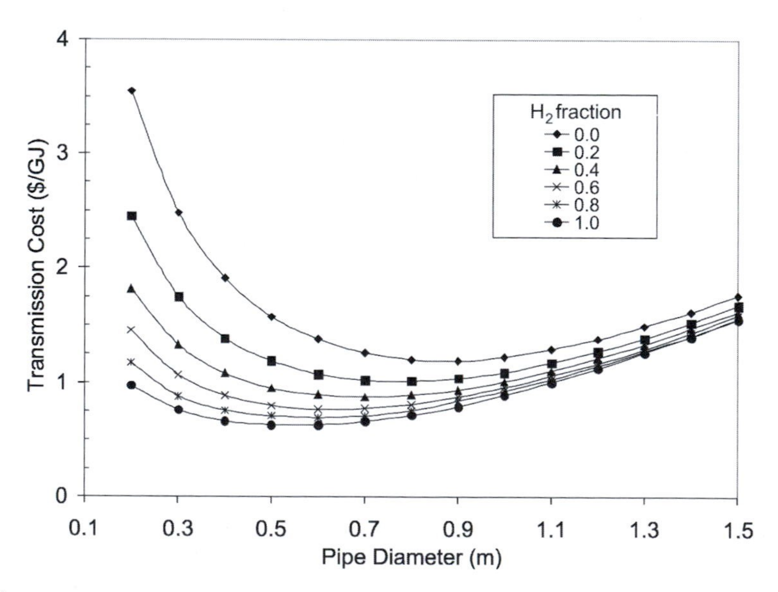

Source: Öney et al (1994)

Figure 10.11 *Variation in the optimal pipe diameter for transmission of natural gas and hydrogen for the given conditions*

truck uses an amount of energy equal to 1.1 per cent of the transported hydrogen according to Wietschel et al (2006).

10.5.3 Hydrogen embrittlement

Hydrogen embrittlement is the process by which various metals, most importantly high-strength steel, become brittle and crack following exposure to hydrogen. Embrittlement involves hydrogen atoms diffusing through the metal and recombining into hydrogen molecules in small voids in the metal matrix, thereby creating pressure from inside the cavity that causes the metal to crack. If steel is exposed to hydrogen at high temperatures, hydrogen will combine with trace carbon in the steel to form tiny pockets of methane. This methane collects in the voids at high pressure and initiates cracks in the steel. This process of *decarburization* of the steel also leads to loss of strength. At present it is not clear how to solve these problems, although there are several possible solutions.

10.6 End-use applications of hydrogen

Hydrogen can be used as a fuel in all applications where fossil fuels are currently used – for ground transportation in cars, truck, buses and railway locomotives; in aircraft; for the generation of electricity; and for space and hot water heating. It can also be used as a reducing agent, in place of coke, in the production of steel. In this section, practical and technical considerations in the uses of hydrogen as an energy carrier are discussed, while environmental and safety-related issues are discussed in section 10.8.

10.6.1 Direct use of hydrogen for surface transportation

Pure hydrogen could be used in fuel cell automobiles if fuel cells fall in cost by the factor of ten or so needed for them to be affordable, and if the vehicles are sufficiently energy efficient that the required volume of hydrogen (probably as a compressed gas) is manageable. Cost, performance and technical challenges related to the use of hydrogen-powered fuel cells in automobiles are discussed at some length in Volume 1 (Chapter 5, subsections 5.4.8, 5.4.10 and 5.4.11). For the convenience of the reader, a simplified version of Table 5.22 from Volume 1 is presented here as Table 10.4. As seen from Table 10.4, advanced hydrogen-fuelled fuel cell vehicles are projected to have almost four times better fuel economy than a current comparable gasoline vehicle and 1.75 times better fuel economy than an advanced gasoline–electric hybrid vehicle. However, as also discussed in Volume 1 (Chapter 5, subsection 5.4.11) and summarized in Table 10.5, direct use of electricity from carbon-free energy sources (such as solar and wind energy) in battery vehicles would be about two and one half times as effective in displacing fossil fuels than using the electricity to make hydrogen that is then used in fuel cells. Thus, optimal use of renewable energy requires maximal reliance on batteries in plug-in hybrid vehicles, with hydrogen used in fuel cells only to provide extended driving range when needed.

As discussed on Volume 1 (Chapter 5, subsection 5.4.8), limited global supplies of Pt are likely to be a significant constraint on the future size of a global fuel cell vehicle fleet. A fleet of 1 billion fuel cell vehicles might be feasible, but not much more. This implies that a strong reliance on auto-sharing schemes so as to reduce the required global stock of vehicles, and much stronger reliance on efficient public transportation and good urban planning so as to minimize the need for automobiles for daily urban travel, will be important. Hydrogen could be used in a modified internal combustion engine (and thus without Pt constraints), but the vehicle efficiency would be much less than using a fuel cell (although about 10–20 per cent more efficient than an engine optimized for gasoline), which would make it difficult to store enough hydrogen onboard to have a driving range comparable to current gasoline and diesel vehicles.

Fuel cells are the ideal energy-conversion devices for use in buses and locomotives, where the need for compact onboard storage of hydrogen is less of a problem (for buses, compressed hydrogen could be stored on the roof of the vehicle). Hydrogen-powered fuel cells could also serve as auxiliary power units in trucks that might use some biofuel for their primary energy requirements. The global demand for Pt would not be excessive (particularly if fuel cells with higher operating temperature are used in locomotives). As

Table 10.4 *Characteristics and performance of current and possible future internal combustion engine (ICE), ICE–battery hybrid and fuel cell–battery hybrid automobiles according to the analysis of Heywood et al (2003)*

Fuel	Gasoline		Diesel		Hydrogen
Year	2001	2020	2020		
Technology	Current	Advanced	Advanced	Advanced	Advanced
Propulsion system	SI ICE	SI hybrid	CI ICE	CI hybrid	FC hybrid
Transmission	Auto	CVT	ACT	CVT	Direct
	Energy and fuel use during operation				
US urban MJ/km	2.82	1.20	1.53	1.03	0.66
US highway MJ/km	2.06	0.91	1.04	0.78	0.51
litres gasoline eq/100km, urban	8.7	3.7	4.7	3.2	2.0
litres gasoline eq/100km, highway	6.4	2.8	3.2	2.4	1.6
miles per gallon, urban	27	63	50	74	115
miles per gallon, highway	37	83	73	97	149
	Lifecycle energy use (MJ/km)				
Operation	2.48	1.07	1.31	0.92	0.59
Fuel production	0.52	0.22	0.18	0.13	0.25
Vehicle manufacture	0.29	0.26	0.26	0.26	0.28
Total	3.29	1.55	1.75	1.31	1.12

Note: FC = fuel cell, SI = spark ignition, CI = compression ignition, ACT = auto-clutch transmission, CVT = continuously variable transmission. Lifecycle energy use is computed assuming driving to be 55 per cent urban, 45 per cent highway. Underlying vehicle characteristics are given in Volume 1, Table 5.22.

discussed in Volume 1 (Chapter 5, subsection 5.7.3), use of fuel cells in locomotives could give a 60 per cent improvement in the locomotive fuel efficiency compared to diesel locomotives.

10.6.2 Use of hydrogen to make methanol from biomass for transportation

If a suitable system for onboard storage of pure hydrogen in cars is not feasible, then methanol could be used as a hydrogen carrier and used either in fuel cells or in an internal combustion engine. As discussed in Volume 1 (Chapter 5, subsection 5.4.8), methanol and other hydrocarbon fuels can be converted to a hydrogen-rich gas with onboard processors. The use of hydrocarbon fuels with fuel cells is less desirable than the direct use of hydrogen, due to the added cost and complexity and reduced efficiency. However, methanol is the least problematic of the various hydrocarbon fuels in technical terms, although it is toxic and soluble in water, which presents environmental hazards.

Methanol can be produced from biomass, but it is richer in both hydrogen and oxygen than biomass, so both must be added, the net reaction being:

$$CH_{1.42}O_{0.64} + 1.79H_2 + 0.18O_2 \rightarrow CH_3OH \quad (10.19)$$

The mass balance of this reaction is such that 0.74 tonnes of biomass and 0.11 tonnes of hydrogen would be used per tonne of methanol produced. If the hydrogen along with the required carbon were supplied by steam reforming of biomass with a water shift reaction, the requirement would be 1.32 tonnes biomass/tonne of methanol plus heat. However, if the hydrogen were supplied externally (ideally produced by electrolysis using electricity from renewable energy sources), only half as much biomass (0.74t/t) would be required. Actual biomass requirements are estimated to be 0.85 and 1.77 tonnes biomass/tonne of methanol with and without external hydrogen, respectively, so one tonne of hydrogen replaces about 8 tonnes of biomass (Overend and Chornet, 1984).

Given HHVs for biomass, methanol and hydrogen of 19.1GJ/t, 22.7GJ/t and 141.9GJ/t, respectively, the

Table 10.5 *Comparison of energy flows and overall efficiency in the utilization of renewable energy for hydrogen produced from solar electricity and used in a fuel cell, and for electricity produced from renewable energy and used in a battery electric vehicle*

FCV with compressed H_2 produced using electricity from PV arrays									
Step	**Sunlight**	**PV array**	**Power conditioning**	**H_2 production**	**H_2 transport**	**H_2 compression**	**Fuel cell**	**Drivetrain**	**Solar energy to wheels**
Efficiency (%)		15	85	80[a]	98[b]	90[c]	51[d]	87[e]	
Remaining energy	1.00	0.1500	0.1275	0.1020	0.1000	0.0900	0.0450	0.0391	3.91%
Battery-powered vehicle using electricity from PV arrays									
Step	**Sunlight**	**PV array**	**Power conditioning**		**DC transmission**		**Battery recharger**	**Drivetrain**	**Solar energy to wheels**
Efficiency (%)		15	85		96.3[f]		95[g]	87	
Remaining energy	1.00	0.1500	0.1275		0.1228		0.1166	0.1015	10.15 per cent

Note: [a]A reasonable target for future system efficiency, including compression to 30atm pressure for subsequent transmission by pipeline (see subsection 10.3.3). [b]Assuming transport a distance of 1000km at a pressure of 30atm, with an energy use equal to 2.1 per cent of the hydrogen energy, based on subsection 10.5.1. [c]The additional energy required for compression from 30atm to 700atm is about 9.6 per cent of the hydrogen energy content based on Figure 10.9. [d]Net efficiency under typical driving conditions, after subtracting energy use by the fuel cell system itself, based on Volume 1, Chapter 5, subsection 5.4.8. [e]Åhman (2001) gives the following efficiencies for the electric drivetrain components: electric motor and control system, 86 per cent today, 89 per cent potential; transmission, 98 per cent. [f]Based on an energy loss of 3.7 per cent for a transmission of 1000km as HVDC, as given in Figure 10.40. [g]Åhman (2001) gives battery recharging efficiencies of 81 per cent for NiMH batteries and 95 per cent for Lithium-ion batteries.

efficiency in producing methanol from biomass alone (at 1.77 tonnes biomass/tonne methanol) is 67 per cent, while the efficiency in producing it from biomass and hydrogen (with the actual inputs given above) is 71 per cent and the ratio of methanol output to biomass input is 1.4. Assuming the biomass input to be produced from solar energy with an efficiency of 1 per cent and the hydrogen to be produced from the sequence, PV cells (15 per cent efficiency), power conditioning (85 per cent efficiency) and electrolysis (80 per cent efficiency), the overall efficiency in producing methanol from solar energy is 1.25 per cent.[5] The efficiency in producing methanol entirely from biomass would be 1.0 per cent × 0.675 = 0.675 per cent.

10.6.3 Hydrogen-fuelled ships

Veldhuis et al (2007) discuss the potential use of liquid hydrogen as a fuel for marine transportation using aero-derivative gas turbines. Because of the low mass of hydrogen, greater payloads could be accepted. Current container ships have engine power ratings of 50–100MW, which is too large to be provided by fuel cells on ships, so the appropriate power source would be a modified internal combustion engine or gas turbine. Liquid fuels are currently used in ships in reciprocating engines or engine/turbine combined cycle to generate electricity with efficiencies that approach those of stationary combined cycle powerplants (50 per cent or better), so there would be little if any energy savings using hydrogen (indeed, when the energy required for liquefaction is accounted for, there would probably be a slight increase in fuel energy requirement per tonne-km of freight transported). However, marine transportation is a disproportionately large source of air pollution (accounting for only 3 per cent of global oil consumption but about 14 per cent of global NO_x emissions and 16 per cent of global SO_x emissions). High-speed cargo catamarans should be technically feasible using hydrogen fuel, thereby providing much shorter delivery times than with current shipping. Propulsion would be provided by aero-derivative gas turbines driving water jets.

10.6.4 Hydrogen-fuelled aircraft

Liquid hydrogen (LH_2) can be used directly as a fuel in commercial aircraft with a number of distinct advantages over jet-A fuel, the special grade of kerosene currently used worldwide for commercial aircraft. The physical characteristics of these fuels are compared in Table 10.1. A substantial amount of work was done in the 1970s and 1980s on exploratory designs of possible hydrogen-fuelled aircraft, and this work has been summarized by Brewer (1991).

Table 10.6 compares some of the characteristics of hydrogen and conventional subsonic aircraft based on the work reviewed by Brewer (1991). Hydrogen fuel weighs less than any of the alternatives, which permits smaller engines and less structural mass, leading to further weight savings. However, the need to insulate the tanks and fuel lines adds additional weight, but the overall take-off weight of a hydrogen-fuelled aircraft is about 25 per cent less than for the corresponding

Table 10.6 *Characteristics of subsonic passenger aircraft fuelled with jet-A fuel or with liquid hydrogen, and the ratios of the two*

Characteristic	Jet-A	LH_2	Jet-A/LH_2
Take-off mass (kg)	237,324	177,675	1.34
Fuel mass (kg)	86,547	27,942	3.10
Empty mass (kg)	110,860	109,817	1.01
Lift/drag ratio	8.65	7.43	1.17
Thrust per engine (Nt)	145,361	127,580	1.14
Energy use (MJ/seat-km)	0.907	0.812	1.12
Take-off distance (m)	2435	1902	1.28
Landing distance (m)	1588	1771	0.90
Production cost (per plane)	26.6	26.9	0.99

Source: Summarized by Brewer (1991) based on design studies carried out in the 1980s

aircraft using jet-A fuel according to these studies. The insulation would consist of vacuum foam between inner and outer walls. The cost of hydrogen-fuelled aircraft is estimated to be comparable to that of a conventional aircraft. Any cost estimates are of course highly uncertain but are driven mainly by the reduced aircraft mass and permitted smaller engines (as represented by the engine thrust data in Table 10.6) for hydrogen-fuelled aircraft. The costs given in Table 10.6 do not include research and development costs.

In order to minimize heat loss from the LH_2 tanks, the surface area of the tanks would need to be minimized. This precludes placing the tanks in the wings, as in current aircraft. Rather, cylindrical tanks would be placed at the front and back of the passenger compartment (which would therefore be separated from the cockpit). This results in a wider diameter fuselage (since not all of the fuselage length can be used to carry passengers) and a smaller wing area than the conventional aircraft, which in turn produces a less aerodynamic aircraft with a 17 per cent penalty in the lift to drag ratio.

The low boiling point of LH_2 (20.3K) and its very high specific heat (9.69J/gm) are such that it can be used as a heat sink with N_2 as an intermediate fluid to cool parts of the aeroplane skin to about 155K. This in turn would induce laminar flow over the aeroplane and dramatically reduce drag. This concept is referred to as 'laminar flow control'. At the same time, the heat absorbed by the LH_2 as it cools the aircraft skin adds energy to the fuel, thereby increasing the amount of energy that is released during combustion. The cold hydrogen can also be used to cool the high-pressure turbine section of the jet engine, with a 5–10 per cent saving in energy use. Today, this is done using air bled from the compressor, which requires the compressor to do more work and results in cooling air discharged into the turbine working fluid, causing some disruption of flow and further loss of efficiency. With LH_2, cooling would occur either indirectly using a second fluid and a heat exchanger, or directly with LH_2.

In addition to insulation around the tank and fuel lines, a continuous vapour barrier would be required so as to prevent any contact between the air and supercooled surfaces in the event that minute cracks in the insulation develop, as otherwise, liquefaction or freezing of the air could occur. The system will have to accommodate thermal expansion and contraction of the tanks and supporting structures as a result of the differences in temperature when the tanks are full and empty, and will have to withstand the shocks of thousands of landings and take-offs over a 15–20 year life.

Another special problem pertains to tank pressurization. With conventional fuel, the fuel tanks are vented such that air can leave or enter the tanks as the plane climbs and ascends. This is not possible with an LH_2 tank because any air entering the tank would freeze, and because some LH_2 would boil off as the pressure inside the tank decreased during ascent. Thus, the tanks would have to be designed to maintain a constant pressure even as the outside atmospheric pressure varies greatly. This would be done through stronger construction, which in turn reduces some of the weight savings due to the low density of hydrogen.

The net result of all of the above considerations (except for laminar flow control) is that the hydrogen-fuelled subsonic jet has a slight fuel efficiency advantage compared to the conventional aircraft.

The CRYOPLANE project, a European consortium of 35 partners from the aviation sector led by Airbus Deutschland, recently made an overall system analysis of hydrogen as an aviation fuel (Airbus, 2003). A range of aircraft categories was considered, from business jets to large long-range aircraft such as the Airbus A380. Per unit of energy, liquid hydrogen has four times the volume of kerosene, so fuel tanks four times as large need to be fitted in, or onto, each aircraft category. The following tank configurations were found to be optimal:

- for business jets and small regional aircraft, the tanks would be located within the fuselage and entirely to the rear of the passenger compartment;
- for regional aircraft up to 100 seats (turboprops and jets) and for short- to medium-range aircraft, tanks would be behind the passenger compartment and on top of the fuselage;
- for long-range and very long-range aircraft, tanks would be located within the fuselage behind the passenger compartment and between the passenger compartment and cockpit.

Locating the fuel tank solely behind the passenger compartment is feasible only when the fuel weight fraction is small, as otherwise the centre of gravity of the aircraft would be shifted too far to the rear.

However, the rear-only tank configuration in a business jet (which is the only practical option for this jet) requires shape changes (or a very large horizontal tail) that increases drag and reduces the maximum lift. Figure 10.12 gives the computed difference in operating empty weight, maximum weight and fuel consumption per passenger-kilometre for LH_2 compared to kerosene-fuelled aircraft for the seven aircraft types considered in the analysis. Unlike the earlier US studies, the analysis in this case indicates increased energy use for LH_2 aircraft, with particularly large penalties for business jets and very-long-range aircraft due to practical restrictions on the placement of the fuel tanks. The technical assessment concluded that hydrogen-fuelled engines will be as energy efficient as kerosene engines, and that conventional turbo engines can be converted to run on hydrogen, although some further research is needed. An aviation-specific safety assessment further concluded that hydrogen-fuelled aircraft will not be less safe than current aircraft, although regulations for airworthiness, ground handling and servicing would need to be adjusted.

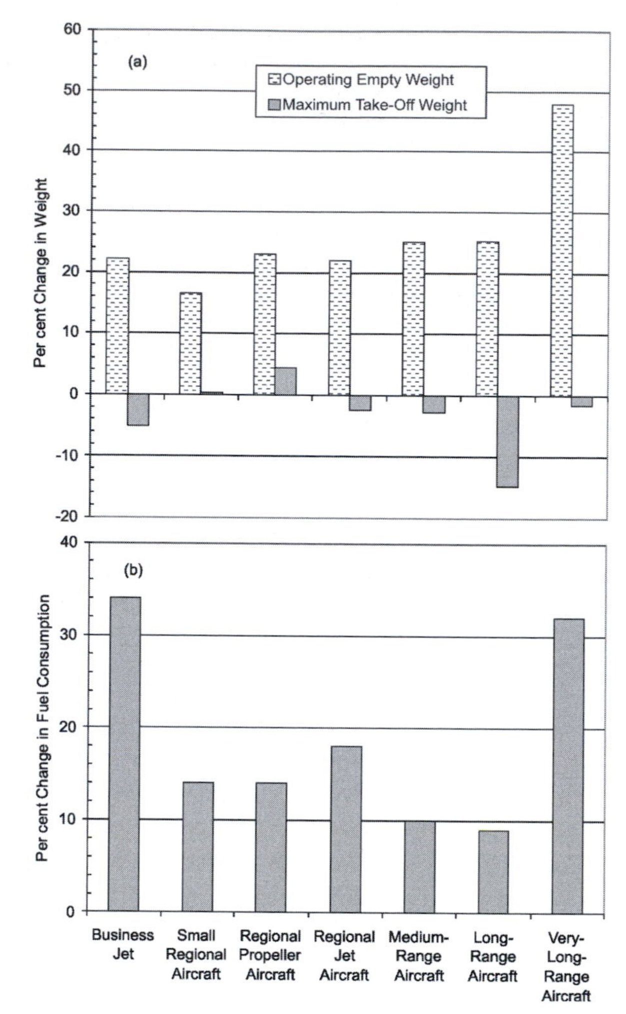

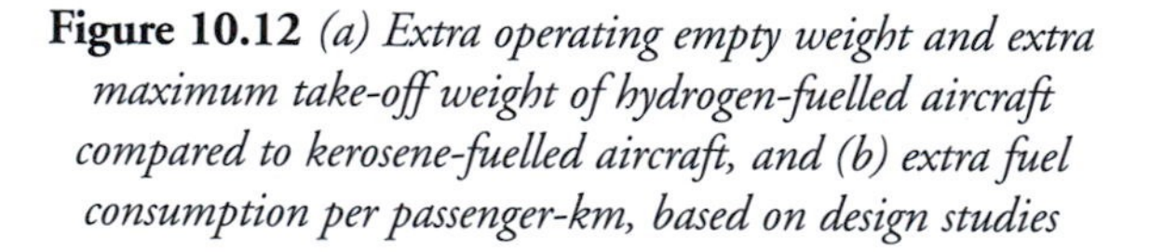

Source: Airbus (2003)

Figure 10.12 *(a) Extra operating empty weight and extra maximum take-off weight of hydrogen-fuelled aircraft compared to kerosene-fuelled aircraft, and (b) extra fuel consumption per passenger-km, based on design studies*

10.6.5 Electricity generation

Electricity and useful heat can be generated from hydrogen using either fuel cells, combined-cycle gas steam turbines, or fuel cell and gas or steam turbine hybrid systems, all of which can also be used with natural gas as the fuel. Table 10.7 gives the reactions that occur in various fuel cells along with the corresponding typical full-load efficiency when operating on pure hydrogen (fuel cells operating on pure hydrogen tend to be more efficient than fuel cells using natural gas as the input fuel, since prior fuel processing to produce a hydrogen-rich gas is not needed). Hydrogen combined-cycle turbines may be more competitive than hydrogen fuel cells at powerplant sizes of one hundred to several hundred megawatts, while fuel cells will probably be more competitive at scales of a few tens of megawatts and less. Electricity-generating efficiencies using fuel cells powered by pure hydrogen should reach 60 per cent. Fuel cell-turbine hybrid system efficiencies at least equal to that expected using natural gas (i.e. 70–75 per cent) should be achievable with pure hydrogen. Fuel cells, particularly those with a higher operating temperature, lend themselves well to cogeneration of electricity, steam and hot water. One can expect overall efficiencies in the cogeneration of heat and electricity of 80–90 per cent, as with large-scale natural gas-based systems.

Generating electricity to produce hydrogen that is later used to regenerate electricity could be worthwhile for one of two reasons: (1) when the electricity source is far removed from the demand centre, with hydrogen transported by pipeline from where electricity is

Table 10.7 *Chemical reactions occurring at the anode and cathode of different fuel cells*

Fuel cell type	Anodic reaction	Cathodic reaction	Electrolyte conducting ion	Efficiency (%)	Operating temperature
AFC	$H_2 + 2OH^- \rightarrow 2H_2O + 2e^-$	$½O_2 + H_2O + 2e^- \rightarrow 2OH^-$	OH^-	50–70	70–130°C
PEMFC	$H_2 \rightarrow 2H^+ + 2e^-$	$½O_2 + 2H^+ + 2e^- \rightarrow H_2O$	H^+	45–55	60–110°C
PAFC	$H_2 \rightarrow 2H^+ + 2e^-$	$½O_2 + 2H^+ + 2e^- \rightarrow H_2O$	H^+	40–55	175–210°C
MCFC	$H_2 + CO_3^{2-} \rightarrow H_2O + CO_2 + 2e^-$ $CO + CO_3^{2-} \rightarrow 2CO_2 + 2e^-$	$CO_2 + ½O_2 + 2e^- \rightarrow CO_3^{2-}$	CO_3^{2-}	50–60	550–650°C
SOFC	$2H_2 + 2O^{2-} \rightarrow 2H_2O + 4e^-$ $2CO + 2O^{2-} \rightarrow 2CO_2 + 4e^-$	$O_2 + 4e^- \rightarrow 2O^{2-}$	O^{2-}	40–72	500–1000°C

Note: AFC = alkaline fuel cell, PEMFC = proton exchange membrane fuel cell (also called solid polymer fuel cell, SPFC), PAFC = phosphoric acid fuel cell, MCFC = molten carbonate fuel cell, SOFC = solid oxide fuel cell.
Source: Harvey (1995) and Edwards et al (2008)

generated to where it is needed, and (2) when renewable electricity exceeds demand, with hydrogen used as a means of storing electricity until it is needed. Hydrogen would need to be compressed prior to compression or storage. In both cases, substantial energy losses would occur. For example, given an electrolysis efficiency of 80 per cent, compressional and transmission energy requirement of 15 per cent (for a transport distance of 3000km) and a fuel cell turbine efficiency of 75 per cent, the round-trip efficiency (electricity to electricity) is only 51 per cent. Thus, if the only mismatch between supply and demand is geographical, HVDC transmission of electricity would be a better option as it would entail smaller losses (about 8 per cent at full capacity for 800kV transmission over a distance of 3000km, including transformer losses at both ends) and would almost certainly cost substantially less.

When hydrogen is used locally to store excess electricity, the round-trip efficiency would still be very low (only the transmission losses would be avoided). However, the energy losses could be greatly reduced if waste heat produced during both electrolysis and electricity generation can be used. Co-production of useful heat and hydrogen would constitute a new form of cogeneration, complementing the later co-production of useful heat and electricity. If both cogeneration efficiencies are 90 per cent and the compressional requirement is 5 per cent, the energy loss during the various conversions is reduced in half (to 23 per cent). Indeed, if the waste heat from both electrolysis and electricity generation can be used, one might be interested in designing for less electricity or hydrogen production but more heat production, since this also reduces the capital costs (through smaller cell area but higher current density). Costs could also be reduced by using the same device as a fuel cell and as an electrolyser. Companies such as Hydrogenics in Canada and Proton Energy in the US are developing reversible PEM fuel cells.[6] Reversible operation requires modifying the electrodes of a non-reversible fuel cell.

10.6.6 Space and water heating

Household furnaces and stoves could be designed to use hydrogen rather than natural gas or heating oil. However, this would require an extensive new infrastructure to distribute hydrogen.

10.6.7 Hydrogen as a reducing agent for industrial processes

Hydrogen can be used in place of coke for the reduction of iron ore – the first step in the production of iron and steel (see Volume 1, Chapter 6, subsection 6.3.1). Hydrogen is used as the sole reducing agent in a steel mill built in Trinidad in 1999 by the German company Outokumpu (Nuber et al, 2006).

10.7 Costs

In this section we discuss costs of producing hydrogen from fossil fuels or electricity, in transmitting and distributing hydrogen and in dispensing hydrogen at fuelling stations for cars and trucks. Several dozen studies of the potential future costs of the hydrogen infrastructure for transportation are reviewed by Agnolucci (2007).

10.7.1 Cost of producing hydrogen

The cost of producing hydrogen depends on the capital cost of the production equipment and the cost of the energy used to make hydrogen. If hydrogen were produced by electrolysis of water using solar-generated electricity in desert areas, the cost of transporting water to the production site would also have to be included. Neglecting O&M costs and water costs, the cost of hydrogen from steam reforming of methane or electrolysis of water can be computed as:

$$C(\$/\mathrm{GJ}) = \frac{C_F(\$/\mathrm{GJ})}{\eta} + \frac{CRF \times C_{CAP}(\$/\mathrm{kWH_2})}{8760(\mathrm{hr/yr})\,CF \times 0.0036(\mathrm{GJ/kWh})} \quad (10.20)$$

where C_F is the fuel (or electricity) cost, η is the fuel (or electricity)-to-hydrogen production efficiency, CRF is the cost recovery factor (see Appendix C), C_{CAP} is the capital cost per kW of H_2 production capacity and CF is the capacity factor (the average output as a fraction of the peak output or capacity).

Figure 10.13 shows the cost of hydrogen produced from steam reforming of methane for natural gas costs of up to \$16/GJ, assuming a capital cost of \$300/kW H_2, 10 per cent interest, a 20-year lifespan, 80 per cent efficiency and a 90 per cent capacity factor. Hydrogen produced by steam reforming of methane ranges from about \$6/GJ to \$21/GJ as the price of natural gas varies from \$4/GJ to \$16/GJ. Figure 10.13 also shows the cost of hydrogen produced by electrolysis for electricity costs up to 8 cents/kWh, assuming an electrolyser capital cost of \$700/kW H_2, 10 per cent interest, a 20-year lifespan, 85 per cent efficiency and capacity factors of 90 per cent and 25 per cent. The 90 per cent capacity factor is included to permit comparison with steam reforming of methane, while the 25 per cent capacity factor corresponds to what could be expected from wind turbines at moderately good sites (see Figure 3.19) or if the electrolysers are sized to use electricity only at times when it is in excess and

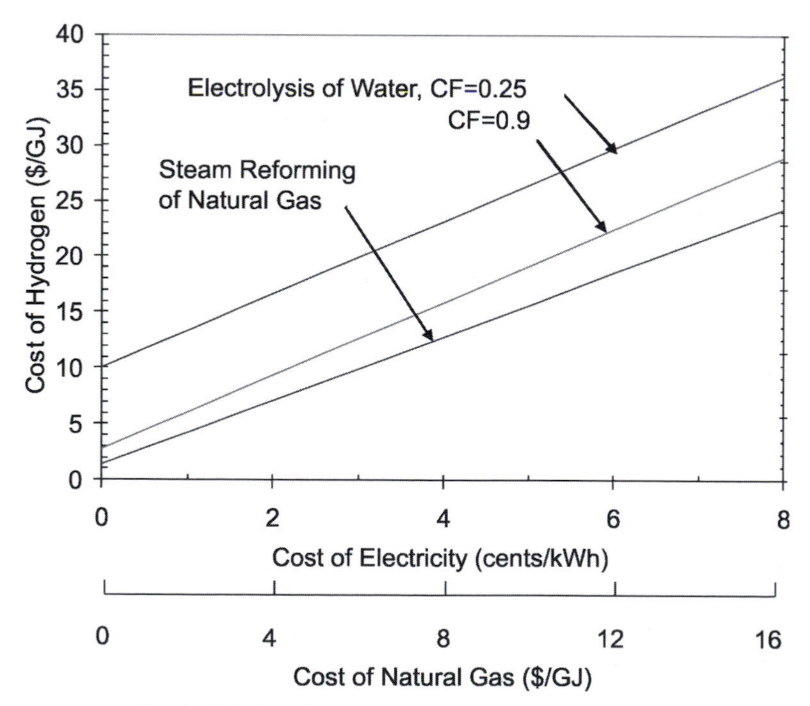

Note: See text for other assumptions and method of calculation.

Figure 10.13 *Cost of hydrogen produced by steam methane reforming as a function of the price of natural gas, and by electrolysis as a function of the cost of electricity*

therefore sold at a discount. Hydrogen from electrolysis ranges from about $13/GJ to $29/GJ as the price of electricity varies from 3 to 8 cents/kWh for a capacity factor of 0.9, and is about $7/GJ more expensive if the capacity factor is 0.25. By comparison, gasoline at $1/litre or $4/gallon corresponds to a cost of about $32/GJ. The total cost of electrolytic hydrogen is dominated by the energy cost rather than by amortization of the capital equipment, so it is worthwhile to accept a low capacity factor if this means purchasing electricity at a deep discount.

Offshore wind farms entail about 50 per cent or more greater capital cost compared to onshore wind farms (including transmission to shore), but can easily have twice the capacity factor due to the greater wind speeds offshore.[7] Thus, the cost of electricity is about the same as from onshore wind energy, but the higher capacity factor reduces the cost of electrolytic hydrogen. If offshore wind farms are dedicated to hydrogen production, then transmission of electricity to shore could occur as DC power, since DC input is required by electrolysers. This would reduce the cost of transmission, as only a single cable would be required for transmission (instead of two). Mathur et al (2008) discuss the economics of using offshore wind to produce hydrogen in India.

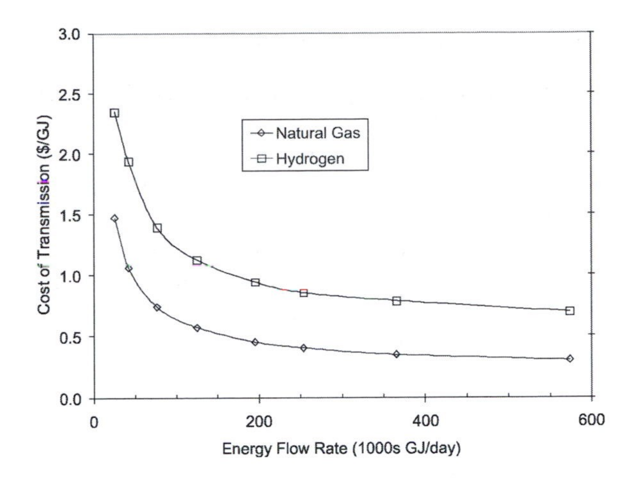

Note: Also shown are costs for transport of natural gas.
Source: Ogden (1999)

Figure 10.14 *Estimated cost of transport of hydrogen by pipeline a distance of 1000km as a function of the flow rate*

10.7.2 Cost of transporting hydrogen

The cost of long-distance transport of hydrogen per unit of energy transported would decrease with the flow rate but would probably be substantially greater than the cost of transporting natural gas at all flow rates. This is illustrated in Figure 10.14, which shows the estimated cost of transporting hydrogen and natural gas 1000km as a function of the flow rate as calculated by Ogden (1999). For a flow rate of 1 million GJ/day (which could serve 10 million cars driven 100km/day with an energy intensity of 1MJ/km), the cost of long-distance transport would be in the order of $1/GJ.

For a more modest initial introduction of hydrogen, with transmission of 100t/day (12,000GJ/day) at a distance of 300km by pipeline, Yang and Ogden (2007) compute a transmission cost of about $6.7/GJ ($0.8/kg). At low rates of market penetration of hydrogen fuel, subsequent distribution by truck as compressed gas or liquefied hydrogen would be less expensive than distribution by pipeline. For example, for distribution of 500kg/day per station to 200 stations within a 25km radius of the pipeline terminus (corresponding to about a 16 per cent market share in US cities), distribution by truck is estimated to cost about $12–15/GJ ($1.5–1.8/kg), compared to $20–40/GJ by pipeline. However, at 100 per cent market penetration of hydrogen vehicles (350 stations at 1800kg/day each within a 25km radius), local distribution by pipeline would be the least expensive option, at around $8–10/GJ.

10.7.3 Cost of refuelling stations

Yang and Ogden (2007) estimated the cost of hydrogen refuelling stations, including land, installation and engineering, O&M and equipment. Equipment costs include compressors or pumps, storage tanks and dispensers. For stations supplying compressed hydrogen to vehicle tanks at 5000psi (345atm), the estimated station cost, amortized over 30 years of hydrogen supply at 15 per cent interest, ranges from about $7.2/GJ ($0.87/kg) at a station size of 500kg/day, to $3.6/GJ (0.44/kg) at a station size of

3000kg/day. Costs for stations delivering liquefied hydrogen are estimated to be about one third less, but the delivered hydrogen would be more costly due to the cost of and losses associated with liquefaction.

10.7.4 Cost of hydrogen delivered to refuelling stations

Based on the analysis of the preceding sections, the costs associated with providing hydrogen to road vehicles at refuelling stations would be:

- for production:
  - $11/GJ for steam methane reforming, assuming a natural gas cost of $8/GJ which, however, is not a viable option due to scarce and/or unreliable natural gas supplies,
  - $36/GJ for electrolysis of water, assuming electrolysers at $700/kW, electricity at 8 cents/kWh and a 25 per cent electrolyser capacity factor,
  - $20/GJ if excess wind-generated electricity is sold a discounted price of 3 cents/kWh, with the same electrolyser cost and capacity factor as above;
- for transmission, $2–8/GJ;
- for distribution, $8–10/GJ;
- for dispensing, $4–8/GJ.

This brings the total cost to $34–61/GJ, which is roughly equivalent to gasoline at $1–2/litre (or $4–8/gallon), before taxes. Given that future hydrogen-powered fuel cell vehicles (FCVs) would travel almost four times further on a given amount of energy then the average of today's vehicles in the US (see Table 10.4), the effective cost of hydrogen in the US (where gasoline costs are among the lowest in the world) would be substantially less than the recent (2008) cost maximum of ~$4/gallon. Compared to future advanced gasoline-powered vehicles, hydrogen-powered FCVs would require about 40 per cent less energy to travel a given distance and so, even after accounting for the greater cost of fuel cell vehicles, could be competitive with gasoline- or diesel-powered vehicles (see below).

The US National Academy of Sciences recently assessed the technical feasibility, costs and barriers to a hydrogen economy (NRC/NAE, 2004). Costs for distribution by pipeline from central hydrogen production facilities (about 1 million kg H_2/day production, serving 2 million cars) were estimated to be about $3.5/GJ with current technology, while costs for distribution by cryogenic truck from middle-size facilities (about 22,000kg/day production serving 40,000 cars/day) were estimated to be about $15/GJ with current technology and $9/GJ with optimistic future technology. Dispensing costs for gaseous hydrogen were estimated to be about $4.5/GJ with current technology (perhaps $3.2/GJ in the future), while dispensing costs for liquid hydrogen were estimated to be $5.2/GJ at present (perhaps $2.5/GJ in the future). Hydrogen produced from biomass costing $2.9/GJ was estimated to cost about $40/GJ (excluding distribution and dispensing), potentially dropping to $30/GJ with future technologies. Even at $30/GJ, such hydrogen would not be competitive with hydrogen from discounted wind-derived electricity, and as shown later, using biomass to produce hydrogen would not be the most effective way of using limited biomass resources.

10.7.5 Competitiveness of hydrogen fuel cell vehicles with gasoline vehicles

As discussed in Volume 1 (Chapter 5, subsection 5.4.11), the additional future cost of a hydrogen FCV compared to an advanced gasoline vehicle is projected to be about $2000. This estimate assumes large-scale production of fuel cells and significant but plausible reductions in Pt loading compared to present fuel cells, leading to a fuel cell cost of $60/kW, compared to $25/kW for internal combustion engines. A $60/kW fuel cell cost is about half what would be expected with large-scale production but using current technology. As noted above, the cost of delivered hydrogen from renewable energy sources could drop to the $30–60/GJ range, whereas gasoline at $1–2/litre (the current range of gasoline costs in the most of the world today) is equivalent to $31–62/GJ. However, a hydrogen FCV would travel about 75 per cent further per unit of fuel energy than an advanced gasoline vehicle (this is equivalent to using 43 per cent less energy to travel a given distance). It is thus possible that savings in operating energy costs of hydrogen FCVs could offset the additional purchase cost.

Figure 10.15 shows the cost of hydrogen at which operating cost savings over a ten-year vehicle lifespan exactly offset the additional purchase cost. Results are shown for gasoline costs of $1.0, $1.5 and $2.0 per litre, and for the following additional assumptions: future cost savings discounted at 10 per cent/year, energy use of 1.07MJ/km and 0.59MJ/km for advanced gasoline and hydrogen vehicles, respectively (taken from Table 10.4), and 20,000km annual driving. From Figure 10.15 it can be seen that:

- for gasoline at $2/litre (about $7.6/gallon and comparable to the cost in Norway), hydrogen at $32/GJ would be competitive with gasoline if the additional vehicle purchase cost is $5000;
- for gasoline at $2/litre, hydrogen at $62/GJ would be competitive with gasoline if the additional vehicle purchase cost is $2000;
- for gasoline at $1/litre, hydrogen at $32/GJ would be competitive with gasoline if the additional vehicle purchase cost is $1000.

10.8 Safety and environmental considerations

10.8.1 Safety issues

Physical properties of hydrogen relevant to its safety as a fuel are given in Table 10.8, and are compared with the corresponding properties for natural gas, gasoline and jet-A fuel. The relative safety of hydrogen and other fuels cannot be assessed from their physical properties alone, but only in the context in which the different fuels would be used.

Hydrogen requires a greater concentration than gasoline before it will burn or explode (the lower flammability and detonability limits), but is exceedingly unlikely to reach the required concentrations except in enclosed unventilated spaces. This is because the diffusivity of hydrogen is an order of magnitude greater than that of gasoline and it is buoyant in air, so the likelihood of it reaching the larger concentration needed to burn or explode is remote. Hydrogen requires much less energy to ignite it than other fuels, but the minimum energy required for ignition is so low that the differences are not meaningful (a spark equivalent to that created by walking across a rug and touching a grounded object will ignite even hydrocarbon fuels in air according to Brewer, 1991). Although its flame temperature falls midway between that of natural gas and gasoline, a hydrogen flame burns out quickly and radiates little heat (IEA, 1999). As a result, objects have to be directly in the flame in order to be severely damaged, and nearby objects are unlikely to catch fire. Calculations indicate that in a vehicle fire involving

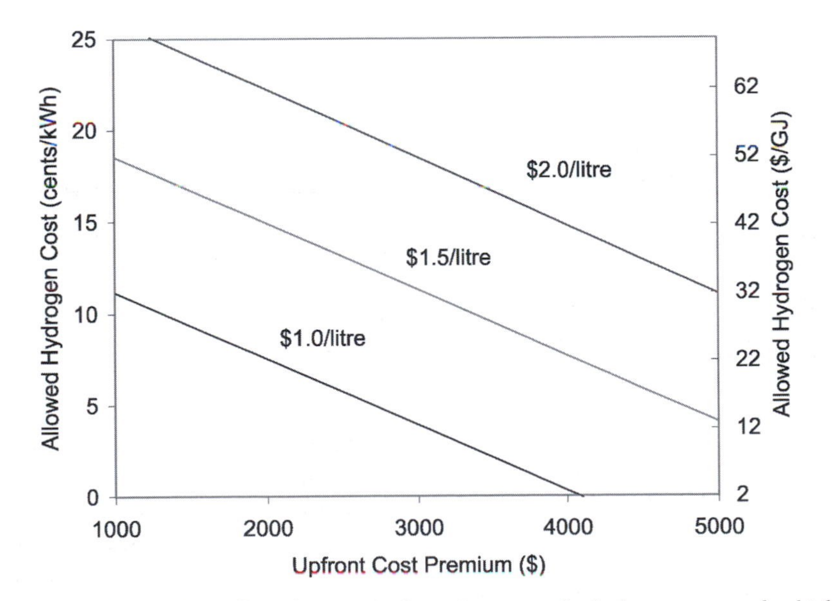

Figure 10.15 *Cost of hydrogen that offsets the increased purchase cost of a hydrogen-powered vehicle over a ten-year operating life, for gasoline at $1.0, $1.5 and $2.0 per litre and for other assumptions given in the text*

Table 10.8 *Safety-related properties of hydrogen, methane or natural gas, gasoline and jet-A fuel*

	Hydrogen	Methane	Gasoline	Jet-A
Flammability limits (% volume)	4.0–75.0	5.3–15.0	1.0–7.6	0.6–4.7
Detonability limits (% volume)	18.3–59.0	6.3–13.5	1.1–3.3	1.1–3.3
Diffusion velocity in air (cm/s)	2.0	0.51	0.17	0.17
Buoyant velocity in air (m/s)	1.2–9.0	0.8–6.0	Non-buoyant	Non-buoyant
Ignition energy at stoichiometric mixture (mJ)	0.02	0.29	0.24	0.25
Ignition energy at lower flammability limit (mJ)	10	20	n.a	
Auto-ignition temperature (°C)	585	540	228–501	>227
Theoretical explosive energy (kg TNT/m^3 gas)	2.02	7.03	44.22	
Flame temperature (°C)	2045	1875	2200	1927
Toxicity	Non-toxic	Non-toxic	Toxic at >500ppm	

Source: Ogden (1999), IEA (1999) and Brewer (1991)

gasoline that would burn for 20–30 minutes, the equivalent hydrogen fire would last 1–2 minutes. The explosive energy of hydrogen is 3.5 times less than that of natural gas and over 20 times less than that of gasoline (Table 10.8). Thus, hydrogen is no more dangerous than natural gas or gasoline, and if its particular properties are taken advantage of, it should be safer. Experimental studies by Tanaka et al (2007) indicate that hydrogen at refuelling stations presents no hazard of exploding if atmospheric concentrations are kept below 15 per cent, which should be easy to do.

Automobiles

As discussed in subsection 10.4.4, the most promising technique for onboard storage of hydrogen in automobiles is as compressed hydrogen (at up to 690atm or 10,000psi). One might be concerned about the durability of such highly pressurized storage tanks in the event of a severe accident. However, the storage tanks would apparently be the strongest part of the car, so the likelihood of the occupants of a car surviving an accident in which the tank could rupture are very small to begin with. Furthermore, the maximum potential energy released by compressed hydrogen increases very slowly with increasing pressure. For example, H_2 at 70atm and 300K can release a theoretical maximum amount of energy of 0.55kWh/kg if it suddenly expands. Increasing the pressure to 1000atm increases the maximum possible energy release by only 10 per cent (Berry et al, 2003b).

However, Shinnar (2003) states that H_2 cars could be easily modified to become an undetectable bomb for a suicide bomber; all that would be required would be for the hydrogen tank to be equipped with a release valve and a delayed detonator. It would, however, be difficult to trigger an explosion. The lower detonability fuel to air ratio for H_2 is 18 per cent or more, which is 1.5–3 times that of natural gas and 6–18 times that of gasoline (Table 10.8). The lower flammability limit is 4 per cent, so for a detonation to occur, H_2 would have to accumulate to 18 per cent concentration in a closed space without ignition. Should an explosion occur, hydrogen has 22 times less explosive energy than an equal volume of gasoline (Table 10.8). Shinnar also notes that propane tanks are stored at much less pressure than proposed for hydrogen storage tanks in cars (300psi versus 6–10,000psi), yet there are strict laws prohibiting the transport of even a small propane tank through all the tunnels in New York. He also notes that, in industry, any technician approaching a pressurized hydrogen tank to check the valves swings a wooden broom so as to detect any potential invisible flame that may come from a leaking valve, as H_2 above 1500psi (about 100atm) is self-igniting. Refuelling stations could also pose a terrorist risk, as the minimum-sized storage tank would have the explosive power of 2000 tonnes of TNT; a terrorist would only need to open a small valve and ignite a small bomb after 20 seconds (Shinnar, 2004). If the risk of terrorism is an issue of concern by the time that hydrogen could play a role in private transportation (probably not before 2025), countermeasures will be needed if hydrogen is to fill that role.

It has been suggested that an odorant could be added to hydrogen to aid in detection of leaks, as both hydrogen and its flame are odourless and invisible.

However, the low molecular weight of hydrogen makes it more mobile than all other gases, and many materials poison fuel cell catalysts (Moy, 2003).

Aircraft

Brewer (1991, Chapter 8) discusses a variety of studies dating back to the 1950s concerning the safety of hydrogen as an aircraft fuel. In the event of a crash in which the passengers can be expected to survive the impact itself, LH_2 tanks are less likely to be ruptured than those of conventional fuels (which are located in the wings) because of their shape and their design to withstand higher pressure than the rest of the fuselage (recall that LH_2 tanks will be designed to maintain the same internal pressure even as the plane climbs to high altitude, which imposes a weight penalty but also provides for enhanced safety). Air is allowed to enter the tanks of conventional fuels, so there is also a fuel–air mixture above the fuel that is susceptible to ignition, but this is not the case with LH_2 tanks because no air would be permitted into the tank. If the tank is ruptured, the hydrogen will evaporate almost immediately and, due to its high buoyancy, will rapidly dissipate into the overlying air. If the spilled fuel is ignited, the duration of the fire would be so brief that it would not heat the fuselage to the point of collapse, as is the case with other fuels. Early tests indicated the following fire durations for spills of 120 litres of various fuels in the presence of an ignition source:

- LH_2, 27 seconds,
- propane, four minutes,
- gasoline, five minutes,
- jet fuel, seven minutes.

As well, the radiant energy emitted by a hydrogen flames is much less than that from other fuels, due to a very small emissivity (0.085 for hydrogen versus 1.0 for hydrocarbon fuels). This in turn reduces the extent to which nearby combustible materials will ignite and will greatly reduce harm to nearby people.

In conclusion, hydrogen as an aircraft fuel is likely to be safer than other fuels in use or contemplated.

10.8.2 Atmospheric impacts

Atmospheric impacts from a hydrogen economy would arise through pollutant emissions (greatly reduced using hydrogen) and potential impacts on atmospheric chemistry of emissions of H_2 in the atmosphere due to leakage from storage and transmission systems, or during use.

Impact on emissions of pollutants

Pollutant emissions at the point of use would be almost completely eliminated for most uses of hydrogen. A potential exception is in the use of hydrogen as a fuel for aircraft. Use of hydrogen in an unmodified engine designed to use kerosene fuel results in increased emissions of nitrogen oxides (NO_x), which in turn contribute to the formation of ozone. However, there are reports of modified engines using hydrogen with NO_x emissions 80–95 per cent less than those of a modern kerosene engine.

Impact of leakage of hydrogen

Hydrogen is a naturally occurring constituent of the atmosphere, with a concentration of around 0.5ppmv. It is produced through photochemical oxidation of methane and other hydrocarbons (about 50 per cent of natural production) and from biogenic processes and combustion (the other 50 per cent). It is removed through reaction with the hydroxyl radical (OH) and biologically through soil organisms, with the latter about three to four times more important than the former. Leakage of hydrogen from a hydrogen energy system would therefore reduce the atmospheric OH concentration, which in turn would increase the concentration of methane, hydrochlorofluorocarbons (HCFCs) and hydrofluorocarbons (HFCs) (all GHGs that are otherwise removed through reaction with OH). Some hydrogen would enter the stratosphere and produce water vapour, where it would serve as a potent GHG and affect stratospheric ozone by cooling the stratosphere. Given a continuing (but decreasing) concentration of chlorine in the stratosphere from the breakdown of chlorofluorocarbons (CFCs) (whose production largely ceased in 2000), this would worsen the depletion of stratospheric ozone. Once the stratospheric concentration of chlorine has returned sufficiently close to the unperturbed concentration (within the next few decades), stratospheric cooling will not induce loss of stratospheric ozone.

There are number of factors to consider in evaluating the impact of leakage from a hydrogen economy on stratospheric ozone. First, H_2 is already released to the atmosphere as a byproduct of the use of fossil fuels, including as a component of car exhaust. If hydrogen is used in fuel cells and efforts are taken to reduce the leakage at all points in the hydrogen energy system, it is possible that total emissions could be comparable to or less than anthropogenic emissions at present. Second, the accumulation of CO_2 in the atmosphere also has a cooling effect on the stratosphere that, like H_2-induced cooling, exacerbates the depletion of stratospheric O_3. The relevant question is whether the switch from a C-based to a H_2-based energy system worsens or alleviates the stratospheric cooling and associated loss of ozone during the next few decades, before stratospheric chlorine levels return to normal.

A number of studies indicate that likely rates of leakage of hydrogen in a hydrogen economy would have a negligible net effect on climate or stratospheric ozone (Kammen and Lipman, 2003; Prather, 2003; Schultz et al, 2003; Trump et al, 2003; Warwick et al, 2004).

Impact of condensation trails

Condensation trails (or 'contrails') from aircraft have a very minor net heating effect on climate, due to their absorption of infrared radiation from the earth's surface exceeding their reflection of solar radiation. Inasmuch as the replacement of kerosene with hydrogen would result in greater emissions of water vapour from aircraft, more extensive contrails can be expected. However, fewer but larger condensation particles would result, presumably due to elimination of emissions of pollutants that serve as condensation nuclei, so the optical depth of the contrails would be smaller. Overall, the impact on radiative heating by contrails of a shift to hydrogen-fuelled aircraft would be negligible.

10.9 Scenarios

Recent studies on how a hydrogen distribution infrastructure might develop, and of how hydrogen-powered fuel cell vehicles might evolve as a series of steps, each of which is a natural progression from the step before, are discussed. We then close with a consideration of if and when the use of renewable energy to produce hydrogen would be the most effective use of renewable energy for transportation.

10.9.1 Steps in the development of a hydrogen distribution infrastructure

Wietschel et al (2006) considered two scenarios for the development of a hydrogen infrastructure in Europe. In Scenario A, hydrogen supplies 20 per cent of stationary and transport energy consumption by 2030, while in Scenario B it supplies 5 per cent. In Scenario A, an infrastructure for hydrogen production, conditioning, distribution and refuelling is built up. In Scenario B, no area-wide infrastructure is built, and hydrogen is used in niche markets and in non-connected systems. It is transported primarily by cryogenic truck. Transportation uses involve public transport and fleets. The required annual average investment amounts to €57 billion for Scenario A and €13 billion for Scenario B (in both scenarios, the largest share of hydrogen is produced by SMR with carbon sequestration). Although this may sound expensive, Wietschel et al (2006) note that the current annual expenditure for new highways in the EU-15 is six times that of Scenario A or 25 times that of Scenario B. Scenario A, while displacing 20 per cent of fossil fuel energy, reduces CO_2 emissions by only 12 per cent, due to the fact that most of the hydrogen is produced from fossil fuels with incomplete sequestration of CO_2.

10.9.2 Steps in the evolution of hydrogen fuel cell vehicles

In the very long term (2050 and later), vehicles running on either hydrogen or biofuels will probably be needed in order to complete the transition to a 100 per cent renewable energy system. Plug-in hybrids could play a role in the transition to hydrogen-fuelled vehicles. Suppes (2006) proposes the following stages:

1. hybrid electric vehicles (HEV);
2. Plug-in hybrid electric vehicles with an extended battery pack adequate for 50km driving (PHEV-50k);
3. PHEV with batteries and reversible fuel cells adequate for 100km driving (PFCHEV-100k);
4. PFCHEV-100k with cheap engines;
5. City battery electric vehicle without engine backup;
6. PFCHEV-100k using mostly regenerative fuel cells;
7. PFCHEV using dual-fuel fuel cells that eliminate the need for an engine.

HEVs are already commercially available, while PHEVs have been presented at auto shows. PHEVs are ideally recharged at night. The first 50km or more each day would be powered by the battery. The PFCHEV (plug-in fuel cell hybrid electric vehicle) is a concept still at the design stage that uses both a battery pack and reversible fuel cell (RFC), along with a standard engine. RFCs are a closed system in which water is electrolysed (using grid electricity) to oxygen and hydrogen under pressure (up to 200atm) and stored. Later, the oxygen and hydrogen recombine to provide power as needed. The RFC would extend the range of the battery pack, while the batteries provide peak power. Use of pure oxygen in the fuel cells doubles the power output compared to using air, so the cost per kW is cut in half; at the end of 2003, a PEM fuel cell using oxygen cost $1100/kW (Suppes, 2006). If, as an electrolyser, the RFC operated for eight hours to generate hydrogen followed by eight hours of use, the cost per unit of energy storage capacity would be about $130/kWh. Storage tanks would cost about $10/kWh for hydrogen and oxygen together, bringing the total cost to $140/kWh. If the cost of PEM fuel cells falls to $60/kW (an entirely reasonable expectation, as discussed in Volume 1, Chapter 5, subsection 5.4.8), the cost of storage capacity would be $17.5/kWh. By comparison, batteries suitable for PHEV-50k vehicles currently cost about $1000/kWh and are projected to drop to $300/kWh according to Figure 5.19 of Volume 1.

A PFCHEV-100k would split a 100km range equally between the battery pack and RFC, still relying on a gasoline (or diesel) engine for travel beyond that range. Stage 4 is based on the fact that, in many cases, the engine would rarely be used, so costs can be reduced by using a very inexpensive engine. Stage 6 represents a significant jump, as the RFC power output would jump from <1.0kW to >15kW as it is transformed from a battery charger to the primary power source. A further significant jump occurs at Stage 7, which uses a fuel cell that runs either on hydrogen (produced electrolytically when the vehicle is parked and connected to the grid) or using methanol (carried on board). Alternatively, Stage 7 could be a FCHEV – a fuel cell relying entirely on onboard hydrogen, with a battery (recharged by the fuel cell) for peak power. In either case, the fuel cell in Stage 7 would operate on air rather than pure oxygen, and so would not be viable until the unit cost using pure O_2 at the optimal pressure has fallen to $100/kW. Each stage in the above evolution would not occur until it can be justified by immediate economic benefits relative to the previous stage. As previously noted (subsection 10.6.1), direct use of renewable electricity to charge batteries in plug-in hybrid vehicles would be over twice as efficient as using renewable electricity to produce hydrogen for use in fuel cell vehicles. Thus, direct use of grid electricity from renewable energy sources for as large a fraction as possible of the total driving is better.

10.9.3 Effectiveness of alternative uses of renewable energy

There are two issues related to the effectiveness of using hydrogen in the transportation sector: (1) the effectiveness of using hydrogen rather than biofuels as a range extender in PHEVs, and (2) the effectiveness of using renewable electricity to produce hydrogen for use in the transportation or industrial sectors compared to using it to displace coal-based electricity during the transition to a completely fossil fuel-free energy system.

Alternative ways to use renewable energy as a range extender in PHEVs

We take it as a given that direct use of renewable electricity to recharge batteries would be far more effective in displacing fossil fuels than using it to produce hydrogen that in turn is used in a fuel cell vehicle. This is because of the substantial losses in producing hydrogen by electrolysis (at 75 per cent efficiency) and in using it in a fuel cell (at 50 per cent efficiency) compared to battery charging (at 95 per cent efficiency), all other steps being comparable for the two choices (see Table 10.5). We also take it as a given that the first priority should be to improve the efficiency of the baseline vehicle. The key questions pertain to the fraction of gasoline or diesel that can be displaced in this way (which depends on the size of battery that can be incorporated into PHEVs at an acceptable cost) and the choice of fuel as a range extender. The two options for a non-fossil fuel range extender are (1) use of biofuels in an ICE or FCV, and (2) use of hydrogen in an ICE or FCV.

In terms of land area ultimately required, hydrogen is the clear winner. The major steps in the chain involving biofuels and their efficiencies are: photosynthesis (~1 per cent), production of biofuels from biomass (~50–60 per cent), use of biofuels in an ICE (~20 per cent) and drivetrain (87 per cent). The result is an efficiency from solar energy to power at the wheels of ~0.1 per cent. If biomethanol is produced and used in a fuel cell with internal reforming of methanol (with a net efficiency of 40–45 per cent), the overall sunlight-to-drivetrain power efficiency would be about 0.25 per cent. If hydrocarbon fuels are produced from a combination of biomass and concentrated solar thermal energy, the overall sunlight to conversion efficiency would be over twice as large (see Chapter 4, subsection 4.3.9), or somewhat better than 0.6 per cent.

In contrast, the major steps in the production and use of hydrogen and their efficiencies would be: sunlight to DC electricity (15–20 per cent), coupling to an electrolyser (95 per cent), electrolysis (80 per cent), compression to 800atm if hydrogen is stored on board at 700atm (85 per cent), use of hydrogen in a fuel cell (50 per cent efficiency) and electric drivetrain (87 per cent). The result is a sunlight-to-wheels efficiency of 4–6 per cent.[8] If solar thermal energy is directly used to produce hydrogen, the sunlight-to-hydrogen efficiency could be as high as 20 per cent, giving an overall sunlight-to-wheels efficiency of 7.4 per cent. If fuel cells cannot be reduced to an acceptable cost (or if they become too costly later due to large increases in the cost of Pt as it becomes scarce), such that hydrogen can only be used in an ICE, the drivetrain efficiency would be about 16 per cent rather than about 43 per cent (0.5 × 0.87), so the sunlight-to-wheels efficiency would only be about 2.7 per cent (and considerably more hydrogen would need to be stored on board, which might not be practical). Nevertheless, use of hydrogen would require far less land than the use of biofuels as a range extender. Furthermore, the hydrogen could be produced from concentrated solar thermal energy in arid regions with minimal water requirements, whereas the biofuels would need to be produced on land having at least minimal biological productivity and would have greater water requirements.

Alternative uses of hydrogen during the transition to a fossil fuel-free energy system

The second question is whether it is better to begin using hydrogen in place of gasoline or diesel fuel as a range extender in PHEVs (assuming that hydrogen vehicles become economically and technically viable) while some electricity is still produced from fossil fuels, or to use renewable energy for the maximum displacement of fossil fuel-based electricity. From Table 10.4, advanced hybrid vehicles would require 1.07MJ/km using gasoline, 0.92MJ/km using diesel and 0.59MJ/km using hydrogen. Thus, 1MJ of hydrogen would displace 1.81MJ of gasoline or 1.56MJ of diesel fuel. Applying markup factors of 1.2 and 1.25 for gasoline and diesel fuel, respectively, and a CO_2 emission factor for petroleum of 20gmC/MJ, the 1MJ of hydrogen used in an advanced vehicle saves about 40gmC. If the hydrogen is produced and compressed at an efficiency of 0.8 × 0.9 (72 per cent overall), then 1.39MJ of electricity are required. If this electricity instead displaces electricity produced from coal at 40 per cent efficiency, then, given a coal CO_2 emission factor of 25gmC/MJ, the avoided CO_2 emission would be 87gmC – over twice the saving. If the electricity used to make hydrogen would have otherwise displaced electricity produced from natural gas at 60 per cent efficiency, the avoided CO_2 emission would be about 32gmC.

In order to compare both options with the effectiveness of using renewable electricity to recharge batteries in a PHEV, we can assume that the baseline vehicle (requiring 1.07MJ gasoline/km) would require 0.13kWh AC electricity per km (or 0.47MJ/km). Thus, the 1.39MJ of electricity used to produce hydrogen can instead displace 2.73MJ of petroleum and avoid an emission of 55gmC.

Thus, the order of preference is:

- use renewable electricity first to displace coal-fired electricity;
- use renewable electricity to recharge batteries in PHEVs;
- use renewable electricity to make hydrogen for use in FCVs (if this becomes a viable option);
- use renewable electricity to displace electricity produced from natural gas.

Practical considerations might lead to several of these options being pursued at the same time. Even if coal supplies electricity to a grid at times, there would be times when excess wind and/or solar electricity are available once these are deployed on a large scale. At these times, displacing coal-fired electricity would not be an option, so the excess electricity could be readily used for recharging batteries. However, making use of wind and solar energy only at times when there is excess supply might not allow production of enough hydrogen to supply the minimum number of refuelling stations and vehicles needed to get started. Dedicated use of some renewable energy sources to produce hydrogen might therefore be needed before coal has been completely eliminated from the electricity system.

10.10 Summary and synthesis

Hydrogen has the potential to serve as an energy currency, replacing fossil fuels in all the ways in which they are used for energy today and, in combination with biomass, replacing them as a chemical feedstock. Hydrogen could be produced by electrolysis of water when solar- and wind-generated electricity are in excess, stored and later used in a fuel cell to produce electricity when there is a shortage of wind or solar power. It could also be transported from distant sunny or windy regions to demand centres. In this way, it could serve to close the spatial and temporal mismatch between intermittent renewable sources of electricity and the demand for electricity. However, much of this gap could also be closed through alternative strategies, such as long-distance HVDC power transmission to link regions of good wind and solar energy with dispatchable hydroelectric facilities and electricity demand centres (as discussed in Chapter 3, subsection 3.11.2), and by using compressed air energy storage supplemented if needed with efficient use of biomass energy for backup power generation. If hydrogen is produced through electrolysis at times of excess renewable power supply and used to produce electricity with fuel cells at times of inadequate renewable power supply, it will be important to perform both steps in a cogeneration mode to the extent possible so as to make use of waste heat and maximize the overall efficiency; otherwise, the round-trip energy losses will amount to about 50 per cent.

Hydrogen could also be used in transportation applications. Its use in automobiles presents major challenges, however, particularly concerning the current high cost of fuel cells and onboard storage, as well as the difficulties of creating a whole new infrastructure to supply hydrogen fuel. Nevertheless, there are indications that the total cost of driving using hydrogen in FCVs could become competitive with gasoline or diesel alternatives, given gasoline at $1/litre ($4/gallon), the projected cost of fuel cells with mass production ($60/kW) and the cost of C-free electricity. Hydrogen-fuelled FCVs are projected to be almost four times as efficient (in terms of km driven per unit of energy in the fuel tank) as current vehicles and about 1.75 times as efficient as advanced gasoline vehicles. However, the overall efficiency in the use of renewably based electricity to power vehicles using hydrogen (produced by electrolysis) would be less than half that with direct use of renewably based electricity to charge batteries. Thus, the optimal solution would be a plug-in hybrid vehicle using onboard hydrogen in a fuel cell only to give extended driving range. Hydrogen is more promising in various niche applications, such as in railway locomotives and to power auxiliary power units in trucks.

Hydrogen could potentially be used in commercial aircraft, although, as with automotive applications, a number of challenges remain. Recent work suggests that the amount of primary energy required to fly a given distance would not be less, and could even be larger, than using hydrocarbon fuels. There are a number of important industrial niche applications for hydrogen in a fossil fuel-free world, such as its use as a reducing agent in the manufacture of iron (in place of coke), in the manufacture of nitrogen fertilizer (in place of natural gas or coal) and in the manufacture of a variety of biomass-based chemicals (in place of petroleum).

The cost of producing hydrogen from renewably based electricity by electrolysis is dominated by the energy cost rather than by the equipment cost, so making use of electricity at times of excess supply (and deeply discounted in cost) would be economically attractive in spite of the lower equipment utilization. Hydrogen would cost $20–35/GJ for electricity at 3–8 cents/kWh with an electrolyser capacity factor of 0.25, while the cost of H_2 dispensed at a fuelling station would be $35–60/GJ. This is comparable to the cost of gasoline at $1–2/litre ($31–62/GJ).

Notes

1 For example, old electric arc furnaces in Japan use 380kWh of electricity and 33Nm^3 of oxygen per tonne of steel produced, while new furnaces use 150kWh of electricity and 45Nm^3 of oxygen (Kato et al, 2005). At 40 per cent electricity generation efficiency, the savings in primary energy is 172.5MJ per extra Nm^3 of oxygen used. Another example is melting of glass. Conventional air-blown combustion melting uses 11MJ/kg-glass. Oxygen-blown combustion reduces this to 6.2MJ/kg while using 0.3Nm^3 O_2/kg, for a savings in primary energy of 16MJ/Nm^3.

2 Generation IV nuclear reactors have also been proposed for direct thermo-chemical splitting of water, as described in subsection 10.3.5.

3 Photon energy in eV times q (1.6×10^{-19} coulombs) gives photon energy in joules; division by Planck's constant, h (6.626×10^{-34}J s), gives the photon

frequency. Wavelength is given by the speed of light (2.999×10^8m/s) divided by frequency.

4 The energy required to transmit CH_4 5000km is 13.6 per cent of the delivered energy, which is greater then 5 × 2.5 per cent because extra hydrogen energy is required to transmit the extra 12.5 per cent needed at the beginning to transmit the 100 units that will be delivered at the end of the pipe. Because of this compounding effect, the extra energy needed to transmit H_2 5000km when *D* is fixed is 4.66 rather than 3.83 times that needed to transmit methane the same distance.

5 That is, 100 per cent × ((22.7GJ/t)/(0.85 × (19.0GJ/t)/0.01) + 0.11 × (141.0GJ/t)/(0.15 × 0.85 × 0.80)).

6 These fuel cells are sometimes referred to as regenerative fuel cells, but this term is also applied to flow batteries (see subsection 3.11.2), which is how it will be used here.

7 For example, as seen from Figure 3.19, the annual average turbine capacity factor at a mean annual wind speed of 6m/s is 0.2, whereas the capacity factor is 0.4 at a mean wind speed of 9m/s.

8 The sunlight-to-wheels efficiency would be about 20 per cent greater in both cases if hybrid vehicles are used, as the average drivetrain efficiencies would be 20 per cent greater than assumed here, but the ratio of the two sunlight-to-wheels efficiencies would be unaffected.

11

Community-Integrated Energy Systems with Renewable Energy

11.1 Introduction

In Volume 1 (Chapter 9), the concept of 'community-integrated energy systems' was introduced. These consist of district heating networks coupled to electricity generation through cogeneration, and of district cooling from centralized electric chillers and/or absorption chillers using waste heat from electricity generation. They allow the aggregation of the heating and cooling loads of a collection of diverse buildings within a given area, the efficient generation of electricity from fuels, the utilization of scattered sources of heat (in addition to or instead of electric powerplants) and the utilization of lower-temperature heat sinks (such as the ground, sewage and lake, river or seawater) for cooling purposes. As discussed in Chapter 9 of Volume 1, substantial energy savings are possible in this way, sometimes with reduced investment costs. In this chapter we extend this concept by considering the integration of district heating and cooling and electricity generation with the use of onsite renewable energy resources. We begin with seasonal underground storage of summer heat and winter coldness.

11.2 Seasonal underground storage of thermal energy

The term 'thermal' energy storage will be used here to refer to storing both heat and coldness, the former by raising the temperature of a storage medium, the latter by lowering the temperature of a medium or by freezing water or some other substance. Diurnal thermal energy storage is already widely used in building heating and cooling systems (see Volume 1, Chapter 4, subsection 4.5.7). Seasonal thermal storage requires larger storage volumes and/or (in the case of heat storage) greater storage temperatures than diurnal storage. A greater volume of storage containers such as water tanks entails greater cost, while a greater storage temperature will lead to greater heat loss. However, the ground itself provides a natural, large-scale medium for storing heat and coldness between summer and winter.

Underground thermal energy storage (UTES) can occur as aquifer thermal energy storage (ATES), as borehole thermal energy storage (BTES) or as cavern thermal energy storage (CTES). In the case of CTES, artificial openings in the rock are created. However, all three involve the pre-existing subsurface rock or sediment as the heat-storage medium, and so will be referred to here as 'natural' UTES. These are in contrast to 'artificial' UTES options, which are: (1) to build a large steel or concrete tank partially or entirely underground, with or without an inner stainless steel lining, and filled with water; or (2) to construct an underground, watertight gravel pit that is saturated with water. A third option is a combination of the two, with a central, high-temperature steel or concrete storage vessel surrounded by a gravel pit or a BTES system, where heat is stored at a lower temperature and extracted with heat pumps. In this way, some of the heat that is lost from the central storage vessel to the surrounding ground is recaptured.

Heat can be stored either as low-temperature heat (<50°C) or as high-temperature heat (>50°C).

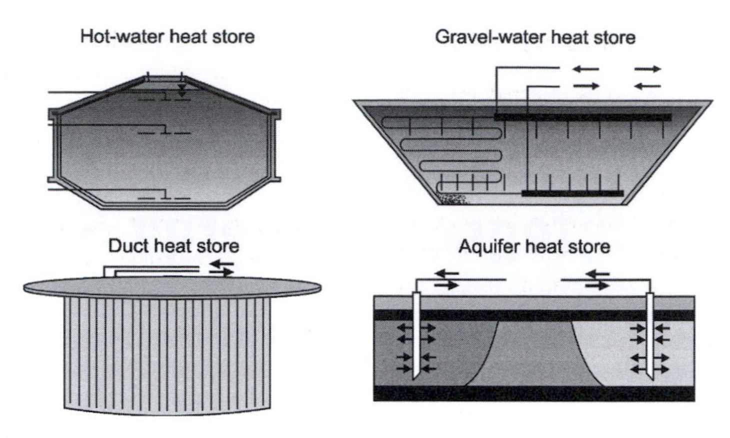

Source: Schmidt et al (2004)

Figure 11.1 *The major types of seasonal underground storage of thermal energy*

Low-temperature storage requires advanced heating systems with a low distribution temperature so that the stored heat can be used, or requires heat pumps (and additional energy input) in order to upgrade the stored heat to higher temperature. High-temperature storage does not require heat pumps, as the heat is already stored at a warm enough temperature to be usable in a district heating system.

The major options are summarized in Figure 11.1 and discussed blow, while technical characteristics of BTES, underground tanks and gravel/water pits are summarized Table 11.1.

Table 11.1 *Characteristics of some alternative options for storing heat underground*

Storage system	BTES	Hot water tank	Gravel/water pit
Storage concept	Hot water circulates through pipes in the ground, to a maximum depth of 150m	Watertight container buried underground	Hot water circulates through pipes in a watertight enclosure around gravel
Construction	U-shaped coaxial plastic pipes with a 1.5–3m separation	Reinforced concrete, steel- or fibreglass-reinforced plastic, or a pit with a cover and lid, stainless steel cover	Pit with waterproof lining, filled with water and gravel without a load-bearing cover
Maximum/minimum volumes	>100,000m^3 due to high lateral heat loss	Maximum 100,000m^3, the largest store designed so far being 28,000m^3	70,000m^3 largest as of 2000
Insulation	Only in the covering layer, 5–10m from the surface	15–30cm on top and at the sides, and also underneath if the pressure can be withstood	As with hot water tank
Storage volume/flat-plate collector area	8–10m^3/m^2	1.5–2.5m^3/m^2	2.5–4m^3/m^2
Cost (€/m^3) at 20,000m^3 storage volume	25	70–80	65–85
Other characteristics	Easily constructed	Container store is costly	With a gravel proportion of 60–70% by volume, the total storage volume is about 50% larger than for a water store

Source: Modified from Eicker (2003)

11.2.1 Aquifer thermal energy storage

In an ATES system, heated water is injected into the ground at a series of injection wells and later withdrawn. The same well(s) can be used for injection and withdrawal, or separate wells can be used for injection and withdrawal. In the latter case, water flows in the aquifer between the injection and withdrawal well. ATES is a *direct* water heat exchange system, in that the water circulated through the buildings is taken directly from the ground. Well depth normally does not exceed 400m. Schaetzle et al (1980, Chapter 7) describe a number of piping arrangements for distributing thermal energy in a community energy system with injection and withdrawal wells. Cool water can be extracted from one well to use for cooling purposes in summer (perhaps with a heat pump). The water is warmed in the process and injected into another well. In winter, the warm water is extracted and used for heating. It is cooled in the process, then injected into the cold well. This is a cyclic process that can be repeated indefinitely while reducing cooling costs by 80 per cent and heating costs by 40 per cent (Dinçer and Rosen, 2002). According to Schaetzle et al (1980), 60 per cent of the surface area of the US is underlain by aquifers adequate for use as thermal energy storage systems, and a similar percentage is expected for other continental areas.

Excess summer heat from a biomass-powered cogeneration system is stored in an aquifer at a depth of 270–390m below the German Reichstag, while winter coldness (for summer air conditioning) is in a second aquifer, at 0–66m depth. The lower aquifer reaches a maximum temperature of 70°C. This temperature drops to about 25°C by the end of the heating season and is upgraded by absorption heat pumps, as a 45°C heat distribution temperature is used. During the winter, the upper aquifer is chilled by ambient cold through three injection wells whenever the ambient temperature is cold enough, reaching a minimum temperature of 10°C. The temperature rises to 30°C by the end of the cooling season.

11.2.2 Borehole thermal energy storage

In a BTES system, U-shaped or coiled pipes are placed in a series of vertical boreholes and a fluid is circulated through the pipe system. Different possible layouts of the pipes and a sample installation are shown in Figure 11.2. The pipe system could serve as the ground coil of a heat pump, if a heat pump will be used to inject and extract heat, or could be connected to a source of hot water (such as a cogeneration powerplant or solar collectors) for

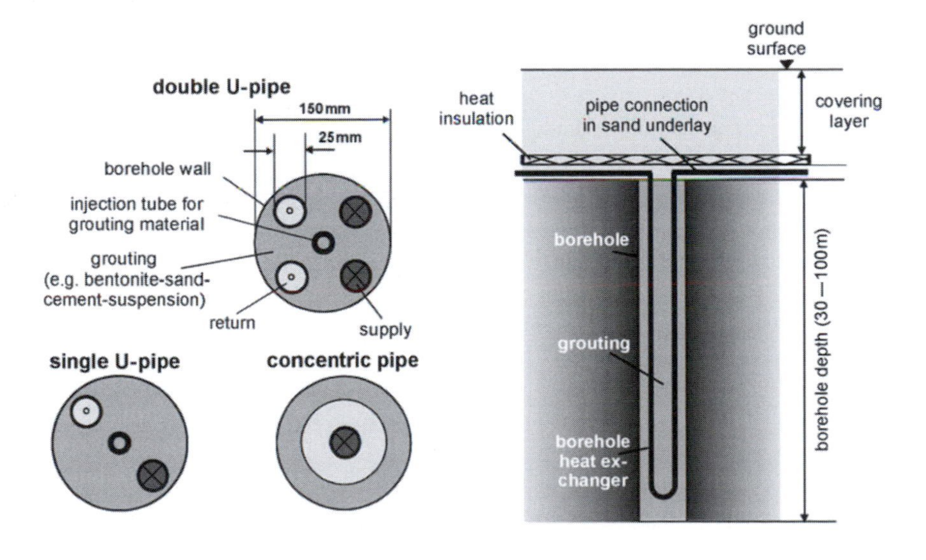

Source: Schmidt et al (2004)

Figure 11.2 *Different arrangements of pipes in a borehole heat exchanger (left) and a sample installation (right)*

unassisted heat transfer. The water circulating through the underground coils is separate from the groundwater, so BTES is referred to as an *indirect* water heat exchanger (in contrast to the direct water exchange of ATES). It is sometimes also referred to as *duct heat storage*. The depth of borehole heat exchangers in BTES systems normally ranges from 30 to 100m (Sanner, 1999). With deep aquifers and deep boreholes (>1000m), the surrounding ground temperature is greater, so the heat loss associated with a given storage temperature is smaller. This blends into the use of geothermal energy for heating purposes, except that heat withdrawn in winter is replaced in summer, thereby overcoming a major disadvantage of geothermal energy – that it is not renewable.

11.2.3 Steel or concrete tanks

Buried stainless steel tanks with a storage volume of up to at least 2000m^3 and concrete tanks (some with a steel lining) with a volume up to 12,000m^3 have been used for seasonal storage of solar thermal energy (Fisch et al, 1998), with storage temperatures varying from 40°C to 90°C seasonally. Bermed tanks (i.e. partially underground with earth piled up against the above-ground portion) can be an economically attractive alternative to complete burial (Rosen, 1998). An advantage of using hot water tanks is that the water can be stratified, with the hottest water on top. Hot water from solar collectors or cogeneration is supplied at the top of the tank and returns to the collectors from the bottom of the tank, while hot water to the heating network is withdrawn from the top of the tank and returns to the bottom of the tank. This results in a greater collector efficiency and greater supply temperatures compared to a well-mixed storage tank. In order to maintain the temperature stratification, diffusing water inlets are needed. For a prototype system built at the University of Calabria (Italy) using 91.2m^2 of vacuum tube solar collectors and a 500m^3 concrete tank with 0.2m of foam insulation, the average collector and storage efficiencies are each about 55 per cent, giving an overall efficiency in the use of solar energy of about 30 per cent (Oliveti et al, 1998). Chung et al (1998) simulated the performance of a solar collector/hot water tank storage system in Korea for heating an office building and greenhouse, for various combinations of collector area and storage volume and for flat-plate and evacuated-tube collectors. For 0.5m of urethane insulation, the storage temperature is greater than 90°C for 4000 hours per year (one year has 8760 hours).

11.2.4 Gravel/water pits

Gravel/water pits are cheaper than concrete/steel storage vessels beyond some minimum size. A recent example of a gravel/water pit for seasonal storage is the one built for a solar housing development in Steinfurt, Germany, in the late 1990s (Pfeil and Koch, 2000). Working from the outside inward, the structure of this pit consists of (1) insulation, (2) aluminium-polyethylene foil, (3) watertight lining, (4) fleece, and (5) the gravel fill. Granulated recycled glass was used as the insulation on the top and sides, while 0.15m thick foam glass plates were used at the bottom. The pit is used for storing summer heat, at temperatures ranging from 90°C (at the start of the heating season) to 30°C (at the end of the heating season). For large-scale systems and a roof-integrated solar collector with a cost of €175/m^2 (which has been reached in Germany once the savings in conventional roofing material is accounted for), the cost of solar thermal energy would be about 12 eurocents/kWh (compared to 8–23 eurocents/kWh from solar combisystems (see Chapter 2, subsection 2.4.6) and 8–12 eurocents/kWh for fossil fuel systems).

11.3 Utilization of renewable energy with district energy systems

The prior existence of a district heating and cooling system in a city makes it much easier to directly utilize renewable energy forms as they become available, and will make it easier to eventually use hydrogen produced from renewable energy sources for heating of individual buildings.

11.3.1 Use of biomass energy for district heating

As noted in Chapter 4 (Figure 4.16), Sweden has been able to switch a large fraction of its building heating requirements to biomass energy (plantation forestry) by switching from fossil fuels to biomass as the fuel for its district heating systems. To directly heat individual buildings with this biomass energy would not have been practical. Similarly, Denmark has been able to utilize biogas from farm manure for heating purposes in some of its district heating systems (Ramage and Scurlock, 1996). By the end of 1997, there were 369 district heating plants in Austria powered by biomass (Faninger, 2000).

Biomass can be used far more efficiently for heating buildings through cogeneration and district heating systems than with dedicated onsite use in pellet boilers. Consider the following three options for the use of 100 units of biomass energy:

1 a pellet boiler at 85 per cent efficiency, supplying 85 units of heat;
2 combustion cogeneration, producing 25 units of electricity that powers onsite electric heat pumps with a COP of 3.0, and 50 units of heat that is distributed with 10 per cent loss, giving a total heat supply of 122.5 units;
3 integrated biomass gasification combined cycle (see Chapter 4, subsection 4.4.3), producing 40 units of electricity and 40 units of heat that are treated in the same way as in case two, giving a total heat supply of 158 units.

Case three provides almost twice as much heat as the direct use of biomass in pellet boilers, and the biomass does not first have to be processed into pellets. However, there can now be excess heat at times, which will diminish the annual savings.[1]

11.3.2 Use of geothermal energy for district heating

Where geothermal energy is available, a district heating network is required in order to distribute the heat to individual buildings. As noted in Chapter 5 (subsection 5.3.1 and Figure 5.4), some municipalities in the Paris region rely largely on geothermal energy to heat buildings through its many local district heating networks.

11.3.3 Solar-assisted district heating

As discussed in subsection 11.2, solar thermal energy, collected during the summer months, has been successfully stored underground and used for heating in winter in several countries. Seasonal thermal energy storage is less expensive at the scale of a small district heating system than at the scale of individual buildings or houses. This is because, as the scale of seasonal energy storage reservoirs increases, the surface-to-volume ratio will decrease. This will decrease the fractional loss of stored thermal energy and, for underground storage in concrete or steel tanks or in gravel pits, decreases the capital cost (which depends largely on the surface area) per unit of storage capacity (unless the size increases to the point where scaffolding needs to be erected during construction).

Figure 11.3 illustrates the basic elements of a solar-assisted district heating system consisting of solar collectors, a short-term storage unit (the buffer tank), seasonal underground heat storage in boreholes, a supplemental boiler and the district heating network. The short-term surface storage tank is used to buffer short-term variations in direct heat supply and demand in winter. The top of the ground storage volume is insulated. Water circulates between the solar heat exchanger and the buffer tank whenever the collector outlet temperature exceeds the water temperature at the bottom of the buffer tank. The return water from the

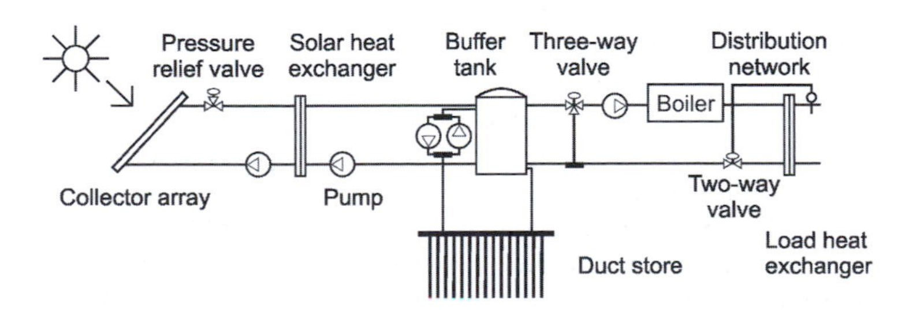

Source: Pahud (2000)

Figure 11.3 *Components and layout of a centralized system for collecting and storing solar thermal energy on a short-term and seasonal basis*

district heating system circulates through the buffer tank when the return-water temperature is less than the water temperature at the top of the tank. The lowest possible flow rate through the district heating system is used at any given time, so as to give the lowest return-water temperature and therefore the maximum possible heat extraction.

European examples

The pioneering work on solar-assisted district heating was carried out in Sweden during the 1970s and 1980s (Dalenbäch, 1990). Solar-assisted district heating systems with storage can be designed such that 32–95 per cent of total annual heating and hot water requirements are provided under German conditions (Lindenberger et al, 2000). By 2003, eight solar-assisted district heating systems had been constructed under the German Solarthermie-2000 programme that began in 1993 to promote solar systems both with and without seasonal storage (Lottner et al, 2000; Schmidt et al, 2004). Technical information and costs are given in Table 11.2. The largest of these systems are in Hamburg (serving 124 single-family homes), Friedrichshafen (serving 570 apartments in eight buildings) and in Neckarsulm (serving six blocks of flats, a school and a commercial centre). The fraction of the heating load supplied by solar energy in these systems ranges from 30 per cent to 62 per cent, but about four years of operation are required to reach the long-term solar fractions, as the ground surrounding the storage tanks is warmed. The cost of solar heat ranges from 16 eurocents/kWh to 42 eurocents/kWh. A number of municipalities and housing companies in Germany have apparently committed themselves to using solar energy by constructing low-energy housing with solar-assisted district heating. Another eight existing and six planned solar-assisted district heating systems in Sweden, Denmark, The Netherlands and Austria are reviewed by Fisch et al (1998). The largest of these, in Denmark, involves 1300 houses, a 70,000m^3 gravel pit for storage and a 30 per cent solar fraction.

Additional information on Danish systems is found in Heller (2000). The European Large-Scale Solar Heating Network maintains a website (main.hvac.chalmers.se/chsp) on solar heating plants with collector areas ranging from 500m^2 to 10,000m^2. The construction of

Table 11.2 *Technical data concerning solar-assisted district heating systems constructed in Germany under the Solarthermie-2000 programme*

Location	First year of operation	Housing type	Heated area (m^2)	Annual heat demand (MWh)	Solar collector area (m^2)	Storage volume (m^3)	Solar fraction	Solar heat cost (€/kWh)
Hamburg	1996	124 single family units	14,800	1610	3000	4500 (hot water)	0.49	0.256
Friedrichshafen	1996	570 apartments in 8 buildings	39,500	4106	5600	12,000 (hot water)	0.47	0.158
Chemnitz	1996	1 office building, hotel and warehouse	4680	573	540 vacuum tubes	8000 (gravel water)	0.30	0.240
Steinfurt	1998	42 apartments in 22 buildings	3800	325	510	1500 (gravelwater)	0.34	0.424
Neckarsulm	1999	6 blocks of flats, school, commercial centre	20,000	1663	2700 (5000, phase II)	20,000 (duct) (63,400, phase II)	0.50	0.172
Attenkirchen	1999	30 apartments	6200	487	800	500 + 9350 (hot water + duct)	0.55	0.170
Rostock	2000	108 apartments	7000	497	1000	20,000 (aquifer)	0.62	0.255
Hannover	2000	106 apartments	7365	694	1350	2750 (hot water)	0.39	0.424

Source: Schmidt et al (2004)

solar-assisted district heating plants has been facilitated by the development of large collectors, with an absorber area of 15m², such that a collector area of 1250m² can be installed within three days. Lessons learned from the Anneberg project in Sweden are presented in Lundh and Dalenbäch (2008). A key finding is that a number of errors were made in the installation of the seasonal solar heating system (but subsequently corrected) due to lack of experience with these systems and their greater complexity (including that of the control systems) compared to conventional heating systems.

North America's first solar-assisted district heating system

The first solar district heating system in North America began construction in 2005 in the town of Okotoks, near Calgary (Alberta, Canada). The project serves 52 small (145m² floor area) detached houses, and consists of a large (250m³) water tank that serves as a short-term thermal energy storage, a field of 144 37m-deep boreholes covered with insulation to RSI 7.04 (R40) for longer-term UTES, a district heat supply loop with a variable speed pump, 2300m² of solar collectors and a high efficiency (90 per cent) backup boiler. The space heating and domestic hot water (DHW) systems are kept separate, as the DHW load is relatively uniform seasonally but requires a higher supply temperature. Separation of the two allows the space heating system to operate at higher efficiency than if the two were combined. The water tank stores heat from the collectors for immediate use, without the need to cycle the heat through the underground heat store. It also allows some daytime heat to be stored and dumped into the BTES overnight, thereby reducing the cost of the BTES by reducing the rate at which it must absorb heat. The decision was made to keep the system as simple as possible in order (1) to reduce costs, (2) to more readily match the ability of installers and maintenance contractors, and (3) to match the operating abilities of homeowners. In order to reduce the required size and hence cost of the solar energy system, the design of the houses was upgraded to meet the Canadian R2000 standard (about a 30 per cent improvement over the heating energy use of houses built to the existing building code).

The BTES can reach 80°C at the end of the summer, but the district heating system distributes heat at 35°C to heating coils inside a regular (for North America) forced air heating system in individual houses. Use of onsite forced air heating systems with such a low-temperature coil required a specially designed heat exchanger and a low flow velocity in order to avoid draughts associated with 'cold heat'. There is also the possibility of increasing the temperature of the short-term storage tank to 55°C in anticipation of forecasted extreme cold, using the backup boiler.

The Okotoks solar system is calculated to provide 95 per cent of the annual space heating load and 50 per cent of the DHW load. The heating loads in the base case are met with a natural gas furnace at 80 per cent efficiency, while heating loads in the upgraded houses and the residual heat loads for the case with solar heat are assumed to be met by a natural gas furnace or a central condensing boiler at 90 per cent efficiency.

Minimizing the distribution temperature

In order to use solar energy in district heating, the required storage and distribution temperatures must be kept as low as possible in order to minimize the heat loss during storage and distribution, while the return temperature should be no more than 40°C if possible so as to permit a relatively high solar collector efficiency (see Chapter 2, subsection 2.4.2). To provide an adequate temperature difference between supply and return hot water flows and a minimal supply temperature, radiant floor heating should be used, or a very efficient heat exchanger in a low-velocity forced air system is needed. The former is used in buildings served by a solar district heating system in Anneberg, Sweden. Solar heat is collected at 60°C using 3000m² of flat-plate collectors and stored at 30–45°C in 60,000m³ of underground rock using an array of 99 65m-deep bores with a 3m spacing (Nordell and Hellström, 2000). Floor heating coils provide heat at 25–32°C. The latter approach is used in the aforementioned Okotoks project in Canada, where district heat at 35°C is supplied to forced air heating systems.

If district heating is to provide domestic hot water, then a somewhat warmer supply temperature (75°C) is needed. In summer, when there is no space heating load, it is not possible to induce a large enough temperature drop to produce return-water no warmer than 40°C. This will diminish the efficiency in collecting solar heat during the summer. In the

Anneberg project, auxiliary electric heaters are used instead to boost the water temperature to at least 55°C for domestic hot water needs, while in the Okotoks project, separate onsite solar thermal collectors are used for DHW.

11.3.4 Hybrid biomass/solar district heating systems

As noted above, biomass energy is being increasingly used as a fuel for district heating systems in some countries. Because the heat load is relatively small during the summer and transitional seasons, the boilers (whether fuelled by biomass or fossil fuel) would operate with frequent on/off operation, which is inherently inefficient. However, this is the time when solar energy is most readily available, so the boilers could be shut down altogether during part of this time. Thus, solar energy complements biomass energy well. The world's first hybrid solar/biomass district heating plant was installed in the village of Deutsch Tschantschendorf in Austria in 1994, and by August 1998, 12 hybrid solar/biomass plants were in operation in Austria with solar collector areas ranging from 225 to 1250m^2 (Faninger, 2000). In these particular cases, a relatively small thermal storage tank was constructed (adequate for three to five days of hot water demand), so the solar contribution to the total heating requirements is rather small: about 15 per cent of the total space plus water heating load.

11.3.5 Wind energy buffer

Wind energy has already become competitive with fossil fuel electricity in regions with good winds (average hub-height wind speed ≥7.5m/s), where electricity can be generated at a cost 5–8 cents/kWh (see Chapter 3, subsection 3.13.1). However, due to the fluctuating and somewhat unpredictable nature of wind energy, it will not be practical to supply more than 20 per cent of total electricity demand with wind energy without some mechanism for storing wind energy. One way to do this is through integration with a district heating and cooling system. If the wind energy component of a local power system is sized such that it sometimes provides more energy than needed, the excess power can be used with heat pumps to supply hot or chilled water in place of fossil fuel or biomass energy, or to charge thermal storage reservoirs. This allows the wind system to be sized to meet a larger fraction of total electricity demand with less waste of excess wind energy. For a scenario in which 50 per cent of Danish electricity demand in 2030 is met by wind, integrated operation of the electricity and heating systems cuts the wasted wind energy potential in half (Redlinger et al, 2002, Chapter 2). This in turn improves the economics of wind energy.

11.3.6 Making the transition to the hydrogen economy

One option to make the final transition away from fossil fuels is to replace fossil fuels with hydrogen fuel in applications where renewable energy cannot be directly used. As discussed in Chapter 10, hydrogen would be produced from a variety of renewable forms of energy (primarily wind and solar energy), in locations where renewable energy is plentiful, and transported largely by pipeline to where it is needed, stored and used when needed.

The switch from fossil fuels to hydrogen for space and water heating will be much easier and less expensive in those communities that are served by district heating systems. In buildings that are individually heated with natural gas, it is likely that a whole new pipeline infrastructure would have to be built in order to utilize hydrogen. In buildings served by district heating systems, the only infrastructure that would need to be rebuilt are the pipelines supplying the central powerplant. Furthermore, if boilers or furnaces need to be modified or replaced in order to use hydrogen fuel, this will be easier and probably less expensive if done at a single, centralized facility rather than in each individual building.

11.4 Summary

Community-integrated energy systems involve centralized production of heat and possibly chilled water that are distributed to individual buildings through district heating and cooling networks. District heating networks can be coupled with large-scale underground storage of heat that is collected from solar thermal collectors during the summer, and used for space heating and hot water requirements during the winter. Heat can also be supplied with biomass (as part of a biomass cogeneration system) or from geothermal heat sources. If both heat and coldness are stored, then heat pumps can be used to recharge the thermal storage reservoirs (or to directly supply heat or coldness to the district heating and cooling networks) during times of excess wind energy. This in turn permits sizing of the wind system to meet a larger fraction of total electricity demand without having to discard as much (or any) electricity generation potential during times of high wind and/or low demand. In the long run, district heating systems with cogeneration will make it easier to make the transition to a hydrogen economy, as a new infrastructure to supply hydrogen to individual buildings would not be needed.

Note

1 The principles with regard to the sizing of fossil fuel cogeneration systems that are discussed in Volume 1, Chapter 9, subsection 9.5.1, and the associated savings are applicable to biomass cogeneration systems.

12

Integrated Scenarios for the Future

This chapter summarizes the key findings of the previous chapters concerning the cost and availability of C-free sources of energy, reviews selected published scenarios involving large-scale use of C-free energy systems and then combines illustrative scenarios for the rapid deployment of C-free energy sources with the demand scenarios for fuels and electricity that were developed in Volume 1 in order to generate scenarios of future CO_2 emissions. It concludes with simulations of the changes in atmospheric CO_2 concentration, global mean temperature and ocean acidity resulting from the emission scenarios, taking into account possible non-linear feedbacks between climate and the carbon cycle.

12.1 Summary of the characteristics of C-free energy sources

The characteristics of various C-free energy sources for the production of electricity, heat and fuels are reviewed here and compared, where appropriate, with those of nuclear energy and fossil fuels. Passive solar energy – in the form of daylighting and passive heating, ventilation and cooling driven by passive ventilation with cool outside air – is the simplest, most reliable and often the least expensive form of renewable energy. It is extensively discussed in Volume 1, where its incorporation into buildings is treated as a reduction in energy demand rather than as energy supply. However, active methods of collecting solar energy (PV modules and solar thermal collectors) are treated as renewable energy supply and so form part of the discussion here.

12.1.1 C-free energy sources for electricity

This section summarizes the cost and performance, land area requirements, and energy return over energy invested (EROEI) of various renewable energy sources of electricity.

Cost and performance

Table 12.1 summarizes the key performance and cost parameters for renewable sources of electricity (wind, PV solar, concentrating solar thermal power, biomass and geothermal energy) and compares these with those of nuclear energy, conventional fossil fuel powerplants and fossil fuel powerplants with CCS. The greatest potential for C-free electricity involves solar and wind, with biomass playing a potentially important supplementary role. Electricity can be generated from sunlight using either PV modules or using mirrors to concentrate sunlight by a factor of 600–2000, producing steam that in turn is used in a steam turbine. PV modules can use either unconcentrated sunlight or sunlight concentrated by a factor of 20–550. Both solar thermal and concentrating PV systems require that mirrors or modules track the sun at all times, and so are applicable only in arid or semi-arid parts of the world (where clouds are rare), but the annual solar irradiance on sun-tracking modules is about 30 per cent greater than for an optimally oriented fixed module. An advantage of concentrating solar thermal power (CSTP) is that heat can be stored overnight, allowing 24-hour-per-day generation of electricity and some ability to vary the electricity production with demand. An advantage of concentrating PV compared to conventional PV is that much smaller quantities of expensive semi-conductor materials are needed, thereby allowing highly efficient multi-junction modules using rare elements to provide a large share of future electricity requirements. Fluorescent dye (also called quantum dot) concentrators, currently under development, offer the possibility of concentrating

Table 12.1 *Costs of C-free electricity and fuels in comparison with fossil fuel-based energy*

	Fuel cost ($/GJ)	Effi-ciency	Capital cost ($/kW)	O&M Fixed (%Inv/yr)	O&M Variable (c/kWh)	Life-span (yrs)	Capacity factor	Cost of electricity (cents/kWh) Capital	Fuel	O&M	Total	Capacity credit (%)
Electricity from renewable energy												
Wind, low cost, onshore	0	0.25	1300	0.7	0.7	25	0.25	4.2	0.0	1.1	5.3	15–30
Wind, high cost, onshore	0	0.25	1900	0.7	0.7	25	0.15	10.3	0.0	1.7	12.0	10–15
Wind, low cost, offshore	0	0.25	1800	0.6	2.1	25	0.4	3.6	0.0	2.4	6.1	20–30
Wind, high cost, offshore	0	0.25	2600	0.6	2.1	25	0.3	7.0	0.0	2.7	9.7	15–25
Solar PV, low cost, today	0	0.15	6000	1.5		25	0.25	19.4	0.0	4.1	23.5	0–65
Solar PV, high cost, today	0	0.15	10,000	1.5		25	0.16	50.6	0.0	10.7	61.3	0–65
Solar PV, low cost, future	0	0.07	1000	1.5		25	0.25	3.2	0.0	0.7	3.9	0–65
Solar PV, high cost, future	0	0.15	2500	1.5		25	0.16	12.7	0.0	2.7	15.3	0–65
CSTP, PT, today	0	0.15	6000	3		30	0.25	17.8	0.0	8.2	26.0	35–85
CSTP, future low	0	0.20	2000	3		30	0.4	3.7	0.0	1.7	5.4	35–85
CSTP, future high	0	0.28	3000	3		30	0.6	3.7	0.0	1.7	5.4	35–85
Geothermal (HDR)	0	0.14	2500	4		50	0.75	2.1	0.0	1.5	3.6	90
Hydropower, large, low	0	0.75	1500	2		100	0.75	0.7	0.0	0.5	1.2	50–80
Hydropower, large, high	0	0.75	3000	2		100	0.75	1.4	0.0	0.9	2.4	50–80
Hydropower, small, low	0	0.75	900	2		100	0.5	0.7	0.0	0.4	1.1	50–80
Hydropower, small, high	0	0.75	1300	2		100	0.5	0.9	0.0	0.6	1.5	50–80
Biomass, low, today	3	0.23	2000	3.5	0.75	20	0.75	2.4	4.7	1.8	9.0	50–90
Biomass, high, today	6	0.32	4000	6.5	1.0	20	0.75	4.9	6.8	5.0	16.6	50–90
Biomass, low, future	2	0.38	1000	2.5	0.5	20	0.75	1.2	1.9	0.9	4.0	50–90
Biomass, high, future	4	0.50	2000	4.5	1.0	20	0.75	2.4	2.9	2.4	7.7	50–90
Electricity from non-renewable energy												
NGCC without C capture, low	6	0.6	600	2.5	0.15	40	0.75	0.5	3.6	0.4	4.5	
NGCC without C capture, high	18	0.6	900	2.5	0.15	40	0.75	0.8	10.8	0.5	12.1	
NGCC with C capture, low	6	0.55	900	2.5	0.3	40	0.75	0.8	3.9	0.6	5.4	92
NGCC with C capture, high	18	0.55	1300	2.5	0.3	40	0.75	1.2	11.8	0.8	13.7	
Coal without C capture, low	2	0.5	1200	3.5	0.2	40	0.75	1.1	1.4	0.8	3.3	
Coal without C capture, high	4	0.5	1500	3.5	0.2	40	0.75	1.3	2.9	1.0	5.2	92
Coal with C capture, low	2	0.45	1500	3.5	0.4	40	0.75	1.3	1.6	1.2	4.1	
Coal with C capture, high	4	0.45	2600	3.5	0.4	40	0.75	2.3	3.2	1.8	7.3	92
Nuclear, low cost	0.64[a]	0.33	4000	1.6	0.05	20	0.75	7.2	0.6	1.0	8.8	
Nuclear, high cost	1.2[a]	0.4	6000	3.3	0.05	20	0.75	10.7	1.2	3.1	15.0	85

Note: CSTP = concentrating solar thermal power, PT = parabolic trough, HDR = hot dry rock, NGCC = natural gas combined cycle. Capital costs for NGCC and coal with carbon sequestration are projected future costs that many might regard as highly optimistic. Nuclear capital costs are based on plausible estimates for plants that might be constructed in the next ten years. Wind and hydro capital costs are current costs, the former excluding discounts of up to 45 per cent for large (>100 turbines) orders. All costs assume financing at 5 per cent interest, except for nuclear energy, where 10 per cent interest is assumed because of greater uncertainty. [a] Contribution of fuel cost to the final cost of electricity in cents/kWh, from Chapter 8, subsection 8.7.3.
Source: Based on estimates presented in previous chapters (some O&M, capacity factors and capacity credits are from GAC (2006)

both direct and diffuse sunlight without the need for a sun-tracking system. There is a large potential to generate electricity using building-integrated PV panels. Costs for both CSTP and PV power should soon be attractive in regions with high solar irradiance.

Wind energy is currently the lowest-cost form of C-free energy that can be exploited on a large scale, and for this reason is experiencing the greatest rate of growth of C-free sources of energy. By combining widely dispersed wind farms with various storage techniques, baseload or dispatchable electricity can be produced from the wind. The most promising parts of the world for wind energy are in middle and high latitudes, and in some low-latitude regions with steady trade winds.

A variety of other renewable energy sources for the generation of electricity – geothermal, hydro and marine – could be locally important, although the first two are not entirely free of GHG emissions and geothermal energy can be replenished only at a very slow rate (through upward conduction of heat from the earth's interior).

Land requirements

Table 12.2 compares the rate at which electricity can be generated per unit of land devoted to biomass plantations, per unit of land flooded by hydroelectric reservoirs, and per unit of land used by solar PV, solar thermal systems, wind turbines and coal and nuclear powerplants. The only land occupied by a wind farm that cannot be used for other purposes is land taken up by turbine foundations and access roads, so wind is by far the most land-efficient form of renewable energy, at

Table 12.2 *Average electricity production per unit of land area using different renewable energy sources, and land area required to supply the global 2005 electricity demand (18,454TWh) indefinitely from renewable energy or over a period of 100 years from coal and nuclear energy*

Energy source	(kWh/yr)/m²	Indefinite supply of 2005 electricity demand
		Direct land requirement
Biomass[a]	1.7–4.2	4–12 × 10^6km²
Hydroelectric power, Brazil	0.4–219	0.1–53 × 10^6km²
Hydroelectric power, Canada	3.5–46	0.4–5.0 × 10^6km²
Solar[b]	45–100	0.2–0.4 × 10^6km²
Wind[c]	3000–4000	5–6 × 10^3km²
Coal (120-400ha/GW @ 0.85CF)	1800–6200	3–10 × 10^3km²
Nuclear (1400ha/GW @ 0.85CF)	520	26 × 10^3km²
	(m²/GWh)	**100 years of 2005 electricity generation**
		Indirect land requirement
Coal at 35% efficiency, surface mining	43–1450	0.08–2.7 × 10^6 km²
Coal at 35% efficiency, underground mining		
– land disturbed by slumping	2.3–510	4-941 × 10^3 km²
– land required to grow wood needed for timber supports	29–262	54–483 × 10^3 km²
– total if additive	31–770	0.06–1.42 × 10^6 km²
Nuclear	70[d]	0.13 × 10^6 km²

Note: [a] Assuming a biomass yield of 150–300GJ/ha/yr and an electricity generation efficiency of 0.4–0.5.
[b] Assuming a mean annual solar irradiance on tilted modules of 200–300W/m², an efficiency of 0.1–0.15 and a land-use fraction of 0.25. [c] Pertains to the 2.3MW turbine of Table 3.6, and is based on the land area taken up by foundations and access roads only, for a mean wind speed at the hub height of about 7.6-8.6m/s (corresponding to a capacity factor of about 0.25–0.40). [d] 40m²/GWh related to mining and milling, an area that will increase as lower grades of ore are mined, and 30m²/GWh related to land that would have been permanently withdrawn according to the proposed (and now cancelled) Yucca Mountain repository for nuclear waste.
Source: Renewable energy, this volume; coal and nuclear, based on Fthenakis and Kim (2009)

3000–4000kWh/m^2/yr. Solar power generation is a distant second, at 45–100kWh/m^2/yr. However, the land area requirements for solar electricity given in Table 12.2 pertain to centralized power generation in desert areas. As noted in Chapter 2 (subsection 2.2.6), a large fraction of current electricity demand could be supplied through building-integrated PV and through PV over parking lots, for which the additional land area is zero. Hydropower spans a wide range of land efficiencies, but assuming that only the most land-efficient schemes are developed in the future, the land efficiency of hydropower will be comparable to that of centralized solar power. Coal powerplants require about 120–400ha per GW of capacity, while nuclear powerplants in the US require about 1400ha/GW.[1] Assuming a capacity factor of 0.85, the annual electricity output per unit of land occupied by a nuclear powerplant and by some coal powerplant is less than that occupied by wind turbines.

Also given in Table 12.2 are two sets of land areas: the land directly required by various types of powerplant to satisfy the global 2005 electricity demand of 18,500TWh, and the indirect requirements for coal and nuclear energy associated with mining and other activities to provide 18,500TWh/yr over a period of 100 years (which is likely to be the maximum feasible duration for the use of these energy sources). The direct land requirements for wind are decidedly less than for nuclear energy and could also be less than for coal. Furthermore, when wind turbines and other non-nuclear powerplants reach the end of their life the same site can be readily used again, but this is not the case of nuclear powerplants (due to the need for decommissioning spanning a period of decades). The indirect land requirement involves stripping off the overburden or removing entire mountain tops in the case of surface coal mining (see online Figures S2.1 and S2.2 of Volume 1), and slumping of land and utilization of land to produce wood for timber supports in the case of underground coal mining. Details can be found in Fthenakis and Kim (2009). The land requirements for nuclear largely involve strip mining of uranium, land set aside to store uranium mining and milling wastes and land permanently set aside in association with nuclear waste repositories. The land requirement given in Table 12.2 for the first two categories probably pertains to the current world average ore grade (about 0.2 per cent) and so will increase dramatically over the course of the next century if nuclear energy is used on a large scale (see Chapter 8, subsection 8.8.1). In any case, the land area impacted by coal and nuclear electricity production over a 100-year time period rivals or exceeds that of many renewable-energy sources. There would also be some small indirect land requirement for renewable energy, associated with land that is transformed during the production of the materials used to manufacture the various powerplants. With end-of-life recycling of materials, this indirect land use will be even smaller and so need not be considered here.

Distribution of concentrating solar thermal and wind electricity potential

Wind energy and CSTP are likely to be the preferred C-free energy sources for providing the bulk of the world's future energy needs, and no major electricity-demand centre is more than 2000–3000km from excellent CSTP or wind resources. Electricity could be transported using HVDC with losses of only 6.2 per cent at full load over a distance of 2000km and 8.7 per cent over a distance of 3000km (see Table 3.20). Figure 12.1 shows the geographical distribution of potential electricity production (GWh/km^2/yr) from wind (top) and CSTP (middle), based on the following assumptions:

- for wind energy, a rotor diameter D of 80m, turbines laid out with a spacing of $7D$ in both directions, a 100m hub height, monthly mean wind power density (WPD) at the hub height computed from monthly mean 100m winds (as obtained on a 1° × 1° grid from the NASA Surface Meteorology and Solar Radiation website discussed in Appendix F) using the relationship between WPD and mean wind speed given in Figure 3.17 for k=2.0, and efficiency at any given mean wind speed as given in Figure 3.18 for the N80-2.5MW turbine;
- for CSTP, annual mean irradiance on a plane that is always pointed directly at the sun (referred to as 'Direct Normal Radiation' or DNR) from the NASA website, a ratio of collector area to ground area of 0.25, and an efficiency in generating electricity of 15 per cent (based on Table 2.15).

The best CSTP regions can produce over 100GWh/km^2/yr, while the best wind regions can produce over 40GWh/km^2/yr based on the areal extent of the wind farms, but over 2000GWh/km^2/yr if only the land area taken up by turbine foundations and access roads is counted. A meaningful way to combine the CSTP and wind energy potential into a single map is plotting the minimum of the CSTP and wind electricity costs. This is shown in the lower panel of Figure 12.1

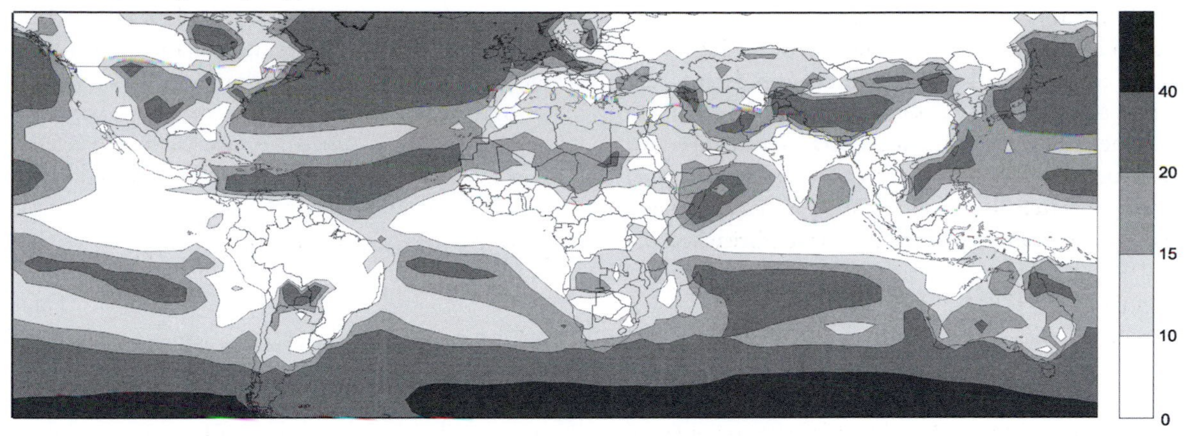

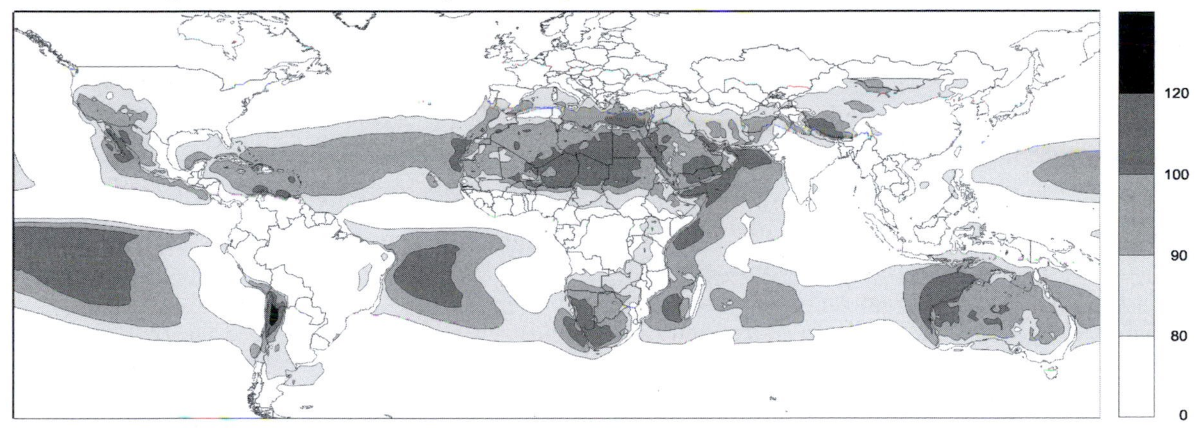

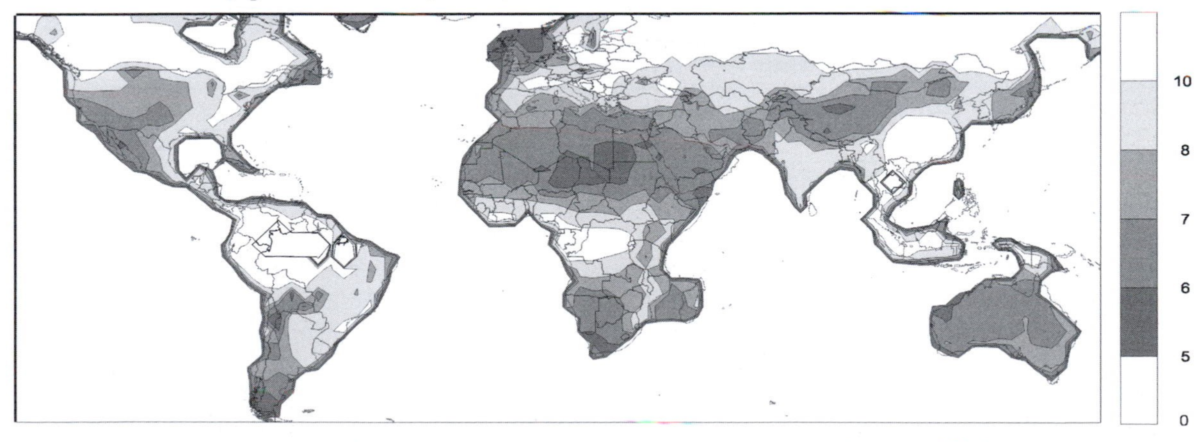

Figure 12.1 *Potential annual electricity production from wind energy (top) and from CSTP (middle), and the minimum of the cost of wind and CTSP electricity (bottom)*

Source: Calculated from NASA 100m wind and direct normal radiation data, as explained in the text

assuming capital costs of wind and CTSP and \$1500/kW and \$3000/kW, respectively, annual O&M costs of 2 per cent and 5 per cent of the wind and CTSP capital cost, respectively, and financing at 5 per cent interest over 20 years. In computing the wind electricity cost, the mean annual capacity factor is obtained by averaging monthly capacity factors that are computed from the monthly mean wind speeds using the relationship shown in Figure 3.45a for the case where the wind farm capacity is set at twice the transmission capacity. For CSTP, the mean annual capacity factor is given by $2DNR(kWh/m^2/yr)/(8760\ I_p)$, where I_p=0.85kW/m^2 is the standardized peak direct irradiance and 8760 is the number of hours in one year. The factor of two conservatively accounts for the effect of thermal energy storage in increasing the annual capacity factor compared to the case with no storage (see Figure 2.37). There are many broad regions where electricity could be produced at a cost of 6–8 cents/kWh, to which modest transmission costs must be added.

Energy return over energy invested

Table 12.3 compares the energy return over energy invested (EROEI) for various renewable energy sources of electricity, as well as for energy from nuclear, coal and natural gas combined cycle (NGCC) powerplants. The EROEI for renewable energy sources is the ratio of lifetime primary energy saved at the alternative fossil fuel powerplant (assumed to be using coal at 40 per cent efficiency) divided by the primary energy required to produce the renewable energy facility. Wind has the most favourable EROEI, followed by biomass cogeneration, CSTP and PV power. The EROEI for nuclear energy depends strongly on the grade of ore used to supply the uranium fuel and so will decrease over time as less favourable uranium ores are utilized. The EROEI for coal and natural gas are based on the energy inputs other than the fossil fuels used at the powerplant. These include energy used to construct the powerplant, in mining and transporting the fuel, and for pollution control. The EROEI is surprisingly low (2.2) for NGCC, and arises because non-fuel energy inputs amount to 25 per cent of the fuel energy input (mostly for constructing the pipelines that supply the natural gas and in operating the pipeline). As seen from Table 12.3, all of the renewable energy options compare highly favourably with coal and natural gas and most already compare highly favourably with nuclear power.

Prospects for nuclear energy and for carbon capture and storage

Nuclear energy is not likely to be able to play a significant role in reducing GHG emissions due to its high cost, the long time (decades) required to develop

Table 12.3 *Comparison of the energy payback time and energy return over energy invested (EROEI) for various sources of electricity*

Electricity source	Energy payback time	EROEI	Reference
Wind in Germany	3–6 months	40–80 if 20-year lifespan	Section 3.15
CSTP in sunny regions			
Parabolic trough	6 months	40 if 20-year lifespan	
Parabolic dish	1.5 years	14 if 20-year lifespan	Subsection 2.3.5
Central tower	2.5 years	8 if 20-year lifespan	
PV in central Europe[a]	1–3 years today < 1 yr, future	8–25-years if 25-year lifespan > 25 if 25-year lifespan	Table 2.6
Biomass cogeneration		36–48 for provision of coppice willow to point of use	Table 4.39
Nuclear		16–18 @ 0.2% grade ore 3–5 @ 0.01% grade ore	Figure 8.23
Coal		5.0 @ 32% powerplant efficiency[b] 6.7 @ 42% powerplant efficiency[b]	Spath et al (1999)
Natural gas		2.2 @ 54% powerplant efficiency[b]	Spath and Mann (2000)

Note: [a] 1700 kWh/m^2/yr solar irradiance [b] LHV basis

a new generation of reactor technologies, unresolved problems related to long-term isolation of radioactive wastes from the biosphere, and the risk of nuclear weapons proliferation and nuclear terrorism (especially if fuel reprocessing were to occur on a large scale). In the absence of reprocessing, the uranium resource thought to be available at a high enough grade to provide an EROEI greater than two from open pit mines and greater than four from underground mines is sufficient for less than 100 years (and possibly much less than 100 years) operation of a nuclear fleet with four times the capacity of the current fleet. Reprocessing of spent fuel would expand the supply by up to a factor of 75, but poses high terrorism and nuclear arms proliferation risks. Thus, nuclear cannot play a large long-term role in reducing GHG emissions. It can delay the last 20 per cent or so of the transition to renewable energy by at most a century, but at the cost of leaving a 1 million-year legacy of nuclear waste for future generations.

CCS requires another one to two decades of research and demonstration projects in order to determine if it would be a viable on a large scale. There are reasons for significant doubts concerning the integrity of storage in deep terrestrial rock layers as well as concerning its impact on potentially useful underground water supplies. Injection in deep ocean water is unacceptable on environmental grounds, but injection in sediments below the sea floor might be a technically and environmentally viable option. However, CCS is likely to be expensive and would require the development of a massive infrastructure if it were to make a significant contribution to reducing CO_2 emissions. The same applies to direct capture of CO_2 from the atmosphere. Neither has any prospect of making a significant contribution before mid-century.

12.1.2 Renewable energy for heat and for liquid and gaseous fuels

Table 12.4 compares the costs associated with providing heat through biomass, solar thermal collectors, natural gas and electric resistance heating. For consistency, all systems are assumed to have an annual capacity factor (average output as a fraction of peak output) of 0.25 and are assumed to be financed at 5 per cent interest over a period of 20 years. Biomass can be used in solid form as a fuel for heating, either as pellets or wood chips. These fuels can often be produced and delivered to users for costs of around \$3–6/GJ, which is less expensive than present and likely future costs of natural gas. The fuel with solar collectors is free but a much greater initial investment is required. The final cost of heat depends strongly on the capacity factor. The combination of low fuel costs compared to natural gas and low investment costs compared to solar thermal makes biomass heating systems very attractive. Electric resistance heating using electricity at 10 cents/kWh is the most expensive option of all.

Table 12.5 compares the cost of transportation fuels made from biomass or from C-free electricity, based on the biomass or electricity cost and conversion efficiency only. Also given are the land areas required per EJ of fuel produced. This table demonstrates the fundamental point that hydrogen from renewable electricity will always be substantially more expensive than hydrogen or liquid fuels made from biomass, but at the same time,

Table 12.4 *Characteristics of different methods of providing heat from solar or biomass energy and comparison with natural gas heating systems in buildings*

	Equipment cost (\$/kW)	Fuel cost (\$/GJ)	Efficiency	Capacity factor	Cost of heat (\$/GJ)
Solar DHW	1000	0	0.6	0.25	10.2
	1500	0	0.4	0.25	15.3
Biomass	75	3	0.9	0.25	3.8
	150	6	0.95	0.25	7.5
Natural gas	50	6	0.9	0.25	6.5
	150	18	0.95	0.25	19.5
Electricity at 10 cents/kWh	50	28	1	0.25	28.3

Note: The cost of heat has been computed assuming that the equipment is financed at 5 per cent/year interest over a 20-year period with the indicated capacity factor and annual O&M costs equal to 2 per cent of the initial investment cost.

Table 12.5 *Characteristics of different fuels that can be produced from biomass energy and of hydrogen produced from wind or solar electricity*

Fuel and source	Cost of electricity or biomass	Conversion efficiency (%)	Cost of fuel ($/GJ)	Land energy yield (GJ/ha/yr)	Land area (1000km^2) per EJ/yr
Ethanol from lignocellulosic biomass	$3–6/GJ	46–54	6–10	200	90–110
Ethanol from sugarcane	$2/GJ	32–45	4–6	869	26–36
H_2 from wind electricity	5–10 cents/kWh	75	19–37	126,000	0.11
H_2 from CSTP energy	5–10 cents/kWh	100	14–28	3600	2.8
H_2 from PV electricity	5–10 cents/kWh	75	19–37	2700	4.9
H_2 from biomass	$3–6/GJ	52–61	5–12	200	80–100

Note: In the case of hydrogen from CSTP energy, high temperature water vapour electrolysis is assumed using waste heat, giving an efficiency based on the electricity input alone of 100 per cent. Land energy yield for sugarcane is a future projection given in Table 4.22, while land energy yields for hydrogen from wind, CSTP or PV are based on the electricity production values given in Table 12.2.

hydrogen from renewable electricity is far more efficient in the use of land than any fuel made from biomass. Among biofuels, sugarcane from Brazilian ethanol is likely to have the smallest land requirements due to high bioenergy yields (>400GJ/ha/yr today, >850 GJ/ha/yr projected for the future according to Table 4.22).[2] Ethanol from sugarcane could therefore be attractive as a long-term liquid fuel for transportation, as long as long-term productivity of the soils can be maintained and its production is restricted in scale so as to preserve important ecosystems (such as the Brazilian cerrado and Amazon rainforest, the latter being threatened by activities displaced by sugarcane plantations rather than directly utilized for sugarcane). Ethanol from fast-growing, hardy and perennial lignocellulosic crops is likely to have much greater land requirements due to the lower yields, which are related to the assumption that these crops will be grown in part on degraded land or surplus agricultural land. The extent to which biofuels can be produced while preserving ecosystems and without competing with land for food production depends strongly on the amount of meat in the average diet, the productivity of agricultural land and the productivity of biomass-for-energy plantations.

Hydrogen fuel can be produced from C-free electricity sources through low-temperature electrolysis. A target system efficiency, including compression to 420atm, is 78 per cent at full load. High-temperature water vapour electrolysis may yield a thermal energy-to-hydrogen efficiency of about 47 per cent (see Chapter 10, subsection 10.3.4), or about 25 per cent solar-to-hydrogen efficiency (compared to 15 per cent for 20 per cent efficient panels and 75 per cent electrolysis efficiency). Even higher solar-to-hydrogen efficiencies, up to 50 per cent, are conceivable with joint PV/thermal panels if all of the excess heat can be used. The waste heat that would accompany concentrating solar thermal generation of electricity can, if applied to high-temperature water vapour electrolysis, reduce the electricity requirement such that the efficiency based on the electricity input only would reach about 100 per cent.

12.2 Summary of recently published low-emission scenarios

Summary information (reference scenario, main measures considered, methods used and major results) on over 30 studies of the potential to reduce greenhouse emissions in various countries or regions of the world is presented in Supplemental Table S12.1.[3] Many of the scenarios – especially those that attempt to determine least-cost supply-side measures – fail to achieve the CO_2 emissions of the magnitude required to limit atmospheric CO_2 concentration to no more than 450ppmv, although some scenarios do achieve 50–100 per cent emission reductions by 2050 or so. Here, we focus on two scenarios, one for the US and one for Europe.

12.2.1 American Solar Energy Society scenario for 2030

The American Solar Energy Society developed a scenario for the US in 2030 with the following key elements: (1) end-use energy efficiency sufficient to reduce primary energy demand in 2030 by 20 per cent compared to business-as-usual, which still leaves total demand 6 per cent above the 2005 level; and (2) a mixture of renewable energy sources (wind, solar, geothermal and biomass) to supply about 50 per cent of total electricity demand.

Table 12.6 gives the savings in CO_2 emissions projected for 2030 and the electrical capacity and electricity supply from various renewable energy sources. Instead of growing from 1684GtC/yr in 2005 to 2235GtC/yr in 2030 (an increase of almost 20 per cent), emissions decrease by 1024GtC/yr – a decrease of 55 per cent compared to the business-as-usual emission in 2030 and 40 per cent compared to the emission in 2005. For each load centre, wind energy was limited to 20 per cent of the total demand, with additional output (if available) exported via additional transmission capacity at a cost of \$620/MW/km) (this is a high cost, based on Figure 3.39). HVDC transmission was not considered in this scenario, so concentrating solar thermal electricity is limited to supplying demand in the American southwest. This and the absence of compressed air energy storage meant that wind energy had to be limited to 20 per cent of total demand. Agricultural residues and lignocellulosic energy crops (such as switchgrass) are assumed to be used for the production of biofuels (with some residues left in the field to control erosion and partly replenish soil nutrients), while forestry residues and municipal organic waste are assumed to be used for the generation of electricity. A total of 133MtC/yr are displaced in this way (58MtC/yr from biofuels, 75MtC/yr from biomass). If all available biomass were used to generate electricity, more CO_2 emissions would be avoided (183MtC/yr) due to the greater effectiveness in using biomass to produce electricity instead of transportation fuels.

Table 12.7 gives the potential electricity generation from concentrating solar thermal power alone in the American southwest. Only land areas with a solar irradiance of 2500kWh/m^2/yr or greater were considered, while land taken up with other uses (including wilderness areas and national parks), land with a slope greater than 1 per cent, or land that was part of a contiguous area of less than 10km^2 was excluded. The technical potential (6876GW and 30,000TWh/yr) vastly exceeds total US electrical capacity (about 1076GW) and annual electrical energy demand (about 4300TWh/yr) in 2005. An analysis using a concentrating solar power deployment model developed by the National Renewable Energy Laboratory indicates that 30GW could be deployed by 2030 with an extension of the 30 per cent investment tax credit, and 80GW by 2030 with a \$35/tCO_2 (\$128/tC) tax.

Table 12.6 *Contribution of different measures to the supply of electricity and to CO_2 emission reduction in the US in 2030 according to the American Solar Energy Society scenario*

Measure	Emissions (MtC/yr)	Electrical capacity (GW)	Average capacity factor	Electricity supply (TWh/yr)	% of grid electricity in 2030[a]
2005 value	1684	800	0.58	4038	
2030 BAU value	2235[b]			5341[b]	
Electricity supply or impact on emissions in 2030					
Efficiency	−688			−1054	−20.0
Wind	−181	245	0.40	860	20.0
Geothermal	−83	50	0.90	395	9.2
Concentrating solar thermal	−63	80	0.43	300	7.0
PV	−63	200	0.17	300	7.0
Biomass	−75	45	0.90	355	8.3
Biofuels	−58				
Total C savings	−1211				51.5
Total renewable capacity or supply		625	0.40	2208	
Net result in 2030	1024			4287	

Note: [a] As a percentage of business-as-usual demand. [b] An increase of 32 per cent.
Source: Kutscher, 2007

Table 12.7 *Potential electricity generation from concentrating solar thermal energy in the American southwest*

State	Area (km^2)	Capacity (GW)	Electricity production (TWh/yr)
Arizona	50,000	2468	5400–10,800
New Mexico	39,400	1940	4250–8500
California	17,900	877	1920–3840
Nevada	14,500	715	1570–3140
Utah	9300	456	1000–2000
Colorado	5400	271	600–1200
Texas	3100	149	325–650
Total	139,600	6876	15,000–30,000
US totals in 2005		1076	4300

Note: Low and high electrical productions are given assuming a capacity factor of 25 per cent (appropriate without thermal storage) and 50 per cent (appropriate with six hours of storage), respectively.
Source: Kutscher (2007)

12.2.2 A renewables-intensive scenario for electricity supply in Europe in 2050

The German Aerospace Center (GAC) developed the Trans-Mediterranean Interconnection for Concentrating Solar Power or TRANS-CSP scenario. This scenario involves a combination of concentrating solar thermal power (CSTP) in southern Europe and imported from the Middle East and North Africa, wind energy (especially in the North Sea region), geothermal energy, hydropower from Norway and other regions, and biomass cogeneration, with a HVDC grid to interconnect these energy sources (see Chapter 3, section 3.12 on the visual impact, cost and energy loss associated with HVDC transmission). Twenty 5GW HVDC transmission lines transmit a total of 700TWh/yr (giving an 80 per cent overall transmission capacity factor) from 20 different locations. In this way, 80 per cent of the projected European electricity demand in 2050 is supplied from renewable energy sources, with the remainder split roughly equally between natural gas and coal. The proposed generation of electricity from different renewable and non-renewable energy sources is shown in Table 12.8, along with capacities and capacity factors. The biomass potential includes 226TWh/yr from agricultural residues, 438TWh/yr from residues from 1.7 million km^2 of forest (with biomass yields for electricity generation ranging from 1.1 to 5.5t/ha/yr) and 100TWh/yr from municipal solid waste. Bioenergy crops specifically for electricity generation are not considered, as these are assumed to instead be dedicated to the transport sector (biofuels). The factor-of-ten growth in electricity supply from biomass requires an average growth rate of 5 per cent/year between 2000 and 2050. The land area used for renewable energy infrastructure in 2050 amounts to 1 per cent of the total land area. The planning and negotiation of international agreements alone for the grid is expected to take 15–20 years. There are already a number of 400–500kV DC interconnections between North Africa and Southern Europe.

Table 12.9 summarizes the investment costs and electricity costs for the various renewable and non-renewable electricity sources in 2000 and as projected for 2020 and 2050 in the TRANS-CSP scenario. The cost given for imported CSTP electricity includes the cost of HVDC transmission. If all of the CSTP plants in North Africa and the Middle East that export power to Europe were cogeneration powerplants, producing electricity and heat that in turn is used to desalinate water through reverse osmosis (as described in Chapter 2, subsection 2.8.1), 75 billion m^3 of water could be desalinated per year as a byproduct of producing electricity for Europe, meeting half of the projected water deficit for the region in 2050.

12.2.3 Development of autonomous renewable energy systems

In the TRANS-CSP scenario presented above, and in the scenario to be presented here, the emphasis is on the large-scale production of electricity from solar and wind

Table 12.8 *Renewable energy electricity scenario for Europe in 2050, and comparison with the situation in 2000*

Energy source	Electricity supply (TWh/yr)			Capacity (GW)		Capacity factor	
	2000	2050	Economic	2000	2050	2000	2050
PV	0.2	153.8	1730	0.2	122.3	0.114	0.144
Local CSP	0.0	111.5		0.0	18.8		0.677
Imported CSTP	0.0	707.5		0.0	102.2		0.790
Wind	23.2	780.7	1520	12.8	276.6	0.207	0.322
Hydro	615.8	748.3	910	190.3	235.7	0.369	0.362
Biomass	49.3	495.4	620	10.8	138.6	0.521	0.408
Geothermal	5.9	200.6	380	0.8	41.2	0.842	0.556
Wave, tidal	0.0	66.7		0.0	16.7		0.456
Total renewable	**694.4**	**3264.5**	**5160**	**214.9**	**952.1**	**0.369**	**0.391**
Natural gas	528.9	431.9		136.9	259.4	0.441	0.190
Oil	195.4	0.0		64.0	0.0	0.349	
Coal	1000.6	362.0		210.4	58.6	0.543	0.705
Nuclear	970.1	0.0		137.8	0.0	0.804	
Total	**3389.4**	**4058.4**		**764**	**1270.1**	**0.506**	**0.365**

Source: Developed by GAC (2006)

Table 12.9 *Investment costs (including discounted decommissioning cost) assumed for various electricity supply technologies in the TRANS-CSP scenario for Europe*

Energy source	Investment cost (€/kW)			Electricity cost (€c/kWh)		
	2000	2020	2050	2000	2020	2050
		Renewable				
PV	5500	1590	910	36.7	8.6	4.8
Local CSTP	3052	4595	4075	17.6	7.5	6.6
Imported CSTP		4200	3500		5.4	4.9
HVDC for imported CSTP		500	450			
Wind	1150	956	832	7.7	6.1	5.2
Hydro	1800	1800	1800	8.4	8.4	8.4
Biomass	2500	1700	1650	6.1	4.3	5.2
Geothermal	13,093	3631	2966	17.5	4.8	5.0
Wave, tidal	3000	2250	2000	8.3	6.2	5.5
		Non-renewable				
Natural gas	450	450	450	4.7	9.9	13.7
Oil	1000	1000	1000	8.1	15.0	12.6
Coal	1150	1150	1150	3.3	5.6	6.0
Nuclear	4000	4500	5250	3.6	4.1	4.9
System average				**4.8**	**6.9**	**6.6**

Note: Costs for biomass, geothermal and CSTP plants increase after 2020 because these plants are assumed to begin to supply peak electricity demand. Coal and natural gas are used largely for cogeneration, while oil and nuclear are phased out by 2050.
Source: GAC (2006)

energy in the parts of the world with the best solar and wind resources, and the transmission of the electricity so generated to the major demand centres by an HVDC backbone grid and its subsequent distribution by the existing AC grid. This will reduce generation costs by increasing the energy production per solar or wind power unit and through economies of scale, thereby offsetting much of the cost of the HVDC transmission while increasing the capacity factor and reliability. However, this assumes the existence of a local distribution grid.

There may be situations where construction of small autonomous renewable energy systems in areas currently remote from the existing grid might be less costly than grid extension and central electricity production. This is already true today in many cases.

The kinds of systems (either existing or hypothetical) that have been studied, and recent examples of published studies and the locations under consideration, are:

- PV/wind/battery – Byrne et al (2007), western China; Katti and Khedkar (2007), India; Dalton et al (2008), subtropical coastal Australia; Nouni et al (2008), India; Bağci (2009), a Hong Kong island;
- PV/wind/fuel cell – Ahmed et al (2008), isolated islands;
- PV/fuel cell/ultra capacitor – Uzunoglu et al (2009), general model;
- Micro-hydro/wind – Ashok (2007), India;
- Micro-hydro/PV/wind/battery – Kenfack et al (2009), equatorial Africa;

Bernal-Austín and Dufo-López (2009) provide references to other analyses of the above systems, made from 1996 to 2008.

12.3 Construction of C-free energy supply scenarios

In this section we construct scenarios of C-free primary energy supply that match the energy demand scenarios developed in Volume 1. The construction of the demand scenarios is first described and summary results are presented.

12.3.1 Energy demand scenarios

The baseline energy demand scenarios (that is, before considering energy efficiency improvements) were developed in the following way:

- The world was divided into ten socio-economic regions,[4] and low and high population and gross domestic product per person (GDP/P) scenarios from 2005 to 2100 were applied to each region.
- Initial (2005) data or estimates of residential and commercial floor space per capita, total distance travelled per capita per year and the proportions of travel by different modes (light-duty vehicles, 2-wheelers, 3-wheelers, buses, mini-buses, rail and air) were determined for each region.
- Growth in GDP/P was used as a driver in a logistic function for increases in residential and commercial floor space and in annual distance travelled per capita (which, when multiplied by population, gives the variation in regional residential and commercial floor space and in total passenger travel).
- Global freight movement (tonne-km per year) was assumed to grow in proportion to the growth in gross world product.
- To account for changes in the structure of the world economy as gross world product increases that are not already accounted for by the logistic growth functions mentioned above, it was assumed that 50 per cent of the industrial and freight transport activity that would have otherwise occurred in 2100 is replaced by an equivalent value added in the services sector, with the additional assumption that the energy intensity (MJ per $ of value added) of the added services activity is a quarter that of the lost industrial and freight activity.

Improvement in energy efficiency and strategies to increase the share of energy-efficient rapid-transit systems were implemented in the above scenarios as follows:

- The initial average energy intensities for each transportation mode in each region were assumed to decrease over time following a logistic function, reaching values by about 2050 that were estimated to be technologically feasible with existing or foreseeable technologies.
- The energy intensities for fuel and electricity use by new and renovated commercial buildings and for fuel use by new and renovated residential buildings in each region were assumed to decrease linearly from the current energy intensities to target energy intensities by a target date (either 2020 or 2050), and an accounting model was used to calculate how the stock average energy intensity changes as existing buildings are renovated and the total building floor area expands (two renovation cycles, between 2005 and 2050 and 2050 and 2095, were considered).
- Energy intensities for residential electricity use decrease over time in wealthy countries (due to

improved end-use efficiency) but were allowed to increase in less wealthy countries as a result of increased provision of electricity services (although the increase is much smaller than if stringent improvements in end-use efficiency were not assumed).

- Energy intensity reductions were also applied to freight, industrial and agricultural energy use.
- Changes in the proportion of total travel by light-duty vehicles (cars, light trucks) versus other modes were considered so as to reflect the impact of alternative forms of urban development.

The assumed reductions in energy intensity in every sector are based on the detailed literature review and analysis presented in the early chapters of Volume 1. Fast (largely by 2020) and slow (by 2050) implementation of energy efficiency measures are considered. Separate energy intensity reductions were applied for fuels and electricity, with the energy intensity reduction factors taking into account shifts from fuels to electricity (through, for example, the development of plug-in hybrid electric light-duty vehicles).

The low and high population scenarios are based on UN low and high projections to 2050 (and extended by the author to 2100), while the GDP/P scenarios allow for a substantial convergence between the lowest-income and highest-income regions (for the low GDP/P scenario, GDP per person is assumed to decrease slightly in North America, reflecting an assumed tradeoff of more leisure time for less material consumption). Figure 12.2 shows the resulting low and high scenarios of global population, global mean income per person and the resulting variation in gross world product when low population and GDP/P are combined and high population and GDP/P are combined. Global population in the low scenario grows from 6.5 billion in 2005 to a peak of 7.64 billion in 2040, then declines to 6.5 billion by 2100, but grows to 10.3 billion by 2100 in the high scenario. Global average GDP/P grows from $8800 in 2005 to $23,800 by 2100 in the low scenario and to $36,800 in the high scenario. The product of these two factors gives a gross world product of $57.2 trillion in 2005 that grows to $154 trillion by 2100 in the low scenario and $379 trillion in the high scenario. Neither the population nor the GDP/P scenarios are intended as predictions; rather, they are meant to span a range of possibilities so as to explore the ultimate consequences in terms of climatic change of high and low scenarios in combination with other assumptions.

Figure 12.3 shows the variation in end-use demand for fuels and for electricity for the low and high scenarios. Results are shown for slow implementation of efficiency measures without the assumed structural change in the economy, and for slow and fast implementation with the structural change. Global demand for fuels peaks at a level 17–33 per cent above the 2005 level in the low scenario and 26–54 per cent above the 2005 level in the high scenario, depending on the rate of improvement in energy efficiency, before beginning to decline. Global electricity demand peaks at 30–40 per cent above the 2005 demand in the low scenario and at about twice the 2005 demand in the high scenario, followed by a slight decline.

The end-use demand for electricity is converted to the amount of electricity that needs to be generated by dividing by the ratio of delivered to generated electricity, which accounts for parasitic loads in the powerplants (that is, electricity consumed by the powerplants themselves) and losses during transmission and distribution. Parasitic losses are rather large for fossil fuel powerplants (6–8 per cent for coal powerplants, 2–3 per cent for natural gas powerplants), and transmission losses are large at present in many developing countries (about 12 per cent on average). Here, it is assumed that the delivered to generated ratio increases from a world average of 0.82 in 2005 to 0.88 by 2040.

12.3.2 Scenarios for the deployment of C-free power

The approach taken here in developing scenarios for the deployment of C-free power is to set ultimate amounts of C-free power capacity for the various C-free energy sources (solar, wind, biomass, hydro and geothermal) such that, collectively, they would be sufficient to supply the entire global demand for fuels and electricity in 2100. These ultimate capacities are approached through a logistic function at rates such that we achieve zero CO_2 emissions by either 2080 or 2120. Two renewable energy pathways are considered: a biomass-intensive pathway, in which all fuel requirements (including liquids for transportation) are met from biomass, and an H_2-intensive scenario, in which many of the fuel

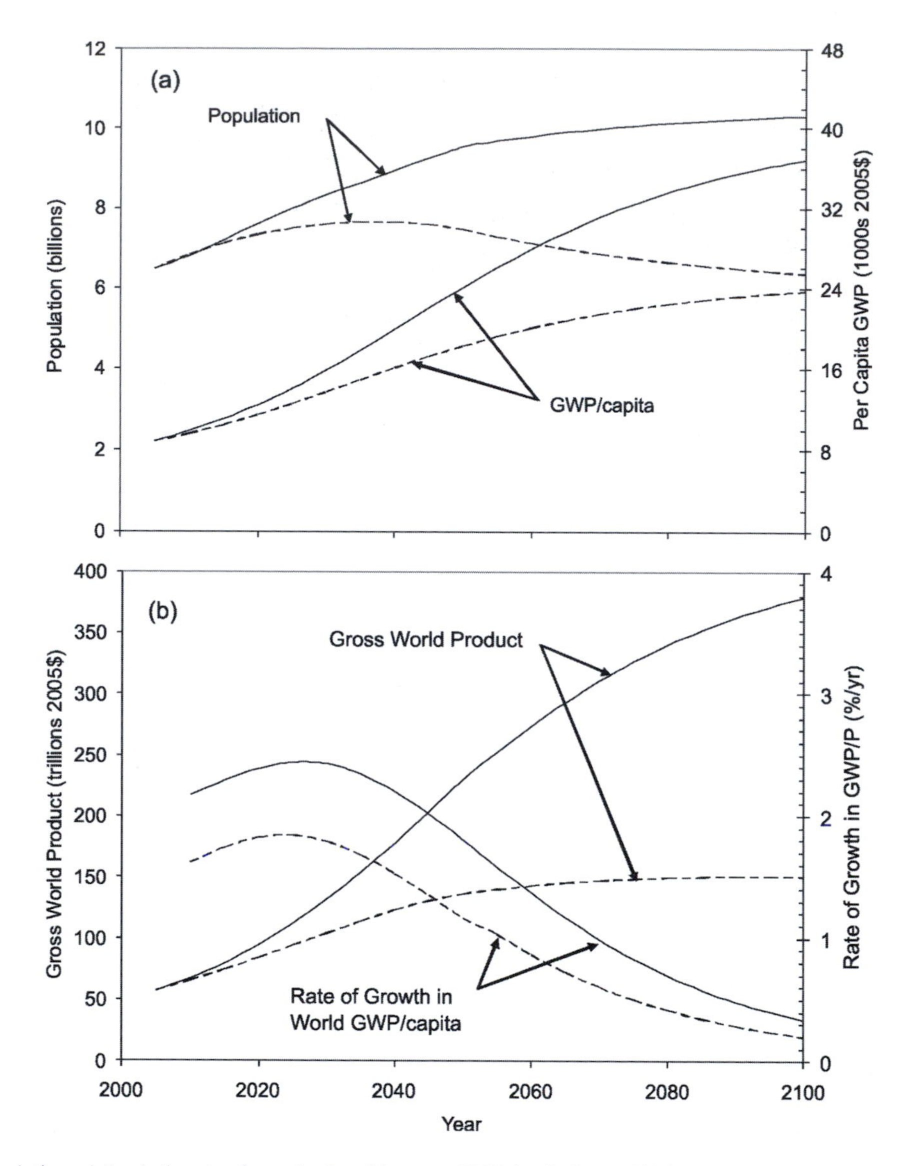

Note: Also shown is the variation in the rate of growth of world average GDP/P for the low and high GDP/P scenarios.

Figure 12.2 *(a) Variation in global population and in global average GDP/P for the high demand scenario (solid lines) and low demand scenario (dashed lines) of Volume 1; (b) variation in gross world product for the combination of high population with high GDP/P growth (solid line) and low population with low GDP/P growth (dashed line)*

requirements (including all of the residual fuel requirements in the transportation sector after penetration of PHEVs) are met with hydrogen and the balance is met with biomass. Altogether then, there are 16 different possible scenarios: four demand scenarios (low and high growth in population × GDP/P, designated as L and H), each with two rates of implementation of efficiency improvements (slow and fast, designated as S and F), two rates of deployment of C-free power (slow and fast, designated as s and f) and two supply-mix scenarios (biomass- and hydrogen-intensive). The different scenario combinations will be denoted here using the letter designators indicated above. For example, LSf is the scenario with low growth

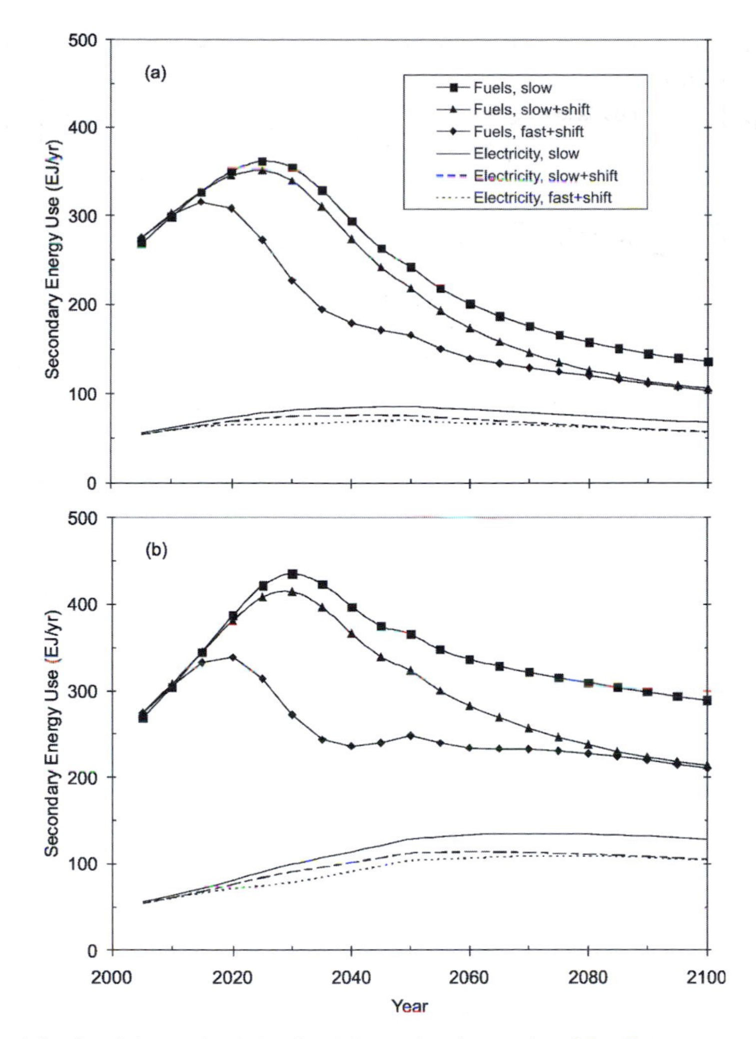

Note: Results are shown with fast (largely by 2020) and slow (largely by 2050) implementation of the efficiency improvements considered here with account taken of illustrative structural shifts in the economy, as well as for slow implementation without structural shifts.

Figure 12.3 *Variation in global demand for fuels and electricity for (a) the low gross world product scenario and (b) the high gross world product scenario*

in world GDP, slow implementation of energy savings measures and fast deployment of renewable energy. One additional scenario, which assumes a strong investment in urban rail-based transit, less total travel per capita and a smaller proportion of total travel by air is considered: LFGf, where 'G' designates 'green' transportation and the specific demand assumptions are as in the Fast+Green demand scenario of Volume 1, subsection 10.4.2.

Potential electric power capacities for various C-free energy sources are assumed to grow following a logistic function as follows:

$$P(t) = \frac{P_u}{1 + ((P_u - P_o)/P_o)e^{-a(t-t_o)}} \tag{12.1}$$

where P_u is the prescribed ultimate capacity, P_o is the capacity in starting year t_o and a is a growth constant.

As hydrogen is currently not used in any sector and biofuels are currently not used for freight or air transport, we set P_o = 0.1EJ/yr in 2015 in order to initialize the logistic curve, but otherwise use data from 2005 for P_o. The P_u values chosen for the low and high energy demand scenarios, and the values of a for slow and fast deployment of renewable energy, are given in Table 12.10. The nuclear energy capacity is computed based on the assumption that all existing nuclear reactors are retired after 40 years of service (using the age distribution for existing reactors shown in Figure 8.27) and that only enough new reactors are built to maintain the 2005 capacity until 2010.

Also given in Table 12.10 are the global mean annual capacity factors assumed for each C-free energy source. The wind capacity is assumed to be mostly located in good wind sites that are far removed from major demand centres, with the wind farms oversized relative to the transmission link (based on the strategy explained in Chapter 3, subsection 3.14.1), so a relatively small wind capacity factor (0.2) is assumed. CSTP, by contrast, is located in desert and semi-arid regions and is assumed to have a sufficiently high concentration factor and sufficient thermal storage capacity that it has an average capacity factor of 0.6. PV is assumed to be largely building-integrated PV and so will be located on average in regions of less solar irradiance than the CSTP plants, so a smaller capacity factor (0.15) is assumed. The total ultimate global capacity limits for PV (3000GW in the low demand scenario, 4000GW in the high demand scenario) combined with a 15 per cent capacity factor result in an annual electricity generation of 4200TWh/yr and 5600TWh/yr for the low and high demand scenarios. These are quite plausible in light of the estimate given in Chapter 2 (Table 2.4) of almost 3100TWh/yr potential electricity generation today from available building roofs and façades in IEA member countries alone.

The ultimate capacities given for the various C-free energy sources represent just one of many possible combinations. The geothermal and biomass electricity generation capacity, for example, could plausibly be several times that shown in Table 12.10. Similarly, wave energy could make a non-negligible contribution to global electricity demand (up to 2000TWh/yr according to Lemonis, 2004). It is also possible that the relative importance of PV and CSTP (3000GW and 6000GW respectively in the LF demand scenario) could be reversed. The assumption here is that PV will be restricted to building-integrated applications to relieve congestion bottlenecks and CSTP will be used in semi-arid regions with thermal energy storage to provide 24 hours of electricity. However, CSTP requires water for effective operation (which could be a limiting factor in arid regions) whereas PV does not (except for very small amounts for occasional cleaning), and CAES provides an alternative means of supplying solar-generated electricity 24 hours per day (see Chapter 2, subsection 2.9.3). Depending on relative future costs and embodied energy requirements, desert PV along with BiPV might be preferable to CSTP. Raugei and Frankl (2009), for example, consider scenarios in which world PV capacity reaches 2400GW or 9000GW by 2050, with a transition from dominance by crystalline silicon at present to dominance by ribbon Si and CdTe by 2025, with about one third the lifecycle embodied energy per kWh of electricity generated as at present and with strong reliance an large-scale CAES. In order to allow the reader to explore alternative scenarios, or to generate his or her own scenarios with more up-to-date performance data that become available after this book is published, the Excel spreadsheets and FORTRAN program used to generate the demand scenarios of Volume 1 and the supply scenarios presented here have been made available with the online supplemental material.

Table 12.10 *Parameters used in the logistic function (Equation (12.1)) to generate scenarios of future C-free electric generating capacity (GW)*

	PV	CSTP	Wind	Biomass	Geothermal	Hydro
Capacity in 2005	5.2	0	59.1	4	8.9	866.8
P_u, low demand	3000	6000	6000	200	100	1200
P_u, high demand	4000	12,000	12,000	200	100	1200
a, slow deployment	0.08	0.08	0.06	0.06	0.06	0.01
a, rapid deployment	0.10	0.10	0.1	0.1	0.1	0.02
Capacity factor	0.16	0.6	0.2	0.6	0.9	0.4

Note: Also given are the capacity factors used to translate capacity into annual generation of electricity.

Table 12.11 gives the allowed ultimate rates of production of biomass energy for the production of fuels in various end-use sectors (passenger ground transportation and agriculture, air travel, freight transport, buildings and industry) for the biomass- and hydrogen-intensive supply scenarios. These production rates times the efficiency in conversion from biomass to biofuels gives the rates of production of biofuels; the assumed conversion efficiencies for transportation fuels increase from 0.46 in 2010 to 0.56 in 2050,[5] and are fixed at 0.9 for building and industrial fuels. The conversion efficiencies account for energy inputs during the production, harvesting, transport and transformation of biomass fuels (which, in the case of solid fuels for buildings and industry, involves only chipping or pelletizing). Also given in Table 12.11 are the allowed ultimate rates of production of H_2 in the hydrogen-intensive scenario. The combined production of biofuels and H_2 is sufficient to displace the projected demand for fossil fuels in these sectors by 2100 or sooner. The biofuel and hydrogen fuel energy requirements for passenger travel in light-duty vehicles take into account the difference in the efficiency with which these fuels can be used in urban and highway driving, which are assumed to account for 60 per cent and 40 per cent, respectively, of total driving (as detailed in Table 10.7 of Volume 1). For air and freight transport, no differences in the end-use efficiency of biofuels and hydrogen are adopted.

The total rate of hydrogen production is divided by an assumed electrolysis plus compression efficiency of 0.75 to give the required generation of electricity (GWh). This is further divided by 0.95 to account for an assumed 5 per cent loss in transporting electricity from where it is generated to where it is used to produce hydrogen (or in transporting hydrogen by pipeline from where it is produced to where it is used). As noted earlier, electricity demand for direct use is divided by a supply efficiency ranging from 0.82 at present to 0.88 in 2040 in order to give the required generation of electricity. The 18 per cent loss at present represents parasitic electricity consumption in fossil fuel power plants and transmission losses. In a future C-free electricity system, the 12 per cent loss assumed here represents average losses in transmission and losses due to the need to temporarily store some portion of the C-free energy supply. Given that storage in hydro-electric reservoirs (excluding pumped-hydro) entails no loss and that thermal energy storage is already included in CSTP electricity production, and given opportunities for end-use demand management, the amount of electricity passing through storage could be rather small (20 per cent or less). An average loss of 35 per cent on 20 per cent stored energy and 5 per cent transmission loss would give an overall loss of 12 per cent.

The annual electricity requirement for production of hydrogen is converted to a required electrical capacity (GW) assuming a capacity factor of 0.6.[6] If the required direct generation of electricity (as provided by the demand scenarios) plus that required to generate hydrogen is less than the generation of C-free electricity (as determined by the logistic function and the parameters given in Table 12.10), then the generation of C-free electricity is reduced. Conversely, any deficit in electricity production is met from fossil fuels.

However, if the C-free electricity supply exceeds the direct electricity demand but is less than the direct plus hydrogen demand, the result will be that some fossil fuels are used to produce electricity that in turn is used to generate hydrogen. To avoid this, hydrogen use by industry is reduced under these circumstances and is assumed to be replaced with biomass (so the hydrogen-intensive scenario moves toward the biomass-intensive scenario). If eliminating all hydrogen use by industry is not sufficient to avoid the electrolytic production of hydrogen from fossil fuels, then some hydrogen in the fossil fuel sector is assumed to be produced directly from fossil fuels. The net result here is that no hydrogen is produced by electrolysis until the global electricity supply is entirely free of fossil fuels.[7]

Figure 12.4 shows the resulting variation of global C-free electricity-generating capacity and electricity generation for the H_2-intensive variant of the LFf scenario. Also shown is the electricity demand for direct use and to generate hydrogen by electrolysis (Figure 12.4c), the natural gas, oil and coal use, the production of biofuels in the hydrogen-intensive scenario and the additional biomass required in the biomass-intensive supply scenario (Figure 12.4d). Figure 12.5 shows the same things for the H_2-intensive HSf scenario. These two figures can be summarized as follows:

For the H_2-intensive scenarios,

- total C-free electricity generation capacity peaks in 2080 at 14.1TW in the LFf scenario and reaches 27.0TW by 2100 in the HSf scenario (by comparison, the total world electricity generating capacity in 2005 was 4.4TW);

Table 12.11 *Parameters used in the logistic function (Equation (12.1)) to generate scenarios for the transition from fossil fuels (for purposes other than to generate electricity) to either biofuels or hydrogen fuel*

	Ground passenger transport and agriculture			Air			Freight			Buildings				Industry			
	B-1	B-2	H	B-1	B-2	H	B-1	B-2	H	Solar	B-1	B-2	H	Solar	B-1	B-2	H
Supply in 2005	1.06	1.06	0	0	0	0	0	0	0	0.9	36.7	36.7	0	0	8.6	8.6	0
P_u, low demand	30	0.10	12	25	0	15	25	0	14	12	31	20	11	5	25	5	20
P_u, high demand	50	0.10	20	45	0	28	60	0	34	12	65	40	30	10	65	15	45
a, slow deployment	0.07	0.005	0.08	0.10	–	0.10	0.10	–	0.10	0.05	0.02	0.02	0.06	0.10	0.0035	0.06	0.08
a, rapid deployment	0.10	0.005	0.10	0.15	–	0.15	0.15	–	0.15	0.10	0.04	0.04	0.08	0.15	0.10	0.10	0.15

Note: Energy terms are in units of EJ/yr. Where current energy use is zero, a small energy use is assumed for 2015 in order to be able to use the logistic function. B-1 and B-2 = biomass required in the biomass-intensive and hydrogen-intensive scenarios, respectively, H is the hydrogen requirement in the hydrogen-intensive scenario, and solar refers to solar thermal heat for space and water heating and low-temperature industrial processes, provided by solar thermal collectors

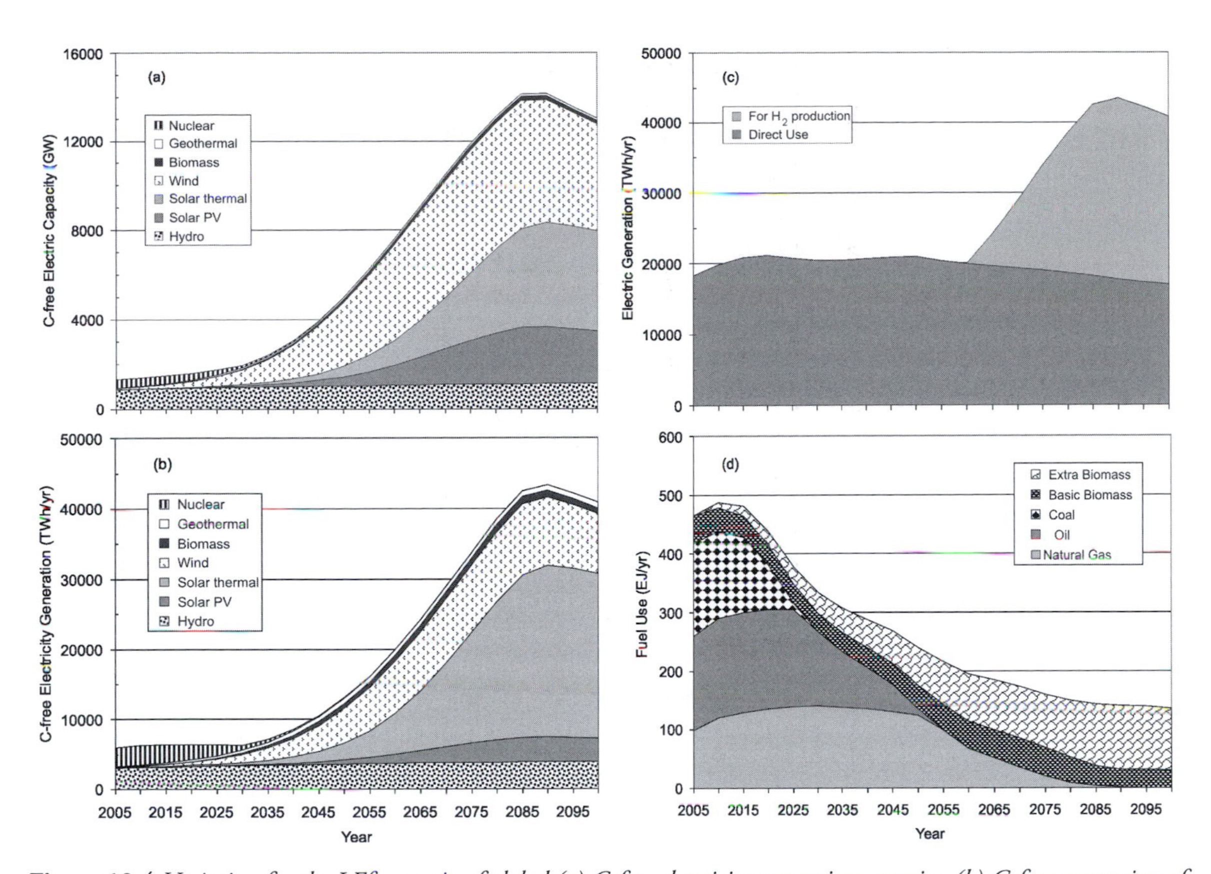

Figure 12.4 *Variation for the LFf scenario of global (a) C-free electricity generating capacity, (b) C-free generation of electricity, (c) electricity demand for direct use and to generate hydrogen by electrolysis, and (d) natural gas, oil and coal demand, as well as the production of biofuels in the hydrogen-intensive scenario and the additional biomass required in the biomass-intensive supply scenario*

- biomass use is essentially unchanged in the LFf scenario (45±10EJ/yr at present, 40EJ/yr in 2100) and approximately doubles (to 98EJ/yr) in the HSf scenario.

For the biomass-intensive scenarios,

- total C-free electricity generation capacity peaks at 7.6TW in 2060 in the LFf scenario and at 13.4TW in 2070 in the HSf scenario;
- biomass use increases to 140EJ/yr and 288EJ/yr for the LFf and HSf scenarios, respectively.

By comparison, most assessments of the global biomass potential place it at 100–200EJ/yr (see Table 4.50), although diets lower in meat than commonly assumed could greatly increase the potential (as seen in Table 4.47).

Figures 12.6 and 12.7 show the rates of installation of new and replacement wind, CSTP and PV capacity for the LFf and HSf scenarios, respectively. Lifespans of 25 years for wind (as in Laxson et al, 2006) and PV and 30 years for CSTP (as in GAC, 2006) are assumed. The average rates of installation of replacement capacity during the final decades of the century are equal to the long-term steady capacities divided by the assumed lifetimes, but the replacement rates fluctuate due to the initially rapid buildup in capacity, which occurs on a timescale that is short compared to the powerplant lifespans. The peak rates of installation of new and/or replacement CSTP and wind capacity are about 190GW/yr and 240GW/yr, respectively, in the LF demand scenarios and about twice that in the HS demand scenarios. These rates apply only to the components that need to be replaced according to the

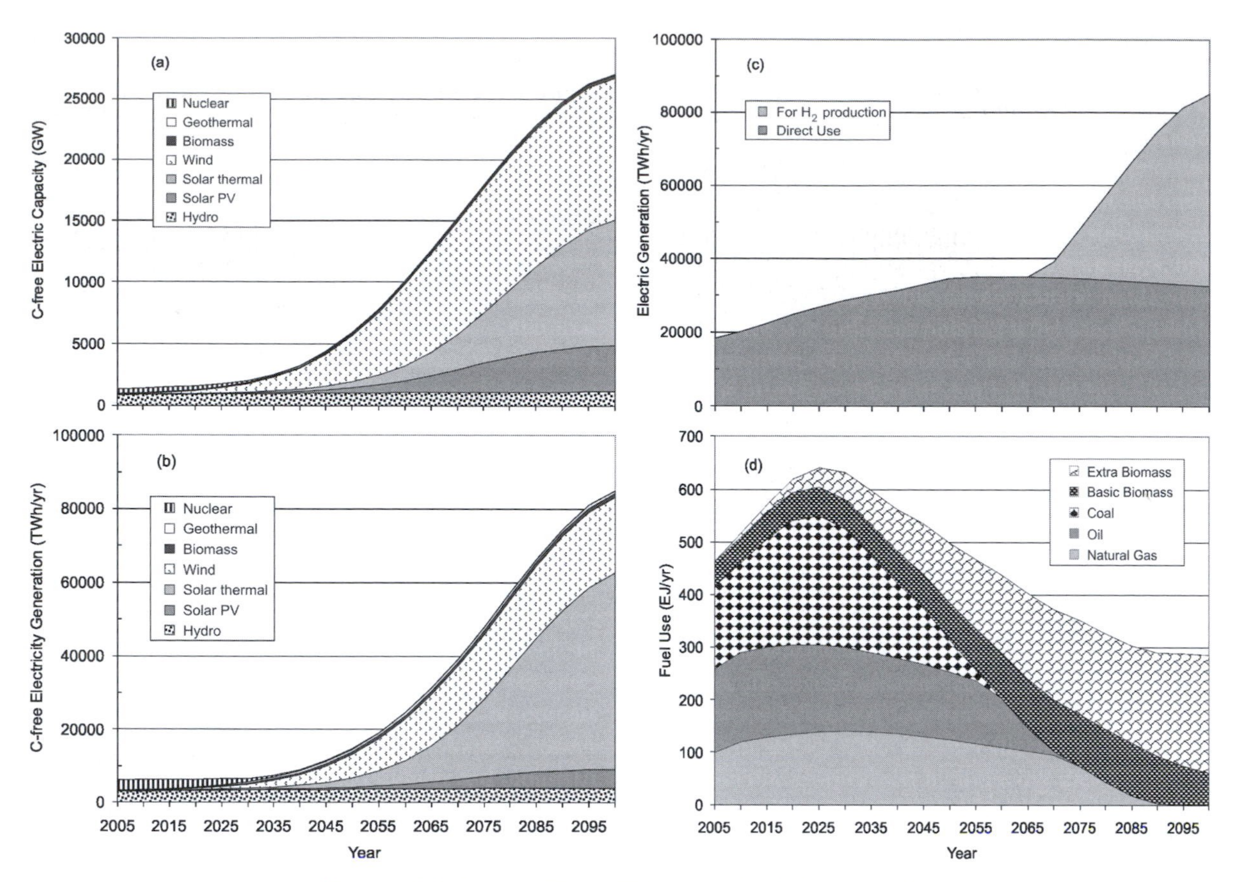

Figure 12.5 *Same as Figure 12.4 except for theHSf scenario*

25- or 30-year lifespans. Concrete foundations are assumed to have twice the lifespan, namely, 50 years for wind turbines and 60 years for CSTP plants (as the CSTP plants will be arid regions, foundation lifetimes could be even longer).

12.3.3 Material and energy flows and net energy yield during the ramp-up

Two important considerations in the deployment of renewable energy systems on the scale envisaged here are the annual material flows and energy requirements for the initial installations and later for replacement of existing facilities as they wear out. Table 12.12 summarizes estimates of the material requirements (tonnes per MW of capacity) for electricity production by wind, solar and (for comparative purposes) nuclear power as given by various studies, as well as for the construction of 800kV HVDC transmission lines. For wind, estimates are given for onshore turbines (which involve substantial amounts of concrete for foundations) and for one floating offshore turbine concept (which requires no concrete but requires substantially more steel than onshore turbines). The material requirements given for HVDC transmission lines assume the use of aluminium wires. Also given is the ratio of automobile mass to engine mechanical power (also t/MW) for a typical US automobile. This is of interest because the world's automobile industry every year produces a mechanical powerplant of about 7000GW (70 million vehicles per year times an average vehicle engine power of about 100kW), which is almost twice as large as the current (2005) total world electricity powerplant of about 4300GW.

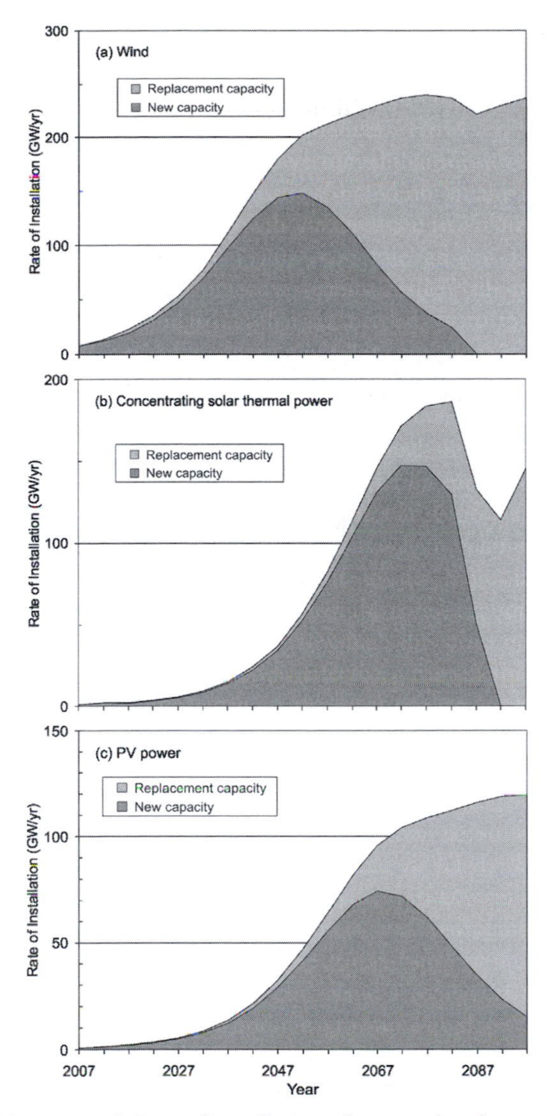

Figure 12.6 *Rate of installation of new and replacement (a) wind capacity, (b) concentrating solar thermal power capacity and (c) PV power capacity for the LFf scenario*

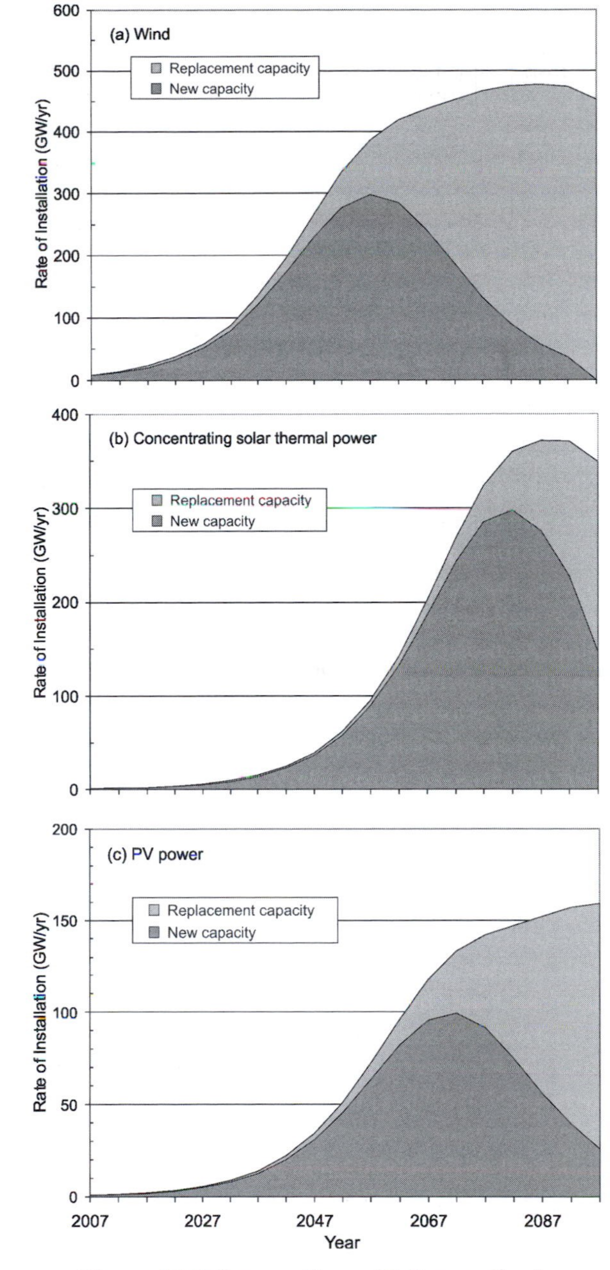

Figure 12.7 *Same as Figure 12.6 except for the HSf scenario*

The information given in Table 12.12 is used along with the rates of installation of new and replacement wind turbines, CSTP and PV power to compute the annual requirements for steel, aluminium, copper and concrete. These are shown separately in Figure 12.8 for CSTP, wind, PV and HVDC transmission. Two sets of material requirements are given for wind, one assuming that all of the installed wind capacity consists of onshore turbines and the other assuming that all of the installed capacity consists of floating offshore turbines.[8] The material requirements for HVDC are computed assuming an average transmission distance of 1500km

Table 12.12 *Material requirements (tonnes per MW of capacity) for electricity production by wind, solar and nuclear power, and ratio of automobile mass to engine mechanical power (also t/MW) for a typical US automobile*

Source of electricity	Concrete	Steel	Al	Cu	Si	FG	Adhesive	OP	Glass	Salt	Other	TOTAL	Total excluding CS	Source
1.5MW onshore wind turbine	1.58	105	0.9	1.9		6.9	1.3					118	116	Lindenberg et al (2008)[a]
4MW onshore wind turbine	2.10	141	1.3	0.8		9.6	1.8				0.6	157	155	Lindenberg et al (2008)[b]
2MW onshore wind turbines	350	119		1.8	0.2	4.4	6.5					481	131	Martínez et al (2009)
Generic wind turbines	360	125	0.4	4.4				9.7				500	140	Voorspools et al (2000)
Adopted here	350	120	1.2	1.2		9	1.8					483	133	
5-MW floating wind turbines[c]	0	324	0.5	11.7		10.4		17.9				364	364	Weinzettel et al (2009)
PV modules only[c]	0		23.0						69.0		18.0	110	110	Perpiñan et al (2009)
PV desert array support structure only	47.4	56	19.4	7.5				5.8			53.5	190	142	Mason et al (2006)
Central tower solar thermal with six hours thermal storage	1147	275	1.0	2.9		4.8		7.5		162		1600	291	Vant-Hull (1991)
Nuclear powerplant	560	60	0.2	0.8							2.7	624	64	Voorspools et al (2000)
500MW pulverized-coal powerplant	158.8	50.7	0.4									210	51	Spath et al (1999)
500MW NGCC powerplant	97.7	31.0	0.2									130	31	Spath and Mann (2000)
HVDC transmission[d]	40	16.3	3.5								0.4	60.2		GAC (2006)
US automobile												13.3		Table 5.22 of Volume 1 (from Heywood et al, 2003)

Note: FG = glass-reinforced plastic, carbon-reinforced plastic or fibreglass. OP = oil and/or plastic. CS = concrete and salt. [a] Also using the total turbine mass of 177.5t, provided by Larry Willey (personal communication, 2009). [b] Also using the total turbine mass of 628.2t, provided by Larry Willey (personal communication, 2009). [c] Used here. [d] Material intensities given here are for 1000km average transmission distance by an 800kV double-dipole DC line.

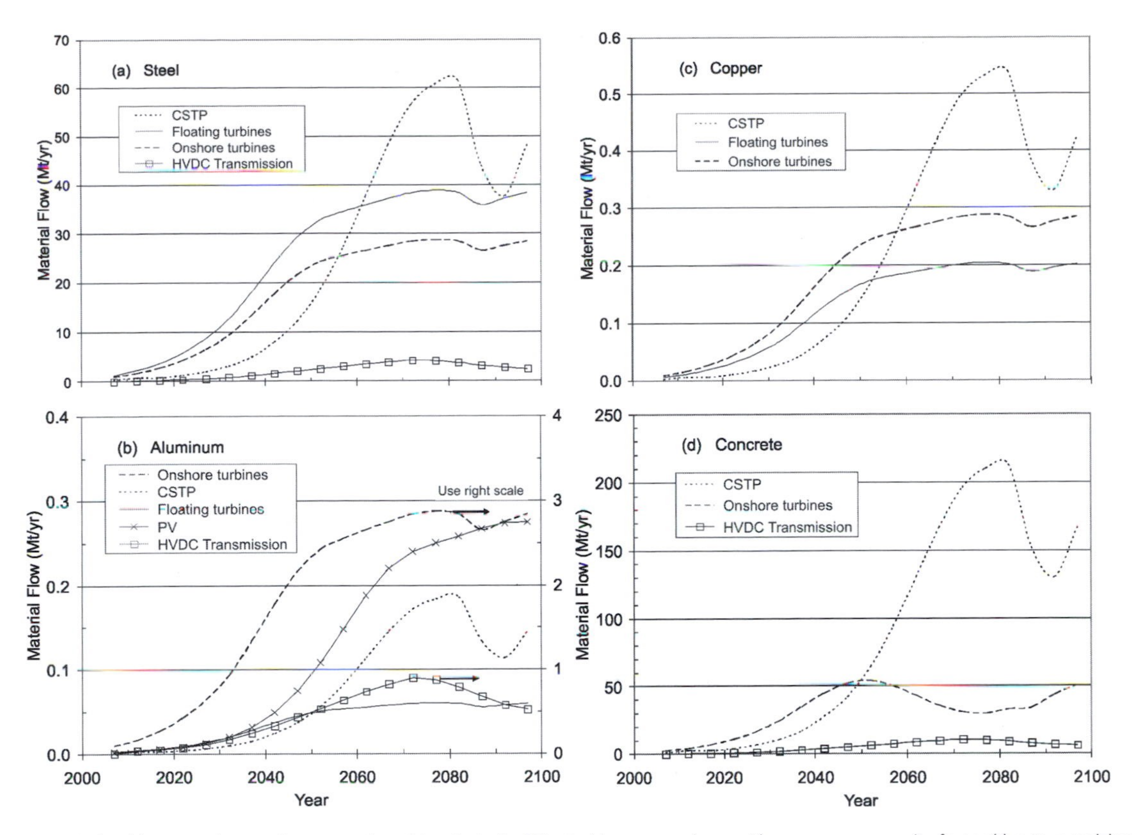

Note: For wind turbines, results are given assuming either that all of the turbines are onshore with an average capacity factor (due to oversizing relative to the transmission links) of only 0.2, or that all of the turbines are floating offshore turbines with an average capacity factor of 0.4 and the required capacity cut in half.

Figure 12.8 *Annual requirements for (a) steel, (b) aluminium, (c) copper and (d) concrete to construct the new or replacement concentrating solar thermal powerplants, wind turbines, PV panels and HVDC transmission lines as required for the LFf scenario*

and are based on the CSTP capacity plus only one third the wind capacity, assuming the wind capacity to be distant onshore wind farms that are oversized by a factor of three on average relative to the transmission capacity (which is consistent with the very low capacity factor assumed for onshore wind farms). The PV capacity is assumed to be BiPV in urban centres, and so does not require long-distance transmission.

Table 12.13 gives the peak material flows for each electricity generation technology and for HVDC transmission lines for the HSf scenario, as well as the total rates of production worldwide of the various commodities, from which it can be seen that the material flows associated with the massive deployment of renewable energy envisaged here amount to about 14 per cent of the 2005 world steel production, 16 per cent of world aluminium production,

Table 12.13 *Peak material flows for the HSf scenario*

	Peak material flows (Mt/yr)				
	Onshore wind turbines	CSTP facilities	HVDC lines	Total here	Global total in 2005
Steel	60	120	8	188	1320
Aluminium	0.6	0.4	1.8	6.1[a]	38
Copper	0.6	1.1		1.7	56
Concrete	107	425	20	514	26,000[b]
Fibreglass	3.3	4.9		6	95[c]
Salt	0	70	0.7	70	

Note: The column 'Global total in 2005' is the total world production of the given commodities for all purposes. [a] Includes PV at 3.7Mt/yr. [b] Assuming the concrete to be 10 per cent cement, such that world concrete production is ten times world cement production. [c] This is the sum of world flat and container glass production in 2004.

3 per cent of world copper production and 2 per cent of world concrete production. The material requirements computed for the LFf scenario are half those shown in Table 12.13. The material (and energy) requirements for onshore and offshore turbines and for CSTP were computed by different people at different times, in different places and with different assumptions, and so should not be compared directly with each other.[9] Rather, the results presented here serve only to indicate the overall magnitude of the material flows in comparison to current global material flows.

The energy flows associated with the deployment of any energy system include the energy used at the construction site (the zero-order input), the energy used in the manufacture of the materials used in the powerplant (the first-order input), the energy used to produce the machinery used to make the materials that go into the energy-producing equipment (a second-order input) and higher-order inputs (as discussed in Appendix B). Here, I calculate and present only the first-order energy inputs (zero-order inputs are very small) but multiply the inputs by a factor of 1.5 in order to crudely account for higher-order inputs. In order to compare the energy inputs with the energy output, an appropriate method of combining fuel and electricity inputs into a single energy input is needed. Energy input to output ratios (and the corresponding energy payback times) presented in earlier chapters for solar and wind sources of energy were generally computed by multiplying electricity inputs by a factor of 2.5 (to give the primary energy required to produce the electricity used in the production of the solar or wind facility) and added to the fuel input, while the electricity output was also multiplied by a factor of 2.5 to give the amount of primary energy saved when solar or wind energy is used in place of fossil fuels to generate electricity.[10] This is inappropriate here because we assume all electricity to be eventually produced from C-free energy sources, while fuels (in the H_2-intensive scenario) consist largely of H_2 produced by electrolysis of water with an efficiency of 0.75. Thus, it is more appropriate to compute the electricity equivalent of the energy used to manufacture wind, CSTP or PV systems and to compare this directly with the electricity output from these systems. The electricity-equivalent input is given by:

$$E_{eq} = 1.5(E_f(\text{GJ/MW}) \times 0.278(\text{MWh/GJ})/\eta_{el} + E_{el}(\text{MWh/MW})) \tag{12.2}$$

where E_f and E_{el} are the direct fuel and electricity inputs to the manufacture of the materials used in the solar or wind power systems, respectively, η_{el} is the electrolysis efficiency, and the factor of 1.5 is meant to account for higher-order inputs.

The energy intensities per MW of wind turbine or CSTP capacity (E_f and E_{el}) are computed here from the material intensities given in Table 12.12 and energy intensities for each of the materials. The energy intensities used here are the future energy intensities given in Chapter 6 of Volume 1, which are summarized in Table 12.14. Materials used in the construction of new powerplants and transmission lines are all assumed

Table 12.14 *Fuel and electricity intensities assumed for primary and secondary materials*

	Primary materials		Secondary materials	
	Fuels (GJ/t)	Electricity (kWh/t)	Fuels (GJ/t)	Electricity (kWh/t)
Steel	12	115	1.2	395
Aluminium	14	11,430	8.4	285
Copper	39	7490	10	2780
Concrete[a]	0.25	6.4		
Fibreglass	2.1	4532	1.9[b]	4079[b]
Salt	7.7[c]		0.77[d]	

Note: [a] Assuming the concrete to be 10 per cent cement. [b] Assumed to be 80 per cent that of primary fibreglass. [c] Derived from Vant-Hull (1991). Could be much less now. [d] I assume that salt from old powerplants can be reused with 10 per cent loss or an energy input for cleaning equal to 10 per cent of the original energy input.
Source: Table 6.37 of Volume 1, except where indicated otherwise

to be primary materials, while materials used in the replacement of powerplants and transmission lines at the end of their lives are all assumed to be secondary (recycled) materials, with the exception of cement (which, it is assumed, cannot be recycled).

In the case of BiPV modules, the energy intensities per m^2 module area are directly specified (combined with the system efficiency, this allows computation of GJ/MW). Alsema et al (2006) projected a 'future' primary energy requirement for multi-crystalline modules with 17 per cent efficiency of 1800MJ/m^2 (compared to 3250MJ/m^2 in then-recent modules), but by 'future' the authors meant only a few years from the time of writing. Jungbluth et al (2008) indicate that, at the time of their writing, 980MJ/m^2 had already been achieved for CdTe modules manufactured in Germany. CdTe modules are expected by some to become the dominant PV technology in the future, but limited Te supplies will probably limit their deployment to about 100GW unless they are used in concentrating PV systems (see Table 2.14). Here, it will be assumed that the energy intensity of 17 per cent efficient PV modules drops from 1800MJ/m^2 in 2010 to 1200MJ/m^2 in 2030. This does not include the energy required for support structures, but neither does it include credits for displaced materials when PV is deployed as BiPV.

Figure 12.9 gives the resulting annual energy requirements for the hydrogen-intensive variant of the LFf scenario. The projected energy requirements for CSTP are quite large. This projection is based on a single rather old (1991) study, and may very well be revised downward if more recent systems have smaller material requirements. Fibreglass and steel are big contributors to the maxima in CSTP energy requirements seen around 2060.

Figure 12.10 compares the energy used to construct or replace solar or wind energy systems in any given year with the electricity produced by all of the units in operation in that year for the LFf scenario (the curves are almost the same for the HSf scenario). This ratio is initially quite large for PV (peaking at almost 30 per cent), then drops to a value of about 6 per cent by the end of the century (at which point new PV is produced only to replace existing PV). The annual energy use to manufacture CSTP plants starts at around 33 per cent of the (initially very small) CSTP energy production, then drops to around 0.3 per cent, while the energy requirement for onshore wind decreases from 5.6 per cent to about 0.6 per cent and that of the floating wind system decreases from 3.5 per cent to about 0.3 per cent. The energy input for the total system (including HVDC transmission lines) peaks at around 9.5 per cent of system output in 2015, then drops to below 1 per cent by the end of the century. These results can be understood in terms of the energy payback times and the lifetimes of individual units in the various systems. Energy payback times based on Equation (12.2) are presented in Table 12.15, and take into account the assumed capacity factors of each system. Using onshore wind as an example, the energy payback time using all recycled materials (except for concrete) with expected future energy intensities is 0.13 years (1.6 months). With an assumed lifespan of 25 years, 4 per cent of the wind powerplant is replaced every year and this requires $0.04 \times 0.13 = 0.0052$ years of output from the entire wind powerplant, or 0.52 per cent of the output.

In spite of the rapid ramp-up of solar and wind energy in the early phases of the scenarios presented here, there is a significant net energy gain from the start. As discussed by Pearce (2008a), there will be a net energy gain during a period of growth if the energy payback time is less than the reciprocal of the percentage growth rate per year.[11] Thus, if an energy supply system is growing by 20 per cent per year, there will be net energy production if the energy payback time is less than five years. This condition is easily satisfied for all the energy systems under consideration here.

12.3.4 Biomass plantation land requirements

The land area required for new biomass plantations is shown in Figure 12.11 for the biomass-intensive variant of the LFs, LSs, HFs and HSs scenarios. The required land areas are computed by first subtracting the biomass energy supplied from sources other than plantations. Present biomass energy supply is about 45EJ/yr (see Figure 4.1), with potential additional supplies of:

- 5–27EJ/yr from primary agricultural wastes (residues);
- 9–25EJ/yr from animal manure;
- 10–16EJ/yr from forestry residues;
- 5EJ/yr from secondary (food processing) agricultural wastes; and
- 1–3EJ/yr from municipal solid waste;

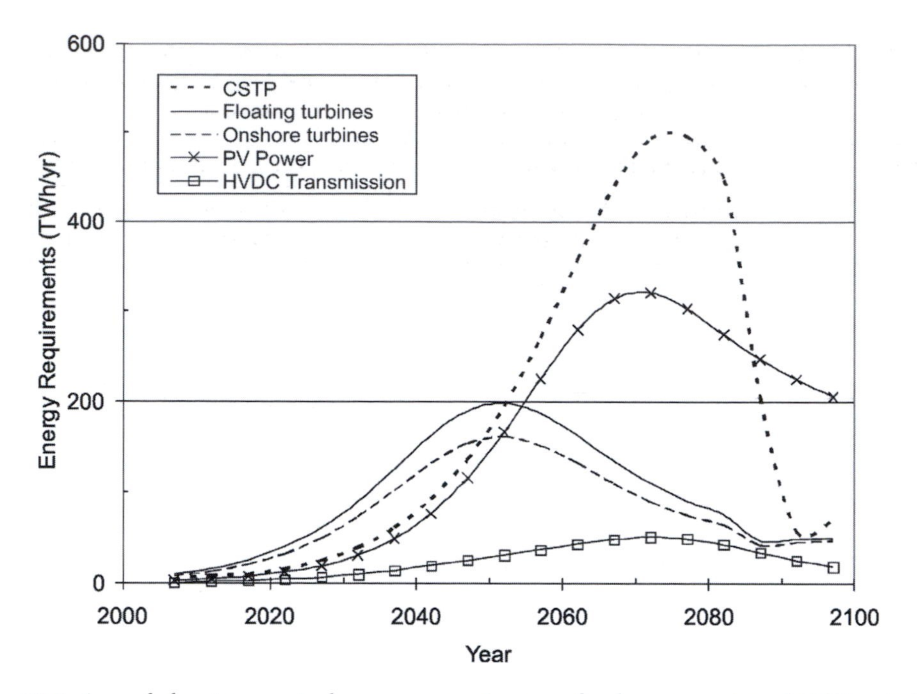

Figure 12.9 *Annual electricity-equivalent energy requirements for the construction of CSTP, wind farms (either entirely as onshore wind farms or entirely as floating offshore wind farms), PV systems and HVDC transmission lines for the LFf scenario*

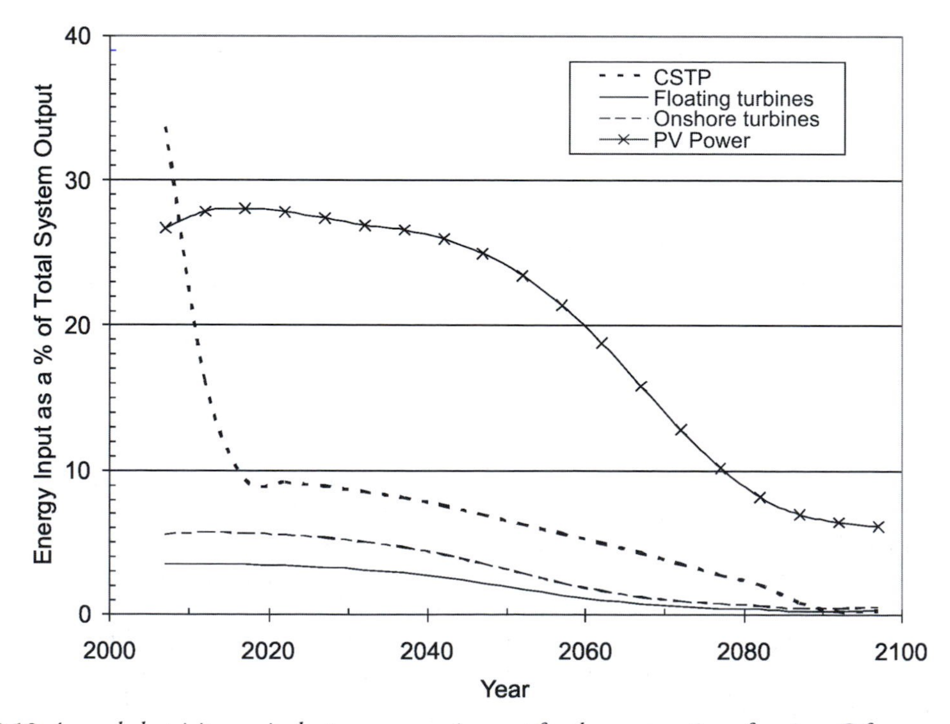

Figure 12.10 *Annual electricity-equivalent energy requirement for the construction of various C-free powerplants as a percentage of the total electricity output from the corresponding type of powerplant for the LFf scenario*

Table 12.15 *Energy payback times for the various solar and wind energy production systems considered here and the percentage of annual energy output needed to replace power generators as they wear out once a steady power capacity is reached*

System	Payback time (years)		Lifespan (years)	Annual energy requirement (%)	
	All primary materials	All secondary materials		All primary materials	All secondary materials
Onshore wind	0.58	0.13	25	2.3	0.51
Floating wind	0.68	0.09	25	2.7	0.38
CSTP	0.63	0.11	30	2.1	0.36
BiPV	2.91	1.08	20	14.5	5.4

Note: The energy payback times are based on expected future energy intensities for various materials, as given in Table 6.37 of Volume 1.

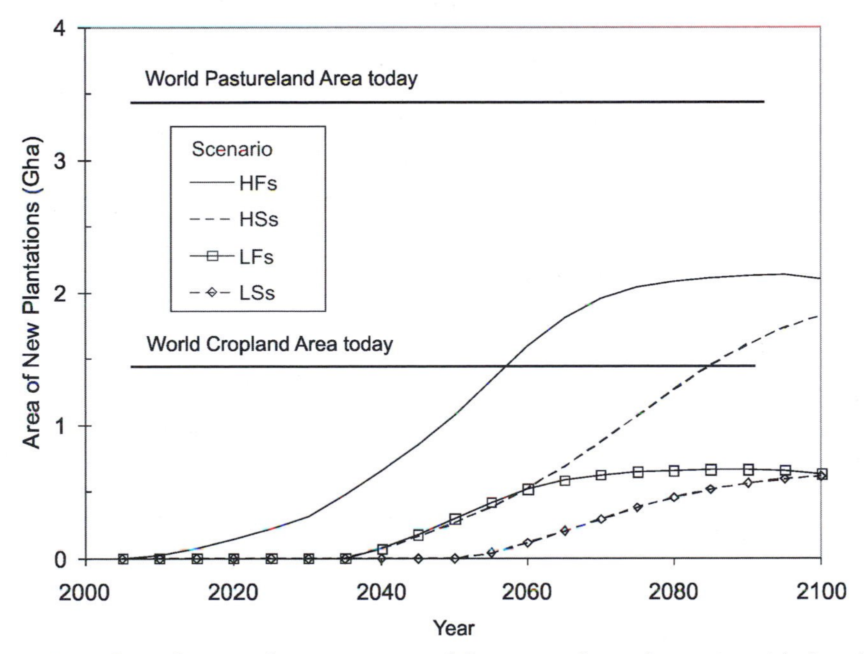

Figure 12.11 *Area of new biomass plantations required for various demand scenarios with slow deployment of C-free energy sources for the biomass-intensive scenarios*

giving a total of 30–76EJ/yr (see Chapter 4, subsection 4.9.5). However, some of the agricultural residues are associated with growing food for animals rather than for direct consumption by people. As the biomass-intensive scenarios require some moderation in meat demand in order for there to be sufficient land available for biomass plantations, consistency requires assuming a rather low supply of bioenergy from animal and primary agricultural wastes. Here, it is assumed that the biomass supply from existing or new non-plantation sources grows from 45EJ/yr in 2005 to 70EJ/yr in 2030 (an increase of only 25EJ/yr).

The remaining biomass energy requirement is assumed to be supplied from plantations with a yield of 100GJ/ha/yr. This may be pessimistic as a global average, as the sugarcane yield in Brazil is more than 400GJ/ha/yr today and is expected to exceed 850GJ/ha/yr in the future. However, most of the bioenergy crops are assumed here to be produced on surplus pastureland. Other workers have assumed plantation yields ranging from 54GJ/ha/yr to 380GJ/ha/yr (as deduced from Table 4.50 using a biomass heating value of 18GJ/t).

As seen from Figure 12.11, the required plantation area for the low-demand scenarios peaks at less than one third of the world's current pastureland area. No new land for biomass plantations is required until 2040 or later in the LF demand scenarios because biomass supplied from various residues and wastes is sufficient until then. However, for the high-demand scenarios the required plantation area reaches about two thirds of the current pastureland area and greatly exceeds the current world cropland area. Thus, there would have to be a significant reduction in land area requirements for crops (through higher yields) and for animal production (through more intensive production systems) in spite of a growing human population and likely adverse climatic changes, or a significant shift away from meat in order to be able to provide the required biomass energy. The required rates of establishment of new bioenergy plantation are about 27Mha/yr around 2050 in the biomass-intensive LFf scenario and 47Mha/yr around 2040 in the biomass-intensive HSf scenario. These rates are higher than in most other biomass-intensive scenarios (see Table 4.50) and are many times the 3Mha/yr rate of establishment of successful biomass plantations today.

12.3.5 Water requirements

In the energy system envisaged here, water would be required for CSTP plants in arid and semi-arid regions, for the production of hydrogen by electrolysis and to meet the water demands of bioenergy crops.

CSTP plants rely on concentrated sunlight to heat steam that is then used in a steam turbine. Water is normally used to cool the condenser of a steam turbine (see Volume 1, Chapter 3, subsection 3.2.1), and the cooling water is itself cooled evaporatively – leading to a loss of water that needs to be replaced. The water requirement will be roughly proportional to the thermal efficiency of the turbine (~40 per cent), which is comparable to that of modern PC powerplants. Thus, the same water requirement (2 litres/kWh) will be assumed here. The electricity production from CSTP plants in the hydrogen-intensive supply scenarios peaks at values ranging from 25,000TWh/yr (for the LFf scenario) to 54,000TWh/yr (for the HSf scenario), and the corresponding water requirements are 60–125 billion m^3/yr – an amount that will certainly not be available in dry regions.[12] About 60 per cent of the peak in global electricity demand is for the production of hydrogen, and even if hydrogen is not produced, water demands for evaporative cooling would still exceed supply in dry regions. Two possibilities are (1) to use air rather than water for cooling the condensers, resulting in a reduction in electricity output by about 10 per cent, or (2) desalinate seawater and pipe it to where it is needed. The energy requirement for state-of-the-art reverse osmosis desalination is 1.6kWh/m^3 (see Volume 1, Chapter 6, section 6.13), so the electrical power required would be 0.36 per cent of the output of the CSTP plants. Additional energy would be required to build the pipelines to supply the water and for dispersing the resulting brine at sea, and for pumping fresh water from the coastal desalination plant to inland CSTP plants (assuming that not all CSTP plants can be located along the coast).[13]

Water is required as a feedstock for the production of hydrogen by electrolysis. Given a heating value for hydrogen of 120.2MJ/kg, the feedstock requirement is 75 million m^3 per EJ of hydrogen produced. The peak hydrogen requirements for the low-slow and high-slow scenarios are 70EJ/yr and 144EJ/yr, respectively, and the correspond to peak feedstock water requirements of 5.2 and 11.0 billion m^3/yr. However, for that portion of the hydrogen that is produced using electricity from CSTP plants (rather than from wind turbines or PV modules), the potential water requirement associated with evaporative cooling of the turbine condensers would be seven to ten times greater than the feedstock requirement.[14] Thus, depending on where the hydrogen is needed and the source of the electricity, fresh water may need to be provided through desalination of seawater.

In the biomass-intensive supply scenarios, the biomass supply peaks at 140–288EJ/yr. According to Table 4.15, 10–95kg of lignocellulosic crop dry matter can be produced per hectare per mm of water transpired. This gives a water requirement of 5.8–55 billion m^3/EJ, or three to four orders of magnitude greater than the feedstock requirement for the production of hydrogen fuel. This water would have to be provided by rain, surface water or shallow groundwater, as pumping groundwater from great depth could consume more energy than is supplied by the biomass[15] (and reliance on groundwater would accelerate the depletion of groundwater supplies). The total water requirement is substantial and it (rather than the supply of land) could be the factor that constrains future bioenergy supply.

12.3.6 Required rates of installation of powerplant manufacturing capacity

Figure 12.12 shows the rate at which new wind, PV and CSTP manufacturing capacity would need to be added in order to achieve the rapid deployment of solar and wind energy envisaged here for the LFf scenario. At the time of most rapid buildup, the equivalent of about seven new factories, each capable of producing 1GW of wind turbines per year, would need to be constructed per year, as well as six new 1GW/yr CSTP factories and three new 1GW/yr PV factories per year. These rates are approximately doubled for the HSf scenario.

As discussed by Laxson et al (2006), the current processes for manufacturing the components of wind turbines are labour-intensive, but do not require specialized fabrication assembly lines (unlike automobile assembly lines, which require many months or years to set up). To increase the rate of production of wind turbines requires producing additional blade moulds, but once the initial plug for a mould has been created, a dozen additional moulds can be produced within a few months.[16] As discussed by Lindenberg et al (2008), segmented moulds could be transported to temporary manufacturing facilities that are established near the site of new large wind farms. Site preparation and pouring of concrete foundations can proceed in parallel with the construction of turbine components, permitting rapid construction of large wind farms.

As discussed in Chapter 2 (subsection 2.2.12), Keshner and Arya (2004) envisage single PV factories producing 2.1–3.6GW/yr of modules, having 100 identical production lines with fully integrated production – an industrial symbiosis. Pearce (2008b) further elaborates on the concept of industrial symbiosis in multi-GW/yr PV manufacturing plants.

12.3.7 CO_2 emissions

Carbon dioxide emissions are computed here by summing the fossil fuel requirements for electricity generation, transportation, buildings and industry, which are obtained by subtracting the available C-free energy sources from the various fuel demands.

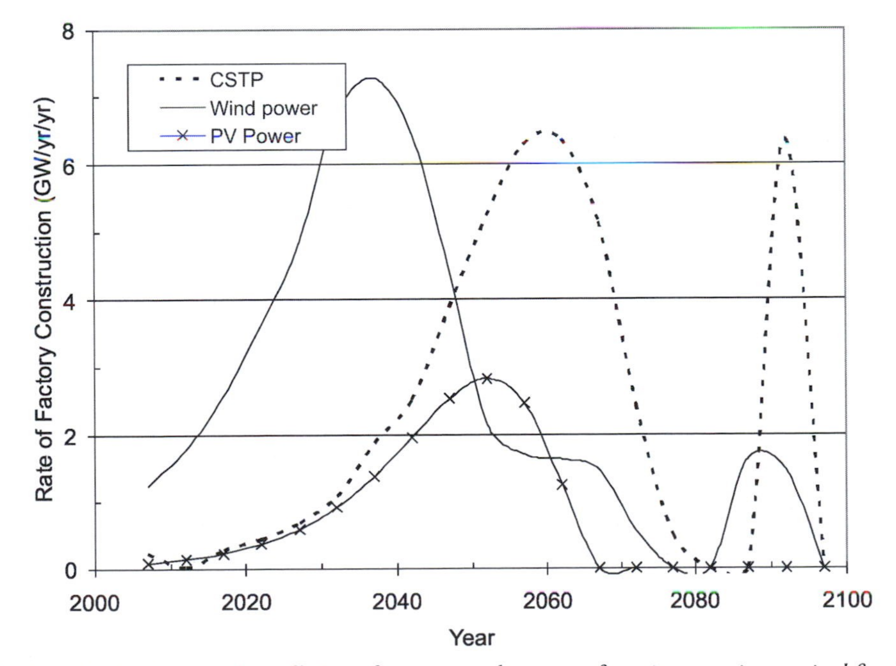

Figure 12.12 *Variation in the rate of installation of new powerplant manufacturing capacity required for the LFf scenario*

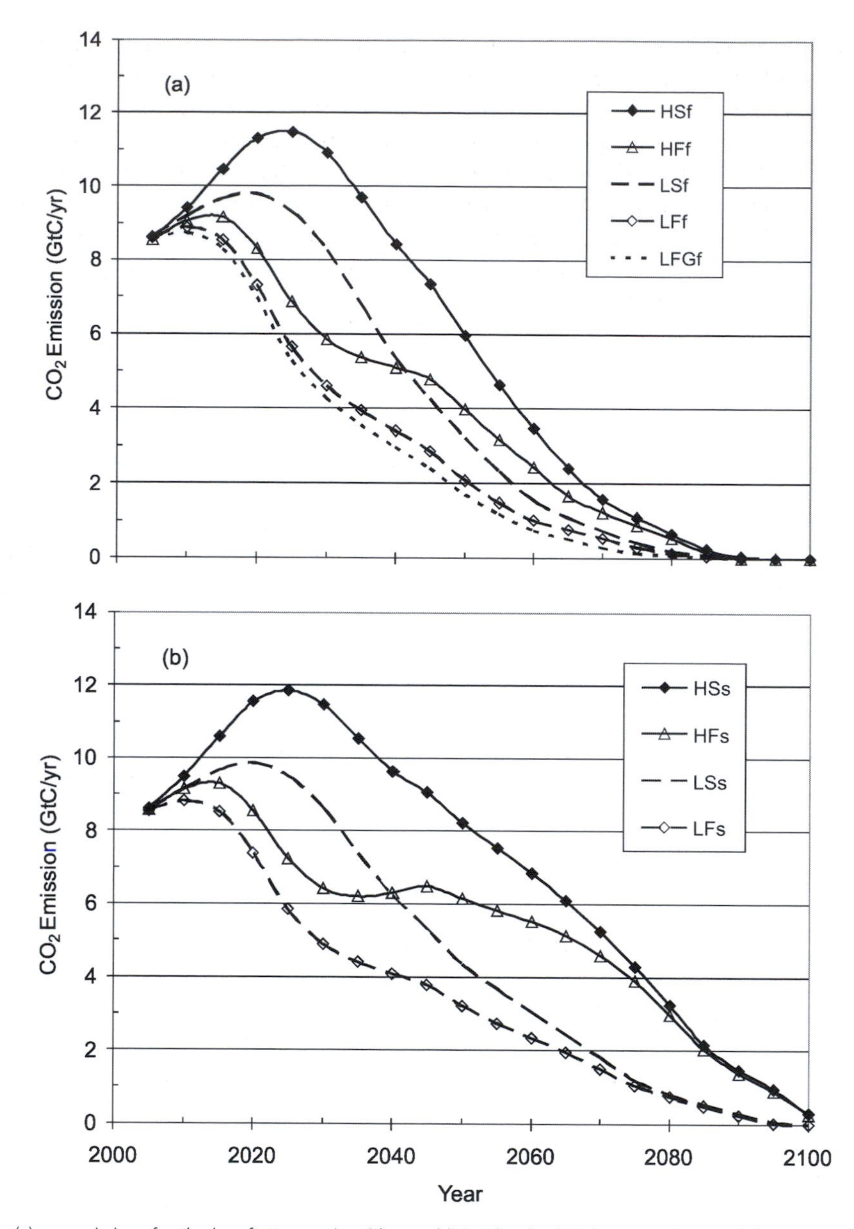

Note: Also shown in (a) are emissions for the low-fast scenario with an additional reduction due to a stronger shift to public transportation.

Figure 12.13 *Variation in fossil fuel CO_2 emissions for the LF, LS, HF and HS demand scenarios for (a) a very rapid transition to C-free energy supplies, and (b) a less rapid transition to C-free energy supplies*

The required fossil fuels are supplied first from natural gas up to the limit of natural gas availability (which peaks around 2030), then from oil up to the limit of oil availability (which peaks in 2015). Any remaining fossil fuel requirement is met from coal. The amounts of natural gas, oil and coal are multiplied by emission factors of 15.7kgC/GJ, 19.7kgC/GJ and 25.1kgC/GJ, respectively, and summed to give the total emission.[17]

Figure 12.13 shows the resulting variation in fossil fuel CO_2 emissions for the various demand scenarios and for the very rapid and less rapid transitions to C-free energy supplies. Results are shown for the H_2-intensive scenarios only, as the emission trajectories are almost identical for the biomass-intensive scenarios (by design). For very rapid deployment of renewable energy, global CO_2 emission peaks at dates ranging from 2010 to 2025 and at values ranging from 8.7 to 11.5GtC/yr. For the less rapid deployment of renewable energy, CO_2 emission also peaks at dates ranging from 2010 to 2025 but at values ranging from 8.8 to 11.9GtC/yr. The main difference between the two rates of deployment in C-free energy is in the rate of decline in global emissions after the peak. By design, global emissions are essentially zero by 2100 or sooner.

It is instructive to derive business-as-usual (BAU) emission scenarios that take into account limited supplies of oil, natural gas and coal, and to compare these with the climate-motivated scenarios developed here. As discussed in Volume 1 (Chapter 2, subsection 2.5.1), peaking in the global supply of oil by 2020 or sooner is a near certainty. This is because rates of discovery of new oil have been steadily declining over the last few decades, so few new large fields remain to be discovered, the rate of decline in the supply of oil from individual utilized oil fields has occurred at progressively faster rates for oil fields peaking progressively later in time, and most oil fields currently in production will soon or have already reached their individual peaks. Data for gas supply are much more uncertain than for oil, but it is likely that gas supply will peak soon after oil supply peaks. Until recently, supplies of coal were thought to be sufficient to last several hundred years. However, estimates of the amount of mineable coal have been steadily falling over the past three decades, and a recent re-evaluation (discussed in Volume 1, subsection 2.5.3) indicates that a third of the ultimately extractable coal may have already been used. Of particular significance is the conclusion of Tao and Li (2007) that Chinese coal supply will peak around 2030, and the conclusion of Croft and Patzek (2009) that coal supply from existing mines in the US will remain comparable to the present rate of supply for only another 20 years before declining, with little prospect for any significant development of new mines.

One way to construct a BAU scenario of future CO_2 emissions is to apply the logistic function (Equation (12.1)) separately to natural gas, oil and coal emissions, and sum the results. To do this, I use the cumulative CO_2 emissions up to 2005 as given by Marland et al (2008), specify the ultimate cumulative emission as some factor of the cumulative emission to 2005 and choose values of the parameter a such that the rate of change in annual emission after 2005 merges smoothly with the trend up to 2005. For the base case BAU scenario, the ultimate cumulative emissions (and hence the ultimate cumulative consumption) are assumed to be three, two and three times cumulative emissions to 2005 for natural gas, oil and coal, respectively. These result in peak supply about now for oil, around 2020 for natural gas and around 2030 for coal, as shown in Figure 12.14.[18] The resulting total fossil fuel CO_2 emission is also shown in Figure 12.14. Surprisingly, global emissions peak around 2020 at about 9GtC/yr. In essence, increasing emissions from growing use of coal do not compensate for falling emissions from oil beginning around 2010 and from natural gas beginning around 2020.

Given uncertainties in projecting the amounts of fossil fuels that are ultimately recoverable, it is important to consider the possibility that the available fossil fuel resource is larger than assumed here. New technologies, spurred by much higher fossil fuel prices, could greatly expand the available fossil fuel supply, causing a break from the previous logistic function consumption pattern. Global CO_2 emissions are therefore compared in Figure 12.15 for three cases: a low case, which is the BAU case considered above, a medium case, in which the amount of remaining available natural gas, oil and coal are all twice as much as assumed in the low case (so remaining natural gas and coal are four times rather than twice the cumulative consumption to date and remaining oil is twice the cumulative consumption to date), and a high case, in which the remaining natural gas and oil are as in the medium case and the remaining coal is three times as much as for the low case (that is, six times cumulative consumption to date). The corresponding CO_2 emission peaks are about 9GtC/yr in 2020, 12GtC/yr in 2040 and 13GtC/yr in 2050. These peaks are in sharp contrast to almost all other BAU scenarios, in which global emissions grow to 20–40GtC/yr by 2100. However, most BAU scenarios do not adequately consider realistic constraints on the supplies of fossil fuels, especially coal, and so give unrealistically high projections of future CO_2 emissions.

Also shown in Figure 12.15 are the global emissions for the low-growth green scenario with fast improvement of energy efficiency and deployment of C-free energy sources (LFGf) and for the high-growth scenario with slow improvement of energy efficiency and deployment of C-free energy sources (HSs). These are the climate policy scenarios developed earlier with

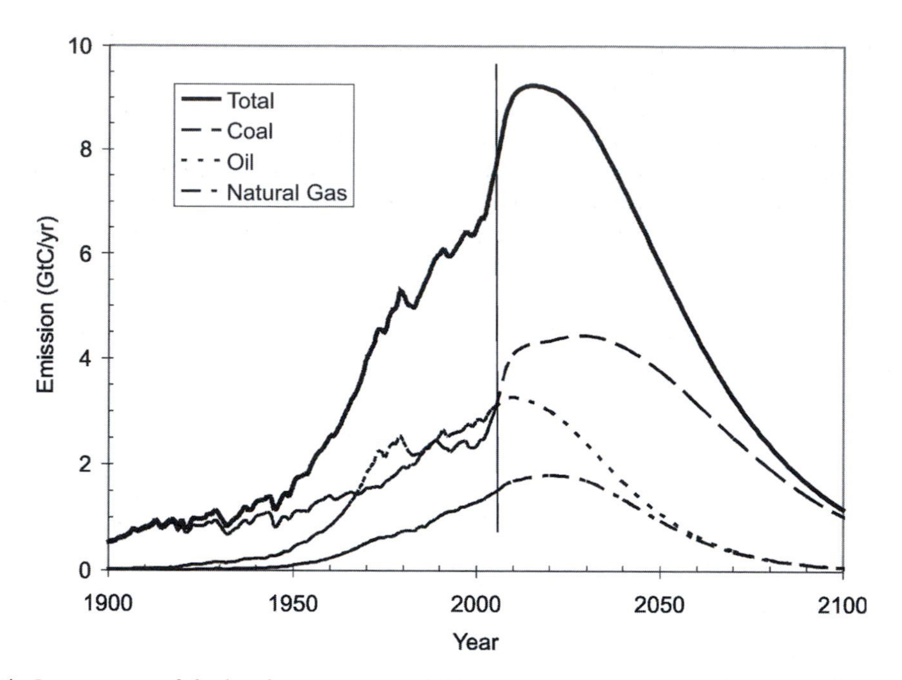

Figure 12.14 *Construction of the low business-as-usual CO_2 emission scenario: global emissions from natural gas, oil and coal based on using estimates in a logistic function of the size of the ultimately extractable resources, and the resulting total global CO_2 emission*

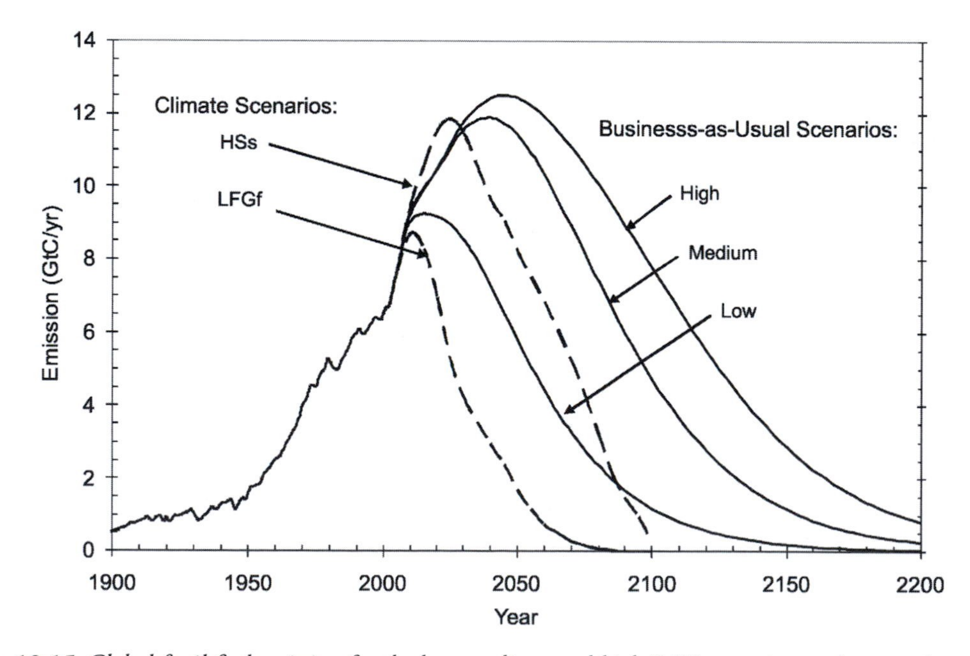

Figure 12.15 *Global fossil fuel emission for the low, medium and high BAU scenarios, and comparison with the lowest (LFGf) and highest (HSs) climate policy scenarios*

the lowest and highest emissions, respectively. Surprisingly, the peak emissions and their timing closely match the peak emissions and timing of the low and high BAU scenarios. This is remarkable because even the HSs scenario has much more rapid and stringent improvements of energy efficiency, and many times more rapid deployment of C-free energy sources than in the vast majority of BAU scenarios. This result implies that very strong efficiency improvements and very rapid deployment of C-free energy sources are required, irrespective of climate concerns, simply to balance energy supply and demand in the future. If the fossil fuel supplies in the low or medium BAU scenarios are close to what we will ultimately be able to use, then rapid energy efficiency and deployment of C-free energy sources will need to be accompanied by much slower economic growth than assumed in most BAU scenarios, and more in line with the lower growth scenarios considered here. The main difference between the climate policy and BAU scenarios is in the rate of decrease in total fossil fuel emissions after the emission peak: in the BAU scenarios this decrease occurs gradually over the following 150–200 years, while in the climate policy scenarios emissions are reduced to zero within 50–90 years of the peak.

In Figure 12.16 the total cumulative CO_2 emissions, from the earliest emissions to the exhaustion of fossil fuel resources in 2300 (or sooner), are given for the BAU and climate policy scenarios. Cumulative emissions range from 1640GtC (for the high BAU scenario) down to 653GtC (for the LFGf scenario). By comparison, cumulative emissions to the end of 2005 were about 311GtC.

12.3.8 Summary

In summary, the major assumptions behind the supply scenarios developed here are as follows:

- As almost no region in the world is more than 3000km from regions of either good winds or semi-arid or arid regions where CSTP is applicable (and most regions are no more than 2000km from such sites), the world's electricity needs can be

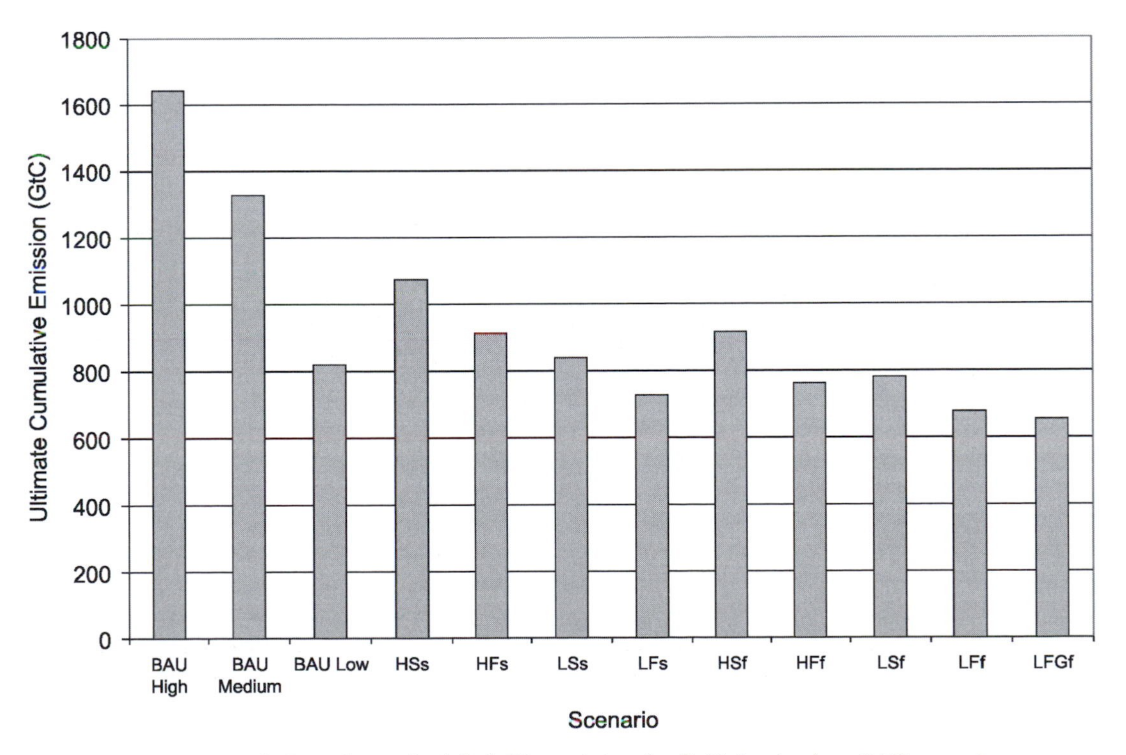

Figure 12.16 *Cumulative fossil fuel CO_2 emission (as GtC) for the three BAU scenarios and for the various climate policy scenarios*

largely met through a combination of dispersed (to reduce variability and increase reliability) large wind farms and CSTP arrays, supplemented by relatively small contributions from hydroelectric power, geothermal power and advanced biomass power generation, and interconnected to each other and to the major electricity demand centres through an HVDC transmission grid.

- Storage at daily and longer timescales needed to meet the remaining variability in power production and to match fluctuating demand can be provided through a combination of hydroelectric reservoirs (where already built or from very limited additional capacity), AA-CAES (which does not require additional fuel) and advanced biomass cogeneration (with thermal energy storage).
- A large portion of the peak in electricity demand (which will invariably be in the summer) can be met through BiPV panels, thereby reducing the requirements for transmission into urban centres from the distant wind and CSTP arrays.
- The future automobile fleet will consist largely of highly efficient PHEVs, thereby shifting 75 per cent of the energy requirements for urban travel from light-duty vehicles to electricity.
- Much of the PHEV fleet will be connected to the grid at any given time, with the batteries in such vehicles serving as a short-term (seconds to hours) buffer for fluctuations in the electricity production from the BiPV powerplant in urban centres.[19]
- District heating and cooling systems with electric heat pumps and the capacity to store excess heat and coldness, ground-source heat pumps for individual buildings and other dispatchable end-use electricity loads will provide other ways of matching the variation in electricity demand and in electricity supply from fluctuating renewable energy sources.
- All existing nuclear powerplants are retired after 40 years of service and no new nuclear powerplants are built after 2010 due to the various problems outlined in Chapter 8.
- CCS is not part of the baseline scenarios developed here because (1) it is not clear that it is viable on a large scale, (2) it is likely to be more expensive than the renewable energy alternatives considered here, and (3) it cannot be deployed as quickly as wind and CSTP systems. There may, however, be a role for CCS in certain niche applications (such as in the production of cement and N fertilizers) and as a means to create a modest C sink through capture and sequestration of CO_2 from the use of solid biomass or in the production of liquid biofuels.

Needless to say, detailed work would need to be carried out at the regional level to determine the optimal mix of the various options adopted here. For simplicity, I have assumed equal ultimate deployment of wind and CSTP capacity, and furthermore, that all the wind capacity is limited to a capacity factor of 0.2 due to oversizing relative to transmission capacity. Some (and perhaps much) of the wind capacity might consist of floating offshore turbines that would be located closer to demand centres than assumed here and probably not oversized, so the capacity factor is likely to be closer to 0.4. Detailed analysis at the regional level would also be required to determine the optimal sizing of thermal storage in CSTP powerplants compared to other options for energy storage, including AA-CAES. It has been assumed here that AA-CAES will develop to the point where it is superior to CAES using gasified biomass as a supplemental fuel. To the extent that a CAES/biomass system is better, the biomass energy requirements will be larger than calculated here.

12.4 Impact on climate and ocean acidification

To assess the impact on global average temperature of a given CO_2 emission scenario requires the use of a coupled climate–carbon cycle computer model. The carbon cycle component of a coupled model computes the rate of removal of emitted CO_2 from the atmosphere by the terrestrial biosphere and oceans. The CO_2 that is not removed accumulates in the atmosphere, leading to an increase in the trapping of radiant heat (infrared radiation) and an increase in temperature. This heat trapping is an example of a *radiative forcing* – an externally imposed alteration in the balance between absorption of solar radiation and emission of infrared radiation to space. The climate component of a coupled model computes the increase in temperature for a given increase in CO_2.

As discussed in Volume 1 (Chapter 1, section 1.2), the key parameter in the climatic response to a given increase in CO_2 is the *climate sensitivity*. This can be loosely defined as the globally averaged temperature warming for a hypothetical *fixed* doubling in the concentration of CO_2 once the climate has had a chance to fully adjust to the higher CO_2 concentration. The climate sensitivity is estimated to range from 1.5–4.5°C. In reality, CO_2 and many other GHGs are increasing together. What matters for the climate is the collective heat trapping by all of the gases together. The climate sensitivity serves as a useful indicator of how responsive the climate is to a change in heat trapping, whatever the combination of GHGs responsible for it. If the climate sensitivity is larger, not only will there be a greater warming for a CO_2 doubling, there will also be a greater warming for any smaller or larger GHG increase.

The increase in temperature caused by an initial increase in CO_2 and other GHGs reduces the subsequent net rate of removal of CO_2 from the atmosphere by the terrestrial biosphere and the oceans. This increases the increase of atmospheric CO_2 concentration compared to what would otherwise occur. That is, there is *positive feedback* between the carbon cycle and the change in climate that amplifies the initial CO_2 increase.

Some positive climate–carbon cycle feedbacks are well understood and will increase in a smooth, predictable manner. An example is the decrease in the ability of the oceans to hold CO_2 due to warming of the ocean surface. However, there is also the possibility of abrupt and large positive feedbacks once uncertain critical temperature thresholds are exceeded. Examples include:

- massive release (2–3GtC/yr) of CO_2 and CH_4 from carbon-rich soils in Siberia known as *yedoma* soils (Khvorostyanov et al, 2008);
- large releases of CO_2 from massive dieback of the Amazon rainforest if a quasi-permanent El Niño-like state develops (Cox et al, 2004); and
- large releases of CH_4 from frozen water-methane deposits known as *clathrates*, which are found in continental shelf sediments worldwide and in permafrost (Harvey and Huang, 1995).

The greater the climate sensitivity, the greater the warming for a given increase in GHGs and the greater the expected and potential additional emissions of CO_2 and CH_4 into the atmosphere from natural sources. Hence, there is a stronger positive climate–carbon cycle feedback if the climate sensitivity is larger, which further amplifies the already stronger warming. We can think of a second climate sensitivity that comes into play on a longer timescale (decades to centuries or even millennia) that accounts for climate–carbon cycle feedbacks and is larger than the short-term climate sensitivity. In the following discussion, the term climate sensitivity refers to the short-term climate sensitivity – the temperature response expected for a fixed doubling of atmospheric CO_2 concentration.

The impact of the scenarios developed in section 12.3 on atmospheric CO_2 and temperature is investigated here using the relatively simple coupled climate–carbon model of Harvey and Huang (2001) and Harvey (2001). In this model, the climate sensitivity can be prescribed by deliberately altering various parameters inside the computer program. There are many parameters and inputs that need to be prescribed in addition to the climate sensitivity and fossil fuel CO_2 emissions, however. These include parameters governing the stimulation of photosynthesis on land by higher atmospheric CO_2 and the increase in plant and soil respiration by warmer temperatures, the emissions of CO_2 from deforestation and other land-related activities, emissions of non-CO_2 GHGs and emissions of aerosol-forming compounds.

Aerosols consist of tiny particles in the atmosphere (such as sulphate, nitrate and organic particles) that mostly have a cooling effect. Their concentrations in the atmosphere have increased greatly due to human-caused emissions of chemicals (called aerosol precursors) that react to form aerosols in the atmosphere, and they could now be suppressing up to half of the heating effect of the cumulative GHG increase so far. However, the lifespan of aerosols in the atmosphere is only a matter of days, compared to years, decades or millennia for GHGs. As most aerosols are produced in association with the burning of fossil fuels, measures to reduce the use of fossil fuels will lead to an immediate reduction in the cooling effect of aerosols, whereas noticeable effects on GHG concentrations will not be seen until much later. Thus, it is possible that the short-term effect of reducing fossil fuel emissions will be an acceleration in the rate of warming. Whether this occurs or not depends on how rapidly fossil fuel emissions are reduced and how strong the aerosol cooling effect is at present (the stronger the effect, the stronger the warming when the suppressing effect of aerosols is reduced). Thus, prior to simulating future temperature changes in response to the emission scenarios developed here, it is important to constrain both the uncertain climate sensitivity and the uncertain aerosol cooling effect.

12.4.1 Constraining the aerosol cooling effect and climate sensitivity

Although the cooling effect of aerosols and the climate sensitivity are both highly uncertain, these two factors are not independent of one another. As explained in Box 12.1, a larger climate sensitivity requires that the present cooling affect due to anthropogenic (human-caused) aerosols also be larger, and vice versa.

Figure 12.17 compares simulations of observed global mean temperature variation with the observed variations from 1856 to August 2009. The model simulations take into account an estimate of the effect of volcanic eruptions and variations in solar luminosity over time, as well as increases in GHGs and aerosols (details are given in Harvey and Kaufmann, 2002). Results are given for the following combinations of climate sensitivity and of the fraction of GHG forcing that is offset by aerosols in 2000:[20]

- 1.0°C climate sensitivity and no aerosol offset;
- 1.5°C climate sensitivity and 15 per cent aerosol offset;
- 2.0°C climate sensitivity and 20 per cent aerosol offset;
- 3.0°C climate sensitivity and 50 per cent aerosol offset;
- 4.5°C climate sensitivity and 50 per cent aerosol offset.

As can be seen from Figure 12.17, the climate model does well in simulating the overall warming from the late 19th century to the present if the climate sensitivity is between 1.5°C and 3.0°C. However, in using the observed temperature variation over the past century to constrain the climate sensitivity, it is assumed that the effective climate sensitivity during the transition from one climate to another is the same as the long-term climate sensitivity.[21] Some coupled three-dimensional atmosphere–ocean climate models show an increase in the effective climate

Box 12.1 Constraining the aerosol cooling effect and climate sensitivity

Changes in the earth's temperature are governed by the difference in the absorption of solar radiation (which has a warming effect) and the emission of infrared radiation to space (which has a cooling effect). For an unchanging climate, these two terms are in balance on an annual and global average basis. An imposed alteration in either term is referred to as a *radiative forcing*. For example, a doubling of the atmospheric CO_2 concentration reduces the outgoing emission of infrared radiation to space by about 4W for every square metre of the earth's surface, so the radiative forcing is about 4W/m^2. Aerosols, in contrast, have a negative forcing (they reduce the available radiative energy by increasing the reflection of solar radiation to space) and so reduce the overall positive radiative forcing that would result from GHG increases alone. That is, they reduce the net radiative forcing.

The temperature response at any given time is proportional to:

$$\text{(net radiative forcing)} \times \text{(climate sensitivity)} \times \text{(a factor to account for the delay effect of the oceans)} \quad (12.3)$$

Thus, given an observed variation of temperature over the last century or so, the greater the climate sensitivity the smaller the net forcing must have been in order to have produced the given warming. If we prescribe too a large climate sensitivity in a climate model when doing a simulation of recent temperature changes, we also need to assume a very large aerosol offset so as not to produce more than the observed warming. However, most of the aerosol forcing is in the northern hemisphere (where the greatest emissions of aerosol precursors occur), so if we have too large a climate sensitivity offset by too large an aerosol cooling effect (compared to reality), we will distort the simulated pattern of temperature change compared to the observed changes. Thus, by considering changes in both global average temperature and the difference between northern and southern hemisphere average temperatures, it is possible to simultaneously constrain the climate sensitivity and present-day aerosol cooling effect. Harvey and Kaufmann (2002) carried out such an exercise and concluded that the climate sensitivity is highly likely to fall between 2°C and 3°C and that the aerosol forcing is likely to be offsetting no more than half of the present-day GHG heating.

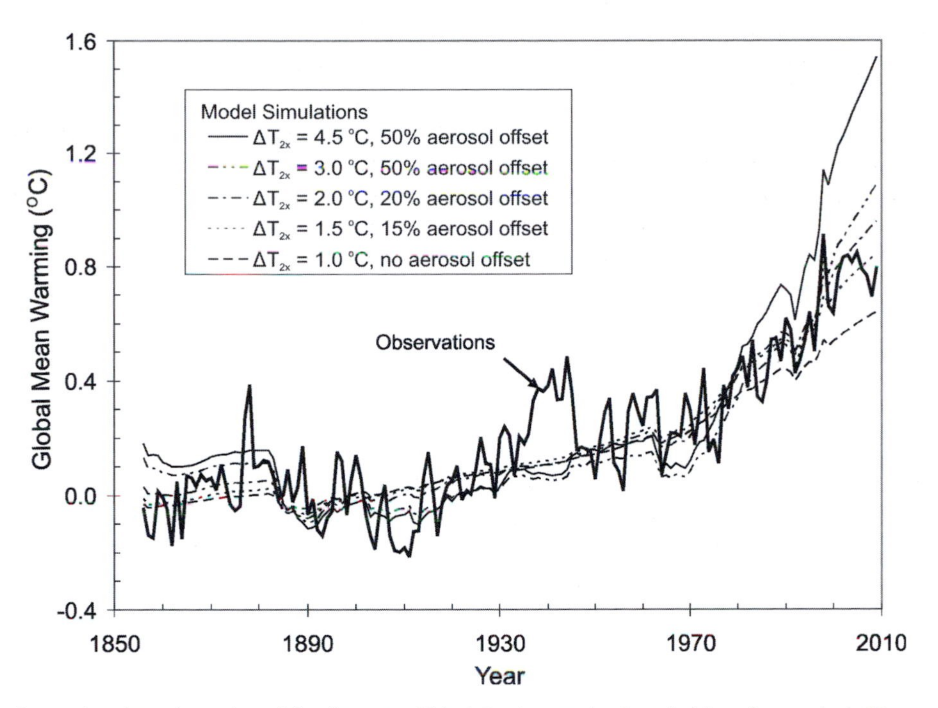

Note: Model results are given for various values of the climate sensitivity (ΔT_{2x}, the warming for a doubling of atmospheric CO_2 concentration) and of the fraction of GHG radiative heating this is offset by aerosols in 2000.

Figure 12.17 *Comparison of observed and model-simulated variation in global mean temperature from 1856 to 2009*

sensitivity over time. As some other approaches yield climate sensitivities of 4.5°C or even larger, climate sensitivities of 1.5°C, 3.0°C and 4.5°C will be considered here in simulations of future climate change.

12.4.2 Simulations of future temperature change

There is a large number of possible combinations of assumptions concerning fossil fuel CO_2 emissions, land use-related CO_2 emissions, emissions of non-CO_2 GHGs and climate sensitivity, not to mention parameters governing the climate–carbon cycle feedback. Here, just a few representative combinations will be considered.[22] The following emission scenarios will be used:

- Scenario 1, high BAU fossil fuel CO_2, land-use CO_2 and non-CO_2 GHG emissions;
- Scenario 2, high BAU fossil fuel CO_2, low land-use CO_2 and non-CO_2 GHG emissions;
- Scenario 3, HSs fossil fuel CO_2 emissions;
- Scenario 4, LFs fossil fuel CO_2 emissions;
- Scenario 5, LFGf fossil fuel CO_2 emissions;
- Scenario 6, LFGf fossil fuel CO_2 emissions with a ramp-up to 1GtC/yr geological or biological sequestration between 2020 and 2050, and held constant thereafter.

Calculations of GHG concentrations and temperature change are performed here using the standard version of the aforementioned climate–carbon cycle model, in which there is only a weak positive climate–carbon cycle feedback due to warming of the ocean surface and a gradual increase in the rate of decomposition of carbon in soils, and using a version that incorporates non-linear feedback relating to thawing of yedoma soils.

Temperature results using the standard climate–carbon cycle model

Figure 12.18 shows atmospheric CO_2 concentration from 1950 to 2250 for the six scenarios given above and the resulting global mean temperature change,

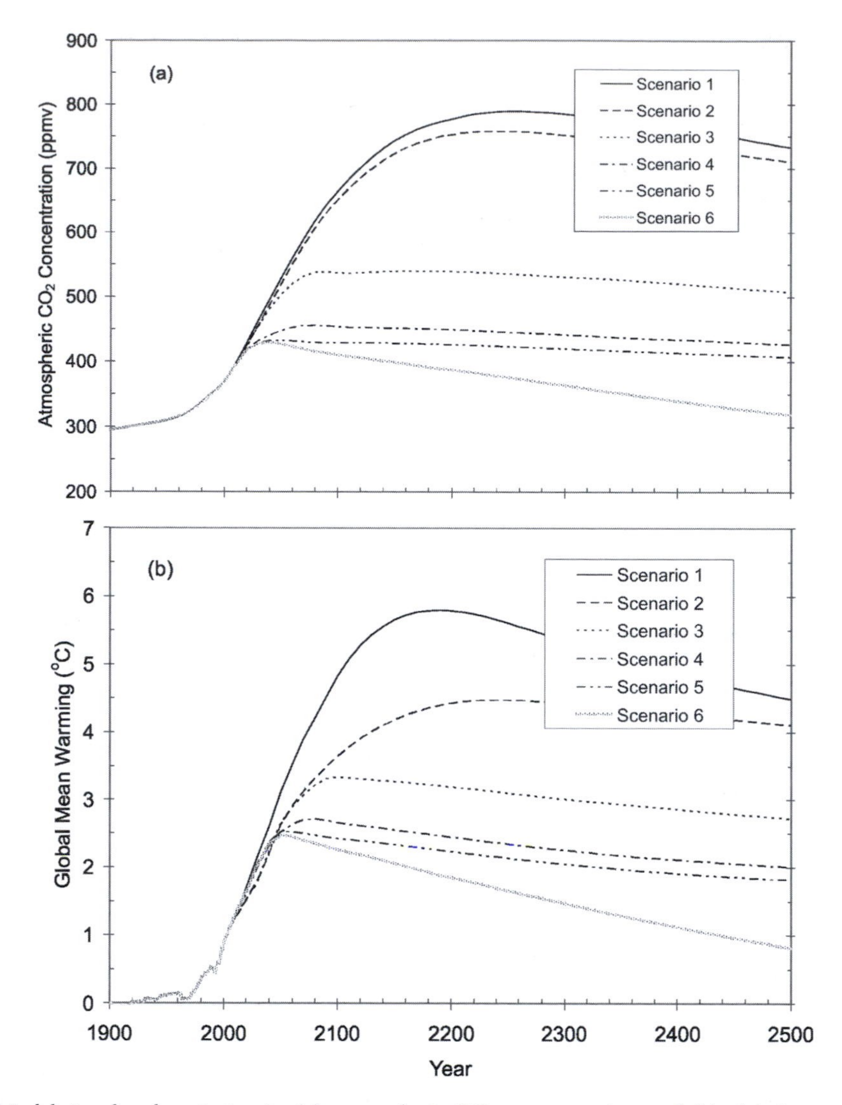

Figure 12.18 *Model-simulated variation in (a) atmospheric CO_2 concentration and (b) global mean warming for the high BAU scenario and the various climate policy scenarios using a climate sensitivity of 3.0°C*

assuming a climate sensitivity of 3°C.[23] In Scenario 1, atmospheric CO_2 concentration rises to about 800ppmv by 2250. In the weakest of the three climate policy scenarios (Scenario 3) CO_2 peaks at about 540ppmv in 2100 and in the strongest climate policy scenario (Scenario 5) it peaks at 433ppmv in 2050, declining to 408ppmv by 2500. For a 3.0°C climate sensitivity, peak global mean warming ranges from 6.0°C (high BAU) to 2.5°C (Scenario 5).[24]

Figure 12.19 compares the global mean warming for the weakest and strongest climate policy scenarios (Scenarios 3 and 5) for climate sensitivities of 1.5°C, 3.0°C and 4.5°C. Recall that the higher climate sensitivities require a stronger aerosol cooling effect so as not to simulate too much warming up to the present. In the strong climate policy scenario, the aerosol concentrations and hence the aerosol cooling effect fall more rapidly than in the weaker climate policy scenario, so there is a few tenths of a degree greater warming by 2030 than for the weaker climate policy scenario. However, after 2050 the temperature trajectories for the two scenarios strongly diverge, with significantly less warming for the stronger emission reduction scenario. For Scenario 5 (with very aggressive emission

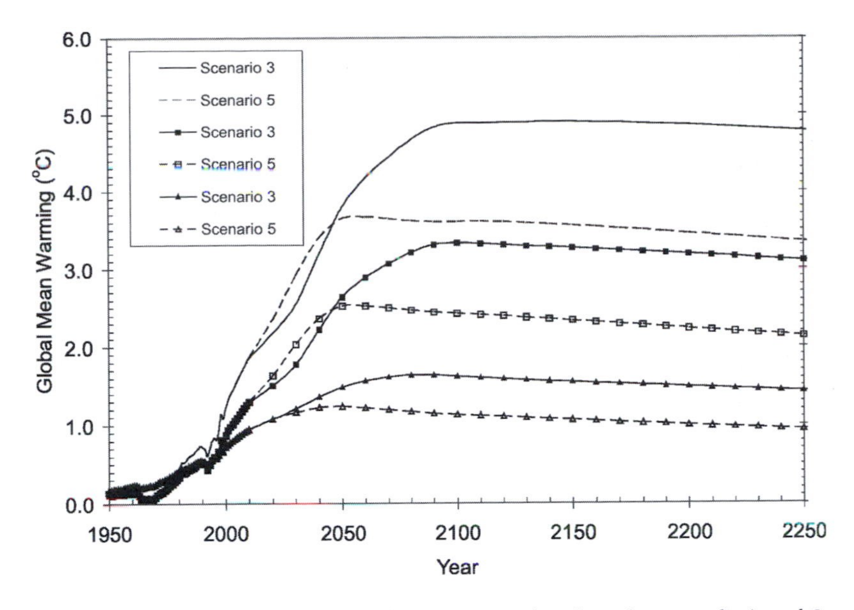

Figure 12.19 *Comparison of global mean warming for Scenario 3 (weakest climate policy) and Scenario 5 (strongest climate policy) for climate sensitivities of 1.5°C (lowest pair), 3.0°C (middle pair) and 4.5°C (upper pair)*

reductions), global mean warming peaks at 1.2–3.7°C, while for Scenario 3 (still aggressive by most standards), global mean warming peaks at 1.6–4.9°C.

In spite of the greater near-term warming with stronger reductions in fossil fuel CO_2 emissions, impacts on ecosystems and food production may not be worse. There are two reasons for this. First, this is because the major aerosols (sulphate and nitrate) are associated with acid deposition. Thus, as aerosol loadings are reduced there will also be a decrease in acid deposition, which will improve the health of forests that are currently adversely affected by acid deposition, in turn improving their ability to tolerate the early stages of global warming. Second, an increase in aerosol concentrations tends to reduce precipitation, so although there will be a faster warming in the near term as aerosol concentrations decrease, there will probably be greater rainfall in some regions due to falling aerosol concentrations. Indeed, the drying in northern China and perhaps also in the Sahel region of Africa during the past two or three decades is probably related to aerosol pollution rather than to recent warming (Zhao et al, 2006; Biasutti and Giannini, 2006).

As seen from Figure 12.19, global mean temperature change as high as 3.8°C could be associated with the most stringent emission reduction considered here (which is perhaps the most stringent considered anywhere, with the CO_2 concentration peaking just below 450ppmv). Even for a 3°C climate sensitivity, the global mean warming reaches 2.5°C. As indicated in Chapter 1, section 1.1 and discussed more fully in Volume 1 (Chapter 1, subsection 1.4.1), a mere 2.5°C warming will probably be highly damaging to many ecosystems and will cause widespread species extinction and water stress in many sensitive regions. If sustained, it also poses a high risk of destabilizing both the Greenland and West Antarctic ice sheets, leading to an eventual sea level rise in excess of 10m. These results reiterate the point made by Harvey (2007) that a CO_2 concentration of 450ppmv constitutes 'dangerous' interference in the climate system, something that is to be avoided under the terms of the United Nations Framework Convention on Climate Change (UNFCCC) that was signed and ratified by almost every country in the world.

Temperature results with climate feedback on yedoma soils

A final simulation is performed to investigate the impact of potential releases of CO_2 and methane from yedoma soils in Siberia. The yedoma soils are estimated to contain about 500Gt of frozen carbon that is particularly susceptible to decomposition if it thaws. Khvorostyanov et al (2008) used a detailed permafrost model to simulate the impact of local warming on the yedoma soils, and they estimate that intense mobilization of carbon (2–3GtC/yr) would begin when the regional warming reaches about 9°C and continue for about 100 years. Inasmuch as the mean annual warming over land at the latitudes (60°N to 70°N) where yedoma soils occur is about twice the global mean warming (see Figure 10.6 of Meehl et al, 2007), this implies that significant carbon release would occur in association with about 4.5°C global mean warming. In the simulations by Khvorostyanov et al (2008), the intense carbon release is triggered by internal heat generation by decomposers, and is irreversible once it starts. Here, it will be assumed that the rate of carbon release increases linearly from 0GtC/yr at 0.8°C global mean warming (i.e. starting from present) to 0.3GtC/yr at 3.3°C global mean warming, and then increases to 3GtC/yr at 4.3°C global mean warming and continues until the cumulative emission reaches 300GtC, then abruptly stops (the reason for choosing the 3.3°C threshold will be evident later). It is further assumed that 25 per cent of the released C is emitted to the atmosphere in the form of CH_4 and the rest as CO_2; the actual proportions as CO_2 and CH_4 would depend on the extent to which the thawed soils are saturated (more extensive saturation would lead to a greater proportion as CH_4). As noted by Walker (2007), some release of methane from yedoma soils is already occurring via thaw lakes that migrate laterally across the landscape (releasing an estimated 0.004GtC/yr of methane at present).

Emissions of CH_4 are of particular concern because the heat trapping of a CH_4 molecule is 26 times that of a CO_2 molecule, and because the methane buildup leads to an increase in its own lifespan in the atmosphere and to increases in tropospheric ozone and stratospheric water vapour, both of which add to the radiative forcing. The emitted CH_4 is oxidized to CO_2 after an average time in the atmosphere (at present) of about 10 years, and the CO_2 so produced is then gradually removed from the atmosphere through the same processes that remove anthropogenic CO_2.

Figure 12.20 shows the impact of global mean temperature for Scenario 3 (HSs) with a climate sensitivity of 3°C. In the absence of yedoma feedback, the global mean warming peaks at about 3.38°C if the climate sensitivity is 3.0°C. However, assuming that carbon release from yedoma soils begins at a warming only 0.08°C smaller unleashes a positive feedback sufficient to push the peak warming to 5.6°C. Figure 12.21 shows the different contributions to the extra radiative forcing caused by emissions of CO_2 and CH_4 from thawing yedoma soils in Scenario 3. The total radiative forcing is more than doubled by the end of the carbon emissions, but the temperature response does not quite double due to the

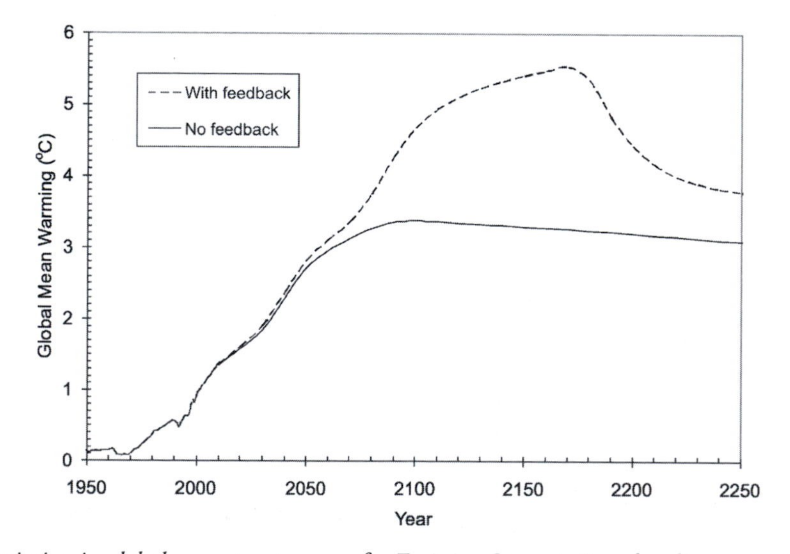

Figure 12.20 *Variation in global mean temperature for Emission Scenario 3 and a climate sensitivity of 3.0°C for cases with no release of carbon from yedoma soils ('No feedback') and when carbon release begins at a global mean warming of 3.3°C ('With feedback')*

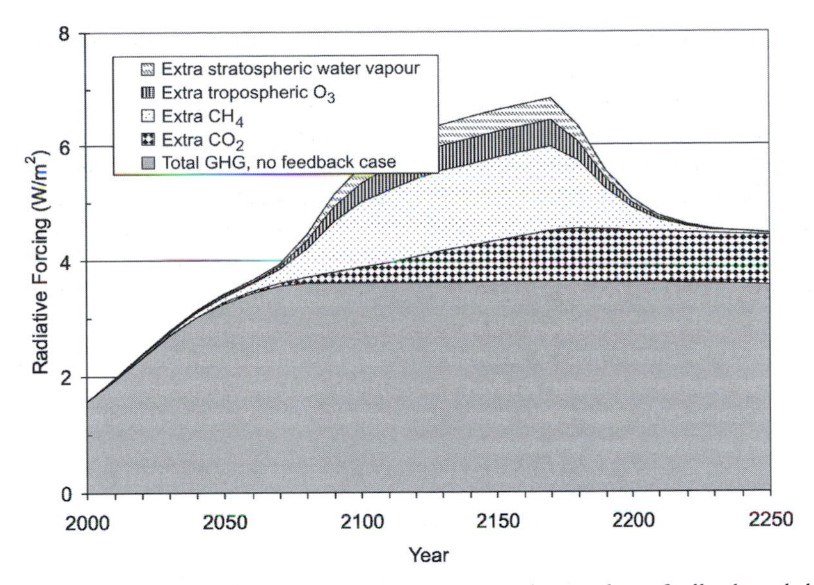

Figure 12.21 *Total radiation forcing from GHGs for Scenario 3 without yedoma feedback and the contribution of different terms to the extra radiative forcing when there is a carbon release from yedoma soils beginning at 3.3°C global mean warming*

delay in warming caused by oceans combined with the rapid decline in ozone and stratospheric water vapour forcings once methane emissions cease.

Scenario 3 is a very aggressive emission reduction scenario, but it is not aggressive enough to avoid a catastrophic positive climate–carbon cycle feedback if the climate sensitivity is 3°C (that is, merely in the mid-point of the accepted uncertainty range) and if the threshold for the beginning of significant carbon release from yedoma soils is 3.3°C global mean warming or less. For Scenarios 4 (LFs) and 5 (LFGf), global mean temperature peaks at less than 3.0°C if the climate sensitivity is no more than 3.0°C, and so would seem to be sufficient to avoid destabilization of yedoma soils (assuming a destabilization threshold of 3.3°C global mean warming or larger). However, there are many other possible positive climate–carbon cycle feedbacks not included here that could push the warming beyond the yedoma destabilization threshold (depending on where the true threshold lies) even for Scenarios 4 and 5 (see, for example, Dorrepaal et al, 2009, and Schuur et al, 2009). The 5.0°C warming reached after carbon release from yedoma soils in turn is likely to be large enough to trigger releases of methane clathrate from continental shelf sediments that in turn could add another 1–2°C global mean warming (Harvey and Huang, 1995). Thus, it is possible that, even with extremely aggressive emission-reduction scenarios beginning within the next decade, global warming could largely slip beyond human control, leading to globally catastrophic consequences.

Impacts on ocean acidity

The absorption of CO_2 from the atmosphere by the oceans causes a decrease in the pH of ocean water and a decrease of the supersaturation of surface water with respect to calcium carbonate – a key structural material of organisms throughout the marine food chain. Figure 12.22 shows the variation in the mean pH of the ocean surface layer for Scenarios 1, 3, 5 and 6. The pH of the ocean surface has already decreased by 0.1 according to the simulation results presented here (and according to simulations with other models too). For the high BAU scenario, the pH falls a total of 0.34 units (from 8.29 to 7.95). For the most stringent climate policy scenario, pH decreases by only a further 0.03 compared to present (decreasing from 8.29 to 8.16).

12.5 A role for carbon sequestration

The climate policy scenarios considered here do not include sequestration of industrial CO_2 because the cost and feasibility of geological carbon sequestration is

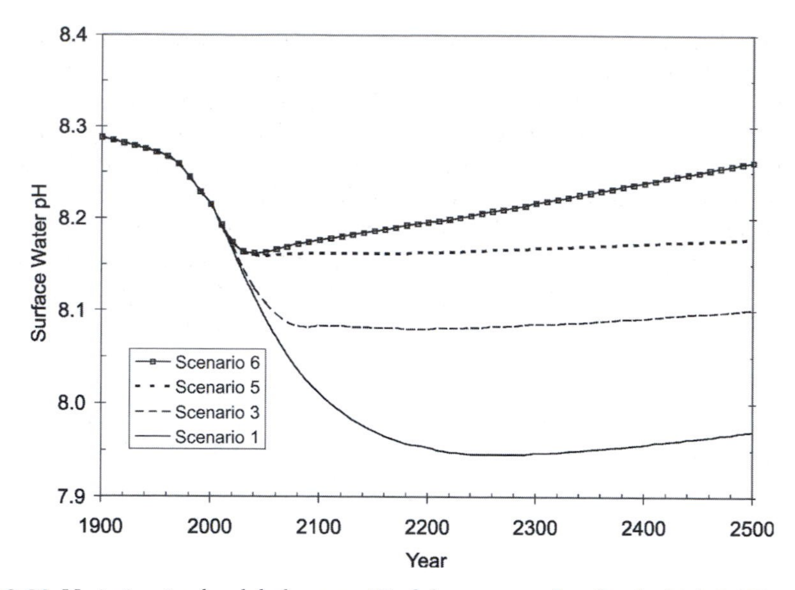

Figure 12.22 *Variation in the global mean pH of the ocean surface for the high BAU scenario and for Climate Policy Scenarios 3, 5 and 6*

unclear, and because wind and concentrating solar thermal power generation provide a faster alternative for eliminating emissions from the electricity generation sector (the sector that accounts for the bulk of potential CO_2 capture and storage). Given the likelihood that the world will substantially overshoot a safe atmospheric CO_2 concentration range, it will almost certainly be necessary to draw down the atmospheric CO_2 concentration from its peak, probably to 350ppmv or lower. The sustainable use of biomass energy is carbon free (except for possible changes in soil carbon), as the carbon released during combustion is simply returning CO_2 to the atmosphere that was removed by photosynthesis during growth of the biomass. If the CO_2 produced during combustion were instead captured and buried, there would be a net removal of CO_2 from the atmosphere – a *negative* emission. Sequestration of bioenergy carbon could therefore play a role in reducing the peak atmospheric CO_2 concentration and in returning the CO_2 concentration to the 350–400ppmv range. The potential role of carbon sequestration in bringing about compliance with the UNFCCC is discussed in some detail in Harvey (2004).

When sequestration of 1GtC/yr per year is added to Scenario 5, the atmospheric CO_2 concentration decreases to 318ppmv by 2500, the warming has decreased from a peak of about 2.5°C to 2.8°C (for a climate sensitivity of 3.0°C), and the pH of ocean surface waters has largely recovered (see Figures 12.18 and 12.22). However, irreversible species losses and ecosystem collapse will have occurred during the intervening 500 years, the magnitude of losses depending on the climate sensitivity and how successful humans are in aiding adaptation by other species. It is not clear whether or not we can avoid irreversibly triggering the eventual collapse of the Greenland and Antarctic ice sheets during this time period. A faster rate of drawdown of atmospheric CO_2 might be possible, reducing the risk of this and other calamities. Sequestration of biomass carbon may also be needed to offset the effect of positive climate–carbon cycle feedbacks that are not included in the standard model simulations. Note, however, that sequestration of (for example) 1Gt of biomass carbon per year would offset the effect of a CO_2 emission due to positive climate–carbon cycle feedbacks of 1GtC/yr, but would not compensate for a similar emission in the form of CH_4 due to the fact that methane is 26 times stronger than CO_2 on a molecule-per-molecule basis.

At present, CCS is being widely promoted as a means to allow continued and even expanded use of coal. However, given substantial uncertainties at present as to how much CO_2 can be safely sequestered geologically, and the likely need to be able to sequester CO_2 from combustion or gasification of bioenergy crops, it would be inadvisable to sequester CO_2 from emission sources that can be eliminated in other ways. Gielen (2003) argues that carbon capture and sequestration should be preferentially applied to steel production rather than to continued coal-based electricity generation because there are comparatively inexpensive renewable energy alternatives to coal-based electricity (as confirmed in this book) but not for steel production.

12.5.1 Implied minimum rate of provision of biomass energy

In Scenarios 3 to 6, fossil fuel emissions decrease to zero between 2080 and 2120. Carbon dioxide for geological sequestration would therefore have to come from use of biofuels or would have to be directly captured from the air. As discussed in Chapter 9 (subsection 9.3.6), capture of CO_2 released from biofuels could be most easily achieved if biomass energy were gasified to produce hydrogen, which would produce a stream of almost pure CO_2 as a byproduct. This CO_2 could be collected, compressed and transported to sequestration sites at relatively little cost.

The amount of biomass energy required in order to generate a CO_2 stream of 1GtC/yr is 25.7EJ/yr (assuming biomass at 18GJ/t to be 70 per cent C). This is small compared to the amount of biomass assumed to be produced for energy purposes in the biomass-intensive scenarios (178–344EJ/yr) and so should be readily available. For the hydrogen-intensive scenarios, only about 40–65EJ/yr of biomass energy are assumed to be provided, so a much larger fraction would need to be combined with CCS in order to achieve a sequestration rate of 1GtC/yr. Given that much of the biomass energy would be from dispersed sources in both sets of scenarios, it is not clear what fraction would be amenable to CO_2 capture and storage. As noted above, we may need to sequester CO_2 at much greater rates in the future. This underlines the importance of limiting fossil fuel emissions as rapidly as possible so as to minimize the magnitude of negative emissions that will probably need to be created in the near future in order to minimize dangerously high global warming (or, at least, to minimize its duration).

It is possible that part of the sequestration requirement identified here could be met by increasing soil carbon (as discussed in Chapter 4, subsection 4.8.3) or through direct capture of atmospheric CO_2 (as discussed in Chapter 9, subsection 9.3.10). Metting et al (2001), while acknowledging that the uncertainties are large and that there are significant gaps in our understanding of the structure and dynamics of soil carbon, indicate a worldwide technical potential to sequester 5.7–8.7GtC/yr in soils for up to 50 years. In practice, 1–2GtC/yr might be achievable if there were sufficient political will to do so. If this is in addition to 1GtC/yr from geological sequestration of biomass C and from direct air capture, the atmospheric CO_2 could be drawn down at a rate of 3–4GtC/yr (1.5–2.0ppmv per year) times a factor of about 0.5 to account for weakening of the pre-existing C sinks due to the reduction of atmospheric CO_2 concentration. The result is a concentration decrease of 75–100ppmv per century.

12.6 Summary

In this chapter, scenarios for the rapid deployment of renewable energy sources have been combined with the aggressive energy efficiency scenarios developed in Volume 1. Rates of increase in C-free energy supply sufficient to eliminate fossil fuel CO_2 emissions between 2080 and 2120 have been postulated, and the rates at which new C-free power supply would need to be added and the annual material flows and energy investments required to build a C-free energy supply system at these rates have been estimated. It has been assumed that wind farms located in regions with good winds, concentrating solar thermal power in arid and semi-arid regions, and building-integrated PV in major demand centres will form the backbone of a future renewable energy system, with transportation fuels consisting either of hydrogen produced from renewable energy sources or liquid fuels produced from biomass. Existing nuclear powerplants are assumed to be retired after a 40-year operating life with no replacement of retired powerplants after 2010.

Subject to these assumptions, it is found that the annual material flows required to build up the renewable energy system are not excessive compared to current material energy flows, and that all supply components of the system quickly become a significant net source of energy. However, water could be a significant limiting factor for the biomass-intensive scenarios. The amount of water required to produce hydrogen in the hydrogen-intensive scenario is small, but if the electricity for hydrogen production by electrolysis is produced from CSTP, the indirect water requirements for the production of hydrogen would be many times larger. Operation of CSTP plants with air rather than water cooling, or desalination of seawater using less than 1 per cent of the energy output of the CSTP plants, are potential solutions. The amount of biomass required in the biomass-intensive supply scenario and the higher-demand scenarios considered here is unlikely to be available unless there is a worldwide shift to diets with low meat consumption.

In the scenarios considered here, CO_2 emissions peak at about 8.4GtC/yr (lowest case) to 11.5GtC/yr (highest case), before declining to zero by 2080–2120. The CO_2 concentration peaks at values of about 430–530ppmv. Even for the lowest emission scenario, global mean warming of 1.2–3.7°C is projected, given a likely climate sensitivity of 1.5–4.5°C. As even the lower part of this range still poses significant risks (such as the eventual collapse of the Greenland and West Antarctic ice sheets) and will probably cause significant damage (widespread collapse of coral reef ecosystems and severe water stress in many regions), the CO_2 concentration will probably have to be drawn back down as rapidly as possible. This can be accomplished through a combination of geological sequestration of CO_2 released during the combustion or gasification of bioenergy crops, biological sequestration in soils (with biochar being particularly promising) and direct capture of CO_2 from the air and its geological sequestration. If 1GtC can be removed from the atmosphere per year beginning in 2050, then the atmospheric CO_2 concentrations, global mean warming and ocean surface water pH can be returned to close to present conditions by 2500 if there are no major releases of methane or other positive climate–carbon cycle feedbacks between now and then. Otherwise, yet larger rates of CO_2 sequestration would be required but it might not be possible to counteract the effect of such feedbacks. In this case, global warming will have slipped beyond human control, with globally catastrophic consequences.

Notes

1 ANL (1981) gives the land area occupied by the nuclear powerplants in the US as 124,000 acres and the nuclear capacity as 36GW in the late 1970s. The very large land area required for nuclear powerplants probably includes a buffer area around the plant with multiple barriers for security purposes.

2 Table 4.10 indicates comparable biofuel yields today for Brazilian sugarcane ethanol or Malaysian palm oil biodiesel.

3 This table will be updated periodically after this book is published.

4 Pacific-Asia OECD, North America, Western Europe, Eastern Europe, former Soviet Union, Latin America, sub-Saharan Africa, Middle East and North Africa, centrally planned Asia and South and Pacific Asia.

5 As indicated in Table 12.5 and in Figure 4.42, these efficiencies are appropriate for the production of Fischer-Tropsch diesel or ethanol from lignocellulosic crops when all energy inputs are considered. Hoogwijk et al (2009) also postulate an increase from 0.4 today to 0.56 by 2050.

6 A capacity factor of 0.6 is appropriate for CSTP with 6 hours of thermal storage, which is desirable when electricity is to be used directly. However, thermal storage might not be needed in CSTP systems that are used to make H_2 by electrolysis of water. However, any production of hydrogen from CSTP would most likely make use of waste heat in high-temperature water vapour electrolysis, which would increase the solar-to-hydrogen efficiency by up to a factor of three. Thus, the procedure used here is still reasonable for obtaining a first estimate of the total C-free power capacity needed to produce hydrogen and to meet direct electricity requirements.

7 In reality, electrolytic hydrogen production could begin sooner in regions that become fossil fuel-free before others. As well, hydrogen could be produced at times of excess C-free electricity supply even though some electricity is generated from fossil fuels at other times.

8 For floating offshore turbines the required capacity and hence rate of installation of capacity has been divided by two in order to reflect the assumed capacity factor of 0.4 instead of 0.2 so that both wind options generate the same amount of electricity. In the case of CSTP, the material flows are for a system with six hours of thermal energy storage, which would correspond to the capacity factor of 0.6 assumed here. A detailed analysis would be required to assess the material requirements for CSTP as a function of the solar field size (which determines the extent to which sunlight is concentrated at the receiver) and the storage system size (which affects the capacity factor).

9 As discussed by Laxson et al (2006), the copper requirements in wind turbines, for example, depend strongly on the type of generator used. High-speed induction generators with steel gear boxes require about 200kg/MW, while direct-drive generators with permanent magnets (which are becoming more popular) require about 1400kg/MW. Permanent magnets also

require use of lanthanide minerals, but it does not appear that the global supply will be a constraint.

10 In some of the studies cited in earlier chapters, a factor of 3.3 (corresponding to 33 per cent electricity generation efficiency) was used instead of a factor of 2.5 (which corresponds to the 40 per cent generation efficiency that is more typical of recently built powerplants).

11 If C is the total existing capacity of a powerplant in GW and h the number of hours per year that it runs, then it produces hC GWh/yr of electricity. If r is the fractional growth rate, then rC is the new capacity that is added. If P is the payback time in years, then $PhrC$ is the energy required to add the increment of capacity rC (assuming that it also runs h hours per year). For there to be net energy production, the energy output from all the pre-existing capacity must exceed that used to add the next increment. Thus, we require that $hC > PhrC$, or $P < 1/r$.

12 For example, total consumptive water use for electricity generation in the US states of Arizona, New Mexico and Colorado is 300 million gallons per day or about 0.4 billion m^3/yr (NETL, 2008, Figure 34), and meeting future water demand for electricity generation is already a matter of concern.

13 Transfer of water by aqueduct from central to southern California requires about 2.5kWh/m^3 according to Volume 1, Table 8.1.

14 Given a hydrogen energy content of 120.2MJ/kg, an electrolysis efficiency based on electricity inputs only of 60–100 per cent and a water requirement of 2 litres/kWh, 67–100kg of water would be required for cooling purposes while producing 1kg of hydrogen from 9kg of water.

15 The electrical energy required to lift water is equal to $gM\Delta z/\eta$, where g is the acceleration due to gravity, M is the mass lifted, Δz is the distance lifted and η is the efficiency of a pump. For water requirements of 100–1000kg per kg of dry matter having a heating value of 18MJ/kg, and assuming the electricity required for pumping to be produced from the biomass itself at an efficiency of 45 per cent, the biomass energy required for pumping alone will equal the energy value of the biomass for water pumped from depths of 660–6600m. Water today is sometimes pumped from depths of 1000m (Pearce, 2004).

16 This is in contrast to nuclear power, where substantial time would be required to ramp up the ability to produce the components needed on a large scale. For example, there is only one facility in the world at present (in Japan) capable of producing the large steel pressure vessels that would be needed for a new wave of nuclear powerplant construction (Schneider et al, 2009).

17 These emission factors are derived from global primary energy and emission data for 2005.

18 A property of the logistic function is that the peak occurs at the point in time when half of the ultimate consumption has occurred. This is borne out by data for oil and coal extraction in individual regions. In the case of natural gas, it is possible that the peak occurs after a larger fraction of ultimate cumulative consumption, followed by more rapid decline after the peak than otherwise.

19 In large dense urban centres in Asia and elsewhere, car ownership rates will of necessity be lower than in the US or Europe due to lack of space, even if high levels of wealth are achieved (such as is the case today in Hong Kong) so PHEVs will be less available to serve as storage for renewable electricity.

20 Aerosol forcing varies in a non-linear manner with emissions of aerosol precursors, again as explained in Harvey and Kaufmann (2002).

21 The climate sensitivity depends on the partial derivatives with respect to temperature of various internal variables that affect net radiation (such as the amount of water vapour in the atmosphere and cloud amounts and properties). The effective climate sensitivity during the transition to a new climate gives the warming that would occur if the partial derivatives observed in a model (the only place where they can be measured) during the early stages of warming were to persist until a new, unchanging climate has been reached. Changes in cloud amounts and locations depend not only on the change in overall temperature, but also on the spatial patterns of temperature change. Because the oceans delay the warming by varying amounts in different locations, the spatial patterns of warming change over time and so do the cloud feedbacks in some models. Thus, the climate could be behaving as if it had one sensitivity for a CO_2 doubling during the early stages of warming, and then move toward a different sensitivity if GHG concentrations are stabilized and regions with delayed warming catch up to the faster-responding regions, thereby altering the cloud feedback. This surely happens to some extent; the question is whether these effects are large or small. For a more complete discussion, see Harvey (2000, Section 11.2).

22 A highly simplified Excel version of the climate–carbon cycle model used here will be available in the online material to accompany Harvey (2010). This model will be driven with energy supply and demand scenarios developed separately for OECD and non-OECD country groups.

23 The CO_2 concentration itself depends only slightly on the climate sensitivity in the standard model.

24 Temperature for the BAU scenario peaks slightly before the peak in CO_2 concentration (in 2190) rather than after the CO_2 concentration peak (in 2260) because the radiative forcing due to non-CO_2 GHGs peaks in 2130.

13

Policy Sketch and Concluding Thoughts

The purpose of this book and of Volume 1 has been to critically assess the technical potential and (where possible) the cost of achieving deep reductions in energy use through improved efficiency, on the one hand, and of rapidly deploying C-free sources of energy, on the other hand. Our focus has been on assessing the prospects of reducing CO_2 emissions rapidly enough that the atmospheric CO_2 concentration will not exceed 450ppmv. Assuming stringent reductions in emissions of other GHGs or their precursors, a CO_2 concentration of 450ppmv gives the climatic equivalent of a CO_2 doubling (from 280ppmv to 560ppmv), which, as discussed here, already poses significant risks and will entail substantial ecological and social losses. However, if CO_2 is allowed to rise above 450ppmv, the losses – both human and ecological – will be greater still.

Our key conclusion is that it is technically feasible and affordable to limit the peak atmospheric CO_2 concentration to 450ppmv through a combination of:

- family planning and other measures sufficient to produce the 'low' population scenario considered here;
- acceptance of a declining rate of growth in world average income per person, rather than continuous exponential growth (something that will probably occur in any case due to declining availability of inexpensive energy and other resources);
- an increase in the rate of improvement in the economic energy intensity from 1.1 per cent/year (the 1965–2005 average) to 2–4 per cent/year until about 2040, and decreasing thereafter; and
- installation of 17–28TW of C-free electricity generation capacity and provision of 40–98EJ/yr of biomass energy by the end of the century (for the H_2-intensive supply scenarios), or installation of 6–11TW of C-free electricity generation capacity and provision of 178–344EJ/yr of biomass energy by mid-century (for the biomass-intensive supply scenarios).

Together, these measures would permit the elimination of fossil fuel CO_2 emissions during the latter quarter of this century. Technological measures – greatly improved energy efficiency and new, C-free energy supplies – will almost certainly not be able to reduce CO_2 emissions rapidly enough if a high population growth scenario is combined with high rates of economic growth throughout the coming century.

Declining rates of economic growth do not mean unemployment or declining human welfare, or entrapment of the poorest countries in their current poverty. The low-growth scenarios considered here assume substantial growth rates throughout the coming century for non-OECD countries, combined with a slight decrease to a slight increase in per capita GDP in OECD countries (depending on the region). Low rates of growth in OECD countries will not lead to unemployment if increasing labour productivity is channelled into increasing leisure time rather than increasing consumption of goods. As discussed in Chapter 11 of Volume 1 (section 11.2), greater leisure time combined with a gradual shift to more compact, vibrant and pedestrian- and transit-oriented cities will do much to increase the quality of life while reducing the proclivity to consume.

In the following pages, broad strategies for achieving rapid deployment of C-free energy sources are outlined, along with a brief consideration of measures to stop further deforestation. Some brief comments on geo-engineering and a final wrap-up are offered.

13.1 Promoting C-free energy

Some C-free energy sources are already economically competitive based on market costs, although in most cases, market costs of C-free energy sources are greater than the market costs of fossil fuels. However, market costs do not reflect a wide variety of social and environmental costs, of which climatic change is only one. These costs are much larger for fossil fuels than for C-free energy sources. To greatly accelerate the deployment of C-free energy sources and to belatedly begin to prepare for the imminent peaking in the supply of oil and later peaking in natural gas and coal supply, deliberate government action is needed. Three overarching policy options are:

- imposition of a gradually increasing C tax, so that some of the climate costs of fossil fuels are reflected in their price;
- removal of all subsidies for fossil fuels (and for nuclear energy); and
- increasing funding of research and development related to a wide array of C-free energy options.

Additional policy options specifically for the electricity sector are:

- developing backbone high-voltage grids (HVDC or HVAC) to permit the linking of scattered regions of high-quality solar and wind resources with the major electricity demand centres;
- promoting an increasing use of C-free energy sources through requirements for minimum fractions of renewable energy (renewable portfolio standards) or through offers by the power utility to purchase C-free electricity at favourable rates (called a fixed feed-in tariff);
- production tax credits, accelerated depreciation and support for green certificates;
- system-wide surcharges to support financial incentive payments and research and development;
- removal of procedural, institutional and regulatory barriers to renewable energy (by, for example, permitting independent power producers to sell to the electricity grid, or permitting net metering of home PV systems); and
- promotion of stakeholder participation (required for public buy-in).

A more complete discussion of these and other options can be found in Aitken (2003), Holm (2005), GAC (2006), Lesser and Su (2008), Barbose et al (2008), Albadi and El-Saadany (2009) and Sovacool (2009). Barbose et al (2008) discuss the approaches used to encourage high PV system performance in 32 prominent PV incentive programmes in the US. Sovacool (2009) reviews and assesses the options for promoting renewable energy (and energy efficiency) based on interviews with 181 energy experts from 93 institutions in 13 countries. A total of 30 different policies were recommended by the interviewed experts, but four policies were repeatedly recommended: eliminating all subsidies for fossil fuels and nuclear energy, pricing electricity more accurately,[1] introduction of fixed feed-in tariffs for renewable energy, and implementation of a system benefits charge to fund public education with regard to efficient use of electricity and to fund concrete measures to use electricity more efficiently. According to Sovacool (2009), fossil fuel and nuclear subsidies in the US amount to about $28 billion/yr. The lack of a carbon tax at present is an additional subsidy to fossil fuels in that future monetizable costs are not included in the current cost of energy. As stressed by Sovacool (2009), pursuing any one of the top four policy recommendations alone will not be particularly effective. For example, removing subsidies without reforming prices still clouds the price signals seen by consumers and will make it difficult for renewable power to compete. Implementing a national feed-in tariff without promoting energy efficiency will cause over-investment in energy supply (and will make it more difficult or impossible to close the gap between demand and C-free energy supply).

The development of bioenergy systems involves complex relationships among a number of different actors. The primary barriers to more widespread use of modern bioenergy systems are primarily non-technical rather than technical. For example, the limited flexibility once energy crops are established makes farmers or landowners reluctant to pursue energy crops, production of bioenergy crops requires a change in business practice and farmers are averse to bearing all of the risk themselves. Strategies for overcoming these and other barriers are discussed by McCormick and Kåberger (2007), Mårtensson and Westerberg (2007), Gan and Yu (2008) and Carlos and Khang (2009). The Biomass Task Force (WGA, 2006) presents specific recommendations for the promotion of biomass energy in the western US. These include providing tax parity among all renewable energy sources, establishing long-term contracts pertaining to the use of biomass from government-owned land, and charging a delivery cost for biomass electricity based on the actual distance transmitted (which is invariably shorter than average) rather than based on the average distance for new power sources (as is currently done). However, as emphasized in earlier chapters, competition between land for food and land for bioenergy crops is likely to be a significant barrier unless future per capita meat consumption is low. Bioenergy should therefore be developed only in step with the freeing of current cropland and pastureland, which will be a relatively slow process. Bioenergy crash programmes must be avoided.

13.2 Research and development

Although great strides have been made in the development of C-free technologies for electricity generation, research and development (R&D) are needed in many areas in order to improve performance and bring down costs. Critical areas include:

- crystalline and thin-film PV modules;
- concentrating solar thermal electricity production;
- advanced adiabatic compressed air underground storage for use with wind and solar energy;
- further improvement in the performance of wind turbines and optimization of the design of offshore wind farms;
- high-voltage DC transmission;
- advanced gasification and cogeneration using biomass fuels;
- biofuels for transportation applications;
- enhanced (hot dry rock) geothermal systems;
- wave, tidal, tidal current and OTEC systems.

It is clear from our discussion of PV power (Chapter 2, section 2.2), concentrating solar thermal power (Chapter 2, section 2.3), wind energy (Chapter 3, sections 3.11 and 3.13) and efficient cogeneration of electricity and heat from biomass (Chapter 4, section 4.4), that these four renewable energy sources are close to the point where they can begin to take over from fossil fuels in the generation of the world's electricity. Wind energy systems are well developed technically and are already highly competitive; concentrating solar thermal electricity is the least expensive of the solar electricity options and, with thermal energy storage, provides electricity 24 hours per day; and PV power, while currently still expensive, is likely to see a factor of two or more reduction in cost during the next decade and to become competitive with peaking fossil fuel-derived electricity. PV electricity is most attractive when PV panels are integrated into the façade or roof of new buildings, as it is then subject to embodied-energy and cost credits from the displacement of conventional building materials, and can provide significant societal-scale benefits through reduced transmission and distribution costs, losses and bottlenecks. By linking dispersed wind farms, concentrating solar thermal arrays and other renewable energy options, and combining these with various storage techniques and dispatchable end-use loads, renewable energy can completely displace fossil fuels for the generation of electricity.

Given the important role that wind energy will play in a C-free electricity system, IEA (2009) recommends substantially increasing the level of R&D funding for wind energy. During the past three decades, OECD funding for wind power has fluctuated between only 1 per cent and 2 per cent of all energy R&D funding. IEA (2009) sees the potential for significant improvements in the capacity factor of wind turbines (by up to 35% of current capacity factors); the development of stronger and lighter materials that would enable larger rotors, lighter nacelles and use of less steel in towers; the development of superconductor technology for lighter and more efficient generators; and significant improvements in all areas related to offshore wind farms. Private wind turbine companies have tended to focus on short-term R&D efforts, where returns on investment are more likely, but according to IEA (2009), long-term fundamental research supported by the public sector is required. The same is surely true of other renewable energy technologies as well.

Given that renewable energy can completely displace fossil fuels for the generation of electricity, the development of 'clean' coal technologies and carbon capture and storage for coal is not recommended. Instead, the overarching policy goal should be to phase out the use of coal altogether as rapidly as possible. As discussed in Volume 1 (Chapter 2, subsection 2.5.2), the mineable coal resource is likely to be much smaller than widely believed, and coal mining itself entails significant negative environmental impacts. CCS technologies are, however, potentially useful in niche applications (such as at cement and fertilizer plants) and for application to biomass CO_2 (so as to create negative emissions). As nuclear energy is not a viable long-term C-free energy source, and it cannot be ramped up fast enough to address the need for early and large reductions in CO_2 emissions, research and development related to nuclear energy should be terminated, with the possible exception of research related to the use of nuclear powerplants to consume discarded plutonium and highly enriched uranium from nuclear weapons as the world moves (hopefully) toward much lower levels of nuclear armaments (if not eventually to complete nuclear disarmament).

In a world of limited funds and limited human scientific resources, choices need to be made while still supporting a diverse portfolio of technologies. Advocates of both nuclear energy and CCS argue that there are no supply-side alternatives to these technologies for reducing CO_2 emissions. The analysis presented in this book shows that this is decidedly not the case.

The technology dynamics literature identifies the possibility of two different future technology paths: the first, a carbon-intensive path, in which gaseous and liquid fuels are made from coal and replace conventional gaseous and liquid fuels as they are depleted (this assumes that the available coal resource is large, an assumption that is questioned in Volume 1, Chapter 2, subsection 2.5.2); and the second, a low-carbon path, in which biomass, solar, wind and other renewable energy sources gradually replace fossil fuels as conventional fuels are depleted. Because initial developments along either path will lead to cost reductions through learning-by-doing, it will be very difficult to switch from one path to another once we have started down one path. This phenomenon is referred to as 'carbon lock-in' and already characterizes the present energy system (Unruh, 2000, 2002).

Thus, if we do not want a carbon-intensive future, or if we do not want to begin travelling a path that will have to be aborted, at great cost, then governments should *not* support research and development of various processes for making gaseous and liquid fuels from coal. If the mineable coal resource is as small as suggested in Volume 1, this path will have to be aborted in any case, so the investment will largely be wasted, additional CO_2 will have been emitted (even if CCS of large point sources can be widely deployed), and the development of a sustainable energy system unnecessarily delayed.

13.3 Preserving forests and enhancing carbon sinks

Deforestation and loss of soil carbon contributed an emission of 1.6±1.1GtC/yr during the 1990s (more recent estimates are not available), compared to a fossil fuel emission of 7.2±0.3GtC/yr during 2000–2005 (Denman et al, 2007). During the 2000–2005 period, the atmospheric CO_2 increase averaged 4.1±0.1GtC/yr. As discussed in Chapter 4 (subsection 4.8.1), it is thought that 0.5–1.5GtC/yr could realistically be absorbed from the atmosphere through reforestation of degraded land or of land in need of erosion protection and flood control, through improved land management and through buildup of soil carbon. The difference between an emission source of about 2GtC/yr and a sink of up to 1.5GtC/yr is almost equal to the current rate of accumulation of CO_2 in the atmosphere. *Thus, with current rates of fossil fuel CO_2 emission, a rapid transition from net deforestation to the estimated maximum achievable rate of biomass accumulation could dramatically slow the increase in atmospheric CO_2 concentration.* Of course, once atmospheric concentration is stabilized the natural sinks (absorption of CO_2 by the oceans and terrestrial biosphere) would themselves weaken, but near-stabilization of atmospheric CO_2 concentration would provide the time for absolute reductions in fossil fuel CO_2 emissions.

As discussed in Nabuurs et al (2007), measures to reduce deforestation and the associated CO_2 emissions can be designed to be compatible with promoting sustainable development. These measures include improving institutional capacity, providing investment capital for investments in sustainable ecotourism and sustainable extractive activities, and implementing appropriate government policies and incentives along with international cooperation. Creating an additional carbon sink of 0.5–1.5GtC/yr requires funds to support reforestation in areas subject to erosion or in need of flood control and to support improvements in land management, as discussed in Chapter 4 (subsection 4.8.1). Kindermann et al (2008) estimate that reducing the global deforestation rate by 50 per cent by 2030 would require funding of $17 billion to $28 billion per year.

13.4 Geo-engineering?

As noted in Chapter 12, it is likely, even with stringent efforts to reduce CO_2 emissions, that the CO_2 concentration will shoot beyond safe and acceptable levels. Even with large negative emissions (1–2GtC/yr), it would take two centuries or more before the CO_2 concentration is brought back down to levels that are likely to be tolerable. In the meantime, the irreversible collapse of the Greenland and/or West Antarctic ice sheets might be triggered or significant release of methane from thawing permafrost soils might begin. In that case, some sort of temporary geo-engineering option – the deliberate manipulation of the earth's radiative balance to offset the effect of higher GHG gas concentrations – might buy us the needed time.

Among the options that have been considered are:

- to place sunshades into earth's orbit in order to block an amount of sunlight sufficient to offset the heating effect of the GHG buildup (which, however, changes over time);
- to spray mist into the atmosphere above the oceans so as to induce the formation of more reflective clouds; and
- to disperse sulphur into the northern polar stratosphere, mimicking the effect of occasional volcanic eruptions, thereby arresting the retreat of Arctic sea ice, the melting of the Greenland ice sheet and the thawing of permafrost.

There are significant uncertainties concerning the effectiveness and side-effects of these and other geo-engineering options, as reviewed by Vaughan and Lenton (forthcoming). Geo-engineering is defensible (if at all) *only* in the context of global efforts to rapidly reduce emissions. There are three primary reasons for this:

1. In the absence of emission reductions leading to at least stabilization of concentrations, the required geo-engineering increases indefinitely, which means that (a) unwanted side-effects increase, and (b) there is a risk of a catastrophically fast warming if the measures should ever stop (for those measures that require continuous application).
2. Geo-engineering does not deal with the problem of ocean acidification.
3. Geo-engineering reduces the incentive to deal directly with the source of the problem (an energy system that requires large use of fossil fuels).

The effort required for some geo-engineering options underlines the importance of addressing the root cause of the problem. For example, as reviewed by Vaughan and Lenton (forthcoming), a sunshade of 4.7million km^2 *in earth's orbit* would be required to offset a mere CO_2 doubling. However, an area of only 84,300km^2 *on the world's land surface* (in desert regions) would need to be covered with 10 per cent efficient PV panels in order to generate an amount of electricity equal to the total world electricity production of about 18,500TWh in 2005. Other options (such as injecting sulphur aerosols into the stratosphere) would require substantially less effort but, like all geo-engineering options, have uncertain side-effects.

In summary, research into some geo-engineering options is certainly justified, but it is too early to say whether or not geo-engineering would be a good idea. Geo-engineering is justifiable, if at all, only as a measure to temporarily mask the slowly decaying CO_2 heating effect as emissions are reduced, and most of its supporters see it only in these terms. It can in no way be justified as a substitute for stringent emissions reduction.

13.5 Concluding thoughts

The approach taken here has been to assume aggressive improvements in energy use over the coming century, to postulate a sufficiently rapid deployment of C-free energy sources that CO_2 emissions are eliminated before the end of this century (fast case) or early in the next century (slow case), and then to estimate the required material and energy flows and the required rates of construction of new PV or wind turbine factories and rates of establishment of new bioenergy plantations. The conclusion is that the required rates of deployment of C-free energy are achievable for the scenario of low growth in world GDP. The required area of new bioenergy plantations under the biomass-intensive scenario would be available for a diet with moderate rather than high (North American) average meat consumption. The high GDP scenario requires about twice the C-free energy as in the low GDP scenario, and almost 2.5 times the land area under the biomass-intensive variant. The hydrogen-intensive scenario requires much less land than the biomass-intensive scenario, and assumes the use of fuel cells in automobiles and trucks. However, material constraints are likely to severely constrain the size of a fuel cell automobile fleet. A truly sustainable future will almost certainly entail very low levels of car ownership in the global average.

Can we afford it? A more relevant question is, do we have sufficient human and physical resources to bring about the required deployment of C-free energy while retrofitting 2.5 per cent of the existing building stock per year to achieve deep savings in energy use? At present, levels of unemployment and under-employment in much of the OECD are hovering around 20 per cent, perhaps more. Similarly, 20–30 per cent of current industrial capacity is idle. We need simply redirect currently unused or underutilized human and industrial capacity to the activities needed to bring about massive reductions in energy requirements and rapid deployment of C-free energy supplies. The limiting factor could very well be the speed with which we can train a workforce skilled in these tasks.

Governments need to set the agenda, at least by setting the overall framework and rules. Careful planning of the logistical requirements associated with the transformation envisaged here is required, especially since the transition must occur quickly (three decades to largely eliminate fossil fuels from the electricity supply in most regions). The situation is analogous to the mobilization that was required of the Allied Forces at the beginning of World War II, and several useful lessons can probably be derived from the study of this mobilization (see, for example, Gilbert and Perl, 2007, concerning the transformation of the US automobile industry to the production of materials needed for the war effort). The big difference between the situation now and at the beginning of World War II is that, in the latter case, there was a clear external and immediate threat. The threat from global warming is not as immediate, nor is it as clear to the general public. To remedy this, climate scientists need to do a better job in communicating their concerns to the public and in countering the well-orchestrated campaign (in the US in particular) to discredit science in general and global warming science in particular (see Mooney, 2005, Monbiot, 2006, Michaels, 2008, Hoggan and Littlemore, 2009, and Schneider, 2010).

The analysis presented here indicates that not only is there a large renewable energy potential, sufficient to power a much more prosperous world than today's, but the transition to renewable energy can be made very rapidly if there is sufficient political will backed up with public support. The obstacles are not technical or financial in nature. The difficult task will be repairing (to the extent that this is possible) the damage done to nature and to the oceans. Elsewhere (Harvey, 2008), I have demonstrated the near-futility of trying to reverse the acidification of the oceans by adding crushed limestone in ocean regions where water that is unsaturated with respect to calcium carbonate is close to the surface. Ultimately, we will need to repair the soil in order to repair the oceans: building up soil carbon is the most effective single mechanism for reducing ocean acidification because it leads to a direct withdrawal of CO_2 from the oceans, whereas the neutralization of ocean acidification caused by inducing the dissolution of limestone in ocean water is partly counteracted by an induced flow of atmospheric CO_2 back into the ocean.

Although we are at or near the peak in the supply of oil, and the availability of natural gas is likely to peak within the next 20–30 years and that of coal by mid-century, this does not come close to implying a future Malthusian disaster of hunger, unemployment, deprivation and the collapse of civilization. Renewable energy, and solar energy in particular, is capable of providing all of the services that are presently provided by fossil fuels. However, it will be possible to meet our needs for energy only if we are highly efficient in the use of energy, particularly with regard to transportation uses – the sector where it will be most difficult to replace fossil fuels. The technologies needed to dramatically reduce our energy use already exist, and most of the technologies needed on the supply side have reached the point where rapid deployment to the needed levels can begin.

However, rapid improvements in energy efficiency and in the deployment of C-free energy supply will not be enough to provide a high probability of avoiding a climate catastrophe. Restraints in economic growth will almost certainly be needed, as only the scenario that combines slow economic growth, rapid improvement in energy efficiency and rapid deployment of C-free energy has a better than 50 per cent chance of averting positive climate–carbon cycle feedbacks that would largely take the global warming issue out of human control, with globally catastrophic consequences. However, the present world leadership is determined that measures to reduce greenhouse gas emissions should not in any way slow down economic growth. As discussed in Volume 1 (Chapter 11, section 11.8), economic growth is at present the policy objective against which all other proposals are judged. What is needed is to move climatic change and reduction of GHG emissions to the place now held by economic growth. All other policies must now be evaluated, at least in part, in terms of how they contribute to the overarching goal of reducing GHG emissions. A price needs to be set on carbon that is initially high enough, and increases at a sufficiently rapid rate, that climate-related targets – such as limiting atmospheric CO_2 concentration to no more than 450ppmv and then drawing down the concentration to 350ppmv or less – are met. How high this price needs to be depends in part on how effective governments are in facilitating the transition to much greater levels of energy efficiency (as discussed in Volume 1) and in facilitating the transition to C-free energy sources (as discussed in this volume). The economy will then adapt, with the economic impact of a high carbon tax strongly dependent on how the revenues are recycled. With economic growth no longer emphasized in government policy, and with both consumers and governments living within their financial means (and moving in the direction of living within the planet's ecological and biophysical means), economic growth will slow. The fundamental problem, as discussed by Jackson (2009, p77), is that there is very little understanding of how to manage an economy that is not growing, or that grows only slowly. Thus, a whole new body of economic analysis and theory is urgently needed.

In summary, then, we need three large transformations: a transformation to vastly greater levels of energy efficiency than at present in all end-use sectors, a rapid deployment of C-free energy sources (primarily wind and solar energy), and a whole new way of thinking that places stabilization of greenhouse gas concentrations at levels below the equivalent of a CO_2 doubling ahead of promotion of economic growth. In the rich countries, that means courageous political leadership to inform the public that we do indeed have to give up something in return for limiting global warming to a level that will still leave much that is beautiful in the world and that provides the basis for a good life for our children and grandchildren. This means reducing our consumption, use and ultimate disposal of material goods; redirecting much of our productive capacity toward public investments in high-quality public transportation and in retrofitting residential and commercial buildings to dramatically reduce their energy consumption; and paying more for energy as we shift to C-free energy sources. In developing countries, economic growth is still needed and is justified, but such growth may indeed have to be slower than in the recent past, so that it is truly sustainable and with far fewer immediate adverse impacts through pollution of air and water. If we do not do this, then we will provoke globally catastrophic consequences (such as several metres sea level rise) that will eventually negate the fruits of the economic growth that we so assiduously promote, and there is a high risk that global warming will slip beyond human control as a result of positive climate–carbon cycle feedbacks that lead to additional GHG emissions that exceed any conceivable counter-acting negative emissions that humans could create after belatedly eliminating fossil fuel emissions.

Note

1 More accurate pricing would involve either elimination of price caps and declining block structures (whereby unit costs decrease with increasing consumption), or introduction of time-of-use rates.

Appendix A

Prefixes and Conversion Factors

The following table gives the prefixes that are used with physical units in the metric system.

The conversion factors between British and metric units encountered in this book are given in Table A.2

Table A.1 *Prefixes used here with physical units in the metric system*

Prefix	Factor	Common examples
Nano	10^{-9}	nm (nanometre)
Micro	10^{-6}	μm (micrometre)
Milli	10^{-3}	mm (millimetre)
Kilo	10^{3}	kW (kilowatt), km (kilometre), kg (kilogram)
Mega	10^{6}	MW (megawatt), MJ (megajoule)
Giga	10^{9}	GW (gigawatt), GJ (gigajoule), Gt (gigatonne)
Tera	10^{12}	TW (terawatt), TJ (terajoule)
Peta	10^{15}	PJ (petajoule)
Exa	10^{18}	EJ (exajoule)

Table A.2 *Conversion factors between and among metric and British units*

British to Metric	Metric to British
Energy and power	
1 British Thermal Unit (Btu) = 1055 joules	1000 joules = 0.9486 Btu
1 calorie = 4.186 joules	
1 therm = 105.5MJ	
1 quad = 1 quadrillion (10^{15}) Btu	
1 kilowatt-hour (kWh) = 1000 watts × 3600 seconds = 3.6 megajoules = 0.0036 gigajoules	
Distance, area, volume, mass and weight	
1 foot = 0.30480 metres	1 metre = 3.28083 feet = 39.37 inches
1 mile = 5280 feet = 1.6093km	1 kilometre (km) = 0.6214 miles
$1ft^2 = 0.092904m^2$	$1m^2 = 10.7638ft^2$
1 acre ($1/640mi^2$) = 4046.87 m^2 = 0.404687ha	1ha = 100m × 100m = 2.47104 acres
$1ft^3$ = 28.32 litres = $0.02832m^3$	$1m^3$ = 1000 litres = $35.3145ft^3$
1 Imperial gallon = 4.546 litres	1 litre = 0.2198 imperial gallons
1 US gallon = 3.785 litres	1 litre = 0.2642 US gallons
1 avoirdupois pound (a unit of weight) corresponds to 0.4536kg on earth. This is what is commonly called 'pound', and is divided into 16 avoirdupois ounces ('ounces'), each corresponding to 28.35gm.	1 kilogram (a unit of mass) corresponds to 2.2046 pounds on earth

Appendix B

Computing the Embodied Energy of Manufactured Materials

The energy embodied in a material is the energy that is required to produce that material. The embodied energy in a tonne of steel, for example, involves the energy used at the steel mill. This is called the *direct* or *zeroth-order* embodied energy in the steel. However, energy is also required in extracting the raw materials used by the steel mill and in bringing them to the mill, and in manufacturing the machinery used by the mill. These would be *first-order* inputs. But the facilities where the equipment used by the mill was manufactured, or where the transportation vehicles were produced, also required inputs from an array of other industries, each of which required energy input. In other words, there are a number of *indirect* energy inputs of first order, second order, third order, and so on. Fully accounting for the embodied energy in a material requires accounting for a very large succession of linkages. This can be done in two ways: by explicitly measuring each of a large number of linkages until it is felt that the excluded higher orders are not important, or through an input–output analysis of the entire economy.

An input–output analysis requires dividing the economy into a number of sectors, which could be in the order of 100. Any one sector could, in principle, have inputs from all the other sectors and its output could be an input to all of the other sectors. All of the possible input–output relationships can be represented by an $N \times N$ matrix, where N is the number of sectors. Through manipulations of this input–output matrix, the full cascade of effects on the embodied energy of products from any one sector can be accounted for. The input–output analysis will identify a large number of linkages between individual sectors that all directly or indirectly contribute to the embodied energy of a given material. For example, the energy path tree accounting for 90 per cent of the total embodied energy in Australian residential buildings involves 592 energy paths, while analysis up to and including 12th order involves over 20 million energy paths (Treloar et al, 2001).

There are two problems with the input–output approach: (1) much of the basic data used to construct the matrix are in terms of the dollar-value (or other currency) of the inputs and outputs, and so have to be converted to energy intensities based on sector-average energy use (since energy data at a more-disaggregated level are usually not available), and (2) different products from a given sector, or different uses of the same product, are not distinguished. However, the energy use at each step in each path is, for that step, a direct energy use – something that can be evaluated in physical terms. The direct energy use at each step of a given path deduced from the input–output matrix is independent of the energy uses deduced for all the other links. Thus, the problems associated with input–output analysis can be mitigated to some extent by replacing the most important energy intensities as deduced from the input–output matrix with the correct intensities as calculated physically. The amounts of inputs into an output at a given link in the input–output analysis as determined by the input–output analysis also need to be replaced with physically determined quantities. This is referred to as a hybrid approach by

Treloar et al (2001), and in their analysis of residential buildings in Australia, 52 per cent of the energy paths in the input–output model were replaced with case-specific process-based energy intensities. For the building analysed by Treloar et al (2001), the unaltered input–output analysis indicated an embodied energy of 834.8GJ. Alteration of a portion of the energy intensities increased this to 1154.4GJ, while alteration of a portion of the energy intensities and product quantities increased the total to 1762.4GJ – a factor of two greater than indicated by the original analysis.

However, to rely entirely on a process-based approach will underestimate the true embodied energy, since interactions beyond some relatively low order will be omitted (that is, there is a truncation error). Lenzen and Dey (2000) find that, in the case of iron and steel in Australia, terminating the analysis after considering only zeroth and first-order terms leads to an estimated energy intensity of 26.05GJ/t, whereas use of an input–output matrix with a process-based analysis for the zeroth and first-order terms leads to an embodied energy of 40.05GJ/t. According to Lenzen and Treloar (2002), going as far as fourth-order inputs for wood-frame and concrete-frame construction still captures only 75 per cent of the total embodied energy; convergence to the full embodied energy is not reached until 12th-order inputs are included. In the case of the PV systems, Crawford et al (2006) find that inclusion of higher-order terms through input–output analysis increases the estimated embodied energy by about 40 per cent.

Apart from these methodological issues, there is a large range in published embodied-energy values for most materials. This is partly due to real differences in the embodied energy of the same material produced in different countries, due to differences in the energy efficiency of the manufacturing plants and in the generation of electricity, and in the distances that raw materials or final products are transported. However, some of the discrepancy is probably due to uncertainties and hence errors in some of the inputs used in the calculations.

Appendix C

Financial Parameters

In this appendix a number of parameters relevant to assessing the costs of alternative investments, or the benefits of measures that reduce energy costs, are introduced. For a more detailed discussion with examples, see Chapter 5 of Masters (2004).

Cost recovery factor and annualized investment cost

The CRF is the amount of money that must be repaid every year to exactly repay an initial load or investment over a period of N years, including interest on the amount still owing at any given time. It depends on the real interest rate expressed as a fraction, *i*, and the repayment time *N* as follows:

$$CRF = i / \left(1 - (1+i)^{-N}\right) \tag{C.1}$$

CRF times the initial investment I_o gives the fixed annual payment needed to exactly pay back the investment plus interest after *N* years. This is referred to as the *annualized investment cost.* If one starts with an initial investment, calculates the interest accumulated the first year based on the amount owing at the start of the year, subtracts $CRF \times I_o$ at the end of the year, then recomputes the annual interest at the start of the next year based on the new balance and continues in this way, one will arrive at exactly zero funds owing after *N* years.

Net present value and net present value factor

The NPV of some income or cost occurring *n* years in the future is the amount of money that would have to be invested today, at an interest rate *i*, in order to have an amount of money equal to that future income or cost after *n* years. A dollar invested now is worth $1 \times (1+i)$ dollars one year from now, $1 \times (1+i)^2$ dollars two years from now, and so on (multiplying by a factor of $1+i$ for each additional year). *X* dollars now is worth $X(1+i)^n$ dollars *n* years in the future, so *X* dollars *n* years from now is worth only $X/(1+i)^n$ dollars today.

For a string of *N* annual costs or savings of equal value spread over *N* years, the NPV of the string is equal to $N \times F_{NPV}$, where F_{NPV} is the *net present value factor* and is given by:

$$F_{NPV} = \frac{(1+i)^N - 1}{(1+i)^N i} \tag{C.2}$$

where *i* is the discount (interest) rate. The NPV factor is equal to the reciprocal of the cost recovery factor (given by C.1).

The NPV of *N* years of savings computed using F_{NPV} is the same as would be obtained if one computed the NPV of each annual savings and added them all up, where the NPV of the savings *X* in year *n* is equal to $X/(1+i)^n$.

Simple payback

If an investment in energy efficiency, for example, entails a first cost of *C* but generates an annual saving in energy costs of *S*, the simple payback time is *C/S.* This is the length of time required for the energy cost savings to pay back the original investment, ignoring interest on the initial investment while it is being paid back.

Simple rate of return

The simple rate of return is the reciprocal of the simple payback period, that is, the annual cost savings divided by the initial investment.

Internal rate of return

If an initial investment I_o generates a string of cost savings over a period of N years that has a NPV S, then the NPV of the investment is:

$$NPV = S - I_o \tag{C.3}$$

The *internal rate of return* (IRR) is the rate of interest that has to be used in calculating the NPV of all the future savings such that the NPV of the investment (given by C.3) is zero. In this case, the cost savings just pay back the initial investment plus interest (at an interest rate equal to the IRR) over the N years. There is no simple general formula for calculating the IRR; rather, it has to be determined through trial and error or through an iterative calculation procedure.

Impact of fuel price escalation different from the general rate of inflation

The interest rate used in the previous formula should always be the real (or nominal) interest rate minus the rate of inflation. This gives the same results as if the real interest were used combined with escalation of future costs and income at the rate of inflation. However, if the future rate of increase in the cost of energy is different from the general rate of inflation (which is used to reduce the effective interest rate), incorrect results will be obtained. Instead, an effective interest or discount rate i' should be used, computed as follows:

$$i' = \frac{i - e}{1 + e} \tag{C.4}$$

where e is the rate of escalation in the price of energy. For projects with a long lifespan relative to the simple rate of return, the simple return of return is a good approximation to the IRR.

Appendix D

Heating Values of Fuels and Energy Equivalents

The heating value of a fuel is the heat that is released when the fuel is combusted. The heat released depends on whether the water vapour produced by combustion is condensed (releasing latent heat) or not. The heating value in the absence of condensation is referred to as the LHV or the net heating value, while the heating value in the presence of condensation is referred to as the HHV or gross heating value. The difference between the two depends on the hydrogen content of the fuel, as this determines how much water vapour is produced in relation to the energy released by combustion. LHV and HHV values, the percentage of hydrogen on a mass basis, and the ratio of LHV to HHV are given in Table D.1 for a variety of fuels.

Table D.1 *LHV, HHV and percentage hydrogen by mass and density of various liquid and gaseous fuels*

	LHV (MJ/kg)	HHV (MJ/kg)	Mass% H	LHV/HHV	Density (kg/L)	Source
			Liquid fuels			
Gasoline (C_5H_{12} to $C_{12}H_{26}$)						
Average $C_{7.14}H_{14.28}$, 85.5% C	44.0	44.4	15.3–16.7	0.99	0.737	Demirbas (2005)
Diesel ($C_{10}H_{22}$ to $C_{15}H_{32}$)						
Average $C_{13.57}H_{27.14}$, 87.0% C	43.0	45.9	15.1–15.5	0.94	0.85	Farrell et al (2006), wikipedia (density)
Kerosene	43.1	46.2		0.92	0.78–0.81	wikipedia
Methanol (CH_3OH) (37.5% C)	19.9	21.1	12.5	0.94		Demirbas (2005)
Ethanol (C_2H_5OH)	26.7	27.6	13	0.97	0.796	Demirbas (2005)
			Natural gas, natural gas liquids and hydrogen			
Methane (CH_4) (75% C)	50.02	55.50	25	0.900	0.7174	Hiete et al (2001)
Ethane (C_2H_6)	47.48	51.87	20	0.915	1.3553	Hiete et al (2001)
Propane (C_3H_8)	46.35	50.34	18.1	0.921	2.0102	Hiete et al (2001)
n-Butane (C_4H_{10})	45.71	49.50	17.2	0.924	2.7030	Hiete et al (2001)
Iso-butane (C_4H_{10})	45.62	49.40	17.2	0.923	2.6912	Hiete et al (2001)
Hydrogen (H_2)	120.2	141.9	100	0.85	0.08988[a] 70.8[b]	

Note: [a] As a gas at STP. Under compression, density would increase in proportion to the pressure. [b] As a liquid at −253°C.

Appendix E

Websites with More Information

International Energy Agency programmes related to C-free energy

PV power systems	www.iea-pvps.org
Solar Heating and Cooling Programme	www.iea-shc.org
District heating and cooling	www.iea-dhc.org
Energy conservation through energy storage	www.iea-eces.org

Solar energy

Solar Millennium	www.mileniosolar.com (German, Spanish)
Solúca Energia S.A.	www.solucar.es
Solargenix Energy	www.solargenix.com
National Renewable Energy Laboratory, US	www.nrel.gov
Solar Paces	www.solarpaces.org
Urban-scale PV systems	www.pvupscale.org
Sandia National Laboratory, US	www.energylan.sandia.gov/sunlab/documents.htm
PV Resources	www.pvresources.com (contains a listing of the largest 600 PV plants in the world, along with photographs)
PV Cool Build	www.pvcoolbuild.com (provides guidance on how to minimize the temperature of building-integrated PV modules)
European PV Industry Association (EPIA)	www.epia.org
European Solar Thermal Industry Federation (ESTIF)	www.estif.org

Wind energy

Associations

World Wind Energy Association	www.wwindea.org
American Wind Energy Association	www.awea.org
Canadian Wind Energy Association	www.canwea.ca
Chinese Wind Energy Association	www.cwea.org.cn (in Chinese only)
COWRIE (Collaborative Offshore Wind Research into the Environment)	www.offshorewindfarms.co.uk
Danish Wind Industry Association	www.windpower.org
European Wind Energy Association	www.ewea.org
Global Wind Energy Council	www.gwec.net
Offshore Wind Energy	www.offshorewindenergy.org

The Danish Wind Industry Association has an excellent online guided tour of wind energy that includes videos. Go to www.windpower.org and look for 'Wind Know-How'.

The European Wind Energy Association has a monthly publication, *Wind Directions*, that can be accessed free online. Papers presented at the annual European Wind Energy Conference are available online from 2004 onward at websites of the form www.ewecYEARinfo, where YEAR = 2004 or later years.

Major turbine manufacturers

See Table 3.2.

Geothermal energy

International Geothermal Association http://iga.igg.cnr.it/index.php
Geo-heat Center, Oregon Institute of Technology http://geoheat.oit.edu
Geothermal Laboratory, Southern Methodist University, Dallas http://smu.edu/geothermal

Appendix F

Software Tools for the Analysis of Renewable Energy

The US NREL (www.nrel.gov) has developed an analysis tool called the Hybrid Optimization Model for Electric Renewables (HOMER). HOMER is a computer model that simplifies the task of evaluating design options for both off-grid and grid-connected power systems for remote, standalone and distributed generation applications. It can look at combinations of PV, wind, batteries, micro-hydro and various engine generators (reciprocating engines, microturbines or fuel cells) powered by a variety of fuels, including biomass. HOMER identifies the least-cost system for particular applications by simulating the hourly performance of different system configurations and ranking them by net present cost.

Natural Resources Canada (NRCan) has developed a series of free tools called RETScreen that is available in 35 languages at www.retscreen.net. It contains Excel-based modules for:

- passive solar air heating;
- active solar space heating;
- solar domestic hot water;
- solar PV;
- wind energy;
- small hydro;
- biomass heating;
- ground-source heat pumps.

Each module contains calculations of the energy production, CO_2 emissions and costs as well as a financial analysis. The software provides links to a global database of climate and solar irradiance data developed by NASA (and directly available at the NASA Surface Meteorology and Solar Energy website http://eosweb.larc.nasa.gov/sse/RETScreen). The website also provides resource maps for renewable energy for individual countries and regions (developed in collaboration with local organizations), as well as a global wind resource map produced by 3TIER based on a ten-year simulation with a global reanalysis model at 5km resolution.

Another useful and free tool is the Global Emission Model for Integrated Systems (GEMIS), developed by the Öko Institute in Germany (www.oeko.de). GEMIS performs a full lifecycle analysis of the energy use and environmental impacts fossil fuel and biomass energy supply systems. Input data required for the analysis are available for 30 countries.

References

ABB (Asea Brown Boveri) (2006) 'ABB opens era of power superhighways', press release, 15 November 2006, www.abb.com/cawp/seitp202/40b621aafd3db79cc12572 25002fcd3d.aspx

Abdel-Dayem, A. M. and Mohamad, M. A. (2001) 'Potential of solar energy utilization in the textile industry – a case study', *Renewable Energy*, vol 23, pp685–694

Abella, M. A. and Chenlo, F. (2004) 'Choosing the right inverter for grid-connected PV systems', *Renewable Energy World*, vol 7, no 2, pp132–146

Accurex Environmental Corporation, Edward, Aul and Associates and E. H. Pechan and Associates (1993) *Emission Factor Documentation for AP-42 Section 1.1, Bituminous and Subbituminous Coal Combustion*, U.S. Environmental Protection Agency, Contract no 68-DO-00120

Achten, W. M. J., Verchot, L., Franken, Y. J., Mathijs, E., Singh, V. P., Aerts, R. and Muys, B. (2008) '*Jatropha* bio-diesel production and use', *Biomass and Bioenergy*, vol 32, pp1063–1084

Adams, E. E., Caulfield, A. J., Herzog, H. J. and Auerbach, D. I. (1997) 'Impacts of reduced pH from ocean CO_2 disposal: Sensitivity of zooplankton mortality to model parameters', *Waste Management*, vol 17, pp375–380

AEC (Architectural Energy Corporation) (2002) *Energy Design Brief: Building Integrated Photovoltaics*, www.archenergy.com

AEP (American Electric Power) (2007) *Interstate Transmission Vision of Wind Integration*, www.aep.com/about/ i765 project/docs/WindTransmissionVisionWhite Paper.pdf

Agnolucci, P. (2007) 'Hydrogen infrastructure for the transport sector', *International Journal of Hydrogen Energy*, vol 32, pp3526–3544

Ahearne, J. F. (1997) 'Radioactive waste: The size of the problem', *Physics Today*, June, pp24–29

Åhman, M. (2001) 'Primary energy efficiency of alternative powertrains in vehicles', *Energy*, vol 26, pp973–989

Ahmed, N. A., Miyatake, M. and Al-Othman, A. K. (2008) 'Power fluctuations suppression of stand-alone hybrid generation combining solar photovoltaic/wind turbine and fuel cell systems', *Energy Conversion and Management*, vol 49, pp2711–2719

Airbus (2003) *Liquid Hydrogen Fuelled Aircraft – System Analysis. Final Technical Report (Publishable Version)*, Airbus Deutschland GmbH (Project Coordinator) Project No GRd1-1999-10014, www.aero-net.org

Aitken, D. W. (2003) *Transitioning to a Renewable Energy Future*, International Solar Energy Society, http:// whitepaper.ises.org

Albadi, M. H. and El-Saadany, E. F. (2009) 'The role of taxation policy and incentives in wind-based distributed generation projects viability: Ontario case study', *Renewable Energy*, vol 34, pp2224–2233

Albright, D. and Feiveson, H. A. (1988) 'Plutonium recycling and the problem of nuclear proliferation', *Annual Review of Energy*, vol 13, pp239–265

Alsema, E. A. and de Wild-Scholten, M. J. (2007) 'Reduction of environmental impacts in crystalline silicon photovoltaic technology: An analysis of driving forces and opportunities', www.ecn.nl

Alsema, E. A., de Wild-Scholten, M. J. and Fthenakis, V. M. (2006) 'Environmental impacts of PV electricity generation – A critical comparison of energy supply options', presented at the 21st European Photovoltaic Solar Energy Conference, Dresden, Germany, 4–8 September, www.ecn.nl

Alvarado, S., Maldonado, P., Barrios, A. and Jauqes, I. (2002) 'Long term energy-related environmental issues of copper production', *Energy*, vol 27, pp183–196

Amigun, B. and von Blottnitz, H. (2007) 'Investigation of scale economies for African biogas installations', *Energy Conversion and Management*, vol 48, pp3090–3094

Andersson, B. A. (2000) 'Materials availability for large-scale thin-film photovoltaics', *Progress in Photovoltaics*, vol 8, pp61–76

Andén, L. (2003) *Solar Installations: Practical Applications for the Built Environment*, James & James, London

Ando, S., Meunchang, S., Thippayarugs, S., Prasersak, P., Matsumoto, N. and Yoneyama, T. (2001) *Sustainability of Sugarcane Production Evaluated Based an N_2 Fixation and Organic Matter Cycle*, JIRCAS Annual Report Topic 3, www.jircas.affrc.go.jp

André, L., Audigane, P., Azaroual, M. and Menjoz, A. (2007) 'Numerical modeling of fluid-rock chemical interactions at the supercritical CO_2-liquid interface during CO_2 injection into a carbonate reservoir, the Dogger aquifer (Paris Basin, France)', *Energy Conversion and Management*, vol 48, pp1782–1797

ANL (Argonne National Laboratory) (1981) *Energy and Land Use*, DOE/EV/10154-5, Argonne National Laboratory, Argonne, Illinois

Anonymous (2004) 'Origin Energy constructing silver solar manufacturing plant', *Renewable Energy World*, vol 7, no 1, p14

Anonymous (2009a) 'First Solar breaks US$1/W barrier', *Renewable Energy World*, vol 12, no 2, p9

Anonymous (2009b) 'AMSC in 10 MW superconducting wind turbine development with US DOE', *Renewable Energy World*, vol 12, no 1, p9

Appleyard, D. (2009) 'Light cycle: Recycling PV materials', *Energy World*, vol 12, no 2, pp28–35

Apps, J. A., Zhang, Y., Zheng, L., Fu, T. and Birkholzer, J. Y. (2009) 'Identification of the thermodynamic controls defining the concentration of hazardous elements in potable ground waters and the potential impact of increasing carbon dioxide partial pressure', *Energy Procedia*, vol 1, pp1917–1924

Archer, C. L. and Jacobson, M. Z. (2002) 'The regularity and spatial distribution of US windpower', *Journal of Geophysical Research – Atmospheres*

Archer, C. L. and Jacobson, M. Z. (2003) 'Spatial and temporal distributions of US winds and windpower at 80 m derived from measurements', *Journal of Geophysical Research–Atmospheres*, vol 108, no D9, 4289, doi: 10.1029/2002JD002076

Archer, C. L. and Jacobson, M. Z. (2005) 'Evaluation of global wind power', *Journal of Geophysical Research–Atmospheres*, vol 110, D12110, doi:10.1029/2004 JD005462

Ardente, F., Beccali, M., Cellura, M. and Lo Brano, V. (2005) 'Life cycle assessment of a solar thermal collector', *Renewable Energy*, vol 30, pp1031–1054

Ardente, F., Beccali, M., Cellura, M. and Mistretta, M. (2008) 'Building energy performance: A LCA case study of kenaf-fibres insulation board', *Energy and Buildings*, vol 40, pp1–10

Aringhoff, R. (2002) 'Proyectos Andasol, Plantas Termosolares de 50 MW', Presentation at the IEA Solar Paces 62nd Exco Meetings Host Country Day

Aringhoff, R. and Brakmann, G. (2003) *Solar Thermal Power 2020: Exploiting the Heat from the Sun to Combat Climate Change*, European Solar Thermal Power Industry Association (ESTIA) and Greenpeace International, Birmingham and Amsterdam

Arrieta, F. R. P., Teixeira, F. N., Yáñez, E., Lora, E. and Castillo, E. (2007) 'Cogeneration potential in the Columbian palm oil industry: Three case studies', *Biomass and Bioenergy*, vol 31, pp503–511

Ashok, S. (2007) 'Optimised model for community-based hybrid energy system', *Renewable Energy*, vol 32, pp1155–1164

Ashwell, T. (2004) *Innovative Design Approaches for Large Wind Turbine Blades: Final Report*, report No SAND2004-0074, Sandia National Laboratories, Albuquerque, New Mexico

Athanasiou, C., Coutelieris, F., Vakouftsi, E., Skoulou, V., Antonakou, E., Marnellos, G. and Zabaniotou, A. (2007) 'From biomass to electricity through integrated gasification/SOFC system-optimization and energy balance', *International Journal of Hydrogen Energy*, vol 32, pp337–342

Atmaca, I. and Yigit, A. (2003) 'Simulation of solar-powered absorption cooling system', *Renewable Energy*, vol 28, pp1277–1293

Aubrey, C. (2008) 'Britannia to rule the waves? UK ambitions for offshore wind', *Renewable Energy World*, vol 11, no 1, pp74–81

Auerbach, D. I., Caulfield, J. A., Adams, E. E. and Herzog, H. J. (1997) 'Impacts of ocean CO_2 disposal on marine life: 1. A toxicological assessment integrating constant-concentration laboratory assay data with variable-concentration field exposure', *Environmental Modeling Assessment*, vol 2, pp333–343

AWEA (American Wind Energy Association) (2009) *Annual Wind Industry Report, Year Ending 2008*, www.awea.org

AWEA/SEIA (American Wind Energy Association/Solar Energy Industries Association) (2009) *Green Power Superhighways, Building a Path to America's Clean Energy Future*, www.awea.org

Bachelet, D., Neilson, R. P., Lenihan, J. M. and Drapek, R. J. (2001) 'Climate change effects on vegetation distribution and carbon budget in the United States', *Ecosystems*, vol 4, pp164–185

Bağci, B. (2009) 'Towards a zero energy island', *Renewable Energy*, vol 34, pp784–789

Bahaj, A. S. (2003) 'Photovoltaic roofing: Issues of design and integration into buildings', *Renewable Energy*, vol 28, pp2195–2204

Bailis, R., Ezzati, M. and Kammen, D. M. (2003) 'Greenhouse gas implications of household energy technology in Kenya', *Environmental Science and Technology*, vol 37, pp2051–2059

Bak, T., Nowotny, J., Rekas, M. and Sorrell, C. C. (2002) 'Photo-electrochemical hydrogen generation from water using solar energy: Materials-related aspects', *International Journal of Hydrogen Energy*, vol 27, pp991–1022

Balaras, C. A., Argiriou, A. A., Michel, E. and Henning, H. M. (2003) 'Recent activities on solar air conditioning', *ASHRAE Transactions*, vol 109, part 1, pp251–260

Balaras, C. A., Grossman, G., Henning, H. M., Ferreira, C. A. I., Podesser, E., Wang, L. and Wiemken, E. (2007) 'Solar air conditioning in Europe – an overview', *Renewable and Sustainable Energy Reviews*, vol 11, pp299–314

Barbier, E. (2002) 'Geothermal energy technology and current status: and overview', *Renewable and Sustainable Energy Reviews*, vol 6, pp3–65

Barbose, G., Wiser, R. and Bolinger, M. (2008) 'Designing PV incentive programs to promote performance: A review of current practice in the US', *Renewable and Sustainable Energy Reviews*, vol 12, pp960–998

Barnett, A. et al (2009) 'Very high efficiency solar cell modules', *Progress in Photovoltaics: Research and Applications*, vol 17, pp75–83

Barry, R. G. and Chorley, R. J. (1982) *Atmosphere, Weather, and Climate*, Methuen, London

Battjes, J. J. (1994) *Global Options for Biofuels from Plantations According to IMAGE Simulations*, Interfacultaire Vakgroep Energie en Milieukunde (IVEM), Rijksuniversiteit Groningen

Beck, P. W. (1999) 'Nuclear energy in the twenty-first century: Examination of a contentious subject', *Annual Review of Energy and Environment*, vol 24, pp113–137

Beck, S. and Haarmeyer, D. (2009) 'The upside in the downturn', *Renewable Energy World*, vol 12, no 2, pp89–97

Bena, B. and Fuller, R. J. (2002) 'Natural convention solar dryer with biomass back-up heater', *Solar Energy*, vol 72, no 1, pp75–83

Benson, S. et al (2005) 'Underground geological storage', in *IPCC Special Report on Carbon Dioxide Capture and Storage*, World Meteorological Organization, Geneva

Berger, A. (1978) 'Long-term variations in daily insolation and Quaternary climatic change', *Journal of Atmospheric Science*, vol 35, pp2362–2367

Bergland, M. and Börjesson, P. (2006) 'Assessment of energy performance in the life-cycle of biogas production', *Biomass and Bioenergy*, vol 30, pp254–266

Bernal-Austín, J. L. and Dufo-López, R. (2009) 'Simulation and optimization of stand-alone hybrid renewable energy systems', *Renewable and Sustainable Energy Reviews*, vol 13, pp2111–2118

Berndes, G. (2002) 'Bioenergy and water – the implications of large-scale bioenergy production for water use and supply', *Global Environmental Change*, vol 12, pp253–271

Berndes, G., Hoogwijk, M. and van den Broek, R. (2003) 'The contribution of biomass in the future global energy supply: A review of 17 studies', *Biomass and Bioenergy*, vol 25, pp1–28

Berry, G. D., Martinez-Frias, J., Espinsoa-Loza, F. and Aceves, S. M. (2003a) 'Hydrogen production', *Encyclopedia of Energy*, Elsevier, vol 3, pp253–265

Berry, G. D., Martinez-Frias, J., Espinsoa-Loza, F. and Aceves, S. M. (2003b) 'Hydrogen storage and transportation', *Encyclopedia of Energy*, Elsevier, vol 3, pp267–281

Bertani, R. (2005) 'World geothermal power generation in the period 2001–2005', *Geothermics*, vol 34, pp651–690

Bhattacharjee, Y. (2007) 'Turning ocean water into rain', *Science*, vol 316, pp1837–1838

BHP Billiton (2004) *Olympic Dam Operations, Environmental Data*, http://hsecreport.bhpbilliton.com/wmc/2004/performance/odo/data/index.htm

Biasutti, M. and Giannini, A. (2006) 'Robust Sahel drying in response to late 20th century forcings', *Geophysical Research Letters*, vol 33, L11706, doi:10.1029/2006GL026067

Birkholzer, J., Zhou, Q. and Tsang, C.-F. (2009) 'Large-scale impact of CO_2 storage in saline aquifers: A sensitivity study of pressure response in stratified systems', *International Journal of Greenhouse Gas Control*, vol 3, pp181–194

Bjorndalen, N., Mustafiz, S. and Islam, M. R. (2003) 'High temperature solar furnace: Current applications and future potential', *Energy Sources*, vol 25, pp153–159

Blanco, J., Malato, S., Fernández-Ibañez, P., Alarcón, Gernjak, W. and Maldonado, M. I. (2009) 'Review of feasible solar energy applications to water processes', *Renewable and Sustainable Energy Reviews*, vol 13, pp1437–1445

Blankinship, S. (2007) 'Post combustion carbon capture', *Power Engineering*, March, p64

Blok, K., Williams, R. H., Katofsky, R. E. and Hendricks, C. A. (1997) 'Hydrogen production from natural gas, sequestration of recovered CO_2 in depleted gas wells and enhanced natural gas recovery', *Energy*, vol 22, pp161–168

Bodansky, D. (2004) *Nuclear Energy: Principles, Practices, and Prospects*, Second Edition, Springer, New York

Bodansky, D. (2006) 'Reprocessing spent nuclear fuel', *Physics Today*, December 2006, pp80–81

Bokhoven, T. P., Van Dam, J. and Kratz, P. (2001) 'Recent experience with large solar thermal systems in the Netherlands', *Solar Energy*, vol 71, pp347–352

Bollinger, M. and Wiser, R. (2008) 'Surpassing expectations: State of the US wind power market', *Renewable Energy World*, vol 11, no 4, pp120–133

Börjesson, P. (1999) 'Environmental effects of energy crop cultivation in Sweden – I: Identification and quantification', *Biomass and Bioenergy*, vol 16, pp137–154

Börjesson, P. and Berglund, M. (2007) 'Environmental systems analysis of biogas systems – Part II: The environmental impact of replacing various reference systems', *Biomass and Bioenergy*, vol 31, pp326–344

Börjesson, P. and Berndes, G. (2006) 'The prospects for willow plantations for wastewater treatment in Sweden', *Biomass and Bioenergy*, vol 30, pp428–438

Bossel, U. (2006) 'Does a hydrogen economy make sense', *Proceedings of the IEEE*, vol 94, pp1826–1837, www.efcf.com

Botha, T. and von Blottnitz, H. (2006) 'A comparison of the environmental benefits of bagasse-derived electricity and fuel ethanol on a life-cycle basis', *Energy Policy*, vol 34, pp2654–2661

Boud, R. (2002) *Status and Research and Development Priorities 2003, Wave and Marine Current Energy*, International Energy Agency, Implementing Agreement on Ocean Energy Systems, www.iea-oceans.org

BP (British Petroleum) (2007) *BP Statistical Review 2007*, www.bp.com

Boyle, G. (1996) 'Solar photovoltaics', in G. Boyle (ed) *Renewable Energy, Power for a Sustainable Future*, Oxford University Press, Oxford

Bradford, T. and Maycock, P. (2007) 'PV market update: Demand grows quickly and supply races to catch up', *Renewable Energy World*, vol 10, no 4, pp60–74

Bradshaw, J., Boreham, C. and La Pedalina, F. (2005) 'Storage retention time of CO_2 in sedimentary basins; examples from petroleum systems', in *Proceedings of the 7th International Conference on Greenhouse Gas Control Technologies (GHGT-7)*, 5–9 September, Vancouver, Canada, vol 1, pp541–550

Breton, S.-P. and Moe, G. (2009) 'Status, plans and technologies for offshore wind turbines in Europe and North America', *Renewable Energy*, vol 34, pp646–654

Brewer, G. D. (1991) *Hydrogen Aircraft Technology*, CRC Press, Boca Raton, Florida

Brewer, P. G., Peltzer, E. T., Friederich, G. and Rehder, G. (2002) 'Experimental determination of the fate of a CO_2 plume in seawater', *Environmental Science and Technology*, vol 36, pp5441–5446

Brovkin, V., Claussen, M., Driesschaert, E., Fichefet, T., Kicklighter, D., Loutre, M. F., Matthews, H. D., Ramankutty, N., Schaeffer, M. and Sokolov, A. (2006) 'Biogeophysical effects of historical land cover changes simulated by six Earth system models of intermediate complexity', *Climate Dynamics*, vol 26, pp587–600

Brown, G. (1996) 'Geothermal energy', in G. Boyle (ed) *Renewable Energy, Power for a Sustainable Future*, Oxford University Press, Oxford

Brown, L. F. (2001) 'A comparative study of fuels for on-board hydrogen production for fuel-cell-powered automobiles', *International Journal of Hydrogen Energy*, vol 26, pp381–397

Brown, M. A., Levine, M. D., Romm, J. P., Rosenfeld, A. H. and Koomey, J. G. (1998) 'Engineering-economic studies of energy technologies to reduce greenhouse emissions: Opportunities and challenges', *Annual Review of Energy and the Environment*, vol 23, pp287–385

BTM Consult (2001) *International Wind Energy Development, World Market Update 2000*, www.btm.dk

Buck, S. (1996) 'Decommissioning nuclear facilties', in P. D. Wilson (ed) *The Nuclear Fuel Cycle, From Ore to Waste*, Oxford University Press, Oxford

Buddhi, D., Sharma, S. D. and Sharma, A. (2003) 'Thermal performance evaluation of a latent heat storage unit for late evening cooking in a solar cooker having three reflectors', *Energy Conversion and Management*, vol 44, pp809–817

Bullough, C., Gatzen, C., Jakiel, C., Koller, M., Nowi, A. and Zunft, S. (2004) 'Advanced adiabatic compressed air energy storage for the integration of wind energy', in *Proceedings of the European Wind Energy Conference, EWEC 2004*, 22–25 November, London, www.2004ewec.info

Burgess, G. and Fernández-Velasco, J. G. (2007) 'Materials, operational energy inputs, and net energy ratio for photobiological hydrogen production', *International Journal of Hydrogen Energy*, vol 32, pp1225–1234

Bustamante, M. M. M., Melillo, J., Conner, D. J., Hardy, Y., Lambin, E., Lotze-Campen, H., Ravindranath, N. H., Searchinger, T., Tschirley, J. and Watson, H. (2009) 'What are the final land limits?', in R. W. Howarth and S. Bringezu (eds) *Biofuels: Environmental Consequences and Interactions with Changing Land Use*, UNESCO and UNEP, http://cip.cornell.edu/biofuels

Butler, D. (2004) 'Nuclear power's new dawn', *Nature*, vol 429, pp238–240

Byrer, C. W. and Guthrie, H. D. (1999) 'Coal deposits: Potential geological sink for sequestering carbon dioxide emissions from power plants', in P. Riemer, B. Eliasson and A. Wokaun (eds) *Greenhouse Gas Control Technologies*, Elsevier Science, New York

Byrne, J., Zhou, A., Shen, B. and Hughes, K. (2007) 'Evaluating the potential of small-scale renewable energy options to meet rural livelihoods: A GIS- and lifecycle cost-based assessment of western China's options', *Energy Policy*, vol 35, pp4391–4401

Caamaño-Martín, E., Laukamp, H., Erge, T., Thornycroft, J., de Moor, H. and Cobben, S. (2008) 'Interaction between photovoltaic distributed generation and electricity networks', *Progress in Photovoltaics: Research and Applications*, vol 16, pp 629–643

Caldeira, K. and Rau, G. H. (2000) 'Accelerating carbonate dissolution to sequester carbon dioxide in the ocean: Geochemical implications', *Geophysical Research Letters*, vol 27, pp225–228

Caldeira, K., Herzog, H. J. and Wickett, M. E. (2001) 'Predicting and evaluating the effectiveness of ocean carbon sequestration by direct injection', First National Conference on Carbon Sequestration, Washington DC, p14–17

Cannell, M. G. R. (2003) 'Carbon sequestration and biomass energy offset: Theoretical, potential and achievable capacities globally, in Europe and the UK', *Biomass and Bioenergy*, vol 24, pp97–116

CA-OWEE (Concerted Action on Offshore Wind Energy in Europe) (2001) *Offshore Wind Energy, Ready to Power a Sustainable Europe, Final Report*, Delft University Wind Energy Research Institute, Duwind, www.offshorewind energy.org

Carlos, R. M. and Khang, D. B. (2009) 'A lifecycle-based success framework for grid-connected biomass energy projects', *Renewable Energy*, vol 34, pp1195–1203

Carpentieri, M., Corti, A. and Lombardi, L. (2005) 'Life cycle assessment (LCA) of an integrated biomass gasification combined cycle (IBGCC) with CO_2 removal', *Energy Conversion and Management*, vol 46, pp1790–1808

Carraretto, C., Macor, A., Mirandola, A., Stoppato, A. and Tonon, S. (2004) 'Biodiesel as alternative fuel: Experimental analysis and energetic evaluations', *Energy*, vol 29, pp2195–2211

Carta, J. A. and Ramírez, P. (2007) 'Analysis of two-component mixture Weibull statistics for estimation of wind speed distributions', *Renewable Energy*, vol 32, pp518–531

Carvalho, J. F. and Sauer, I. L. (2009) 'Does Brazil need new nuclear power plants?', *Energy Policy*, vol 37, pp1580–1584

Cavallo, A. J. (1995) 'High-capacity factor wind energy systems', *Journal of Solar Energy Engineering*, vol 117, pp137–143

Cavallo, A. J. (2007) 'Controllable and affordable utility-scale electricity from intermittent wind resources and compressed air energy storage (CAES)', *Energy*, vol 32, pp120–127

Charalambous, P. G., Maidment, G. G., Kalogirou, S. A. and Yiakoumwtti, K. (2007) 'Photovoltaic thermal (PV/T) collectors: A review', *Applied Thermal Engineering*, vol 27, pp275–286

Charlier, R. H. (2003) 'Sustainable co-generation from the tides: A review', *Renewable and Sustainable Energy Reviews*, vol 7, pp187–213

Chiu, Y.-W., Walseth, B. and Suh, S. (2009) 'Water embodied in bioethanol in the United States', *Environmental Science and Technology*, vol 43, pp2688–2692

Cho, A. (2007) 'Catalyzing the emergence of a practical biorefinery', *Science*, vol 315, pp795

Chow, T. T., Hand, J. W. and Strachan, P. A. (2003) 'Building-integrated photovoltaic and thermal applications in a subtropical hotel building', *Applied Thermal Engineering*, vol 23, pp2035–2049

Chung, M., Park, J. and Yoon, H. (1998) 'Simulation of a central solar heating system with seasonal storage in Korea', *Solar Energy*, vol 64, pp163–178

CISAC (Committee on International Security and Arms Control) (1994) *Management and disposition of excess weapons plutonium*, US National Academy Press, Washington

Clément, A., McCullen, P., Falcão, A., Fiorentino, A., Gardner, F., Hammarlund, K., Lemonis, G., Lewis, T., Nielsen, K., Petroncini, S., Pontes, M. T., Schild, P., Sjöström, B., Sørensen, H. C. and Thorpe, T. (2002) 'Wave energy in Europe: Current status and perspectives', *Renewable and Sustainable Energy Reviews*, vol 6, pp405–431

Cleveland, C. J. (2005) 'Net energy from the extraction of oil and gas in the United States', *Energy*, vol 30, pp769–782

Cleveland, C. J. and Ruth, M. (1995) 'Interconnections between depletion of minerals and fuels: The case of copper production in the United States', *Energy Sources*, vol 18, pp355–373

Clifton-Brown, J. C., Bruer, J. and Jones, M. B. (2007) 'Carbon mitigation by the energy crop, *Miscanthus*', *Global Change Biology*, vol 13, pp2296–2307

Cochran, T. B. and Paine, C. E. (1995) *The Amount of Plutonium and Highly-Enriched Uranium Needed for Pure Fission Nuclear Weapons*, Natural Resources Defense Council, New York

Cooper, M. (2009) *The Economics of Nuclear Reactors: Renaissance or Relapse?*, Institute for Energy and the Environment, Vermont Law School, www.neimagazine.com/journals/Power/NEI/August_2009/attachments/Cooper%20Report%20on%20Nuclear%20Economics%20FINAL.pdf

Corbière-Nicollier, T., Laban, B. G., Lundquist, L., Leterrier, Y., Månson, J. A. E. and Jolliet, O. (2001) 'Life cycle assessment of biofibres replacing glass fibres as reinforcement in plastics', *Resources, Conservation and Recycling*, vol 33, pp267–287

Corti, A., Fiaschi, D. and Lombardi, L. (2004) 'Carbon dioxide removal in power generation using membrane technology', *Energy*, vol 29, pp2025–2043

Cowan, J. (2008) 'Wind chill: Losing the PR battle over wind power', *The National Post*, 7 November, www.nationalpost.com/most_popular/story.html?id=941547

Cowie, A. L. and Gardner, W. D. (2007) 'Competition for the biomass resource: Greenhouse impacts and implications for renewable energy incentive schemes', *Biomass and Energy*, vol 31, pp601–607

Cowie, A. L., Smith, P. and Johnson, D. (2006) 'Does soil carbon loss in biomass production systems negate the greenhouse benefits of bioenergy?', *Mitigation and Adaptation Strategies for Global Change*, vol 11, pp979–1002

Cox, P. M., Betts, R. A., Collins, M., Harris, P. P., Huntingford, C. and Jones, C. D. (2004) 'Amazonian forest dieback under climate-carbon cycle projections for the 21st century', *Theoretical Applied Climatology*, vol 78, pp137–156

Crawford, R. H. and Treloar, G. J. (2004) 'Net energy analysis of solar and conventional domestic hot water systems in Melbourne, Australia', *Solar Energy*, vol 76, pp159–163

Crawford, R. H., Treloar, G. J., Fuller, R. J. and Bazilian, M. (2006) 'Life-cycle energy analysis of building integrated photovoltaic systems (BiPVs) with heat recovery unit', *Renewable and Sustainable Energy Reviews*, vol 10, pp559–575

Croft, G. D. and Patzek, T. W. (2009) 'Potential for coal-to-liquids conversion in the U.S.-resource base', *Natural Resources Research*, vol 18, pp173–180

Currie, M. J., Mapel, J. K., Heidel, T. D., Goffri, S. and Baldo, M. A. (2008) 'High-efficiency organic solar concentrators for photovoltaics', *Science*, vol 321, pp226–228

Curtright, A. E. and Apt, J. (2008) 'The character of power output from utility-scale photovoltaic systems', *Progress in Photovoltaics: Research and Applications*, vol 16, pp241–247

Cyranoski, D. (2009) 'Beijing's windy bet', *Nature*, vol 457, pp372–374

Czisch, G. and Giebel, G. (2000) 'A comparison of intra- and extra-European options for an energy supply with wind power', presented at Wind Power for the 21st Century, Kassel, September, www.iset.uni-kassel.de

Dahowski, R. T. and Dooley, J. J. (2004) 'Carbon management strategies for US electricity generation capacity: A vintage-based approach', *Energy*, vol 29, pp1589–1598

Dai, Y. J., Huang, H. B. and Wang, R. Z. (2003) 'Case study of solar chimney power plants in Northwestern regions of China', *Renewable Energy*, vol 28, pp1295–1304

Dalenbäch, J.-O. (1990) *Central Solar Heating Plants with Seasonal Storage – A Status Report*, IEA Solar Heating and Cooling Programme, Task VII, www.iea-shc.org

Dalton, G. J., Lockington, D. A. and Baldock, T. E. (2008) 'Feasibility analysis of stand-alone renewable energy supply options for a large hotel', *Renewable Energy*, vol 33, pp1475–1490

Damen, K., Faaij, A., van Bergen, F., Gale, J. and Lysen, E. (2005) 'Identification of early opportunities for CO_2 sequestration – worldwide screening for CO_2-EOR and CO_2-ECBM projects', *Energy*, vol 30, pp1931–1952

Damen, K., van Troost, M., Faaij, A. and Turkenburg, W. (2006) 'A comparison of electricity and hydrogen production systems with CO_2 capture and storage. Part A: Review and selection of promising conversion and capture technologies', *Progress in Energy and Combustion Science*, vol 32, pp215–246

Das, D. and Veziroğlu, T. N. (2001) 'Hydrogen production by biological processes: A survey of the literature', *International Journal of Hydrogen Energy*, vol 26, pp13–28

Davies, P. A. (2005) 'A solar cooling system for greenhouse food production in hot climates', *Solar Energy*, vol 79, pp661–668

Davies, P. A. and Paton, C. (2005) 'The seawater greenhouse in the United Arab Emirates: Thermal modeling and evaluation of design options', *Desalinization*, vol 173, pp103–111

Davison, J. (2007) 'Performance and costs of power plants with capture and storage of CO_2', *Energy*, vol 32, pp1163–1176

Day, D., Evans, R. J., Lee, J. W. and Reicosky, D. (2005) 'Economical CO_2, SO_x, and NO_x capture from fossil-fuel utilization with combined renewable hydrogen production and large-scale carbon sequestration', *Energy*, vol 30, pp2558–2579

DeCarolis, J. F. and Keith, D. W. (2006) 'The economics of large-scale wind power in a carbon constrained world', *Energy Policy*, vol 34, pp395–410

del Cañizo, C., del Coso, G. and Sinke, W. C. (2009) 'Crystalline silicon solar module technology: Towards the 1 € per watt-peak goal', *Progress in Photovoltaics: Research and Applications*, vol 17, pp199–209

DeLucchi, M. (1991) *Emissions of Greenhouse Gases from the Use of Transportation Fuel and Electricity*, vol 2, Appendices, ANL/ESD/TM-22, Argonne National Laboratory, Argonne, Illinois

de Mira, R. and Kroeze, C. (2006) 'Greenhouse gas emissions from willow-based electricity: A scenario analysis for Portugal and The Netherlands', *Energy Policy*, vol 34, pp1367–1377

Demirbas, A. (2005) 'Biodiesel production from vegetable oils via catalytic and non-catalytic supercritical methanol transesterification methods', *Progress in Energy and Combustion Science*, vol 31, pp466–487

Denholm, P. (2006) 'Improving the technical, environmental and social performance of wind energy systems using biomass-based energy storage', *Renewable Energy*, vol 31, pp1355–1370

Denholm, P. (2007) *The Technical Potential of Solar Water Heating to Reduce Fossil Fuel Use and Greenhouse Gas Emissions in the United States*, Technical Report NREL/TP-640-41157, National Renewable Energy Laboratory, Golden, Colorado

Denholm, P. and Kulcinski, G. L. (2004) 'Life cycle energy requirements and greenhouse gas emissions from large scale energy storage systems', *Energy Conversion and Management*, vol 45, pp2153–2172

Denholm, P. and Margolis, R. M. (2007a) 'Evaluating the limits of solar photovoltaics (PV) in traditional electric power systems', *Energy Policy*, vol 35, pp2852–2861

Denholm, P. and Margolis, R. M. (2007b) 'Evaluating the limits of solar photovoltaics (PV) in electric power systems utilizing energy storage and other enabling technologies', *Energy Policy*, vol 35, pp4424–4433

Denholm, P. and Short, W. (2006) *Documentation of the WinDS Base Case Data, Version AEO 2006 (1)*, National Renewable Energy Laboratory, Golden, Colorado

Denholm, P., Kulcinski, G. L. and Holloway, T. (2005) 'Emissions and energy efficiency assessment of baseload wind energy systems', *Environmental Science and Technology*, vol 39, pp1903–1911

Denholm, P., Margolis, R. and Milford, J. (2008) *Production Cost Modeling for High Levels of Photovoltaics Penetration*, Technical Report NREL/TP-581-42305, National Renewable Energy Laboratory, Golden, Colorado

Denman, K. L. et al (2007) 'Couplings between changes in the climate system and biogeochemistry', in S. Solomon, D. Qin, M. Manning, Z. Chen, M. Marquis, K. B. Averyt, M. Tignor and H. L. Miller (eds) *Climate Change 2007: The Physical Science Basis. Contribution of Working*

Group I to the Fourth Assessment Report of the Intergovernmental Panel on Climate Change, Cambridge University Press, Cambridge and New York

Dersch, J., Geyer, M., Herrmann, U., Jones, S. A., Kelly, B., Kistner, R., Ortmanns, W., Pitz-Paal, R. and Price, H. (2004) 'Trough integration into power plants – a study on the performance and economy of integrated solar combined cycle systems', *Energy*, vol 29, pp947–959

Detzel, A., Krüger, M. and Ostermayer, A. (2006) 'Assessment of bio-based packaging materials', in J. Dewulf and H. van Langenhove (eds) *Renewables-Based Technology*, John Wiley and Sons, Chichester

Deutch, J., Moniz, E. J., Ansolabehere, S., Driscoll, M., Gray, P. E., Holdren, J. P., Joskow, P. L., Lester, R. K. and Todreas, N. E. (2003) *The Future of Nuclear Power*, Massachusetts Institute of Technology, Cambridge, http://web.mit.edu/nuclearpower

de Vries, E. (2008) 'Float on, floating offshore opens up the deep', *Renewable Energy World*, vol 11, no 2, pp46–54

de Vries, E. (2009a) 'REPower systems, less is more offshore', *Renewable Energy World*, vol 12, no 2, pp53–56

de Vries, E. (2009b) 'Wind technology trends: Why small steps matter', *Renewable Energy World*, vol 12, no 4

de Wild-Scholten, M., Alsema, E. A., ter Horst, E. W., Bächler, M. and Fthenakis, V. M. (2006) 'A cost and environmental impact comparison of grid-connected rooftop and ground-based PV systems', presented at the 21st European Photovoltaic Solar Energy Conference, Dresden, Germany, 4–8 September, www.ecn.nl

Diab, Y. and Achard, G. (1999) 'Energy concepts for utilization of solar energy in small and medium cities: The case of Chambéry', *Energy Conversion and Management*, vol 40, pp1555–1568

Dimitriou, I. and Aronsson, P. (2005) 'Willows for energy and phytoremediation in Sweden', *Unasylva*, vol 56, no 221, pp47–50

Dinçer, I. and Rosen, M. A. (2002) *Thermal Energy Storage, Systems and Applications*, John Wiley, New York

Doctor, R., Palmer, A., Coleman, D., Davison, J., Hendriks, C., Kaarstad, O., Ozaki, M. and Austell, M. (2005) 'Transport of CO_2', in *IPCC Special Report on Carbon Dioxide Capture and Storage*, World Meteorological Organization, Geneva

Dominguez-Faus, R., Powers, S. E., Burken, J G. and Alvarez, P. J. (2009) 'The water footprint of biofuels: A drink or drive issue?', *Environmental Science and Technology*, vol 43, pp3005–3010

Dornburg, V., Lewandowski, I. and Patel, M. (2004) 'Comparing the land requirements, energy savings, and greenhouse gas emission reductions of biobased polymers and bioenergy: An analysis and system extension of life-cycle assessment studies', *Journal of Industrial Ecology*, vol 7, pp93–116

Dornburg, V., Termeer, G. and Faaij, A. P. C. (2005) 'Economic and greenhouse gas emission analysis of bioenergy production using multi-product crops – case studies for the Netherlands and Poland', *Biomass and Bioenergy*, vol 28, pp454–474

Dorrepaal, E., Toet, S., van Logtestijn, R. S. P., Swart, E., van de Weg, M. J., Callaghan, T. V. and Aerts, R. (2009) 'Carbon respiration from subsurface peat accelerated by climate warming in the subarctic', *Nature*, vol 460, pp616–620

dos Santos, M. A., Rosa, L. P., Sikar, B., Sikar, E. and dos Santos, E. O. (2006) 'Gross greenhouse gas fluxes from hydro-power reservoir compared to thermal plants', *Energy Policy*, vol 34, pp481–488

Drake, B. and Hubacek, K. (2007) 'What to expect from a greater geographic dispersion of wind farms? A risk portfolio approach', *Energy Policy*, vol 35, pp3999–4008

Drange, H., Alendal, G. and Johannessen, O. M. (2001) 'Ocean release of fossil fuel CO_2: A case study', *Geophysical Research Letters*, vol 28, pp2637–2640

Du, Y. and Parsons, J. E. (2009) *Update on the Cost of Nuclear Power*, Center for Energy and Environmental Policy Research, Massachusetts Institute of Technology, Cambridge, MA

Duan, L., Lin, R., Deng, S., Jin, H. and Cai, R. (2004) 'A novel IGCC system with steam injected H_2/O_2 cycle and CO_2 recovery', *Energy Conversion and Management*, vol 45, pp797–809

Duchemin, E., Lucotte, M., St-Louis, V. and Canuel, R. (2002) 'Hydroelectric reservoirs as an anthropogenic source of greenhouse gases', *World Resource Review*, vol 14

Dutton, A. G., Bleijs, J. A. M., Dienhart, H., Falchetta, M., Hug, W., Prischich, D. and Ruddell, A. J. (2000) 'Experience in the design, sizing and implementation of autonomous wind-powered hydrogen production systems', *International Journal of Hydrogen Energy*, vol 25, pp705–722

Dyer, J. A. and Desjardins, R. L. (2006) 'Carbon dioxide emissions associated with the manufacturing of tractors and farm machinery in Canada', *Biosystems Engineering*, vol 93, pp107–118

Dwivedi, P. and Alavalapati, J. R. R. (2009) 'Economic feasibility of electricity production from energy plantations on present community-managed forestlands in Madhya Pradesh, India', *Energy Policy*, vol 37, pp352–360

EC (European Commission) (2007) *Concentrating Solar Power, from Research to Implementation*, EC, Brussels, www.solarpaces.org

Eccles, J. K., Pratson, L., Newell, R. G. and Jackson, R. B. (2009) 'Physical and economic potential of geological CO_2 storage in saline aquifers', *Environmental Science and Technology*, vol 43, pp1962–1969

Eck, M., Zarza, E., Eickhoff, M., Rheinländer, J. and Valenzuela, L. (2003) 'Applied research concerning the direct steam generation in parabolic troughs', *Solar Energy*, vol 74, pp341–351

EcoHeatCool (2006) *EcoHeatCool Workpackage 1, The European Heat Market, Final Report*, Euroheat and Power, Brussels, www.euroheat.org/ecoheatcool/

Edwards, P. P., Kuznetsov, V. L., David, W. I. F. and Brandon, N. P. (2008) 'Hydrogen and fuel cells: Towards a sustainable energy future', *Energy Policy*, vol 36, pp4356–4362

EER (Emerging Energy Research) (2009) *Global Concentrated Solar Power Markets and Strategies 2009–2020*, www.emerging-energy.com

Egger, C., Öhlinger, C. and Dell, G. (2003) 'Pellets: Clean, convenient and carbon-neutral', *Renewable Energy World*, vol 6, no 5, pp82–89

Eicker, U. (2003) *Solar Technologies for Buildings*, John Wiley, Chichester

Eiffert, P. and Kiss, G. J. (2000) *Building-Integrated Photovoltaic Designs for Commercial and Institutional Structures: A Sourcebook for Architects*, NREL/BK-520-25272, National Renewable Energy Laboratory, Golden, Colorado

Eiffert, P. and Task-7 Members (2002) *Building Integrated Photovoltaic Power Systems: Guidelines for Economic Evaluation*, International Energy Agency, Photovoltaic Power Systems Programme, Task 7, Paris, www.iea-pvps.org

Elliott, D. (1996) 'Tidal power', in G. Boyle (ed) *Renewable Energy: Power for a Sustainable Future*, Oxford University Press, Oxford

Enibe, S. O. (2002) 'Performance of a natural circulation solar air heating system with phase change material energy storage', *Renewable Energy*, vol 27, pp69–86

Enslin, J. H. R., Jansen, C. P. J. and Bauer, P. (2004) 'In store for the future? Interconnection and energy storage for offshore wind farms', *Renewable Energy World*, vol 7, no 1, pp104–113

EPA (Environmental Protection Agency) (2009) *Federal Requirements Under the Underground Injection Control (UIC) Program for Carbon Dioxide (CO_2) Geologic Sequestration Wells; Notice of Data Availability and Request for Comment*, EPA-HQ-OW-2008-0390; FRL-RIN 2040-AE98, http://edocket.access.gpo.gov/2008/E8-27738.htm

EPIA (European Photovoltaic Industry Association) (2009) *Global Market Outlook for Photovoltaics Until 2013*, www.epia.org

EPRI (Electric Power Research Institute) (2002) *Deterring Terrorism – Aircraft Crash Impact Analyses Demonstrate Nuclear Power Plant's Structural Strength*, Nuclear Energy Institute, www.nei.org

Epstein, M., Bertocchi, R. and Karni, J. (2004) 'Solar fixation of atmospheric nitrogen', *Journal of Solar Energy Engineering*, vol 126, pp626–632

ERCOT (Energy Reliability Council of Texas) (2008) *Competitive Renewable Energy Zones (CREZ) Transmission Optimization Study*, www.ercot.com/news/press_releases/2008/nr04-02-08

Erdle, E., Gross, J. and Meyringer, V. (1986) 'Possibilities for hydrogen production by combination of a solar thermal central receiver system and high temperature electrolysis of steam', *Solar Thermal Central Receiver Systems Proceeding of the Third International Workshop*, pp727–736

ESTIF (European Solar Thermal Industry Federation) (2003) *Sun in Action II – A solar thermal strategy for Europe. Volume 1: Market Overview, Perspectives and Strategy for Growth*, ESTIF, Brussels, www.estif.org

ESTIF (2007) *Solar Thermal Action Plan for Europe*, ESTIF, Brussels, www.estif.org

Everett, B. (1996) 'Solar thermal energy', in G. Boyle (ed) *Renewable Energy, Power for a Sustainable Future*, Oxford University Press, Oxford

EWEA (European Wind Energy Association) (2005) *Large Scale Integration of Wind Energy in the European Power Supply: Analysis, Issues, and Recommendations*, EWEA, Brussels, www.ewea.org

Ewing, R. C. and von Hippel, F. N. (2009) 'Nuclear waste management in the United States – starting over', *Science*, vol 325, pp151–152

Ezzati, M., Mbinda, B. M. and Kammen, D. M. (2000) 'Comparison of emissions and residential exposure from traditional and improved cookstoves in Kenya', *Environmental Science and Technology*, vol 34, pp578–583

Faaij, A. P. C. (2006) 'Bio-energy in Europe: Changing technology choices', *Energy Policy*, vol 34, pp322–342

Faaij, A., Meuleman, B., Turkenburg, W., van Wijk, A., Bauen, A., Rosillo-Calle, F. and Hill, D. (1998) 'Externalities of biomass based electricity production compared with power generation from coal in The Netherlands', *Biomass and Bioenergy*, vol 14, pp125–147

Fairless, D. (2007) 'The little shrub that could – maybe', *Nature*, vol 449, pp652–655

Faninger, G. (2000) 'Combined solar biomass district heating in Austria', *Solar Energy*, vol 69, pp425–435

FAO (Food and Agriculture Organization) (2002) *Tropical Forest Plantation Areas 1995 Dataset*, FAO Working Paper FP/18, Food and Agriculture Organization, Rome

FAO (2005) *Global Forest Resources Assessment 2005: Progress towards Sustainable Forest Management*, FAO Forestry Paper 147. Food and Agriculture Organization, Rome

Fargione, J., Hill, J., Tilman, D., Polasky, S. and Hawthorne, P. (2008) 'Land clearing and the biofuel carbon debt', *Science*, vol 319, pp1235–1238

Farrell, A. E., Plevin, R. J., Turner, B. T., Jones, A. D., O'Hare, M. and Kammen, D. M. (2006) 'Ethanol can contribute to energy and environmental goals', *Science*, vol 311, pp506–508

Fearnside, P. M. (2002) 'Greenhouse gas emissions from a hydroelectric reservoir (Brazil's Tucuruí Dam) and the energy policy implications', *Water, Air, and Soil Pollution*, vol 133, no 1–4, pp69–96

Fearnside, P. M. (2004) 'Greenhouse gas emissions from hydroelectric dams: Controversies provide a springboard for rethinking a supposedly "clean" energy source, an editorial comment', *Climatic Change*, vol 66, pp1–8

Fearnside, P. M. (2005a) 'Brazil's Samuel dam: Lessons for hydroelectric policy and the environment in Amazonia', *Environmental Management*, vol 35, pp1–19

Fearnside, P. M. (2005b) 'Do hydroelectric dams mitigate global warming? The case of Brazil's Curuá-Una dam', *Mitigation and Adaptation Strategies for Global Change*, vol 10, pp675–691

Feddema, J., Oleson, K., Bonan, G., Mearns, L., Washington, W., Meehl, G. and Nychka, D. (2005) 'A comparison of a GCM response to historical anthropogenic land cover change and model sensitivity to uncertainty in present-day land cover representations', *Climate Dynamics*, vol 25, pp581–609

Feltrin, A. and Freundlich, A. (2008) 'Material considerations for terawatt level deployment of photovoltaics', *Renewable Energy*, vol 33, pp180–185

Fernandes, S. D., Trautmann, N. M., Streets, D. G., Roden, C. A. and Bond, T. C. (2007) 'Global biofuel use, 1850–2000', *Global Biogeochemical Cycles*, vol 21, GB2019

Fiedler, F. (2004) 'The state of the art of small-scale pellet-based heating systems and relevant regulations in Sweden, Austria and Germany', *Renewable and Sustainable Energy Reviews*, vol 8, pp201–221

Firestone, J., Kempton, W. and Krueger, A. (2009) 'Public acceptance of offshore wind power projects in the USA', *Wind Energy*, vol 12, pp183–202

Fisch, M. N., Guigas, M. and Dalenbäck, J. O. (1998) 'A review of large-scale solar heating systems in Europe', *Solar Energy*, vol 63, pp355–366

Fischer, G. and Schrattenholzer, L. (2001) 'Global bioenergy potentials through 2050', *Biomass and Bioenergy*, vol 20, pp151–159

Fischer, G., Teixeira, E., Hizsnyik, E. T. and van Velthuizen, H. (2009) 'Land use dynamics and sugarcane production', in P. Zuurbier and J. van de Vooren (eds) *Sugarcane Ethanol, Contributions to Climate Change Mitigation and the Environment*, Wageningen Academic Publishers, Wageningen

Flamant, G., Ferriere, A., Laplaze, D. and Monity, C. (1999) 'Solar processing of materials: Opportunities and new frontiers', *Solar Energy*, vol 66, pp117–132

Florides, G. A., Tassou, S. A., Kalogirou, S. A. and Wrobel, L. C. (2002) 'Review of solar and low energy cooling technologies for buildings', *Renewable and Sustainable Energy Reviews*, vol 6, pp557–572

Fowles, M. (2007) 'Black carbon sequestration as an alternative to bioenergy', *Biomass and Bioenergy*, vol 31, pp426–432

Fricker, H. W. (2004) 'Regenerative thermal storage in atmospheric air system solar power plants', *Energy*, vol 29, pp871–881

Frogatt, A. (2005) *Nuclear Reactor Hazards*, Nuclear Issues Paper No 2, Heinrich Böll Foundation, Berlin, www.boell.de/nuclear

Fthenakis, V. (2000) 'End-of-life management and recycling of PV modules', *Energy Policy*, vol 28, pp1051–1058

Fthenakis, V. and Alsema, E. (2006) 'Photovoltaics energy payback times, greenhouse gas emissions and external costs: 2004–early 2005 status', *Progress in Photovoltaics: Research and Applications*, vol 14, pp 275–280

Fthenakis, V. and Kim, H. C. (2007) 'Greenhouse-gas emissions from solar electric and nuclear power: A life-cycle study', *Energy Policy*, vol 35, pp2549–2557

Fthenakis, V. and Kim, H. C. (2009) 'Land use and electricity generation: A life-cycle analysis', *Renewable and Sustainable Energy Reviews*, vol 13, pp1465–1474

Fthenakis, V. and Zweibel, K. (2003) 'CdTe PV: Real and perceived EHS risks', Conference Paper NREL/CP-520-33561, www.nrel.gov

Fthenakis, V. C., Wang, W. and Kim, H. C. (2009a) 'Life cycle inventory analysis of the production of metals used in photovoltaics', *Renewable and Sustainable Energy Reviews*, vol 13, pp493–517

Fthenakis, V. C., Mason, J. E. and Zweibel, K. (2009b) 'The technical, geographical, and economic feasibility for solar energy to supply the energy needs of the US', *Energy Policy*, vol 37, pp387–399

GAC (German Aerospace Center) (2006) *Trans-Mediterranean Interconnection for Concentrating Solar Power, Final Report*, GAC, www.dlr.de/tt/trans-csp

Gad, H. E., Hamed, A. M. and El-Sharkawy, I. I. (2001) 'Application of a solar desiccant/collector system for water recovery from atmospheric air', *Renewable Energy*, vol 22, pp541–556

Gale, J., Bradshaw, J., Chen, Z., Garg, A., Gomez, D., Rogner, H. H., Simbeck, D., Williams, R., Toth, F. and van Vuren, D. (2005) 'Sources of CO_2', in *IPCC Special Report on Carbon Dioxide Capture and Storage*, Cambridge University Press, Cambridge

Gallagher, S. J., Norton, B. and Eames, P. C. (2007) 'Quantum dot solar concentrators: Electrical conversion efficiencies and comparative factors of fabricated devices', *Solar Energy*, vol 81, pp813–821

Gan, L. and Yu, J. (2008) 'Bioenergy transition in rural China: Policy options and co-benefits', *Energy Policy*, vol 36, pp531–540

García, M., Vera, J. A., Marroyo, L. Lorenzo, E. and Pérez, M. (2009) 'Solar-tracking PV plants in Navarra: A 10

MW assessment', *Progress in Photovoltaics: Research and Applications*, vol 17, pp337–346

García-Rodríquez, L. (2003) 'Renewable energy applications in desalination: State of the art', *Solar Energy*, vol 75, pp381–393

Garwin, R. L. (2007) 'Plutonium recycle in the US nuclear power system?', presentation to AAAS Symposium, 17 February, www.fas.org/RLG

Gasol, C. M., Gabarrell, X., Anton, A., Rigola, M., Carrasco, J., Ciria, P., Solano, M. L. and Rieradevall, J. (2007) 'Life cycle assessment of a *Brassica carinata* bioenergy cropping system in southern Europe', *Biomass and Bioenergy*, vol 31, pp543–555

Gaunt, J. L. and Lehmann, J. (2008) 'Energy balance and emissions associated with biochar sequestration and pyrolysis bioenergy production', *Environmental Science and Technology*, vol 42, pp4152–4158

Gee, R., Cohen, G. and Greenwood, K. (2003) 'Operation and preliminary performance of the Duke solar power roof: A roof-integrated solar cooling and heating system', in *2003 International Solar Energy Conference*, Kohala Coast, Hawaii, 15–18 March, pp295–300

Geiser, K. (2001) *Materials Matter: Towards a Sustainable Materials Policy*, MIT Press, Cambridge, MA

Ghany, N. A. A., Kumugai, N., Meguro, S., Asami, K. and Hashimoto, K. (2002) 'Oxygen evolution anodes composed of anodically deposited Mn-Mo-Fe oxides for seawater electrolysis', *Electrochimica Acta*, vol 48, pp21–28

Ghosh, P. C., Emonts, B., Janßen, H., Mergel, J. and Stolten, D. (2003) 'Ten years of operational experience with a hydrogen-based renewable energy supply system', *Solar Energy*, vol 75, pp469–478

Gibbins, J. and Chalmers, H. (2008) 'Preparing for global rollout: A "developed country first" demonstration programme for rapid CCS deployment', *Energy Policy*, vol 36, pp501–507

Gielen, D. (2003) 'CO_2 removal in the iron and steel industry', *Energy Conversion and Management*, vol 44, pp1027–1037

Gilbert, R. and Perl, A. (2007) *Transport Revolutions: Moving People and Freight Without Oil*, Earthscan, London

Goldemberg, J., Coelho, S. T. and Guardabassi, P. (2008) 'The sustainability of ethanol production from sugarcane', *Energy Policy*, vol 36, pp2086–2097

Goossens, D. and Kerschaever, E. V. (1999) 'Aeolian dust deposition on photovoltaic solar cells: The effects of wind velocity and airborne dust concentration on cell performance', *Solar Energy*, vol 66, pp277–289

Gordon, J. M. and Ng, K. C. (2000) 'High-efficiency solar cooling', *Solar Energy*, vol 68, pp23–31

Green, J. A. S. (2007) *Aluminum Recycling and Processing for Energy Conservation and Sustainability*, ASM International, Materials Park, Ohio

Green, M. A. (2000) 'Photovoltaics: Technology overview', *Energy Policy*, vol 28, pp989–998

Green, M. A. (2001) 'Photovoltaic physics and devices', in J. Gordon (ed) *Solar Energy: The State of the Art, ISES Position Papers*, James & James, London

Green, M. A. (2003) 'Crystalline and thin-film silicon solar cells: State of the art and future potential', *Solar Energy*, vol 74, pp181–192

Green, M. A. (2006) 'Consolidation of thin-film photovoltaic technology: The coming decade of opportunity', *Progress in Photovoltaics: Research and Applications*, vol 14, pp383–392

Green, M. A., Emery, K., Hishikawa, Y. and Warta, W. (2009) 'Solar cell efficiency tables (version 34)', *Progress in Photovoltaics: Research and Applications*, vol 17, pp320–326

Greenblatt, J. B., Succar, S., Denkenberger, D. C., Williams, R. H. and Socolow, R. H. (2007) 'Baseload wind energy: Modeling the competition between gas turbines and compressed air energy storage for supplemental generation', *Energy Policy*, vol 35, pp1474–1492

Greenpeace (2005) *Wind Force 12: A Blueprint to Achieve 12% of the World's Electricity from Wind Power by 2020*, Global Wind Energy Council, www.gwec.org and www.ewea.org

Greijer, H., Karlson, L., Lindquist, S. and Hagfeldt, A. (2001) 'Environmental aspects of electricity generation from a nanocrystalline dye sensitized solar cell system', *Renewable Energy*, vol 23, pp27–39

Gross, R., Heptonstall, P., Leach, M., Skea, J., Anderson, D. and Green, T. (2007) 'The UK energy research centre review of the costs and impacts of intermittency', in G. Boyle (ed) *Renewable Energy and the Grid: The Challenge of Variability*, Earthscan, London

Guiney, B. and Henkel, T. (2003) 'Solar thermal for cooling, heating, and power generation', *Renewable Energy World*, vol 6, no 2, pp93–99

Gunter, W. D., Gentzis, T., Rottenfusser, B. A. and Richardson, R. J. H. (1997) 'Deep coalbed methane in Alberta, Canada: A fuel resource with the potential of zero greenhouse gas emissions', *Energy Conversion and Management*, vol 38, ppS217–S222

Gustavsson, L. and Karlsson, A. (2002) 'A system perspective on the heating of detached houses', *Energy Policy*, vol 30, pp553–574

Gustavsson, L. and Madlener, R. (2003) 'CO_2 mitigation costs of large-scale bioenergy technologies in competitive electricity markets', *Energy*, vol 28, pp1405–1425

Gutschner, M. and Task-7 Members (2001) *Potential for Building Integrated Photovoltaics*, International Energy Agency, Photovoltaic Power Systems Programme, Task 7, Paris, www.iea-pvps.org.

GWEC (Global Wind Energy Council) (2006) *Global Wind Energy Outlook 2006*, www.gwec.net

GWEC (2010) 'Global wind power boom continues despite economic woes', news release of 3 February 2010, annex

Hafner, S. (2003) 'Trends in maize, rice, and wheat yields for 188 nations over the past 40 years: A prevalence of linear growth', *Agriculture Ecosystems and Environment*, vol 97, pp275–283

Hallenbeck, P. C. and Benemann, J. R. (2002) 'Biological hydrogen production: Fundamentals and limiting processes', *International Journal of Hydrogen Energy*, vol 27, pp1185–1193

Hamelinck, C. N. and Faaij, A. P. C. (2006) 'Outlook for advanced biofuels', *Energy Policy*, vol 34, pp3268–3283

Hamelinck, C. N., van Hooijdonk, G. and Faaij, A. P. C. (2005a) 'Ethanol from lignocellulosic biomass: Techno-economic performance in short-, middle- and long-term', *Biomass and Bioenergy*, vol 28, pp384–410

Hamelinck, C. N., Suurs, R. A. A. and Faaij, A. P. C. (2005b) 'International bioenergy transport costs and energy balance', *Biomass and Energy*, vol 29, pp114–134

Hansson, A. and Bryngelsson, M. (2009) 'Expert opinions on carbon dioxide capture and storage – A framing of uncertainties and possibilities', *Energy Policy*, vol 37, pp2273–2282

Harvey, L. D. D. (1995) 'Solar-hydrogen electricity generation in the context of global CO_2 emission reduction', *Climatic Change*, vol 29, pp53–89

Harvey, L. D. D. (2000) *Global Warming: The Hard Science*, Prentice Hall, Harlow

Harvey, L. D. D. (2001) 'A quasi-one-dimensional coupled climate-carbon cycle model, Part II: The carbon cycle component', *Journal of Geophysical Research – Oceans*, vol 106, pp22,355–22,372

Harvey, L. D. D. (2003) 'Impact of deep-ocean carbon sequestration on atmospheric CO_2 and on surface-water chemistry', *Geophysical Research Letters*, vol 30, doi:10.1029/2002GLO16224

Harvey, L. D. D. (2004) 'Declining temporal effectiveness of carbon sequestration: Implications for compliance with the United Nations Framework Convention on Climate Change', *Climatic Change*, vol 63, pp259–290

Harvey, L. D. D. (2007) 'Dangerous anthropogenic interference, dangerous climatic change, and harmful climatic change: Non-trivial distinctions with significant policy implications', *Climatic Change*, vol 82, pp1–25

Harvey, L. D. D. (2008) 'Mitigating the atmospheric CO_2 increase and ocean acidification by adding limestone powder to upwelling regions', *Journal of Geophysical Research*, vol 113, C04028, doi:10.1029/2007 JC004373

Harvey, L. D. D. (2010) 'Climate and climate system modelling', in J. Wainwright and M. Mulligan (editors) *Environmental Modelling, Finding Simplicity in Complexity, 2nd Edition*, John Wiley, Hoboken, NJ

Harvey, L. D. D. and Huang, Z. (1995) 'Evaluation of the potential impact of methane clathrate destabilization on future global warming', *Journal of Geophysical Research*, vol 100, pp2905–2926

Harvey, L. D. D. and Huang, Z. (2001) 'A quasi-one-dimensional coupled climate-carbon cycle model, Part 1: Description and behavior of the climate component', *Journal of Geophysical Research – Oceans*, vol 106, pp22,339–22,353

Harvey, L. D. D. and Kaufmann, R. (2002) 'Simultaneously constraining climate sensitivity and aerosol radiative forcing', *Journal of Climate*, vol 15, pp2837–2861

Hasnain, S. M. (1998) 'Review on sustainable thermal energy storage technologies, Part I: Heat storage materials and techniques', *Energy Conversion and Management*, vol 39, pp1127–1138

Hastings, S. R. and Mørck, O. (2000) *Solar Air Systems: A Design Handbook*, James & James, London

Haszeldine, R. S. (2006) 'Deep geological CO_2 storage: Principles reviewed, and prospecting for bio-energy disposal sites', *Mitigation and Adaptation Strategies for Global Change*, vol 11, pp377–401

Hattenbach, R. P., Wilson, M. and Brown, K. R. (1999) 'Capture of carbon dioxide from coal combustion and its utilization for enhanced oil recovery', in P. Riemer, B. Eliasson and A. Wokaun (eds) *Greenhouse Gas Control Technologies*, Elsevier Science, New York

Haugan, P. M. and Alendal, G. (2005) 'Turbulent diffusion and transport from a CO_2 lake in the deep ocean', *Journal of Geophysical Research-Oceans*, vol 110, C09S14, doi:10.1029/2004JC002583

Haugen, H. S. and Eide, L. I. (1996) 'CO_2 capture and disposal: The realism of large-scale scenarios', *Energy Conversion and Management*, vol 37, pp1061–1066

Hay, J. E. (1986) 'Calculation of solar irradiances for inclined surfaces: Validation of selected hourly and daily models', *Atmosphere-Ocean*, vol 24, pp16–41

Hazell, P. and Wood, S. (2008) 'Drivers of change in global agriculture', *Philosophical Transactions of the Royal Society B*, vol 363, pp495–515

Hecker, S. S. (2006) 'Towards a comprehensive safeguards system: Keeping fissile materials out of terrorists' hands', *Annals of the American Academy of Political and Social Science*, vol 607, pp121–132

Hegazy, A. A. (2001) 'Effects of dust accumulation on solar transmittance through glass covers of plate-type collectors', *Renewable Energy*, vol 22, pp525–540

Hegedus, S. (2006) 'Thin film solar modules: The low cost, high throughput and versatile alternative to Si wafers', *Progress in Photovoltaics: Research and Applications*, vol 14, pp393–411

Hektor, E. and Berntsson, T. (2007) 'Future CO_2 removal from pulp mills – Process integration consequences', *Energy Conversion and Management*, vol 48, pp3025–3033

Heller, A. (2000) '15 years of R&D in central solar heating in Denmark', *Solar Energy*, vol 69, pp437–447

Heller, M. C., Keoleian, G. A. and Volk, T. A. (2003) 'Life cycle assessment of a willow bioenergy cropping system', *Biomass and Bioenergy*, vol 25, pp147–165

Henderson-Sellers, A. and Robinson, P. J. (1986) *Contemporary Climatology*, Longman, Harlow

Henning, H. M. (ed) (2004) *Solar-assisted Air Conditioning in Buildings – A Handbook for Planners*, Springer-Verlag Wien, Vienna

Henning, H. M. (2007) 'Solar assisted air conditioning of buildings – an overview', *Thermal Engineering*, vol 27, pp1734–1749

Hermann, B. G., Blok, K. and Patel, M. K. (2007) 'Producing bio-based bulk chemicals using industrial biotechnology saves energy and combats climate change', *Environmental Science and Technology*, vol 41, pp7915–7921

Hertwich, E. G. and Zhang, X. (2009) 'Concentrating-solar biomass gasification process for a 3rd generation biofuel', *Environmental Science and Technology*, vol 43, pp4207–4212

Herzog, H. J. (2001) 'What future for carbon capture and sequestration?', *Environmental Science and Technology*, vol 35, pp148–153

Hesketh, K. W. (1996) 'Power reactors', in P. D. Wilson (ed) *The Nuclear Fuel Cycle: From Ore to Waste*, Oxford University Press, Oxford

Hestnes, A. G. (1999) 'Building integration of solar energy systems', *Solar Energy*, vol 67, pp181–187

Heywood, J. B., Weiss, M. A., Schafer, A., Bassene, S. A. and Natarajan, V. K. (2003) *The Performance of Future ICE and Fuel Cell Powered Vehicles and their Potential Fleet Impact*, MIT LFEE 2003-004 RP, www.lfee.mit.edu/publications

Hiete, M., Berner, U. and Richter, O. (2001) 'Calculation of global carbon dioxide emissions: Review of emission factors and a new approach taking fuel quality into consideration', *Global Biogeochemical Cycles*, vol 15, pp169–181

Himmel, M. E., Ding, S. Y., Johnson, D. K., Adney, W. S., Nimlos, M. R., Brady, J. W. and Foust, T. D. (2007) 'Biomass recalcitrance: Engineering plants and enzymes for biofuels production', *Science*, vol 315, pp804–807

Hinshelwood, E. (2000) 'Community funded wind power – the missing link in UK wind farm development?', *Wind Engineering*, vol 24, pp299–305

Hirst, E. (2002) 'Integrating wind output with bulk power operations and wholesale electricity markets', *Wind Energy*, vol 5, pp19–36

Hoffert, M. I. and Potter, S. D. (1997) 'Energy supply', in R. G. Watts (ed) *Engineering Response to Global Climate Change: Planning a Research and Development Agenda*, Lewis Publishers, Boca Raton

Hoggan, J. and Littlemore, R. (2009) *Climate Cover Up: The Crusade to Deny Global Warming*, Greystone Books, Vancouver

Högselius, P. (2009) 'Spent nuclear fuel policies in historical perspective: An international comparison', *Energy Policy*, vol 37, pp254–263

Holloway, S. (2001) 'Storage of fossil fuel-derived carbon dioxide beneath the surface of the Earth', *Annual Review of Energy and the Environment*, vol 26, pp145–166

Holloway, S. (2005) 'Underground sequestration of carbon dioxide – a viable greenhouse gas mitigation option', *Energy*, vol 30, pp2318–2333

Holloway, S., Pearce, J. M., Hards, V. L., Ohsumi, T. and Gale, J. (2007) 'Natural emissions of CO_2 from the geosphere and their bearing on the geological storage of carbon dioxide', *Energy*, vol 32, pp1194–1201

Holm, D. (2005) *Renewable Energy Future for the Developing World*, International Solar Energy Society, http://whitepaper.ises.org

Hooda, N. and Rawat, V. R. S. (2006) 'Role of bio-energy plantations for carbon-dioxide mitigation with special reference to India', *Mitigation and Adaptation Strategies for Global Change*, vol 11, pp445–467

Hoogwijk, M., Faaij, A., van den Broek, R., Berndes, G., Gielen, D. and Turkenburg, W. (2003) 'Exploration of the ranges of the global potential of biomass for energy', *Biomass and Bioenergy*, vol 25, pp119–133

Hoogwijk, M., de Vries, B. and Turkenburg, W. (2004) 'Assessment of the global and regional geographical, technical and economic potential of onshore wind energy', *Energy Economics*, vol 26, pp889–919

Hoogwijk, M., Faaij, A., Eickhout, B., de Vries, B. and Turkenburg, W. (2005) 'Potential of biomass energy out to 2100, for four IPCC SRES land-use scenarios', *Biomass and Bioenergy*, vol 29, pp225–257

Hoogwijk, M., van Vuuren, D., de Vries, B. and Turkenburg, W. (2007) 'Exploring the impact on cost and electricity production of high penetration levels of intermittent electricity in OECD Europe and the USA, results for wind energy', *Energy*, vol 32, pp1381–1402

Hoogwijk, M., Faaij, A., de Vries, B. and Turkenburg, W. (2009) 'Exploration of regional and global cost-supply curves of biomass energy from short-rotation crops at abandoned cropland and rest land under four IPSS SRES land-use scenarios', *Biomass and Bioenergy*, vol 33, pp26–43

Horn, M., Führing, H. and Rheinländer, J. (2004) 'Economic analysis of integrated solar combined cycle power plants. A sample case: The economic feasibility of an ISCCS power plant in Egypt', *Energy*, vol 29, pp935–945

Hosseini, R., Soltani, M. and Valizadeh, G. (2005) 'Technical and economic assessment of the integrated solar combined cycle power plants in Iran', *Renewable Energy*, vol 30, pp1541–1555

House, K. Z., Schrag, D. P., Harvey, C. F. and Lackner, K. S. (2006) 'Permanent carbon dioxide storage in deep-sea sediments', *Proceedings National Academy of Sciences*, vol 103, pp12,291–12,295

Howarth, R. W. and Bringezu, S. (eds) (2009) *Biofuels: Environmental Consequences and Interactions with Changing Land Use*, Scientific Committee on Problems of the Environment (SCOPE), UNESCO and UNEP, http://cip.cornell.edu/biofuels.

Hu, S. Y. and Cheng, J. H. (2007) 'Performance evaluation of pairing between sites and wind turbines', *Renewable Energy*, vol 32, pp1934–1947

Huld, T., Šúri, M. and Dunlop, E. D. (2008a) 'Geographical variation of the conversion efficiency of crystalline silicon photovoltaic modules in Europe', *Progress in Photovoltaics: Research and Applications*, vol 16, pp 595–607

Huld, T., Šúri, M. and Dunlop, E. D. (2008b) 'Comparison of potential solar electricity output from fixed-inclined and two-axis tracking photovoltaic modules in Europe', *Progress in Photovoltaics: Research and Applications*, vol 16, pp47–59

Hultquist, G., Szakálos, P., Graham, M. J., Sproule, G. I. and Wikmark, G. (2008) 'Detection of hydrogen in corrosion of copper in pure water', presented at the International Corrosion Congress, Las Vegas, Nevada

Hultquist, G. et al (2009) 'Water corrodes copper', *Catal Letters*, DOI 10.1007/s10562-009-0113-x

Huntley, M. E. and Redalje, D. G. (2007) 'CO_2 mitigation and renewable oil from photosynthetic microbes: A new appraisal', *Mitigation and Adaptation Strategies for Global Change*, vol 12, pp573–608

Hurley, B., Hughes, P. and Giebel, G. (2007) 'Reliable power, wind variability and offshore grids in Europe', in G. Boyle (ed) *Renewable Energy and the Grid: The Challenge of Variability*, Earthscan, London, pp181–199

Ibrahim, H., Ilinca, A. and Perron, J. (2008) 'Energy storage systems – Characteristics and comparisons', *Renewable and Sustainable Energy Reviews*, vol 12, pp1221–1250

IEA (International Energy Agency) (1999) *Hydrogen – Today and Tomorrow*, International Energy Agency, Paris, www.ieagreen.org.uk

IEA (2001a) *Carbon Dioxide Capture from Power Stations*, International Energy Agency, Paris, www.ieagreen.org.uk

IEA (2001b) *Carbon Dioxide Disposal from Power Stations*, International Energy Agency, Paris, www.ieagreen.org.uk

IEA (2001c) *Ocean Storage of CO_2*, International Energy Agency, Paris, www.ieagreen.org.uk

IEA (2002) *Reclaiming Heat From Shower Water*, CADDET Demo 48, International Energy Agency, Paris, www.caddet-ee.org

IEA (2003) *Renewables for Power Generation: Status and Prospects*, 2003 Edition, International Energy Agency, Paris

IEA (2006a) *Energy Technology Perspectives 2006: In Support of the G8 Plan of Action*, International Energy Agency, Paris

IEA (2006b) *Offshore Wind Experiences*, International Energy Agency, Paris

IEA (2006c) *Renewable Energy: RD&D Priorities, Insights from IEA Technology Programmes*, International Energy Agency, Paris

IEA (2006d) *Review and Analysis of Ocean Energy Systems Developments and Supporting Policies*, International Energy Agency, Paris

IEA (2008) *Trends in Photovoltaic Applications: Survey Report of Selected IEA Countries Between 1992 and 2007*, Photovoltaic Power Systems Programme, Report IEA-PVPS T1-17:2008, International Energy Agency, Paris

IEA (2009) *Technology Roadmap, Wind Energy*, International Energy Agency, Paris

Infield, D. and Watson, S. (2007) 'Planning for variability in the longer term: The challenge of a truly sustainable energy system', in G. Boyle (ed) *Renewable Energy and the Grid: The Challenge of Variability*, Earthscan, London

Inui, Y., Matsumae, T., Koga, H. and Nishiura, K. (2005) 'High performance SOFC/GT combined power generation system with CO_2 recovery by oxygen combustion method', *Energy Conversion and Management*, vol 46, pp1837–1847

IPCC (Intergovernmental Panel on Climate Change) (2005) *IPCC Special Report on Carbon Dioxide Capture and Storage*, Cambridge University Press, Cambridge, www.ipcc.ch

Ito, M., Kato, K., Komoto, K., Kichimi, T. and Kurokawa, K. (2008) 'A comparative study on cost and life-cycle analysis for 100 MW very large-scale PV (VLS-PV) systems in deserts using m-Si, a-Si, CdTe, and CIS modules', *Progress in Photovoltaics: Research and Applications*, vol 16, pp17–30

Ivy, J. (2004) *Summary of Electrolytic Hydrogen Production: Milestone Completion Report*, NREL/MP-560-36734, National Renewable Energy Laboratory, Golden, Colorado

Jackson, F. (2008) 'Why renewable energy infrastructure has become an exciting investment option', *Renewable Energy World*, vol 11, no 4, pp29–35

Jackson, T. (2009) *Prosperity Without Growth: Economics for a Finite Planet*, Earthscan, London

Jahn, U. and Nasse, W. (2004) 'Operational performance of grid-connected PV systems on buildings in Germany', *Progress in Photovoltaics: Research and Applications*, vol 12, pp441–448

James, R. R., DiFazio, S. P., Brunner, A. M. and Strauss, S. H. (1998) 'Environmental effects of genetically engineered woody biomass crops', *Biomass and Bioenergy*, vol 14, pp403–414

Janjai, S. and Tung, P. (2005) 'Performance of a solar dryer using hot air from roof-integrated solar collectors for drying herbs and spices', *Renewable Energy*, vol 30, pp2085–2095

Jasinskas, A., Zaltauskas, A. and Kryzeviciene, A. (2008) 'The investigation of growing and using tall perennial grasses as energy crops', *Biomass and Bioenergy*, vol 32, pp981–987

Jayaraman, K. S. (2007) 'India's carbon dioxide trap', *Nature*, vol 445, p350

Jobert, A., Laborgne, P. and Mimler, S. (2007) 'Local acceptance of wind energy: Factors of success identified in French and German case studies', *Energy Policy*, vol 35, pp2751–2760

Johansson, M. and Laike, T. (2007) 'Intention to respond to local wind turbines: The role of attitudes and visual perception', *Wind Energy*, vol 10, pp435–451

Johansson, T. B., Kelly, H., Reddy, A. K. N. and Williams, R. H. (1993) 'A renewables-intensive global energy scenario', in T. B. Johansson, H. Kelly, A. K. N. Reddy and R. H. Williams (eds) *Renewable Energy, Sources for Fuels and Electricity*, Island Press, Washington, DC

Jones, J. (2005) 'Getting warmer, solar thermal systems on the up', *Renewable Energy World*, vol 8, no 4, pp124–133

Jones (1996) ? in P. D. Wilson (ed) *The Nuclear Fuel Cycle, From Ore to Waste*, Oxford University Press, Oxford

Jungbluth, N., Tuchschmid, M. and de Wild-Scholten, M. (2008) *Life Cycle Assessment of Photovoltaics: Update of ecoinvetn data v2.0*, www.esu-services.ch

Junginger, M., Faaij, A. and Turkenburg, W. C. (2005) 'Global experience curves for wind farms', *Energy Policy*, vol 33, pp133–150

Junginger, M., de Visser, E., Hjort-Gregersen, K., Koornneef, J., Raven, R., Faaij, A. and Turkenburg, W. C. (2006) 'Technological learning in bioenergy systems', *Energy Policy*, vol 34, pp4024–4041

Kagel, A. (2008) *The State of Geothermal Technology. Part II: Surface Technology*, Geothermal Energy Association, Washington, DC, www.geo-energy.org

Kalogirou, S. (1998) 'Use of parabolic trough solar energy collectors for sea-water desalination', *Applied Energy*, vol 60, pp65–88

Kalogirou, S. (2003) 'The potential of solar industrial process heat application', *Applied Energy*, vol 76, pp337–361

Kalogirou, S. (2005) 'Seawater desalination using renewable energy sources', *Progress in Energy and Combustion Science*, vol 31, pp242–281

Kalogirou, S. (2009) *Solar Energy Engineering Processes and Systems*, Academic Press, Elsevier Science, Burlington, MA

Kalogirou, S. and Tripanagnostopoulos, Y. (2006) 'Hybrid PV/T solar systems for domestic hot water and electricity production', *Energy Conversion and Management*, vol 47, pp3368–3382

Kalogirou, S. and Tripanagnostopoulos, Y. (2007) 'Industrial application of PV/T solar energy systems', *Applied Thermal Engineering*, vol 27, pp1259–1270

Kammen, D. M. and Lipman, T. E. (2003) 'Assessing the future hydrogen economy', *Science*, vol 302, p226

Kammen, D. M., Bailis, R. and Herzog, A. V. (2001) *Clean Energy for Development and Economic Growth: Biomass and Other Renewable Energy Options to Meet Energy Development Needs in Poor Nations*, Policy Discussion Paper for the United Nations Development Programme, Environmentally Sustainable Development Group (ESDG) and the Climate Change Clean Development Mechanism (CDM) for Distribution at the Seventh Conference of Parties to the UNFCCC, Marrakech, Morocco

Karagiorgas, M., Botzios, A. and Tsoutsos, T. (2001) 'Industrial solar thermal applications in Greece: Economic evaluation, quality requirements and case studies', *Renewable and Sustainable Energy Reviews*, vol 5, pp157–173

Karekezi, S. and Kithyoma, W. (2004) 'Part II: Renewables and Rural Energy in sub-Saharan Africa – An Overview', AFREPREN/FWD Secretariat, www.afrepren.org/Pubs/bkchapters/rets/part2a.pdf

Karellas, S., Karl, J. and Kakaras, E. (2008) 'An innovative biomass gasification process and its coupling with microturbine and fuel cell systems', *Energy*, vol 33, pp284–291

Kartha, S. and Larson, E. D. (2000) *Bioenergy Primer: Modernized Biomass Energy for Sustainable Development*, United Nations Development Programme, New York

Kartha, S. and Leach, G. (2001) 'Using modern bioenergy to reduce rural poverty', report to the Shell Foundation, Stockholm Environment Institute, Stockholm

Kashiwa, B. A. and Kashiwa, C. B. (2008) 'The solar cyclone: A solar chimney for harvesting atmospheric water', *Energy*, vol 33, pp331–339

Kato, T., Kubota, M., Kobayashi, N. and Suzuoki, Y. (2005) 'Effective utilization of by-product oxygen from electrolysis hydrogen production', *Energy*, vol 30, pp2580–2595

Katti, P. K. and Khedkar, M. K. (2007) 'Alternative energy facilities based on site matching and generation unit sizing for remote area power supply', *Renewable Energy*, vol 32, pp1346–1362

Kaushika, N. D. and Sumathy, K. (2003) 'Solar transparent insulation materials: A review', *Renewable and Sustainable Energy Reviews*, vol 7, pp317–351

Kazimi, M. S. and Todreas, N. E. (1999) 'Nuclear power economic performance: Challenges and opportunities', *Annual Review of Energy and Environment*, vol 24, pp139–171

Keith, D. W., DeCarolis, J. F., Denkenberger, D. C., Lenschow, D. H., Malyshev, S. L., Pacala, S. and Rasch, P. J. (2004) 'The influence of large-scale wind power on global climate', *Proceedings National Academy of Science*, vol 101, no 46, pp16,115–16,120

Kempton, W. and Tomić J. (2005) 'Vehicle-to-grid power implementation: From stabilizing the grid to supporting large-scale renewable energy', *Journal of Power Sources*, vol 144, pp280–294

Kempton, W., Archer, C. L., Dhanju, A. and Garvine, R. W. (2007) 'Large CO_2 reductions via offshore wind power matched to inherent storage in end-uses', *Geophysical Research Letters*, vol 34, no L02817, doi:10.1029/2006GL 028016

Kenfack, J., Neirac, F. P., Tatietse, T. T., Mayer, D., Fogue, M. and Lejeune, A. (2009) 'Microhydro-PV-hybrid system: Sizing a small hydro-PV-hybrid system for rural electrification in developing countries', *Renewable Energy*, vol 34, pp2259–2263

Keoleian, G. A., Spatari, S. and Beal, R. (1997) *Life Cycle Design of a Fuel Tank System*, US Environmental Protection Agency, EPA/600/SR-97/118

Keshner, M. S. and Arya, R. (2004) *Study of Potential Cost Reductions Resulting from Super-Large-Scale Manufacturing of PV Modules*, NREL/SR-520-36846, National Renewable Energy Laboratory, Golden, Colorado

Khan, F. I., Hawboldt, K. and Iqbal, M. T. (2005) 'Life cycle analysis of wind-fuel cell integrated system', *Renewable Energy*, vol 30, pp157–177

Khan, J. and Bhuyan, G. S. (2009) *Ocean Energy: Global Technology Development and Status*, IEA-OES Document T0104, IEA, www.iea-oceans.org/_fich/6/ANNEX_1_Doc_T0104.pdf

Khanna, M., Dhungana, B. and Clifton-Brown, J. (2008) 'Costs of producing miscanthus and switchgrass for bioenergy in Illinois', *Biomass and Bioenergy*, vol 32, pp482–493

Kheshgi, H. S. and Archer, D. E. (1999) 'Modelling the evasion of CO_2 injected into the deep ocean', in P. Riemer, B. Eliasson and A. Wokaun (eds) *Greenhouse Gas Control Technologies*, Elsevier Science, New York

Kheshgi, H. S., Prince, R. C. and Marland, G. (2000) 'The potential of biomass fuels in the context of global climate change: Focus on transportation fuels', *Annual Review of Energy and the Environment*, vol 25, pp199–244

Khvorostyanov, D. V., Ciais, P., Krinner, G. and Zimov, S. A. (2008) 'Vulnerability of east Siberia's frozen carbon stores to future warming', *Geophysical Research Letters*, vol 35, L10703, doi:10.1029/2008GL033639

Kim, S. and Dale, B. E. (2004) 'Allocation procedure in ethanol production system from corn grain. I. System expansion', *International Journal of Life Cycle Assessments*, vol 7, pp237–243

Kim, S. and Dale, B. E. (2004) 'Cumulative energy and global warming impact from the production of biomass for biobased products', *Journal of Industrial Ecology*, vol 7, no 3–4, pp147–233

Kim, S. and Dale, B. E. (2005a) 'Environmental aspects of ethanol derived from no-tilled corn grain: Nonrenewable energy consumption and greenhouse has emissions', *Biomass and Bioenergy*, vol 28, pp475–489

Kim, S. and Dale, B. E. (2005b) 'Life cycle assessment of various cropping systems utilized for producing biofuels: Bioethanol and biodiesel', *Biomass and Bioenergy*, vol 29, pp426–439

Kindermann, G., Obersteiner, M., Sohngen, B., Sathaye, J., Andrasko, K., Rametsteiner, E., Schlamadinger, B., Wunder, S. and Beach, R. (2008) 'Global cost estimates of reducing carbon emissions through avoided deforestation', *Proceedings of the National Academy of Sciences*, vol 105, pp10,302–10,307

Kinsey, G. S., Herbet, P., Barbour, K. E., Krut, D. D., Cotal, H. L. and Sherif, R. A. (2008) 'Concentrator multijunction solar cell characteristics under variable intensity and temperature', *Progress in Photovoltaics: Research and Applications*, vol 16, pp503–508

Kintisch, E. (2008) 'The greening of synfuels', *Science*, vol 320, pp306–308

Klee, R. J. and Graedel, T. E. (2004) 'Elemental cycles: A status report on human or natural dominance', *Annual Review of Environment and Resources*, vol 29, pp69–107

Klemeš, J., Cockerill, T., Bulatov, I., Shackley, S. and Gough, C. (2006) 'Engineering feasibility of carbon dioxide capture and storage', in S. Shackley and C. Gough (eds) *Carbon Capture and its Storage, An Integrated Assessment*, Ashgate, Hampshire, pp43–85

Koh, L. P. and Wilcove, D. S. (2007) 'Cashing in palm oil for conservation', *Nature*, vol 448, pp993–994

Kolb, G. J. and Diver, R. B. (2008) *Screening Analysis of Solar Thermochemical Hydrogen Concepts*, Sandia Report SAND2008-1900, Sandia National Laboratories, Albuquerque, New Mexico

Kosnik, L. (2008) 'The potential of water power in the fight against global warming in the US', *Energy Policy*, vol 36, pp3252–3265

Kreider, J. F. and Kreith, F. (1981) *Solar Energy Handbook*, McGraw Hill, New York

Kreusch, J., Neumann, W., Appel, D. and Diehl, P. (2006) *Nuclear Fuel Cycle*, Nuclear Issues Paper No 3, Heinrich Böll Foundation, Berlin, www.boell.de/nuclear

Kristmannsdóttir, H. and Ármannsson, H. (2003) 'Environmental aspects of geothermal energy utilization', *Geothermics*, vol 32, pp451–461

Krohn, S., Morthorst, P. E. and Awerbuch, S. (2009) *The Economics of Wind Energy: A Report by the European Wind Energy Association*, European Wind Energy Association, Brussels, www.ewea.org

Kuckshinrichs, W., Zapp, P. and Poganietz, W.-R. (2007) 'CO_2 emissions of global metal-industries: The case of copper', *Applied Energy*, vol 84, pp842–852

Kumar, A. and Kandpal, T. C. (2005) 'Solar drying and CO_2 emissions mitigation: Potential for selected cash crops in India', *Solar Energy*, vol 78, pp321–329

Kumar, R., Adhikari, R. S., Garg, H. P. and Kumar, A. (2001) 'Thermal performance of a solar pressure cooker based on evacuated tube solar collector', *Applied Thermal Engineering*, vol 21, pp1699–1706

Kuo, C.-T., Shin, H.-Y., Hong, H.-F., Wu, C.-H., Lee, C.-D., Lung, I.-T. and Hsu, Y.-T. (2009) 'Development of the high concentration III-V photovoltaic system at INER, Taiwan', *Renewable Energy*, vol 35, pp1931–1933

Kurokawa, K. (ed) (2003) *Energy From the Desert – Feasibility of Very Large Scale Photovoltaic Power (VLS-PV) Systems*, James & James, London

Kurtz, S. (2008a) 'A comparison of theoretical efficiencies of multi-junction concentrator solar cells', *Progress in Photovoltaics: Research and Applications*, vol 16, pp537–546

Kurtz, S. (2008b) *Opportunities and Challenges for Development of a Mature Concentrating Photovoltaic Power Industry*, National Renewable Energy Laboratory, Golden, Colorado

Kutscher, C. F. (ed) (2007) *Tackling Climate Change in the U.S.: Potential Carbon Emission Reductions from Energy Efficiency and Renewable Energy by 2030*, American Solar Energy Society, www.ases.org/climatechange

Labrecque, M. and Teodorescu, T. I. (2001) 'Influence of plantation site and wastewater sludge fertilization on the performance and foliar nutrient status of two willow species grown under SRIC in southern Quebec (Canada)', *Forest Ecology and Management*, vol 150, pp223–239

Lamp, P., Costa, A., Ziegler, F., Collares Pereira, M., Farinha Mendes, J., Ojer, J. P., Conde, A. G. and Granados, C. (1998) 'Solar assisted absorption cooling with optimized utilization of solar energy', in *Natural Working Fluids '98, IIR – Gustav Lorentzen Conference*, 2–5 June, Oslo, International Institute of Refrigeration, Paris, pp530–538.

Langdon, C., Takahashi, T., Sweeney, C., Chipman, D., Goddard, J., Marubini, F., Aceves, H., Barnett, H. and Atkinson, M. (2000) 'Effect of calcium carbonate saturation state on the calcification rate of an experimental coral reef', *Global Biogeochemical Cycles*, vol 14, pp639–654

Larson, E. D. (1993) 'Technology for electricity and fuels from biomass', *Annual Review of Energy and Environment*, vol 18, pp567–630

Larson, E. D. (2006) 'A review of life-cycle analysis studies on liquid biofuel systems for the transport sector', *Energy for Sustainable Development*, vol 10, pp109–126

Lasher, S., McKenney, K., Sinha, J., Ahluwalia, R., Hua, T. and Peng, J. K. (2009) *Technical Assessment of Compressed Hydrogen Storage Tank Systems for Automotive Applications*, Report to the United States Department of Energy, Office of Energy Efficiency and Renewable Energy, Hydrogen, Fuel Cells and Infrastructure Technology Program, Part 1. TIAX LLC, Cambridge, Massachusetts, Federal Grant Number DE-FC36-04GO14203

Laxson, A., Hand, M. M. and Blair, N. (2006) *High Wind Penetration Impact on U.S. Wind Manufacturing Capacity and Critical Resources*, Technical Report NREL/TP-500-40482, National Renewable Energy Laboratory, Golden, Colorado

Lechón, Y., Cabal, H., de la Rúa, C., Caldés, N., Santamaría, M. and Sáez, R. (2009) 'Energy and greenhouse gas emission savings of biofuels in Spain's transport fuel: The adoption of the EU policy on biofuels', *Biomass and Bioenergy*, vol 33, pp920–932

Ledford, H. (2006) 'Making it up as you go along', *Nature*, vol 444, pp677–678

Lee, J., Cho, H., Choi, B., Sung, J., Lee, S. and Shin, M. (2000) 'Life cycle analysis of tractors', *International Journal of Life Cycle Analysis*, vol 5, pp205–208

Lee, J. W. and Li, R. (2003) 'Integration of fossil energy systems with CO_2 sequestration through NH_4HCO_3 production', *Energy Conversion and Management*, vol 44, pp1535–1546

Lehmann, J., Gaunt, J. and Rondon, M. (2006) 'Bio-char sequestration in terrestrial ecosystems – A review', *Mitigation and Adaptation Strategies for Global Change*, vol 11, pp403–427

Lemonis, G. (2004) 'Wave and tidal energy conversion', *Encyclopedia of Energy*, Elsevier, vol 6, pp385–396

Lemus, R. and Lal, R. (2005) 'Bioenergy crops and carbon sequestration', *Critical Reviews in Plant Sciences*, vol 24, pp1–21

Lenzen, M. (1999) 'Greenhouse gas analysis of solar-thermal electricity generation', *Solar Energy*, vol 65, pp353–368

Lenzen, M. (2006) *Life-Cycle Energy Balance and Greenhouse Gas Emissions of Nuclear Energy in Australia*, University of Sydney, www.isa.org.usyd.edu.au/publications/documents/ISA_Nuclear_Report.pdf

Lenzen, M. (2008) 'Life cycle energy and greenhouse gas emissions of nuclear energy: A review', *Energy Conversion and Management*, vol 49, pp2178–2199

Lenzen, M. and Dey, D. (2000) 'Truncation error in embodied energy analysis of basic iron and steel products', *Energy*, vol 25, pp577–585

Lenzen, M. and Munksgaard, J. (2002) 'Energy and CO_2 life-cycle analyses of wind turbines – review and applications', *Renewable Energy*, vol 26, pp339–362

Lenzen, M. and Treloar, G. (2002) 'Embodied energy in buildings: Wood versus concrete – reply to Börjesson and Gustavsson', *Energy Policy*, vol 30, pp249–255

Lesser, J. A. and Su, X. (2008) 'Design of an economically efficient feed-in tariff structure for renewable energy development', *Energy Policy*, vol 36, pp981–990

Lew, D. J., Williams, R. H., Shaoxiong, X. and Shihui, Z. (1998) 'Large-scale baseload wind power in China', *Natural Resources Forum*, vol 22, pp165–184

Li, Q., Wu, Z. and Li, X. (2009) 'Prediction of CO_2 leakage during sequestration into marine sedimentary strata', *Energy Conversion and Management*, vol 50, pp503–509

Liebig, M. A., Johnson, H. A., Hanson, J. D. and Frank, A. B. (2005) 'Soil carbon under switchgrass stands and cultivated cropland', *Biomass and Bioenergy*, vol 28, pp347–354

Lind, L. R. and Wang, M. Q. (2004) 'A product-nonspecific framework for evaluating the potential for biomass-based products to displace fossil fuels', *Journal of Industrial Ecology*, vol 7, pp17–32

Lindenberg, S., Smith, B., O'Dell, K., DeMeo, E. and Ram, B. (2008) *20% Wind Energy by 2030: Increasing Wind Energy's Contribution to U.S. Electricity Supply*, US Department of Energy, Washington, DC

Lindenberger, D., Bruckner, T., Groscurth, H. M. and Kümmel, R. (2000) 'Optimization of solar district heating systems: Seasonal storage, heat pumps, and cogeneration', *Energy*, vol 25, pp591–608

Londo, M., Dekker, J. and ter Keurs, W. (2005) 'Willow short-rotation coppice for energy and breeding birds: An exploration of potentials in relation to management', *Biomass and Bioenergy*, vol 28, pp281–293

Long, S. P., Ainsworth, E. A., Leakey, D. B., Nösberger, J. and Ort, D. R. (2006) 'Food for thought: Lower-than-expected crop yield stimulation with rising CO_2 concentrations', *Science*, vol 312, pp1918–1921

Lotspeich, C. and van Holde, D. (2002) 'Flow batteries: Has really large battery storage come of age?' in *Proceedings of the 2002 ACEEE Summer Study on Energy Efficiency in Buildings*, vol 3, American Council for an Energy Efficient Economy, Washington, DC

Lottner, V., Schulz, M. E. and Hahne, E. (2000) 'Solar-assisted district heating plants: Status of the German programme Solarthermie-2000', *Solar Energy*, vol 69, pp449–459

Lund, H. and Kempton, W. (2008) 'Integration of renewable energy into the transport and electricity sectors through V2G', *Energy Policy*, vol 36, pp3578–3587

Lund, H. and Münster, E. (2006) 'Integrated energy systems and local energy markets', *Energy Policy*, vol 34, pp1152–1160

Lund, H. and Ostergard, P. A. (2000) 'Electric grid and heat planning scenarios with centralized and distributed sources of conventional, CHP and wind generation', *Energy*, vol 25, pp299–312

Lund, J. W., Freeston, D. H. and Boyd, T. L. (2005) 'Direct application of geothermal energy: 2005 Worldwide review', *Geothermics*, vol 34, pp691–727

Lundh, M. and Dalenbäch, J.-O. (2008) 'Swedish solar heating residential area with seasonal storage in rock: Initial evaluation', *Renewable Energy*, vol 33, pp703–711

Luque, A., Sala, G. and Luque-Heredia, I. (2006) 'Photovoltaic concentration at the onset of its commercial development', *Progress in Photovoltaics: Research and Applications*, vol 14, pp413–428

Lutz, A. E., Bradshaw, R. W., Keller, J. O. and Witmer, D. E. (2003) 'Thermodynamic analysis of hydrogen production by steam reforming', *International Journal of Hydrogen Energy*, vol 28, pp159–167

Lutz, J. D., Klein, G., Springer, D. and Howard, B. D. (2002) 'Residential hot water distribution systems: Roundtable session', in *Proceedings of the 2002 ACEEE Summer Study on Energy Efficiency in Buildings*, vol 1, American Council for an Energy Efficient Economy, Washington, DC

Macedo, I. G., Seabra, J. E. A. and Silva, J. E. A. R. (2008) 'Green house gases emission in the production and use of ethanol from sugarcane in Brazil: The 2005/2006 averages and a prediction for 2020', *Biomass and Bioenergy*, vol 32, pp582–595

Maeng, H., Lund, H. and Hvelplund, F. (1999) 'Biogas plants in Denmark: Technological and economic developments', *Applied Energy*, vol 64, pp195–206

Magelli, F., Boucher, K., Hsiaotao, T. B., Melin, S. and Bonoli, A. (2009) 'An environmental impact assessment of exported wood pellets from Canada to Europe', *Biomass and Bioenergy*, vol 33, pp434–441

Makareviciene, V. and Janulis, P. (2003) 'Environmental effect of rapeseed oil ethyl ester', *Renewable Energy*, vol 28, pp2395–2403

Makhijani, A. (2007) Carbon-free and Nuclear-Free: A Roadmap for U.S. Energy Policy, Institute for Energy and Environmental Research, Takoma Park, Maryland. Available from www.ieer.org

Mancini, T., Heller, P., Butler, B., Osborn, B., Schiel, W., Goldberg, V., Buck, R., Diver, R., Andraka, C. and Moreno, J. (2003) 'Dish-Stirling systems: An overview of development and status', *Journal of Solar Energy Engineering*, vol 125, pp135–151

Marbe, Å., Harvey, S. and Berntsson, T. (2004) 'Biofuel gasification combined heat and power – new implementation opportunities resulting from combined supply of process steam and district heating', *Energy*, vol 29, pp1117–1137

Marbe, Å., Harvey, S. and Berntsson, T. (2006) 'Technical, environmental and economic analysis of co-firing of gasified biofuel in a natural gas combined cycle (NGCC) combined heat and power (CHP) plant', *Energy*, vol 31, pp1614–1631

Marland, G., West, T. O., Schlamadinger, B. and Canella, L. (2003) 'Managing soil organic carbon in agriculture: The net effect on greenhouse gas emissions', *Tellus*, vol 55, pp613–621

Marris, E. (2006) 'Nuclear reincarnation', *Nature*, vol 441, pp796–797

Marsden, W. (2007) *Stupid to the Last Drop: How Alberta is Bringing Environmental Armagedddon to Canada (and Doesn't Seem to Care)*, Vintage Canada, Toronto

Mårtensson, K. and Westerberg, K. (2007) 'How to transform local energy systems towards bioenergy? Three strategy models for transformation', *Energy Policy*, vol 35, pp6095–6105

Martinelli, L. A. and Filoso, S. (2007) 'Polluting effects of Brazil's sugar-ethanol industry', *Nature*, vol 445, p364

Martínez, E., Sanz, F., Pellegrini, S., Jiménez, E. and Blanco, J. (2009) 'Life cycle assessment of a multi-megawatt wind turbine', *Renewable Energy*, vol 34, pp667–673

Mason, J. E., Fthenakis, V., Hansen, T. and Kim, H. C. (2006) 'Energy payback and life-cycle CO_2 emissions of the BOS in an optimized 3.5 MW PV installations', *Progress in Photovoltaics: Research and Applications*, vol 14, pp179–190

Mason, J. E., Fthenakis, V., Zweibel, K., Hansen, T. and Nikolakakis, T. (2008) 'Coupling PV and CAES power plants to transform intermittent PV electricity into a dispatchable electricity source', *Progress in Photovoltaics: Research and Applications*, vol 16, pp649–668

Masters, G. M. (2004) *Renewable and Efficient Electric Power Systems*, Wiley-Interscience, Hoboekn, NJ

Mathur, J., Agarwal, N., Swaroop, R. and Shah, N. (2008) 'Economics of producing hydrogen as transportation fuel using offshore wind energy systems', *Energy Policy*, vol 36, pp1212–1222

Matthews, R. W. (2001) 'Modelling of energy and carbon budgets of wood fuel coppice systems', *Biomass and Bioenergy*, vol 21, pp1–19

Maycock, P. and Bradford, T. (2006) 'PV market update: Soaring demand continues despite predictions of slowdown', *Renewable Energy World*, vol 9, no 4, pp68–80

McConnell, R. D. (2002) 'Assessment of the dye-sensitized solar cell', *Renewable and Sustainable Energy Reviews*, vol 6, pp271–295

McCormick, K. and Kåberger, T. (2007) 'Key barriers for bioenergy in Europe: Economic conditions, know-how and institutional capacity, and supply chain co-ordination', *Biomass and Bioenergy*, vol 31, pp443–452

McElroy, M. B., Lu, X., Nielsen, C. P. and Wang, Y. (2009) 'Potential for wind-generated electricity in China', *Science*, vol 325, pp1378–1380

McHenry, M. P. (2009) 'Agricultural bio-char production, renewable energy generation and farm carbon sequestration in Western Australia: Certainty, uncertainty and risk', *Agriculture, Ecocsystems and Environment*, vol 129, pp1–7

Meehl, G. A. et al (2007) 'Global climate projections', in S. Solomon, D. Qin, M. Manning, Z. Chen, M. Marquis, K. B. Averyt, M. Tignor and H. L. Miller (eds) *Climate Change 2007: The Physical Science Basis. Contribution of Working Group I to the Fourth Assessment Report of the Intergovernmental Panel on Climate Change*, Cambridge University Press, Cambridge and New York, pp747–845

Mehling, H., Cabeza, L. F., Hippeli, S. and Hiebler, S. (2003) 'PCM-module to improve hot water heat stores with stratification', *Renewable Energy*, vol 28, pp699–711

Meng, K. C., Williams, R. H. and Celia, M. A. (2007) 'Opportunities for low-cost CO_2 storage demonstration projects in China', *Energy Policy*, vol 35, pp2368–2378

Mertens, S. and de Vries, E. (2008) 'Small wind turbines: Driving performance', *Renewable Energy World*, vol 11, no 3, pp130–139

Metlay, D. (2000) 'From tin roof to torn wet blanket: Predicting and observing groundwater movement at a proposed nuclear waste site', in D. Sarewitz, R. A. Pielke Jr and R. Byerley Jr (eds) *Prediction: Science, Decision Making, and the Future of Nature*, Island Press, Covelo, California

Metting, F. B., Smith, J. L., Amthor, J. S. and Isaurralde, R. C. (2001) 'Science needs and new technology for increasing soil carbon sequestration', *Climatic Change*, vol 51, pp11–34

Michaels, D. (2008) *Doubt is Their Product*, Oxford University Press, Oxford

Midilli, A., Dogru, M., Akay, G. and Howarth, C. R. (2002) 'Hydrogen production from sewage sludge via a fixed bed gasifier product gas', *International Journal of Hydrogen Energy*, vol 27, pp1035–1041

Milborrow, D. (2007) 'Wind power on the grid', in G. Boyle (ed) *Renewable Energy and the Grid: The Challenge of Variability*, Earthscan, London

Mills, D. and Keepin, B. (1994) 'Baseload solar power: Near-term prospects for load-following solar thermal electricity', *Energy Policy*, vol 22, pp841–857

Mills, D. and Morrison, G. L. (2003) 'Optimisation of minimum backup solar water heating system', *Solar Energy*, vol 74, pp505–511.

Mills, D. R. (2001) 'Solar thermal electricity', in J. Gordon (ed) *Solar Energy: The State of the Art, ISES Position Papers*, James & James, London

Mink, G., Horvath, L., Evseev, E. G. and Kudish, A. I. (1998) 'Design parameters performance testing and analysis of a double-glazed, air-blown solar still with thermal energy recycle', *Solar Energy*, vol 64, no 4–6, pp265–277

MIT (Massachusetts Institute of Technology) (2003) *The Future of Nuclear Power: An Interdisciplinary MIT Study*, MIT, Cambridge, MA

MIT (2006) *The Future of Geothermal Energy: Impact of Enhanced Geothermal Systems (EGS) on the United States in the 21st Century*, MIT, Cambridge, MA

MIT (2007) *The Future of Coal: An Interdisciplinary MIT Study*, MIT, Cambridge, MA

MIT (2009) *Update of the MIT 2003 The Future of Nuclear Power: An Interdisciplinary MIT Study*, MIT, Cambridge, MA

Mock, J. E., Tester, J. W. and Wright, P. M. (1997) 'Geothermal energy from the Earth: Its potential as an environmentally sustainable resource', *Annual Review of Energy and Environment*, vol 22, pp305–356

Mohr, N. J., Schermer, J. J., Huijbregts, M. A. J., Meijer, A. and Reijnders, L. (2007) 'Life cycle assessment of thin film GaAs and GsInP/GaAs solar modules', *Progress in Photovoltaics: Research and Applications*, vol 15, pp163–179

Möllersten, K., Yan J. and Moreira, J. R. (2003) 'Potential market niches for biomass energy with CO_2 capture and storage – Opportunities for energy supply with negative CO_2 emissions', *Biomass and Bioenergy*, vol 25, pp273–285

Möllersten, K., Gao, L., Yan, J. and Obersteiner, M. (2004) 'Efficient energy systems with CO_2 capture and storage from renewable biomass in pulp and paper mills', *Renewable Energy*, vol 29, pp1583–1598

Momirlan, M. and Veziroğlu, T. N. (2002) 'Current status of hydrogen energy', *Renewable and Sustainable Energy Reviews*, vol 6, pp141–179

Monbiot, G. (2006) *Heat: How to Stop the Planet Burning*, Allen Lane, London

Mondol, J. D., Yohanis, Y. G. and Norton, B. (2007) 'The effect of low insolation conditions and inverter oversizing on the long-term performance of a grid-connected photovoltaic system', *Progress in Photovoltaics: Research and Applications*, vol 15, pp353–368

Mondol, J. D., Yohanis, Y. G. and Norton, B. (2009) 'Optimising the economic viability of grid-connected photovoltaic systems', *Applied Energy*, vol 86, pp985–999

Mooney, C. (2005) *The Republican War on Science*, Basic Books, New York

Moore, L. M. and Post, H. N. (2008) 'Five years of operating experience at a large, utility-scale photovoltaic generating plant', *Progress in Photovoltaics: Research and Applications*, vol 16, pp249–259

Moreira, J. R. (2006) 'Global biomass energy potential', *Mitigation and Adaptation Strategies for Global Change*, vol 11, pp313–342

Moreira, J. R. and Goldemberg, J. (1999) 'The alcohol program', *Energy Policy*, vol 27, pp229–245

Morrison, G. L. (2001) 'Solar water heating', in J. Gordon (ed) *Solar Energy: The State of the Art, ISES Position Papers*, James & James, London

Morrison, G. L., Budihardjo, I. and Behnia, M. (2004) 'Water-in-glass evacuated tube solar water heaters', *Solar Energy*, vol 76, pp135–140

Morton, D. C., DeFries, R. S., Shimabukuro, Y. E., Anderson, L. O., Arai, E., Espirito-Santo, F., Freitas, R. and Morisette, J. (2006) 'Cropland expansion changes deforestation dynamics in the southern Brazilian Amazon', *Proceedings National Academy of Sciences USA*, vol 103, pp14,637–14,641

Moy, R. (2003) 'Liability and the hydrogen economy', *Science*, vol 301, p47

Mudd, G. M. and Diesendorf, M. (2008) 'Sustainability of uranium mining and milling: Toward quantifying resources and eco-efficiency', *Environmental Science and Technology*, vol 42, pp2624–2640

Mueller-Langer, F., Tzimas, E., Kaltschmitt, M. and Peteves, S. (2007) 'Techno-economic assessment of hydrogen production processes for the hydrogen economy for the short and medium term', *International Journal of Hydrogen Energy*, vol 32, pp3797–3810

Murray, J. P. (1999) 'Aluminum production using high-temperature solar process heat', *Solar Energy*, vol 66, pp133–142

Murray, J. P. and Steinfeld, A. (1999) 'Clean metals production using high-temperature solar process heat', in B. Mishra (ed) *Proceedings of the 1999 TMS Annual Meeting/EPD Congress 1999*, San Diego, 28 February–4 March, pp817–830

Musial, W. (2006) 'Country review 2006, USA', *IEA-OES Annual Report 2006*, www.iea-oceans.org

Nabuurs, G. J. et al (2007) 'Forestry', in B. Metz, O. R. Davidson, P. R. Bosch, R. Dave and L. A. Meyer (eds) *Climate Change 2007: Mitigation*, Contribution of Working Group III to the Fourth Assessment Report of the Intergovernmental Panel on Climate Change, Cambridge University Press, Cambridge and New York

Nakagami, H., Ishihara, O., Sakai, K. and Tanaka, A. (2003) 'Performance of residential PV system under actual field conditions in western part of Japan', in *2003 International Solar Energy Conference*, Kohala Coast, Hawaii, 15–18 March, pp491–498

Nakicenovic, N. and Swart, S. (eds) (2000) *Emission Scenarios*, Special Report of the Intergovernment Panel on Climate Change, Cambridge University Press, Cambridge

Nakicenovic, N., Grübler, A. and MacDonald, A. (1998) *Global Energy Perspectives: International Institute for Applied Systems Analysis/World Energy Council*, Cambridge University Press, Cambridge

Nandwani, S. S. (2007) 'Design, construction and study of a hybrid solar food processor in the climate of Costa Rica', *Renewable Energy*, vol 32, pp427–441

Nassauer, O. (2005) *Nuclear Energy and Proliferation, Nuclear Issues Paper No 4,* Heinrich Böll Foundation, Berlin, www.boell.de/nuclear

Navarte, L. and Lorenzo, E. (2008) 'Tracking and ground cover ratio', *Progress in Photovoltaics: Research and Applications,* vol 16, pp703–714

NDA (Nuclear Decommissioning Authority) (2006) *NDA Strategy*, www.nda.gov.uk

NEA/IAEA (Nuclear Energy Agency/International Atomic Energy Agency) (2008) *Uranium 2007: Resources, Production and Demand*, OECD Publishing, Paris

Nemet, G. F. (2006) 'Beyond the learning curve: Factors influencing cost reductions in photovoltaics', *Energy Policy*, vol 34, pp3218–3232

Nepstad, D. C., Stickler, C. M. and Almeida, O. T. (2006) 'Globalization of the Amazon soy and beef industries: Opportunities for conservation', *Conservation Biology*, vol 20, pp1595–1603

NETL (National Energy Technology Laboratory) (2007) *Carbon Sequestration Atlas of the United States and Canada*, www.netl.doe.gov/technologies/carbon_seq/refshelf/atlas/index.html

NETL (2008) *Estimating Freshwater Needs to Meet Future Thermo-electric Generation Requirements, 2008 Update*, DOE/NETL-400/2008/1339

Ni, M., Leung, M. K. H. and Leung, D. Y. C. (2007) 'Energy and exergy analysis of hydrogen production by solid oxide elctrolyzer plant', *International Journal of Hydrogen Energy*, vol 32, pp4648–4660

Nilsson, S. and Schopfhauser, W. (1995) 'The carbon-sequestration potential of a global afforestation program', *Climatic Change*, vol 30, pp267–293

Niven, R. K. (2005) 'Ethanol in gasoline: Environmental impacts and sustainability review article', *Renewable and Sustainable Energy Reviews*, vol 9, pp535–555

Nordell, B. and Hellström, G. (2000) 'High temperature solar heated seasonal storage system for low temperature heating of buildings', *Solar Energy*, vol 69, pp511–523

Nordmann, T. and Clavadetscher, L. (2003) 'Understanding temperature effects on PV system performance', IEA PVPS Task 2 Activity 2.4, Paper 7P–B3-14, 3rd PV World Conference (WCPEC) in Osaka, Japan, May 2003

Norgate, T. E. and Rankin, W. J. (2000) 'Life cycle assessment of copper and nickel production', in *Proceedings of MINPREX 2000*, Australian Institute of Mining and Metallurgy, Melbourne

Norgate, T. E., Jahanshahi, S. and Rankin, W. J. (2007) 'Assessing the environmental impact of metal production processes', *Journal of Cleaner Production*, vol 15, pp838–848

Noufi, R. and Zweibel, K. (2006) *High-Efficiency CdTe and CIGS Thin-Film Solar Cells: Highlights and Challenges*, Conference Paper NREL/CP-520-39894, www.nrel.gov

Nouni, M. R., Mullick, S. C. and Kandpal, T. C. (2008) 'Providing electricity access to remote areas in India: An approach towards identifying potential areas for decentralized electricity supply', *Renewable and Sustainable Energy Reviews*, vol 12, pp1187–1220

Nowotny, J., Bak, T., Nowotny, M. K. and Sheppard, L. R. (2007) 'Titanium oxide for solar hydrogen I. Functional properties', *International Journal of Hydrogen Energy*, vol 32, pp2609–2629

NRC/NAE (National Research Council and National Academy of Engineering) (2004) *The Hydrogen Economy: Opportunities, Costs, Barriers, and R&D Needs*, National Academies Press, Washington, DC

NREL (National Renewable Energy Laboratory) (1998) *A Look Back at the US Department of Energy's Aquatic Species Program: Biodiesel from Algae*, NREL/TP-580-24190, www.nrel.gov.

Ntziachristos, L., Kouridis, C., Samaras, Z. and Pattas, K. (2005) 'A wind-power fuel-cell hybrid system study on the non-interconnected Aegean islands grid', *Renewable Energy*, vol 30, pp1471–1487

Nuber, D., Eichberger, H. and Rollinger, B. (2006) 'Circored fine ore direct reduction', *Millennium Steel 2006*, www.millennium-steel.com, pp37–40

Obernberger, I., Biedermann, F., Widmann, W. and Riedl, R. (1997) 'Concentrations of inorganic elements in biomass fuels and recovery in the different ash fractions', *Biomass and Bioenergy*, vol 12, pp211–224

Odeh, N. A. and Cockerill, T. T. (2008) 'Life cycle GHG assessment of fossil fuel power plants with carbon capture and storage', *Energy Policy*, vol 36, pp367–380

Odeh, S. D., Behnia, M. and Morrison, G. L. (2003) 'Performance evaluation of solar thermal electric generation systems', *Energy Conversion and Management*, vol 44, pp2425–2443

Ogawa, M., Okimori, Y. and Takahashi, F. (2006) 'Carbon sequestration by carbonization of biomass and forestation: Three case studies', *Mitigation and Adaptation Strategies for Global Change*, vol 11, pp429–444

Ogden, J. M. (1999) 'Prospects for building a hydrogen energy infrastructure', *Annual Review of Energy and the Environment*, vol 24, pp227–279

Oliveti, G., Arcuri, N. and Ruffolo, S. (1998) 'First experimental results from a prototype plant for the interseasonal storage of solar energy for the winter heating of buildings', *Solar Energy*, vol 62, pp281–290

Omer, S. A., Wilson, R. and Riffat, S. B. (2003) 'Monitoring results of two examples of building integrated PV (BiPV) systems in the UK', *Renewable Energy*, vol 28, pp1387–1399

O'Neill, M. J., McDanal, A. J. and Jaster, P. A. (2002) 'Development of terrestrial concentrator modules using high-efficiency multi-junction solar cells', 29th IEEE Photovoltaic Specialists Conference, New Orleans, May

Öney, F., Veziroğlu, T. N. and Dülger, Z. (1994) 'Evaluation of pipeline transportation of hydrogen and natural gas mixtures', *International Journal of Hydrogen Energy*, vol 19, pp813–822

Ordorica-Garcia, G., Douglas, P., Croiset, E. and Zheng, L. (2006) 'Technoeconomic evaluation of IGCC power plants for CO_2 avoidance', *Energy Conversion and Management*, vol 47, pp2250–2259

Othman, M. Y. H., Sopian, K., Yatim, B. and Daud, W. R. W. (2006) 'Development of advanced solar assisted drying systems', *Renewable Energy*, vol 31, pp703–709

Overend, R. P. and Chornet, E. (1984) 'Hydrogen and upgrading of biomass', in *Hydrogen Energy Progress V, Proceedings of the Fifth World Hydrogen Energy Conference*, Toronto, Canada, 15–20 July, pp1773–1784

Ozaki, M., Ohsumi, T. and Masuda, S. (1999) 'Dilution of released CO_2 in mid ocean depth by moving ship', in P. Riemer, B. Eliasson and A. Wokaun (eds) *Greenhouse Gas Control Technologies*, Elsevier Science, New York

Padin, J., Veziroğlu, T. N. and Shahin, A. (2000) 'Hybrid solar high-temperature hydrogen production system', *International Journal of Hydrogen Energy*, vol 25, pp295–317

Pahkala, K., Aalto, M., Isolahti, M., Poikola, J. and Jauhiainen, L. (2008) 'Large-scale energy grass farming for power plants – A case study from Ostrobothnia, Finland', *Biomass and Bioenergy*, vol 32, pp1009–1015

Pahud, D. (2000) 'Central solar heating plants with seasonal duct storage and short-term water storage: Design guidelines obtained by dynamic system simulations', *Solar Energy*, vol 69, pp495–509

Paish, O. (2002) 'Small hydro power: Technology and current status', *Renewable and Sustainable Energy Reviews*, vol 6, pp537–556

Pangavhane, D. R., Sawhney, R. L. and Sarsavadia, P. N. (2002) 'Design, development and performance testing of a new natural convection solar dryer', *Energy*, vol 27, pp579–590

Park, B. B., Yanai, R. D., Sahm, J. M., Lee, D. K. and Abrahamson, L. P. (2005) 'Wood ash effects on plant and soil in a willow bioenergy plantation', *Biomass and Bioenergy*, vol 28, pp355–365

Parawira, W., Read, J. S., Mattiasson, B. and Björnsson, L. (2008) 'Energy production from agricultural residues: High methane yields in pilot-scale two-stage anaerobic digestion', *Biomass and Bioenergy*, vol 32, pp44–50

Parrish, D. J. and Fike, J. H. (2005) 'The biology and agronomy of switchgrass for biofuels', *Critical Reviews in Plant Sciences*, vol 24, pp423–459

Patel, N. V. and Philip, S. K. (2000) 'Performance evaluation of three solar concentrating cookers', *Renewable Energy*, vol 20, pp347–355

Patzek, T. W. (2004) 'Thermodynamics of the corn-ethanol biofuel cycle', *Critical Reviews in Plant Sciences*, vol 23, pp519–567

Patzek, T. W. and Pimentael, D. (2005) 'Thermodynamics of energy production from biomass', *Critical Reviews in Plant Sciences*, vol 24, pp327–364

Paul, B. and Andrews, J. (2008) 'Optimal coupling of PV arrays to PEM electrolysers in solar-hydrogen systems for remote area power supply', *International Journal of Hydrogen Energy*, vol 33, pp490–498

Paulsen, K. and Hensel, F. (2007) 'Design of an autarkic water and energy supply driven by renewable energy using commercial available components', *Desalination*, vol 203, pp455–462

Pearce, F. (2004) 'Asian farmers suck continent dry', *New Scientist*, 28 August, pp6–7

Pearce, J. M. (2008a) 'Thermodynamic limitations to nuclear energy deployment as a greenhouse gas mitigation technology', *International Journal of Nuclear Governance, Economy and Ecology*, vol 2, pp113–130

Pearce, J. M. (2008b) 'Industrial symbiosis of very large-scale photovoltaic manufacturing', *Renewable Energy*, vol 33, pp1101–1108

Peharz, G. and Dimroth, F. (2005) 'Energy payback time of the high-concentration PV system FLATCON®', *Progress in Photovoltaics: Research and Applications*, vol 13, pp627–634

Pehnt, M., Oeser, M. and Swider, D. J. (2008) 'Consequential environmental system analysis of expected offshore wind electricity production in Germany', *Energy*, vol 33, pp747–759

Pelland, S. and Abboud, I. (2008) 'Comparing photovoltaic capacity value metrics: A case study for the City of Toronto', *Progress in Photovoltaics: Research and Applications*, vol 16, pp715–724

Perez, R., Margolis, R., Kmiecik, M., Schwab, M. and Perez, M. (2006) 'Update: Effective load carrying capability of photovoltaics in the United States, NREL/CP-620-40068', presented at American Solar Energy Conference 2006, Denver CO, 8–13 July, www.nrel.gov/pv/pdfs/40068.pdf

Perpiñan, O., Lorenzo, E., Castro, M. A. and Eyras, R. (2009) 'Energy payback times of grid connected PV systems: Comparison betweem tracking and fixed systems', *Progress in Photovoltaics: Research and Applications*, vol 17, pp137–147

Pervaiz, M. and Sain, M. M. (2003) 'Carbon storage potential in natural fiber composites', *Resources Conservation and Recycling*, vol 39, pp325–340

Pfeil, M. and Koch, H. (2000) 'High performance–low cost seasonal gravel/water storage pit', *Solar Energy*, vol 69, pp461–467

Phani, G., Tulloch, G., Vittorio, D. and Skryabin, I. (2001) 'Titania solar cells: New photovoltaic technology', *Renewable Energy*, vol 22, pp303–309

Pimentel, D. (2004) 'Livestock production and energy use', in *Encyclopedia of Energy*, vol 3, Elsevier

Pimentel, D. and Patzek, T. W. (2005) 'Ethanol production using corn, switchgrass, and wood; biodiesel production using soybean and sunflower', *Natural Resources and Research*, vol 14, pp65–76

Pimentel, D., Marklein, A., Toth, M. A., Karpoff, M., Paul, G. S., McCormack, R., Kyriazis, J. and Krueger, T. (2008) 'Biofuel impacts on world food supply: Use of fossil fuel, land, and water resources', *Energies*, vol 1, pp41–78

Popovski, K. and Popovska-Vasilevska, S. (2003) 'Prospects and problems for geothermal use in agriculture in Europe', *Geothermics*, vol 32, pp545–555

Pottier, J. D. and Blondin, E. (1995) 'Mass storage of hydrogen', in *Hydrogen Energy Systems, Utilization of Hydrogen and Future Aspects*, NATO ASI Series E-295, Kluwer, Dordrecht

Prasad, D. and Snow, M. (2005) *Designing with Solar Power: A Source Book for Building Integrated Photovoltaics (BiPV)*, Earthscan, London

Prather, M. J. (2003) 'An environmental experiment with H_2?', *Science*, vol 302, pp581–582

Price, H., Lüpfert, E., Kearney, D., Zarza, E., Cohen, G., Gree, R. and Mahoney, R. (2002) 'Advances in parabolic trough solar power technology', *Journal of Solar Energy Engineering*, vol 124, pp109–125

Price, H., Mehos, M., Kutscher, C. and Blair, N. (2007) 'Current and future economics of parabolic trough technology', in Proceedings of Energy Sustainability 2007, 27–30 June, Long Beach, California

Prueksakorn, K. and Gheewala, S. H. (2008) 'Full chain energy analysis of biodiesel from *Jatropha curcas* L. in Thailand', *Environmental Science and Technology*, vol 42, pp3388–3393

Quaschning, V. (2004) 'Solar thermal water heating', *Renewable Energy: World*, vol 7, no 2, pp95–99

Ramage, J. (1996) 'Hydroelectricity', in G. Boyle (ed) *Renewable Energy: Power for a Sustainable Future*, Oxford University Press, Oxford

Ramage, J. and Scurlock, J. (1996) 'Biomass', in G. Boyle (ed) *Renewable Energy: Power for a Sustainable Future*, Oxford University Press, Oxford

Rand, B. P., Genoe, J., Heremans, P. and Poortmans, J. (2007) 'Solar cells utilizing small molecular weight organic semiconductors', *Progress in Photovoltaics: Research and Applications*, vol 15, pp659–676

Rasi, S., Veijanen, A. and Rintala, J. (2007) 'Trace compounds of biogas from different biogas production plants', *Energy*, vol 32, pp1375–1380

Rau, G. H. and Caldeira, K. (1999) 'Enhanced carbonate dissolution: A means of sequestering waste CO_2 as ocean bicarbonate', *Energy Conversion and Management*, vol 40, pp1803–1813

Rau, G. H., Knauss, K. G., Langer, W. H. and Caldeira, K. (2007) 'Reducing energy-related CO_2 emissions using accelerated weathering of limestone', *Energy*, vol 32, pp1471–1477

Raugei, M. and Frankl, P. (2009) 'Life cycle impacts and costs of photovoltaic systems: Current state of the art and future outlooks', *Energy*, vol 34, pp393–399

Raugei, M., Bargigli, S. and Ulgiati, S. (2007) 'Life cycle assessment and energy pay-back time of advanced photovoltaic modules: CdTe and CIS compared to poly-Si', *Energy*, vol 32, pp1310–1318

Raven, R. P. J. M. and Gregersen, K. H. (2007) 'Biogas plants in Denmark: Successes and setbacks', *Renewable and Sustainable Energy Reviews*, vol 11, pp116–132

Ravindranath, N. H. and Hall, D. O. (1996) 'Estimates of feasible production of short rotation tropical forestry plantations', *Energy for Sustainable Development*, vol 2, pp14–20

Ravindranath, N. H., Balachandra, P., Dasappa, S. and Rao, K. U. (2006) 'Bioenergy technologies for carbon abatement', *Biomass and Bioenergy*, vol 30, pp826–837

Raymer, A. K. P. (2006) 'A comparison of avoided greenhouse gas emissions when using different kinds of wood energy', *Biomass and Bioenergy*, vol 30, pp605–617

Redlinger, R. Y., Andersen, P. D. and Morthorst, P. E. (2002) *Wind Energy in the 21st Century: Economics, Policy, Technology and the Changing Electricity Industry*, Palgrave, Basingstoke

Reijnders, L. (2006) 'Conditions for the sustainability of biomass based fuel use', *Energy Policy*, vol 34, pp863–876

Reiner, D. M. and Herzog, H. J. (2004) 'Developing a set of regulatory analogs for carbon sequestration', *Energy*, vol 29, pp1561–1570

REN21 (Renewable Energy Policy Network for the 21st Century) (2005) *Renewables 2005 Global Status Report*, Worldwatch Institute, Washington, DC

REN21 (2007) *Renewables 2007 Global Status Report*, Worldwatch Institute, Washington, DC, www.ren21.net

REN21 (2009) *Renewable 2009 Update Global Status Report*, Worldwatch Institute, Washington, DC, www.ren21.net

Rhodes, J. S. and Keith, D. W. (2005) 'Engineering economic analysis of biomass IGCC with carbon capture and storage', *Biomass and Bioenergy*, vol 29, pp440–450

Richards, K. R. and Stokes, C. (2004) 'A review of forest carbon sequestration cost studies', *Climatic Change*, vol 63, pp1–48

Riebesell, U., Zondervan, I., Rost, B., Tortell, P. D., Zeebe, R. and Morel, F. M. M. (2000) 'Reduced calcification of marine plankton in response to increased atmospheric CO_2', *Nature*, vol 407, pp364–367

Righelato, R. and Spracklen, D. V. (2007) 'Carbon mitigation by biofuels or by saving and restoring forests?', *Science*, vol 317, p902

Rodrigues, M., Faaij, A. P. C. and Walter, A. (2003) 'Techno-economic analysis of co-fired biomass integrated gasification/combined cycle systems with inclusion of economies of scale', *Energy*, vol 28, pp1229–1258

Rosa, L. P., Santos, M. A., Matvienko, B., Santos, E. O. and Sikar, E. (2004) 'Greenhouse Gas Emissions from Hydroelectric Reservoirs in Tropical Regions', *Climatic Change*, vol 66, pp9–21

Rosen, J., Tietze-Stöckinger, I. and Rentz, O. (2007) 'Model-based analysis of effects from large-scale wind production', *Energy*, vol 32, pp575–583

Rosen, M. A. (1998) 'The use of berms in thermal energy storage systems: Energy-economic analysis', *Solar Energy*, vol 63, pp69–78

Rosenkranz, G. (2006) *Nuclear Power – Myth and Reality. The risks and prospects of nuclear power*, Nuclear Issues Paper No 1, Heinrich Böll Foundation, Berlin, www.boell.de/nuclear

Rössing (2004) *2004 Report to Stakeholders*, Rössing Uranium Limited, www.rossing.com

Rotty, R. M., Perry, A. M. and Resier, D. B. (1975) *Net Energy from Nuclear Power*, Institute for Energy Analysis, IEA-75-3, Oak Ridge Associated Universities, Oak Rdige, Tennessee

Roy, S. B., Pacala, S. W. and Walko, R. L. (2004) 'Can large wind farms affect local meteorology?', *Journal of Geophysical Research*, vol 109, D1901, doi:10.1029/2004JD004763

Rubin, E. S., Yeh, S., Antes, M., Berkenpas, M. and Davison, J. (2007) 'Use of experience curves to estimate the future cost of power plants with CO_2 capture', *International Journal of Greenhouse Gas Control*, vol 1, pp188–197

Rytter, L. (2002) 'Nutrient content in stems of hybrid aspen as affected by tree age and tree size, and nutrient removal with harvest', *Biomass and Bioenergy*, vol 23, pp13–25

(S&T)^2 Consultants Inc. (2005a) *Ethanol GHG Emissions Using GHGenius, An Update*, prepared for Natural Resources Canada, www.ghgenius.ca

(S&T)[2] Consultants Inc. (2005b) *Documentation for Natural Resources Canada's GHGenius Model 3.0*, www.ghgenius.ca

(S&T)[2] Consultants Inc. (2005c) *Biodiesel GHG Emissions Using GHGenius, An Update*, prepared for Natural Resources Canada, www.ghgenius.ca

Sagar, A. D. and Kartha, S. (2007) 'Bioenergy and sustainable development?', *Annual Review of Environment and Resources*, vol 32, pp131–167

Sailor, D. J., Smith, M. and Hart, M. (2008) 'Climate change implications for wind power resources in the Northwest United States', *Renewable Energy*, vol 33, pp2393–2406

Sailor, W. C., Bodansky, D., Braun, C., Fetter, S. and van der Zwaan, B. (2000) 'A nuclear solution to climate change?', *Science*, vol 288, pp1177–1178

Saitoh, H., Hamada, Y., Kubota, H., Nakamura, M., Ochifuji, K., Yokoyama, S. and Nagano, K. (2003) 'Field experiments and analyses on a hybrid solar collector', *Applied Thermal Engineering*, vol 23, pp2089–2105

Samaniego, J., Alija, F., Sanz, S., Valmaseda, C. and Frechoso, F. (2008) 'Economic and technical analysis of a hybrid wind fuel cell energy system', *Renewable Energy*, vol 33, pp839–845

Samson, R., Mani, S., Boddey, R., Sokhansanj, S., Quesada, D., Urquiaga, S., Reis, V. and Ho Lem, C. (2005) 'The potential of C4 perennial grasses for developing a global BIOHEAT industry', *Critical Reviews in Plant Sciences*, vol 24, pp461–495

Sanderson, K. (2006) 'A field in ferment', *Nature*, vol 444, pp673–676

Sanderson, M. A., Adler, P. R., Boateng, A. A., Casler, M. D. and Sarath, G. (2006) 'Switchgrass as a biofuels feedstock in the USA', *Canadian Journal of Plant Science*, vol 86, pp1315–1325

Sandnes, B. and Rekstad, J. (2002) 'A photovoltaic/thermal (PV/T) collector with a polymer absorber plate. Experimental study and analytical model', *Solar Energy*, vol 72, pp63–73

Sanner, B. (1999) *High Temperature Underground Thermal Energy Storage, State-of-the-art and Prospects*, Lenz-Verlag-Giessen, Giessen

Saouter, E., van Hoof, G., Stalmans, M. and Brunskill, A. (2006) 'Oleochemical and petrochemical surfactants', in J. Dewulf and H. van Langenhove (eds) *Renewables-Based Technology*, John Wiley & Sons, Chichester

Sarenbo, S. (2009) 'Wood ash dilemma-reduced quality due to poor combustion performance', *Biomass and Bioenergy*, vol 33, pp1212–1220

Sartori, F., Lal, R., Ebinger, M. H. and Parrish, D. J. (2006) 'Potential soil carbon sequestration and CO_2 offset by dedicated energy crops in the USA', *Critical Reviews in Plant Sciences*, vol 25, pp441–472

Saxe, M. and Alvfors, P. (2007) 'Advantages of integration with industry for electrolytic hydrogen production', *Energy*, vol 32, pp42–50

Schaeffer, M., Eickhout, B., Hoogwijk, M., Strengers, B., van Vuuren, D., Leemans, R. and Opsteegh, T. (2006) 'CO_2 and albedo climate impacts of extratropical carbon and biomass plantations', *Global Biogeochemical Cycles*, vol 20, GB2020, doi:10.1029/2005GB002581

Schaetzle, W. J., Brett, C. E., Grubbs, D. M. and Seppanen, M. S. (1980) *Thermal Energy Storage in Aquifers, Design and Applications*, Pergamon Press, New York

Schenk, N. J., Moll, H. C., Potting, J. and Benders, R. M. J. (2007) 'Wind energy, electricity, and hydrogen in the Netherlands', *Energy*, vol 32, pp1960–1971

Scheuermann, K., Boleyn, D., Lilly, P. and Miller, S. (2002) 'Measured performance of California buydown program residential PV systems', in *Proceedings of the 2002 ACEEE Summer Study on Energy Efficiency in Buildings*, Vol 1, American Council for an Energy Efficient Economy, Washington, DC, pp273–285

Schimel, D. (2006) 'Climate change and crop yields: Beyond Cassandra', *Science*, vol 312, pp1889–1890

Schlesinger, W. H. (2000) 'Carbon sequestration in soils: Some cautions amidst optimism', *Agriculture Ecosystems and Environment*, vol 82, pp121–127

Schmidt, T., Mangold, D. and Muller-Steinhagen, H. (2004) 'Central solar heating plants with seasonal storage in Germany', *Solar Energy*, vol 76, pp165–174

Schmitz, M. (2009) 'Salt-free solar: CSP tower using air', *Renewable Energy World*, vol 12, no 1, pp51–52

Schneider, E. A. and Sailor, W. C. (2008) 'Long-term uranium supply estimates', *Nuclear Technology*, vol 162, pp379–387

Schneider, M., Thomas, S., Froggatt, A. and Koplow, D. (2009) *The World Nuclear Industry Status Report 2009, With Particular Emphasis on Economic Issues*, German Federal Ministry of the Environment, Nature Conservation and Reactor Safety, www.bmu.de/english/nuclear_safety/downloads/doc/44832.php

Schneider, S. H. (2010) *Science As a Contact Sport: Inside the Battle to Save Earth's Climate*, National Geographic, Washington, DC

Schoen, T. J. N. (2001) 'Building-integrated PV installations in the Netherlands: Examples and operational experiences', *Solar Energy*, vol 70, pp467–477

Schramek, P. and Mills, D. R. (2003) 'Multi-tower solar array', *Solar Energy*, vol 75, pp249–260

Schramek, P. and Mills, D. R. (2004) 'Heliostats for maximum ground coverage', *Energy*, vol 29, pp701–713

Schultz, M. G., Diehl, T., Brasseur, G. P. and Zittel, W. (2003) 'Air pollution and climate-forcing impacts of a global hydrogen economy', *Science*, vol 302, pp624–627

Schuur, E. A. G., Vogel, J. G., Crummer, K. G., Lee, H., Sickman, J. O. and Osterkamp, T. E. (2009) 'The effect of permafrost thaw on old carbon release and net carbon exchange from tundra', *Nature*, vol 459, pp556–559

Scott, D. A. and Dean, T. J. (2006) 'Energy trade-offs between intensive biomass utilization, site productivity loss, and ameliorative treatments in loblolly pine plantations', *Biomass and Bioenergy*, vol 30, pp1001–1010

Searchinger, T., Heimlich, R., Houghton, R. A., Dong, F., Elobeid, A., Fabiosa, J., Tokgoz, S., Hayes, D. and Yu, T. H. (2008) 'Use of US croplands for biofuels increases greenhouse gases through emissions from land use change', *Science*, vol 319, pp1238–1240

Sellers, W. D. (1965) *Physical Climatology*, University of Chicago Press, Chicago

Service, R. F. (2007a) 'Biofuel researchers prepare to reap a new harvest', *Science*, vol 315, pp1488–1491

Service, R. F. (2007b) 'Rethinking Mother Nature's choices', *Science*, vol 315, p793

Service, R. F. (2009) 'ExxonMobil fuels Venter's efforts to run vehicles on algae-based oil', *Science*, vol 325, p379

Shanmugam, V. and Natarajan, E. (2006) 'Experimental investigation of forced convection and desiccant integrated solar dryer', *Renewable Energy*, vol 31, pp1239–1251

Sharma, A., Chen, C. R. and Lan, N. V. (2009) 'Solar-energy drying systems: A review', *Renewable and Sustainable Energy Reviews*, vol 13, pp1185–1210

Sharma, S. D., Buddhi, D., Sawhney, R. L. and Sharma, A. (2000) 'Design development and performance evaluation of a latent heat storage unit for evening cooking in a solar cooker', *Energy Conversion and Management*, vol 41, pp1497–1508

Sheehan, J., Camobreco, V., Duffield, J., Graboski, M. and Shapouri, H. (1998) *An Overview of Biodiesel and Petroleum Diesel Life Cycles*, National Renewable Energy Laboratory, Golden, Colorado

Sheehan, J., Aden, A., Paustian, K., Killian, K., Brenner, J., Walsh, M. and Nelson, R. (2004) 'Energy and environmental aspects of using corn stover for fuel ethanol', *Journal of Industrial Ecology*, vol 7, pp117–146

Shi, J. Q. and Durucan, S. (2004) 'A numerical simulation study of the Allison Unit CO_2-ECBM pilot: The effect of matrix shrinkage and swelling on ECBM production and CO_2 injectivity', in *Proceedings of the 7th International Conference on Greenhouse Gas Control Technologies, Vol 1 Peer-Reviewed Papers and Plenary Presentations,* British Columbia, 5–9 September, pp431–442

Shinnar, R. (2003) 'The hydrogen economy, fuel cells, and electric cars', *Technology on Society*, vol 25, pp455–476

Shinnar, R. (2004) 'The mirage of the H_2 economy', *Clean Technologies and Environmental Policy*, vol 6, pp223–226

Shultis, J. K. and Faw, R. E. (2008) *Fundamentals of Nuclear Science and Engineering*, CRC Press, Boca Raton

Shum, K. L. and Watanabe, C. (2008) 'Towards a local learning (innovation) model of solar photovoltaic deployment', *Energy Policy*, vol 36, pp508–521

Shuster, E. (2008) *Estimating Freshwater Needs to Meet Future Thermoelectric Generation Requirements,* US Department of Energy, National Energy Technology Laboratory, DOE/NETL-400/2008/1339

Sick, F. and Erge, T. (eds) (1996) *Photovoltaics in Buildings: A Design Handbook for Architects and Engineers*, International Energy Agency, Paris

Sims, R. E. H. (2002) 'The brilliance of bioenergy: Small projects using biomass', *Renewable Energy World*, vol 5, no 1, pp57–63

Sims, R. (2010) 'Heat and power generation by gasification and combustion', in B. Singh (ed), *Industrial Crops and Uses*, Commonwealth Agricultural Bureau International (CABI), Wallingford, UK (in press)

Sims, R. E. H., Hastings, A., Schlamadinger, B., Taylor, G. and Smith P. (2006) 'Energy crops: Current status and future prospects', *Global Change Biology*, vol 12, pp2054–2076

Sistla, P. (2006) 'Country review 2006, India', *IEA-OES Annual Report 2006*, www.iea-oceans.org

Slade, A. and Garboushian, V. (2005) '27.6% efficient silicon concentrator cell for mass production', Technical Digest, 15th International Photovoltaic Science and Engineering Conference, Shanghai, October

Slootweg, H. and de Vries, E. (2003) 'Inside wind turbines: Fixed vs. variable speed', *Renewable Energy World*, vol 6, no 1, pp30–40

Smeets, E. M. W., Faaij, A. P. C., Lewandowski, I. M. and Turkenburg, W. C. (2006) 'A bottom-up assessment and review of global bio-energy potentials to 2050', *Progress in Energy and Combustion Science*, vol 33, pp56–106

Smeets, E. M. W., Junginger, M., Faaij, A., Walter, A., Dolzan, P. and Turkenburg, W. C. (2008) 'The sustainability of Brazilian ethanol: An assessment of the possibilities of certified production', *Biomass and Bioenergy*, vol 32, pp781–813

Smeets, E. M. W., Lewandowski, I. M. and Faaij, A. P. C. (2009) 'The economical and environmental performance of miscanthus and switchgrass production and supply chains in a European setting', *Renewable and Sustainable Energy Reviews*, vol 13, pp1230–1245

Smith, K. R., Uma, R., Kishore, V. V. N., Zhang, J., Joshi, V. and Khalil, M. A. K. (2000) 'Greenhouse implications of household stoves: An analysis for India', *Annual Review of Energy and the Environment*, vol 25, pp741–763

Snyder, B. and Kaiser, M. J. (2009) 'Ecological and economic cost–benefit analysis of offshore wind energy', *Renewable Energy*, vol 34, pp1567–1578

Söker, H., Rehfeldt, K., Santjer, F., Strack, M. and Schreiber, M. (2000) *Offshore Wind Energy in the North Sea: Technical Possibilities and Ecological Considerations – A Study for Greenpeace*, www.greenpeace.org

Solúcar (2006) *10MW Solar Thermal Power Plant for Southern Spain, Final Technical Progress Report,* NNES-1999-356. Available from http://ec.europa.eu/energy/res/sectors/doc/csp/ps10_final_report.pdf

Sørensen, B. (1999) *Long-term scenarios for global energy demand and supply: Four global greenhouse mitigation scenarios*, Roskilde University, Institute 2, Energy and Environment Group, Denmark

Sovacool, B. K. (2008) 'Valuing the greenhouse gas emissions from nuclear power: A critical survey', *Energy Policy*, vol 36, pp2940–2953

Sovacool, B. K. (2009) 'The importance of comprehensiveness in renewable electricity and energy-efficiency policy', *Energy Policy*, vol 37, pp1529–1541

Sparovek, G. and Schnug, E. (2001) 'Temporal erosion-induced soil degradation and yield loss', *Soil Science Society of America Journal*, vol 65, pp1479–1486

Spath, P. L. and Mann, M. K. (2000) *Life Cycle Assessment of a Natural Gas Combined-Cycle Power Generation System*, Technical Report NREL/TP-570-27715, National Renewable Energy Laboratory, Golden, Colorado

Spath, P. L., Mann, M. K. and Kerr, D. R. (1999) *Life Cycle Assessment of Coal-fired Power Production*, Technical Report NREL/TP-570-25119, National Renewable Energy Laboratory, Golden, Colorado

Starr, F. (2007) 'Flexibility of fossil fuel plant in a renewable energy system: Possible implications for the UK', in G. Boyle (ed) *Renewable Energy and the Grid: The Challenge of Variability*, Earthscan, London

Steinfeld, A. (2002) 'Solar hydrogen production via a two-step water splitting thermochemical cycle based on Zn/ZnO redox reactions', *International Journal of Hydrogen Energy*, vol 27, pp611–619

Steinfeld, A. and Thompson, G. (1994) 'Solar combined thermochemical processes for CO_2 mitigation in the iron, cement, and syngas industries', *Energy*, vol 19, pp1077–1081

Stephanopoulos, G. (2007) 'Challenges in engineering microbes for biofuels production', *Science*, vol 315, pp801–804

Stevens, S. H., Kuuskraa, V. A., Spector, D. and Riemer, P. (1999) 'CO_2 sequestration in deep coal seams: Pilot results and wordwide potential', in P. Riemer, B. Eliasson and A. Wokaun (eds) *Greenhouse Gas Control Technologies*, Elsevier Science, New York

Stone, K. W., Garboushian, V., Boehm, R., Hurt, R., Gray, A. and Hayden, H. (2006) 'Analysis of five years of field performance of the Amonix high concentration PV system', *Powergen 2006*, April, Las Vegas, Nevada, www.amonix.com

Stone, R. (2007) 'Can palm oil plantations come clean?', *Science*, vol 317, p1491

Stoppato, A. (2008) 'Life cycle assessment of photovoltaic electricity generation', *Energy*, vol 33, pp224–232

Straatman, P. J. T. and van Sark, W. G. J. H. M. (2008) 'A new hybrid ocean thermal energy conversion–offshore solar pond (OTEC-OSP) design: A cost optimization approach', *Solar Energy*, vol 82, pp520–527

Strengers, B. J., van Minnen, J. G. and Eickhout, B. (2008) 'The role of carbon plantations in mitigating climate change: potentials and costs', *Climatic Change*, vol 88, pp343–366

Styles, D. and Jones, M. B. (2007) 'Energy crops in Ireland: Quantifying the potential life-cycle greenhouse gas reductions of energy-crop electricity', *Biomass and Bioenergy*, vol 31, pp759–772

Succar, S. and Williams, R. H. (2008) *Compressed Air Energy Storage: Theory, Resources, and Applications for Wind Power*, Princeton Environment Institute, Princeton University, Princeton, New Jersey

Suppes, G. J. (2006) 'Roles of plug-in hybrid electric vehicles in the transition to the hydrogen economy', *International Journal of Hydrogen Energy*, vol 31, pp353–360

Swanson, R. M. (2006) 'A vision for crystalline silicon photovoltaics', *Progress in Photovoltaics: Research and Applications*, vol 14, pp443–453

Swedish Energy Agency (2008) *Energy in Sweden 2008*, Swedish Energy Agency, Eskilstuna, www.stem.se

Swisher, J. and Wilson, D. (1993) 'Renewable energy potentials', *Energy*, vol 18, pp437–459

Syed, A., Maidment, G. G., Missenden, J. F. and Tozer, R. M. (2002) 'An efficiency comparison of solar cooling schemes', *ASHRAE Transactions*, vol 108 (Part 1), pp877–886

Szakálos, P., Hultquist, G. and Wikmark, G. (2007) 'Corrosion of copper by water', *Electrochemical and Solid-State Letters*, vol 10, C63–C67

Takeshita, T. and Yamaji, K. (2008) 'Important roles of Fischer-Tropsch synfuels in the global energy future', *Energy Policy*, vol 36, pp2773–2784

Tamburri, M. N., Peltzer, E. T., Friederich, G. E., Aya, I., Yamane, K. and Brewer, P. G. (2000) 'A field study of the effects of CO_2 ocean disposal on mobile deep-sea animals', *Marine Chemistry*, vol 72, pp95–101

Tanaka, T., Azuma, T., Evans, J. A., Cronin, P. M., Johnson, D. M. and Cleaver, R. P. (2007) 'Experimental study on hydrogen explosions in a full-scale hydrogen filling station model', *International Journal of Hydrogen Energy*, vol 32, pp2162–2170

Tao, Z. and Li, M. (2007) 'What is the limit of Chinese coal supplies? A STELLA model of Hubbert peak', *Energy Policy*, vol 35, pp3145–3154

Taylor, M. A. (2007) *The State of Geothermal Technology. Part I: Subsurface Technology*, Geothermal Energy Association, Washington, DC, www.geo-energy.org

Teske, S. (2004) 'Solar thermal power 2020: A fine future for solar thermal electricity', *Renewable Energy World*, vol 7, no 1, pp120–124

Thambimuthu, K., Soltanieh, M. and Abanades, J. C. (2005) 'Capture of CO_2', in B. Metz, O. Davidson, H. C. de Coninck, M. Loos and L. A. Meyer (eds) *IPCC Special Report on Carbon Dioxide Capture and Storage*, Cambridge University Press, Cambridge

Tharakan, P. J., Volk, T. A., Lindsey, C. A., Abrahamson, L. P. and White, E. H. (2005) 'Evaluating the impact of three incentive programs on co-firing willow biomass with coal in New York State', *Energy Policy*, vol 33, pp337–347

Thomas, R. and Fordham, M. (eds) (2001) *Photovoltaics and Architecture*, E & FN Spon, London

Thomas, S. (2005) *The Economics of Nuclear Power*, Nuclear Issues Paper No 5, Heinrich Böll Foundation, Berlin, www.boell.de/nuclear

Thompson, S. L. and Barron, E. J. (1981) 'Comparison of Cretaceous and present Earth albedos: Implications for the cause of paleoclimates', *Journal of Geology*, vol 89, pp143–167

Thomson, G. (2009) *Burying Carbon Dioxide in Underground Saline Aquifers: Political Folly or Climate Change Fix?*, Program on Water Issues, Munk Centre for International Studies, University of Toronto, www.powi.ca.

Thür, A., Furbo, S. and Shah, L. J. (2006) 'Energy savings for solar heating systems', *Solar Energy*, vol 80, pp1463–1474

Tijmensen, M. and van den Broek, R. (2004) 'Clean power from farm waste: International experiences with anaerobic digestion of farm manure', *Renewable Energy World*, vol 7, no 2, pp124–131

Tilman, D., Hill, J. and Lehman, C. (2006) 'Carbon-negative biofuels from low-input high-diversity grassland biomass', *Science*, vol 314, pp1598–1600

Tingzhen, M., Wei, L., Guoling, X., Yanbin, X., Xuhu, G. and Yuan, P. (2008) 'Numerical simulation of the solar chimney power plant systems coupled with turbine', *Renewable Energy*, vol 33, pp897–905

Toke, D. (2005) 'Explaining wind power planning outcomes: Some findings from a study in England and Wales', *Energy Policy*, vol 33, pp1527–1539

Tollefson, J. (2008) 'Not your father's biofuels', *Nature*, vol 451, pp880–883

Topic, M., Brecl, K. and Sites, J. (2007) 'Effective efficiency of PV modules under field conditions', *Progress in Photovoltaics: Research and Applications*, vol 15, pp19–26

Treloar, G. J., Love, P. E. D. and Holt, G. D. (2001) 'Using national input–output data for embodied energy analysis of individual residential buildings', *Construction Management and Economics*, vol 19, pp49–61

Tremeac, B. and Meunier, F. (2009) 'Life cycle analysis of 4.5 MW and 250 W wind turbines', *Renewable and Sustainable Energy Reviews*, vol 13, pp2104–2110

Trevisani, L., Fabbri, M., Negrini, F. and Ribani, P. L. (2007) 'Advanced energy recovery systems from liquid hydrogen', *Energy Conversion and Management*, vol 48, pp146–154

Tripanagnostopoulos, Y., Souliotis, M., Battisti, R. and Corrado, A. (2006) 'Performance, cost and life-cycle assessment study of hybrid PVT/air solar systems', *Progress in Photovoltaics: Research and Applications*, vol 14, pp65–76

Troncoso, E. and Newborough, M. (2007) 'Implementation and control of electrolyzers to achieve high penetrations of renewable power', *International Journal of Hydrogen Energy*, vol 32, pp2253–2268

Tsang, C. F., Birkholzer, J. and Rutqvist, J. (2008) 'A comparative review of hydrologic issues involved in injection storage of CO_2 and in injection disposal of liquid waste', *Journal of Environmental Geology*, vol 54, pp1723–1737

Tsoutsos, T., Gekas, V. and Marketaki, K. (2003) 'Technical and economical evaluation of solar thermal power generation', *Renewable Energy*, vol 28, pp873–886

Tuck, G., Glendining, M. J., Smith, P., House, J. I. and Wattenbach, M. (2006) 'The potential distribution of bioenergy crops in Europe under present and future climate', *Biomass and Bioenergy*, vol 30, pp183–197

Tuskan, G. A. (1998) 'Short-rotation wood crop supply systems in the United States: What do we know and what do we need to know?' *Biomass and Bioenergy*, vol 14, pp307–315

Twidell, J. (2003) 'Technology fundamentals: Wind turbines', *Renewable Energy News*, vol 6, no 3, pp102–111

Tzimas, E. and Peteves, S. D. (2005) 'The impact of carbon sequestration on the production cost of electricity and hydrogen from coal and natural-gas technologies in Europe in the medium term', *Energy*, vol 30, pp2672–2689

Tzimas, E., Mercier, A., Cormos, C. C. and Peteves, S. D. (2007) 'Trade-off in emissions of acid gas pollutants and of carbon dioxide in fossil fuel plants with carbon capture', *Energy Policy*, vol 35, pp3991–3998

UN (United Nations) (2008) *2005 Energy Statistics Yearbook*, Department of Economic and Social Affairs, United Nations, New York

Unruh, G. C. (2000) 'Understanding carbon lock-in', *Energy Policy*, vol 28, pp817–830

Unruh, G. C. (2002) 'Escaping carbon lock-in', *Energy Policy*, vol 30, pp317–325

Upson, P. C. (1996) 'Isotopic enrichment of uranium', in P. D. Wilson (ed) *The Nuclear Fuel Cycle: From Ore to Waste*, Oxford University Press, Oxford

US CSP (United States Concentrating Solar Power) (2002) *Status of Major Project Opportunities*, presentation at the 2002 Berlin Solar Paces CSP Conference

US DOE (United States Department of Energy) (2006) *Multi-Year Program Plan 2007–2011*, Solar Energy Technologies Program, US Department of Energy, Washington DC, www1.eere.energy.gov/solar

US EIA (United States Energy Information Administration) (2007) *Solar Explained: Photovoltaics and Electricity*, available from http://tonto.eia.doe.gov/energyexplained/index.cfm?page=solar_photovoltaics

Uzunoglu, M., Onar, O. C. and Alam, M. S. (2009) 'Modeling, control and simulation of a PV/FC/UC based hybrid generation system for stand-alone applications', *Renewable Energy*, vol 34, pp509–520

van Dam, J., Faaij, A. P. C., Lewandowski, I. and Fischer, G. (2007) 'Biomass production potentials in Central and Eastern Europe under different scenarios', *Biomass and Bioenergy*, vol 31, pp345–366 van den Wall Bake, J. D., Junginger, M., Faaij, A., Poot, T. and Walter, A. (2009) 'Explaining the experience curve: Cost reductions of Brazilian ethanol from sugarcane', *Biomass and Bioenergy*, vol 3, pp644–658

van Hulle, F. et al (2009) *Integrating Wind, Developing Europe's Power Market for the Large-Scale Integration of Wind Power*, Intelligent Energy-Europe Programme, TradeWind, http://www.trade-wind.eu

van Leeuwen, J. W. S. (2007) *Nuclear Power: The Energy Balance*, Ceedata Consultancy, Chaarn, Netherlands, www.stormsmith.nl

van Mierlo, B. and Oudshoff, B. (1999) *Literature Survey and Analysis of Non-Technical Problems for the Introduction of Building Integrated Photovoltaic Systems*, International Energy Agency, Photovoltaic Power Systems Programme, Task 7, Paris, www.iea-pvps.org.

van Sark, W. G. J. H. M., Alsema, E. A., Junginger, H. M., de Moor, H. H. C. and Schaeffer, G. J. (2008) 'Accuracy of progress ratios determined from experience curves: The case of crystalline silicon photovoltaic module technology development', *Progress in Photovoltaics: Research and Applications*, vol 16, pp441–453

Vannoni, C., Battisti, R. and Drigo, S. (2008) *Potential for Solar Heat in Industrial Processes*, International Energy Agency, Solar Heating and Cooling Programme Task 33, Paris

Vant-Hull, L. L. (1991) 'Solar thermal electricity: An environmentally benign and viable alternative', in *Proceedings of the World Clean Energy Conference*, 4–7 November, Geneva, World Energy Coalition for the Global Energy Charter, pp350–356

Vattenfall (2007) *The Nuclear Fuel Cycle*, Vattenfall AB, Stockholm, www.vattenfall.com

Vaughan, N. E. and Lenton, T. M. (forthcoming) *Climatic Change*, in press

Veldhuis, I. J. S., Richardson, R. N. and Stone, H. B. J. (2007) 'Hydrogen fuel in a marine environment', *International Journal of Hydrogen Energy*, vol 32, pp2553–2566

Verlinden, P. J., Lewandowski, A., Bingham, C., Kinsey, G. S., Sherif, R. A. and Lasich, J. B. (2006) 'Performance and reliability of multijunction III-V modules for concentrator dish and central receiver applications', in *Proceedings of the 4th World Conference on Photovoltaic Energy Conversion*, Waikoloa, Hawaii, pp592–597

Volk, T. A., Abrahamson, L. P., Nowak, C. A., Smart, L. B., Tharakan, P. J. and White, E. H. (2006) 'The development of short-rotation willow in the northeastern United States for bioenergy and bioproducts, agroforestry and phytoremediation', *Biomass and Bioenergy*, vol 30, pp715–727

von Hippel, F. N. (2001) 'Plutonium and reprocessing of spent nuclear fuel', *Science*, vol 293, pp2397–2398

von Hippel, F. N., Bunn, M., Diakov, A., Goldston, R., Katsuta, T., Ramana, M. V., Rogner, H. H., Suzuki, T. and Yu, S. (2010) 'Nuclear Energy', in *Global Energy Assessment*, International Institute for Applied Systems Analysis, Laxenburg, Austria

Voorspools, K. R. and D'haeseleer, W. D. (2006) 'An analytical formula for the capacity credit of wind power', *Renewable Energy*, vol 31, pp45–54

Voorspools, K. R., Brouwers, E. A. and D'haeseleer, W. D. (2000) 'Energy content and indirect greenhouse gas emissions embedded in "emission-free" power plants: Results for the Low Countries', *Applied Energy*, vol 67, pp307–330

Wadia, C., Alivisatos, A. P. and Kammen, D. M. (2009) 'Materials availability expands the opportunity for large-scale photovoltaics deployment', *Environmental Science and Technology*, vol 43, pp2072–2077

Wagner, H. J. and Pick, E. (2004) 'Energy yield ratio and cumulative energy demand for wind energy converters', *Energy*, vol 29, pp2289–2295

Walker, G. (2007) 'A world melting from the top down', *Nature*, vol 446, pp718–721

Wan, Y. H. and Bucaneg, D. (2002) 'Short-term power fluctuations of large wind power plants', *Journal of Solar Energy Engineering*, vol 124, pp427–431

Wang, M., Wu, M. and Huo, H. (2007) 'Life-cycle energy and greenhouse gas emission impacts of different corn ethanol plant types', *Environmental Research Letters*, vol 2, April–June, p024001

Wardle, D. A., Nilsson, M. C. and Zackrisson, O. (2008) 'Fire-derived charcoal causes loss of forest humus', *Science*, vol 320, p629

Warwick, N. J., Bekki, S., Nisbet, E. G. and Pyle, J. A. (2004) 'Impact of a hydrogen economy on the stratosphere and troposphere studied in a 2-D model', *Geophysical Research Letters*, vol 31, ppL05107

Watson, R. T., Noble, I. R., Bolin, B., Ravindranath, N. H., Verardo, D. J. and Dokken, D. J. (eds) (2000) *Land Use, Land-use Change, and Forestry, Special Report of the Intergovernmental Panel on Climate Change*, Cambridge University Press, Cambridge

WCD (World Commission on Dams) (2000) *Dams and Development: A New Framework for Decision-Making*, Earthscan, London, www.dams.org

WEC (World Energy Council) (2007) *2007 Survey of Energy Resources*, World Energy Council, London, www.world energy.org

Weidema, B. P. (2000) 'Avoiding co-product allocation in life-cycle assessment', *Journal of Industrial Ecology*, vol 4, pp11–33

Weinert, J. X. and Lipman, T. E. (2006) *An Assessment of the Near-Term Costs of Hydrogen Refueling Stations and Station Components*, Institute of Transportation Studies, University of California, Davis

Weinzettel, J., Reenaas, M., Solli, C. and Hertwich, E. G. (2009) 'Life cycle assessment of a floating offshore wind turbine', *Renewable Energy*, vol 34, pp742–747

Weiss, W. (2003) *Solar Heating Systems for Houses: A Design Handbook for Solar Combisystems*, James & James, London

Weiss, W. and Rommel, M. (2008) *Process Heat Collectors, State of the Art within Task 33/IV*, International Energy Agency, Solar Heating and Cooling Task 33, Solar Heat for Industrial Process, Paris, www.iea-shc.org

Weiss, W., Bergmann, I. and Stelzer, R. (2009) *Solar Heat Worldwide: Markets and Contribution to the Energy Supply 2007*, International Energy Agency, Solar Heating and Cooling Programme, Paris, www.iea-shc.org

Wentzel, M. and Pouris, A. (2007) 'The development impact of solar cookers: A review of solar cooking impact research in South Africa', *Energy Policy*, vol 35, pp1909–1919

Wenzel, E. (2008) 'Mining Lithium from geothermal "lemonade"', *Green Tech*, vol 28, February, http://news.cnet.com/8301-11128_3-9881869-54.html

Werder, M. and Steinfeld, A. (2000) 'Life cycle assessment of the conventional and solar thermal production of zinc and synthesis gas', *Energy*, vol 25, pp395–409

West, T. O. and Marland, G. (2002) 'A synthesis of carbon sequestration, carbon emissions, and net carbon flux in agriculture: Comparing tillage practices in the United States', *Agriculture, Ecosystems and Environment*, vol 91, pp217–232

WGA (Western Governors' Association) (2006) *Clean and Diversified Energy Initiative, Biomass Task Force Report*, www.westgov.org/wga/initiatives/cdeac/Biomass-full.pdf

White, S. W. and Kulcinski, G. L. (2000) 'Birth to death analysis of the energy payback ratio and CO_2 gas emission rates from coal, fission, wind, and DT-fusion electrical power plants', *Fusion Engineering and Design*, vol 48, pp473–481

Wicke, B., Dornburg, V., Junginger, M. and Faaij, A. (2008) 'Different palm oil production systems for energy purposes and their greenhouse gas implications', *Biomass and Bioenergy*, vol 32, pp1322–1337

Wiemken, E., Beyer, H. G., Heydenreich, W. and Kiefer, K. (2001) 'Power characteristics of PV ensembles: Experiences from the combined power production of 100 grid connected PV systems distributed over the area of Germany', *Solar Energy*, vol 70, no 6, pp513–518

Wietschel, M., Hasenauer, U. and de Groot, A. (2006) 'Development of European hydrogen infrastructure scenarios – CO_2 reduction potential and infrastructure investment', *Energy Policy*, vol 34, pp1284–1298

Wihersaari, M. (2005) 'Greenhouse gas emissions from final harvest chip production in Finland', *Biomass and Energy*, vol 28, pp435–443

Wilf, M. (2004) 'Fundamentals of RO-NF Technology', presented at the International Conference on Desalination Costing, Lemesos, Cyprus, 6–8 December, www.medrc.org

Williams, R. H. (1998) 'Fuel decarbonization for fuel cell applications and sequestration of the separated CO_2', in R. U. Ayres and P. M. Weaver (eds) *Ecorestructuring: Implications for Sustainable Development*, United Nations University Press, Tokyo

Williams, R. H., Bunn, M., Consonni, S., Gunter, W., Holloway, S., Moore, R. and Simbeck, D. (2000) 'Advanced energy supply technologies', in *World Energy Assessment: Energy and the Challenge of Sustainability*, United Nations Development Programme, New York

Wilson, J. R. and Burgh, G. (2008) *Energizing Our Future, Rational Choices for the 21st Century*, Wiley-Interscience, Hoboken, New Jersey

Wilson, P. D. (1996a) 'Basic principles', in P. D. Wilson (ed) *The Nuclear Fuel Cycle: From Ore to Waste*, Oxford University Press, Oxford

Wilson, P. D. (ed) (1996b) *The Nuclear Fuel Cycle: From Ore to Waste*, Oxford University Press, Oxford

Winfield, M., Jamison, A., Wong, R. and Czajkowski, P. (2006) *Nuclear Power in Canada: An Examination of Risks, Impacts and Sustainability*, Pembina Institute, Calgary

Wirsenius, S. (2000) 'Human use of land and organic materials', PhD Thesis, Chalmers University of Technology, Göteborg

WNA (World Nuclear Association) (2006) *Energy Analysis of Power Systems*, www.world-nuclear.org

Wolf-Gladrow, D. A., Riebesell, U., Buckhardt, S. and Bijma, J. (1999) 'Direct effects of CO_2 concentration on growth and isotopic composition of marine plankton', *Tellus*, vol 51B, pp461–476

Wolfson, R. (1993) *Nuclear Choices: A Citizen's Guide to Nuclear Technology*, MIT Press, Cambridge, MA

Wolsink, M. (2000) 'Wind power and the NIMBY-myth: Institutional capacity and the limited significance of public support', *Renewable Energy*, vol 21, pp49–64

Wolsink, M. (2007a) 'Planning of renewable schemes: Deliberative and fair decision-making on landscape issues instead of reproachful accusations of non-cooperation', *Energy Policy*, vol 35, pp2692–2704

Wolsink, M. (2007b) 'Wind power implementation: The nature of public attitudes: Equity and fairness instead of "backyard motives"', *Renewable and Sustainable Energy Reviews*, vol 11, pp1188–1207

Wong, S., Foy, C., Gunter, B. and Jack, T. (1999) 'Injection of CO_2 for enhanced energy recovery: Coalbed methane versus oil recovery', in P. Riemer, B. Eliasson and A. Wokaun (eds) *Greenhouse Gas Control Technologies*, Elsevier Science, New York

Wood, E. (2003) 'The US offshore wind market: Can it stay on course?', *Renewable Energy World*, vol 6, no 3, pp50–58

Wood, H. G., Glaser, A. and Kemp, R. S. (2008) 'The gas centrifuge and nuclear weapons proliferation', *Physics Today*, September, pp40–45

Wüstenhagen, R., Wolsink, M. and Burer, M. J. (2007) 'Social acceptance of renewable energy innovation: An introduction to the concept', *Energy Policy*, vol 35, pp2683–2691

WWEA (World Wind Energy Association) (2009) *World Wind Energy Report 2008*, www.wwindea.org

WWI (Worldwatch Institute) (2006) *Biofuels for Transportation: Global Potential and Implications for Sustainable Agriculture and Energy in the 21st Century*, WWI, Washington, DC

Wyman, C. E. (1999) 'Biomass ethanol: Technical progress, opportunities, and commercial challenges', *Annual Review of Energy and the Environment*, vol 24, pp189–226

Yamada, N., Hoshi, A. and Ikegami, Y. (2009) 'Performance simulation of solar-boosted ocean thermal energy conversion plant', *Renewable Energy*, vol 34, pp1752–1758

Yamashita, K. and Barreto, L. (2005) 'Energyplexes for the 21st century: Coal gasification for co-producing hydrogen, electricity and liquid fuels', *Energy*, vol 30, pp2453–2473

Yang, C. and Ogden, J. (2007) 'Determining the lowest-cost hydrogen delivery mode', *International Journal of Hydrogen Energy*, vol 32, pp268–286

Yang, M. (2007) 'Climate change drives wind turbines', *Energy Policy*, vol 35, pp6546–6548

Yattara, A., Zhu, Y. and Ali, M. M. (2003) 'Comparison between solar single-effect and single-effect double-lift absorption machines (Part I)', *Applied Thermal Engineering*, vol 3, pp1981–1992

Yemshanov, D. and McKenney, D. (2008) 'Fast-growing poplar plantations as a bioenergy supply source for Canada', *Biomass and Bioenergy*, vol 32, pp185–197

Yildiz, B. and Kazimi, M. S. (2006) 'Efficiency of hydrogen production systems using alternative nuclear energy systems', *International Journal of Hydrogen Energy*, vol 31, pp77–92

Young, L. and Pian, C. C. P. (2003) 'High-temperature, air-blown gasification of dairy-farm wastes for energy production', *Energy*, vol 28, pp655–672

Yu, J. and Chen, L. X. L. (2008) 'The greenhouse gas emissions and fossil energy requirement of bioplastics from cradle to gate of a biomass refinery', *Environmental Science and Technology*, vol 42, pp6961–6966

Zaaijer, M. and Henderson, A. (2003) 'Offshore update: A global look at offshore wind energy', *Renewable Energy News*, vol 6, no 4, pp102–119

Zalesny, R. S., Wiese, A. H., Bauer, E. O. and Riememschneider, D. E. (2009) '*Ex situ* growth and biomass of *Populus* bioenergy crops irrigated and fertilized with landfill leachate', *Biomass and Bioenergy*, vol 33, pp62–69

Zarza, E., Rojas, M. E., González, L., Caballero, J. M. and Rueda, F. (2006) 'INDITEP: The first pre-commercial DSG solar power plant', *Solar Energy*, vol 80, pp1270–1276

Zeman, F. (2007) 'Energy and material balance of CO_2 capture from ambient air', *Environmental Science and Technology*, vol 41, pp7558–7563

Zhai, X. Q. and Wang, R. Z. (2008) 'Experiences on solar heating and cooling in China', *Renewable and Sustainable Energy Reviews*, vol 12, pp1110–1128

Zhang, L. and Sun, K. (2007) 'Tidal current energy developments in China', in *IEA Ocean Energy Systems Newsletter*, May, www.iea-oceans.org

Zhang, N. and Lior, N. (2006) 'A novel near-zero CO_2 emission thermal cycle with LNG cryogenic exergy utilization', *Energy*, vol 31, pp1666–1679

Zhang, X. R. and Yamaguchi, H. (2008) 'An experimental study on evacuated tube solar collector using supercritical CO_2', *Applied Thermal Engineering*, vol 28, pp1225–1233

Zhao, C., Tie, X. and Yunping, L. (2006) 'A possible positive feedback of reduction of precipitation and increase in aerosols over eastern central China', *Geophysical Research Letters*, vol 33, L11814, doi:10.1029/2006GL025959

Zhao, Y., Hao, L.-S. and Wang, Y.-P. (2009) 'Development strategies for wind power industry in Jiangsu Province, China: Based on the evaluation of resource capacity', *Energy Policy*, vol 37, pp1736–1744

Zhou, G. Y., Morris, J. D., Yan, J. H., Yu, Z. Y. and Peng, S. L. (2002) 'Hydrological impacts of reafforestation with eucalypts and indigenous species: A case study in southern China', *Forest Ecology and Management*, vol 167, pp209–222

Zhou, S., Zhang, X. and Liu, J. (2009) 'The trend of small hydropower development in China', *Renewable Energy*, vol 34, pp1078–1083

Zinn, Y. L., Resck, D. V. S. and da Silva, J. E. (2002) 'Soil organic carbon as affected by afforestation with *Eucalyptus* and *Pinus* in the *Cerrado* region of Brazil', *Forest Ecology and Management*, vol 166, pp285–294

Zondag, H. A. (2008) 'Flat-plate PV-thermal collectors and systems: A review', *Renewable and Sustainable Energy Reviews*, vol 12, pp891–959

Zweibel, K. (2005) *The Terawatt Challenge for Thin-Film PV*, NREL Technical Report, www.nrel.gov

Index